Advanced Electronic Packaging

IEEE Press
445 Hoes Lane
Piscataway, NJ 08854

IEEE Press Editorial Board
Mohamed E. El-Hawary, *Editor in Chief*

M. Akay	T. G. Croda	M. S. Newman
J. B. Anderson	R. J. Herrick	F. M. B. Periera
R. J. Baker	S. V. Kartalopoulos	C. Singh
J. E. Brewer	M. Montrose	G. Zobrist

Kenneth Moore, *Director of IEEE Book and Information Services (BIS)*
Catherine Faduska, *Acquisitions Editor*
Jeanne Audino, *Project Editor*

IEEE Component, Packaging, and Manufacturing Technology Society, Sponsor
CPMT-S Liaison to IEEE Press, Joe E. Brewer

Advanced Electronic Packaging

Second Edition

Edited by
Richard K. Ulrich

William D. Brown

IEEE Press Series on Microelectronic Systems
STUART K. TEWKSBURY, Series Editor

A JOHN WILEY & SONS, INC., PUBLICATION

Copyright © 2006 by the Institute of Electrical and Electronics Engineers, Inc. All rights reserved.

Published by John Wiley & Sons, Inc. Published simultaneously in Canada.

No part of this publication may be reproduced, stored in a retrieval system, or transmitted in any form or by any means, electronic, mechanical, photocopying, recording, scanning, or otherwise, except as permitted under Section 107 or 108 of the 1976 United States Copyright Act, without either the prior written permission of the Publisher, or authorization through payment of the appropriate per-copy fee to the Copyright Clearance Center, Inc., 222 Rosewood Drive, Danvers, MA 01923, (978) 750-8400, fax (978) 750-4470, or on the web at www.copyright.com. Requests to the Publisher for permission should be addressed to the Permissions Department, John Wiley & Sons, Inc., 111 River Street, Hoboken, NJ 07030, (201) 748-6011, fax (201) 748-6008, or online at http://www.wiley.com/go/permission.

Limit of Liability/Disclaimer of Warranty: While the publisher and author have used their best efforts in preparing this book, they make no representations or warranties with respect to the accuracy or completeness of the contents of this book and specifically disclaim any implied warranties of merchantability or fitness for a particular purpose. No warranty may be created or extended by sales representatives or written sales materials. The advice and strategies contained herein may not be suitable for your situation. You should consult with a professional where appropriate. Neither the publisher nor author shall be liable for any loss of profit or any other commercial damages, including but not limited to special, incidental, consequential, or other damages.

For general information on our other products and services or for technical support, please contact our Customer Care Department within the United States at (800) 762-2974, outside the United States at (317) 572-3993 or fax (317) 572-4002.

Wiley also publishes its books in a variety of electronic formats. Some content that appears in print, may not be available in electronic formats. For more information about Wiley products, visit our web site at www.wiley.com.

Library of Congress Cataloging-in-Publication Data is available.

ISBN-13 978-0-471-46609-X
ISBN-10 0-471-46609-3

Printed in the United States of America

10 9 8 7 6 5 4 3 2 1

Contents

Preface to the Second Edition xv
Contributors xvii
Acronyms xix

1. Introduction and Overview of Microelectronic Packaging 1

W. D. Brown

1.1 Introduction 1
1.2 Functions of an Electronic Package 2
1.3 Packaging Hierarchy 3
    1.3.1 Die Attach 5
    1.3.2 First-Level Interconnection 5
    1.3.3 Package Lid and Pin Sealing 7
    1.3.4 Second-Level Interconnection 7
1.4 Brief History of Microelectronic Packaging Technology 8
1.5 Driving Forces on Packaging Technology 19
    1.5.1 Manufacturing Costs 20
    1.5.2 Manufacturability Costs 20
    1.5.3 Size and Weight 20
    1.5.4 Electrical Design 21
    1.5.5 Thermal Design 21
    1.5.6 Mechanical Design 21
    1.5.7 Manufacturability 22
    1.5.8 Testability 22
    1.5.9 Reliability 23
    1.5.10 Serviceability 23
    1.5.11 Material Selection 23
1.6 Summary 24
    References 25
    Exercises 26

2. Materials for Microelectronic Packaging 29

W. D. Brown and Richard Ulrich

2.1 Introduction 29
2.2 Some Important Packaging Material Properties 29
    2.2.1 Mechanical Properties 30
    2.2.2 Moisture Penetration 30
    2.2.3 Interfacial Adhesion 31
    2.2.4 Electrical Properties 31
    2.2.5 Thermal Properties 33
    2.2.6 Chemical Properties 34
    2.2.7 System Reliability 34
2.3 Ceramics in Packaging 35
    2.3.1 Alumina (Al_2O_3) 37
    2.3.2 Beryllia (BeO) 39
    2.3.3 Aluminum Nitride (AlN) 40
    2.3.4 Silicon Carbide (SiC) 41
    2.3.5 Boron Nitride (BN) 41
    2.3.6 Glass–Ceramics 42
2.4 Polymers in Packaging 43
    2.4.1 Fundamentals of Polymers 43
    2.4.2 Thermoplastic and Thermosetting Polymers 45
    2.4.3 Effects of Water and Solvents on Polymers 46
    2.4.4 Some Polymer Properties of Interest 47
    2.4.5 Primary Classes of Polymers Used in Microelectronics 50
    2.4.6 First-Level Packaging Applications of Polymers 55
2.5 Metals in Packaging 57
    2.5.1 Die Bonding 57
    2.5.2 Chip to Package or Substrate 58
    2.5.3 Package Construction 64
2.6 Materials Used in High-density Interconnect Substrates 66
    2.6.1 Laminate Substrates 67
    2.6.2 Ceramic Substrates 70
    2.6.3 Deposited Thin-Film Substrates 71
2.7 Summary 73
    References 73
    Exercises 75

3. Processing Technologies   77

H. A. Naseem and Susan Burkett

- 3.1 Introduction   77
- 3.2 Thin-Film Deposition   77
  - 3.2.1 Vacuum Facts   78
  - 3.2.2 Vacuum Pumps   79
  - 3.2.3 Evaporation   81
  - 3.2.4 Sputtering   84
  - 3.2.5 Chemical Vapor Deposition   88
  - 3.2.6 Plating   89
- 3.3 Patterning   93
  - 3.3.1 Photolithography   93
  - 3.3.2 Etching   96
- 3.4 Metal-to-Metal Joining   98
  - 3.4.1 Solid-State Bonding   98
  - 3.4.2 Soldering and Brazing   101
- 3.5 Summary   102
- References   102
- Exercises   103

4. Organic Printed Circuit Board Materials and Processes   105

Richard C. Snogren

- 4.1 Introduction   105
- 4.2 Common Issues for All PCB Layer Constructions   106
  - 4.2.1 Data Formats and Specifications   106
  - 4.2.2 Computer-Aided Manufacturing and Tooling   107
  - 4.2.3 Panelization   108
  - 4.2.4 Laminate Materials   109
  - 4.2.5 Manufacturing Tolerance Overview   111
- 4.3 PCB Process Flow   112
  - 4.3.1 Manufacture of Inner Layers   112
  - 4.3.2 Manufacture of MLB Structure and Outer Layers   118
  - 4.3.3 Electrical Test   124
  - 4.3.4 Visual and Dimensional Inspection   124
  - 4.3.5 Contract Review   125
  - 4.3.6 Microsection Analysis   125
- 4.4 Dielectric Materials   127
  - 4.4.1 Dielectric Material Drivers   127
  - 4.4.2 Dielectric Material Constructions and Process Considerations   128
- 4.5 Surface Finishes   133
- 4.6 Advanced PCB Structures   134
  - 4.6.1 High-density Interconnection (HDI) or Microvia   134
- 4.7 Specifications and Standards   141
  - 4.7.1 The IPC, a Brief History   141
  - 4.7.2 Relevant Standards to Organic Printed Circuit Boards   141
- 4.8 Key Terms   142
- References   145
- Exercises   145

5. Ceramic Substrates   149

Aicha A. R. Elshabini and Fred D. Barlow III

- 5.1 Ceramics in Electronic Packaging   149
  - 5.1.1 Introduction and Background   149
  - 5.1.2 Functions of Ceramic Substrates   149
  - 5.1.3 Ceramic Advantages   150
  - 5.1.4 Ceramic Compositions   150
  - 5.1.5 Ceramic Substrate Manufacturing   151
- 5.2 Electrical Properties of Ceramic Substrates   152
- 5.3 Mechanical Properties of Ceramic Substrates   153
- 5.4 Physical Properties of Ceramic Substrates   154
- 5.5 Design Rules   154
- 5.6 Thin Film on Ceramics   155
  - 5.6.1 Introduction and Background   155
  - 5.6.2 Deposition Techniques   155
  - 5.6.3 Thin-Film Substrate Properties   156
- 5.7 Thick Films on Ceramics   156
  - 5.7.1 Introduction and Background   156
  - 5.7.2 Screen Preparation and Inspection   157
  - 5.7.3 Screen-Printing Process   158
  - 5.7.4 Substrate Cleaning and Process Environment   159
  - 5.7.5 Thick-Film Formulations   159
  - 5.7.6 Heat Treatment Processes for Pastes   160
  - 5.7.7 Thick-Film Metallizations   161
  - 5.7.8 Thick-Film Dielectrics   162
  - 5.7.9 Thick-Film Resistors   162

- 5.8 Low-Temperature Cofired Ceramics (LTCC) 163
  - 5.8.1 LTCC Technology 163
  - 5.8.2 Tape Handling and Cleanroom Environment 166
  - 5.8.3 Via Formation 167
  - 5.8.4 Via Fill 169
  - 5.8.5 Screen-Printing Considerations for Tape Materials 171
  - 5.8.6 Inspection 172
  - 5.8.7 Tape Layer Collation 172
  - 5.8.8 Lamination 173
  - 5.8.9 Firing 175
  - 5.8.10 Postprocessing 177
  - 5.8.11 Design Considerations 178
  - 5.8.12 Shrinkage Prediction and Control 179
- 5.9 HTCC Fabrication Process 180
  - 5.9.1 HTCC Process 180
  - 5.9.2 Multilayer ALN 180
- 5.10 High-Current Substrates 180
  - 5.10.1 Direct Bonded Copper Process 181
  - 5.10.2 Active Metal Brazing (AMB) 182
- 5.11 Summary 182
  - References 183
  - Exercises 184

6. Electrical Considerations, Modeling, and Simulation 187

S. S. Ang and L. W. Schaper

- 6.1 Introduction 187
  - 6.1.1 When Is a Wire Not a Wire? 187
  - 6.1.2 Packaging Electrical Functions 187
- 6.2 Fundamental Considerations 188
  - 6.2.1 Resistance 189
  - 6.2.2 Self and Mutual Inductance 194
  - 6.2.3 Capacitance 200
  - 6.2.4 Parameter Extraction Programs 202
- 6.3 Signal Integrity and Modeling 202
  - 6.3.1 Digital Signal Representation and Spectrum 203
  - 6.3.2 Driver and Receiver Models 205
  - 6.3.3 RC Delay 207
- 6.4 Transmission Lines 212
  - 6.4.1 Microstrip Transmission Lines 215
  - 6.4.2 Termination Reflections 217
  - 6.4.3 Signal Line Losses and Skin Effect 223
  - 6.4.4 Net Topology 224
- 6.5 Coupled Noise or Crosstalk 226
- 6.6 Power and Ground 230
  - 6.6.1 Dynamic Power Distribution 231
  - 6.6.2 Power System Impedance 231
  - 6.6.3 Resonance of Decoupling Capacitance 232
  - 6.6.4 Power Distribution Modeling 232
  - 6.6.5 Switching Noise 233
- 6.7 Overall Packaged IC Models and Simulation 236
  - 6.7.1 Simulation 237
- 6.8 Time-Domain Reflectometry 238
- 6.9 Summary 242
  - References 242
  - Exercises 243

7. Thermal Considerations 247

Rick J. Couvillion

- 7.1 Introduction 247
  - 7.1.1 Heat Sources 247
  - 7.1.2 Approaches to Heat Removal 249
  - 7.1.3 Failure Modes 250
- 7.2 Heat Transfer Fundamentals 251
  - 7.2.1 Heat Transfer Rate Equations 251
  - 7.2.2 Transient Thermal Response of Components 255
  - 7.2.3 Conduction in Various Shapes 257
  - 7.2.4 Overall Resistance 264
  - 7.2.5 Forced Convection Heat Transfer 268
  - 7.2.6 Natural or Free Convection Heat Transfer 276
- 7.3 Air Cooling 282
- 7.4 Liquid Cooling 282
  - 7.4.1 Single-Phase Liquid Cooling 282
  - 7.4.2 Two-Phase Liquid Cooling 282

- 7.5 Advanced Cooling Methods 286
  - 7.5.1 Heat Pipes 286
  - 7.5.2 Thermoelectric Cooling 287
  - 7.5.3 Microchannel Cooling 288
- 7.6 Computer-Aided Modeling 289
  - 7.6.1 Solids Modeling 289
  - 7.6.2 Computational Fluid Dynamics 290
  - 7.6.3 Levels of Decoupling 290
  - 7.6.4 Typical Results 290
- 7.7 Summary 290
- References 292
- Appendix: Thermophysical Properties for Heat Transfer Calculations 292
- Exercises 297

8. Mechanical Design Considerations 299

William F. Schmidt

- 8.1 Introduction 299
- 8.2 Deformation and Strain 299
- 8.3 Stress 303
- 8.4 Constitutive Relations 307
  - 8.4.1 Elastic Material 307
  - 8.4.2 Plastic Material 309
  - 8.4.3 Creep 310
- 8.5 Simplified Forms 311
  - 8.5.1 Plane Stress and Plane Strain 311
  - 8.5.2 Beams 312
- 8.6 Failure Theories 317
  - 8.6.1 Static Failure 318
  - 8.6.2 Fracture Mechanics 320
  - 8.6.3 Fatigue 321
- 8.7 Analytical Determination of Stress 323
  - 8.7.1 Bi-Material Assembly–Axial Effects 323
  - 8.7.2 Bi-Material Assembly–Bending Effects 328
  - 8.7.3 Peeling Stress 329
  - 8.7.4 Tri-Material Assembly 331
- 8.8 Numerical Formulations 335
  - 8.8.1 Finite-Element Method 335
  - 8.8.2 Commercial Codes 338
  - 8.8.3 Limitations and Hazards 340
- 8.9 Summary 341
- References 341
- Bibliography 341
- Exercises 342

9. Discrete and Embedded Passive Devices 349

Richard Ulrich

- 9.1 Introduction 349
- 9.2 Passives in Modern Electronic Systems 350
- 9.3 Definitions and Configurations of Passives 354
- 9.4 Film-Based Passives 356
- 9.5 Resistors 358
  - 9.5.1 Design Equations 358
  - 9.5.2 Sizing Embedded Resistors 360
  - 9.5.3 Materials for Resistors 361
- 9.6 Capacitors 363
  - 9.6.1 Paraelectrics and Ferroelectrics 365
  - 9.6.2 Sizing Dielectric Areas 367
  - 9.6.3 Dielectric Materials Used in Capacitors 369
- 9.7 Inductors 371
- 9.8 Electrical Characteristics of Passives 372
  - 9.8.1 Modeling Ideal Passives 373
  - 9.8.2 Modeling Real Capacitors 374
  - 9.8.3 Differences in Parasitics Between Discrete and Embedded Capacitors 375
  - 9.8.4 Modeling Real Inductors 377
  - 9.8.5 Modeling Real Resistors 379
- 9.9 Issues in Embedding Passives 379
  - 9.9.1 Reasons for Embedding Passives 379
  - 9.9.2 Problems with Embedding Passive Devices 380
- 9.10 Decoupling Capacitors 381
  - 9.10.1 Decoupling Issues 381
  - 9.10.2 Decoupling with Discrete Capacitors 382
  - 9.10.3 Decoupling with Embedded Capacitors 383
- 9.11 Future of Passives 384
- References 385
- Exercises 386

10. Electronic Package Assembly 389

Tarak A. Railkar and Robert W. Warren

- 10.1 Introduction 389

10.2	Facilities 389				10.8.1	Wafer Bumping 419
	10.2.1	Cleanroom Requirements 389			10.8.2	Fluxing 422
	10.2.2	Electrostatic Discharge Requirements 391		10.9	Package Sealing/Encapsulation/Coating 425	
	10.2.3	Moisture Sensitivity Level (MSL) Requirements 392			10.9.1	Hermetic Package Sealing 426
	10.2.4	Reflow Temperatures 393			10.9.2	Hermetic Package Testing 426
10.3	Component Handling 393				10.9.3	Nonhermetic Encapsulation 427

10.2 Facilities 389
    10.2.1 Cleanroom Requirements 389
    10.2.2 Electrostatic Discharge Requirements 391
    10.2.3 Moisture Sensitivity Level (MSL) Requirements 392
    10.2.4 Reflow Temperatures 393
10.3 Component Handling 393
    10.3.1 Shipping 393
    10.3.2 Storage 393
    10.3.3 Handling/Processing 394
10.4 Surface-Mount Technology (SMT) Assembly 395
    10.4.1 Solder Printing and Related Defects 395
    10.4.2 Component Placement 397
    10.4.3 Solder Reflow 398
    10.4.4 Cleaning 399
10.5 Wafer Preparation 399
    10.5.1 Wafer Probing 399
    10.5.2 Wafer Mounting 401
    10.5.3 Wafer Backgrinding/Thinning 401
    10.5.4 Wafer Sawing 402
    10.5.5 Wafer Scribing 403
    10.5.6 Equipment 404
10.6 Die Attachment 405
    10.6.1 Epoxy 405
    10.6.2 Thermoplastics and Thermosets 406
    10.6.3 Solder 407
    10.6.4 Rework 407
    10.6.5 Die-Attach Equipment 408
10.7 Wirebonding 409
    10.7.1 Thermocompression Wirebonding 409
    10.7.2 Ultrasonic Wirebonding 409
    10.7.3 Thermosonic Wirebonding 410
    10.7.4 Ribbon Bonding 410
    10.7.5 Ball Bonding 410
    10.7.6 Wedge Bonding 411
    10.7.7 Wirebond Testing 411
    10.7.8 Tape-Automated Bonding 414
    10.7.9 Plasma Surface Treatment 415
10.8 Flip-Chip 417
    10.8.1 Wafer Bumping 419
    10.8.2 Fluxing 422
10.9 Package Sealing/Encapsulation/Coating 425
    10.9.1 Hermetic Package Sealing 426
    10.9.2 Hermetic Package Testing 426
    10.9.3 Nonhermetic Encapsulation 427
10.10 Package-Level Processes 429
    10.10.1 Lead Trim, Form, and Singulation 430
    10.10.2 Solder Ball Attach and Singulation 430
    10.10.3 Marking 430
10.11 State-of-the-Art Technologies 430
    10.11.1 3D and Stacked Die 430
    10.11.2 Radio Frequency (RF) Modules 431
    10.11.3 Microelectromechanical Systems (MEMS) and Microoptoelectromechanical Systems (MOEMS) 432
    10.11.4 Nanotechnology 434
10.12 Summary 435
References 435
Exercises 436

11. Design Considerations   437

J. P. Parkerson and L. W. Schaper

11.1 Introduction 437
11.2 Packaging and the Electronic System 437
    11.2.1 Packaging Functions 437
    11.2.2 System and Packaging Metrics 438
    11.2.3 System Constraints and Trade-Offs 440
    11.2.4 System Partitioning 442
11.3 Trade-Offs Among Packaging Functions 445
    11.3.1 Signal Wiring 445
    11.3.2 Power Distribution 452
    11.3.3 Thermal Management 455
    11.3.4 Interconnect Testing 456
11.4 Trade-Off Design Example 458
11.5 Product Development Cycle 460
    11.5.1 Traditional and Modified Product Cycles 461

x Contents

- 11.5.2 Market Analysis and Product Specification 463
- 11.5.3 Block Diagram and Partitioning 464
- 11.5.4 Technology Selection 464
- 11.5.5 ASIC/PCB/MCM Design 465
- 11.5.6 Thermal/Mechanical Design 466
- 11.5.7 Test Program Development 466
- 11.5.8 Manufacturing Tooling 467
- 11.5.9 Fabrication/Assembly 467
- 11.5.10 Characterization 467
- 11.5.11 Qualification 467
- 11.5.12 Product Introduction 468

11.6 Design Concepts 468
- 11.6.1 Component Overview 469
- 11.6.2 Schematic Overview 471
- 11.6.3 Design Viewpoint 474
- 11.6.4 Back Annotation 475
- 11.6.5 Simulation and Evaluation 476

11.7 PCB/MCM Board Design Process 477
- 11.7.1 PCB Design Flow 477
- 11.7.2 Librarian 477
- 11.7.3 Package 479
- 11.7.4 Layout 479
- 11.7.5 Fablink 481
- 11.7.6 Summary of Design Concepts 484

11.8 Summary 484
References 484
Bibliography 485
Exercises 485

12. Radio Frequency and Microwave Packaging 487

Fred Barlow and Aicha Elshabini

12.1 Introduction and Background 487
- 12.1.1 Nature of High-Frequency Circuits 487
- 12.1.2 Applications of High-Frequency Circuits 488
- 12.1.3 Basic Concepts 490

12.2 Transmission Lines 494
- 12.2.1 Transmission Line Modes 495
- 12.2.2 System-Level Transmission Lines 496
- 12.2.3 Planar Transmission Lines 499
- 12.2.4 Discontinuities 505

12.3 High-Frequency Circuit Implementation 510
- 12.3.1 Material Considerations 510
- 12.3.2 Microwave Monolithic Integrated Circuits 513
- 12.3.3 MIC Technologies 513

12.4 Lumped-Element Components 515
- 12.4.1 Capacitors 515
- 12.4.2 Inductors 516
- 12.4.3 Resistors and Terminations 518

12.5 Distributed Components 518
- 12.5.1 Impedance-Matching Devices 518
- 12.5.2 Filters 519
- 12.5.3 Power Dividers 520
- 12.5.4 Couplers 522

12.6 Simulation and Circuit Layout 523
12.7 Measurement and Testing 525
12.8 Frequency-Domain Measurements 525
- 12.8.1 Measurement Systems 525
- 12.8.2 Probing Hardware and Connectors 526

12.9 Time-Domain Measurements 527
12.10 Design Example 528
12.11 Summary 531
References 531
Exercises 535

13. Power Electronics Packaging 537

Alexander B. Lostetter and Kraig Olejniczak

13.1 Introduction 537
13.2 Semiconductor Power Device Technology 537
- 13.2.1 Ideal and Nonideal Power Switching 538
- 13.2.2 Power Diodes 540

- 13.2.3 Thyristors 541
- 13.2.4 Power Bipolar Junction Transistors 542
- 13.2.5 Power MOSFETs 542
- 13.2.6 Insulated Gate Bipolar Transistors 542
- 13.2.7 Static Induction Transistors (SITs) 543
- 13.2.8 Silicon Carbide Semiconductor Devices 543

13.3 Commercially Available Power Packages 547
- 13.3.1 Discrete Power Device Packages 547
- 13.3.2 Multichip Power Modules (MCPMs) and Completely Integrated Solutions 548
- 13.3.3 Thermal Performance of Commercial Packages [53–58] 552

13.4 Power Packaging Design Methodology 561
- 13.4.1 Overall System Design Philosophies 561
- 13.4.2 Substrate Selection 563
- 13.4.3 Baseplate and Heat Spreader Selection 565
- 13.4.4 Die-Attach Methods [62–64] 565
- 13.4.5 Wirebonding [65] 570
- 13.4.6 Thermal Design 573
- 13.4.7 Electromagnetic Interference (EMI) and Electromagnetic Compliance (EMC) 576
- 13.4.8 High-Temperature Power Electronics 576

13.5 Summary 577
References 577
Exercises 579

14. Multichip and Three-Dimensional Packaging 583

James Lyke

14.1 Introduction 583
- 14.1.1 Brief History of Multichip Packaging 583
- 14.1.2 Motivations for Multichip Packaging 585

14.2 Packaging Hierarchy and Taxonomy 588
- 14.2.1 Hierarchy 588
- 14.2.2 Anatomy of an MCM 588
- 14.2.3 Planar MCM Approaches 591

14.3 Three-Dimensional Systems 599
- 14.3.1 Defining Characteristics of 3D Systems 599
- 14.3.2 Die and Package Stacks 602
- 14.3.3 MCM Stacks 605
- 14.3.4 Folding Approaches 607

14.4 Options in Multichip Packaging 608
- 14.4.1 Yield/Known Good Die 608
- 14.4.2 Process Compatibility 609
- 14.4.3 Density Metrics in 2D and 3D Packaging 609
- 14.4.4 Wiring Density 609
- 14.4.5 Input/Output 610
- 14.4.6 Electrical Performance and Substrate Selection 613
- 14.4.7 Thermal Management 613
- 14.4.8 Testability 615
- 14.4.9 System in a Package Versus System on a Chip 615

14.5 Emerging Trends in Density Scaling 615
- 14.5.1 Method 1: For Regular and/or Low-Pincount Assemblies 617
- 14.5.2 Method 2: For Moderately Complex Pincount Assemblies 618
- 14.5.3 Method 3: For Moderately Complex Pincount Assemblies 619
- 14.5.4 Issues in Ultradense Packaging 619

14.6 Summary 621
References 622
Exercises 623

15. Packaging of MEMS and MOEMS: Challenges and a Case Study 625

Ajay P. Malshe, Volkan Ozguz and John Patrick O'Connor

15.1 Introduction 625
15.2 Background 625
- 15.2.1 Mixed Signals, Mixed Domains, and Mixed Scales Packaging: Toward the Next-Generation

           Application-Specific
           Integrated Systems   626
    15.2.2 Microelectromechanical
           Systems   626
15.3 Challenges in Mems Integration
     628
     15.3.1 Release and Stiction   630
     15.3.2 Dicing   631
     15.3.3 Die Handling   631
     15.3.4 Wafer-Level Encapsulation
            631
     15.3.5 Stress   632
     15.3.6 Outgassing   632
     15.3.7 Testing   633
     15.3.8 State-of-the-Art in MEMS
            Packaging   633
     15.3.9 Future Directions   635
15.4 Packaging Considerations and
     Guidelines Related to the Digital
     Micromirror Device   636
     15.4.1 Introduction and Background
            to MOEMS and Particularly
            DMD Devices   636
     15.4.2 Parameters Influencing DMD
            Packaging   637
     15.4.3 DMD Package Design   640
     15.4.4 DMD Hermetic Package
            Assembly   646
15.5 Future Packaging Challenges
     647
     References   648
     Exercises   650

16. Reliability Considerations   651

Richard Ulrich

16.1 Introduction   651
     16.1.1 Definitions   651
     16.1.2 Patterns of Failure
            653
     16.1.3 Coverage in This Chapter
            654
16.2 Failure Mechanisms   655
     16.2.1 Corrosion   656
     16.2.2 Mechanical Stress   659
     16.2.3 Electrical Stress   660
     16.2.4 Techniques for Failure
            Analysis   660
16.3 Accelerated Testing   661
     16.3.1 Accelerated Environmental
            Testing   663

     16.3.2 Electrostatic Discharge
            Accelerated Testing   666
     16.3.3 Other Accelerated Tests
            666
     16.3.4 Test Structures   667
16.4 Reliability Metrology   668
     16.4.1 Failure Rate, MTBF, and FITs
            668
     16.4.2 Reliability Functions   668
     16.4.3 Weibull Distribution   674
     16.4.4 Normal Distribution   677
     16.4.5 Failure Distributions and the
            Bathtub Curve   680
16.5 Failure Statistics for Microelectronic
     Systems   681
     16.5.1 Predicting Failure in
            Components That Have
            Multiple Failure Modes
            683
16.6 Industrial Practice of Reliability
     Science for Microelectronics
     684
     Bibliography   684
     Exercises   684

17. Cost Evaluation and Analysis   691

Terry R. Collins, Scott J. Mason, and Heather Nachtmann

17.1 Introduction   691
17.2 Product Cost   691
     17.2.1 Direct Costs   692
     17.2.2 Indirect Costs   692
     17.2.3 Traditional Volume-Based
            Costing   692
     17.2.4 Activity-Based Costing
            694
17.3 Break-even Analysis   696
     17.3.1 Linear Break-even Analysis
            696
     17.3.2 Piecewise Linear Break-even
            Analysis   698
17.4 Learning Curve Relationships
     698
     17.4.1 Determining Exponent
            Values for Improvement Rates
            700
     17.4.2 Learning Curve Examples
            702
17.5 Forecasting Models   703

	17.5.1	Mean-Squared Error (MSE) 704
	17.5.2	Mean Absolute Deviation (MAD) 704
	17.5.3	Mean Percentage Error (MPE) 704
	17.5.4	Mean Absolute Percentage Error (MAPE) 705
	17.5.5	Moving Average 705
	17.5.6	Forecasting Sales Based on Historical Data 706
	17.5.7	Exponential Smoothing 707
	17.5.8	Least-Squares Regression 712
17.6	Comparative Analysis 714	
	17.6.1	Capital Project Selection and Evaluation 715
	17.6.2	Replacement Analysis 716
17.7	Sensitivity Analysis 717	
	17.7.1	Single-Parameter Sensitivity Analysis 718
	17.7.2	Optimistic–Pessimistic Sensitivity Analysis 719
17.8	Summary 720	
	References 721	
	Exercises 721	

18. Analytical Techniques for Materials Characterization 725

Emily A. Clark, Ingrid Fritsch, Seifollah Nasrazadani, and Charles S. Henry

18.1	Overview 725	
18.2	X-Ray Diffraction 725	
	18.2.1	Summary 728
	18.2.2	Basic Principles 728
	18.2.3	Instrumentation 729
	18.2.4	Practical Considerations and Applications 731
18.3	Raman Spectroscopy 734	
	18.3.1	Summary 734
	18.3.2	Basic Principles 735
	18.3.3	Instrumentation 735
	18.3.4	Practical Considerations and Applications 736
18.4	Scanning Probe Microscopy 740	
	18.4.1	Summary 740
	18.4.2	STM Principles and Instrumentation 740
	18.4.3	SFM Principles and Instrumentation 741
	18.4.4	Practical Considerations and Applications 742
18.5	Scanning Electron Microscopy and Energy Dispersive X-ray Spectroscopy 744	
	18.5.1	Summary 744
	18.5.2	Basic Principles 745
	18.5.3	Instrumentation 746
	18.5.4	Practical Considerations and Applications 748
18.6	Confocal Microscopy 750	
	18.6.1	Summary 750
	18.6.2	Basic Principles 750
	18.6.3	Instrumentation 750
	18.6.4	Practical Considerations and Applications 751
18.7	Auger Electron Spectroscopy 752	
	18.7.1	Summary 752
	18.7.2	Basic Principles 753
	18.7.3	Instrumentation 757
	18.7.4	Practical Considerations and Applications 759
18.8	X-ray Photoelectron Spectroscopy 766	
	18.8.1	Summary 766
	18.8.2	Basic Principles 766
	18.8.3	Instrumentation 769
	18.8.4	Practical Considerations and Applications 770
18.9	Secondary Ion Mass Spectrometry 775	
	18.9.1	Summary 775
	18.9.2	Basic Principles 776
	18.9.3	Instrumentation 778
	18.9.4	Practical Considerations and Applications 782
	References 786	
	Exercises 790	

Index 793

Preface to the Second Edition

When the first edition of this text was written back in 1997, there was already a rich history of teaching and research in electronic packaging at the University of Arkansas. As its editor, Dr. William Brown, described in the preface to that book, there was such a concentration of expertise available locally across many facets of packaging that it seemed important to set it on paper as both a tool for us to use in teaching our Advanced Electronic Packaging courses as well as for the rest of the academic community. The resulting work, *Advanced Electronic Packaging—With Emphasis on Multichip Modules* proved to be a major success as both a textbook and as a desk reference for the practicing engineer.

Since that time, microelectronic packaging, as well as the research and teaching programs at U of A, have evolved and expanded. The growth of the university's programs has, we hope, paralleled the changes in the industry as a whole, since they are reflected in this new edition. We still maintain a strong interdisciplinary teaching program in packaging in the form of two consecutive graduate-level courses. As before, experts in the various topics teach from 1 to 3 weeks each semester, enabling the students to be exposed to the latest developments through the shortest possible path. These professors represent the departments of Electrical, Mechanical, Chemical and Industrial Engineering as well as Chemistry, Physics and, new since the first edition, Microelectronics-Photonics.

The changes have come about in the number of topics covered, with several new ones added, and the way that the old topics are covered, a natural consequence of both experience in teaching and the evolution of the sciences involved. Specifically, the byline from the first edition, "With Emphasis on Multichip Modules," is gone, since that term is no longer in widespread use. Of course, the concept is not gone; it has just morphed into other names such as "System in a Package" and "3D Packaging," which are both covered in their new forms in this new edition. Organic and ceramic substrates are now separate chapters, and there are new chapters on integrated passives, RF packaging, assembly, and cost evaluation. Some organizational modifications were also made, such as moving processing technologies closer to the front so that the student has an understanding of how to make the objects when they are later described in detail. More industrial experts from outside the university have been enlisted in order to bring a more pragmatic flavor to the coverage.

This was written from the start to be a teaching text, with many examples and exercises. As we continue to teach the courses during the writing phase, the students are providing continuous feedback on some of the chapters as the material is consolidated. The editors and chapter authors hope that this edition will provide a valuable teaching and reference tool in a vital and fast-moving area of technology.

Fayetteville, Arkansas
September, 2005

RICHARD K. ULRICH
WILLIAM D. BROWN

Contributors

EDITORS

DR. RICHARD ULRICH
Department of Chemical Engineering
University of Arkansas
Fayetteville, Arkansas

DR. WILLIAM D. BROWN
Department of Electrical Engineering
University of Arkansas
Fayetteville, Arkansas

CONTRIBUTORS

DR. SIMON ANG
Department of Electrical Engineering
University of Arkansas
Fayetteville, Arkansas

DR. FRED BARLOW
Department of Electrical Engineering
University of Arkansas
Fayetteville, Arkansas

DR. WILLIAM D. BROWN
Department of Electrical Engineering
3217 Bell Engineering Center
University of Arkansas
Fayetteville, Arkansas

DR. SUSAN BURKETT
Department of Electrical Engineering
University of Arkansas
Fayetteville, Arkansas

EMILY CLARK
Department of Chemistry and Biochemistry
University of Arkansas
Fayetteville, Arkansas

DR. TERRY R. COLLINS
Department of Industrial Engineering
Department
Texas Tech University
Lubbock, Texas

DR. RICK J. COUVILLION
Department of Mechanical Engineering
University of Arkansas
Fayetteville, Arkansas

DR. AICHA ELSHABINI
Department of Electrical Engineering
University of Arkansas
Fayetteville, Arkansas

DR. INGRID FRITSCH
Department of Chemistry and Biochemistry
University of Arkansas
Fayetteville, Arkansas

DR. CHARLES HENRY
Department of Chemistry
Colorado State University
Fort Collins, Colorado

DR. ALEXANDER B. LOSTETTER
Arkansas Power Electronics International, Inc.
Fayetteville, Arkansas

JAMES LYKE
AFRL/VSSE
Kirtland AFB, New Maxico

DR. AJAY P. MALSHE
Department of Mechanical Engineering
University of Arkansas
Fayetteville, Arkansas

Dr. Scott J. Mason
Department of Industrial Engineering
University of Arkansas
Fayetteville, Arkansas

Dr. Heather Nachtmann
Department of Industrial Engineering
University of Arkansas
Fayetteville, Arkansas

Dr. Hameed A. Naseem
Department of Electrical Engineering
University of Arkansas
Fayetteville, Arkansas

Dr. Seifollah Nasrazadani
Materials Science and Engineering Department
University of North Texas
Denton, Texas

John Patrick O'Connor
Texas Instruments Digital Imaging
Plano, Texas

Dr. Kraig J. Olejniczak
Valparaiso University
Valparaiso, Indiana

Dr. Volkan Ozguz
Chief Technology Officer
Irvine Sensors Corporation
Costa Mesa California

Dr. James Patrick Parkerson
Department of Computer Science and Computer Engineering
University of Arkansas
Fayetteville, Arkansas

Dr. Tarak A. Railkar
Skyworks Solutions, Inc.
Irvine California
currently with:
Texas Instruments, Inc.
Dallas, Texas

Dr. Leonard Schaper
Department of Electrical Engineering
University of Arkansas
Fayetteville, Arkansas

Dr. William F. Schmidt
Department of Mechanical Engineering
University of Arkansas
Fayetteville, Arkansas

Richard C. Snogren
Bristlecone LLC
Littleton, Colorado

Dr. Richard Ulrich
Department of Chemical Engineering
University of Arkansas
Fayetteville, Arkansas

Robert Warren
Skyworks Solutions, Inc.
Irvine, California

Acronyms

ABC	activity-based costing
AC	alternating current
ADP	atmospheric downstream plasma
AES	auger electron spectroscopy
ALIVH	any layer interstitial via hole
AlN	aluminum nitride
ALU	arithmetic logic unit
AMB	active metal brazing
ANSI	American National Standards Institute
AOI	automated optical inspection
APCVD	atmospheric pressure chemical vapor deposition
AQL	acceptable quality level
ARAES	angle-resolved auger electron spectroscopy
ASCII	Standard American Code for Information Interchange
ASIC	application-specific integrated circuit
ASM	American Society for Metals *or* Association for Systems Management
ASME	American Society of Mechanical Engineers
ASTM	American Society for Testing and Measurements
ATC	assembly test chip
ATCM	advanced thermal conduction module
ATPG	automatic test pattern generation
ATT	American Telephone and Telegraph
AW	artwork
BBUL	bumpless build-up layer
BCB	benzocyclobutene
BDD	binary decision diagram
BE	backscattered electron
BECW	buried engineering change wires
BGA	ball grid array
BiCMOS	bipolar complementary metal–oxide–semiconductor
BILBO	built-in logic block organization
BIST	built-in self-test
BJT	bipolar junction transistor
BLM	ball limiting metallurgy
BSCCO	bismuth–strontium–calcium–copper–oxide
BSE	backscattered electron
BT	bismaleinide triazine
C4	controlled-collapse chip connect
CAD	computer-aided design
CAE	computer-aided engineering
CAF	conductive anodic filament

CAM	computer-aided manufacturing
CBIC	cell-based ASIC
CBGA	ceramic ball grid array
CCD	charge-coupled device
CCGA	ceramic column grid array
CDR	critical design review
CerDIP	Ceramic dual-in-line package
CET	conducted emissions test
CF	coupling fault
CF-4	tetrafluoride
CFD	computational fluid dynamics
CHA	concentric hemispherical analyzer
CIC	copper–invar–copper
CINDAS	Center for Information and Numerical Data Analysis and Synthesis
CISC	complex instruction set computing
CLA	centerline average
CM	contract manufacturer
CMA	cylindrical mirror analyzer
CMOS	complementary metal–oxide–semiconductor
CNC	computer numerically controlled
COB	chip on board
COF	chip on flex
CPLD	complex programmable logic device
CPU	central processor unit
CQFP	ceramic quad flat pack
CSP	chip-scale package *or* chip size package
CSTP	circular self-test path
CTE	coefficient of thermal expansion
CTL	computational tree logic
CVD	chemical vapor deposition
DARPA	defense advanced research projects agency
DBC	direct-bonded copper
DC	direct current
DCA	direct chip attach
DDQ	design data questionnaire
DDR–LVCMOS	double data rate–low-voltage complimentary metal–oxide–semiconductor
DF	dissipation factor
DFT	design for testability *or* design for test
DGEBA	diglycidyl ether of bispenol-A
DI	deionized
DIP	dual-in-line package
DLC	diamondlike carbon
DLP	digital light processing
DM	direct metallization
DNP	distance to neutral point
DRAM	dynamic random-access memory
DRC	design rule check
DSP	digital signal processing (or processor)
DUT	device under test
EBCDIC	extended binary coded decimal interchange code
EC	engineering change

ECL	emitter coupled logic
ECPI	elastomeric conductive polymer interconnection
ECR	electron cyclotron resonance
EDA	electronic digital analyzer *or* electronic differential analyzer *or* electronic design automation
EDAX	energy dispersive X-ray
EDDM	electronic design data model
EDS	energy dispersive spectroscopy
EDTA	ethylene diamine tetraacetic acid
EDX	energy-dispersive X-ray spectroscopy
EFO	electronic flame-off
EIA	Electronic Industry Association
EIAJ	Electronic Industry Association of Japan
ELO	epitaxial liftoff
EMC	electromagnetic compliance *or* electromagnetic compatibility
EMI	electromagnetic interference
EMS	electronic manufacturing service
ENIG	electroless nickel immersion gold
ESCA	electron spectroscopy for chemical analysis
ESD	electrostatic discharge
ESL	equivalent series inductance
ESR	equivalent series resistance
FE	finite element
FEA	finite-element analysis
FEM	finite-element method
FEP	fluorinated ethylenepropolgene
FIB	focused ion beam
FIT	failure in time
FOM	figure of merit
FPGA	field-programmable gate array
FSM	finite state machine
FWHM	full width at half maximum
GALPAT	Galloping patern
GPS	Global Positioning System
GSG	ground–signal–ground
GTO	gate turn-off
GUI	graphics user interface
HASL	hot air solder level
HAST	highly accelerated stress test
HBT	heterojunction bipolar transistor
HDI	high-density interconnect
HDL	hardware description language
HDS	heat dissipation surface
HEMT	high-electron-mobility transistor
HEPA	high-efficiency particulate air
HgBCCO	mercury–barium–calcium–copper–oxide
HIC	humidity indicator card
HiDEC	high-density electronics center
HIPP	highly integrated packaging and processing
HOPG	highly oriented polytropic graphite

HTCC	high-temperature cofired ceramic
HTMOS	high-temperature metal–oxide–semiconductor
HTSC	high-temperature superconductor
IBAD	ion-beam-assisted deposition
IC	integrated circuit
ICPMS	inductively coupled plasma mass spectroscopy
IEDM	International Electron Devices Meeting
IEEE	Institute of Electrical and Electronics Engineers
IEPS	International Electronics Packaging Society
IGBT	insulated gate bipolar transistor
ILB	inner lead bonding
IMAPS	International Microelectronics and Packaging Society
IMPS	interconnected Mesh Power System
IMS	insulated metal substrate
I/O	input/output
IPC	Institute for Interconnecting and Packaging Electronic Circuits
IPN	interpenetrating polymer network
IR	infrared
IRPS	International Reliability Physics Symposium
ISHM	International Society of Hybrid Microelectronics
ISO	International Organization for Standardization
ITRS	International Technology Roadmap for Semiconductors
JCPDS	Joint Committee on Powder Diffraction Standards
JEDEC	Joint Electron Device Engineering Council
JESD	Joint Electronics Standard Description
JFET	junction field-effect transistor
JTAG	joint testability action group
KGD	known good die
LAN	local area network
LCCC	leaded ceramic chip carrier
LCM	large core memory *or* least common multiple
LDI	laser direct imaging
LDMS	laser desorption mass spectroscopy
LED	light-emitting diode
LEED	low-energy electron diffraction
LFSR	linear feedback shift register
LGA	land grid array
LMI	laminated microinterconnect
LMIG	liquid metal ion gun
LNA	low-noise amplifier
LOC	lead on chip
LPCVD	low-pressure chemical vapor deposition
LPI	liquid photoimageable
LRRM	line reflect reflect match
LRU	least replaceable unit
LS	low-power Schottky (with reference to TTL)
LSI	large-scale integration
LSSD	least sensitive scan design
LTCC	low-temperature cofired ceramic
LVS	layout versus schematic

MAD	mean absolute deviation
MAPE	mean absolute percentage error
MBB	moisture barrier bags
MBGA	metal ball grid array
MCM	Multichip module
MCM-C	Multichip module—ceramic
MCM-D/C	multichip module—deposited/ceramic
MCM-D	multichip module—deposited
MCM-L	multichip module—laminate
MCM-Si	multichip module—silicon
MCP	multichip package *or* multicomponent package
MCPM	multichip power module
MDL	minimum detection level
MEMS	microelectromechanical system
MESFET	metal–semiconductor field-effect transistor
MFM	magnetic force microscope
MGA	masked gate array (as it applies to ASIC)
MIC	microwave integrated circuit
MIPS	millions of instructions per second
MLB	multilayer board
MLC	multilayer ceramic
MLTF	multilayer thin film
MMC	metal matrix composite
MMIC	microwave monolithic integrated circuit
MMU	memory management unit
MOCVD	metal organic chemical vapor deposition
MOEMS	microoptoelectromechanical systems
MOS	metal–oxide–semiconductor
MOSFET	metal–oxide–semiconductor field-effect transistor
MPE	mean percentage error
MPECVD	microwave power enhanced chemical vapor deposition
MRP	mass resolving power
MSE	mean square error
MSI	medium-scale integration
MSL	moisture sensitivity level
MTBF	mean time between failures *or* mean time before failure
MTTF	mean time to failure
NASA	National Aeronautics and Space Administration
NCREPT	National Center for Reliable Electric Power Transmission
NEC	Nippon Electric Company
NEMI	National Electronics Manufacturing Initiative
NHE	normal hydrogen electrode
NIST	National Institute of Standards and Technology
NMOS	N-channel metal–oxide–semiconductor
NPSF	Neighborhood Pattern Sensitive Fault
NRE	nonrecurring engineering
NTRS	National Technology Roadmap for Semiconductors
NTT	Nippon Telephone and Telegraph
OEM	original equipment manufacturer
OLB	outer lead bonding
OSP	organic solder preservative

PAC	power amplifier contorl
PBGA	plastic ball grid array
PC	personal comuter
PCB	printed circuit board
PCMCIA	Personal Computer Memory Card International Association
PCR	polyerase chain reactions
PCTFE	polychlorotrifluoroethylene
PDA	personal digital assistant
PDE	partial differential equation
PDMS	poly(dimethylsiloxane)
PDR	preliminary design review
PDS	power distribution system
PECVD	plasma-enhanced chemical vapor deposition
PFBCC	polymeric film-based chip carrier
PGA	pin grid array
PHEMT	P-high electron mobility transistor
PLCC	plastic leaded chip carrier
PLD	programmable logic device
PMMA	poly(methyl methacrylate)
PMT	photomultiplier tube
PO	purchase order
PODEM	path oriented decision making
ppm	parts per million
PPP	polypropylene
PPQ	polyphenylquinoxaline
PQFP	plastic quad flatpack
PSG	phosphosilicate glass
PSPD	position-sensitive photodiode
PTF	polymer thick film
PTFE	polytetrafluoroethylene
PTH	plated through hole
PVC	polyvinyl chloride
PVD	physical vapor deposition
PWB	printed wiring board
PWM	pulse width modulation
QFP	quad flatpack
QMA	quadrapole mass analysis
RAM	random-access memory
RCC	resin-coated copper
RCG	Roadmap Coordinating Group
REFDES	reference designation
RF	radio frequency
RFA	retarding field analyzer
RH	relative humidity
RIE	reactive ion etching
RIM	reactive ion milling
RISC	reduced instruction set computing
ROM	read-only memory
RSF	relative sensitivity factor
RTL	register transfer language
SAF	stuck-at-fault

SAM	scanning auger microscopy
SCI	serial communication interface
SDR–LVMOS	single data rate—low-voltage complementary metal–oxide–semiconductor
SE	secondary electrons
SECM	Scanning electrochemical microscopy
SEM	Scanning electron microscope
SERS	Surface-enhanced Raman spectroscopy
SFM	Scanning force microscope
SHE	standard hydrogen electrode
SIA	Semiconductor Industry Association
SIMS	secondary ion mass spectroscopy
SIP	system in a package
SISO	Serial in–serial out
SIT	static induction transistor
SLCC	Stackable leadless chip carrier
SLICC	slightly larger than integrated circuit carrier
SLT	solid logic technology
SM	solids modeling
SMA	subminiature type A
SMD	surface-mount device
SMP	surface-mount package
SMT	surface-mount technology *or* surface-mount transistor
SNA	scalar network analyzer
SNMS	sputtered neutral mass spectroscopy
SOAC	system on a chip
SOAP	system on a package
SOC	system on a chip
SOG	spin on glass
SOI	silicon on insulator
SOIC	small outline integrated circuit
SOLT	short open load thru
SOP	system on a package *or* small outline package
SOT	small outline transistor
SPM	scanning probe microscopy
SRAM	static random-access memory
SRC	Semiconductor Research Corporation
SRF	self-resonant frequency
SSI	small-scale integration
SSMS	spark source mass spectroscopy
STD	simple to design
STM	scanning tunneling microscopy
TAB	tape-automated bonding
TBCCO	thallium–barium–calcium–copper–oxide
TBGA	tape ball grid array
TCC	temperature coefficient of capacitance
TCE	thermal coefficient of expansion
TCK	test clock
TCM	thermal conduction module
TCR	temperature coefficient of resistance
TDI	test data input
TDO	test date output

TDR	time-domain reflectometry	
TDT	time-domain transmission	
TE	transverse electric	
TEM	transverse electromagnetic *or* transmission electron microscopy	
TEOS	tetraethosysilane	
TF	thin film *or* transition fault	
TFM	thick-film multilayer	
TM	transverse magnetic	
TO	transistor outline	
TOF	time of flight	
TRAM	transmit–receive antenna module	
TRL	transmission reflect line	
TSOP	thin small outline package	
TTL	transistor–transistor logic	
TTRAN	tape transfer	
TWG	Technology Working Group	
UBM	under-bump metallurgy	
UHV	ultra high vacuum	
UPH	units per hour	
UTIC	Universal Test Interface Code	
UUT	unit under test	
UV	ultraviolet	
VBC	volume-based costing	
VCR	voltage coefficient of resistance	
VCSEL	vertical cavity surface emitting laser	
VHDL	Verilog Hardware Description Language	
VHSIC	very high speed integrated circuit	
VLSI	very large scale integration	
VME	VersaModule Eurocard	
VNA	vector network analyzer	
VSWR	voltage standing-wave ratio	
WDS	wavelength dispersive spectroscopy	
WLAN	wireless local area network	
WL-CSP	wafer-level–chip-scale packaging	
WLP	wafer-level packaging	
WLS	wafer-level stacking	
WSI	wafer-scale integration	
WSP	wafer-scale package	
XPS	X-ray photoelectron spectroscopy	
XRD	X-ray diffractometry	
YBCO	yttrium–barium–copper–oxide	
YSZ	yttrium-stabilized zirconia	

Chapter 1

Introduction and Overview of Microelectronic Packaging

W. D. BROWN

1.1 INTRODUCTION

An exact date for the advent of *electronic packaging* is difficult, if not impossible, to establish due to the diversity of opinion on what constitutes an electronic package. However, there is no doubt that, in general, nearly all structures that are involved in the generation, transmission, and utilization of electrical signals are packaged in some manner, even if the packaging method and material are somewhat primitive. For example, the insulation surrounding a wire is technically a package. Thus, an *electronic package* might be defined as that portion of an electronic structure that serves to protect an electronic/electrical element from its environment and the environment from the electronic/electrical element. However, in addition to providing encapsulation for environmental protection, a package must also allow for complete testing of the packaged device and a high-yield method of assembly to the next level of integration [1–3].

With the commercialization of the silicon transistor in the 1950s, electronic packaging technology development took on a new level of intensity resulting in the proliferation of standard electronic packages, a few of which are shown in Figure 1.1. However, for several years, the performance of packaged electronic components was limited by the component itself and was impacted little, if any, by packaging technology. At some later date, parasitics associated with the package housing the component began to adversely affect the performance of the device. Consequently, packaging of electronic components, in particular integrated circuits (ICs), became the focus of an intense developmental effort and continues to challenge the microelectronics industry today [4]. Essentially, the point has been reached where advancement in IC performance now drives packaging technology. No matter what approach is taken to package an IC, the primary objective is to realize maximum performance from the IC, and, thereby, electronic systems, by minimizing the impact of the package. It is the intent of this book to discuss all aspects of electronic packaging that impact the performance of electronic devices that they house.

This book addresses all aspects of electronic packaging beginning with an introduction and overview of packaging including definitions, functions, and classifications of microelectronics packaging. Subsequent chapters provide a thorough treatment of technical subject

---

Advanced Electronic Packaging, Second Edition, Edited by Richard K. Ulrich and William D. Brown
Copyright © 2006 the Institute of Electrical and Electronics Engineers, Inc.

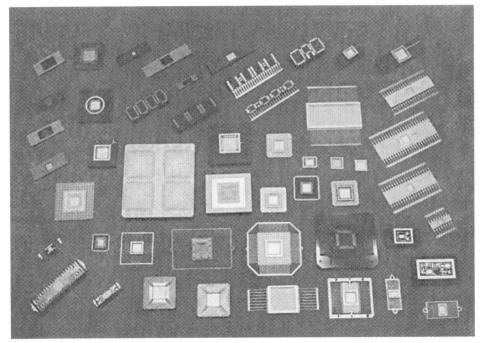

Figure 1.1 Photograph of a number of standard electronic packages (courtesy of Kyocera).

matter important to both the package designer engaged in developing custom packages and the design engineer who must choose from the available standardized packages. The fundamental topics covered include packaging materials and applications, processing technologies, electrical fundamentals, design considerations, thermal considerations, mechanical considerations, package assembly, reliability, and cost evaluation and analysis. Other chapters address electrical modeling and simulation, discrete and integrated passive devices, radio frequency (RF) and microwave packaging, power electronics packaging, multichip and three-dimensional (3D) packaging, and microelectromechanical system (MEMS) and optoelectronic packaging. The final chapter discusses analytical techniques for materials characterization, a subject of considerable interest since materials, and particularly, new materials, have played such an important role in the evolution of electronic packaging technology.

1.2 FUNCTIONS OF AN ELECTRONIC PACKAGE

In the previous section, a very simple definition of an electronic package was given. If we restrict the definition of an electronic package to the housing and interconnection of integrated circuits (also referred to as ICs, silicon chips, chips, or die) to form an electronic system, then we can restrict the discussion to a subset of the multitude of electronic packages that would otherwise have to be considered. For this reduced set of packages, the functions that the package must provide include a structure to physically support the chip, a physical housing to protect the chip from the environment, an adequate means of removing heat generated by the chips or system, electrical connections to allow signal and power access to and from the chip, and a wiring structure to provide interconnection between the chips

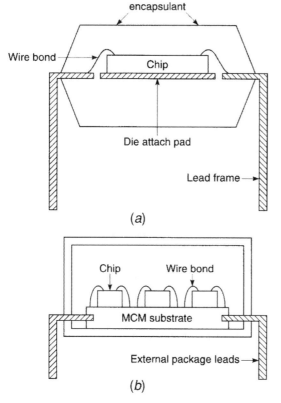

Figure 1.2 (*a*) Single- and (*b*) multichip packages illustrating signal distribution, heat dissipation, power distribution, and circuit support and protection (signal and power distribution are accomplished through leads and wire bonds, heat dissipation is accomplished through leads and chip support, and support and protection are accomplished through the lead frame, substrate, and external package).

of an electronic system [5]. Thus, the package must provide for:

- Signal distribution
- Heat dissipation
- Power distribution
- Circuit support and protection

Figure 1.2 illustrates these various functions for both single-chip and multichip packaging schemes.

In addition to providing the four basic requirements listed above, an electronic package must also function at its designed performance level while still allowing for a product that is high quality, reliable, serviceable, and economical [5]. The rearrangement or addition of functional features, such as upgrading the memory of a computer, is also a desirable feature of an electronic packaging technology.

1.3 PACKAGING HIERARCHY

Typical electronic systems are made up of several layers or levels of packaging, and each level of packaging has distinctive types of interconnection devices associated with it. One way in which this hierarchy of interconnection levels can be divided is as follows:

Level 0   Gate-to-gate interconnections on a monolithic silicon chip.

Level 1   Packaging of silicon chips into dual-in-line packages (DIPs), small outline integrated circuit (SOICs), chip carriers, multichip packages, and so on, and the chip-level interconnects that join the chip to the lead frames. Occasionally, this level is skipped when tape-automated bonding (TAB) or chip on-board (COB) technologies are utilized.

Level 2   Printed wiring board (PWB), also referred to as a printed circuit board (PCB), level of interconnection. Printed conductor paths connect the device leads of components to PWBs and to the electrical edge connectors for off-the-board interconnection.

Level 3   Connections between PWBs. This may include PWB-to-PWB interconnections or card-to-motherboard interconnections.

Level 4   Connections between two subassemblies. For example, a rack or frame may hold several shelves of subassemblies that must be connected together to make up a complete system.

Level 5   Connections between physically separate systems such as host computer to terminals, computer to printer, and so on.

The various levels of interconnection are illustrated in Figure 1.3.

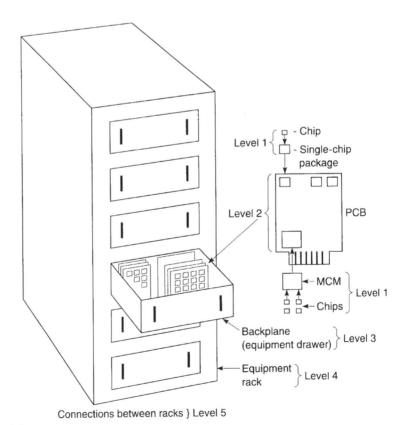

Figure 1.3 Electronic packaging hierarchy.

Gate-to-gate interconnections categorized as level 0 are formed during IC fabrication. They will not be discussed in this text. Persons interested in learning more about this level of interconnection should consult one of the many books on IC fabrication. Likewise, levels 3 through 5 are not pertinent to IC packaging, so they will not be discussed either. Although not considered a part of the packaging hierarchy, die attach (or die bonding) is a crucial step in packaging technology, so it is discussed prior to addressing levels 1 and 2.

1.3.1 Die Attach

An IC must be mounted on a substrate or metal lead frame by a die attach material, which permits heat conduction while assuring mechanical stability. The die attach may also provide for electrical grounding if the material used to secure the chip is sufficiently conductive. The three primary types of die attach materials used are soldering (both soft and hard solders can be used), which is also referred to as eutectic bonding, metal-filled polymers (epoxies), and metal-filled glasses. Hard solders, such as a silicon–gold alloy, can present reliability problems due to mismatched thermal expansion coefficients. Soft solders generally do not suffer from this problem because they exhibit plastic flow, which prevents chips from seeing high stresses. However, they can result in failure due to fatigue and creep.

Metal-filled epoxies and polyimides are another alternative for die bonding. As a general rule, they are less expensive, produce lower stress levels, and require significantly lower processing temperatures. Metal loading, generally silver, yields acceptable values of electrical and thermal conductivity, although normal thermal cycling during use can cause drift in these parameters over time. Polyimides require higher processing temperatures for complete curing but also exhibit a higher stability at high temperatures than materials containing metal fillers.

Glass adhesives have also been used for die attach. The primary problems with glasses are the excessively high processing temperatures and the possible presence of oxidizing agents, which can corrode the IC. Silver-loaded glass adhesives have been used in ceramic packages.

1.3.2 First-Level Interconnection

First-level packaging (or interconnection) refers to the technology required to get electrical signals into and out of a single transistor or IC; in other words, the connections required between the bonding pads on the IC and the pins of the package. This is generally accomplished by wire bonding, flip-chip bonding, or TAB, all of which are illustrated in Figure 1.4.

1.3.2.1 Wire Bonding

Wire bonding is the oldest method used for first-level interconnection and is still the dominant method used today, particularly for chips with a moderate number of inputs/outputs (I/O) (i.e., 200). This technique involves connecting gold or aluminum wires between the chip bonding pads, located around the periphery of the chip, and contact points on the package. Although this process has been automated for many years, it is still time consuming because each wire, requiring two bonding operations, must be attached individually. Limitations of wire bonding include the requirement for minimum spacing between adjacent bonding sites to provide sufficient room for the bonding tool, the number of bonding pads

6 Chapter 1 Introduction and Overview of Microelectronic Packaging

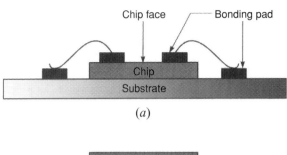

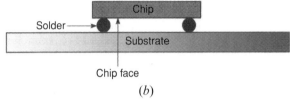

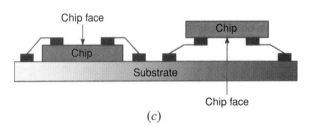

Figure 1.4 Illustrations of (*a*) wire, (*b*) flip-chip, and (*c*) tape-automated bonding.

that can be located around the periphery of the chip, signal delay, and crosstalk between adjacent wires.

1.3.2.2 Flip-Chip Bonding

In this interconnection technology, the chip is mounted upside down onto a carrier, module, or PWB. Electrical connection is made via solder bumps much like those used in TAB, which is discussed in the next section. The solder bumps are located over the surface of the chip in a somewhat random pattern or an array so that periphery limitation, such as that encountered in wire bonding, does not limit the I/O capability. The I/O density is primarily limited by the minimum distance between adjacent bonding pads on the chip and the amount of chip area that can be dedicated to interconnection. Additionally, the interconnect distance between chip and package is minimized since bumps can essentially be located anywhere on the chip. Although this technique is attractive for use in multichip packaging technology because chips can be located very close together, fatigue of solder joints due to thermal expansion mismatch of the chip–bond–substrate, heat removal from the back of the chip, and difficulty inspecting the solder joints after the chip has been attached to the substrate offer special challenges to the packaging specialist.

1.3.2.3 Tape-Automated Bonding

Tape-automated bonding, developed in the early 1970s, is a third method for accomplishing first-level interconnection. It is most often used with chip carriers or PWBs. In this technique,

ICs are first mounted on a flexible polymer tape, such as polyimide, containing repeated, flat, wide copper interconnection patterns formed lithographically from a metal laminate. Each pad on the IC is aligned to a metal interconnection stripe on the tape and attachment is effected by thermocompression bonding. All bonds are formed to the IC at the same time by a process called gang bonding. This is referred to as inner lead bonding (ILB). At this point, the IC can be tested and burned-in, allowing poorly bonded or defective chips to be eliminated prior to packaging. Good chips are then outer lead bonded (OLB) to lead frames or PWBs using the thermocompression process. The shape of the interconnections and the use of copper with its lower resistivity results in low inductance and low resistance that minimizes signal distortion. However, TAB requires the use of complex metallurgy, multilayer solder bumps either on the tape or IC or both in order to effect a bond. Generally, the bumps utilize gold or copper as the primary constituent, along with titanium or tungsten as a diffusion barrier to prevent alloying. Finally, a TAB tape can only be used for a chip and package that matches its interconnection pattern. Thus, each TAB tape is, in effect, a custom tape. However, TAB will likely continue to gain popularity as a method for bonding chips directly to a PWB and for large chips with high I/O requirements.

1.3.3 Package Lid and Pin Sealing

Lid and pin sealing generally utilize materials that are similar or identical to those used elsewhere in packaging technology. Low-melting-temperature glasses are used for sealing both the lid and I/O pins in hermetic packages. Required properties of these sealing glasses include (1) the ability to form a hermetic seal, (2) temperatures required to effect the seal must be compatible with the other materials in the package, (3) must adhere well to the materials being sealed, (4) must have a thermal coefficient of expansion compatible with the materials being sealed, and (5) the sealing glass must be electrical insulating. The primary problems encountered with glass seals are low strength and brittleness.

Metal hermetic lid seals are effected by low-melting-temperature brazing, alloy soldering, or welding. Solders used for lid sealing are dictated by the temperature hierarchy of the processes that precede and follow the sealing operation, the required seal strength, and cost. It is less desirable than brazing because of its lower strength, tendency to become embrittled by intermetallic formation, and the use of fluxes. Brazing, which consists of reflowing a preform of a eutectic, for example, Au–Sn (80–20), yields a stronger, more corrosion-resistant seal, although it is a little more difficult to perform consistently. On the other hand, welding is still the most popular method of realizing high-reliability hermetic seals because of the high yield and the fact that the high-current pulses used to effect the weld produce only local heating.

1.3.4 Second-Level Interconnection

As noted previously, level 2 interconnection refers to the electrical connection of an IC to a circuit board, the most common one being a conventional PWB. Following level 1 interconnection, single IC chips normally undergo encapsulation in either plastic or ceramic-based packages prior to connection to a PWB. In other cases, the IC is bonded directly to a PWB either by a die attach/wire bonding scheme or using TAB. The chip is then protected by a "glob-top" encapsulant such as a topical epoxy or a silicone. This interconnection technique, referred to as chip on board (COB), has the advantages of reduced PWB area and cost because of the elimination of an extrinsic package. For multichip packages, second-level interconnection is the connection between a package containing more than one chip, which are interconnected via an imbedded conductor network in a substrate, and a PWB. However, the substrate on which the chips are mounted prior to being inserted into a package actually qualifies as a PWB in the broadest sense of the definition of a PWB.

1.4 BRIEF HISTORY OF MICROELECTRONIC PACKAGING TECHNOLOGY

It is probably impossible to determine the exact date that electronic packaging began to be viewed as an engineering and science technology. However, microelectronics packaging technology began in earnest in response to the discovery of the transistor in the late 1940s and has continued to evolve to serve the increasing complexity and performance of ICs since that time. Early transistors were of the alloy structure and were housed in plastic packages, providing little in the way of protection for the device. However, once the military became interested in these new devices for high-reliability applications, the need for hermeticity to prevent transistor gain degradation and junction leakage current due to contamination and moisture led to the development of the metal transistor outline (TO) packages shown in Figure 1.5. These packages consist of a gold-plated metal base containing external leads (generally three or four leads for discrete transistors), commonly referred to as a header, and a metal lid that is sealed to the header by welding in an inert atmosphere, such as nitrogen or argon, to ensure that moisture and other contaminants are excluded from the ambient in which the electronic device must operate.

With the development of silicon planar technology, electronic packages were developed to accommodate the large number of I/O leads of ICs. Initially, ICs were merely packaged in higher pin count versions of the TO can, which quickly became inadequate to handle the more complex and higher I/O integrated circuits. Consequently, the 1960s saw a rapid proliferation of new packages for ICs. Because of cost considerations, the lack of standards for package design, and difficulty in mounting some packages onto PWBs, only

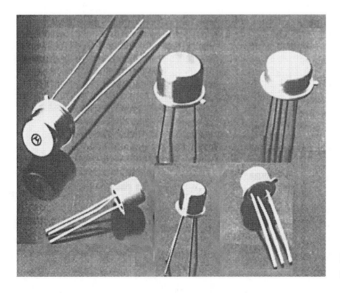

Figure 1.5 Photograph of a selection of TO packages.

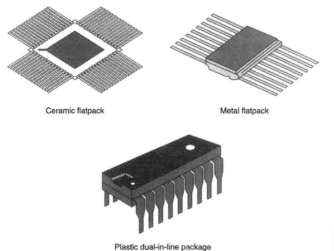

Figure 1.6 Dual-in-line and flatpack packages.

the "flatpack" and dual-in-line package (DIP) survived. Examples of these two major types of IC packages are shown in Figure 1.6. It should be noted that the flatpack leads, which extend from all four sides, are essentially planar with the package, while those of the DIP are perpendicular to the body of the package and exit on two sides. This has some implications in terms of mounting to PWBs. In particular, the DIP is ideally suited for insertion mounting onto PWBs via plated through holes (PTHs) by automatic insertion machines and wave soldering, whereas the flatpack has to be mounted using special methods and tools. Consequently, the DIP became the primary package for ICs and, along with through-hole PWBs, has long dominated the electronics assembly market.

Packaging costs also received considerable attention during the 1960s while the industry was deciding which package or packages would become standard. One of the early attempts to develop a low-cost, hermetic package resulted in the CerDIP, which was a DIP constructed of two pieces of sandwiched ceramic with the leads protruding from between the slabs of ceramic as shown in Figure 1.7. The two pieces of ceramic were held together using a low-melting-temperature glass, which also acted as a seal, thereby providing hermeticity. Unfortunately, the glass used originally outgassed moisture, which created reliability problems. The development of vitreous sealing glasses, which do not outgas moisture, coupled with performing the sealing operation in a nitrogen ambient reduced the reliability problem to a tolerable level.

Driven by the need to further reduce packaging costs, the industry pursued fully automated manufacturing of plastic DIPs. The result of this effort was a low-cost plastic package, which was transfer molded around an IC chip that had previously been die and wire bonded to a lead frame. Unfortunately, plastics are not hermetic because of their high permeability to moisture and poor adhesion to the metal leads, which provides a path for water vapor to access the IC. Furthermore, the resins and fillers either initially contained undesirable contaminants (e.g., Cl and Na) or they were polymerization by-products of the plastics, leading to degradation of the IC. These problems were eventually resolved by improving the encapsulant applied to the chip prior to plastic packaging and improvement in the properties of the plastics themselves.

10   Chapter 1   Introduction and Overview of Microelectronic Packaging

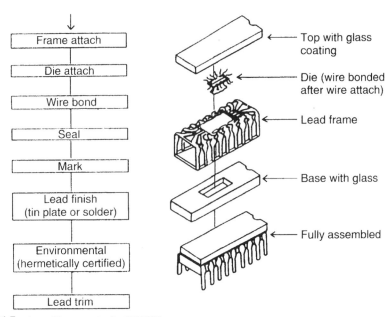

Figure 1.7   Assembly sequence for CERDIP.

The 1970s and 1980s saw the development of several types of IC packages, including surface-mount packages (SMPs), in response to a need for higher density PWBs. When mounted on a PWB, the SMP's leads do not penetrate the PWB like those of through-hole-mounted packages. Thus, they can be mounted on the side of the PWB containing conductor traces. Consequently, SMPs can be mounted on both sides of a PWB. Mounting of SMPs to PWBs is accomplished by reflow solder technology, which gave new life to the flatpack, actually the first surface-mount package. Also, small outline packages (SOPs), which resemble miniature versions of DIPs, as shown in Figure 1.8, were developed for use in surface-mount technology (SMT). The two types of small outline packages are the small outline transistor (SOT) and the small outline IC (SOIC) shown in Figure 1.9. This same period saw the development of chip carriers and quadpacks (or quad flatpacks). The chip carrier is available in both leadless and leaded versions, as well as with plastic and

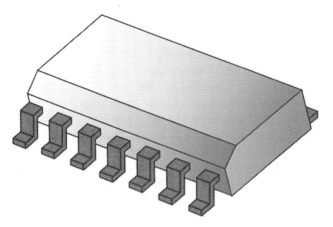

Figure 1.8   Small outline package (SOP).

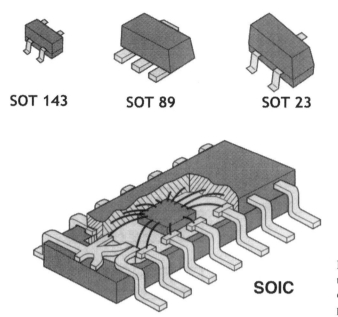

Figure 1.9 Small outline transistor (SOT) and small outline integrated circuit (SOIC) packages.

ceramic bodies (see Fig. 1.10). These type packages conform very closely to the size of the ICs they contain and have leads on all four sides. The quadpack, one of the earliest plastic surface-mount IC packages, comes in a variety of sizes and lead configurations and also has leads on all four sides as shown in Figure 1.11. The terminal pitches vary from 0.3 to 1.0 mm (10 to 50 pins/cm^2). Thus, the quadpack (I/O ≈ 300) is generally used when the chip I/O requirements are greater than can be addressed using a DIP (I/O ≈ 64 with 10 pins/cm^2).

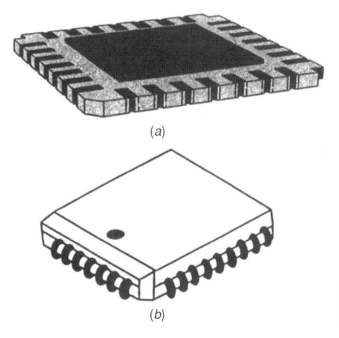

Figure 1.10 (*a*) Ceramic leadless chip carrier and (*b*) plastic leaded chip carrier.

12 Chapter 1 Introduction and Overview of Microelectronic Packaging

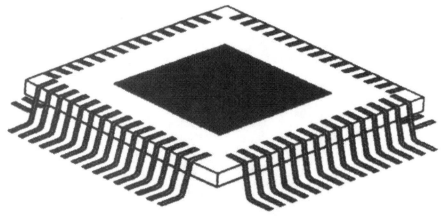

Figure 1.11 Quad flatpack (or quadpack).

The ultimate goal for high-performance electronic systems is generally to pack devices as close together as possible in order to minimize circuit path length. In response to this need, the early 1990s saw the emergence of both pin grid array (PGA) and ball grid array (BGA) packages as a replacement for quad flatpacks (QFPs), primarily because of their high I/O density (the terminals are arrayed on part or all of the bottom of the package), minimum footprint, and shorter electrical paths, which means that they have better electrical performance. For QFPs, lead counts higher than 200 require lead spacings of 0.5 mm, and for 300 leads, the spacing approaches 0.3 mm. Unfortunately, as spacings become tighter, the yield falls exponentially with lead spacing. For I/Os greater than 250, the PGA and BGA have an advantage over the QFP in that they always occupy less space than a QFP. However, PGA and BGA construction is inherently more expensive than that of the QFP because of costs associated with the component carrier substrate. The primary terminal pitches of the BGA are 1.27 and 1.5 mm, yielding a mounting density of 40 to 60 pins/cm^2.

The BGA package evolved from flip-chip technology, also referred to as controlled-collapse chip connect (C4), pioneered by IBM for ICs [6]. Thus, the BGA package can be identified by the solder bumps on the bottom of the package. The solder bumps can be arranged in a uniform full-matrix array (i.e., over the entire bottom surface), a staggered full

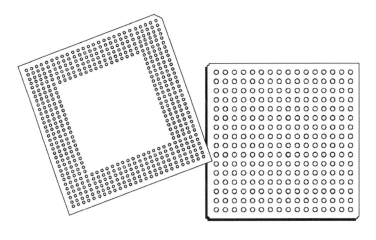

Figure 1.12
Examples of full matrix and perimeter array BGA I/O solder bumps.

1.4 Brief History of Microelectronic Packaging Technology   13

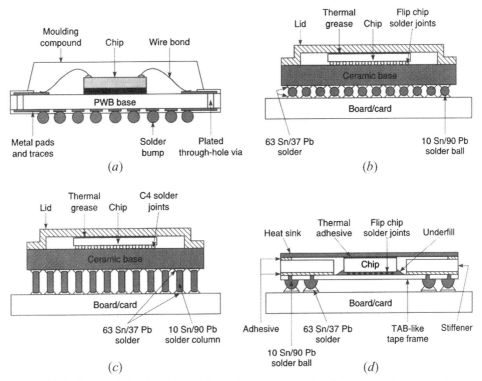

Figure 1.13 Conceptual sketch of (*a*) plastic ball grid array, (*b*) ceramic ball grid array, (*c*) ceramic column grid array, and (*d*) tape ball grid array.

array, or around the perimeter in a multiple number of rows (see Fig. 1.12). No matter how the bumps are arranged, the result is a smaller footprint than that of conventional packages. As noted previously, for a given amount of real estate, the BGA package provides more I/O than QFPs. Thus, the BGA package is considered the package of choice for high-density and high-I/O ICs.

As is the case with other packages, BGAs have been broadly classified according to the type and form of the die carrier substrate material. Thus, the packages have been classified as CBGAs (ceramic), CCGAs (ceramic column), PBGAs (plastic), MBGAs (metal), and TBGAs (tape). These packages, some of which are conceptually illustrated in Figure 1.13, can house either a single chip or multiple chips.

As noted previously, one of the big advantages of BGA packages is that they offer high I/O. Typical I/O for several types of BGAs are 400 I/O PBGAs, 736 I/O TBGAs, and 625 I/O CBGAs, although BGA packages are available with more than 2000 leads. Ceramic column BGAs are available with I/Os greater than 1000. Current and planned designs have solder-bump-array pitches in the range of 40 to 150 μm. The solder bumps are generally of tin–lead or tin–lead–silver composition. Thus, the BGA is a leadless package that is not susceptible to bent or skewed leads, which means that it can be easily handled. However, there are some aspects of BGA packages of concern. In particular, solder joint defects, warpage during reflow, a large variation in solder ball size, the inability to visually inspect the solder joints, reduced resistance to thermal cycling, and problems associated with rework. These can be overcome with good design, process development, and process control

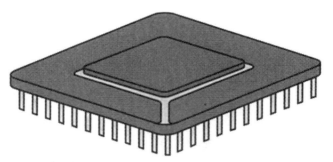

Figure 1.14 Pin grid array package.

so that the resulting yield makes inspection and rework of minor importance. Considering all the pluses and minuses, BGA still appears to be the surface-mount package of the future for both single-chip and multichip packaging. Similarly, the PGA package (I/O ≈ 600), illustrated in Figure 1.14, is used when the I/O requirement is higher than that provided by quadpacks.

There always has been and will continue to be motivation to pack more electronic functionality and higher speed performance into a smaller volume of space. Packaging of ICs is one area that offers attractive benefits for reducing size and improving performance by either eliminating the package or reducing the size to the point where it takes up very little more space than the IC. Elimination of the package [i.e., direct chip attach (DCA)] still presents some problems where full functionality and reliability testing are concerned. In other words, the ability to test and burn-in bare die has not yet reached the quality and reliability levels comparable to the same die in a package. Consequently, packaging ICs for testing and burn-in is still very attractive. In the late 1990s, the BGA concept was applied to a packaging technology referred to as chip-scale packaging (CSP) by reducing the terminal pitch to 1.0 mm or less for a mounting density greater than 100 pins/cm^2 [7, 8]. CSP contributed significantly to a reduction in the size, weight, and performance of products such as the cellular phone.

CSPs are essentially "packages" that ruggedize the IC for ease of handling, testing, and assembly. Thus, CSPs are a viable substitute for "known good die" (KGD) if low-cost test and burn-in methods are not available for bare die. CSPs, also referred to as slightly larger than IC carrier (SLICC), are generally defined as packages that are equal to or smaller than 1.2 times the bare die size. Microball grid array, miniball grid array, and micro-SMT packages fit this definition since they are of minimum size and employ direct surface mounting instead of wire bonds. In fact, in excess of 60 different chip-scale package types have been proposed over the years. The primary difference in the various types is the material layers, which serve as compliant members, space transformers, and mechanical protection, between the silicon and the bump array. These material layers serve to categorize the CSPs into tape carrier (flexible laminate), resin mold, ceramic carrier, silicon-based, rigid laminate, lead-on-chip (LOC), and lead frame types.

CSPs are designed to be flip-chip mounted using conventional equipment and solder reflow. Essentially, CSPs take advantage of the attributes of a flip chip in a surface-mountable package. Thus, CSPs offer a method for subjecting ICs to full functional and reliability testing using a packaging technology while essentially maintaining the size and performance of bare die.

1.4 Brief History of Microelectronic Packaging Technology

The 1980s also marked the turning point in the way electronic engineers viewed IC packaging technology. As noted previously, for many years the electronics industry had been concentrating on increasing the performance of ICs (i.e., more circuitry/silicon area operating at higher speeds) with little consideration of the fact that ICs in an electronic system must communicate with each other through the packages that contain them. As a result of the trend toward higher circuit densities and operating speeds on a chip, the following effects became important considerations for packaging engineers:

- I/O requirements increased sharply.
- Signal transition time between chips became a factor limiting system speed.
- Signal integrity between silicon chips degraded.
- Power requirements per chip increased.
- A problem with heat dissipation was created.

All of these factors forced electronic packaging technology into the spotlight, resulting in a reconsideration of how ICs were being packaged. From this reconsideration evolved multichip packaging, for example, multichip module (MCM) packaging technology [5], examples of which are shown in Figure 1.15. Although the basic concept of a multichip module was not new (hybrid circuits had been around for nearly 50 years), in the late 1980s and throughout the 1990s interest was renewed in mounting a multiple number of ICs in a single package in order to take advantage of inherently shorter interconnection distances between ICs. Thus, the development of MCM technology became an industry effort to push the performance of electronic systems to higher and higher levels.

The simplest definition of an MCM is that of a single electronic package containing more than one IC. The ICs are interconnected through a substrate. Based on this simple definition, an MCM combines high-performance ICs with a custom-designed common substrate structure, which provides mechanical support for the chips and multiple layers of conductors to interconnect them. This arrangement takes better advantage of the performance of the ICs than does interconnecting individually packaged ICs because the interconnect length is much shorter. The really unique feature of MCMs is the complex substrate structure, which is fabricated using multilayer ceramics, polymers, silicon, metals, glass-ceramics, laminates, and the like. Thus, multichip modules are not really new. They have been in existence since the first multichip hybrid circuit was fabricated. Conventional PWBs utilizing chip on board (COB), a technique where ICs are mounted and wire bonded directly to the board (direct chip attach), have also existed for some time. However, if packaging efficiency (also called silicon density), defined as the percentage of area on an interconnecting substrate that is occupied by silicon, is the guideline used to define an MCM, then many hybrid and COB structures with less than 30% silicon density do not qualify as MCMs. The fundamental (or basic) intent of MCM technology is to provide an extremely dense conductor matrix for the interconnection of bare IC chips. Consequently, some companies designated their MCM products as high-density interconnect (HDI) modules and others dropped the MCM designation for multichip packages (MCP).

MCM technology, still a viable technology today, played a major role in the evolution of multichip packaging in the 1990s, which includes three-dimensional (3D) stacking of ICs within a single package [9–13]. Although chips can be stacked physically without connecting them to each other to save board space, present development efforts focus on interconnected, stacked chips. One of the first commercial efforts to stack chips within a single package mated flash memory with static random-access memory (SRAM). However,

16 Chapter 1 Introduction and Overview of Microelectronic Packaging

Figure 1.15 Packaged MCMs (courtesy of nCHIP).

several patents issued in the 1980s described and illustrated chip stacking approaches to multichip packaging. As a rule, multichip packaging can take many forms but is simply the packaging of more than one IC in a single package, no matter the details of the IC interconnection or whether or not they are electrically interconnected. In fact, another approach to multichip packaging is referred to as "few-chip packaging" [14]. This generally refers to the placing of 2 to 5 ICs on a laminate substrate in a BGA package that looks very similar to a single-chip package.

The industry desire to reduce product size, weight, and cost [e.g., cell phones, personal digital assistants (PDAs), and other hand-held applications] while providing extra performance (i.e., shorter interconnects that lower capacitance and inductance, reducing crosstalk, and lower power consumption) and increasing functionality, has driven 3D packaging of both chips and packages. Figures 14.16 and 14.17 illustrate some approaches to chip stacking and interconnection, respectively. Chip stacking offers the additional advantage of assembling die of dissimilar geometries and the use of mixed technologies.

Chip stacking may be the proper choice when high-yielding ICs are available in bare die form. However, chip stacking has resulted in the need to thin chips to 100, 75, 50, 20 or even 10 µm and below, so that a multiple number of them can be mounted in a conventional package without increasing the vertical profile [15, 16]. Even so, chip stacking is usually limited to 2 or 3 chips, yielding silicon efficiencies of approximately 140 and 220%, respectively, but as many as 8 chips in a single package has been achieved. The goal appears to be a maximum package height of 1.2 mm for two-die stacks and 1.4 mm for three-die stacks. This packaging approach has been used as an alternative to system on a chip (SOC) [17–20], which has been made possible partly because of the improvement in bare die testing [21–23]. The result is essentially a system on a package (SOP) [24] or a system in a package (SIP) [25]. Interestingly, the success with chip thinning has even spawned an effort to develop thinner packages, referred to as ultrathin or paper-thin packages, for single-chip packaging. The development of such packages also makes package stacking more attractive.

Thinning of chips improves their thermal performance, allows for mounting on irregular surfaces, relieves stress in the chip when mounted, and produces a more mechanically reliable device. However, thinning chips also creates a whole new set of challenges because of the wafer thinning process. These include induced stresses, microcracking, and the creation of defects in the chips. Additionally, dicing and handling of very thin chips during assembly can be problematic. Furthermore, design software, thermal management, and a single chip failure in a stacked system can all contribute to higher cost for a system.

Figure 1.16 shows approaches to package stacking. Any approach to package stacking results in silicon efficiencies in excess of 100%. Package stacking is fairly straightforward and is preferable for lower yielding devices or where chips require burn-in. One merely needs to develop techniques for connecting the packages together mechanically and electrically. This can be accomplished by simply soldering the leads of the packages together (Fig. 1.16a) or by the use of an interposer such as a flexible PWB (Fig. 1.16b). Package stacking has the advantages of using existing technology, it can have a CSP footprint, and it allows reasonably easy integration of different processes and materials. The biggest disadvantage of package stacking is the height of the resulting unit.

The most recent approach to the packaging of chips is referred to as wafer-level packaging (WLP) or wafer-scale packaging (WSP), which began in the late 1990s [26–30]. In WLP, the die and package are manufactured and tested on the wafer prior to separation of the packaged devices by dicing in the flip-chip fashion. Consequently, the packages are really

18 Chapter 1 Introduction and Overview of Microelectronic Packaging

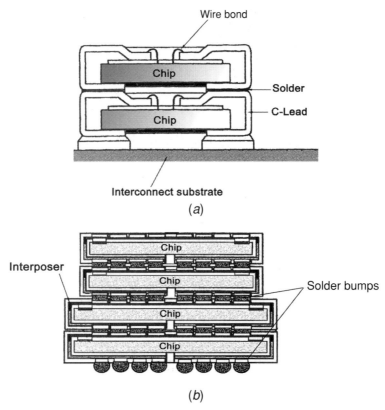

Figure 1.16 Three-dimensional package stacking.

chip size as opposed to chip scale, which has produced the name wafer-scale/chip-scale packages (WS–CSP) or wafer-level/chip-scale packaging (WL–CSP).

The motivation for WLP (WSP) is that peripherally designed die can be transformed, using a thin-film technology, into a standard bump or ball footprint that is compatible with current PCB layout rules, device test practices, and assembly practice. Originally, wafer-level redistribution was intended to physically redistribute perimeter bonding pads to an area array for flip chip technology.

In addition to the package size, some other benefits of WLP are lowest cost per I/O since all interconnections are formed at the wafer level at the same time, lowest testing cost since it can be done at the wafer level, lowest burn-in cost because it is done at the wafer level, elimination of underfilling, and enhanced electrical performance because of the short interconnections.

The next logical step in packaging is presently under development and is referred to as wafer-level stacking (WLS) to yield stacked die that are packaged at the wafer level. Wafers containing ICs are interconnected using a thinly polished wafer (a few tens of microns thick) containing through-wafer vias. Figure 1.17 provides an illustration of wafer-level stacking. An active wafer can be joined to a passive interposer or another active wafer using any convenient joining method. The top wafer is then thinned to expose through-hole contacts. The process is repeated to add more wafers to the stack. Upon completion of the wafer-level packaging process, individual packages containing stacked die are obtained by singulation.

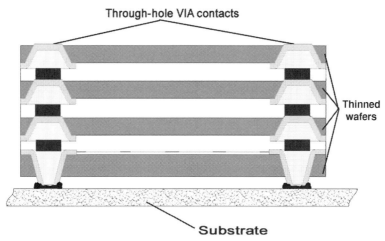

Figure 1.17 Wafer-level stacking to yield wafer-level packaging of stacked die.

There appears to be little doubt that the impact of packaging on future electronic systems will continue to increase with time. In fact, the trend will continue to be toward a more integrated approach to semiconductor packaging and system design. Consequently, the boundary between semiconductor fabrication and packaging will blur, and packaging will, by necessity, become an integral part of system design.

As packaging technology progresses, there are many problems that must be addressed. Some of these are organic materials for substrates that address problems inherent to organic substrates such as moisture absorption, dielectric loss, warpage, limited minimum interconnect dimensions, and so forth. Lower loss dielectrics are also critically important to higher frequency system operation. Given that lead-free soldering probably will be mandated worldwide at some point in time, development of lower cost materials with lower processing temperatures must be developed [31, 32]. Then, for new materials, reliability issues must also be addressed. As minimum chip geometries continue to shrink into the nanometer range, higher pin count, higher wiring density, and passive device integration will provide significant challenges to packaging technology. If these challenges can be and are met, then thermal management will continue to provide challenges for the researcher. Finally, the emerging need for packaging of MEMS, optoelectronics, and nanodevices ensure that packaging technology will remain fertile ground for research and development into the foreseeable future.

1.5 DRIVING FORCES ON PACKAGING TECHNOLOGY

Historically, packaging has always been a substantial fraction of the price of an IC (10 to 50%), and, consequently, reducing packaging costs while maintaining reliability and performance has been the focus of packaging engineers for many years. During this time, IC technology has transitioned from small-scale integration (SSI) in the 1960s to submicron minimum dimension very large scale integration (VLSI) at the present time. Since packaging technology has not enjoyed anywhere near the performance advancement of ICs over the past 30 years, electronic system performance has become increasingly limited by IC packages. Thus, design decisions facing packaging engineers today are becoming increasingly driven by system performance for a market that is still very cost conscious.

Given that cost and performance are the primary concerns in electronic packaging, it is important to examine the factors that relate performance and cost to packaging technology choices. Some factors, which must be considered, are manufacturability, reliability, serviceability, size, weight, signal integrity, mechanical stability, and power consumption with its accompanying heat dissipation problem. In general, packaging costs are driven by materials and fabrication requirements associated with actual manufacturing and by testing and rework associated with manufacturability. In the case of multichip packaging, manufacturing costs include the cost of the IC chips, generally referred to as "known good die." On the other hand, performance is a function of electrical, thermal, and mechanical design constraints, material selection, and fabrication limitations. A few of these cost–performance factors, treated in detail in later chapters, are briefly discussed below.

1.5.1 Manufacturing Costs

As noted previously, manufacturing costs include all the materials and fabrication steps required to produce each packaged IC including the cost of the IC itself. The materials include not only the actual material that goes into the package but also materials such as chemicals required to process the material, which ultimately are part of the package. Manufacturing costs vary widely for packaging and interconnection elements.

1.5.2 Manufacturability Costs

Manufacturability costs, because they are generally associated with how well the product was designed for manufacturability, are incurred as a result of testing, reworking when possible, retesting after rework, and fabrication yield loss. Since testing is, at the present time, the only way to ensure full functionality and high reliability for both single-chip and multichip packaged parts, packaging technology should provide for full testing of the finished product. Furthermore, since the final yield can be increased above initial yield by reworking some of the parts that fail initial testing, it is important that packaging technology also allow for an inexpensive and easy rework process. Although chips usually cannot be repaired, this is not necessarily the case for multichip packaging. Reworking multichip packages, which generally refers to the replacement of a chip or chips within the package is an important cost issue because rework costs are generally high. Consequently, high value chips should be thoroughly tested prior to mounting in a multichip package.

1.5.3 Size and Weight

Because of the many environments in which packaged electronic systems must operate, size and weight are often considered to be performance goals. Restrictions placed on these two physical parameters may be the result of personal preferences or system requirements. For example, weight may be a critical factor if the system is to operate in a satellite or a portable computer. Size (either area or volume or some combination of the two) may be critical in applications such as calculators and cellular phones. Most important, specifying these parameters can impact other design aspects of the electronic package such as material specifications and cooling requirements.

1.5.4 Electrical Design

As noted previously, the on-chip switching speeds of ICs are continuously increasing. Furthermore, noise margins are generally decreasing at the same time. Unfortunately, chip I/O count and interconnection speed have not kept pace so that packaging interconnects now play a dominant and limiting role in determining overall system performance. Each lead from the chip to the package and each package lead to the outside world has some parasitic capacitance, resistance, and inductance that limits switching speed, distorts the shape of signals passing through it, and serves as a source of electrical noise. These leads are also a source of reliability problems. Likewise, the pattern of metal and dielectric that forms the circuitry between chips and from chips to the outside world of an MCP contribute to the degradation of electrical performance. Consequently, some electrical design factors that must be considered include signal lead length (short parallel runs to minimize mutual inductance and crosstalk, and short runs near ground planes to minimize capacitive loading), use of matched impedances to avoid signal reflection, low ground resistance for minimum power supply voltage drop, and power supply spiking caused by signal lines switching simultaneously. All of these factors are functions of geometries and materials.

1.5.5 Thermal Design

The primary objective of thermal design is to remove heat from the junctions of ICs to ensure that they operate properly and to avoid triggering temperature-activated failure mechanisms. This is generally accomplished by conducting the heat away from the chips and into a gas or liquid coolant. Although the power dissipation per gate of ICs has decreased in recent years, the power dissipation per chip has increased during the same time since power per gate scales linearly with feature size, while on-chip circuit power density increases as the square of the feature size reduction ratio. Even complimentary metal–oxide–semiconductor (CMOS) circuit densities and operating frequencies are becoming great enough that thermal design cannot be ignored when packaging these chips. Thermal issues are even more significant in MCP design because the heat density is generally high as a result of closely spaced, high-density, high-performance chips. Consequently, thermal design of MCPs must consider such factors as (1) how heat is to be removed from the IC (through the substrate or directly off the backside of the chip), (2) whether to use forced air or liquid cooling (forced air is generally much cheaper and the system is more likely to survive a fan failure than a pump failure), (3) the use of a high thermal conductivity substrate (high thermal conductivity usually implies a high dielectric constant, diamond being an exception) or thermal vias that can consume a large area of the substrate, thereby reducing interconnect capacity, and (4) stresses induced in the chips and substrate due to mismatches in thermal coefficients of expansion (TCEs). All of these considerations impact performance, cost, and reliability.

1.5.6 Mechanical Design

In the previous section, it was noted that mismatches in thermal coefficients of expansion cause stresses to be induced in ICs, high-density interconnect substrates, and packages of an electronic system as the temperature changes. Such stresses can be very localized, for example, under a small portion of a chip, or universal such as across an entire layer of an MCP substrate. Thus, the mechanical design aspect of electronic packaging technology is, in general, closely related to changes in temperature. Another mechanical property that may need to be considered is stiffness, characterized by the tensile modulus (E), which is important in areas such as chip attach where three materials (chip, substrate or package, and adhesive) form two interfaces. Thermal stresses increase with increasing E and decreasing thickness of the adhesive layer. Thus, for example, a thin layer of high E adhesive material should only be used with large-area chips if the TCE of the substrate or package closely matches that of the chip.

1.5.7 Manufacturability

No matter how much effort and exactness goes into the design of an electronic package, ultimately high-quality packages must be manufacturable in sufficient quantities to allow for competitive pricing. Successful manufacturing depends on many factors such as the availability of materials, process technologies, and automated fabrication equipment at acceptable costs, cost of piece parts, manufacturing yield, manufacturing cycle time, and repairability. Design specifications must not overly challenge fabrication technology because the limitations of fabrication processes and equipment for any microelectronic technology essentially establish the ultimate capability of the technology. Although zero-defect manufacturing is possible, it is not realistic because of high costs and possible loss of competitive edge due to conservative dimensions, tolerances, materials, and process choices that would all impact performance in a negative way. Thus, the alternative is to balance the above-noted variables against loss of products due to defects, which must be screened out by testing.

Although MCPs offer significant benefits in terms of greater densities, smaller size, and better performance, implementing this technology requires significant changes in design and manufacturing methodologies, fabrication processes, and equipment. Manufacturers have, to some degree, avoided making these changes by "pushing" conventional technologies to achieve higher density/performance levels. MCP technology will continue to offer challenges to manufacturing that are somewhat unique and, thus, will continue to require the development of new materials, new processing technologies, and specialized fabrication equipment.

1.5.8 Testability

A primary motivation for testing is to find and eliminate manufacturing-induced defects. Other reasons for testing include the need to prove a new design, to verify that manufacturing processes are under control, and to predict product performance under normal operating conditions. Generally, testing can be divided into two broad categories: in-process and stress testing. In-process testing is nondestructive and is used to screen for defects that occur as a result of manufacturing and prevent the product from ever operating properly. Such testing can be performed at almost any stage of the fabrication process. At every testing point, factors to consider include testing costs, investment in the product to the point of test, and the scrap and possible repair costs. On the other hand, stress testing, which is often destructive to the product, is most frequently performed on the finished product to evaluate long-term reliability. Thus, stress testing finds product defects that do not initially prevent the product from operating, but do so later. The results of both types of testing are used as feedback to improve both product design and manufacturing.

MCPs must be subjected to the same test requirements as the single-chip packages they replace. However, MCPs present unique challenges in testing compared to single-chip packages because of their unique structure. The usual approach to MCP testing is to perform separate testing of the substrate on which the chips are to be mounted and interconnected, the ICs prior to mounting on the substrate, and the assembled module prior to sealing in case rework is required. Since the completed MCP is a complex electrical system that must be tested as a single unit, electrical testability must be integrated into the MCP package design. At the present time, not all of the problems associated with assembled MCP testing have been solved.

1.5.9 Reliability

In recent years, highly reliable products have become commonplace in the electronics industry. Consequently, sustained success in this market has depended on a company's ability to provide superior products that are reliable. Reliability assurance is tied very closely to thorough product testing. However, consistently high product reliability requires the interaction of product design, manufacturing, and testing. When properly performed and integrated, these three functions produce a reliable product. At the system level, reliability is highly dependent on the failure rate of its component parts. Thus, in the case of MCPs, reliability of the final substrate is the combined result of the reliabilities of each of its components, pointing out the need to thoroughly in-process test the substrate, ICs, and assembled substrate, followed by stress testing of functional substrates.

1.5.10 Serviceability

Serviceability of a product refers to the demand placed on it by the need to replace failed components. For MCPs containing expensive ICs, this capability may be important when failure of a module is a result of chip failure. Given that rework costs are lower than the price of a new module, the design of the module must allow for chip replacement. Factors that impact this capability are chip spacing, method and material used to attach the die, and method and material used for electrical interconnects. In many cases, the design of this capability into the product may be in complete opposition to the performance criteria, for example, when close spacing of the ICs is extremely important.

1.5.11 Material Selection

Packaging materials play critical roles in the proper functioning of a packaged electronic system [33, 34]. For example, metals provide the means for conducting signals throughout the system via thin-film conductors, wires, contacts, vias, and the like. On the other hand, insulating materials are used to prevent loss of signal currents by confining them to the metal paths. Other materials are used to provide physical and structural support. Finally, there are materials whose primary function is to protect the system from the environment.

The packaging industry is concerned with the electrical, mechanical, thermal, chemical, and physical performance of all materials that are used in electronic packages. Table 1.1 lists some specific material properties in each of these areas that impact the fabrication, performance, and reliability of electronic packages. Because of the wide variation in these properties, material selection for electronic packaging technology is not an easy task. As is generally the case throughout the microelectronics industry, the final choice of a material for a specific application most likely results from a series of compromises or trade-offs aimed at achieving and/or optimizing one or more performance criteria. For interconnect

Table 1.1 Material Properties of Importance in Packaging Technology

Electrical properties	Thermal properties	Mechanical properties	Physical properties	Chemical properties
Dielectric constant	Coefficient of thermal expansion (CTE) (ppm/°C)	Young's modulus (GPa of kpsi)	Microstructure (grain size)	Metal oxidation
Loss tangent (tan δ)	Decomposition temperature	Poisson's ratio	Flatness and planarization	Metal migration
Resistivity (Ω-cm) volume surface	Melting point	Stress (dyn/cm^2) Shear strength (MPa)	Viscosity (poise)	Reactivity
Dielectric strength (V/cm)	Glass transition temperature (T_g)	Curing temperature	Hermeticity	Adhesion
Temperature coefficient of resistance (Ω-cm/°C)	Thermal conductivity (W/m · K)	Glass transition temperature (T_g)	Melting point	Toxicity
	Shrinkage	Dimensional stability	Eutectic temperature	Environmental
	Curing temperature	Tensile strength (GPa)	Density (g/cm^3)	
	Thermal stability	Flexural strength (GPa)	Glass transition temperature (T_g)	
	Temperature coefficient of resistance (Ω-cm/°C)	Adhesion strength	Hardness (Brinell)	
		Peel strength		
		Ductility		
		Maleability		
		Interface energy		

substrates, the selection of materials can be a particularly challenging task because of their complex multilayered structure.

1.6 SUMMARY

Because of the diversity of opinion on what constitutes an electronic package, an exact date for its origin is not possible. However, the beginning of modern electronic packaging can probably be dated around 1950, shortly after the discovery of the transistor. Since that time, a myriad of electronic packages and packaging materials have evolved. In all cases, the package must provide a structure to physically support the electronic device and protect it from the environment, a means of removing heat generated by the device, and electrical connections to and from the device. Over the years, chip packaging has evolved from simple transistor outline (TO) packages that generally hold a single transistor chip to dual-in-line packages (DIPs) for ICs to quadflatpacks (QFPs) to chip-scale packages (CSPs). The most recent electronic packaging technologies, referred to as multichip packaging (MCP), house an electronic system by interconnecting a multiple number of ICs within a single packaging structure. This can be done either in a planar fashion, such as the case for multichip modules (MCMs), or in the vertical direction by chip stacking. Chip stacking can be accomplished by either stacking single-chip packages or by stacking a multiple number of chips in a single package, or a combination of the two. More recently, wafer-level packaging, which involves creating the package while chips are still in wafer form and then separating them by dicing, is gaining popularity. Future packaging likely will involve wafer-level packaging of stacked wafers, resulting in stacked chips interconnected by vias formed through the material used to physically separate the wafers. Thus, system packaging at the wafer level will be possible and will permit the mixing of different technologies in a single package.

REFERENCES

1. R. R. TUMMALA AND E. J. RYMASZEWSKI, eds., *Microelectronics Packaging Handbook*, New York: Van Nostrand Reinhold, 1989.
2. D. P. SERAPHIM, R. LASKY, AND C.-Y. LI, *Principles of Electronic Packaging*, New York: McGraw-Hill, 1989.
3. T. L. LANDERS, W. D. BROWN, E. W. FANT, E. M. MALSTROM, AND N. M. SCHMITT, *Electronics Manufacturing Processes*, Englewood Cliffs, NJ: Prentice Hall, 1994.
4. J. ADAM, C. S. CHANG, J. J. STANKUS, M. K. IYER, AND W. T. CHEN, Addressing Packaging Challenges, *IEEE Circuits Devices Mag.*, p. 40, July 2002.
5. W. D. BROWN, ed., *Advanced Electronic Packaging: With Emphasis on Multichip Modules*, Piscataway, NJ: IEEE Press, 1999.
6. G. PHILLIPS, Flip-Chip Packaging Moves into the Mainstream, http://www.reed-electronics.com/semiconductor/index.asp?layout=articlePrint&articleID=CA239576, September 3, 2003.
7. J. FJEISTAD, Trends in Chip-Scale Packaging, *HDI*, p. 22, January 2000.
8. C. MITCHELL, Taking Chip-Scale Packaging to New Heights, http://www.elecdesign.com/Articles/Index.cfm?ArticleID=2880, September 16, 2003.
9. Y. W. HEO, A. YOSHIDA, AND R. GROOVER, Advances in 3D Packaging—Trends and Technologies for Multichip Die and Package Stack, Proceedings of the VLSI Packaging Workshop of Japan, Kyoto, Japan, November 12, 2002.
10. K. TAKAHASHI, K. TANIDA, M. UMEMOTO, M. KOJIMA, M. ISHINO, AND M. BONKOHARA, 3-Dimensional Chip Stacking for High-Density Electronic Packaging, Abs. 598, 204th Meeting, The Electrochemical Society, 2003.
11. G. REED, 3-D Packaging—Nowhere to Go But Up, http://www.reed-electronics.com/semiconductor/index.asp?layout=articlePrint&articleID=CA319211, September 19, 2003.
12. J. C. DEMMIN, Stacked CSPs: Issues and Results, http://www.reed-electronics.com/semiconductor/index.asp?layout=articlePrint &articleID=CA302689, September 19, 2003.
13. L. SMITH AND T. TESSIER, Stacked Chip-Scale Packages: They're Not Just for Cell Phones Anymore, http://www.chipscalereview.com/issues/0701/f4_01.html, September 16, 2003.
14. K. RINEBOLD, Few-Chip Packaging—An MCM Renaissance, *HDI*, p. 18, October 2000.
15. W. J. KRONINGER, F. HECHT, G. LANG, F. MARIANI, S. GEYER, AND L. SCHNEIDER, Time for Change in Pre-Assembly? The Challenge of Thin Chips, Proceedings of the 2001 Electronic Components and Technology Conference (0-7803-7038-4/01).
16. G. KLINK, M. FEIL, F. ANSORGE, R. ASCHENBRENNER, AND H. REICHL, Innovative Packaging Concepts for Ultra Thin Integrated Circuits, Proceedings of the 2001 Electronic Components and Technology Conference (0-7803-7038-4/01).
17. W. F. SHUTLER, Examining Technology Options for System on a Package, *Electronics Packaging and Production*, p. 32, September 2000.
18. J. BALIGA, Packaging Provides Viable Alternatives to SOC, *Semiconductor International*, p. 169, July 2000.
19. J. A. BRINTON, SOC Isn't Cutting It Yet. Is Multi-Chip Package a Better Answer Today? *Semiconductor Business News*, January 28, 2000.
20. J. WOIDA, The Challenge of System on a Chip, *Electronic Packaging and Production*, p. 18, May 2000.
21. J. RATES, KGD: A State of the Art Report, *Adv. Packaging*, p. 30, September 1999.
22. C. BEDDINGFIELD, W. BALLOULI, F. CARNEY, AND R. NAIR, Wafer-Level KGD, *Adv. Packaging*, p. 26, September 1999.
23. M. A. FRY, J. D. KLINE, J. L. PRINCE, AND G. A. TANEL, In-Line KGD Test Speeds Flip Chip Assembly, *Electronic Packaging and Production*, p. 36, February 2001.
24. W. F. SHUTLER, A. PAROLO, S. OGGIONI, AND C. DALL'ARA, Examining Technology Options for System on a Package, *Electronic Packaging and Production*, p. 32, September 2000.
25. C. TRUZZI AND S. LERNER, Broadening the Platforms for System-in-Package Solutions, *Solid-State Technol.*, p. 115, November 2000.
26. T. DISTEFANO, Wafer-Level Packaging Is Driving the Convergence of Fab and Assembly, http://www.chipscalereview.com/issues/0702/wafer_level.html, August 12, 2003.
27. C. LINDER, W. RIETZLER, AND H. AUER, Wafer-Level Packaging: Making 300 mm a Reality, http://www.chipscalereview.com/issues/0702/f9_01.html, August 12, 2003.
28. M. MASUMOTO, K. MASUMOTO, N. SADAKATA, A. KUROSAKA, T. SUZUKI, M. M. INABA, T. INOUE, M. KAIZU, T. OHMINATO, AND M. I. INABA, Wafer-level Chip Scale Package, http://www.fujikura.co.jp/gihou/gihou30e/30e_18.html, September 9, 2003.
29. Author Unknown, Chip and Wafer-Scale Packaging or Chip-on-Board, http://web.pb.izm.fhg.de/hdi/010_technologies/030_wlp/, September 16, 2003.

30. Author Unknown, Wafer Level Packaging, http://web.pb.izm.fhg.de/hdi/010_technologies/030_wlp/, September 16, 2003.
31. G. REED, Packaging According to Lead-Free Mandates, http://www.reed-electronics.com/semiconductor/index.asp?layout=article&stt=000&articleid=CA312528, August 21, 2003.
32. J. J. WIMER, 3-D Chip Scale with Lead-Free Processes, http://www.reed-electronics.com/semiconductor/index.asp?layout=articlePrint&articleID=CA326072, October 3, 2003.
33. D. R. FEARS AND S. THOMAS, Emerging Materials Challenges in Microelectronics Packaging, *MRS Bull.*, p. 68, January 2003.
34. V. H. OZGUZ AND J. YAMAGUCHI, Materials for 3D Packaging of Electronic and Optoelectronic Systems, *MRS Bull.*, p. 35, January 2003.

EXERCISES

1.1. What do you think is meant by the phrase, "the point has been reached where advancement in integrated circuit performance now drives packaging technology"?

1.2. What are the four basic functions that an electronic package must provide? In addition to these functions, give four other desirable characteristics of an electronic package.

1.3. Define the six levels of packaging hierarchy.

1.4. Which level of the packaging hierarchy is relevant to MCMs and why?

1.5. What are the three primary types of die attach materials?

1.6. What are the three methods used to provide connections between an IC and the package that holds it?

1.7. Lid and pin sealing material must possess five important properties. What are they?

1.8. What three features of the dual-in-line package (DIP) helped it survive the package evolution and become the primary package used today?

1.9. What package feature distinguishes a flatpack from a DIP?

1.10. Describe the construction of a CerDIP.

1.11. What is the most attractive feature of a surface-mount package (SMP) and why is it important?

1.12. From an electrical performance point of view, what three advantages do PGA and BGA packages offer?

1.13. There are five broad classifications of BGA packages according to the form of the die carrier substrate material. What are they?

1.14. The BGA package is not susceptible to bent or skewed leads, but some aspects of BGA packages are of concern. What are they?

1.15. What five effects were created by the trend in the IC industry toward higher circuit densities and operating speeds that gave impetus to the development of MCM technology?

1.16. CSPs are generally defined as packages that are equal to or smaller than _____ times the base die size.

1.17. Thinning of ICs for 3D chip stacking has both positive and negative aspects. What are four positive aspects? What are five negative aspects?

1.18. What are six benefits of WLP?

1.19. What are some of the factors that relate performance and cost to packaging technology choices? Mention at least eight different factors.

1.20. Distinguish between manufacturing costs and manufacturability costs associated with packaging.

1.21. Leads from an IC to a package and from the package to the outside world deleteriously impact electrical signals in three ways. What are they?

1.22. What three electrical design factors must be considered in order to minimize the deleterious effects (from previous exercise) that packaging interconnects have on electrical signals?

1.23. Distinguish between in-process and stress testing of electronics packages. What factors must be considered at every testing point in the fabrication sequence?

1.24. What are some of the problems that must be addressed as packaging technology progresses? Consider at least ten problems.

Chapter 2

Materials for Microelectronic Packaging

W. D. BROWN AND RICHARD ULRICH

2.1 INTRODUCTION

The materials that make up an electronic package perform several functions, the most obvious being the conduction of signals by metals, confinement of signals to conduction paths by electrical insulation, physical support for the various components in the system, and protection from the environment [1–8]. All of the materials that make up the package are crucial to its functioning properly and reliably, and must be carefully selected. There exists no single packaging solution for all electronic systems, but some choices are clearly better than others from the point of view of performance and cost.

This chapter is devoted to the properties, processing, and interactions of materials used in microelectronic packaging. The first section presents a general discussion of important properties of packaging materials, followed by three sections addressing ceramics, polymers (or "plastics"), and metals in general. In the final section, materials used for substrates, dielectrics, metallizations, and packaging in the three fundamental approaches to high-density interconnect (HDI) substrate technology are discussed. The objective is to provide data and performance information on some of the materials commonly encountered in modern electronic packaging technologies. For more in-depth discussions of a wider range of materials used in electronic packaging, the reader is referred to the many materials handbooks that are available.

2.2 SOME IMPORTANT PACKAGING MATERIAL PROPERTIES

Since all measures of packaging density increase at every level in electronics, the selection of materials is increasingly important. Materials for these applications fall into three major categories: metals, polymers, and ceramics. Several basic properties of materials used in packaging are critical to the evolution of high-performance packaging [9–11]. These are discussed in the following subsections with particular regard paid to the differences between these three main material types.

Advanced Electronic Packaging, Second Edition, Edited by Richard K. Ulrich and William D. Brown
Copyright © 2006 the Institute of Electrical and Electronics Engineers, Inc.

2.2.1 Mechanical Properties

The strength and stiffness of packaging materials are primarily important for providing the mechanical support and physical protection for components in the system. This topic is covered extensively in Chapter 8, but a brief introduction will be provided here.

The stress inflicted on a material is defined as the force divided by the area over which is it applied. If the force is perpendicular to the material's surface, it is called normal stress; if it is parallel to the surface, it is shear stress. Inflicting a stress on any material will cause it to deform and, even if the amount of deformation is not visible to the unaided eye, it may be enough to damage the very small structures present in modern electronic systems. This deformation is called strain and can either be elastic, in which case the material bounces back to its original shape after the stress is removed, or plastic, in which case the strain is so large that the material is permanently deformed even after the stress is removed. The ratio of stress applied to the resulting strain is a measure of the stiffness of the material, and goes by the name Young's modulus. Ultimate properties describe the limits that the material can withstand before physically failing, as in ultimate stress and ultimate strain. Residual forces would never be purposely built into an electronic assembly, but stresses (and the resulting strains) commonly arise from mismatch of the coefficient of thermal expansion (CTE) values between two or more materials that are tightly joined together. For instance, most polymers have a much higher CTE than most metals and, when a polymer board with thin metal conductors is heated, the expansion of the thicker board could cause the metal lines to break.

Ceramic materials tend to have very high Young's moduli or, in other words, they are very stiff. However, their ultimate strain tends to be low, compared to metals and plastic, so that they are prone to cracking if deformed by such forces as thermal mismatch or impact. Metals are strong compared to polymers but are also generally heavier and more expensive. Polymers have less ultimate strength than the other two classes of materials but higher ultimate strain, enabling them to tolerate considerably more deformation before failing.

2.2.2 Moisture Penetration

For an electronics packaging engineer, controlling the penetration of moisture into electronics packages is very important to reliability performance because the presence of water in an electronic system is almost always detrimental. Moisture can cause metals to corrode, polymers to swell, interfaces to delaminate, and some materials to lose mechanical strength and insulating ability. Polymers alone cannot prevent the penetration of water into a system; external changes in humidity are transmitted through millimeters of almost any polymer in time scales of less than a day. Only *hermetic* enclosures, which completely surround the assembly with metal and ceramic, can truly exclude water. Any path, even very small, through polymeric materials will allow moisture to intrude. Metals and ceramics both have such low moisture diffusivities that even submillimeter-thick layers will effectively exclude water permanently, if there are no cracks, pinholes, or entry paths along the material interfaces.

Some polymers are better than others in this regard, but none are truly hermetic. For example, fluorocarbon polymers repel water relatively well, while epoxies and polyimides (Kapton) readily absorb up to a couple of percent by weight of water. The actual water flux into a plastic package is proportional to the diffusivity of moisture through that material and the driving force for diffusion (the difference in relative humidity inside and out) and inversely proportional to the thickness of the material. Still, the diffusivity of water through plastic is approximately 10^8 and 10^{15} higher than the diffusivity of water through ceramic and metal, respectively. Although increasing the thickness of the polymer barrier increases the time required for water to penetrate it, the increased thickness does not add significantly to the water penetration time, but it does increase the path length for heat dissipation, so this is not usually a viable approach.

Moisture can also penetrate a package along interfaces. For example, in a plastic package, an interface exists between the plastic and the input/output (I/O) leads. If the adhesion of the plastic to the metal leads is poor, moisture can migrate to the interior of the package along this interface. To prevent such migration, adhesion between the two materials is very important.

2.2.3 Interfacial Adhesion

Electronic packages, for the most part, consist of multiple layers of dissimilar materials, thereby ensuring that several interfaces exist within any package. The integrity of these interfaces has a significant impact on the reliability of the packaging system. CTE mismatch and moisture penetration are major reasons for material delamination. In the previous section, the importance of interfacial adhesion between materials used to package integrated circuits (ICs) was noted. Adhesion results from some combination of mechanical, chemical, and electrical interaction between the two materials comprising the interface region. Surface energies of materials are a key factor in determining adhesion quality. Chemically, reactions can form chemical bonds between the constituent materials resulting in a third material that acts as an adhesion layer. Microscopically, the roughness of surfaces, which are to be joined, contribute to the mechanical strength of the interface. Electrostatically, surface charge can influence the adhesion between two materials.

Surface preparation prior to joining of the materials has a significant impact on the quality of the resulting bond, as does the selection of the process to create the interface. For example, precleaning of the metal surfaces and use of the proper temperature are critical steps in the formation of a reliable, low-resistance solder joint. In molding operations, agents are often added to the epoxies to enhance release of the package from the mold, which further complicates interface adhesion.

2.2.4 Electrical Properties

There are two electrical properties of interest in common packaging materials: resistivity (ρ in Ω-cm and dielectric constant (k or, less commonly, ε_r, dimensionless). Other electrical properties of interest might include dissipation factor and magnetic susceptibility, depending on the application. There are two main and opposite functions of electrical materials in packaging: conducting current and preventing current from leaking from one conductor to another. Also, materials are used as dielectrics in capacitors and as magnetic field concentrators in inductors, both of which can be embedded or discrete.

Probably the best known electrical property of a material is its resistivity in Ω-cm. It is an intrinsic property—one that is not a function of the size or mass of the material—and is related to the resistance of an object by:

$$\text{Resistance} = \text{resistivity}(\text{conduction length/conduction area}) = \rho(L/Wh) \qquad (2.1)$$

where R is resistance in ohms, ρ is resistivity in Ω-cm, L is the length of the material, h is the thickness of the material, and W is the width of the material.

Electrical conductivity is the reciprocal of electrical resistivity and is expressed as $1/\Omega$-cm or mho/cm, sometimes called Siemens/cm. Conductivity is related to the number of charge carriers, the charge per carrier, and carrier mobility, and is, in general, a function of the temperature and crystal structure of the material. Material resistivity is important for conductors, which readily transfer electrical charges in a circuit, and dielectrics (nonconductors), which prevent charge transfer. If a dielectric material is used solely as an insulator to prevent charge transfer, its dielectric strength is important. Dielectric strength is the value of the externally applied electric field (in volts/distance) at which breakdown or failure of a dielectric occurs, usually resulting in permanent damage to the dielectric. It should be noted that, although a material's ability to insulate is directly proportional to its thickness, surface area, porosity, and defects also influence its insulating characteristics.

Dielectric materials are those whose resistivity is so high that they are, effectively, insulators to the point that voltages common on the circuit board will not result in significant current through them. They have two applications: insulating conductors from one another and as the dielectric layer in capacitors. Requirements for the first application are fairly obvious: The material needs to be resistive enough and thick enough to prevent significant current flow between adjacent conductors that have a different voltage between them. The second application involves a much more exacting set of electrical properties. The most important of these is the relative dielectric constant k, often referred to simply as the dielectric constant, which can be understood in terms of a capacitor, a device for storing electrical charge. A capacitor consists of a negative and positive electrode separated by a dielectric material [12]. With a voltage applied between the two electrodes, the negative electrode stores charge, and charge is removed from the positive electrode. The amount of stored charge is dependent on, among other things, the dielectric material between the electrodes. Dielectrics do not transport electric charges, but they are not inert insulators because an externally applied field (voltage/distance) can displace the electronic and ionic charges from their normal positions in the internal structure of the material. The charges return to their normal positions when the electric field is removed. Referring back to the application of dielectrics as insulating layers, their dielectric constant can be important here as well, since it affects the way high-frequency signals are transmitted down signal lines. This is discussed in considerable detail in Chapter 6 and 9.

The charge Q (expressed as coulombs, or A-s), which is held by a capacitor, is proportional to the voltage V across the capacitor, and the capacitance C expressed in farads:

$$Q = CV \tag{2.2}$$

The capacitance further depends on the relative dielectric constant k and the geometry of the capacitor. For a parallel-plate capacitor:

$$C = \varepsilon_0 k A / h \tag{2.3}$$

where C is capacitance in farads, k is the dielectric constant, ε_0 is the permittivity of a vacuum (8.85×10^{-12} F/m), A is the electrode area, and h is the separation of the electrodes. Thus, the quantity of electrical charge stored in a capacitor at a given voltage is proportional to the dielectric constant. It should be noted that, in general, the dielectric constant is not really a constant at all; it is a function of temperature, film thickness, frequency, voltage, age, and other factors as well.

Thin-film dielectrics used in the semiconductor and discrete capacitor industry are either organic or inorganic materials, and the organics are usually polymers. Many polymers exhibit excellent electrical, thermal, and mechanical properties at low cost and, consequently, are popular materials in microelectronic packaging. Materials selected for use as dielectric layers in thin-film multilayer structures should, in general, exhibit a low dielectric constant (<4) to avoid signal distortion at high frequencies, good mechanical properties, a low coefficient of thermal expansion, close to that of silicon for chip carriers or to copper for circuit boards, low water and solvent absorption, thermal stability (up to 400°C), and good adhesion. Furthermore, the materials should be reliable, relatively inexpensive, and easily processed.

Polymers are very attractive for packaging applications from an electrical performance point of view. They generally are excellent insulators because both their bulk and surface resistivities are high. Polymer response to an applied voltage involves both dipolar and ionic responses, but they generally possess sufficient breakdown strength for packaging applications. However, moisture absorption reduces both the electrical resistivity and the breakdown strength. The alternating current (AC) properties of a polymer can be more important than its direct current (DC) resistance because of how it affects the propagation of high-frequency signals. The dielectric constant is a complex number composed of an in-phase and an out-of-phase component. The in-phase component is related to charge storage and the out-of-phase component results in power loss. The ratio of loss to storage is $\tan \delta$, which is referred to as the loss tangent or dissipation factor. For plastics, both parts of the dielectric constant and the loss tangent are functions of temperature, water content, impurities, degree of cure, and the frequency of the applied field.

The capacitance in a package resulting from the dielectric properties of polymers presents problems for electrical signal propagation. These problems include *RC* and *LC* time delays and signal distortion. The whole dielectric spectrum of an insulating material is of concern to packaging engineers because it impacts on package encapsulation, chip interconnections, and multichip interconnections. An advantage of polymers is their low dielectric constant and loss tangent factor. For example, if LC delay limits signal speed, the signal speed can be increased by using an insulating material of lower dielectric constant because signal propagation delay time is roughly equal to the square root of the dielectric constant of the insulating material.

2.2.5 Thermal Properties

Electrical engineers are most familiar with energy in the form of flowing electrons or trapped charge, but heat is nothing but another form of energy, stored as the vibrational energy of atoms in a solid or translational energy of atoms and molecules in liquids or gases. In other words, heat is manifested as the kinetic energy of atoms and molecules, and temperature is a measure of the density of that energy. As temperature is increased by the addition of heat, the molecules of a solid vibrate more strongly until their energy breaks the bonds between them, causing either melting or thermal degradation of the material. In the same manner, increasing the temperature of a liquid causes its molecules to break the weak bonds that hold a liquid together, resulting in vaporization. Thermal concepts such as these are explored further in Chapter 7.

This concept of heat and temperature also explains the coefficient of thermal expansion. As the molecules vibrate more strongly, they tend to move a larger distance about their average position, causing the material to effectively increase in size. The fractional increase in dimension per degree of temperature increase is the CTE, measured in ppm/°C. The glass transition temperature of a polymer is a phase transition that is weaker than that of better-known phase transitions such as melting or vaporization. At this temperature, the long strands of polymer molecules are vibrating so strongly that they break the weak bonds between the individual strands, resulting in a sudden increase in CTE of the polymer. It is usually desirable to operate well below this temperature to prevent the thermal stresses that can result from large CTEs. Polymers generally show much larger CTEs than metals and ceramics. Glass transition temperatures are seen mainly in thermoplastic polymers, those that can be melted and reformed, more weakly in thermoset polymers such as epoxies, and not at all in metals and ceramics.

A given temperature difference across a material results in the conduction of heat from high to low temperature. The constant of proportionality between these two quantities is the thermal conductivity in W/mK. Polymer materials have a thermal conductivity that is orders of magnitude lower than that of metals, with ceramics in between. The low conductivity of polymers, a very common material in electronic packaging, is problematic to maintaining acceptable temperatures in modern high-power systems.

2.2.6 Chemical Properties

Essentially all materials used by engineers suffer from deterioration of some kind, although usually at significantly different rates, depending on the source of the degradation. Chemical attack can come from acids, salts, gases, and even pure water. Basically, the ability of a material to resist chemical corrosion due to its environment is of primary importance. Since corrosion is frequently irregular in attack, it is very difficult to quantify. Consequently, the packaging engineer must be aware of the chemical interaction of materials used in packaging, such as polyimide and copper or the interdiffusion of metals, and take steps to minimize, if not eliminate, the possibility of chemical activity between materials that would lead to reduced reliability.

2.2.7 System Reliability

In simple terms, reliability is a measure of how well and for how long something is able to perform its intended function. Although operation forever without failure is the goal, the effort to ensure failure-free operation and the effort required to get a failed system back into operation after a failure are important considerations of reliability. The reliability of electronic systems begins with the design phase of the electronic components and eventually progresses to how well the engineering of the package is performed. Although package engineering involves many parameters, one of the more important ones is the management of heat from both internal and external sources. One of the very basic facts of electronic reliability is that reliability degrades with increasing temperature. In fact, accelerated life testing, that is, testing at elevated temperatures, is routinely performed on electronic components to determine the mean time to failure (MTTF). These topics are discussed in more detail in Chapter 16.

As noted previously, another important consideration is moisture. Moisture can be handled if hermetic packaging is utilized. However, this is accomplished at a significant increase in packaging cost. Molded plastic packages greatly reduce the cost of packaging ICs, but at the risk of reduced reliability, although dramatic improvements in moisture penetration of molded plastics have been made in the past few years. Coupled with passivation coating of ICs, molded plastic packaging has reached a point where its reliability is quite good.

Finally, the packaging engineer must consider the impact of shock and vibration on packaged electronics. For example, the flexing of printed circuit boards, or other interconnect substrates, under vibrational and/or shock forces can produce opens in conductor lines, cracks in dielectric layers, and other effects that can lead to premature failure of the product. These potential problem areas can be compounded if the packaging engineer reduces line widths, dielectric thicknesses, and the like to accommodate a requirement for reduction in the size and/or weight of the electronic system. The bottom line is that electronic package engineering plays a very important role in establishing the reliability of an electronic system.

The reliability of an IC package, whether it be a single-chip or multichip package, is only as good as the stability of the materials from which it is formed. The materials must be able to undergo temperature cycling during assembly and while in use without suffering degradation. Likewise, exposure to chemicals must not adversely affect the performance of packaging materials. Since the end goal is to ensure that the ICs in the package operate reliably over their designated lifetime, factors such as thermal stress cycling, moisture and other impurity intrusion, dielectric stress, interface delamination, overheating, and so forth, must be considered when selecting packaging materials.

2.3 CERAMICS IN PACKAGING

Ceramics are metal–nonmetal compounds that are attractive for electronic packaging applications because of their unique electrical, chemical, mechanical, and physical properties [13–16]. Some of the ones commonly encountered in electronics are silicon nitride (Si_3N_4), aluminum oxide (Al_2O_3), silicon dioxide (SiO_2), aluminum nitride (AlN), silicon carbide (SiC), magnesium oxide (MgO), tungsten carbide (WC), boron nitride (BN), and beryllium oxide (BeO). They consist of metallic and nonmetallic elements bonded together primarily by ionic and/or covalent bonds. Typically, they are hard and brittle with low toughness and ductility. In general, they are electrically and thermally insulating because of the lack of conduction electrons. Ceramic materials normally have high melting temperatures (see Table 2.1) and high chemical stability because of their strong bonds. Since all solid materials that are not metallic or organic are broadly referred to as ceramics, they may be amorphous, polycrystalline, or crystalline. Thus, glasses are a subgroup of ceramics and the term *glass–ceramics* commonly heard in the packaging industry is redundant, even though glasses and crystalline ceramics differ sufficiently for a distinction between them to be justified.

Table 2.1 Melting Temperatures of Some Ceramic Materials

Material	Melting Temperature (°C)
SiO_2	1715
Si_3N_4	1900
Quartz	1938
Al_2O_3	2050
AlN	2677
BeO	2725
MgO	2798
WC	2850
BN	3000
SiC	3100

36 Chapter 2 Materials for Microelectronic Packaging

The use of ceramics in electronics dates from the earliest vacuum tubes where they were used in envelopes and spacers. Likewise, from the very earliest stages of the evolution of microelectronics in the 1950s, ceramics have played a substantial role, not only as materials but also by providing special processing techniques, such as green tape, metallization, sealing, joining, and thick-film printing. As noted in Chapter 1, early efforts to provide a hermetic environment for ICs led to the development of the transistor outline (TO) metal package generally called a "TO can." It consisted of a gold-plated metal base containing space to mount the chip and up to 14 I/O leads. Electrical connection between the chip and package leads is accomplished by aluminum wire bonding. A nickel-based metal lid is welded to the base in a controlled ambient, such as dry nitrogen, to provide hermeticity. The rapid evolution of ICs soon made the TO package unsuitable because of its limited I/O capability. This led Fairchild to develop a rectangular ceramic package, in the 1960s, with two rows of leads on opposite sides of the structure. Because of the two rows of parallel leads, this structure was referred to by the name dual-in-line package, or DIP. There were two methods used to fabricate these packages. Figure 2.1 illustrates a lamination process in which a ceramic green sheet (or tape) is formed by doctor blading a slurry of alumina, glass, and organic binders. The green tape is then metallized with a tungsten paste by screen printing. The desired number of layers is then laminated by firing at approximately 1500°C, followed by nickel plating of the exposed tungsten metallization. The lid and external leads are then attached by brazing and exposed metal surfaces are gold plated.

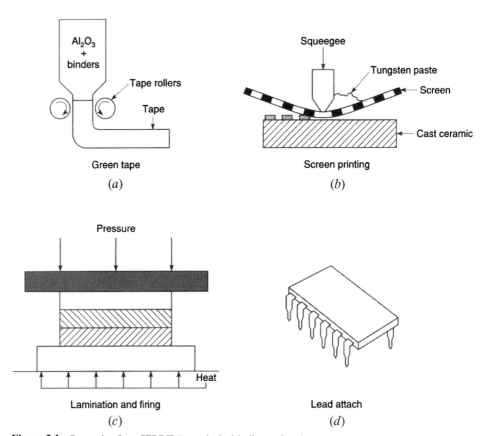

Figure 2.1 Processing for a CERDIP (ceramic dual-in-line package).

At about the same time, IBM developed a ceramic package called the solid logic technology package, or SLT. The base of this package, made of dry-pressed alumina, was square and contained screen-printed and fired (~800°C) conductors and resistors. I/O pins exited through the bottom and were soldered to the conductor lines. Semiconductor chips were then soldered in place and a lid was epoxy sealed over the module. These processes are explained in more detail in Chapter 5.

Modern ceramics find application in the packaging of discrete semiconductor devices, single integrated circuits, multiple integrated circuits, hybrid circuits, and circuit boards for high-performance packaging. They can be a constituent part or the only material used in substrates, circuit components and conductors, sealing compounds, and package covers or lids. One of their more attractive properties is that packages made entirely of ceramic, or ceramic in combination with metals, can be made hermetic. The particular application of a ceramic material in electronics packaging, as is the case for other type materials, dictates which of its properties are important. Generally, there is no material that possesses all the requirements for a given application. Thus, trade-off decisions must be made with regard to the various material properties to achieve optimization with respect to application requirements. Table 2.2 lists a number of ceramic materials along with some of their more important properties. Some of the ones most commonly used, or which have been targeted for use in electronic packaging, particularly high-density interconnect substrates, are discussed in the following subsections.

2.3.1 Alumina (Al_2O_3)

Alumina or aluminum oxide (Al_2O_3) based ceramics are the most commonly used ceramics in electronic packaging. They evolved because manufacturing facilities were already available since alumina had been used for some time in electron tubes, they were stronger than glasses that were being used as substrates, their electrical insulating properties were better than glasses, and their thermal conductivity was acceptable. It should be noted that sapphire is a single-crystal form of alumina and is used in the microelectronics industry, primarily as an insulating substrate.

Alumina compositions normally encountered range from 90 to 99+ weight percent purity with the other constituents being oxides of silicon (SiO_2), magnesium (MgO), and calcium (CaO). These oxides, in combination with some amount of powdered alumina, form a glass-phase material in the area between the grains of alumina, thereby bonding them together into a solid material after firing at high temperatures (1500 to 1600°C). The composition of this glass-phase grain boundary area is important for the adherence of thick-film materials to alumina when it is used as a substrate. Table 2.3 gives some properties of high-alumina-content ceramics.

To some degree, alumina represents an optimal combination of properties for many applications in electronic packaging: electrical resistivity, dielectric loss, and physical strength (high strength-to-weight ratio) are excellent. It is chemically and thermally stable and is the lowest cost ceramic material. Because the thermal conductivity is 20 times that of most oxides, it is attractive for high-power circuits. The flexural strength (modulus) is higher (2 to 4 times) than that of most oxide ceramics. However, thermal conductivity degrades very rapidly with weight percent of the glassy phase. In fact, it degrades to 50% of its maximum value for a weight percent of alumina equal to 90.

The powders, of grain size 1 to 3 μm, can be hot pressed, extruded, injection molded, or prepared in slurries for slip casting or tape casting. Tape casting utilizes a slurry that

Table 2.2 Properties of Some Ceramic Materials[a]

Material	Dielectric constant	Dissipation factor ($\tan \delta$)	Electrical resistivity (Ω-cm)	Thermal expansion (ppm/°C)	Thermal conductivity (W/m K)	Flex strain (MPa)	Density (kg/cm^3)
Alumina	4.5–10	0.0004–0.001	>14^{14}	6.5–7.2	22–40	300–385	3.75–4.0
AlN	8.5–10	0.001	>10^{14}	2.7–4.6	100–260	280–320	3.2
BeO	6.5–8.9	<0.001	>10^{15}	6.3–9.0	260–300	170–240	2.95
BN	4.1–5.4	0.10	>10^{14}	2.6–8.6	55–600	110	2.2–3.1
SiC	20–45	0.05	>10^{14}	2.8–4.6	70–270	450	3.0–3.2
Si$_3$N$_4$	5–10	—	>10^{14}	2.3–3.2	25–35	255–690	2.4–3.4
SiO$_2$	3.8	0.004	>10^{14}	0.5	1.6	50	2.2
Si	12	—	10^5	2.6	120–150	690	2.33
Glass	5.7–7.2	0.006	>10^{14}	9.2	2	50	2.9
Cordierite	4.5	0.400	10^6–10^{12}	2.5	2.5	70	2.7
Forsterite	6.2	0.50	10^{10}–10^{12}	9.8	3.3	170	2.9
Mullite	6.2–6.8	0.02	>10^{14}	4.0–4.9	5.0–10	140	3.1
Steatite	5.7	0.1	>10^{12}	4.2	2.5	170	2.7
Glass–Ceramic	4.5–8.5	0.002	>10^{13}	2.5–6.5	0.8–2.3	150–240	2.9

[a]Ranges of values (resulting from different phases, method of preparation, etc.) are provided when available.

Table 2.3 Properties of Some High-Alumina Content Ceramics at Room Temperature

%Al_2O_3	Dissipation constant	Dissipation factor (tan δ)	Bulk resistivity (Ω-cm)	Coefficient of thermal expansion (ppm/°C)	Flex strength (MPa)	Maximum service temperature (°C)	Thermal conductivity (W/m K)
85	8.2–8.5	0.0010–0.0014	>10^{14}	5.3–6.5	298	1400	7.5–13.0
90	8.8–10.0	0.0006–0.001	>10^{14}	6.2–6.8	339	1500	12–13
96	9.0–9.5	0.0003	>10^{14}	6.6–6.7	360	1700	21–25
99.5	9.5–9.8	0.0001	>10^{14}	6.9–7.1	381	1750	29–35
99.9	9.9–10.3	0.0003	>10^{15}	7.0–7.2	394	1900	37–45

consists of powder and organic-based binders, dispersants, deflocculants, and solvents. The slurry is drawn under a knife edge (doctor blading) onto a material from which it can be easily removed, usually polymer or glass. The result upon drying is a flexible thin "green" (unfired) sheet. The green sheet can be cut, punched, or otherwise shaped. After firing at high temperatures, the resulting material can be used as a substrate in electronic packaging. The surface finish depends on the method of manufacture, but normal processing yields a flatness of 3 to 25 µm/in. Lapping/ polishing can produce a surface flatness of about 2µm/in. and a surface roughness of about 100×. However, voids are still present after polishing and present problems that can be overcome by glazing with materials that are essentially inert or coating with another material such as polyimide.

Properties of this material, which limit its use and performance, are its relatively high thermal expansion, intermediate between silicon and copper, which prevents its use with large integrated circuits, and its moderately high dielectric constant, which becomes a problem for high-speed circuits. It is rarely used apart from being bonded to metals. However, refractory metals, such as tungsten and molymanganese (Mo-Mn), must be used for conductors because of the high firing temperature required for alumina. In some applications, electroplating of copper or nickel over the refractory metal is performed to lower the conductor resistivity.

2.3.2 Beryllia (BeO)

Beryllium oxide, or beryllia, has been around for many years and is, in many ways, superior to alumina, particularly with regard to thermal conductivity. Beryllia materials are formed in much the same manner as alumina-based materials. Unfortunately, beryllia powder is very toxic. So, the primary reason for its limited use is the health risk posed by airborne particles that impacts machining processes needed to shape it. Consequently, the toxicity factor requires that it be processed in facilities built to deal with this problem. Its cost is higher, about 10 times the cost of alumina, in part, because there are a limited number of suppliers willing to deal with this material.

For many years, beryllia was considered to be the only ceramic material with a high thermal conductivity. In fact, it has a thermal conductivity about 10 times higher than that of alumina-based materials, but decreases fairly significantly with temperature. However, it still finds use as a heat sink for TO packages, a heat spreader for high-power transistors, and a substrate for microwave devices. Its thermal expansion is comparable to alumina and its dielectric constant is lower than alumina. It is considered to have good dielectric properties for packaging. Generally, beryllia-based material is formed by dry pressing or tape casting. Sputtered or evaporated thin-film metals are used for microstrip circuitry. Cofired and postfired metallization systems have also been developed for use with beryllia. However, because of its toxicity, its use is generally restricted to applications where thermal management is of primary concern.

2.3.3 Aluminum Nitride (AlN)

Aluminum nitride is a relatively new packaging material that evolved during the 1980s [17–20]. It represents an alternative to BeO without the toxicity concerns. In general, its electrical properties (dielectric constant, loss tangent, resistivity, dielectric strength) and mechanical properties (bending strength, thermal expansion) are comparable to those of alumina and beryllia. Its thermal conductivity is slightly better than that of BeO in the temperature range where ICs operate. Although its thermal conductivity decreases with increasing temperature, the degradation is less severe than that of BeO. However, its thermal conductivity, which is a sensitive function of density, is significantly reduced by very low levels of contaminants, especially oxygen. On the other hand, the value of thermal conductivity of pure AlN material approaches that of aluminum. The dielectric constant of aluminum nitride, around 9, is comparable to alumina, so it has similar problems when used in high-speed circuits. Fortunately, both the dielectric constant and loss tangent vary only slightly with frequency.

Aluminum nitride, like most other ceramic packaging materials, has a thermal coefficient of expansion that increases with temperature. In magnitude, the CTE of AlN is comparable to, but slightly higher than, that of silicon and tracts with temperature in a similar fashion. Thus, it is attractive as a material for high-density interconnect substrates, which contain large-area ICs. AlN-based materials are formed using either low-temperature pressureless sintering or hot-pressing methods only slightly different from those used to form alumina- and beryllia-based materials. AlN powders are sintered at relatively high temperatures (1650°C) in an inert atmosphere, such as nitrogen/hydrogen, in order to prevent conversion of the Al to Al_2O_3. CaO and Y_2O_3 are also used during sintering because of their ability to scavenge oxygen and form yttrium aluminum oxide phases during densification of AlN. Formation by hot pressing, which utilizes binders and plasticizers, requires AlN powder of extremely low oxygen content for the reason noted above.

The surface roughness of as-fired AlN is worse than that of alumina because of a larger grain size than tape-cast alumina, which means larger voids. As-fired or ground material is available with a surface roughness of better than 35 μin. However, parts that have been lapped to a surface finish of less than 5 μin. are available. In general, conventional thick-film pastes cannot be used with AlN because of poor adhesion and other reasons. Efforts to develop new thick-film pastes for AlN has led to products based on Cu, RuO_2, and MnO_2. Similarly, NiCr/Au and Ti/Pt/Au thin-film metallization systems have been studied. Also, tungsten shows some promise for cofired metallurgy.

There is a reliability concern about the possibility of hydrolysis with long-term exposure to humidity at high temperatures. The susceptibility to hydrolysis, however, is a function of the surface chemistry, which can be altered by chemical or thermal treatments, producing relatively stable surfaces. Additionally, aluminum nitride reacts with alkaline solutions, which are often used in cleaning and plating operations in electronics.

2.3.4 Silicon Carbide (SiC)

Silicon carbide is another material that has been around for many years and has been used previously by the electronics industry [21]. Traditionally, it has been used because of its polycrystalline semiconducting phase. Only since Hitachi developed an electrically insulating phase in the early 1980s with a relatively high thermal conductivity has silicon carbide attracted significant attention as an insulating substrate material. In fact, in its pure form it is not a very good insulator, but small amounts of materials such as BeO and B_2O_3 can be used to increase its resistivity such that it can be considered electrically insulating. SiC combined with a small amount of BeO is called Hitaceram. When mixed, BeO segregates in the grain boundaries between the SiC grains so the grains of SiC remain very pure (see Fig. 2.2). However, this results in a dielectric strength that is lower than that of other ceramic materials. Hitaceram has a thermal conductivity about twice that of SiC (270 versus 140 W/mK), which is actually better than that of aluminum metal. In addition to the high thermal conductivity, SiC has an excellent CTE match to silicon. Unfortunately, it has a dielectric constant of 40 or greater at 1 MHz and a high dielectric loss factor, which presents problems for high-speed circuits. Other electrical properties are also poor.

The formation of solid SiC is accomplished by hot pressing (sintering) at temperatures in excess of 1900°C. Unfortunately, as with other ceramics, the surface finish of the raw material is a potential problem. However, the surface can be polished to an excellent finish. At the present time, compatible cofired metallization and acceptable processing technology do not exist to make SiC a viable ceramic substrate for packaging. However, there is some indication that cofired tungsten may be a potential metallurgy for cofired packaging systems.

2.3.5 Boron Nitride (BN)

Boron nitride exhibits three allotropic forms, but h-BN, a soft, graphitelike, hexagonal structure, and c-BN, a hard, diamondlike, cubic form, are the most commercially important

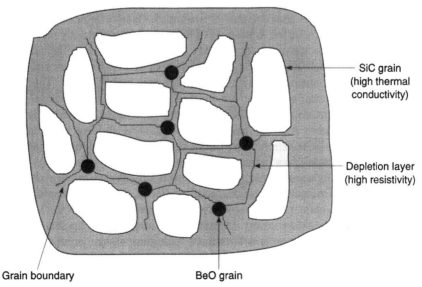

Figure 2.2 Structure of Hitaceram, a mixture of SiC and BeO.

[22, 23]. In bulk form, c-BN is formed from h-BN by subjecting it to high pressure and temperature. The c-phase, exhibiting a thermal conductivity of 600 W/m K, has been deposited in thin-film form via chemical vapor deposition (CVD). Many of the properties of hexagonal BN are anisotropic because of its layer structure. Furthermore, it is soft, weak, susceptible to moisture, and hard to metallize. Nevertheless, it is finding electronic applications as an additive to improve the properties of other ceramics (Al_2O_3/BN, AlN/BN, SiC/BN), as a heat sink material, and as a dielectric. It has a low dielectric constant and CTE.

2.3.6 Glass–Ceramics

Glass–ceramics are formed by adding glass to materials, such as alumina, in a slurry that then can be used in forming green sheet [24]. The controlled crystallization of glass in a slurry or green sheet is an alternative method of producing ceramics of high crystallinity during firing. The crystalline phase has a higher melting point than the glass from which it is made and is stable as long as the original processing temperature is not exceeded. The properties of glass–ceramics are somewhat tailorable by varying the percentage of glass versus ceramic. Glass–ceramic material technology originated with the Corning Glass Works in the 1950s and yielded well-known products such as Corningware.

With a dielectric constant of around 5, glass–ceramics are one of the best high-performance dielectric materials. Their properties actually approach those of an ideal substrate material for electronic packaging except for their low thermal conductivity. In addition to their low dielectric constant, they have a good thermal match with silicon and are very compatible with gold and copper metallization. Table 2.4 lists some of the more common glasses, ceramics, and glass–ceramic materials along with their dielectric constants at 1 MHz and room temperature. It should be noted that borosilicate glass has a dielectric constant of about 4 and has been used in the industry since about 1970 in multilayer thin-film packages because of its excellent thermal expansion match to silicon.

Cofired glass–ceramic technology was introduced by IBM in the late 1970s. It consisted of vitreous densification of cordierite, a magnesium aluminum silicate, and spodumene glass powders by crystallization below 1000°C (i.e., below the glass melting and deformation temperature) into cordierite and spodumene glass–ceramics. This process was combined

Table 2.4 Dielectric Constant of Some Glasses, Ceramics, and Glass + Ceramics at 1 MHz

Material	Dielectric constant
Glasses	
Borate	3.2
Silica	3.8
Borosillicate	4.0
Lead Borosilicate	7.0
Ceramics	
Cordierite	4.5
Mullite	6.7
Alumina	9.8
Glass + Ceramic	
Borosilicate + silica	3.9
Borosilicate + alumina	5.6
Lead borosilicate + alumina	7.8

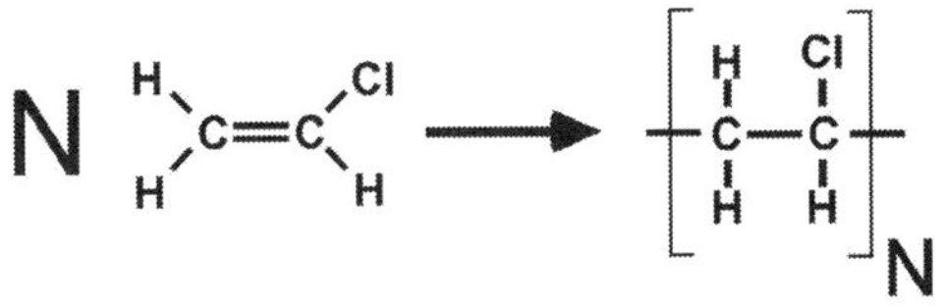

Figure 2.3 Vinlyl chloride monomer turned into PVC polymer.

with gold, silver, or copper to form multilayer substrates. New mixtures, formed by the addition of high thermal conductivity materials such as beryllia, silicon nitride, and diamond to glass, have also been explored in an attempt to further enhance the thermal conductivity of glass–ceramics.

2.4 POLYMERS IN PACKAGING

Polymeric materials, or *plastics*, are used in all aspects of electronic systems, such as package molding, passivation coatings, glob tops for chip-on-board (COB), die-attach adhesives, and printed circuit board (PCB) substrates [25–29]. Although polymers do not have the high hermeticity of ceramics or the strength of metals, they are very important in packaging due to their ease of processing, flexibility, insulating properties, and low cost.

2.4.1 Fundamentals of Polymers

A polymeric material is one that contains many chemically bonded units that form a chainlike structure [30–32]. The backbone of almost all polymers used in electronic packaging, and elsewhere for that matter, is carbon, although silicon alternating with oxygen is the linear structure in silicone polymers. The term *degree of polymerization* refers to the number of repeating units in the average molecule, which can exceed tens of thousands. Multiplication by the molecular weight of each repeating unit gives the average molecular weight for the polymer, which can exceed a million. The molecular architecture, or chain structure, is defined by the basic linear connection scheme exhibited by the polymer. Figure 2.3 shows the conversion of a monomer, chloroethylene (also known as vinyl chloride) into polyvinyl chloride, or PVC. While this double-bond opening mechanism is one of several common polymerization chemistries, it does show the general concept of a linear polymer backbone that is many repeat units long, in this case N, and the idea of pendant groups hanging off of that backbone, in this case Cl.

Polymers can be linear, branched, or cross-linked, as shown in Figure 2.4. In a linear polymer, the monomer is only connected to two other monomers if it is in the backbone or to

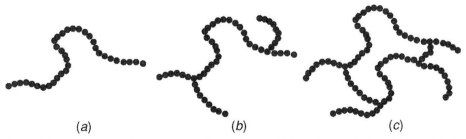

Figure 2.4 Three major chain architectures of homopolymers: (*a*) linear, (*b*) branched, and (*c*) cross-linked.

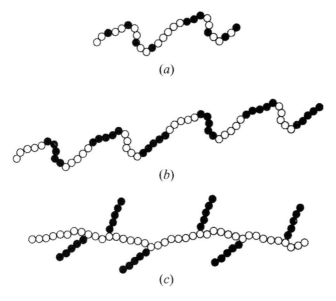

Figure 2.5 Possible monomer arrangements for copolymers: (*a*) random, (*b*) block, and (*c*) graft.

one if it is on the end. However, in branched and cross-linked polymers, some of the repeat units are connected to more than two other units at the branch or cross-link point. Cross-linked polymers have connections between individual polymer backbone molecules, resulting in a harder, more solvent-resistant, and unmeltable solid. Mixing component monomers by copolymerization can be used to create an infinite number of polymer alloys, as shown in Figure 2.5. The open circles are one type of momomer and the filled circles another.

An infinite number of variations in the basic linear monomer's connection scheme is possible. As shown in Figure 2.4, polymers can branch at points of triple functionality introduced at random points along the linear chain backbone, and side branches can grow off the main branches so that the polymer looks like a tree. Finally, cross-linked network polymers result from branches on polymer chains reaching from chain to chain, connecting them together in three dimensions to form one giant molecule. For instance, a bowling ball is cross-linked rubber and is one big 16-lb molecule! The glass transition temperature and brittleness of a polymer increase with the cross-link density. The glass transition temperature (T_g) is the temperature above which the individual polymer molecules can wiggle independently, so above T_g a polymer has a softer and more rubbery texture due to the relative motion of the individual strands. It also has a significantly higher CTE. Cross-linking is the process by which natural rubber is vulcanized into a stronger, stiffer polymer but, once cross-linked, it cannot be melted and reformed by molding, rolling, extrusion, and the like. Cross-linked polymers exhibit little or no T_g behavior since the chains are less free to move relative to one another.

There are also numerous possible arrangements of the units that make up the polymer. For example, the repeat units of a homopolymer are all identical, while copolymers are composed of more than one kind of monomer. Random copolymers are composed of monomers in no definite sequence of units. Regular copolymers contain a regular alternating sequence of two monomer units. The properties of both random and regular copolymers are usually quite different from those of the related homopolymers. Block copolymers are linear chains that contain long contiguous blocks of two (or more) repeating units combined in the chains. They retain most of the physical characteristics of the two homopolymers. The

Table 2.5 Dissociation Energy for Bond Types Present in Polymers

Bond type	Dissociation energy (kcal/mol)
Primary covalent	50–100
Ionic	10–20
Hydrogen bond	3–7
Dipole interaction	1.5–3
van der Waals	0.5–2

spatial arrangement of the units along the backbone can be random or ordered. Branched copolymers can have backbones that are a homopolymer with different grafted side chains.

What is clear at this point is that an infinite number of different polymer systems can be constructed from one or two monomers by varying the chain length, side branching, primary sequencing along the backbone, cross-linking, spatial arrangements, and composition. However, the chemical and physical characteristics of the resulting material are also related to the basic building blocks, the monomeric units, and the forces between them. These basic monomeric units are held together by various types of bonding forces present in polymers: (1) primary covalent, (2) ionic, (3) hydrogen bond, (4) dipole interaction, and (5) van der Waals forces. Primary forces hold the backbone together and provide cross-linking, while the other four are collectively known as secondary forces and, as shown in Table 2.5, are far weaker than covalent bonding and provide attraction between the various strands of polymer but are much weaker than covalent cross-linking.

2.4.2 Thermoplastic and Thermosetting Polymers

Thermoplastics and thermosets have already been defined with regard to cross-linking between polymer chains. From Table 2.5, it is obvious that the covalent bond is the strongest and the last one to be broken as the thermal energy of the material is raised. For linear and branched polymers, secondary bonds act as weak cross-links and hold the individual polymer chains together with each other.

Thermoplastics are high-molecular-weight polymers that can be melted and reformed any number of times because they do not have covalent cross-links between the separate polymer molecules. For temperatures below the glass transition temperature, they are stiff because the molecules are held in place with one another by secondary forces that act as weak cross-links. As the temperature is raised, a point is reached ($T > T_g$) where the individual molecules gain enough kinetic energy that they overcome some of the intermolecular secondary forces, causing them to wiggle independent of one another. This has two major effects: The material becomes much softer and the CTE increases. As the temperature is raised further, a point is reached where the energy of the system is sufficient to overcome the remaining secondary bonds and the polymer melts into a liquid. Since the viscosity of the resulting liquid is very high due to molecular entanglement, molding requires a large input of pressure energy to force the material into a mold. Table 2.6 gives properties of several thermoplastic polymers.

Thermosetting polymers are polymers that form covalent cross-links during final processing, usually referred to as *curing*. Chemical reactions are used to form covalent bonds between different molecules resulting in a more rigid, chemically resistant and higher T_g

Table 2.6 Properties of Some Thermoplastic Polymers

Material	Dielectric constant	Bulk resistivity (Ω-cm)	Loss tangent	Heat distortion temperature (°C)	Linear expansion (ppm/°C)	Tensile strength (psi)
Silicone polyimide	3.0	10^{15}–10^{17}	0.007	300–475	40	300
Polystyrene	2.4–3.1	10^{18}	0.0002	180	4	4000
Polyetheylene	2.3	10^{16}	0.0005	—	9.5	3000
Fluorocarbon	2.1	10^{18}	0.0003	250	5.5	600

material than a thermoplastic. The material is supplied to the molder in non-cross-linked form so it can be melted and formed into the desired shape. Covalent cross-links are usually formed in response to either an added cross-linking chemical or by heat or both. Once this conversion occurs, the further addition of heat does not cause the material to melt but can degrade the material since the cross-links require about the same energy to be broken as do the bonds along the backbone. When the dissociation energy of the primary covalent bonds is reached, the bonds of both the chains and cross-links fail, and the polymer degrades into a variety of organic gases. For most polymers, this happens around 250 to 350°C. Epoxies are the most used thermosets in electronic packaging, by far, since they make up the majority of FR4 circuit board materials and molded IC packaging, but polyimides and benzo-cyclobutane (BCB) are also cross-linked. Table 2.7 gives properties of several classes of thermosetting polymers.

2.4.3 Effects of Water and Solvents on Polymers

Polymer materials have porous structures that allow the absorption of water molecules or ions to a greater extent and much faster than through metals or ceramics [33]. Chemical solvents can penetrate in between the molecules of a thermoplastic polymer, causing them to separate enough to dissolve the polymer. On the other hand, cross-linked polymers generally will swell somewhat in solvents due to solvent penetration, but the solvent cannot separate primary covalent cross-linking bonds. Polymers that are lightly cross-linked will swell much more than those with extensive cross-linking. In other words, highly cross-linked polymer

Table 2.7 Property Values for Some Classes of Thermosetting Plastics

Material	Dielectric constant	Dissipation factor (tan δ)	Resistivity (Ω-cm)	Water absorption (%)	Heat deflection temperature (°F)	Tensile strength (psi)	CTE (ppm/°C)
Epoxies	4.6–5.0	0.01	3.8–9×10^{15}	0.04–0.12	250–400	15K–30K	1.7–2.2
Phenolics	6–10	0.1–0.7	10^{13}–10^{14}	0.5–0.7	340–500	7K–11K	0.88–2.5
Alkyds	4.6–4.7	0.02	10^{13}–10^{14}	0.07–0.08	350–400	3K–6K	2–3
Polyester	4.5	0.05	10^{14}	0.5	—	9K	2
Polyimide	3.4	0.01	10^{18}	2.9	680	17K	2.8
Polyurethane	3.0	0.02–0.075	10^{13}–10^{15}	0.11–1.1	190	1K–5K	12–25
Benzo-cyclobutene	2.7	0.0008	10^{19}	0.2	350	—	52

materials tend to be more resistant to water absorption. Ones, such as epoxies, polyamides, and polyvinyl alcohols, contain polar groups and absorb large amounts of water compared to nonpolar materials, such as fluorocarbons and hydrocarbons.

2.4.4 Some Polymer Properties of Interest

Polymers exist either as semicrystalline or amorphous materials; fully crystalline polymers can only be made from a few backbone compositions and then only in very small quantities. The amorphous materials do not exhibit the crystalline phase either as a solid or liquid, so they are similar to glasses and are often called *glassy polymers*. On the other hand, semicrystalline polymers contain both amorphous and crystalline phases. In these, a single polymer molecule extends through both amorphous and crystalline regions, and one molecule can go through several of each. Higher crystallinity can be achieved by solidifying the liquid polymer very slowly, giving enough time for individual polymer molecules to move to low-energy crystal sites. Some electrical, thermal, and mechanical properties of polymer classes are given in Table 2.8.

2.4.4.1 Dielectric Constant

For applications as interconnect substrates, the dielectric constant should be as low as possible, less than 5 at least, and constant over a wide frequency range [34–37]. If k varies with frequency, then some portions of a signal will travel faster than others through an interconnect and distortion will occur. Likewise, the dielectric loss tangent, tan δ, ideally would be low (<0.01) and frequency independent to minimize loss of signal energy. For many of the polymers used in electronic packaging, such as FR4 at about 4.6, the dielectric constant can be considered reasonably constant with frequency up to the low GHz. Fluoropolymers, such as Teflon-based materials, have the lowest k of any common substrate material at around 2 to 2.5 and the lowest dissipation factors but are considerably more expensive than epoxy-based substrates due to processing complications. Basically, the more polar and the more crystalline a material is, the higher its dielectric constant. Few polymers have k's higher than about 8, which is several times lower than that achievable from amorphous ceramic oxides and up to three orders of magnitude lower than that of highly crystalline ferroelectric ceramics.

2.4.4.2 Coefficient of Thermal Expansion

Differences in CTE of the materials used in electronic packaging are a primary source of mechanical stress. As is true for most materials, the CTE of polymers depends on many

Table 2.8 Ranges of Property Values for Some Polymer Classes

Polymer class	Dielectric constant	T_g (°C)	TCE In-plane (ppm/°C)	TCE out-of-plane (ppm/°C)	Modulus (GPa)	Water uptake (%)	Planarization (%)	Shrinkage (%)
PIs	2.5–2.8	300–400+	2–60	60–100	1.9–9.0	0.25–4.0	0.05–0.5	10–60
Fluoropolymers	1.9–2.6	160–320	90–300	90–300	1.4–1.6	0.01–0.1	Poor	None
BCBs	2.6–2.8	310–350+	65	65	3.3	0.25	Up to 0.95	Little
PPQs	2.8–3.0	360	40	40	3.45	0.1	0.2–0.7	20–35
PIQs	3.2–3.4	300–400+	3–58	—	3.8	0.8–1.0	0.4–0.5	0–35

factors, such as temperature, degree of cure, thermal cycling schedule, direction of measurement [38], chain orientation, and so forth in a complex fashion. In general, the CTE of polymers is considerably higher than those of common metals and ceramics. The CTE of thermoplastics is somewhat higher above their glass transition temperature. If thick films are desired, the polymer should have a low elastic modulus to offset stresses inflicted on it by CTE mismatch.

2.4.4.3 Glass Transition Temperature (T_g)

As noted previously, T_g is the temperature (or more correctly, the temperature range) above which the molecular chains of an amorphous or *glassy* thermoplastic start to wiggle independently due to thermal kinetic energy. Fully crystalline thermoplastic polymers—and there are very few of these—show almost no T_g, while semicrystalline thermoplastics show a weak T_g, usually in proportion to the fraction of amorphous phase. Highly cross-linked polymers may show little discernable T_g behavior. The glass transition temperature is important because several important properties change as T exceeds T_g. For example, above T_g the CTE increases, the elastic modulus decreases, and ultimate tensile strain increases.

2.4.4.4 Thermal Stability

There is no question that a dielectric must be stable and not outgas under thermal cycling such as might occur during subsequent processing, die attach, and rework. Physical changes, such as shrinkage due to solvent loss, can lead to delamination due to mechanical stresses and/or degradation of other properties such as adhesion, glass transition temperature, elastic modulus, and so forth due to increased cross-linking of the polymer [39]. Furthermore, outgassing can result in contamination of other materials in a multimaterial structure.

2.4.4.5 Mechanical Properties

In general, mechanical properties of a polymer, or any electronic packaging material for that matter, are important for the long-term reliability of the product. However, during the fabrication of the product, mechanical properties are also extremely important. Polymers are somewhat unique among packaging materials because mechanical stresses produce a response that is both elastic (like a solid) and viscous (like a liquid). The response can vary from tough to brittle to rubbery to flowing depending on the temperature and T_g. Thus, the response depends on such things as cross-linking density, chain orientation, and fillers. Some of the properties, which are usually specified in order to mechanically characterize a material, are Young's modulus, tensile strength, elongation, yield strength, ultimate strength, elastic modulus, and tensile modulus.

Mechanical failures of polymer films, resulting from stress in the films, are manifested in delamination of the film, cracking, crazing, or deformation zone formation and result in stress relief [36]. Delamination occurs when the adhesive strength of the material is exceeded, and cracking, crazing, and deformation void formation occur when the cohesive strength of the material is exceeded. Often these failures occur at points where the stress has been concentrated by foreign particles, vias, or corners and edges of metal lines. They also occur because of fatigue and weakness created by certain types of postdeposition processing.

2.4.4.6 Adhesion

Delamination in a multilayer system occurs if either the cohesive or adhesive strength of the layer material is exceeded [40–47]. Often, the adhesive strength of a material is weaker than its cohesive strength. Although delamination can be totally destructive, partial delamination can eventually lead to failure by allowing contaminants to "leak in" and cause corrosion on metals or further swelling of polymers. The adhesion strength of the interface between two joined materials depends on the properties of both materials and the manner in which the adhesive bond is created. Sometimes, the adhesion performance of one or both materials is enhanced by the use of an adhesion promoter, such as silane coupling agents or reactive metal layers. In microelectronics, polymers are required to adhere well to themselves, metals, ceramics, and glasses. Equally important is the adhesion of metals deposited onto polymer material. In general, good adhesion can be achieved between a polymer and a second material if their surfaces are free from contamination, if they wet each other well, and if the polymer is a thermoset. However, a wide range of adhesive behavior is exhibited by polymers interfaced with other materials.

2.4.4.7 Water Absorption

Unfortunately, polymers have a propensity for absorbing moisture, leading to undesirable swelling and changes in adhesion, dielectric constant, and stress in some situations. Under the right conditions, the opportunity for corrosion is enhanced, the polymer can blister, and hydrolytic molecular breakdown of the polymer can happen. Thus, it is critically important to properly cure polymers and to control the environment to which they are exposed.

2.4.4.8 Planarization

The ability to perform fine-line photolithography in both semiconductor manufacturing and high-density electronic packaging depends critically on the flatness and uniformity of the surface to be patterned. The ability to produce a flat surface over features is referred to as planarization. Planarity is defined as the percent of difference between the feature height before and after coating, divided by the original feature height. Thus, from Figure 2.6,

$$\text{Degree of planarization} = (1 - X_1/X_2)100\% \qquad (2.4)$$

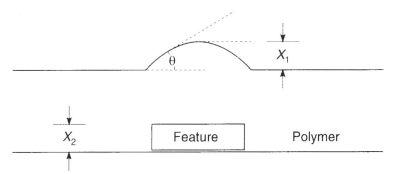

Figure 2.6 Planarization parameters for polymer overcoats.

The ability of a material, such as a polymer, to produce a planar surface when it is applied to a surface containing features such as metal interconnects, is determined by its viscosity, solids content, shrinkage, and flow during cure. High solid content, low viscosity solutions yield the best degree of planarization (>95% is achievable). However, a sufficient thickness of polymer must be deposited to ensure that the film has the required mechanical properties. Polymer films also shrink since they contain solvents, which are driven off during curing. Any shrinkage degrades the planarization. Finally, thermoset polymers can flow during curing resulting in increased planarization.

2.4.5 Primary Classes of Polymers Used in Microelectronics

Polymers are used in several key applications in microelectronic fabrication such as photoresists, intermetallic dielectrics, and packaging [42–44]. Within these major categories, specific applications include die attach, package molding, tape-automated bonding (TAB) substrates, PCBs, glob top encapsulants, chip passivation, interdielectric layers for multi-level metal ICs, and dielectric layers for interconnect substrates. In this section, the use of polymers in post-IC fabrication, particularly transfer-molded IC packaging, is addressed. The specific classification of plastic materials used in these applications include epoxies, silicones, and polyimides.

2.4.5.1 Epoxy

An epoxy system consists primarily of a resin and a hardener. The resin is the base organic binder material and holds all of the constituent materials together. It also imparts many of the inherent characteristics, such as glass transition temperature, moisture absorption, chemical inertness, and shrinkage on cure, to the epoxy compound. Hardeners or curing agents commonly of the catalytic, amine, or acid anhydride type, affect the viscosity and reactivity of the compound, establish the dominant chemical bonding, and control the cross-linking in the mixture. Hundreds of compounds have been found that will perform as hardeners, and they all produce different rates of curing, different reaction products, and different product properties. Figure 2.7 shows the chemical structure of the basic epoxy group and the two most common forms of epoxy. The benzene structure, which is an integral building block of two epoxy types, is also illustrated in the figure.

Accelerators are used to catalyze cross-linking of the epoxy when heat is applied during the molding process, thereby reducing the set time. The molding operation also requires the use of inert fillers or extenders with epoxy resins and curing agents in order to extend or enhance specific properties for a molding compound. Fillers reduce compound cost, increase hardness, decrease stresses due to a lower CTE, retard moisture absorption, and increase thermal conductivity. On the other hand, they can increase the density of the composite, increase the dielectric constant, lower the flexural and tensile strength, and reduce the ability of the material to flow during molding. Consequently, the type and amount of filler used must be given careful consideration. Materials that can be used as fillers are fused silica, quartz, glass, talc, metals, and clay. For present day IC packaging applications, inert silicas, such as fused quartz (SiO_2), are in common use.

Other less significant additives to epoxy compounds are flame retardants, mold release agents, and colorants. The flame retardant, usually a Br compound, is necessary to meet product safety requirements for plastic electronic modules. The mold release agent, normally

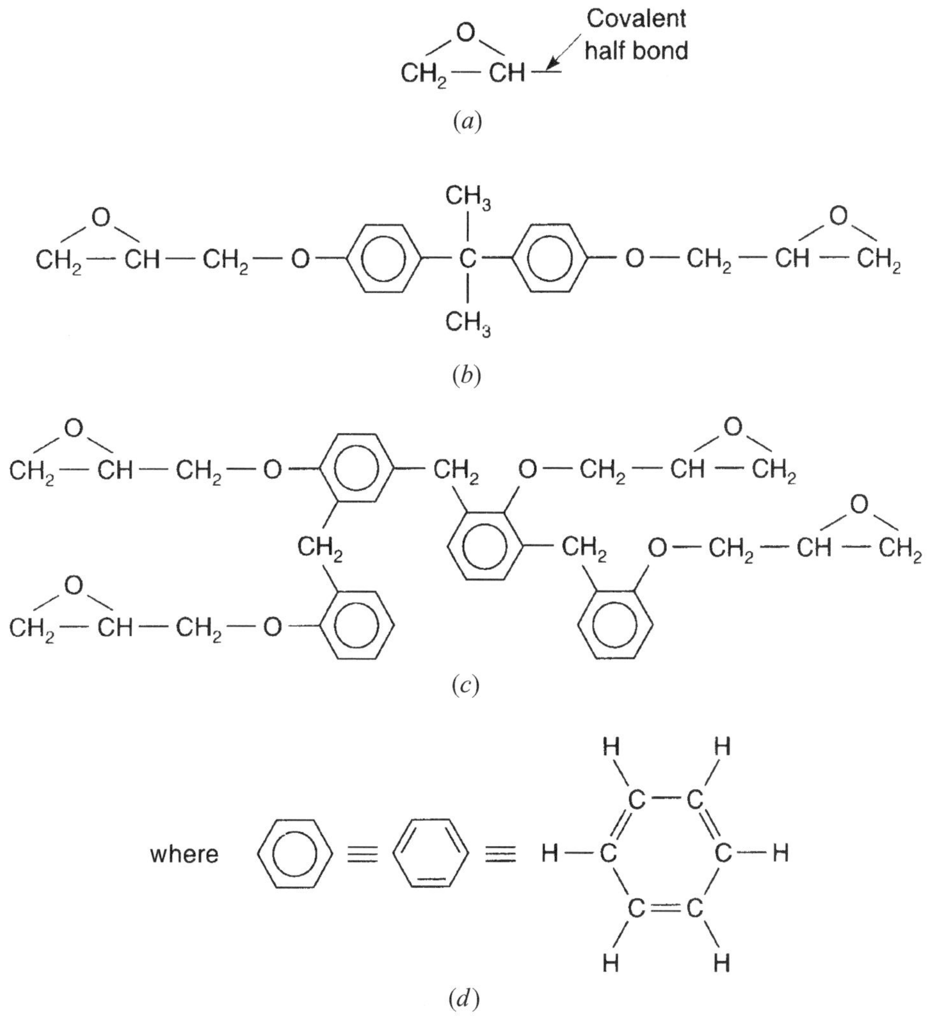

Figure 2.7 Chemical structures of (a) epoxide group, (b) diglycidyl ether of bisphenol-A (DGEBA), (c) epoxy novolar, and (d) the benzene structure.

a synthetic ester wax or a palm tree wax called carnauba, is used to facilitate the release of the package from the mold cavity. A colorant, either lamp-black or an organic dye, is added for its aesthetic effect. Molding plastic is often black because it's a cheap, easy color.

In the early years of plastic IC packaging, the excellent molding characteristics of phenolics, bisphenol epoxies, and silicones were used. Unfortunately, each of these compound types proved to have deficiencies that resulted in module failure during both stress testing and under operation in the field. For example, ICs in phenolic packages suffered from aluminum corrosion when subjected to temperature–humidity stress testing. Additionally, phenolic can emit ammonia and ionic contaminants during curing and readily absorb moisture.

In the 1960s, molding compounds, based on diglycidyl ether of bisphenol-A (DGEBA), a product of the reaction of epichlorohydrin (made from propylene and chlorine) and bisphenol-A, were used for molding packages. This compound contains chloride as a contaminant, an H_2O soluble ion, which, when in contact with moisture, vigorously attacks aluminum. Since the chloride can be diluted by the addition of certain materials to the base resin, efforts to reduce hydrolyzable chloride have been successful in reducing the levels to better than 400 ppm. Packages molded from bisphenol-A resin-based compounds also suffer from a failure mode referred to as *windowing* because they have a fairly low glass transition temperature (T_g) (100 to 120°C). Above T_g, the CTE increases dramatically. Consequently, during temperature cycling tests, which normally have a maximum temperature of 125°C, the expansion rate increases. The stress on bond wires can be enhanced to the point that the wires are pulled from their bond pads. The term "windowing" comes from the fact that, when cool, the bond wire makes contact with the bond pad. However, when heated, the wire is pulled away from the bond pad creating an open. The temperature range that alternately produces an open and contact is referred to as the temperature window. This problem, more than anything else, spelled the end for bisphenol-based molding compounds.

From the previous discussion, it is obvious that a molding compound must have a sufficiently high T_g so that it is not exceeded either during stress screening or operation in the field. Low moisture uptake is also a very critical property. These requirements led to the development of semiconductor grade novolac resins, produced by the reaction of phenol and formaldehyde in the presence of acid, in the 1970s. They have high cross-linking density, which implies a low moisture absorption rate, are solids at room temperature, have a T_g of around 150°C, and are thermally stable. Epoxy novolacs are produced from epichlorohydrin using the same reaction process as for the bisphenols. Novolac-based molding compounds for semiconductor packaging generally consist of cresol novolac epoxy resin, a curing agent, an accelerator, a flame retardant, an inert filler, a mold release agent, and a colorant. Generally, a molding compound consists of about 25% resin and 72% inert filler with the other constituents contributing the remaining 3%. Epoxy novolac compounds developed for semiconductor packaging are the primary molding compounds used today and are also used in glob top encapsulation and the construction of PCBs.

2.4.5.2 Silicone

Silicones are a class of generally soft polymers based on a chain of repeating silicon–oxygen bonds with hydrocarbon side groups attached to the silicon atoms. The silicon–oxygen bond is somewhat stronger than the carbon–carbon bonds found in most organic resins, and, consequently, their properties and behavior depart from conventional polymers. They have a wider useful temperature range, reasonable water resistance, oxidative stability, and lack ionic contaminants. The T_g is low for these materials, although they can maintain their useful characteristics up to 350 to 500°C and remain fairly stable and usable at temperatures down to −80°C, unlike most other polymers. Silicones have been used in the electronics industry for encapsulation since the 1960s, but they never have enjoyed the popularity of epoxies even though they exhibit low-dielectric constants, high heat resistance, low moisture absorption, and good low-temperature properties.

Molding compounds are made by combining a silicone resin with fillers, such as fused silica, carbon black, or titanium oxide. Fused silica enhances mechanical and rheological properties, while carbon black and titanium oxide are used as radiation screens. Silicone has the problem of a CTE, which is about double that of most polymers, but the low modulus usually more than makes up for it. Figure 2.8 shows the chemical structure of the repeating unit of silicone.

There are several processes used to manufacture silicones. Two examples are the reaction of silicon and methyl chloride to form methyl silicone and the reaction of silicon and

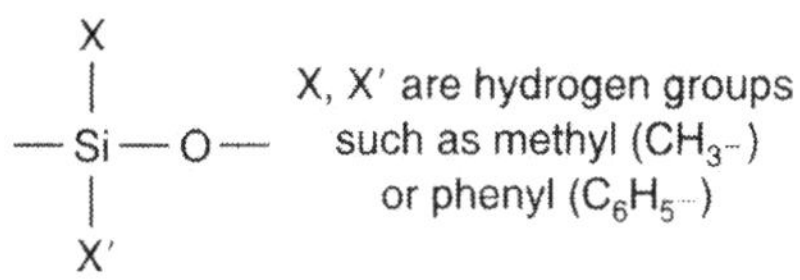

```
    X
    |
 —Si—O—        X, X' are hydrogen groups
    |          such as methyl (CH₃–)
    X'         or phenyl (C₆H₅–)
```

Figure 2.8 Basic chemical repeating unit of silicone.

chlorobenzene to form phenyl silicone. Commercially, the primary source of silicone polymers is the Rochow process in which a heated bed of copper and pure silicon is subjected to alkyl or aryl monohalide. The organohalosiloxane complexes resulting from this reaction are purified and hydrolyzed into a blend of cyclic siloxane oligomers, which are ring-opened into linear polymers later, and linear hydroxy end-blocked siloxane polymers. Resins with viscosities that vary over a 4 order of magnitude range can be produced, although not all viscosities can be used in encapsulation processes.

Commercial encapsulation silicones are available in one- and two-part systems and can be cured via amine, oxime, alcoxy, and acetoxy condensations. Carboxylate curing is not desirable since one by-product is carboxylic acid, which can corrode circuit metallization. Alcoxy cure, although a slower process, is preferred for circuit encapsulation because the by-products are alcohols.

2.4.5.3 Polyimide

Polyimide and polyimide copolymers have been available since the 1920s and have found widespread use in the electronics industry as chip carriers, laminates, film insulation, flexible circuits and cables, moldings, fiber reinforcements, tapes, and circuit board coatings. They are the primary material of choice as an organic dielectric because they possess most of the ideal characteristics and, in general, they are well characterized in terms of dielectric and electrical performance in microelectronic applications.

Polyimides are cyclic-chain polymers characterized by the imide functionality, a cyclic secondary amine bound to two carbonyl groups shown in Figure 2.9. Synthesis of polyimides occurs in two stages. First, polyamic acid is formed by a polycondensation reaction of an acid dianhydride with a difunctional base. Then, the polyamic acid is converted to polyimide by a high-temperature operation (250 to 350°C), which removes the solvent and water (see Fig. 2.10). The conditions of this curing or imidization step determine the electrical, mechanical, and thermal properties of the resulting polyimide. The isolation of the first-stage product, polyamic acid, allows end users of polyimides to apply it in polyamic acid form and then convert it to polyimide during a subsequent high-temperature operation. After complete curing, the polyimide is insoluble in virtually all common solvents.

Polyimides have excellent thermal stability and are resistant to temperatures up to 350°C for extended periods of time. In general, they have a low dielectric constant, typically 3.5 or

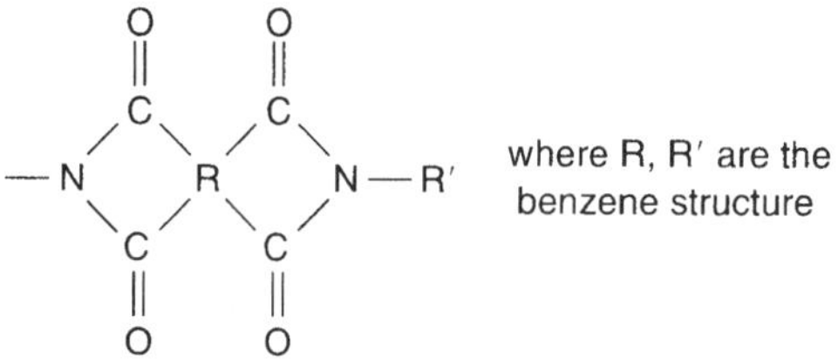

Figure 2.9 Imide functionality of polyimide.

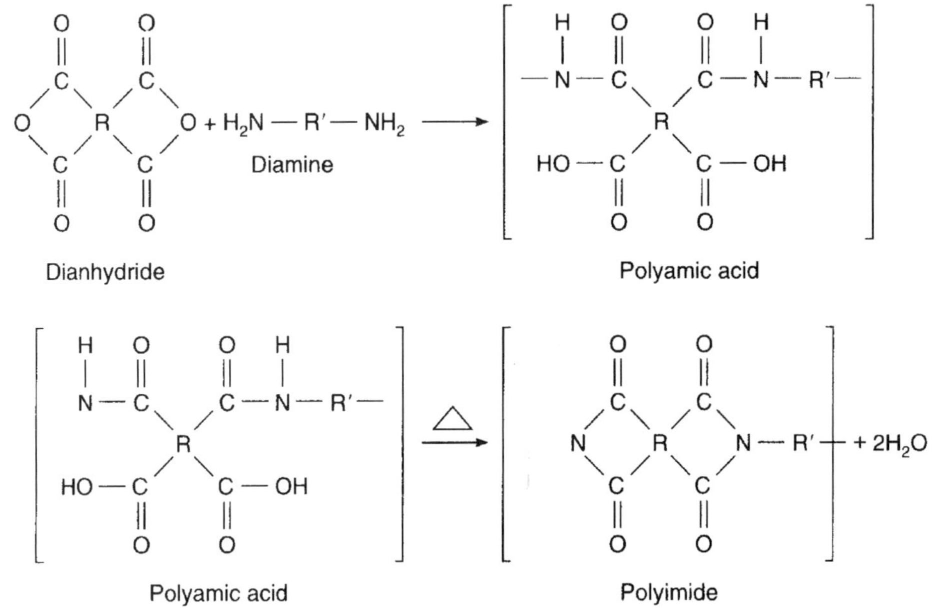

Figure 2.10 Chemical reaction by which polyimide can be created.

less, and a dielectric strength and volume resistivity comparable to silicon dioxide, which makes them attractive for use in high-performance circuitry. Unfortunately, the dielectric properties of polyimides can be a function of film thickness, cure conditions, and moisture content. Less desirable features are a high CTE, althouh they have less residual stresses than conventional dielectrics, and low thermal conductivity. However, they have excellent solvent resistance, although they are attacked by alkalies and organic acids. Since they are based on the imide ring, they do suffer from some problems with hydrolytic instability. The fact that they generally adhere well to metals and ceramics, coupled with the single most important property of polymides, the ability to planarize underlying topology, makes them particularly attractive for interlevel dielectric applications in multilevel metal structures. Patterning can be accomplished by wet etching or dry etching, using an oxygen plasma, reactive ion etching (RIE), or reactive ion milling (RIM).

As noted previously, the conventional approach to the use of polyimides in IC fabrication was to spin coat the polyamic acid precursor of polyimide and then subject the material to a sufficiently high temperature to cause conversion to polyimide. Later, preimidized material in solution form was made available to IC manufacturers. The processing of these materials is similar to that of conventional polyimides with the exception that a significantly lower curing temperature (250°C) is required. The resulting polyimide films are soluble in many common solvents. In addition to the low bake temperature, preimidized material has the advantages of less shrinkage, less outgassing, and an easy rework capability.

A specialized type of polyimide is finding increased application in microelectronic fabrication, combining the photosensitive properties of photoresists with the electrical and mechanical properties of conventional polyimides [45–48]. They contain photoreactive groups that cross-link with adjacent polymer chains when exposed to ultraviolet (UV) light. The exposed and unexposed portions of the film then exhibit different solubilities and can be developed like conventional negative photoresists. Prior to their introduction, photoresist had to be applied over the polyimide, defined, and developed using standard photolithographic techniques in order to pattern the polyimide for etching. Thus, the availability of photosensitive polyimides has reduced the complexity of polyimide patterning by the addition of photoreactive groups to the polyimide, which reduces the processing steps and increases throughput. However, photosensitive polyimides do have lower resolution than photoresists because of shrinkage in film thickness during heat treatment.

2.4.6 First-Level Packaging Applications of Polymers

Plastic packaging of single chips simply involves physically and electrically attaching a chip onto a lead frame, followed by encapsulation in a dense and rigid plastic using transfer molding. The plastic provides protection for the IC and defines the shape of the outer surface of the package. The popularity of plastic packages is evident from the fact that they account for over 80% of the packaged ICs sold today. The economic impact of plastic on the IC packaging industry has been such that the cost of plastic packaging a given part is about one-fourth the cost of hermetic ceramic packaging.

Discrete transistor encapsulation in plastic dates from the late 1950s. A very simple packaging technique, referred to as "glob top," used a ceramic header to hold a wire-bonded chip that was covered with a "glob" of cured epoxy resin [49]. A second technique, dating from the same time period, utilized a metal frame to support the chip and provide for electrical connections. Following wire bonding of the transistor to the metal frame, the structure was covered by injection molding plastic around the chip, wires, and metal frame to form the package. This package was the forerunner to all plastic IC packages used today.

The DIP evolved from the flatpack because of economic pressures to develop an IC package whose terminals (generally limited to about 64) could be inserted through a PCB and wave soldered into place. Although originally constructed from ceramic, glass, and metal, the need to reduce the cost of packaging ICs quickly led to the development of the molded plastic DIP. The molded plastic DIP was the most popular IC package when plated through-hole mounting on PCBs was common. With the advent of surface mount, other plastic packages have evolved to replace the DIP.

During the 1980s, the use of solder pastes and reflow soldering of PCBs gave new life to variations on the original flatpack packaging concept. The plastic small outline package (SOP), previously shown in Figure 1.8, is actually constructed like a DIP except that the leads are designed for surface mounting. As the need for I/O connections increased, plastic leaded chip carriers (PLCCs) were developed that have pins on all four sides. By reducing the width of the pins and their spacing (pitch), pin counts as high as 400 are possible. Reducing the lead pitch also allowed the development of a quad flatpack with a lead count of 200. Pin grid arrays (PGAs) are another form of first-level packages. They differ from other packages in that they utilize a circuitized plastic substrate with connecting pins spaced either in multiple rows around the perimeter of the substrate, or uniformly over the entire bottom surface of the substrate, yielding pin counts in excess of 300 for single-chip packages and over 800 for multichip packages. These are high-cost packages and present problems in routing of printed wiring board (PWB) conductor lines to all the pins. Nevertheless, for all of these packages, the basic construction is the same as that used for plastic DIPs, which consists of a metal lead frame or circuitized substrate, a chip, chip bonding material, chip bonding wire, and plastic encapsulation material. Figure 2.11 shows illustrations of the most common plastic packages.

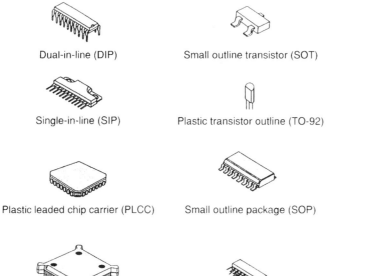

Figure 2.11 Some common plastic packages.

2.4.6.1 Plastic Molding

Thermosetting polymers are used for plastic IC packaging. In the early days of plastic packaging, the most widely used materials were epoxy-based resins, although silicones, for high-temperature applications, and unsaturated polyesters, which cure very rapidly, were used to some degree (see Table 2.9 for a comparison of the properties of these materials). Because epoxies react with a variety of curing agents, resin systems can be formulated with a wide range of properties, such as high mechanical strength, low cure shrinkage (1 to 2%) yielding low residual stress, good wetting and adherence, fast curing, low moisture permeability, good dimensional stability, reasonable end-use temperatures of 125 to 150°C, good dielectric properties, and a wide range of viscosities. Thus, they are attractive from a processing point of view, but their reliability performance is less than desired.

Transfer molding is the real workhorse of the IC plastic packaging industry, which, in fact, is really nothing more than compression molding. A second plastic packaging technique is called radial spread coating, a process like glob top, used in TAB encapsulation. In general, this approach yields rather thick coatings. Finally, reaction-injection molding is accomplished by rapid in-mold polymerization of two to three streams of reactive molding

Table 2.9 Property Value Ranges for Molding Epoxies, Silicones, and Polyesters

Material	Dielectric constant	Dissipation factor ($\tan \delta$)	Resistivity (Ω-cm)	Water Absorption (%)	CTE (ppm/°C)	Thermal conductivity (W/m k)	Linear Shrinkage (%)
Epoxy	3.2–5.0	0.01–0.03	10^{15}–10^{16}	0.04–0.2	9.4–31	0.25–0.87	0.3–0.5
Silicone	2.7–3.7	0.001–0.003	10^{14}–10^{15}	0.12–0.15	15.6–22	0.22–0.45	0.2–0.4
Polyester	3.1–4.7	0.0016–0.03	10^{14}–10^{15}	0.3–1.4	11.1–44.4	0.16–0.58	3.0

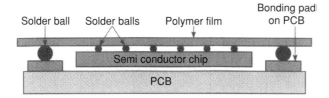

Figure 2.12 Polymer film-based chip carrier (PFBCC).

components. The reactants undergo turbulent mixing and are then conveyed to the mold cavity.

2.4.6.2 Thin-Film Packaging

In place of a molded plastic chip carrier, a polymeric film-based chip carrier (PFBCC) can be employed as shown in Figure 2.12. The thin-film carrier, based on polyimides (e.g., Kapton) or polyesters (e.g., Mylar), is flexible, which has contributed to its use in cables and TAB. These materials also have a low dielectric constant, which contributes to better electrical performance. TAB chip carriers have a lead frame formed on the polymeric film carrier by metal deposition, photolithographic definition, and etching. PFBCC is merely a variation or extension of TAB. In TAB, the film carrier acts as temporary support until the chip is excised and mounted to a substrate via the metal I/O leads. In the PFBCC, the circuitry is supported by the thin-film carrier so that the chip is not excised from the film. Other materials are being considered for application in PFBCCs, but they must meet the requirements for interconnect metallization and chip attachment.

2.5 METALS IN PACKAGING

The number of different metals and combinations of metals used in the electronics industry is quite large. However, the following discussion will be restricted to those metals that find application in level 1 and 2 of the packaging hierarchy. In particular, metals used to provide signal paths from the chip to the package, metals used to provide interconnection for multichip packages, multichip modules, and package construction including lead frames.

2.5.1 Die Bonding

For cases where the backside of an IC must be electrically connected to the outside world, or if heat transfer from the IC to the package is of primary concern, a metal-based adhesive can be used to attach an IC to a package or substrate. In the early history of IC packaging, a gold–silicon eutectic (6% Si–94% Au) was used to mount silicon chips to gold-plated package bases, and then electrical connectivity was provided by wire bonding. Other eutectics that have found application as die bonding material include gold–tin (Au–Sn) and gold–germanium (Au–Ge). Since these are hard solders, they have high flow stresses and, thus, offer excellent fatigue and creep resistance. On the other hand, they create high stresses in the chip due to thermal mismatch with the material on which the chip is mounted. Softer solders, such as lead–silver–indium (Pb–Ag–In) and lead–tin (Pb–Sn), reduce stress in chips.

More recently, silver-filled glasses have found application as die bond adhesives. These materials exhibit excellent thermal stability and normally yield a void-free interface between

Chapter 2 Materials for Microelectronic Packaging

Table 2.10 Properties of Some Die-Attach Materials

Material	Resistivity (Ω-cm)	Dielectric constant	Dissipation factor (tan δ)	Shear strength (MPa)	Density (kg/m^3)	Thermal conductivity (W/m K)	TCE (ppm/°C)	Maximum temperature (°C)
Silicone	10^{13}–10^{15}	2.9–4.0	0.001–0.002	—	7.9	6.4–7.5	262	—
Polyurethane	3×10^{10}	6.0–8.5	0.05–0.06	15.5	1.4–2.0	1.9–4.6	90–450	—
Epoxy novolac	10^{13}–10^{16}	3.5	0.016	26.2	—	—	—	—
Epoxy phenolic	6×10^{14}	3.4	0.32	—	—	25–75	33	—
Epoxy bisphenol-A	10^{14}–10^{16}	3.2–3.8	0.013–0.024	—	—	—	—	—
AuSn	4×10^{-7}	—	—	185	14,520	251	16	280
AuSi	8×10^{-6}	—	—	—	1,568	293	10–12	370
SnIn50	10^{-7}	—	—	—	—	—	—	117
AgIn90	8×10^{-8}	—	—	—	—	—	—	114

the chip and substrate. However, the presence of solvents and binders, coupled with high curing temperatures, create some special processing challenges. Epoxy and polyimide materials filled with metals have found application in plastic packaging. Outgassing and poor thermal stability have hindered their acceptance for high-reliability packaging. Silicone gels are also receiving some attention for this application. Polymer die bonding is attractive because it is faster than eutectic bonding. The properties of a few die-attach materials are given in Table 2.10.

2.5.2 Chip to Package or Substrate

As noted previously, connections made between ICs and packages or substrates is usually accomplished using wire, tape-automated, or flip-chip bonding. Each of these chip-to-package (or substrate) interconnect methods have unique material requirements and are discussed separately in the following sections. Figure 2.13 illustrates the distinctive features of each chip-to-package (or substrate) interconnect method.

2.5.2.1 Wire Bonding

The early history of microelectronics saw the development of ultrasonic wire bonding for use with aluminum wire to provide electrical connections from a chip to its package. Actually, pure aluminum is difficult to draw into a wire because it is too soft. To toughen the aluminum, silicon (1%) is added. Unfortunately, 1% silicon exceeds the solubility of silicon in aluminum at room temperature. This can lead to silicon precipitation, forming a second silicon phase, which causes hardening of the wire. Magnesium can also be added to aluminum to allow it to be drawn into wire. Approximately the same amount of magnesium is needed (0.5 to 1%) to strengthen aluminum sufficiently for bonding wire. The 1% magnesium concentration in aluminum is below the equilibrium solid solubility at room temperature, so second-phase magnesium is not a problem. Furthermore, the resistance to degradation of 1% Mg–Al wire at elevated temperatures is much better than that of 1% Si–Al wire.

Gold wire also finds use in microelectronics packaging when it is alloyed with materials such as beryllium and copper. Beryllium-doped wire, is about 10 times stronger than

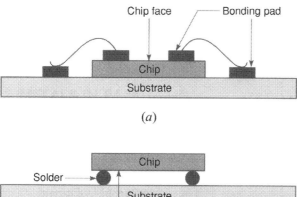
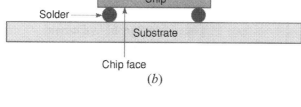
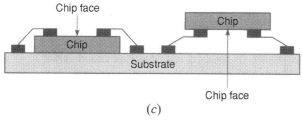

Figure 2.13 Illustrations of (*a*) wire, (*b*) flip-chip, and (*c*) tape-automated bonding.

copper-doped wire, which is an advantage in automated thermocompression bonding. Unfortunately, the price of gold has prompted a search for other bonding wire materials. To date, focus has been directed at silver, copper, and palladium for both thermocompression and thermosonic bonding.

Low cost and resistance to sweep during plastic encapsulation are the advantages of using copper wire for wire bonding. Copper wire bonding must be performed in an inert ambient to prevent the formation of copper oxide, which degrades the quality of the wire bond. Also, since copper is harder than gold or aluminum, metals onto which copper wire is bonded must be harder.

Some properties of aluminum, gold, and copper bonding wire are given in Table 2.11. The metallizations used for bonding pads on chips, substrates, and packages to which wire bonds must be attached include aluminum, gold, silver, nickel, and copper. Silver is usually

Table 2.11 Properties of Bonding Wire

Material	Thermal conductivity (W/m K)	Melting point (°C)	Electrical resistivity (Ω-m)	TCR[a] (Ω-m/°C)	Elastic modulus (GPa)	TCE (1/K)	Hardness (Brinell)	Elongation (%)
Al	237	660	2.7×10^{-8}	4.3×10^{-11}	35	4.6×10^{-5}	17	50
Au	319	1065	2.3×10^{-8}	4×10^{-11}	77⁰	1.4×10^{-5}	18.5	4
Cu	403	1085	1.7×10^{-8}	6.8×10^{-11}	13⁰	1.6×10^{-5}	37	51

[a]TCR = temperature coefficient of resistance.

the metal plated onto lead frames and alloyed with platinum or palladium in thick-film hybrid circuit conductor pastes.

2.5.2.2 Tape-Automated Bonding

Tape-automated bonding (TAB) is a method of providing electrical interconnection between specially prepared silicon chips and second-level packaging by bonding ICs to finely etched metal leads using thermocompression bonding, as illustrated in Figure 2.14. The leads usually are supported by polymeric film (50 to 125 μm thick), but a support film is not a requirement. The conductor metal is usually either "rolled and annealed" or electroplated copper (20 to 75 μm thick), although aluminum and rolled nickel can be used. The two approaches to forming the copper TAB interconnections yield materials with different microstructure and mechanical properties, which impacts bond formation. Electroplated copper often requires thin layers of sputtered chromium and copper to promote adhesion of the plated thick copper to the polymer. A barrier layer metal, such as tin, gold, or gold–nickel, is generally electroplated to the copper to promote solderability of the TAB lead to the IC and substrate.

Attachment of the TAB leads to the IC involves the use of aluminum, gold, copper, or solder (95Pb–5Sn) bumps on either the TAB tape or the IC, as illustrated in Figure 2.15. Gold is by far the most popular bump metal. The interface between the bump and the chip pads normally has a special metallurgy consisting of adhesion, bonding, and barrier layers. Barrier metals of palladium, titanium–tungsten, platinum, copper, or nickel, covered with a thin layer of gold, which acts as the bonding layer, are used between the aluminum bonding pads on the chip and a gold bump. Metals used to assure adhesion to both the aluminum

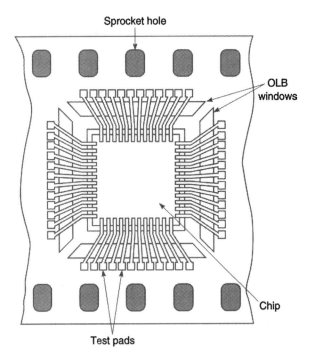

Figure 2.14 Illustration of an IC mounted on a TAB tape.

2.5 Metals in Packaging 61

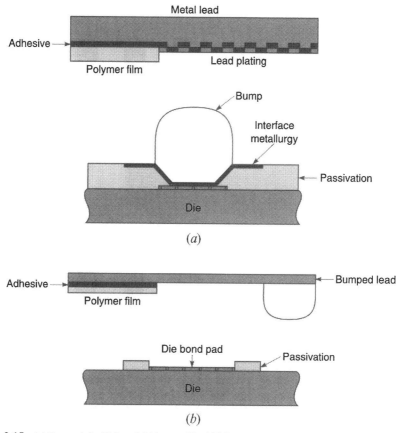

Figure 2.15 (*a*) Bumped die TAB and (*b*) bumped lead TAB.

bond pads and silicon dioxide on the chip can be titanium, chromium, or aluminum. If the chip is to be soldered to the TAB tape, gold or copper bumps can be tin-plated. Connection between the TAB lead and the substrate typically uses gold–tin solder bumps. Recently, formation of the bumps using electroless nickel plating has been studied. The polymer tape can be polyimide, polyester (Mylar), or epoxy/fiberglass.

2.5.2.3 Flip-Chip Bonding

Flip-chip bonding is a first-level chip to package connection technology that has its roots in the controlled-collapse chip connection (C4) process developed by IBM. In this bonding technique, ICs are directly bonded to pads on a substrate in a face-down orientation, providing the shortest interconnection path from IC to substrate. Most ICs that are flip-chip bonded must undergo a process called bumping, which is very similar to the bumping process for TAB. In this process, multiple plating operations are used to place a specific thickness (bump) of a tin–lead alloy on each bonding pad of the IC [50–52]. The solder bumps are generally 100 to 250 μm in diameter and 50 to 200 μm high. The locations of the solder bumps match the locations of bonding pads on the substrate, as shown in Figure 2.16.

62 Chapter 2 Materials for Microelectronic Packaging

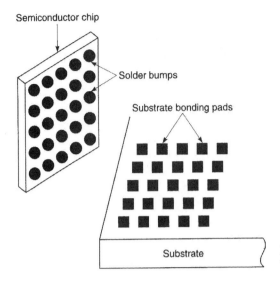

Figure 2.16 Illustration of solder bump and bonding pad alignment in flip-chip bonding.

The bumps used for flip-chip bonding are Pb–Sn or Pb–In solder of various compositions, depending on desired characteristics of things such as ease of fabrication, reflow temperature, mechanical performance, and corrosion resistance. The metallurgy of the 95Pb–5Sn C4 solder bump is shown in Figure 2.17. The metals between the solder and IC bonding pad promote adhesion, ensure wettability by the solder, and provide a barrier between the solder and the underlying IC. Note that the bottom layer of metal must adhere well to both aluminum and the dielectric material, which forms the solder dam. Both chrome (Cr) and titanium (Ti) are excellent barrier materials, but they are not solderable. Solderable metals, which can also function as barriers, include silver (Ag), copper (Cu), and nickel (Ni). Thus, the barrier metallurgy in Figure 2.17 includes a thin transition layer of Cr–Cu and a thick layer of Cu. The gold (Au) layer serves to prevent oxidation and has excellent

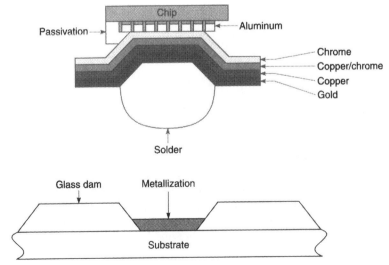

Figure 2.17 Metallurgy of a solder bump used in flip-chip bonding.

Table 2.12 Utilization and Properties of Materials Used in Flip-Chip Bonding

Materials	Function
Cr, Ti, Pt, Al, Ti–W	Adhesion
Cu, Pd, W, Pt, Ni, Ti–W	Barrier
Au, Ti	Bonding
Au, Au–Cu, solder–Cu	Bump
Cu	Leads
Sn, Au, Au–Ni	Plating on leads

		Bump material	
Property	Gold	Copper	95Pb–5Sn
Tensile strength (MPa)	152–213	207–276	20
Elongation (%)	3.5–6.3	2–3	25
Young's modulus (GPa)	77	83	24
Resistivity ($\Omega \times 10^{-8}$)	2.0–2.5	1.75–20	20
Thermal condition (W/m K)	318	380	36
TCE (ppm/°C)	15	17	29

solderability. Table 2.12 presents a summary of some of the materials utilized in flip-chip bonding, where they are commonly used, and some properties of materials routinely used to fabricate the bump.

Another approach to flip-chip bonding utilizes conductive polymers whose conductivity is anisotropic. In other words, these polymers only conduct current in a direction perpendicular to the surface of the IC. They are formulated by controlling the dispersion of metallic particles in the material so that sufficient particles are present to only allow conduction perpendicular to the plane of the IC surface. Commonly used metal particles include Au, Ag, and Ni. The advantages of this method of chip connect are that patterning and reflow are not required and inductance is negligible. However, alignment of chip and substrate bonding pads is critical.

2.5.3 Package Construction

Military interest in the transistor led to the development of metal transistor packages because of the need for a highly controlled and reliable operating environment. The transistor outline (TO) package, also called a round header package, was the first metal package developed. Other common types of metal packages are the flatpack (or butterfly), monolithic, and platform. These four metal transistor packages are illustrated in Figure 2.18. Metal-body packages are considered to be mature products and are still used in high-reliability applications in which hermiticity is required.

2.5.3.1 Base

The base of a metal package (header) is most often made of an alloy of steel and may be either nickel or gold plated. It may have a copper heat spreader brazed to it with a buffer layer of nickel or gold brazed to the heat spreader. The buffer metal layer must be compatible with the metal on the back side of the chip. The package pins are usually copper core steel but may be silver plated if gold is used for the wire bonding. The packaged devices may be

64 Chapter 2 Materials for Microelectronic Packaging

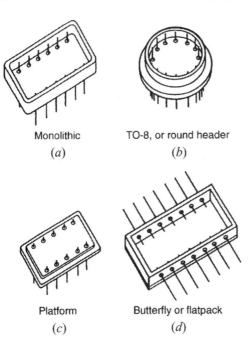

Figure 2.18 Common types of metal packages.

plated with gold, silver, or tin, or the leads may be solder dipped to enhance solderability following a long shelf life. In some cases, the base of the package is constructed of copper or aluminum or their alloys.

2.5.3.2 Lead Frames

The lead frame of an electronic package serves to mechanically support the chip during the assembly of the package and provides electrical connection between the chip and the outside world, as illustrated in Figure 2.19. They are made by stamping or chemically etching strips of nickel–iron, clad strip, or copper-based alloys. Many alloys exist within these categories. The most widely used metal for lead frames is alloy 42 (42% nickel–58% iron) because its thermal expansion coefficient is close to that of silicon. Table 2.13 lists several lead frame material compositions along with some of their properties. Another iron alloy is Kovar, which is composed of Fe–Ni–Co. Clad materials include copper-clad stainless steel. The copper alloys include Cu–Fe, Cu–Cr, Cu–Ni–S i, and Cu–Sn. Properties, such as electrical resistivity, thermal conductivity, solderability, coefficient of thermal expansion, and lead bend fatigue, all vary with composition. In the 1960s, some lead frames were completely gold or silver plated. In recent years, however, costs have reduced the plating of lead frames to silver on the chip attach platform and bonding finger ends.

2.5.3.3 Lids

The lids of TO packages are usually a nickel-plated steel alloy of the same composition as the base. Lid sealing can be accomplished by resistance welding, seam welding, and cold welding. Brazing is performed using either Au–Sn, In–Cu–Ag, or Cu–Ag. Solder sealing is accomplished using Pb–Sn or glass solders. The lids of ceramic packages are generally constructed of a steel alloy and, in some cases, are gold plated. The lid sealing material is generally either soft or hard solder to ensure hermiticity.

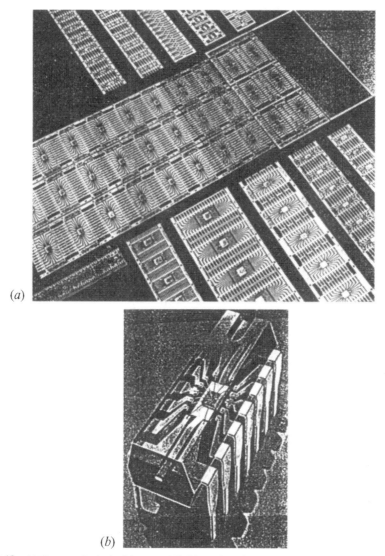

Figure 2.19 (a) Photograph of a lead frame and (b) illustration of a chip wire bonded to a lead frame.

Table 2.13 Lead Frame Base Materials and Some of Their Properties

Material	Elastic Modulus (GPa)	CTE (ppm/°C)	Thermal Conductivity (W/m K)	Electrical Resistivity (Ω-cm)
42Ni–58Fe	142	4.5	16	70
50Ni–50Fe	142	10.0	100	50
29Ni–17Co–54Fe	138	5.5	40	50
1.0Zr–99Cu	117	17.5	380	2.0
2.35Fe–0.03P–0.12Zn–Cu	118	17.5	260	2.5
0.6Fe–0.2P–0.4Mg–Cu	117	18	320	2.2

2.6 MATERIALS USED IN HIGH-DENSITY INTERCONNECT SUBSTRATES

Many of the materials used in high-density interconnect (HDI) substrate packaging are common to all packaging technologies. So often, differences may only be in how the materials are applied/deposited, which affects their microstructure and resultant properties, how they are defined, and the minimum geometries possible for a given processing technology. Since level 1 interconnection technology (chip to package) and level 2 interconnection technology (package to interconnect substrate, e.g., PWB) are discussed elsewhere, this section will be restricted to the materials used to fabricate HDI substrates to which chips are attached for purposes of interconnection with each other. Thus, the major interconnect substrate component parts that are addressed are the substrate, dielectric, and metallization.

An ideal substrate material should combine all of the following general characteristics [53–57]:

High mechanical strength and good thermal shock capability

High thermal conductivity and a CTE equal to that of silicon

Low dielectric constant and loss factor

High dielectric strength and resistivity

Surface perfection and ability to maintain tight dimensional control

Nontoxicity

Atmospheric, chemical, and thermal stability under subsequent processing and use conditions

Cofirable with Au, Ag, or Cu

Inexpensive

However, the most critical requirements placed on the substrate material for use in HDI substrates are mechanical integrity (Young's modulus) and thermal conductivity.

The dielectric material is the one key element that drives HDI substrate technology because a low dielectric constant material as the insulating layer between multiple metal layers is the only hope for achieving the high packing density required by multichip packaging while maintaining a characteristic 50-Ω impedance [58, 59]. An ideal dielectric candidate should have:

Low dielectric constant (2 to 4)

Excellent adhesion to substrate materials and metallization

Good mechanical properties

Thermal stability

High resistance to chemicals

Resistance to moisture absorption

Low residual stress

High degree of planarization

Additionally, they should be capable of being applied in thick coatings (10 µm or better) and be etchable and ionically clean.

The major factors that contribute to the metal conductor choice for HDI substrates include:

Good adhesion to the substrate, dielectric, and other metals

Low resistivity/high electrical conductivity

Good corrosion resistance

Chemical inertness

Good solderability

Can be electrolytically or chemically plated

Forms an inert native oxide

Inexpensive

A final, and perhaps most important, factor is that the metal be well understood by packaging engineers [60].

2.6.1 Laminate Substrates

Since the early 1970s, ICs have been die and wire bonded directly to printed wiring boards (PWBs) using what has been defined as chip-on-board or COB, technology. In simple terms, an HDI laminate substrate is a PWB to which bare ICs are mounted and through which they are electrically connected to each other and other electronic components using line widths and spaces somewhat smaller than those of conventional PWBs. Recently developed die-attach technologies, that are finding more frequent use in HDI laminate substrates, are TAB and flip chip. Thus, an HDI laminate substrate simply refers to a multichip package that utilizes unpackaged chips on a laminate substrate. Of the three primary types of HDI substrates, the laminate substrate is the least dense in terms of interconnect density, even though high-resolution printed wiring board technology is used in their fabrication. They are, however, the least expensive at about \$1/in.^{2}. Although HDI substrate and PWB technologies have obvious connections to each other in a qualitative sense, the miniaturized features of HDI laminate substrates lead to technologies quite distinctive from PWBs.

2.6.1.1 Substrates

The primary base material of HDI laminate substrates, a special category of PWBs, is a polymer, often an epoxy, flame-retardant phenolic, polyfunctional epoxy, or polyimide resin, which means that it is a dielectric. Of these, epoxies and polyimides are the most commonly used. Standard epoxies have relatively low values of T_g and, therefore, high values of expansion so they suffer from copper delamination at temperatures used for soldering. For applications where material stability is required at higher temperatures, materials such as bismaleimide triazine (BT resin) and polyimides with high glass transition temperatures can be used. The polymer is usually reinforced with woven glass fibers [e-glass, paper, s-glass, quartz, aramid (Kevlar), etc.] to provide mechanical rigidity once several layers are laminated (see Table 2.14). A very common laminate material, consisting of epoxy and e-glass, is referred to as FR-4. The reinforcement material generally has a lower dielectric constant than the resin, so its presence does reduce the dielectric constant of the composite structure below that of the resin. However, a substantial reduction in the laminate dielectric

Table 2.14 Properties of Fiber Materials Used to Reinforce PCBs

Material	Tensile strength (Kg/mm)	Elongation (maximum) (%)	Thermal conductivity (W/m K)	CTE (ppm/°C)	Dielectric constant (1 MHz)	Dissipation factor (1 MHz)	Young's modulus (kg/mm)
E-glass	350	4.9	0.9	5.0	5.9	0.0012	7500
S-glass	480	5.5	0.9	2.9	4.5	0.003	8500
Quartz	200	5.1	1.2	0.50	3.4	0.0002	7500
Aramid	410	4.5	0.5	60	4.05	0.001	12,800

constant requires the use of a different, low dielectric constant resin. Development efforts are constantly targeting lower dielectric constant polyimide materials.

If laminates are not reinforced, they are quite flexible and are referred to as "flex" circuits. The dielectric material from which flexible substrates are made include Kapton (a polyimide), PVC, Mylar (a polyester terephthalate), and Nomex (a random fiber aramid). Interconnects are provided by thin, flexible copper foil. Locations on the flexible PWB, where discrete components are to be mounted, are often reinforced.

Combining flex laminates with rigid laminates in a substrate yields a "rigid-flex" structure. Often the center or main portion of an HDI laminate substrate is rigid and the flexible portions appear as appendages, which emerge from the main part of the board.

Unlike other HDI technologies, the laminate material serves as a mechanical support and provides isolation for the electrical signals. Other materials receiving some attention include aramids and fluoropolymers because some of their properties are tailorable. Electrical properties of interest are:

Dielectric constant   Loss tangent (or dissipation factor)
Dielectric strength   Dielectric breakdown
Insulation resistance Current handling limitations

Similarly, mechanical and physical properties, which are important to the proper functioning of HDI laminate substrates, include:

Thickness   Flexural strength
Bow and twist   Thermal conductivity
Coefficient of thermal expansion Dimensional stability
Chemical resistance   Fungal resistance
Water absorption   Thermal stress
Copper surface and peel strength Flammability

Table 2.15 provides information on some materials commonly used for HDI laminate substrates, along with ranges of property values of these materials.

2.6.1.2 Dielectrics

As noted in the previous section, the dielectric in HDI laminate substrate technology is the same material that forms the substrate if a PWB is constructed using conventional lamination procedures. Alternatively, a reinforced polyimide prepreg material can be coated with a dielectric material, such as unreinforced aliphatic resin, to form dielectric layers between layers of copper. In this approach, the interconnections are sequentially built up on a conventional PWB using alternating layers of dielectric and copper. The advantages of this approach are enhanced interconnect density and low cost.

2.6 Materials Used in High-Density Interconnect Substrates   69

Table 2.15 Some Materials Used in Laminate Substrates and Ranges of Property Values of Such Substrate Materials

Laminate substrates	Property values
FR4 (epoxy + e-glass)	Glass transition temperature, T_g (°C): 75–260
Polyimode + Kevlar	Thermal conductivity (W/m K): 0.16–0.6
Polyimide + e-glass	Tensile modulus (10^6 psi): 0.2–4.4
Epoxy + e-glass	Tensile strength (10^3 psi): 20–60
Teflon + e-glass	Dielectric constant (1 MHz): 2.1–5.5
Epoxy + aramid	Dissipation factor (1 MHz): 0.0002–0.02
Epoxy + fused silica	Volume resistivity (Ω-cm): 10^7–10^{14}
Polyimide film	X-Y plane thermal expansion (ppm/°C): −5.0–55
Polyester film	Z axis thermal expansion (ppm/°C): 24–400
Polymide + fused silica	
Fused silica fabric	
Epoxy + Kevlar	

2.6.1.3 Metallization

The most common metallization used in HDI laminate substrates is copper, although aluminum and polymer thick films (carbon-loaded or carbon/silver-loaded) are also used in special applications. Table 2.16 gives properties of some interconnect materials used in HDI laminate substrates, as well as in other electronic packaging technologies. The copper is applied either by electroplating or rolling and annealing copper foil. If surface mount or

Table 2.16 Properties of Electronic Package Conductor Materials

Metal	Resistivity (Ω-cm)	Thermal conductivity (W/m K)	Thermal expansion coefficient (ppm/°C)	Melting temperature (°C)
Aluminum	2.65–4.3	240–247	23–25	660
Copper	1.67	390–420	17–20	1064–1083
Gold	2.2–2.35	297	14–14.2	1065
Silver	1.6	420	20	960–962
Tungsten	5.5	160–200	4.5	3415
Molybdenum	5.2	146	5.0	2610–2625
Platinum	10.6	71	9.0	1772–1777
Nickel	6.8–10.8	92	13.3–13.5	1455
Palladium	10.8	70–92	12–13.3	1550–1552
Chromium	13–20	66	6.3–6.5	1875–1900
Invar	46–80	11	1.5–3.1	1425–1500
Kovar	50	17	5.3	1450
Silver palladium	20	150	14	1145
Gold palladium	30	130	10	1350
Au–20%Sn	16	57	15.9	280
Pb–5%Sn	19	63	29	310
20%Cu–W	2.4	248	7.0	1083
20%Cu–Mo	2.5	197	7.2	1083
Titanium	5.5	22	9.0	1665
Tantalum	15.6	58	6.5	2980

through-hole components are to be mounted on the PWB, the copper pads where the device is to be mounted must be plated with Sn–Pb solder or nickel–gold. If bare chips are to be mounted to the PWB for wire, tape-automated, or flip-chip bonding, then the bonding pads must be coated with gold or Sn–Pb solder.

2.6.2 Ceramic Substrates

High-density interconnect ceramic substrate technology is an outgrowth of traditional hybrid circuit technology in which thick-film pastes were used to print conductors on ceramic substrates. The three ceramic-based structures that can be considered HDI ceramic substrate technologies are thick-film multilayer (TFM), *high-temperature cofired ceramic* (HTCC), and *low-temperature cofired ceramic* (LTCC). TFM is the only technology of the three to require a substrate platform for mechanical support. The ceramic used in HTCC and LTCC technologies serves as the interlevel dielectric, substrate, and package body. The cofired ceramic modules can also be classified as laminate structures since several layers of dielectric ceramic material, containing interconnection patterns, are bonded together in a parallel fashion to form the completed HDI substrate using heat and/or pressure. Although the HDI ceramic substrate technology is a higher performance technology than HDI laminate substrates, it is significantly more expensive at approximately \$10/in.^{2}.

2.6.2.1 Substrates

Ceramics, such as aluminum nitride, beryllia, alumina, mullite, mullite + glass, alumina + borosilicate, alumina + lead borosilicate, and cordierite glass–ceramic, are all candidates for application in HDI ceramic substrate technology. Other materials presently being investigated are boron nitride and silicon carbide. However, the most commonly used substrate material is alumina containing various amounts of glass. A 96% alumina–4% glass mixture is used for the majority of thick-film circuits because of cost and performance characteristics. HTCC technology utilizes 94 to 96% alumina. For other applications, the alumina content can vary from 40 to 99+%. Common substrate materials for LTCC are 55% alumina–45% lead borosilicate glass, 45% cordierite–55% borosilicate glass, and 35% silica–65% borosilicate glass. The percentages are by weight. Fosterite and Steatite have also found application in TFM circuits because their dielectric constants are lower than that of alumina.

2.6.2.2 Dielectrics

In cofired structures, the ceramic materials that provide mechanical support for the circuit (substrate) also provide for electrical isolation (dielectric) required between the several layers of interconnect metal. Thus, the dielectric materials used in cofired technology were discussed in the previous section. For thick-film multilayer structures, commonly used dielectric materials for crossovers, insulation between conductors, capacitor dielectric layers, encapsulation, and hermetic seals include low-melting-point glasses, mixtures of glass and ceramic, and crystallizable material such as magnesium aluminum silicate (cordierite).

2.6.2.3 Metallization

The most commonly encountered metals are tungsten, molybdenum, nickel, molymanganese, copper, silver, gold, platinum, and aluminum. Multiple layered metallization structures include titanium–palladium–gold, chromium–gold, chromium–copper, tantalum nitride–chromium–gold, and silver–palladium. Table 2.16 gives important properties for some of these metals.

2.6.3 Deposited Thin-Film Substrates

Although all three HDI substrate technologies offer enhanced performance over conventional packaging techniques, deposited thin-film substrate technology is unique because its manufacturing methods are different than those of the other two HDI interconnect substrate technologies. Furthermore, it provides the highest interconnect density, or alternatively, the highest packaging efficiency. Thus, it has the potential to offer the highest signal propagation speed. Since it is primarily based on conventional semiconductor processing technology, the materials used are identical or very similar to those developed for and used to manufacture ICs. However, the technology is expensive at the present time, costing about \$100/in.^{2}.

2.6.3.1 Substrates

Materials that are used or are being explored for application as deposited thin-film substrates are silicon, ceramics such as alumina, mullite, aluminum nitride, silicon carbide, glass, glass–ceramics, and beryllia, metals such as aluminum, copper, copper–invar–copper, steel, tungsten, copper-clad molybdenum, and molybdenum, diamond-coated substrates, aluminum matrix reinforced with silicon carbide composites, and glass–ceramic. Unlike the other HDI substrate technologies, deposited thin-film substrate technology requires a separate package body to house the substrate and to make electrical and mechanical connections to the PCB.

Silicon, one of the earliest choices for substrates, has the advantages of a relatively high thermal conductivity, an extremely smooth surface, which is important for fine-line lithography, and the capability for incorporating decoupling capacitors into the substrate. Unfortunately, silicon is a brittle material, so it cannot be used as a package body. Furthermore, it is not very strong, so it will bend when coated with an organic material of significantly different CTE. Of course, if silicon chips are used with a silicon substrate, thermal expansion differences do not exist and attachment failures are not a problem.

Some of the other candidate materials for use as substrates in deposited thin-film substrate technology, such as alumina, aluminum nitride, and silicon carbide, are discussed elsewhere in this chapter.

2.6.3.2 Dielectrics

The most common dielectric materials are silicon dioxide, silicon nitride, glass, polyimides, polyquinoline, Teflons, triazine, bisbenzo-cylobutenes, benzo-cyclobutene (BCB), and polyphenylquinoxaline (PPQ). Silicon dioxide was one of the first dielectrics to be used in deposited thin-film substrate technology, in particular, in multichip modules—deposited (MCM-D) because it was well understood and is easy to deposit. It has a moderate dielectric constant that, when combined with an aluminum interconnect of 10 μm width and a silicon dioxide thickness of 10μm yields a characteristic impedance of 50 Ω. However, silicon dioxide does not work well with copper metallization, so aluminum is generally used, resulting in a relatively low conductivity. Silicon dioxide also absorbs moisture, which results in a change in dielectric constant. Properties of some types of thin-film dielectric materials

Table 2.17 Properties of Some Thin-Film Interconnect Substrate Dielectric Type of Materials

Material	Dielectric constant	TCE (ppm/°C)	Moisture absorption	Glass Transition temperature (°C)	Young's modulus (kpsi)	Degree of planarization (%)
Standard polyimides	3.4–3.8	20–50	1–3.5	310	350–1000	25–35
Low stress polyimides	2.9–3.4	3–6	0.5–2.0	>400	1250	10–30
Acetylene terminated polyimides	2.8–3.2	38	0.8–3.0	225	400	90–95
Silicon polyimides	2.8–3.5	6–15	0.8–0.9	<300	100–250	25–30
BCBA	2.6–2.7	35–70	0.3–0.5	350	350	90–94
PPQA	2.7–2.8	50–60	0.9–1.0	360	—	—

are given in Table 2.17. Of these, polyimides are the most widely used interlayer dielectric because of their low dielectric constants. Table 2.18 gives properties of some commercially available polyimides. They can be applied using a variety of techniques, but are normally applied by spinning. Typical thicknesses range from 5 to 25 μm and can be applied either as a single coating or by multiple applications. Polyimides can be used with both aluminum and copper metallization, although metal barrier/adhesion layers are generally required between copper and polyimide to prevent corrosion of and promote adhesion to the copper. One of the most serious problems with many polyimides is moisture absorption, which can be significant and which changes the dielectric constant substantially. Benzo-cyclobutene is a good alternative to polyimides since it has a lower dielectric constant and much lower moisture absorption. It also exhibits excellent planarization, unlike many polyimides.

2.6.3.3 Metallization

The most common metals used in deposited thin-film substrate technology are aluminum, gold, and sputtered or plated copper. Barrier layer metals include chromium, titanium, and nickel. Aluminum is relatively cheap, readily available, and compatible with polyimide and other dielectrics. However, it is difficult to solder, cannot be plated, and has a lower conductivity than copper. On the other hand, copper does not suffer from the limitations of aluminum, although it degrades when in contact with uncured polyimide, as noted previously. The need for barrier metals to promote adhesion and prevent corrosion adds to the

Table 2.18 Properties of Some Commercial Polyimides

Material	Dielectric constant	Dielectric loss (tan δ)	CTE (ppm/°C)	T_g (°C)	Flex modulus (GPa)	Tensile strength (MPa)
Dupont PI 2555	3.3	0.002	40	>320	2.4	133
Hitachi PIQ-13	3.4	0.002	50	>350	—	133
Dow BCB-13005	2.7	0.002	50	>350	3.3	85

fabrication costs. Since the metallization is created using integrated circuit techniques, a higher interconnect density is possible (i.e., narrower lines) with deposited thin-film substrate technology than with ceramic and laminate substrate technologies, which, in turn, reduces the required number of interconnect levels. However, narrower lines result in high DC line resistance with corresponding propagation delays. Some important properties of these metal are given in Table 2.16.

2.7 SUMMARY

The packaging of electronics involves a fairly large variety of materials that must protect and ensure that the system inside the package operates within specifications and reliably over an extended period of time. Materials must be selected that allow electrical signals to be routed from point to point within the system with a minimum of loss in amplitude and minimal distortion. This involves the selection of materials for substrates, dielectrics, and interconnections that are compatible with each other and suitable for the amplitude and frequency of the system's electrical signals. The three primary classifications of such packaging materials are ceramics, polymers, and metals. When choosing materials for packaging applications, the packaging engineer must consider properties such as mechanical strength, moisture penetration, interfacial adhesion, dielectric constant and loss, and thermal characteristics in order to ensure that the system operates reliably over its expected life. For information on a much larger number of packaging materials than is provided in this chapter, the reader is referred to the Microelectronics Packaging Materials Database, a product of the Semiconductor Research Corporation (SRC)–funded research conducted at Purdue University by the Center for Information and Numerical Data Analysis and Synthesis (CINDAS). It contains data on thermal, mechanical, electrical, and physical properties of selected microelectronics packaging materials [61].

REFERENCES

1. D. P. SERAPHIM, R. LASKY, AND C.-Y. LI, *Principles of Electronic Packaging*, New York: McGraw-Hill, 1989.
2. D. A. DOANE AND P. D. FRANZON, eds., *Multichip Module Technologies and Alternatives: The Basics*, New York: Van Nostrand Reinhold, 1993.
3. G. MESSNER, I. TURLIK, J. W. BALDE, AND P. E. GARROU, *Thin Film Multichip Modules*, Reston, VA: ISHM, 1992.
4. R. R. TUMMALA AND E. J. RYMASZEWSKI, eds., *Microelectronics Packaging Handbook*, New York: Van Nostrand Reinhold, 1989.
5. P. GARROU AND A. KNUDSEN, Aluminum Nitride for Microelectronics, *Advancing Microelectronics*, Vol. 21, No. 1, p. 6, 1994.
6. *Electronic Materials Handbook, Vol. 1: Packaging*, Cleveland Ohio: ASM International, 1989.
7. T. L. LANDERS, W. D. BROWN, E. W. FANT, E. M. MALSTROM, AND N. M. SCHMITT, *Electronics Manufacturing Processes*, Englewood Cliffs, NJ: Prentice-Hall, 1994, p. 100.
8. F. RODRIGUEZ, *Principles of Polymer Systems*, 3rd ed., New York: Hemisphere, 1989.
9. E. D. FEIT AND C. W. WILKINS, *Polymer Materials for Electronic Applications*, Washington, DC: American Chemical Society, 1982.
10. W. R. BRATSCHUN, A. J. MOUNTVALA, AND A. G. PINCUS, *Uses of Ceramics in Microelectronics: A Survey*, Washington, DC: National Aeronautics and Space Administration, 1971.
11. B. C. H. STEELE, *Electronic Ceramics*, New York: Elsevier Applied Science, 1991.
12. F. N. SINNADURAI, *Handbook of Microelectronics Packaging and Interconnection Technologies*, Scotland: Electrochemical, 1985.
13. T. G. TESSIER, G. M. ADEMA, AND I. TURLIK, Polymer Dielectric Options for Thin Film Packaging Applications, Proceedings of the 39th Electronic Components Conference, 1989, pp. 127–134.
14. R. J. JENSEN, Polyimides as Interlayer Dielectrics for High-Performance Interconnections of Integrated Circuits, *Polymers for High Technology*, American Chemical Society, 1987, pp. 466–481.
15. T. G. TESSIER, I. TURLIK, G. M. ADEMA, D. SIVAN, E. K. YUNG, AND M. J. BERRY, Process Considerations in Fabricating Thin Film Multichip Modules, Technical Report TR89-45, MCNC, 1989.
16. G. GESCHWIND AND R. M. CLARY, Multichip Modules: An Overview, *PC FAB*, pp. 28–38, November 1990.
17. D. WILSON, H. D. STENZENBERGER, AND P. M. HERGENROTHER, *Polyimides*, New York: Chapman and Hall, 1990.

18. S. L. ROSEN, *Fundamental Principles of Polymeric Materials*, New York: Wiley, 1993.
19. W. F. SMITH, *Principles of Materials Science and Engineering*: 2nd ed., New York: McGraw-Hill, 1990.
20. R. D. ROSSI, Polyimides, reprint from *Engineered Materials Handbook, Vol. 3: Adhesives and Sealants*, Cleveland, Ohio: ASM International, 1991.
21. H. GRIGORIEW AND J. LECIEJEWICZ, X-ray and Electron Microscopy Study of Amorphous Boron Nitride Films, *Thin Solid Films*, Vol. 127, pp. L75–L79, 1989.
22. A. LIPP, K. A. SCHWETZ, AND K. HUNOLD, Hexagonal Boron Nitride: Fabrication, Properties and Applications, *J. European Ceramic Soc.*, Vol. 5, pp. 3–9, 1989.
23. R. F. DAVIS, Diamond and Silicon Carbide Thin Films: Present Status and Potential as Wide Band Gap Semiconducting Materials, *Int. J. Mat. Product Technol.*, Vol. 4, No. 2, pp. 81–103, 1989.
24. N. L. D. SOMASIRI, R. L. D. ZENNER, AND J. C. HOUGE, A Process for Surface Texturing of Kapton Polyimide to Improve Adhesion to Metals, *IEEE Trans. Components, Hybrids, and Manufact. Technol.*, Vol. 14, No. 4, pp. 798–801, 1991.
25. G. LEHMAN-LAMER, D. B. HOY, AND K. M. MIDDO, New Multilayer Polyimide Technology Teams with Multilayer Ceramics to Form MultiChip Modules, *Hybrid Circuit Technol.*, pp. 21–26, October 1990.
26. R. C. BUCHANAN, ed., *Ceramic Materials for Electronics*, New York: Marcel Dekker, 1986.
27. L. M. LEVINSON, ed., *Electronic Ceramics: Properties, Devices, and Applications*, New York: Marcel Dekker, 1988.
28. J. A. KING, ed., *Materials Handbook for Hybrid Microelectronics*, Boston: Artech House, 1988.
29. D. D. MARCHANT AND T. E. NEMECEK, Aluminum Nitride: Preparation, Processing, and Properties, *Adv. Ceramics*, Vol. 26, pp. 52–81, 1989.
30. T. J. BUCK, Substrates for High Density Packaging, Proc. of the National Electronic Packaging and Production Conference, pp. 650–659, 1990.
31. E. S. DETTMER, H. K. CHARLES, S. J. MOBLEY, AND B. M. ROMEMESKO, Hybrid Design and Processing Using Aluminum Nitride Substrates, Proc. of the Intl. Sym. on Microelectronics, 1988, pp. 545–553.
32. D. D. DENTON, D. R. DAY, D. F. PRIORE, AND S. D. SENTURIA, Moisture Diffusion in Polyimide Films in Integrated Circuits, *J. Electronic Mat.*, Vol. 14, No. 2, pp. 119–136, 1985.
33. D. BURDEAUX, P. TOWNSEND, AND J. CARR, Benzocyclobutene (BCB) Dielectrics for the Fabrication of High Density, Thin Film Multichip Modules, *J. Electronic Mat.*, Vol. 19, No. 12, pp. 1357–1366, 1990.
34. K. K. CHAKRAVORTY, C. P. CHEN, J. M. CECH, M. H. TANIELIAN, AND P. L. YOUNG, High-Density Interconnection Using Photosensitive Polyimide and Electroplated Copper Conductor Lines, *IEEE Trans. Components, Hybrids, Manufact. Technol.*, Vol. 13, No. 1, pp. 200–206, 1990.
35. H. TAKASAGO, K. ADACHI, AND M. TAKADA, A Copper/Polyimide Metal-Base Packaging Technology, *J. Electronic Mat.*, Vol. 18, No. 2, pp. 319–326, 1989.
36. C. FEGER, M. M. KHOJASTEH, AND J. E. MCGRATH, eds., *Polyimides: Materials, Chemistry, and Characterization*, Amsterdam: Elsevier Science, 1989.
37. Y.-H. KIM, J. KIM, G. F. WALKER, C. FEGER, AND S. P. KOWALCZYK, Adhesion and Interface Investigation of Polyimide on Metals, *J. Adhesion Sci. Technol.*, Vol. 2, No. 2, pp. 95–105, 1988.
38. D. A. SCOLA AND J. H. VONTELL, High Temperature Polyimides, Chemistry and Properties, *Polymer Composites*, Vol. 9, No. 6, pp. 443–452, 1988.
39. T. COBB, Organics or Inorganics—Which Dielectric, *Advanced Packaging*, pp. 35–39, Fall 1993.
40. T. RUCKER, V. MURALI, R. SHUKLA, AND H. NEUHAUS, Polyimide for Multichip Modules: Materials and Process Challenges, *Mat. Res. Symp. Proc.*, Vol. 264, pp. 71–82, 1992.
41. K. C. CHUANG, R. D. VANNUCCI, AND B. W. MOORE, Effects of a Noncoplanar Diphenyldiamine on the Processing and Properties of Addition Polyimides, NASA Tech Briefs, LEW-15399, Lewis Research Center Houston.
42. T. G. TESSIER, G. M. ADEMA, AND I. TURLIK, Polymer Dielectric Options for Thin Film Packaging Applications, Proc. 39th Electronic Components Conference, pp. 127–134, 1989.
43. K. K. CHAKRAVORTY, C. P. CHIEN, J. M. CECH, L. B. BRANSON, J. M. ATENCIO, T. M. WHITE, L. S. LATHROP, B. W. AKER, M. H. TANIELIAN, AND P. L. YOUNG, High Density Interconnection Photosensitive Polyimide and Electroplated Copper Conductor Lines, Proc. 39th Electronic Components Conference, pp. 135–142, 1989.
44. A. W. LIN, Evaluation of Polyimides as Dielectric Materials for Multichip Packages with Multilevel Interconnection Structure, Proc. 39th Electronic Components Conference, pp. 148–154, 1989.
45. B. T. MERRIMAN, J. D. CRAIG, A. E. NADER, D. L. GOFF, M. T. POTTIGER, AND W. J. LAUTENBERGER, New Low Coefficient of Thermal Expansion Polyimide for Inorganic Substrates, Proc. 39th Electronic Components Conference, pp. 155–159, 1989.
46. P. V. NAGARKAR, P. C. SEARSON, F. BELLUCI, M. G. ALLEN, AND R. M. LATANISION, Interfacial Interactions Affecting Polyimide Reliability, Proc. 39th Electronic Components Conference, pp. 160–166, 1989.
47. N. SASHIDA, T. HIRANO, AND A. TOKOH, Photosensitive Polyimides with Excellent Adhesive Property for Integrated Circuit Devices, Proc. 39th Electronic Components Conference, pp. 167–170, 1989.
48. T. BANBA, E. TAKEUCHI, A. TOKOH, AND T. TAKEDA, Positive Working Photosensitive Polymers for Semiconductor Surface Coating, Proc. 41st Electronic Components Conference, pp. 564–567, 1991.
49. P. G. JOBE, C. PUGLISI, J. MCMAHON, AND R. D. ROSSI, Acetylene-Terminated Low-Stress Polyimide Oligomer for Interlayer Dielectric Applications, Proc. 41st

Electronic Components Conference, Atlanta, pp. 568–571, 1991.
50. W. VOLKSEN, D. Y. YOON, AND J. HEDRICK, Polyamic Alkyl Esters: Versatile Polyimide Precursors for Improved Dielectric Coatings, Proc. 41st Electronic Components Conference, pp. 572–579, 1991.
51. T. F. REDMOND, C. PRASAD, AND G. A. WALKER, Polyimide Copper Thin Film Redistribution on Glass Ceramic/Copper Multilevel Substrates, Proc. 41st Electronic Components Conference, pp. 689–692, 1991.
52. A. SASAKI AND Y. SHIMADA, Electrical Design Technology for Low Dielectric Constant Multilayer Ceramic Substrate, Proc. 41st Electronic Components Conference, pp. 719–726, 1991.
53. P. B. CHINOY AND J. TAJADOD, Processing and Microwave Characterization of Multilevel Interconnects Using Benzocyclobutene Dielectric, *IEEE Trans. Components, Hybrids, Manufact. Technol.*, Vol. 16, No. 7, pp. 714–719, 1993.
54. G. K. H. SCHAMMLER, V. GLAW, AND G. CHMIEL, Comparison of the Metallization of Chemically and Laser-Etched Structures in BPDA-PDA Polyimide, *IEEE Trans. Components, Hybrids, Manufact. Technol.*, Vol. 16, No. 7, pp. 720–723, 1993.
55. J. M. CECH, A. F. BURNETT, AND C.-P. CHIEN, Reliability of Passivated Copper Multichip Module Structures Embedded in Polyimide, *IEEE Trans. Components, Hybrids, Manufact. Technol.*, Vol. 16, No. 7, pp. 752–758, 1993.
56. D. W. SWANSON, COB Encapsulants, *Advanced Packaging*, pp. 12–16, October 1993.
57. M. E. O'DAY AND G. L. LEATHERMAN, Static Fatigue of Aluminum Nitride Packaging Materials, *Intl. J. Microcircuits Electronic Packaging*, Vol. 16, No. 1, pp. 41–48, 1993.
58. N. B. NGUYEN, Using Advanced Substrate Materials with Hybrid Packaging Techniques for Ultrahigh-Power ICs, *Solid State Technol.*, pp. 59–62, February 1993.
59. R. ISCOFF, Application Dictates MCM Substrate Choice, *Semiconductor International*, p. 54, September 1991.
60. B. C. FOSTER, F. J. BACHNER, E. S. TORMEY, M. A. OCCHIONERO, AND P. A. WHITE, Advanced Ceramic Substrates for Multichip Modules with Multilevel Thin Film Interconnects, *IEEE Trans. Components, Hybrids, Manufact. Technol.*, Vol. 14, No. 4, pp. 784–789, 1991.
61. Microelectronics Packaging Materials Database, available from CINDAS LLC, P.O. Box 3814, West Lafayette, IN 47996-3814 (www.cindasdata.com).

EXERCISES

2.1. Electronic packaging materials perform several important functions. What are they?

2.2. The selection of materials and manufacturing processes for packaging of ICs is important because they impact both the fabrication process and operation of the packaged IC. Give five parameters impacted by these selections.

2.3. Give the seven basic properties of materials used in packaging that are critical to the evolution of high-performance packaging.

2.4. Why is mechanical strength of packaging materials so important?

2.5. Why is moisture penetration of concern in electronic packaging?

2.6. Which of the two types of polymers, *fluorocarbon* or *epoxy*, repels water fairly well?

2.7. How does water flux (F) into a plastic electronic package depend on the driving force (D) and the thickness (t) of the package? On what does D depend?

2.8. If moisture penetration of plastic packages is a problem, why not just increase the thickness of the package to eliminate the problem?

2.9. What factors influence the quality of the I/O interface of electronic packages and allow moisture penetration to occur (i.e., what factors contribute to poor adhesion)?

2.10. To what are the dielectric constant and loss tangent (or dissipation factor) of plastics sensitive? Give at least five.

2.11. What two thermal properties of materials are of importance to packaging and why?

2.12. What are some of the characteristics of ceramic materials?

2.13. Describe the composition of alumina-based ceramics.

2.14. What properties of alumina limit its usefulness in electronic packaging?

2.15. What are the more common uses of beryllia in electronics?

76 Chapter 2 Materials for Microelectronic Packaging

2.16. Why is AlN an attractive substrate material for use in an HDI deposited thin-film substrate?

2.17. What are the attractive properties of SiC with regard to its use in electronic packaging? What are its primary limitations?

2.18. What are glass–ceramics? Why are they attractive for electronics?

2.19. Define the term *plastics*. Where are plastics used in electronic packaging?

2.20. What are the three basic chain architectures of polymers?

2.21. Define the term *glass transition temperature* as it applies to a polymer.

2.22. What are the five types of bonding forces present in polymers?

2.23. Define the terms *thermoplastic* and *thermoset* as they apply to polymers.

2.24. Give nine properties of polymers of interest to packaging specialists?

2.25. Given that an HDI substrate uses 1-μm-thick copper as the first-level interconnect metal, what is the height of the feature in a polymer dielectric layer over the copper interconnect if the degree of planarization is 90%?

2.26. What are the commonly used additives with epoxy compounds (i.e., resin + hardeners) and what is their purpose?

2.27. Define the term *elastomer* and give an example of a common elastomer.

2.28. What properties and/or behavior separate silicones from conventional polymers?

2.29. What are the attractive properties of polyimides with regard to application in the electronics industry? What are their less desirable properties?

2.30. What are three commonly used methods of providing electrical connections between ICs and packages?

2.31. Which metals are commonly used in wire bonding?

2.32. What materials are usually used to form bumps for flip-chip bonding?

2.33. Describe the composition of conductive polymers that are used in flip-chip bonding. How do they provide electrical conduction in only one direction?

2.34. What functions does a lead frame perform in an electronic package?

2.35. What are the characteristics of an ideal HDI substrate material?

2.36. What are the characteristics of an ideal dielectric candidate for use in an HDI substrate?

2.37. What are the factors to consider when selecting a metal for HDI substrates?

2.38. What are the three primary classifications of HDI substrates?

2.39. Describe the construction of an HDI laminate substrate.

2.40. What are the three ceramic-based structures that can be considered HDI ceramic substrate technologies?

Chapter 3

Processing Technologies

H. A. NASEEM AND SUSAN BURKETT

3.1 INTRODUCTION

Thick or thin films of conductor metals are encountered in many places in microelectronic packaging. For example, screen-printed thick films of conductor traces are utilized in the manufacture of multilayer ceramic (MLC) substrates. In this chapter, some of the thin-film deposition techniques most commonly used in packaging are discussed. In particular, evaporation, sputtering, and electro- and electroless plating techniques, which are used in the fabrication of corrosion-resistant metal pads on MLC packages, tape-automated bonding (TAB)-metallized tape, solder bumps and substrate pads in flip-chip (C4) technology, and plated conductor (Cu) foils and plated through holes in multilayer printed circuit boards (PCB), are described. These techniques are often used with photolithography and etching in order to obtain metal traces, pads, or bumps in desired regions. Photolithography and etching processes used in patterning metal traces are treated in later sections of this chapter. Finally, some metal-to-metal joining techniques, such as bonding and soldering, are briefly described.

3.2 THIN-FILM DEPOSITION

Vapor-phase deposition of thin films can be performed via physical vapor deposition (PVD) or chemical vapor deposition (CVD) techniques. In PVD processes, the depositing species are turned into atomic or molecular form, which then condense on the substrate, as well as on the chamber walls. There is no chemical reaction either in the gas phase or at the depositing surface. Evaporation and sputtering are examples of PVD. On the other hand, in CVD, gaseous species react at the depositing surface to form a thin film. There are several inherent advantages to CVD, such as good step coverage and conformal coating, which are not available in PVD, but CVD of metals is still in its infancy and needs considerable development. Currently, most thin-film metallizations utilize evaporation and sputtering technologies, which are described in some detail in the following sections. Chemical vapor deposition is also described with an emphasis on the process that is used for typical growth of dielectric thin films or metals, such as Ta, Ti, and W. Metallorganic compounds can be used to deposit Al or Cu by the process of metalorganic CVD (MOCVD), but these processes are not fully developed. Plating techniques are most commonly used with circuit

---

Advanced Electronic Packaging, Second Edition, Edited by Richard K. Ulrich and William D. Brown
Copyright © 2006 the Institute of Electrical and Electronics Engineers, Inc.

boards and MCMs. In many cases, however, a seed layer is deposited using PVD before plating. Plating is discussed in Section 3.2.6.

3.2.1 Vacuum Facts

The forced removal of gases from a container is referred to as evacuation. At a given temperature, gas molecules within a container are in a state of random motion, striking each other and the walls of the container. The latter is measured as the pressure in the container. Boltzmann statistics adequately describe such a system in equilibrium. As the gas molecules are removed from a fixed volume, its pressure decreases. In traditional measurement, pressure is given in units of Torr (named after the Italian pioneer Torricelli). One Torr is equivalent to 1 mm of Hg. The SI unit of pressure is the pascal, denoted as Pa, named after the French scientist by the same name. One Pa is equivalent to a force per unit area of one newton per meter squared. The cgs unit of dyn/cm^2 is called a bar. In what follows, the traditional unit of Torr is be used; 1 Torr equals 133 Pa. The standard (or normal) pressure is 760 Torr or 1.01×10^5 Pa, which is the nominal pressure on the surface of Earth. Typically, pressure values in the range of 10^{-15} Torr are found approximately 10,000 miles above Earth's surface. It is interesting to note that a normal human being can create a vacuum of about 740 Torr in a small volume by his own force of suction; an octopus can deliver up to 700 Torr. A pressure in the range of 10^{-7} Torr (called high vacuum) is required for evaporation of metal films.

Mean free path is a measure of how far a gas molecule or evaporated atomic species can travel on an average without colliding with another. It is given by

$$L = kT/(p s^2 \sqrt{2}) \tag{3.1}$$

where T is the temperature in kelvin, s is the molecular size in meters, p is the pressure in Pa, and k is the Boltzmann constant. Thus, at a pressure of 10^{-6} Torr, and for a typical size of a molecule of 0.3 nm, the mean free path is roughly 400 m. A source-to-substrate distance of 30 cm, which is typical of evaporation systems, is adequate to ensure a collision-free path for metal atoms to travel to the substrate (line-of-sight deposition).

Residual gases, especially water vapor, in the deposition chamber also strike the surface of the growing film and may react with the metal atoms and become incorporated into it. The reactivity of the impinging residual gas atom determines its sticking coefficient; a sticking coefficient of 1 represents 100% incorporation, whereas, a value of 0 means complete reevaporation or desorption from the growth surface. For example, during the deposition of Al, Cr, or Ti, the sticking coefficient for oxygen-bearing species in the chamber can be taken as 1 because of the strong reactivity of these metals to oxygen. On the other hand, the sticking coefficient of noble gases, such Ar or He, can be taken as nearly 0 for obvious reasons.

The rate at which residual gas molecules strike any surface within the chamber is given by

$$\frac{dn}{dt} = \frac{p}{\sqrt{2mkT}} \; \mathrm{m}^{-2}\,\mathrm{s}^{-1} \tag{3.2}$$

where m is the mass of the impinging molecule, T is temperature in kelvin, k is the Boltzmann constant, and p is the partial pressure of the molecules.

If the deposition rate of metal atoms and the sticking coefficient of the reactive species are known, the composition of the deposited film can be estimated. For example, let us say that Ti is being evaporated under a base pressure of 1×10^{-5} Torr at a rate of 1 nm/s. This corresponds to an impingement rate of Ti of 5.7×10^{15} cm^{-2} s^{-1}. Assuming that the pressure in the chamber is predominantly due to the oxygen-carrying species (e.g., water vapor), the O impingement rate is 4.8×10^{15} cm^{-2} s^{-1}. Taking the sticking coefficient to be 1 (which is very justifiable), the deposited film will not be Ti, but TiO.

3.2.2 Vacuum Pumps

Evacuation of a metallization system is generally achieved in two steps. In the first step, the system is pumped from atmospheric pressure to the 10-mTorr range using a mechanical rotary vane pump. Direct-drive, dual-stage, oil-sealed pumps are normally used. Figure 3.1 shows the schematic of such a pump. Basically, it consists of two eccentrically mounted rotors having spring-loaded vanes that, upon rotation, pull in and compress a certain amount of gas from the chamber and push it out into the oil reservoir and, thus, out to the exhaust. High-grade turbine oil, which is vacuum distilled to remove all high vapor pressure components, is used in these pumps. This oil not only seals the pumping mechanism from the atmosphere but also serves as a coolant and lubricant. Dual-stage pumps can easily pump a system down to the mTorr range, whereas single-stage pumps are limited to 25 mTorr. Direct-drive pumps offer much higher rotor speeds (1725 rpm) and, hence, much shorter pumping times, as compared to ones with belts and pulleys (rotor speed of 300 to 600 rpm).

For a rotary vane pump, the maximum throughput is given as the volume of free air displaced by the pump at atmospheric pressure. The pumping speed decreases as the pressure in the chamber decreases, as shown in Figure 3.2. It can be seen that the throughput is fairly constant above 10 mTorr but falls off rapidly as the pressure decreases below 1 mTorr. The mechanical pump must not be used to pump the metallization system below 10 mTorr for extended periods of time (more than 10 min) otherwise oil vapors will backstream into the clean chamber. The pump-down time can be estimated from the formula:

$$t = (V/S) \ln(P_0/P) \tag{3.3}$$

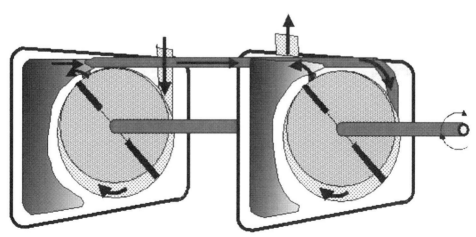

Figure 3.1 Schematic of a two-stage rotary vane pump showing the eccentric rotor shaft and spring-loaded vanes.

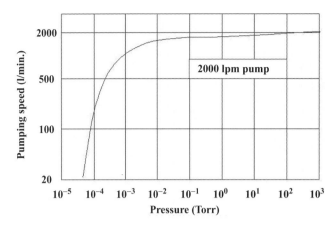

Figure 3.2 Typical plot of pumping speed versus pressure for a two-stage rotary vane pump.

where V is the chamber volume, S is the average pump speed, P_0 is the starting atmospheric pressure, and P is the desired pressure. In a well- designed system, it should not take more than 15 min to bring the chamber pressure down from atmospheric pressure to about 10 mTorr.

High vacuum is achieved in the deposition chamber by pumping it down using a second pump. There are three types of high-vacuum pumps that can be used in metallization systems: a diffusion pump, a turbomolecular pump, or a cryogenic pump. Diffusion pumps work on the principle of transferring momentum from high-speed oil vapor molecules to the gas molecules such that they can be removed from the chamber. Figure 3.3 shows a cross-sectional view of a diffusion pump. The bottom of the pump contains an electric heater that brings the pumping oil to the boiling point at about 50 mTorr. This low pressure is achieved in the diffusion pump using a mechanical pump. The oil vapor is forced up the central columns of the jet assembly where it is directed downward in the form of jets of heavy oil molecules. When these oil molecules strike a gas molecule belonging to the chamber, it is forced down. Normally there are three stages of oil jets that compress and move the chamber gas toward the bottom from where they are pumped out using a mechanical pump. Diffusion pumps can easily pump a system down to the 10^{-7} Torr range if the system does not outgas profusely above this pressure. Its pumping speed remains constant in the 10^{-3} to 10^{-8} Torr range. An estimation of the pumping time is, therefore, straightforward.

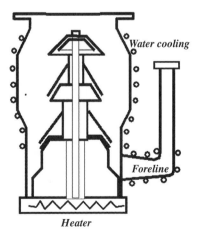

Figure 3.3 Cross-sectional view of a typical diffusion pump.

Hydrocarbon contamination of a metallization system, arising from the diffusion pump, can be completely avoided by using a turbomolecular or cryogenic pump. In fact, these two pumps have become the workhorse of metallization technology in the industry. However, they will not be described in this chapter. Rather, a basic understanding of their operation is assigned as an exercise.

3.2.3 Evaporation

Evaporation is the simplest technique for depositing a thin film. In this process, vapors are produced from a material located in a source, which is heated using any one of several methods. For example, when Al or Au is heated, it first melts. Upon supplying further thermal energy, the metal atoms gain sufficient kinetic energy to dislodge from the melt surface and launch in all possible directions away from the source. When these atoms strike a relatively cooler surface, for example, the substrate or the walls of the deposition chamber, they condense to form a thin film of the material being evaporated, Al or Au in this example. Since most materials, including metals, vaporize at relatively high temperatures (500 to 2000°C), reaction with ambient gas is of major concern during evaporation. For example, if Al is evaporated in an ambient containing oxygen or moisture, the deposited film is most likely to be aluminum oxide rather than pure aluminum. In fact, reactive evaporation in a controlled gaseous ambient is used to deposit nitrides and oxides of various metals such as Ti, Ta, Cr, and Al. Thus, an evaporation system designed to deposit metal films is necessarily equipped with an evacuation system to remove all reactive gases from the chamber. Another advantage of evaporating in a vacuum chamber is that the vaporized metal atoms do not collide with other atoms before reaching the substrate. This assures high deposition rates in desired line-of-sight areas and no deposition in others. Line-of-sight deposition also allows the use of shadow masks for selective metallization.

3.2.3.1 Film Deposition

The deposition of a thin film onto a substrate is believed to take place by the following sequence of events. The evaporated material, in the form of an atomic or molecular beam, arrives at the surface where it is adsorbed physically (physisorbed). Depending on the incident particle energy, as well as the energy provided by the substrate, the incident atom may be desorbed from the surface or be chemically adsorbed (chemisorption). For both physisorption and chemisorption, the atoms have sufficient surface mobility to move about the growth surface. At this stage, atoms combine to form nuclei, which can then form and dissociate randomly all over the surface. When the size of these nuclei grows above a critical size (a diameter at which the increase in surface energy due to the increased size is more than balanced by a decrease in the volume energy), they become stable and start to grow. When these nuclei bump into each other, they coalesce to form a continuous film. Depending on the deposition conditions, such as substrate temperature, rate of evaporation, and the particular material being evaporated, there will be a certain degree of surface roughness. This is sketched in Figure 3.4. The deposited film normally is polycrystalline with random or slightly preferential orientation.

Thermal evaporation sources, as well as electron beam sources, can be treated as point sources. Thus, if a 15-cm-diameter substrate is positioned about 30 cm away from a point source, it is clear that the deposited film will not be uniform across the substrate. In general, the mass deposited per unit area is given by

$$R_d = (M_e/r^2)\cos\theta\cos\phi \qquad (3.4)$$

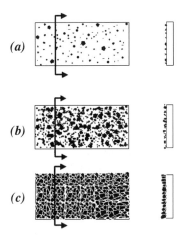

Figure 3.4 Illustration of nucleation and growth phenomena in thin films.

where M_e is the total mass of the evaporated area, and r, θ, and ϕ are shown in Figure 3.5. This is known as the cosine law. In order to achieve acceptable uniformity on large areas, therefore, it becomes necessary to rotate the substrate.

Figure 3.6 shows a sketch of an evaporator system with its components properly marked. A substrate heater is often required for good-quality (low stress, large grain size, etc.) films. Substrate rotation is incorporated for thickness uniformity. Thickness and deposition rates are also measured and controlled in most sophisticated systems. Multiple hearth e-beam sources, as well as multifilament systems, are available for co-deposition to form alloys or sequential deposition for multilayered structures. Load-lock chambers are used to reduce the pump-down time of the deposition chamber. Since water vapor is difficult to pump, not exposing the deposition chamber to ambient air improves the pumping speed. Liquid nitrogen traps are also employed to condense out moisture within the evaporator. The trap also helps to prevent back streaming of pump oil vapors into the chamber. Glass bell jar chambers can be used for Al and Au depositions, but water-cooled, stainless steel chambers are used when refractory metals, such as W, Mo, and Ta, are evaporated

3.2.3.2 Evaporation Sources

There are two types of evaporation sources used in metallization applications. These are (1) thermal evaporation sources and (2) electron beam sources. In the case of thermal

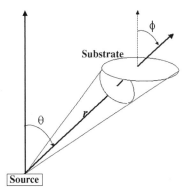

Figure 3.5 Diagram defining angles and distances used in the derivation of deposition rate for evaporation from a source of small area.

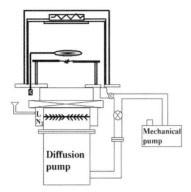

Figure 3.6 Sketch of an evaporator system showing its main components.

evaporation, a high electric current at a low voltage is passed through a refractory metal filament heater. These filament heaters come in a variety of shapes and forms, some of which are shown in Figure 3.7. Figures 3.7a and b show a spiral and a basket filament, respectively. These are made from multistrand tungsten (W) wire and are well suited for aluminum evaporation. Pieces of Al wire, cut up into small candy sticks, are hung from the spiral filament. Upon melting, surface tension pulls the molten Al onto the W filament where it evaporates very rapidly. Aluminum pellets can be used with basket filaments. Although W filaments work well for aluminum evaporation, tungsten reacts with Al, becomes brittle, and the filament breaks after multiple uses.

Figures 3.7c and d show a dimple boat and crucible, respectively. These can be made of Mo, W, or Ta. For gold or Cu evaporation, Mo or Ta boat filaments work very well and can be reused without any problems. These boats are also available with an insulating barrier coating of aluminum oxide to prevent reaction with the filament and to prevent current from shorting through the thick molten conductor layer being evaporated. Some of the high-melting-point materials, such as chromium, can be evaporated rather easily using Cr-plated W wire filaments. Extremely high melting and boiling point materials, such as tungsten, molybdenum, and tantalum, cannot be evaporated using thermal evaporation. However, electron beam evaporation can be successfully utilized for such applications.

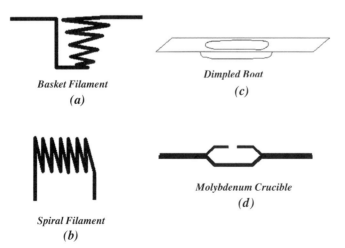

Figure 3.7 Various sources for thermal evaporation of materials by resistance heating.

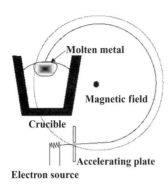

Figure 3.8 Sketch of a single-hearth electron beam evaporation source.

The electron beam has become the source of choice for reproducible and high rate deposition of metal films. One of the most appealing features of this source is its ability to deposit refractory metals such as W, Mo, Ta, Ti, and so forth, which, owing to their high melting and boiling points, cannot be evaporated by conventional thermal means. In this source, an intense electron beam (a beam current of 1 A) is accelerated across several kilovolts of potential. The beam is then directed toward the metal target contained in a water-cooled crucible. Almost all of the kinetic energy (several kilowatts) is converted into heat, locally, in the spot where the beam hits. Electrostatic fields are used to scan the beam across larger surfaces for better uniformity and control. Figure 3.8 shows a sketch of a single-hearth e-beam source; multiple-hearth sources are used for sequential evaporation of metals without breaking the vacuum. The electron-emitting filament is shielded from the evaporating metal using a permanent magnet to turn the beam around by 270°. One of the disadvantages of e-beam evaporation is the emission of X rays from the source target, which causes radiation-induced defects in many active devices. However, this fact is probably not pertinent to the problem at hand, that is, metallization for packaging applications.

3.2.4 Sputtering

Sputtering falls into the category of PVD processes because the deposition material of choice (target material) is physically removed from the target and deposited onto the substrate. Electrically, a potential gradient is established and ions, typically Ar^+, are removed from the target surface (cathode) and then condense on the substrate (anode) to form a film. This method is used extensively in the integrated circuit industry for thin-film deposition. It is an attractive process for depositing materials, such as high-purity metals, that possess the same composition as the target. When metal oxide, nitride, or carbide films are desired, a reactive gas (oxygen, nitrogen, or methane) is introduced to the gas flow in combination with argon. This process is referred to as *reactive* sputtering.

Sputtering is especially useful for the deposition of alloys due to the ease of acquiring an alloy target with the desired combination of elements. Deposition of alloys by evaporation, for example, is very difficult if the alloy materials have significantly different melting temperatures. In the case of sputtering, alloy composition will be maintained regardless of differing material properties of the alloy constituents. Sputtering does result in the incorporation of a very small amount of argon into the film, and heating of the substrate can also occur. In terms of step coverage, sputtering provides a moderate ability to cover surface topology. The coverage is better than what is observed for evaporation but not as good as the conformal films that result from CVD processes.

3.2.4.1 Direct Current Sputtering

Direct current (DC) sputtering, sometimes referred to as DC diode sputtering, is useful for depositing metallic films because the target is conducting. In DC sputtering, however, a low ion current density results in a relatively low sputtering rate, and therefore, magnetron sputtering has become a very common method for thin-film deposition due to the increased sputtering rate observed for magnetron sputtering (see Section 3.2.4.3). Diode discharges are widely used for plasma-enhanced deposition and reactive ion etching (RIE). For DC sputtering, the target material is conducting and, therefore, easy to deposit under DC conditions.

Typical base pressure for the vacuum system is in the range of 1×10^{-6} to 1×10^{-9} Torr. A gas inlet allows argon flow into the chamber, and sputtering pressures are typically in the 20-mTorr region. The target material is typically composed of hot pressed sintered disks and may be water cooled depending on the specific equipment type or application. A voltage is applied and a plasma, or glow discharge, is established between the parallel electrodes. The target is bombarded by energetic positively charged argon ions, and, as a result, target atoms are transported to the substrate material. The use of a shield between the target and electrode allows the cleaning of surface contamination from the target material before beginning the sputter deposition. Electrical connections can also be reversed to allow short-duration sputtering from the substrate to remove substrate contamination. A schematic of a typical sputtering system is shown in Figure 3.9. Sputtered target particles lose their initial energy as the number of collisions is increased. Therefore, the average kinetic energy of the sputtered material is dependent on the sputtering pressure.

The sputtered material arrives and condenses on the substrate as loosely bonded atoms. The rate of diffusion is dependent on the substrate material and the substrate temperature. Nuclei are formed and growth proceeds by diffusion. Many times substrate heating is employed to encourage diffusion processes. The rate of sputter deposition is determined by the sputter yield (S); S is the ratio of the number of target atoms ejected to the number of ions incident. Sputter yield depends on a number of factors such as target material, mass of bombarding ions, and the energy of bombarding ions. Since, typically, the target material is a specific desired material and the mass of bombarding ions will be the mass of argon ions, the variable under control is the energy of bombarding ions. There is a minimum energy for an ion to produce sputtering, and, after that threshold energy is achieved, S increases as ion energy increases. Sputter yield also varies with the angle of incidence of the ions. As the angle of incidence increases, atoms near the surface have a higher chance of being sputtered until the angle begins to approach 90° and then the ions have a higher probability of being reflected.

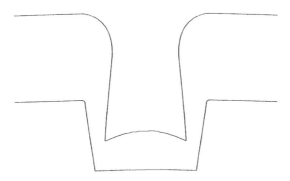

Figure 3.9 Schematic of a sputter deposition system.

3.2.4.2 Radio Frequency Diode Sputtering

For the deposition of insulating materials, where the target is nonconducting, a radio frequency (RF) generator is necessary. RF power applied between the electrodes enhances the plasma by sweeping the electrons back and forth between the cathode and the anode, increasing the lifetime of the individual electrons and, therefore, the probability of ionization. RF diode sputtering can be used to deposit nonconducting, conducting, and semiconducting materials. An RF diode vacuum system is arranged similarly to the DC diode system with the exception of the target power supply. The electrodes reverse the cathode–anode position on each half-cycle. Most RF power supplies operate at 13.56 MHz and, at these high frequencies, there is no charge accumulation on the target due to the short cycle time on each electrode.

An RF diode sputtering system can be operated at lower pressures than those for DC diode discharges. This is due to the fact that fewer electrons are lost and the ionization efficiency of the argon gas is increased. Typical operating pressures for RF sputtering are 1 to 15 mTorr. A disadvantage to this technique is the fact that the power supply is more complicated than a DC power supply and requires a matching network to operate efficiently. RF power can also be used with magnetron sputtering, as well as diode configurations.

3.2.4.3 Magnetron Sputtering

Magnetron sputtering systems have the addition of magnetic fields near the cathode in a similar arrangement to the diode sputtering systems. A magnetic field is applied at right angles to the electric field in these systems. This causes the electrons to follow spiral paths and, therefore, causes a confinement of electrons to the area around the target surface. This increase in electron density increases the probability of ionization of argon atoms, which also ultimately increases the sputtering deposition rates for this process. Without confinement, as in diode sputtering, electrons are lost to the chamber walls and sputter deposition rates are reduced. The use of a magnetron also allows the formation of a plasma at lower chamber pressures, typically 10^{-5} to 10^{-3} Torr.

Magnetrons vary in design with planar and cylindrical magnetron sources being the most common. Although they have different geometries, they operate under the same principles. In microelectronic production, large-diameter silicon wafers are difficult to place on the walls of a chamber possessing cylindrical geometry. Planar magnetrons are more widely used because of the geometry supplied with commonly used planar targets. The magnetic field created by a planar magnetron is toroidal and results in a high plasma density in this region. This area resembles a *racetrack* due to the ringlike nature and erodes the target more quickly in this area because of the higher sputtering rates that occur here.

3.2.4.4 Film Growth

Processing parameters influence the structure and properties of a sputter-deposited film. Ion bombardment, substrate temperature, sputtering pressure, target voltage, and current density are examples of processing parameters under operator control. A model has been developed, the zone model, that predicts characteristics of thin films that are formed by evaporation or sputtering based on the substrate temperature and ion energy. For low substrate temperature and low ion energy, the deposited film tends to be amorphous. In general, grain size and overall crystallinity in the film increases as the substrate temperature increases, the pressure decreases, or the ion energy increases. Intrinsic and thermal stresses may be incorporated in sputter-deposited films. Thermal stresses occur if there is a difference in thermal expansion

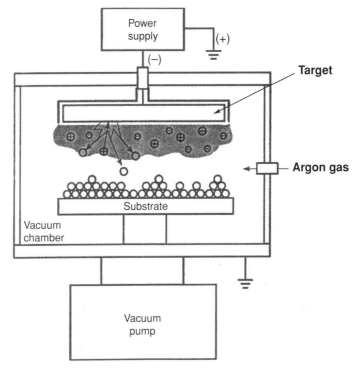

Figure 3.10 Cross section of a thin film deposited into a via opening by sputtering.

coefficients of the film and substrate, while intrinsic stress depends on the nature of film growth and is based on sputter conditions (e.g., argon gas pressure, substrate temperature, ion energy).

Integrated circuits have developed to incredible levels of complexity. Smaller circuit geometries demand higher aspect ratio (depth/diameter) vias to interconnect adjacent layers. Step coverage of thin films deposited by PVD is very important. In the case of sputtering, step coverage is improved when compared to evaporation due to the higher gas pressures and increased energy of depositing species that allow greater distribution of these sputtered species from the increased number of collisions. Typical step coverage of a via opening by sputter deposition is shown in cross section in Figure 3.10. It is clear from this figure that deposition into higher aspect ratio vias can be a problem due to the higher deposition rate at the top surface and corners compared to the via side walls. However, this step coverage can be improved by heating of the substrate to increase surface diffusion.

Other sputtering techniques exist that may be used depending on a specific application and the deposited thin-film requirements. Pulsed sputtering combines the advantages of DC and RF sputtering by using a single power supply in a frequency range of 10 to 100 kHz and is very useful for reactive sputtering to produce insulating films. Electron cyclotron resonance (ECR) microwave plasma sources can be used to increase sputter deposition rates. Diode systems may be enhanced with a second cathode (triode sputtering), which serves as a secondary source of electrons to increase ionization in the plasma and also results in higher deposition rates. Separate ion beam sources may be used to generate Ar^+ ions, independent of the ions bombarding the target, and this is referred to as ion beam sputtering, or ion-beam-assisted deposition (IBAD). This separate source of ions provides improved control over film growth.

3.2.5 Chemical Vapor Deposition

Chemical vapor deposition (CVD) processes rely on chemical reactions rather than physical reactions to promote the growth of thin films. Reactive gases are passed over a heated substrate in a furnace for thermal CVD. It is a method that is well suited to growing a wide variety of materials, and CVD-deposited films exhibit much improved step coverage compared to films deposited by PVD methods. Typically, the reactor is composed of a tube furnace and a vacuum pumping system if CVD is performed below atmospheric pressure. Reactor geometries affect deposition and film parameters, so it is important to fully characterize the specific reactor equipment in each laboratory. Vacuum systems, such as the ones commonly used in sputtering or etch processes, may be used in plasma-enhanced CVD processing (see Section 3.2.5.3).

One of the most important reasons for using CVD is the reduced temperature at which it can be accomplished compared to high-temperature furnace processes. When an integrated circuit has gone through metallization processes in the fabrication sequence, the metals generally will not tolerate high-temperature furnace operations. In the case of deposition of oxide or nitride passivation layers after metallization, for example, CVD provides a lower temperature alternative process. One disadvantage, however, is that oxide films deposited by CVD are inherently less dense than ones deposited by thermal oxidation.

CVD Reactors Principal requirements of CVD reactors are to provide a uniform supply of gaseous reactants to the substrate surface and to supply activation energy for the reaction. There are several types of reactors in use including:

1. *Horizontal Reactors* These reactors are similar to furnace systems used in oxidation or diffusion processing. They are essentially tube furnaces that include an inclined heated susceptor for uniform gas flow. The susceptor is the substrate holder. In the past few years, tube furnaces have been designed to allow vertical chambers that have several advantages over horizontal tubes. A reduced particle count is one of the most important advantages.

2. *Vacuum Reactors* These reactors consist of a general vacuum system that includes inlets for gas flow, vacuum pumping, vacuum exhaust, and a power supply to internal electrodes. This type of reactor is commonly used for plasma-enhanced CVD. Due to the toxic nature of some CVD gases, the exhaust from vacuum pumps should be treated for safety reasons.

3. *Continuous-Feed Flow Reactors* In this reactor, substrates are placed on a conveyor belt that moves underneath gas nozzles and is typically used in atmospheric pressure CVD (APCVD) processing.

Different reactor types may also be termed hot- or cold-walled reactors.

1. *Hot-Wall Reactor* The walls of the reactor chamber are heated directly by resistive heating elements and the heat is transmitted to the substrates. This has the advantage of uniform temperature distribution and convection effects. However, film will deposit on the walls of the chamber and may become a source of particles in the film.

2. *Cold-Wall Reactor* In this reactor, the substrates are heated to a much higher temperature than the walls of the chamber by heating the susceptor, and, therefore, film deposition on the chamber walls is reduced.

3.2.5.1 Low-Pressure CVD

Low-pressure CVD (LPCVD) is performed under vacuum in either horizontal or vertical reactors at pressures in the range of 0.1 to 1.0 Torr. Either hot- or cold-walled reactors can be used. Due to the requirement for vacuum, vacuum system design and maintenance are considerations in choosing this particular process. These systems also require strict safety protocols because the exhaust gases are reactive and usually toxic or flammable. Step coverage for LPCVD-deposited films is good due to the conformal nature of CVD films.

Materials commonly deposited by LPCVD are polycrystalline silicon and dielectric films. Most LPCVD polysilicon is done with silane (SiH_4) in furnaces at temperatures ranging from 575 to 650°C. Typical polysilicon deposition rates are 100 to 1000 Å/min. Various chemistries can be used for deposition of oxide in LPCVD reactors, such as silane and oxygen, dichlorosilane ($SiCl_2H_2$) and nitrous oxide (N_2O), and tetraethoxysilane (TEOS). Silicon nitride is most commonly deposited from a mixture of dichlorosilane and ammonia (NH_3).

3.2.5.2 Atmospheric Pressure CVD

Atmospheric pressure CVD is performed at atmospheric pressure, thus no vacuum system is required. Films that are deposited by this method have the advantage of a high deposition rate, but the trade-offs are particulate problems and relatively poor uniformity. This process is more commonly used for depositing thick dielectrics because the deposition rates exceed 1000 Å/min. Wafers move along a conveyor belt that is heated to temperatures of approximately 350°C. Gas nozzles supply the reactant and diluent gases for the reaction. Oxide and nitride films may be deposited by APCVD. Another material of interest is phosphosilicate glass (PSG), which is a silicon dioxide film containing phosphorus. This material softens and reflows at moderate temperatures and can promote planarization of surfaces.

3.2.5.3 Plasma-Enhanced CVD

An efficient method for fragmenting gas molecules into reactive parts is by the formation of a plasma. Therefore, vacuum chambers equipped with electrode systems for generating a plasma may be used to increase the reactivity of this deposition process. This increased reactivity by the plasma allows deposition to be performed at lower temperatures than those required by LPCVD. Typical deposition temperatures of most materials in a plasma-enhanced CVD (PECVD) system do not exceed 400°C. The PECVD process is useful for depositing films for integrated circuits (IC) that cannot tolerate typical CVD temperatures because of the presence of metallization on the IC, for example. Substrates are immersed in the plasma generated by a parallel-plate electrode system. Due to the ability to provide ion bombardment of the surface, species diffuse along the surface. Thus, PECVD is known as a good process for filling small feature sizes. High-density plasmas can also be generated with electron cyclotron resonance (ECR) sources to further lower substrate processing temperatures.

3.2.6 Plating

Plating is another very popular and important technique for depositing high-quality conductor films and foils used extensively in the packaging of single integrated circuit chips, in conventional circuit boards, and in multichip module packaging. Both electro- and electroless plating are used in the electronic industry. A few examples are mentioned here, details of which are to be found in appropriate sections dealing with them. Copper foil used in multilayer circuit board fabrication is mass produced in the form of rolls using electroplating technology. In plated through hole technology, copper is first deposited electrolessly and then by electroplating into high-aspect-ratio through holes. Tape-automated bonding (TAB) strips are also made by plating or vacuum coating conductor and solder metals sequentially on a polymeric supporting film. Similarly, electronic connectors encountered in various forms and shapes, such as package pins in through holes, pad-to-pad connectors in surface mounting, or spring-to-pin connectors, usually consist of a base metal structure (made of copper, phosphorus–bronze, beryllium–copper, iron–nickel alloys, or nickel) plated with a hard coating of gold, followed by a flash (100 to 200-nm-thick plating) of soft gold. For all copper-based connectors, a layer of Ni is plated onto the copper before the Au to prevent Cu oxidation and intermixing of Cu and Au. For multichip modules constructed of Cu/polyimide or Cu/SiO$_2$, plated copper is used industrywide because of its cost effectiveness.

Both electro- and electroless plating involve the deposition of metal onto a desired surface from an electrolyte solution containing a salt of the metal. In its simplest form, the metal deposition process can be thought of as a reduction of metal cations, M^{n+}, at the substrate surface according to the reaction

$$M^n + n\,e^- \longrightarrow M^o \tag{3.5}$$

Such a reduction reaction must be accompanied by a corresponding oxidation reaction to supply the requisite electrons. In electroless plating, an appropriate reducing agent, added to the plating bath, provides such a reaction. Whether a particular reducing agent will oxidize to provide the necessary electrons for the reduction of metal at the substrate depends on the redox potentials for the electrochemical reactions involved. Additive agents and pH control may be needed to tip the reaction in favor of the plating reaction. In electroplating, metal reduction from the electrolyte takes place at the cathode, whereas the oxidation reaction proceeds at the anode. An overpotential is provided by the applied electric field for the desired reaction to take place. In either case, once the metal atom is abstracted by the substrate surface, the process follows the nucleation and growth mechanisms previously discussed in relation to vapor-phase deposition of metals.

3.2.6.1 Electroless Plating

Electroless plating is commonly used for copper plating in PCB fabrication on insulating epoxy–glass laminates in additive fine-line processes and as the seed layer (for later electroplating) inside plated through holes (PTH) for both additive and subtractive processes. Electroless plating of many other metals, such as Ag, Au, Pd, Ni, and Cr is also used extensively for technological or ornamental applications. Copper plating, however, is emphasized here.

Although electroless copper is used primarily on nonconducting surfaces, the reduction reaction required to deposit Cu must occur initially on a metallic surface. The insulating surface, therefore, must be activated (catalyzed) first to provide reducing metal seeds on its surface. Palladium and tin alloys are created by an autocatalytic reduction of metal ions from solutions. The substrate to be plated is sequentially dipped in SnCl$_2$ and PdCl$_2$ solutions. This forms Pd nuclei on the insulating surface. In another technique, the surface is treated with a colloid of Pt/Sn metallic particles. The surface charge on the metal particles

Table 3.1 Typical Electroless Copper Plating Bath Composition

Component	Concentration (g/L)
Copper sulfate (pentahydrate)	15
Formaldehyde	10
Ethylene diamine tetraacetic acid (EDTA)	20
Sodium potassium tartrate	10
Sodium hydroxide	15

is negative, which keeps them from agglomeration and helps them deposit on positively charged sites on the insulator.

Initial copper nucleation on Pd nuclei is believed to occur because of the catalysis of the dissociation of formaldehyde into formate ions and hydrogen. The excellent ability of hydrogen abstraction by Pd causes the catalysis. Pt, Ni, and even Cu, which adsorb hydrogen, will support electroless plating.

Once the surface is covered with deposited Cu, the reduction reaction continues. However, the reducing agent must be chosen such that the reaction only occurs at the substrate surface and not throughout the solution; the latter will cause precipitation of the metal throughout the bath (bath decomposition). In fact, bulk precipitation may occur due to local changes in the pH value. Complexants are added to control this problem by reducing the concentration of free metal ions below the value necessary to trigger the breakup of the metal complex in the bath. It also allows the pH of the plating bath to be increased, which makes Cu deposition thermodynamically more favorable. Bath decomposition may also occur due to the presence of dust or metallic particulate. This can be avoided by adding stabilizers, such as oxygen, which are adsorbed on the particulate surfaces to prevent heterogeneous reduction on these surfaces. During plating, the pH of the bath changes unless buffers are added. Carboxylic acids and organic amines are used as buffers for acidic and alkaline baths, respectively. The addition of complexant may depress the deposition excessively. Therefore, exaltants or accelerators, which are thought to increase the oxidation rate, are sometimes added to the bath.

The composition of an alkaline electroless copper bath is given in Table 3.1. It consists of copper sulfate as the electrolyte, formaldehyde as the reducing agent, sodium hydroxide for alkalinity, ethylene diamine tetraacetic acid (EDTA) as a complexant and pH buffer, and sodium potassium tartrate (Rochelle salt) as a complexant.

3.2.6.2 Electroplating

Electroplating has been around since the early nineteenth century when silver was electroplated. Other metals that were plated during the early twentieth century include Cu, Ni, Au, and Cr. Electroplating technology has come a long way since then in terms of mass production of extremely tightly quality-controlled films and foils for microelectronic applications. In this section, Cu is treated as an example despite the fact that Au, Cr, and Ni plating are by no means technologically less important.

Faraday's laws of electrolysis predicts the weight of electrodeposited material at the electrode in terms of the current passed through the electrolyte as

$$M = \frac{QA}{qNn} = \frac{ItA}{Fn} \tag{3.6}$$

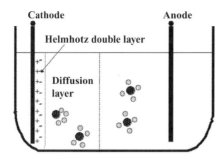

Figure 3.11 Schematic representation of the plating mechanism.

where M is the mass of deposited metal in grams, Q is the total charge in coulombs (equal to I time t where I is the constant current in amperes and t is the duration of the plating), A is the atomic weight of the metal, F is the Faraday constant (equal to the product of electronic charge, q, and Avogadro's number, N), and n is the number of electrons transferred per metal ion. This expression assumes 100% current efficiency in plating the required metal. There may, however, be other reactions, such as hydrogen evolution, that reduce the efficiency.

Figure 3.11 represents an electrolytic plating bath with a certain electrolyte in it. Before an external voltage is applied across the cathode and anode, a certain potential exists between the metal cathode (on which metal is to be plated) and its ions in the electrolyte. This potential is normally reported with respect to the standard hydrogen electrode (SHE), which is defined to be 0 V. Because of this potential, ions or dipolar complexes line up along the metal electrode as shown in the figure. This is known as the Helmholtz double layer. When an external voltage is applied, the cell potential changes from its equilibrium value by a voltage called the overpotential, which determines the rate at which the plating reaction proceeds.

The region adjacent to the Helmholtz region, where the ion concentration is less than the equilibrium concentration, is called the diffusion region. A concentration gradient exists across this region due to the reduction of metal at the cathode, which forces the metal complexes to diffuse toward the cathode. The diffusion region is usually thousands of times larger than the double layer. With forced agitation, the diffusion layer can almost be eliminated.

Plating can be done either potentiostatically or galvanostatically. In the former, the applied voltage is kept constant, whereas in the latter the current is kept constant. The advantage of constant current plating is that it can be easily timed for a certain amount of metal deposition. The constant current or voltage in the two techniques can be pulsed between on and off to achieve what is known as pulse plating. This method gives much denser, more uniform, and smoother depositions.

There are several electroplating solutions used for copper plating, but the one used most widely in microelectronic packaging is the copper sulfate/sulfuric acid bath. Copper sulfate is the source for cupric ions, whereas, sulfuric acid provides high conductivity for the bath. There are two main types of acid copper baths. The *high-speed* bath consists of a high concentration of copper ($\sim$100g/L) and a low concentration of acid ($\sim$30g/L), and is operated at high currents (hundreds of mA/cm^2). Conversely, the *high-throw* bath has a low concentration of Cu ($\sim$15g/L) and a high concentration of acid ($\sim$200g/L), and is used with low current densities ($\sim$30 mA/cm^2). The high-speed bath has low conductivity and, therefore, does not deposit uniformly inside high-aspect-ratio through holes; the high-throw bath is used for such applications. There are some organic additives that are used to improve film quality. Brighteners improve the grain size and microstructure from columnar-like to more uniform grains.

3.3 PATTERNING

In the previous section, it was shown how various substrates, such as ceramic chip carriers, TAB tape, or via-drilled, glass–epoxy laminates, can be metallized. Metallized surfaces are further processed to form metal traces using photolithography and etching techniques. This is known as subtractive metallization. Sometimes photolithography precedes the metallization step, resulting in a patterned metal deposit without the need for etching. This is called additive metallization.

3.3.1 Photolithography

Photolithography is a technique by which photoimaging methods are used to transfer the image of a mask onto a substrate. It utilizes a photosensitive material called photoresist, which undergoes chemical changes upon exposure to intense ultraviolet light, becoming either more (or less) soluble in certain solvents and, thus, leaving a positive (or negative) replica of the mask on the substrate. A resist that leaves a positive image of the mask is called positive photoresist and the other is called negative photoresist. These photoresists consist primarily of long-chain polymers and a certain percentage of sensitizers. Negative photoresist has cyclized polyisoprene (rubber) as the resin and bis-aryl diazide as the sensitizer. Upon exposure to ultraviolet (UV) light, cross-linking of the polymer takes place, making it even more difficult to dissolve or etch. Positive resists have orthoquinone diazides as the resin and an almost equal amount of a sensitizer. When exposed to UV light, the bonds in the resin break, turning it into carboxylic acid, which can be easily dissolved in alkaline developers. The name *resist* indicates that the regions of the substrate that are protected by the resist will not be attacked by etchants that remove the photoresist from unprotected regions. This is illustrated schematically in Figure 3.12.

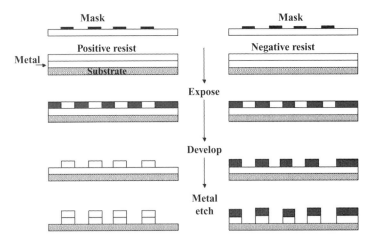

Figure 3.12 Steps involved in the patterning of metal traces using positive and negative photoresists.

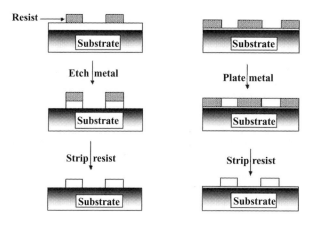

Figure 3.13 Illustrations of the additive and subtractive techniques used to pattern interconnects on substrates.

3.3.1.1 Patterning Schemes

Figure 3.13 depicts a method of transferring a pattern from a mask onto a copper-plated substrate. The copper film could have been deposited on a glass–epoxy laminate with through holes by an electroless and electroplating combination, or a Cu foil could have been laminated first, followed by plating to metallize the through holes. This method of patterning a circuit board is referred to as the subtractive technique. Another method of fabricating Cu traces on circuit boards is called the additive technique. In this case, the pattern is transferred onto the bare substrate (glass–epoxy laminate, e.g.), and then electroless plating is used to build up the copper traces. These are illustrated in Figure 3.13.

3.3.1.2 Process Steps

Photo-definable materials are available in liquid as well as thin solid sheet form. Liquid photoresists are either spray- or spin-coated onto the substrate. This process causes the solvent to evaporate, leaving a gel-like film on the substrate. On the other hand, dry-film photoresists do not have any solvents in them. Instead, they contain some thermoset polymers. The photoresist film is hot roller-coated directly onto the substrate. Liquid photoresists are commonly used in IC fabrication and on silicon substrate multichip modules (MCMs), whereas dry films are used extensively in multilayer circuit board and thick-film screen fabrication where the requirement on linewidth is not very demanding.

3.3.1.3 Resist Application

As noted previously, liquid resists are applied to substrates using spin or spray techniques. In the spin-coating technique, the substrate is held on a chuck capable of high-speed rotation by a vacuum. The chuck rotates at about 500 rpm while a measured amount of resist is dispensed onto the substrate. After a few seconds, the chuck speed is increased to around 3000 rpm and maintained there for 20 to 30 s. The spinner is then turned off. Depending on the viscosity of the resist and the spin speed, a certain thickness of photoresist remains on the substrate. Usually, an edge bead is formed around the periphery of the substrate, which is removed using a solvent. The photoresist contains residual solvents, which are removed by a soft bake process at about 90°C for 30 min in air.

Dry-film photoresist is supplied between two protective sheets; a top polyester support sheet and a bottom polyolefin cover sheet. The bottom layer is removed before laminating the dry film to a substrate using a hot roller. The heating causes the resist to flow and follow the surface morphology. Sometimes adhesion promoters are added to the dry film for improved adhesion.

3.3.1.4 Resist Exposure

The resist-coated substrate is placed in an aligner that ensures that the pattern on the mask will be transferred onto each substrate in an identical fashion. This is even more important for double-sided laminates, which must register properly with all other laminates, so that plated through holes will connect the correct conductor lines. In sophisticated aligners (those used for IC fabrication), alignment patterns on the substrate and the mask are aligned using micromanipulators. Currently, these costly aligners are being used for MCM-D (MCM—deposited) technology. (This is overkill since minimum linewidths of 10 to 15 μm are all that is required in MCM-D applications.) In circuit board fabrication, proper alignment can be achieved using alignment holes in the boards, as well as in the mask, which are then placed over registration pins in the aligner. The mask may be placed in contact with the resist-coated surface (contact printing), or slightly removed from it (proximity printing). In contact printing, copies of the master mask are used since the masks are damaged due to repeated contact with substrates. Usually, the master copy is made of glass, whereas the copies are disposable emulsion masks. Proximity printing avoids mask damage at the cost of a slight loss in resolution. Exposure time depends on the resist sensitivity, spectral distribution of the UV lamp, and its intensity. A dry-film resist-coated substrate requires a waiting period of about 30 min before developing in order to allow the polymerization process to be completed. Certain dyes are added to dry photoresist that cause the color of the exposed region to change, thus creating a negative image of the mask in the exposed resist. The top protective polyester film is removed before the film is developed.

3.3.1.5 Resist Developing

A photoresist supplier usually provides an appropriate developer and a developing recipe. A negative resist developer is a xylene-based solvent that dissolves unexposed resist but leaves the exposed and, hence, cross-linked resist intact. Positive resist developer is an alkaline aqueous-based liquid that easily removes the exposed resist where the resin has turned to a carboxylic acid group. Following resist development, the substrates are rinsed and inspected for completion of the developing process before etching the underlying (e.g., metal) film. Etch chemistries and recipes are discussed in Section 3.3.2. First, however, photoresist stripping, which is performed after the etch step, is described.

3.3.1.6 Resist Stripping

In some multilayer circuit board fabrication, the remaining hardened photoresist is not removed. Rather, it is imbedded into the multilayer structure. The adhesion and out-gassing of such photoresists must meet more stringent reliability requirements. Photoresist manufacturers supply the stripping solution and its recipe. Normally, strong oxidizers, such as freshly mixed hydrogen peroxide/sulfuric acid can be used to remove cross-linked negative photoresist, but the same mixture also etches copper. Nophenol-based strippers can be used with metals. These contain sulfonic acid and other proprietary solvents. Positive resists can be stripped in organic solvents, such as acetone, or those supplied by the resist manufacturer.

3.3.2 Etching

Etching a material means removing it from areas where it is not wanted by reacting it with liquids or gases to form soluble or volatile products, respectively. When etching is performed in a liquid, it is called wet etching, and when it is done in a gaseous or vapor phase, it is called dry etching. Thus, during etching of Cu in sulfuric acid, copper reacts to form copper sulfate, which is soluble in the aqueous solution being used and is, therefore, easily removed. Similarly, when gold is etched in a CCl_4 plasma (dry etching), volatile gold chloride radicals are formed and are swiftly removed from the area being etched by the process pump. One of the important requirements of a good etchant is that it does not react with the resist masking layer, or the etching selectivity between the two should be very large. Currently, wet etching is used almost exclusively in circuit board manufacturing because (1) the capital investment is quite low, (2) the minimum linewidth requirement is not stringent (upwards of 25μm), and (3) a dry etching process for copper does not exist. Even for etching of Au, high selectivity wet etchants are available.

3.3.2.1 Wet Etching

In order to etch a pattern onto a substrate photolithographically, a patterned substrate is immersed in an etchant solution for a predetermined time. Various agitation schemes (such as magnetic stirring, air bubbling, liquid recirculation, etc.) may be utilized to improve the etch uniformity across the substrate. The etch rate may be a function of certain ionic or complexed radical concentrations that can decrease with continued etching. This results in continually decreasing etch rates. Buffers are added to solve this problem so that substrate-to-substrate and lot-to-lot reproducibility is within specifications.

Wet etching normally yields extremely good metal-to-mask selectivities. For example, the use of a ferric chloride or sodium bisulfate/sulfuric acid mixture does not adversely affect negative photoresist while etching copper at a fairly high rate. On the other hand, if a hydrogen peroxide/sulfuric acid mixture is used for copper etching, it will strip the photoresist as well. Thus, the proper choice of etchant is necessary for photoengraving of patterns by wet etching.

Wet etching is normally isotropic. That is, it etches in the lateral direction, under the mask, as it etches vertically through the film thickness. This is shown in Figure 3.14a. This

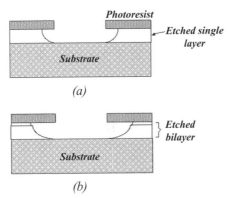

Figure 3.14 Isotropic wet etching of (a) a single-layer and (b) a bilayer thin film.

Table 3.2 Etchants for Some Selected Metals Encountered in Packaging Technology

Metal	Etchant
Gold (Au)	Aqua regia ($3HNO_3/1HCl$)
Aluminum (Al)	$85H_3PO_4/5CH_3COOH/5HNO_3/5H_2O$
Aluminum (Al)	$1NaOH/5H_2O$
Chromium (Cr)	$2KMnO_4/1NaOH$
Chromium (Cr)	$1HCl/1H_2O$
Copper (Cu)	$1HNO_3/1H_2O$
Copper (Cu)	$2CuCl/1HCl$
Nickel (Ni)	$FeCl_3$
Titanium (Ti)	$1H_2SO_4/1H_2O$

causes the final width of a line of thickness, d, to be approximately $2d$ smaller than the width of the mask line. As long as the mask is designed to take this artifact into account, it will not cause any problems. However, the problem can be exacerbated in situations where a double layer has to be etched. If the etch rate of the top layer is higher than the bottom layer in the same etchant, there will be considerable undercutting. This is shown in Figure 3.14b.

Etching of a metal is, in fact, an oxidation reaction whereby a neutral metal atom loses one or more electrons (oxidation), becomes ionized, and moves into the solution. It may form complexes within the solution, but that is beside the point. The electrons lost by the metal ion must be collected by radicals provided by the etchant solution (reduction). Furthermore, the flow of electron(s) from the metal being etched to the radicals must be energetically favored. From an electrochemical perspective, half-cell potentials of reduction and oxidation reactions of the radicals involved can be used to predict if etching will take place or not. Consider using a ferric chloride solution to etch copper. If the reaction goes to completion, then

$$Cu^0 \longrightarrow Cu^{2+} + 2e^- \tag{3.7}$$

and

$$2Fe^{3+} + 2e^- \longrightarrow 2Fe^{2+} \tag{3.8}$$

The overall reaction is thus

$$Cu^0 + 2Fe^{3+} \longrightarrow Cu^{2+} + 2Fe^{2+} \tag{3.9}$$

The half-cell potential for the first oxidation reaction is -0.34 V, whereas it is 0.77 V for the second. Therefore, for the overall reaction, the change in potential is 0.43 V. This positive potential indicates that the etching reaction is thermodynamically favorable.

There are other wet etchants for copper as well. Table 3.2 gives recipes for etching a few of the metals commonly encountered in packaging technology.

3.3.2.2 Reactive Ion Etching

Etching by dry methods is performed to avoid the isotropic nature of wet etching. Dry etching is done under vacuum by generating a plasma with the reactive gaseous species flowing into the chamber. Ions bombard the substrate and remove the targeted material that reacts with the etch species. This chemical etch performed under vacuum is referred to as reactive ion etching (RIE). It is extremely useful for etching a wide variety of materials and the resulting etch profile is anisotropic. This is due to a highly directional ion beam, which creates a preference for the base of the opening compared to the side walls.

Parallel-plate electrode systems are commonly used in a low-pressure vacuum system. Chlorine-based plasmas, as well as fluorine-based, are used to etch materials such as silicon, silicon dioxide, and aluminum. One problem in chlorine-based chemical reactions is the degradation of photoresist by chlorine in this process. Another problem is residual film post-RIE processing. Another general problem is the potential for electrical damage to devices. Postetch cleaning and annealing schemes can be used to minimize etch damage.

3.4 METAL-TO-METAL JOINING

In electronic packaging, situations are often encountered where one metal is to be joined to another metal. Examples are bond pads on the chip-to-bond pads on the package, multilayer ceramic package output pads to pins and leads, package-brazed ring-to-metal cap, and package pins to plated through holes or surface-mount pads on second-level package boards. There are three ways by which such metal-to-metal contacts can be made reliably: (1) solid-state bonding, (2) soldering and brazing, and (3) fusion welding. An example of fusion welding, is electric arc welding, which is used to hermetically seal ceramic packages by joining a metal cap to a brazed metal ring on the package. This method is normally used for larger packages and for multichip modules. Brazing is more commonly used to attach metal lids to packages because of the much lower temperature involved and the reworkability of the process. This method is not discussed further here. Solid-state bonding and soldering/brazing techniques, along with examples, are discussed in the following sections.

3.4.1 Solid-State Bonding

Metal pads on a chip can be electrically connected to the package pads by stitching conducting flat or round wire between them. Commonly used techniques are referred to as wire bonding or tape-automated bonding. In either of the techniques, the conductor wire is pressed against the metal pad, which may be elevated to high temperatures (but much below the melting temperatures of the metals involved) to promote solid-state interdiffusion of metals. Ultrasonic rubbing of the wire is also used to improve adhesion. Yet, the interdiffusion should not be so extensive as to form voids at the interface, nor should it form intermetallics that are resistive and brittle.

3.4.1.1 Wire Bonding

The metal pads on most ICs are aluminum, whereas the bond pads on most ceramic packages and metal lead frames (for plastic packages) are plated with multiple layers of metal and a final flash coating of soft gold. In general, the wire material must have the following properties: (1) highly conductive, (2) bondable to Al and Au, (3) corrosion resistance, (4) strong, (5) drawable in fine-wire form, and (6) ductile. Both Al and Au satisfy these conditions and are used extensively for wire bonding. There are two types of wire bonders, which are routinely used in microelectronics packaging. These are briefly described here.

Thermocompression Bonding In thermocompression bonding, both temperature and pressure are used to make a metal-to-metal bond. If either pressure or temperature alone is used, the process will not result in a bond. The chip carrier with the die already attached to it is placed on a heated block. The chip temperature should be about 220 to 250°C. A 1-mil (25 µm) diameter gold wire is passed through a ceramic capillary, which may or may not be heated. The end of the gold wire is made into a perfect ball using a hydrogen torch or an electric arc. The compression head is aligned with the pad using x-y micromanipulators. The capillary is brought down on the metal pad, which causes the gold ball to strike the bond pad, be flattened, and bonded to it. The capillary is then pulled back and moved to the package lead bond pad where it is lowered and the Au wire is pressed onto the lead pad to form the second (a stitch) wire bond. The capillary is then pulled back a little distance before a clamp is actuated, which grabs hold of the wire and pulls it. The wire breaks at the weak point (the heel of the wire bond) leaving a tailless bond. All these steps are illustrated in Figure 3.15. One of the attractive features of this wire bonding technique is that, following formation of the first bond the capillary head can be moved in any direction to form the second bond.

Thermocompression bonding to gold pads using gold wire produces extremely high reliability bonds. A gold wire to an aluminum pad bond is not nearly as good. In fact, at elevated temperatures, such bonds are very unreliable due to the formation of various Al–Au intermetallic compounds. One such intermetallic is commonly referred to as the

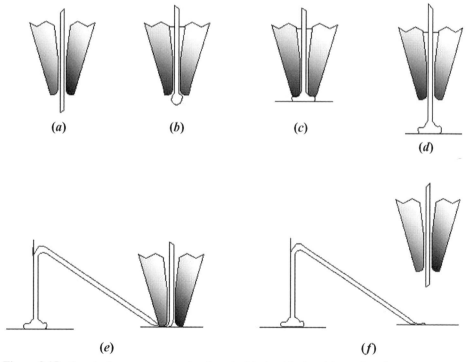

Figure 3.15 Steps in thermocompression bonding of gold wire: (*a*) wire sticks out from the capillary, (*b*) hydrogen torch makes ball at the end of the wire, (*c*) the ball is compressed against the heated bond pad to form a nail head, (*d*) the capillary is moved away to form the loop, (*e*) the second bond is made, and (*f*) the wire is broken off by pulling.

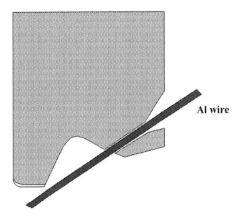

Figure 3.16 Sketch of the wedge used for ultrasonic bonding of aluminum wires.

"purple plaque." In the past, it has been identified as the cause of extremely poor wire bond reliability.

Ultrasonic Wedge Bonding In a wedge bonder, ultrasonic agitation provides the energy to deform the wire and induce diffusion to yield a strong bond. It uses Al–Si (1% Si) wire of about 2 mils diameter. The wire passes underneath a wedge-shaped tool, which has a hole in the back and through which the wire passes as shown in Figure 3.16. The wedge is brought into contact with the bond pad and pressed against it while the ultrasonic source is turned on. A good bond is formed, although not as good as a thermocompression bond. The wedge is then moved to the lead bond pad region of the package where it comes down a second time to form a similar bond, except that, in this case, the wedge is tilted slightly so that the wire becomes deformed at the heel of the bond. The wire is unreeled a little and a clamp is activated to pull on the wire to break it at the weakest point. One of the disadvantages with this type of bonder is the inability to turn in any direction for the second bond.

3.4.1.2 Tape-Automated Bonding

The wire bonding process is very slow, requiring operator vigilance even though modern bonders have image recognition capabilities and can be programmed to perform wire bonding at a very fast rate. Still, the ideal situation is where all the connections are made at the same time. This is accomplished using tape-automated bonding (TAB). In TAB technology, flexible leads are photoengraved from either vacuum-deposited or plated copper layers on polyimide or polyester films. The inner pads are connected to all the bond pads on the chip at one time using thermocompression "gang bonding". This is shown in Figure 3.17. The

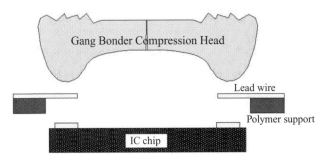

Figure 3.17 Schematic showing the "gang bonding" of a TAB lead strip to a chip.

outer pads can also be gang bonded to the package, or they may be soldered. Likewise, the inner leads can be bonded using soldering. In this case, appropriate solder metals must be vacuum deposited or plated onto the Cu.

The Al bond pads on the chip are "bumped" by depositing gold on them. A thin layer of Ti–W or Cr may be deposited to prevent interdiffusion between the two. Cu TAB leads may also be coated with Au–Ni where Ni prevents interdiffusion between the Au and Cu. If, on the other hand, TAB leads are to be soldered to the pads, they are coated with Sn in order to eutectically solder them to Au-coated pads. The eutectic temperature for the Au–Sn system (28% Sn) is 280°C, slightly higher than the thermocompression gang bonding temperature. More about soldering is presented in the next section.

3.4.2 Soldering and Brazing

Soldering and brazing techniques are methods of joining two metals using an alloy that melts at a temperature lower than either of the constituents, wets both surfaces well, and flows easily. Brazing is also called hard soldering since the only difference between the two is the temperature at which the joining occurs. Brazing is usually accomplished at high temperatures ($>400°C$) where wetting, as well as diffusion mechanisms, are considered important in providing strength to the joint. In soldered joints, only wetting is believed to be involved. Soldering is used in the die-attach (back-side bonding) process if good thermal contact to the substrate is required. It is also used in flip-chip die attach, also known as the controlled-collapse chip connection (C4) process. Brazing, on the other hand, is used for attaching pins to multilayer ceramic (MLC) PGAs and for metal lid joining in hermetic sealed packages. The flip-chip solder process and pin brazing for pin grid arrays (PGAs) are described in the following sections.

3.4.2.1 Controlled-Collapse Chip Connection (C4)

Wire bonding and TAB were treated in the previous section as the first-level interconnection between an IC chip and its package. Wire bonding is a sequential process and, therefore, becomes prohibitively expensive as the number of I/Os on the chip increases. TAB solves this problem by gang bonding. Both these schemes require bond pads around the periphery of the chip, which strongly limits the ability to increase packing density of electronic systems. IBM developed an area array chip to package interconnect scheme in the early 1960s, known as flip-chip bonding, that resolves this issue. There are two flip-chip techniques, the beam lead and C4 processes. Only the C4 technique is discussed here.

In the flip-chip process, the contact bond pads are spread over the entire chip surface. Adhesion promoter/diffusion barrier thin-film coatings of Cr/Cr–Cu/Cu/Au are deposited onto the Al contact pads. The Cu layer is usually very thick, but the Au layer is thin and serves to prevent oxidation of the Cu. This interfacial metallurgy is necessary because Al does not wet Pb–Sn solder, whereas Cu does. Cr improves adhesion between Al and thick Cu and prevents intermetallic formation. The Pb and Sn films are then deposited to several mils thickness. The Pb-to-Sn ratio is kept in the lead-rich region (i.e., above the eutectic composition of 67:33). After metal evaporation through a shadow mask, the solder is reflowed in a belt furnace in a hydrogen ambient to form the solder bumps. Nonwetability of glass and surface tension result in well-formed and well-mixed bumps. The ceramic chip carrier is metallized with Cr/Cu/Cr thin films and photoetched to form connecting

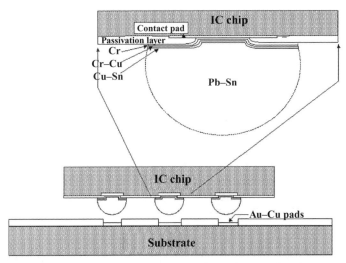

Figure 3.18 Schematic of a flip-chip solder bump and the chip attachment to the substrate.

lines and pads for solder bump attach. The Cr is stripped from the bump pad areas and a thin gold layer is deposited to improve soldering. The chip is aligned and the solder is reflowed in the presence of a proper flux in a furnace to form the C4 contacts. Figure 3.18 shows a schematic of a flip-chip solder bump and a chip attachment to a substrate.

3.4.2.2 Pin Brazing

In MLC packages, the package pins are brazed to pads on the back of the substrate. These pads are made of screen-printed Mo that are formed during the high-temperature firing of the screen-printed green stack. A 5-μm-thick electroless nickel layer is then plated on the pads, followed by a thin layer of similarly deposited gold. These pads are then screen printed with Au–Sn paste of eutectic composition. The kovar (Co–Ni–Fe alloy) pins, plated with Ni and Au, are held in a fixture in contact with the Au–Sn screen-printed pads and brazed at 390°C.

3.5 SUMMARY

In summary, some techniques and processing technologies used in electronic packaging were briefly discussed in this chapter. This included a description of metallization techniques used in printing circuit boards, MCM-D packaging, TAB, solder bumps, and so forth in some detail. These techniques included evaporation, sputtering, and plating. Also, techniques used to fabricate conductor traces, that is, photolithography and etching, were described. Finally, brief accounts of wire bonding, TAB, flip-chip attachment, and brazing were presented in this chapter.

REFERENCES

1. T. L. LANDERS, W. D. BROWN, E. W. FANT, E. M. MALSTROM, AND N. M. SCHMITT, *Electronics Manufacturing Processes*, Englewood Cliffs, NJ: Prentice Hall, 1994.
2. D. P. SERAPHIM, R. LASKY, AND C. Y. LI, eds., *Principles of Electronic Packaging*, New York: McGraw-Hill, 1989.
3. R. A. COLCLASER, *Microelectronics: Processing and Device Design*, New York: Wiley, 1980.
4. W. S. RUSKA, *Microelectronic Processing*, New York: McGraw-Hill, 1987.
5. S. M. SZE, *Semiconductor Devices—Physics and Technology*, New York: Wiley, 1985.

6. S. P. MURARKA AND M. C. PECKERAR, *Electronic Materials—Science and Technology*, San Diego: Academic, 1989.
7. W. WESTWOOD, in *Sputter Deposition*, H. Tompkins, ed., AVS, 2003.
8. J. A. THORNTON, Structure-Zone Models of Thin Films, *Proc. SPIE*, Vol. 821, pp. 95–103, 1987.

EXERCISES

3.1. Calculate the mean free path of the residual gas in a vacuum chamber evacuated to pressures of 10^{-5}, 10^{-6}, and 10^{-7} Torr. Assume that the residual gas is water (H_2O), which is almost spherical, and has a diameter of 0.13 nm. (Note that 1 Torr equals 133.3 Pa.)

3.2. In a vacuum evaporation system, Al is being evaporated at a rate of 2 nm/s. What is the impingement rate of Al atoms at the substrate? What impingement rates are expected for the residual H_2O molecules at the substrate for base pressures of 10^{-5}, 10^{-6}, and 10^{-7} Torr. Assuming a sticking coefficient of 1 (i.e., every H_2O molecule striking the substrate surface leaves an oxygen atom to be incorporated into the growing film), what will be the atomic percent oxygen in the Al film for the three pressures?

3.3. Gold is to be evaporated from a molybdenum boat onto a substrate situated at a distance of 20 cm. What amount of gold (in grams) should be loaded into the boat, assuming that it will be completely evaporated, in order to achieve a film thickness of 500 nm on the substrate. The density of gold is 19.3 g/cm^3.

3.4. A copper-plating bath is operated at a current density of 25 mA/cm^2 for 20 min. What thickness of plated copper is obtained if the plating efficiency is assumed to be 100%. The density of copper is 8.96 g/cm^3.

Chapter 4

Organic Printed Circuit Board Materials and Processes

RICHARD C. SNOGREN

4.1 INTRODUCTION

Printed circuit boards (PCBs), printed wiring boards (PWBs), or just plain circuit boards have been around in some form or another for over half a century. They use multiple layers of etched copper circuitry to interconnect microprocessors, passive devices, and other electronic components to create a functional product subassembly and have evolved over the past 50+ years from a simple etched copper pattern on one side of a dielectric substrate comprised of paper impregnated with phenolic resin, known as XXXP or triple XP, to the modern, vertically connected stacks of over 30 layers. If we look back from then to where we are today, there appears to be a natural progression:

- Single-sided XXXP.
- Double-sided XXXP with side-to-side connections made by soldering wires (Z wires).
- Double-sided XXXP with side-to-side connections made by soldering brass eyelets procured from the New England shoe industry.
- Evolution of new dielectrics, epoxy resins, glass reinforcements.
- Plated through hole came in the late 1940s early 1950s. This was pivotal to the technology enabling the development of multilayer boards (MLBs)
- Multilayer boards.
- Surface mount.
- High-density interconnection (HDI) and microvias.
- More and improved dielectrics for high-speed and high-frequency applications.
- Embedding or integrating passive devices, that is, resistors and capacitors, directly into the MLB structure.

This technical evolution came with increasing performance and reliability and decreasing cost. In the beginning, copper conductor widths were in the range of 3 to 4 mm

---

Advanced Electronic Packaging, Second Edition, Edited by Richard K. Ulrich and William D. Brown
Copyright © 2006 the Institute of Electrical and Electronics Engineers, Inc.

with huge variations. Today traces of 0.125 mm are commonplace and 0.05 mm is on the horizon. Correspondingly, via diameters have shrunk from over 1 mm to less than 0.25 mm in commonplace systems and down to 0.10 mm in specialized applications. This trend is true of all of the features of PCBs. PCBs are commonly known as level 2 interconnections, where level 0 is gate-to-gate interconnections on the silicon chip, and level 1 is the chip-to-board connection. As the density of interconnections increases on the silicon, there must be a corresponding increase at the other levels as well. It is increased functionality on silicon that drives the density of the interconnection on the PCB.

Printed circuit board manufacture is an inherently parallel process, with the various layers manufactured and tested separately. These layers generally consist of a polymer-based support with the X and Y conductors made from etched Cu foil. The known good layers are then brought together for lamination into a stack and vertical interconnection. This chapter describes the process through that point, while Chapter 10 describes the addition of parts to the board to complete a functional subassembly, a process called assembly. The organization of this chapter follows:

- Common input data and processing flow of basic PCB fabrication.
- Current commercialized dielectric materials with their properties, applications, sources and links for more detailed information. These links are critical today to enable the engineer to keep current on material changes and advancements. In recent years, electronic materials technology has changed so rapidly that printed literature, in the form of catalogs and data sheets, may be obsolete before it reaches your desk. Therefore, maintaining Web links to the sources enables the engineer to be current at all times.
- Surface finishes.
- Advanced PCB structures including:
  HDI (high-density interconnection)
  Microvias
  Buried and blind vias
- Key terms and definitions.
- Questions and exercises.

This is the classic organic printed circuit board process and there are many variations. For example, the process sequence may vary by manufacturer or equipment platform, and there are literally hundreds, perhaps thousands, of combinations of polymers, reinforcement, finished thickness, tolerances, and copper types and thickness. Some notable variations may be mentioned; however, to include them all would require more pages than the entire text of this book.

4.2 COMMON ISSUES FOR ALL PCB LAYER CONSTRUCTIONS

4.2.1 Data Formats and Specifications

Customer inputs are the data that the PCB manufacturer receives from the designer. PCB designs are generated using electronic PCB computer-aided design (CAD) systems. The data may be "unintelligent" or "intelligent." The traditional and old Gerber format is *unintelligent data* and expresses only the physical characteristics of the board as coordinates and vectors, including the positions of all traces, vias, contact pads, and the like. *Intelligent data* parametrically relates the schematic, parts list and associated placement, footprint and connectivity information to the physical PCB. ODB++ is a proprietary open database format that is rising in popularity. The Institute for Interconnecting and Packaging Electronic Circuits (IPC), a trade association representing electronic circuit design and manufacturing, offers, at no charge, a similar format called GEN CAM—Off Shoot.

The primary driver for industry standards and specifications is the IPC, a North American trade association connecting the electronic industries. A specification tree posted on ipc.org is useful in understanding the broad range of standards available, and other sets of standards have been developed by the government [1, 2]. Standards, in general, should be classified and specified based on performance needs and can be broken down into three classes:

Class 1, the lowest level, is for general electronic products, including consumer products, some computers, and computer peripherals.

Class 2 is for dedicated service electronics, which includes communications equipment, sophisticated business machines, instruments and military equipment where performance and extended life are required, but for which uninterrupted service is not critical.

Class 3 is for high-reliability electronic products, including commercial, military, and life-supporting products where continued performance or performance on demand is critical and equipment downtime cannot be tolerated.

4.2.2 Computer-Aided Manufacturing and Tooling

The word *tooling* conjures up visions of hardware, but in PCB manufacturing it is a generic name for various bits of machine instructions necessary for every step for a given end product board. These usually include the process traveler, photolithography masks and other artwork (AW), as well as programs for drilling, routing, automated optical inspection (AOI), and final electrical test. Once in the hands of the PCB tooling engineer, often called front-end engineer or computer-aided manufacturing (CAM) engineer, the input data is processed through the CAM system to create the required tooling.

Photo tools use a silver halide photographic emulsion on a 0.175-mm Mylar film. The image created by the CAD and CAM systems is transferred to the film using a laser plotter with pixel sizes ranging from 12.5 to 1.5 µm depending on the equipment platform and the resolution required. Cleanliness control is critical in all of the photographic and image transfer processes, and they must be run in a class 10,000, or better, clean room with tight temperature and humidity controls. Laser direct imaging (LDI) eliminates photo tools and will be discussed in Section 4.3.

Engineering or fabrication drawings do not accompany the job through the manufacturing process. The tooling engineer analyzes the drawings and all of the requirements of the product and reduces these to clear work instructions that are compatible with the applicable manufacturing operations. In addition, the tooling engineer analyzes and processes the inputted CAD data, using the CAM system. This analysis includes a check similar to the designer's design rules check, commonly referred to as a manufacturing rules check or design for manufacturing. This checking process verifies the manufacturability of the product within predefined rules for trace widths, spacing, registration, plated through hole aspect ratio, dielectric thickness, stack-up, layer-to-layer registration, and minimum annular ring, hole-size to pad-size relationship. If the manufacturing rules check is satisfactory, the job proceeds. If there are issues of manufacturability, the customer is contacted and a resolution is developed. Etch compensation of the artwork is required to assure that the specified line

108 Chapter 4 Organic Printed Circuit Board Materials and Processes

width and spacing are achieved. This is especially critical for controlled impedance circuits. The basis of this is described in Section 4.3.

Due to the instability of the thin-core laminate materials, primarily shrinkage, artwork is scaled to accommodate these known changes in dimensions, as described in Section 4.2. Scale factors are developed from the PCB manufacturer's experience but vary by the construction and materials of the laminate used, as well as the type and orientation of the copper image. There is also an X and Y variation due to the specific type of reinforcement and the nature of the warp and fill of the fabric.

4.2.3 Panelization

Circuit board manufacturers do not build boards, they build panels in standard sizes that consists of multiple boards. The most common panel size is 46 × 61 cm (18 × 24 in.) with a 2.5-cm border all around to accommodate tooling holes, plating thieves, test coupons, and other manufacturing necessities. Many other panel sizes are used at the choice of the manufacturer, with the sizes trending upwards for greater manufacturing economy. The boards then can be any shape desired but must have a separation distance of 1.5 to 2.5 mm. Figures 4.1 shows a twelve-up panel.

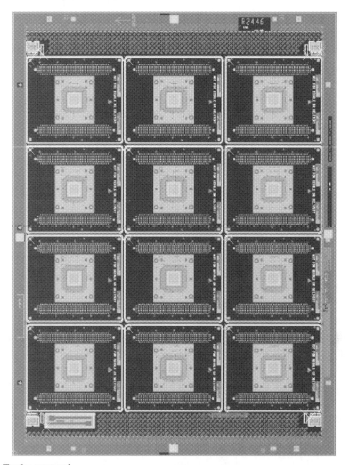

Figure 4.1 Twelve-up panel.

4.2 Common Issues for All PCB Layer Constructions   109

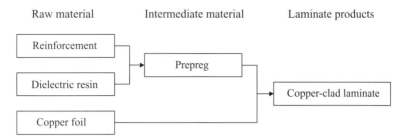

Figure 4.2 Copper-clad laminate material supply chain.

4.2.4 Laminate Materials

4.2.4.1 Composites and Product Choices

Laminates are comprised of a resin matrix of polymeric dielectric, usually reinforced with glass fabric or fibers and clad with copper foil on one or both sides. The raw material supply chain is shown in Figure 4.2. Considering that there are 8 basic dielectric polymers with at least 6 variations of each, depending on the source, and up to 40 reinforcements that can produce laminate in over 300 specified thicknesses and tolerances with 6 common copper thicknesses of 3 grades available on one or both sides of the laminate, that equates to over 20 million possible combinations of materials in the laminate. Conservatively, if one queried all of the laminate manufacturers and tabulated their total product offerings, there would be well in excess of 1000 products to choose from. In this text, unless otherwise noted, we will refer to laminate in the general terms of common epoxy resin reinforced with glass fabric.

The industry specification for rigid multilayer board materials is IPC-4101 [3] and includes laminate and prepreg requirements with specific sheets to generically identify commercially available products and their properties.

The glass fabric, Figure 4.3, is a weave made from what is known as *e-glass* for electronic applications. It has been available since the 1940s and is made from strands

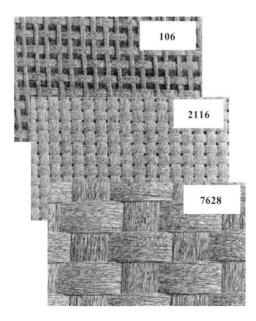

Figure 4.3 Three common glass styles.

110 Chapter 4 Organic Printed Circuit Board Materials and Processes

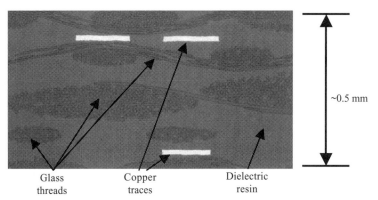

Figure 4.4 Glass–resin matrix cross section.

of continuous glass filaments that are plied and twisted into yarns and woven into fabric. The filaments are treated with silane to promote adhesion and are sold to the laminators in rolls approximately a meter wide. The laminator dissolves the resin, which consists of specific polymer formulations with cross-linking agents added, in solvent, and impregnates the fabric. This process, called treating, results in a dry and partially cured (polymerized) resin–fabric composite known as prepreg since it is preimpregnated with partially cured polymer. One or more plies of prepreg are clad with copper foil by laminating under pressure at the curing temperature of the resin, which completes the cure at the same time. Prepreg is also a product and is used by the PCB manufacturer in the multilayer lamination process. Figure 4.4 is a cross section from a multilayer PCB and shows the matrix of copper traces, glass threads, and dielectric resin.

A primary element of PCB construction is the stackup. This is the layer-by-layer, core-by-core, and material-by-material buildup of the board. Some stackups are specifically defined on the drawing, others give flexibility to the manufacturer and only critical features are specified. In this discussion, we will review only the basic types of construction, not detail materials and geometries used to control impedance and other signal characteristics.

4.2.4.2 Dimensional Stability

The three basic materials and their coefficients of thermal expansion (CTE) are Cu at 17 ppm/°C, glass at about 3, and polymer resin from 55 to 100. CTEs for resin systems vary considerably with the particular polymeric structure, curing mechanism, and thermoplastic characteristics. The main purpose of adding the glass cloth to the polymer used to make prepreg is to reduce its overall CTE down to near that of Cu. Figure 4.5 illustrates how stresses can build into the final assembly due to curing shrinkage of the polymer. Epoxy laminates, for example, are cured at approximately 185°C, and, as it reaches that temperature,

Low stress at cure initiation (<185°C)

Cure occurs at 185°C

High stress after being cooled to ambient:

Figure 4.5 Stress induced during laminate manufacture.

Table 4.1 Laminate Shrinkage over 61-cm (24-in.) Span

Dielectric thickness (μm)	Copper thickness (μm)	Epoxy glass shrinkage (μm)	Polyimide glass shrinkage (μm)
125	17/17	50	75
125	34/34	100	150
125	68/68	200	300
125	103/103	450	675
75	103/103	600	900

the resin is still a viscous liquid, so there are virtually no stresses in the composite at this time. As the resin begins to cross-link or cure, it becomes solid and shrinks slightly due to the cross-linking mechanism increasing the density of the polymer. Further, solvents present in the resin from the previous treating process are also given off, resulting in additional shrinkage. Although stresses have started to build, they are still relatively low at the high temperature. As the composite cools down to ambient temperature, if the CTE of the laminate is higher than the Cu, it goes into additional tension while the copper goes into compression. The thicker the copper, the less it will compress, creating higher stress. When the copper is etched away to form the circuit patterns, some of the stress is relieved and the overall composite shrinks. Table 4.1 illustrates the degree of final shrinkage as a function of the polymer, dielectric thickness, and copper thickness. Fortunately, the shrinkage is fairly predictable for a given source of material and construction and therefore, the artwork scaling procedure enables the PCB manufacturer to have reasonably good control of layer-to-layer registration in the final MLB structure. However, there are limits and this is one of many manufacturing tolerances that must be considered in the PCB design process.

4.2.5 Manufacturing Tolerance Overview

The following is a brief overview of manufacturing tolerances and their general magnitude, as related to PCB manufacture. Tables 4.2 and 4.3 illustrate the major tolerances for images and holes, respectively. These numbers are not absolute, and they will vary with manufacturer, material, and equipment, but they represent the tolerances for standard epoxy

Table 4.2 Manufacturing Tolerances of Image-Related Processes

Image-related processes	Manufacturing tolerance (± μm)
CAD inputs and conversion	Virtually perfect
Laser plotter accuracy	12
Mylar film stability	25
Inner layer side to side	25
Postetch laminate shrinkage (w/scaling)	75
Postetch tooling hole punching	25
Lamination pinning error	25
Laminate instability during press cure	75
Outer-layer photo-tool punching	25
Outer layer side to side	25
Total	**±312 μm**

Table 4.3 Manufacturing Tolerances of Hole-Related Processes

Hole-related processes	Manufacturing tolerance (± μm)
Panel position on driller	50
Drill machine positional accuracy	18
Drill bit size	Negligible
Spindle accuracy and run-out	Negligible
Drill wander and splay	50
Total	**±118 μm**

fiber-glass constructions for up to eight layers. Do not be too alarmed by the sum of these tolerances as everything does not go in one direction at the same time. If it did, we would not be able to manufacture PCBs as we know them today. If you apply a simple root-mean-square (rms) analysis to these numbers, you will see that the total registration tolerance, the sum of holes and images, comes out to about ±196 μm. This is not inconsistent with the manufacturing allowances for high-density or reduced producibility products as published in industry design and fabrication standards [4, 5]. The essence of this is that since pads are intended to encircle holes with some margin, commonly referred to as minimum annular ring, the PCB design must make pads larger than holes at least equivalent to the sum of two times the minimum annular ring plus the manufacturing allowance or tolerance. For a conventional MLB, the referenced IPC manufacturing allowance is 200 μm. To this add 25 μm per side for annular ring, and the pad should be 250 μm larger than the drilled hole.

4.3 PCB PROCESS FLOW

Figures 4.6, 4.7, and 4.8 show the process flow for single-sided, double-sided, and multilayer PCBs. Inputs are provided to the board manufacturer by the PCB designer in Gerber, ODB++, or GEN CAM formats and processed through the PCB manufacturers CAM systems to produce tooling (photo tools, drill programs, route programs, and AOI and electrical test programs). In the following text, we will discuss the classic multilayer board process. To see the process in single- or double-sided PCB terms, simply ignore those process steps not appearing on the given process flowchart.

4.3.1 Manufacture of Inner Layers

The appropriate copper-clad laminate core material is taken from stock and identified by job number and layer number and, with their written work instructions, are moved to the imaging area. Dry film photoresist is a thin (25- or 50-μm) photo-sensitive material that has a slight thermoplastic characteristic so that it adheres to copper when heat and pressure are applied with a hot roll laminator. These are generally negative working, meaning that a positive image on the photo tool will transfer as a negative image on the resist. It is now ready to receive the image projected through the artwork. The artwork is laser plotted at full size on clear plastic sheets and visually checked. It is then precision punched with an optically aligned system to enable accurate layer-to-layer registration by pins, sometimes retractable, in the glass exposure frame outside of the layer material edges. The exposure unit is closed and a vacuum drawn to press it all together. Once the unit is evacuated, a separate vacuum is pulled on precision grooves in the top and bottom glass plates, which

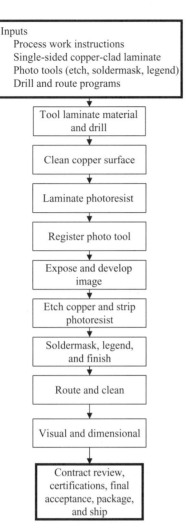

Figure 4.6 Single-sided PCB process flow.

firmly secures each of the artwork to the opposing glass plates in registration to each other, and the unit is ready to expose (Fig. 4.9). The top and bottom artwork tools are usually registered to each other within ± 17 to 25 μm.

Laser direct imaging (LDI) is an image transfer process that is gaining in use since it eliminates the need for the photo tools by plotting the image directly on the photoresist. Special laser plotters and resists are required for LDI and, as with any new manufacturing technology, the equipment and process materials are very expensive. With time, these high costs will begin to decrease and the use of LDI will increase. However, for the purposes of this text we will work with conventional photo tools. When exposed to light from either type of source, the resist polymerizes, becoming harder and chemically resistant. The unexposed resist is easily washed off or "developed" using a mild caustic solution. The pH is closely controlled, and dissolved solids are removed from the developer solution. The layer is new ready to be immersed or sprayed in a solution that will etch away the Cu in the areas not covered by patterned resist. Regardless of the etch chemistry used, alkaline or acid, a certain amount of undercut occurs, and traces are not tidy rectangular cross sections but are always

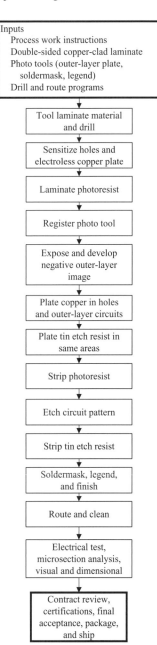

Figure 4.7 Double-sided PCB process flow.

trapezoidal, as shown in Figure 4.10. Line width and space between lines are measured at the foot or base of the trace, the intersection between the copper and the dielectric. The artwork is compensated slightly to assure good line definition at the foot and allow for some degree of overetch. Compensation depends on the copper thickness and results in an etched line width tolerance of $+10/-10$ μm for 17-μm-thick copper to $+25/-25$ μm for 68-μm-thick copper. The compensation permits a slight overetch that ensures removal of all copper flash at the foot. The final tolerance given in Figure 4.10 includes the process tolerances of the artwork plotting and photo tool developing, printing and layer image developing, and

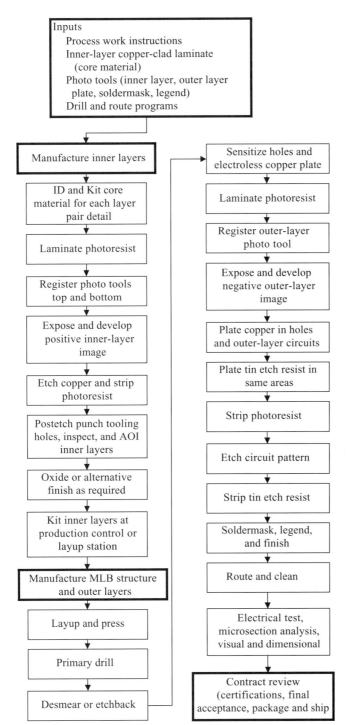

Figure 4.8 Multilayer PCB process flow.

Figure 4.9 Exposure unit with glass exposure frame and artwork installed.

the etching processes, including both chemistry and equipment process variations. When controlled impedance traces are required, specific traces are measured on each layer to verify that the process is keeping the trace dimension in tolerance.

The layers are now inspected by automated optical inspection, which uses cameras to scan the layer detail that is illuminated by either laser or large-field light, and the reflected image is compared to the actual design data. Breaks or opens in circuits and copper bridges or shorts can be detected. Verification stations are used to manually confirm the reported defect and correct it if possible. Minor excess copper can be manually removed but opens cannot easily be reworked. Some false readings are caused by particulate contamination, but repetitive errors can quickly be traced upstream in the process and the root cause identified and corrected. Further, AOI systems can collect statistical data to feed into continuous process improvement systems. Verification of controlled impedance trace dimensions is also performed at the AOI station. Some AOI equipment can automatically measure these traces. Otherwise, optical measurements are made and recorded manually.

Holes used to pin etched layer pairs together in registration during the layup and pressing process are punched after etching and become primary reference markers for the

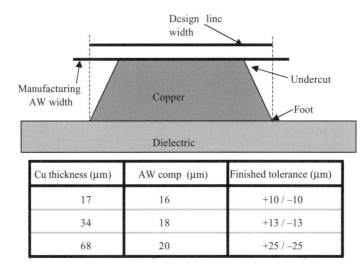

Cu thickness (μm)	AW comp (μm)	Finished tolerance (μm)
17	16	+10 / −10
34	18	+13 / −13
68	20	+25 / −25

Figure 4.10 Undercut and etch compensation of artwork.

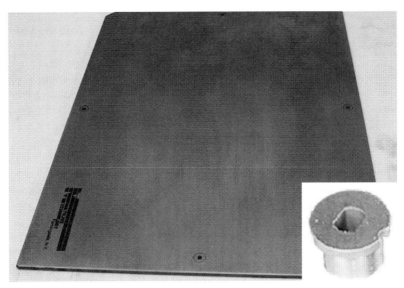

Figure 4.11 Lamination plate with slotted bushings.

mechanical processes of drilling to the inner-layer image. For a given pair of inner layers, the top and bottom inner-layer artwork is reasonably well registered in the glass plates prior to exposure. At this time there are no tooling holes in the core material because they are not required for pinning the artwork to the inner-layer core material. Further, as discussed earlier, there may be a great deal of residual stress in the thin-core dielectrics, which is relieved after etch, resulting in movement of the etched copper image. To minimize error from this shrinkage, two perimeter targets are part of the artwork and etched in the layer. These targets are used in the postetch punch system to optically align the four-hole punch for a best fit, distributing any postetch error evenly across the panel with minimum error in the center of the panel. The four tooling holes are punched close to the center axis of both X and Y on all sides of the panel, approximately 2 cm in from the edge, as shown in Figure 4.11, and the layup stack is shown in Figure 4.12. The holes are slotted slightly in the direction of their respective axis to allow for movement of the laminate during subsequent

Figure 4.12 MLB layup, prepreg, and inner-layer cores.

processes and assures that the error is uniformly distributed with minimum error in the center of the panel. The center of the panel is commonly referred to as the "sweet spot" with respect to registration.

To assure optimum adhesion between the internal copper circuitry and the prepreg used in the final pressing operation, some treatment or finish must be applied to the copper surface before layup. Double-treated copper cores can be purchased where the surface is already treated and no subsequent treatment is required. Most often, an oxide or alternative oxide finish is applied just prior to layup. After this process, the layers are ready to be kitted for layup and pressing.

Individual layer pairs are manufactured as separate details with their own work instructions and process flow sheets. They are typically released to manufacturing at the same time but may be processed as separate orders. To continue the process into the MLB structure, the appropriate layers must converge and be kitted appropriately for layup and pressing. This may be done at a formal production control staging station, or they may accumulate right in the layup department. All of the process flow sheets are compiled with the new MLB work instructions and flow sheets for the outer layer process. After pressing, the process is very similar to that used for a double-sided board.

4.3.2 Manufacture of MLB Structure and Outer Layers

4.3.2.1 Layup and Drilling

Prepreg and inner-layer cores are also layed-up on pins in special precision ground steel plates using slotted holes. The slot punched in the inner layer is longer than the pin diameter, allowing the layer to move about the center axis during the lamination cycle, thereby distributing dimensional error about the center of the panel. Again, the panel center is the sweet spot where alignment is guaranteed. The layer details are carefully placed over these pins. The work instructions and process flow sheet includes layup and stackup information. This shows the orientation of the layers, overall thickness without prepreg, before press and after press, and critical thickness tolerance between layers. The latter is used in the case of controlled impedance. The overall thickness without prepreg is measured to verify that all of the cores are correct. The overall thickness out of the press is the final verification that the layup and process is correct. Although the lot-to-lot variation in the flow characteristics of prepreg is minor, it is sufficient that it has to be monitored closely by the layup technicians. The technicians do not have the latitude of changing prepreg types, that is, resin systems, yet they are required to adjust the glass style and resin content to achieve the required thickness. In these cases, the process documentation is changed to reflect the actual build for future reference.

Pressing, bonding, and lamination are synonymous terms. The MLB composite is pressed at high temperature and pressure in a vacuum. The temperature is dictated by the resin system; epoxy cures at about 185°C and polyimide at about 250°C. Press cycles have controlled temperature and pressure ramp rates, both up and down, to assure the optimum properties of the composite. Some resins require a postcure, which may be accomplished in an oven or in the press. Vacuum pressing is almost universally used today because it assures that entrapped air is removed from the composite, resin flow is maximized, the final MLB composite is free from voids, and the dielectric properties are optimized.

Primary drilling is the formation of holes that will subsequently be plated through. These plated through holes (PTHs) are the backbone of Z connectivity for the MLB circuit matrix. Holes are drilled through copper pads in order to connect them vertically, so pad stack registration is critical to interconnection reliability. Advances in mechanical via formation by drilling have been very significant in recent years. As recent as the late 1980s, a 0.35-mm bit was near state of the art, and 0.65 mm was a typical size. Today, 0.25 mm is near average and 0.10 mm is near state of the art. Speeds and drill rates vary by the bit design, type of material being drilled, and the drill machine. For a 0.25-mm bit in a polyimide glass board, the speed is 136,000 rpm at a drill rate of 132 cm/min while the same material, a 0.10-mm-diameter bit has a speed of 160,000 rpm at 69 cm/min. Without going into further detail of the drilling process, it is noteworthy that this is an extremely sophisticated and technical operation. Even though the capital cost of drill machines has increased over the years, the cost per hole drilled has decreased significantly in the same time period. Equipment maintenance is critical as is the skill level required by drilling technicians. There are 245 standard PCB drill bit sizes ranging from 0.10 to 6.6 mm, and the typical PCB manufacturer stocks about half of these sizes. The cost of bits ranges from approximately $1.00 to $12.00 each, consequently, drill bit inventory cost can be high. The number of hits per bit is determined by the stackup and specifications and can be as low as 100, although more typically around 1000, and in less critical applications over 2500. Cost to repoint bits is in the range of $0.20 to $0.30 each, with only 2 or 3 repoints available per bit.

The panels are registered to the drill machine, the appropriate drill bits are loaded, the program is loaded, and the drilling begins. Outer-layer alignment holes are also drilled at this time and will be used to register the outer-layer artwork to the primary drill pattern. Since they are drilled in the same setup as the primary drill pattern, their positional tolerance to the primary pattern is very good. Backup and entry materials influence the hole quality as well as drilling costs. Their primary function is burr reduction, improved bit cooling, registration, lubrication, and minimizing drill splay.

The major quality issues with drilling are rough holes, damaged copper pads on the inner layers, and wedge voids. The latter is a separation between the prepreg and the copper pad caused by the drill process. This is most common in high-layer-count polyimide boards with holes less than 0.30 mm. All of these conditions may affect the uniformity and reliability of the plating in the hole and the metallurgical bond needed between that plating and the inner-layer interconnecting pads. Conductive anodic filament (CAF) growth, although not necessarily caused by rough drilling, can be aggravated by it. Considerable work has been done on this topic in recent years. Where there is a bias voltage between closely spaced holes or between holes and very close internal traces, and in the presence of humidity and given time, conductive copper filaments have been shown to grow, creating shorts and failures. CAF-resistant laminates are also being offered by the material suppliers.

4.3.2.2 Desmear or Etchback

Drilling creates heat that may soften the resin in the structure, causing it to smear onto the sides of the freshly cut surface of the copper in the internal pads. Although the smear may be extremely thin and will be plated over, it is a contaminate on the copper and must be removed to assure a reliable metallurgical bond to the plating. Strong oxidizers, such as hot concentrated potassium permanganate solution, are used to desmear or etchback the hole prior to plating. Plasma is a more friendly process and is accomplished using specialized vacuum chambers where a fluorinated gas may be combined with nitrogen and/or oxygen and maintained at a sufficiently low pressure to enable ionization by powered radio frequency (RF) electrodes. The panels are placed between the electrodes, and the ionized gas flows through the chamber from side to side on periodic cycles to assure that the holes are uniformly processed. The ions attack and remove only the organic materials, therefore, the

Solder | Copper plating | Electroless copper interface | Copper foil 34 μm thick

Figure 4.13 Cross section of a desmeared PTH to pad interface after solder float.

resin is removed without affecting the copper and/or the glass. A subsequent wet chemical process, referred to as glass etch, is used to remove the short glass fibers remaining in the hole.

Figure 4.13 shows a portion of one side of a vertical cross section of a desmeared PTH after a solder float for 10 s at 288°C. The electroless copper interface is approximately 0.75 μm thick and is highlighted by a microetch process used in the preparation of the microsection. This will be discussed more in the plating section. Note that the hole wall is smooth, and there is no evidence of anything but the electroless copper between the plating and the inner pad.

Figure 4.14 shows resin and glass etched back from the drilled surface of the copper in the hole. There is evidence of wicking, Cu plating going into a hole in the resin, but it is in the acceptable range. The basis for etchback goes back many years to the early development of MLBs and PTHs, circa the mid-1960s when poor quality drilling, low T_g laminates, resin smear, and plating were more of a problem than they are today. In the 1960s, T_g, the glass

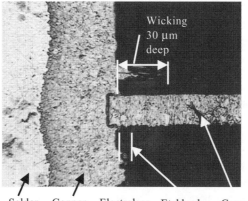

Solder | Copper plating | Electroless copper interface | Etchback 7 μm deep | Copper foil 17 μm thick

Figure 4.14 Cross section of an etched back PTH to pad interface after thermal stress in liquid solder.

transition temperature, was not part of our PCB vocabulary. Low T_g relates to high Z axis CTE and excessive stresses on PTHs during soldering and normal life thermal cycling. Low T_g also means softer materials that are more susceptible to resin smear, especially when the drill bits and the drill processes are not that great. The notion of etchback came about primarily as a method to assure that all of the resin smear was removed, and to provide a three-point contact between the plating and the copper foil. This indeed worked. Early processes used concentrated sulfuric or fluorosulfonic acid to remove resin and hydrofluoric acid to remove the glass.

Today there are three primary industry specifications for high-reliability PCBs. These are MIL-PRF-55110, MIL-PRF-31032, and IPC 6012 Class 3. Although MIL-PRF-55110 is technically obsolete since 1997, it is still in use as of this writing. These specifications all have common elements. First, they are all used to specify high-reliability PCBs for critical applications including military, aerospace and satellites, and commercial applications where failures would be catastrophic. Second, they all give an option for the use of etchback. In MIL-PRF-55110, if the PCB manufacturer is certified with etchback then he may use etchback or not depending on the specific drawing requirements. A manufacturer can be certified without etchback; however, then, that manufacturer cannot build products requiring etchback. The point of this discussion is that with or without etchback, these accepted industry standards are used for manufacture, acceptance, and certification of high-reliability PCBs. This might give rise to the following question: Is etchback really necessary for reliability?

4.3.2.3 Sensitize Holes and Apply Electroless Copper Plate

Electroplating does not adhere to nonmetals without some form of treatment of the nonmetal surface. In this PCB instance, the nonmetal is the dielectric that is plastic (resins, polymers, dielectrics) and glass or other organic reinforcements. The most common and original process, which has been in service over 35 years, is electroless copper using autocatalytic palladium–tin (Pd–Sn) as an activator for the dielectric surface. Most of the industry calls the activator the catalyst. It involves softening the plastic surface with solvents and surfactants, followed by treatment with the catalyst. The catalyst is a palladium–tin colloidal chemistry that adheres to the plastic surfaces as well as the copper surfaces due to very small forces not unlike van der Waals' forces. The tin is subsequently reduced and removed, leaving tiny palladium particles stuck to plastic and copper surfaces. This is followed by the electroless (no applied current) deposition of copper on the palladium particles resulting in a very thin adherent copper coating. The overall process for deposition of electroless copper in the hole is called "Dep." This coating is typically in the range of 0.75 to 2.0 μm. It also covers all of the copper in the holes and on the surface of the panel. The result is that we now have an electrically conductive surface that we can electrolytically plate. Panel plating is the process of plating the entire panel, including the holes, followed by a print and etch, similar to an inner-layer etch. Since this process has limited use for today's PCB manufactures, the following discussions will cover the pattern plating process, where only the holes and the outer-layer circuit patterns are plated.

Direct metallization (DM) processes are slowly gaining a position as a replacement for electroless copper. These go by trade names of Black Hole (Hunt Chemical), Shadow (Electrochemicals), Crimson (Shipley), Neopact (Atotech), and Phoenix (MacDermid). These are conveyorized processes with high throughput and relatively low recurring costs, the need for the electroless copper is eliminated and only the plastic surface is treated.

Carbon is used to render the dielectric surfaces conductive. There are still many skeptics about DM reliability in the MLB applications and hence adoption is slow.

4.3.2.4 Pattern Cu Using Plate-up Through a Photoresist

Electroplating Cu through openings in a resist is very similar to the process described for inner layers in Section 4.3.1 except that 50-μm resist is used and the photo tool is a positive image. The thicker resist stands up better to the plating process, and it is thick enough that the plated up copper will not "mushroom" over the top as it plates up. Unlike the inner layer use of dry film resist as an etch resist, this time it is to resist the plating of copper in defined areas; hence it is a plating resist. Automatic alignment systems are becoming available for outer-layer printing. Four perimeter tooling holes are drilled at primary drill; the artwork has corresponding fiducials. Top and bottom artwork is registered to the glass plates and held by vacuum the same as with inner layers. The panel is inserted and the plates closed. The cameras register the best-fit of the panel and artwork fiducials. Manipulators move the panel to the best-fit position with respect to the tooling holes and the artwork fiducials. Accuracies of ± 20 μm are reported with semiautomatic and automatic systems.

4.3.2.5 Plate Copper in Holes and Outer-Layer Circuits

Copper is the conductor, whether plated or foil. Depositing copper in the PTH of MLBs completes the connectivity of the wiring matrix by adding the Z direction interconnects. Plating processes vary in size and through-put from small tanks that can run one or two 46×61 cm panels at a time to larger automatic lines that can handle 16 or more panels at once. Like all other segments of the PCB manufacturing process, copper quality is critical. Ductility, tensile strength, metallurgical bonding to the foil, and uniformity in the hole and on the surface are a few issues associated with the copper used. None of these have been show stoppers in the evolution of reliable PCBs, but problems have occurred and been solved over time such that today's PCB PTHs are robust and reliable. The normal minimum copper plating thickness is 25 μm with allowance to go down to 20 μm in isolated areas. Many in the high-reliability community live with the notion that more is better. Copper plating thickness is determined by destructive analysis of metallurgical microsections cut vertically through the PTH. Industry standards are used to control the microsectioning process, measurement, and analysis.

4.3.2.6 Plate Tin Etch Resist in the Same Areas and Strip Resist

If electrodeposited tin plating is used as an etch resist, it is plated in line with the copper plating and must cover both the external circuits as well as the inside of the plated holes. Any form of contamination that may accumulate in the holes will prevent the tin from depositing and result in an etch-out condition. This results in scrap. Tin etch resist is a consumable part of the process since this plating is not delivered as part of the product. The plating resist is stripped by the same process earlier described for the inner-layers etch resist. The area under the resist will be bare copper that will be subsequently removed by etching.

4.3.2.7 Etch the Circuit Pattern

The tin plate is the etch resist. Although the same etch chemistry may be used for the outer layer as well as the inner layers, typically, different etchants and separate etch lines are used.

Acid etch (cupric chloride) is used for inner layers and alkaline etch is used for outer layers. The tin plate is stripped using proprietary stripping solutions. These usually are composed of a mild solution of 20% nitric acid with ferric ions and antitarnish agents.

4.3.2.8 Soldermask, Legend, and Finish

Prior to the soldermask application, outer layers are visually inspected to assure all circuits are present with no opens and/or shorts. Cleanliness testing is also performed according to industry standards, prior to soldermask, to assure that no inorganic residues remain on the board. Panels are precleaned with a mild acidic cleaner, rinsed, and dried prior to moving to the soldermask application process. Today, virtually all soldermask is LPI (liquid photoimageable). It may be applied by curtain coating, spraying, or screen printing. All of these processes are good and all of them have their benefits and disadvantages. The object of the coating process is to apply a uniform coating across both sides of the panel with perfect fill at the foot of all of the copper features. The imaging is transferred photographically after the application is dried tack free. For smaller lot production, the semiautomatic screen printing application is attractive. Both sides of the panel are coated simultaneously. The critical parameters of squeegee material, angle, pressure, and travel rate are all controlled. Following coating, the panels are tack dried in either a box oven or conveyorized infrared oven. Outer-layer panels are registered with soldermask photo tools, similar to the outer-layer image process previously described, and exposed.

The legend consists of the various printing and other line artwork that is applied to the board to indicate surface-mount component placement, identification numbers, manufacturing dates, and so on. This is commonly screen printed but direct imaging techniques are being developed for the legend application. Ink jet technology has been demonstrated to improve the resolution of the legend characters as well as a more efficient process since it is software driven. The screening process requires the generation of artwork for the image and transferring the image to the screen, preparation of the screen with the appropriate emulsion, and cleaning screens between images.

Final metal finishes are specified on the design documentation and may include hot air solder level (HASL), reflowed tin–lead plate, electroless nickel immersion gold (ENIG), immersion tin, immersion silver, electrolytic nickel and hard gold or soft gold, and organic solder preservative (OSP). All of these finishes, except HASL and reflowed tin–lead plate, are very thin and provide a flat and planar surface that is compatible with and friendly to high-density surface-mount technology (SMT). HASL and reflowed tin–lead plate produce slightly nonuniform surface finishes. HASL has limited shelf life (less than a year) due to very thin areas that result in the formation of a ternary alloy (copper/tin/lead) that ultimately becomes nonsolderable. Reflowed tin–lead minimizes this issue due to a thicker and more uniform solder coating that is less susceptible to the effects of the ternary alloy.

4.3.2.9 Secondary Drill, Route, and Clean

The almost completed manufacturing panel now goes back to the drill room for secondary drill and route. These are basically machining processes, however, the same type of numerically controlled equipment that is used for primary drilling is used for this operation. Special carbide router bits are used, and route programs created by the tooling engineer drive the machines. Some holes are not to be plated through and, for that reason, are drilled after the plating process. If the nonplated holes are very small, less than about 2 mm, they will be drilled at primary drill and tented over by the plating resist and prevented from being plated. However, if they are larger holes, the resist will not tent consistently and the holes are drilled after plating. Conveniently, this is done at the same time as the final route process. The secondary drilling and routing are done on the same machine. Test coupons will be routed at this same time as well. The final step is cleaning to remove all residual dust from the parts once they are singulated. This is either a batch or conveyorized process involving washing in a mild detergent and rinsing with deionized water followed by warm air drying.

4.3.3 Electrical Test

Only the simplest single- and double-sided boards do not need to be individually electrically tested prior to shipment. Since boards are becoming increasingly complex, virtually all of them are tested these days. *Bed-of-nails fixtures* with *golden boards* were the early test methodology. This expression comes from the fact that test fixtures were manufactured using spring-loaded pins (pogo pins) to interface between the tester and the PCB. The testers had a single test head with contacts on 2.5-mm centers. Fixtures were made from clear acrylic plastic plates separated by spacers approximately 8 cm long. The bottom plate had holes on 2.5-mm centers to align with the grid of the tester head. The top plate was drilled to match the physical pads on the board that were to be tested. Since the pins were 1.3 mm in diameter, the density of the boards testable was limited. However, keep in mind at that time boards were typically layed out on a 2.5-mm pitch and migrating to a 1.25-mm pitch was a major move. The term "golden board" comes from the era when a board, judged to be good or correct, is used as a standard to compare to other boards. The testers were actually self-programmed from the golden board by reading in its nets. Production boards were tested in the same way and compared to this one.

As of this writing we have two fundamental test methods—bed of nails (fixture testing) and flying probe (fixtureless testing). Both of these methods involve programming the tester from a netlist. The bed of nails utilizes a unique dedicated test fixture that physically contacts all of the end points of the nets simultaneously and provides a hard-wire connection between the board and the test set. Unlike the original spring-loaded pogo pins, the test heads (two of them, one top and one bottom) each have spring-loaded contacts. Fixtures are made to 0.5-mm pitch using very short, fine, and stiff "piano wire" pins. Test pitches down to 0.25 mm are achieved by staggering the pins. Programs are created using the design netlist and is relatively easy. The test is very fast and accurate in verification of continuity of traces and isolation nets. Rates up to 8000 points/second are common. Manufacturing the test fixtures is an expensive proposition, but when amortized over large quantities of boards, the recurring cost is very small. Conversely, the flying probe test technology contacts only 4 to 16 points on the board at a time. It is programmed to contact all end points and verify continuity. Similarly, a probe contacts a given net and simultaneously probes adjacent nets to verify isolation. The setup for flying probes is very fast and inexpensive, however, the recurring costs are high as the test run times are longer (10 to 40 points/second), especially for very dense boards with high net counts. Most manufacturers have both types of equipment, and trade-offs are made based on the cost and volume of product to be tested. When quantities are small, flying probe is the choice; when quantities are high, then bed of nails takes over.

4.3.4 Visual and Dimensional Inspection

The products are now submitted for final visual and dimensional checking. Critical dimensions are verified and overall visual inspections are performed. Many tools are used for this final check including stereo microscopes, *XY* coordinatographs, halo lights, micrometers, pin gages, and various other hand tools. Depending on the lot size and customer-specific requirements, inspections may be 100% of all PCBs submitted, or they may be on an acceptable quality level (AQL), where the sample size is statistically determined for the particular lot size. Where industry standards are specified on the engineering drawing, the acceptance criteria is given in IPC-6012 A [5] and includes 44 very specific features and characteristics to be inspected. Sampling plans are also given as a function of the three PCB performance classes.

4.3.5 Contract Review

Contract review is a term that has evolved from the ISO 9002 standard certifications and is another expression for what the PCB industry formerly called "final inspection." As customer's products have become more complex, performance requirements have increased and so has the process of specifying and ordering PCBs. Today, the contract for PCBs goes far beyond what was formerly considered in a purchase order (PO) for boards. The contract actually includes all of the language of the PO plus the associated specifications and data that is provided by the customer for the manufacture of the boards. Hence, final inspection is more than merely looking at the boards and confirming that they meet the drawing, that is, they are pretty and the dimensions are within tolerance. The organizational function performing contract review must review in depth 100% of the requirements of the contract and certify with documented evidence that the contract has been met. This may be as simple as a certification that all dimensions and primary specification requirements are met. However, it may also include certificates of conformance for raw materials used in the construction of the board; process traceability; records and verification of the drilling sequence, how many hits were made per drill bit, and on which holes; copies of all process documents, either delivered with the product or retained in a system where they are retrievable on demand by the customer for periods ranging from 4 to 16 years. The latter includes all associated records and microsections. Packaging is also not to be overlooked in this scenario since there are occasions where packaging requirements exceed the individual raw board costs.

4.3.6 Microsection Analysis

Microsectioning is a metallurgical technique of preparing, cross sectioning, and examining a plated through hole with a metallograph at magnifications from $50\times$ to $500\times$. The technique is used for in-process control as well as satisfying end product acceptance and specification requirements. Methods and acceptance criteria are given in References 1, 2, 5, and 6. Microsections are performed on representative test strips manufactured on the panel with the product. Industry design standards establish the details of the coupon design and layout [7]. Figure 4.15 shows such a coupon area on a manufacturing panel. They are tested in an "as-received" or as-manufactured condition without added stress, or after thermal stress. Thermal stress is a standard test where a small section is fluxed and floated in molten solder for 10 s at 288°C (Fig. 4.16). This is an extreme test to simulate the effects of subsequent soldering operations, and has become the accepted minimum standard for the constructional integrity and durability of PCBs. Coupons are then mounted and potted in an acrylic or epoxy compound, ground and polished, and then given a mild etch to bring out the grain structure of the copper and highlight the interface between the pad and plating. Several critical features are examined by microsection including: layer-to-layer registration,

126 Chapter 4 Organic Printed Circuit Board Materials and Processes

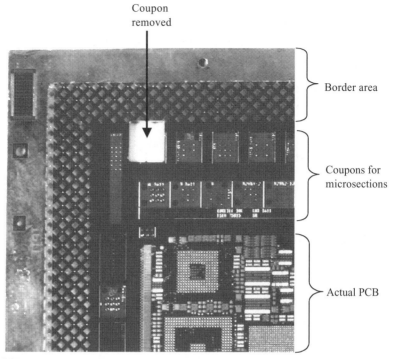

Figure 4.15 One corner of a panel showing coupons for microsections. Note that one coupon has been removed for analysis.

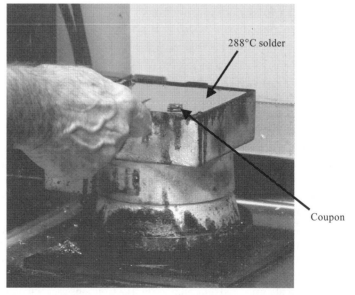

Figure 4.16 Test coupon being solder floated (thermal stress test) for 10 s at 288°C.

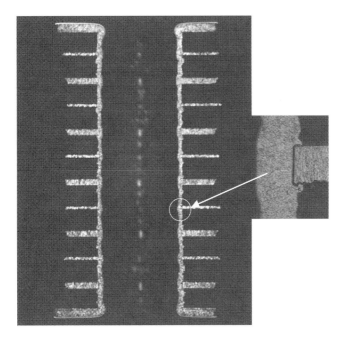

Figure 4.17 Microsection representing a high-reliability PCB for space craft application. Note the mixed copper foil thicknesses and etchback.

minimum annular ring, plating thickness and integrity, plating to internal copper pad bond quality, copper foil thickness, dielectric thickness, and overall laminate quality. Figure 4.17 shows a high-layer-count microsection exhibiting acceptable features. In this instance, the microsection represents a board actually flying on NASA's Mars Reconnaissance Orbiter, designed and manufactured by Lockheed Martin Astronautics Corporation.

4.4 DIELECTRIC MATERIALS

Materials were discussed briefly in Section 4.2.4.1. Notably, there are over a thousand commercial choices of raw materials for PCBs today with literally millions of possible combinations of polymer, reinforcement, and copper resulting in a wide variety of chemical, physical, and electrical properties as well as wide variations in cost. Although the basic epoxies (FR-4, GF, and GFGs) represent the largest market for PCBs, their contribution to high numbers of possibilities is relatively small.

4.4.1 Dielectric Material Drivers

The main drivers are electrical and mechanical characteristics, manufacturability and cost. The order of these depends on the market and application. For many markets, cost is the biggest driver and performance may be traded-off. For others, performance is the primary driver and cost is not an issue. Manufacturability affects cost as well as quality and reliability.

Electrical and mechanical characteristics relate directly to performance. The electrical properties of most interest are the dielectric constant (k) and dissipation factor (DF). Stability of these materials and their respective electrical properties with varying temperature and humidity is a factor. Low k and low DF are properties necessary for high-speed designs. PTFE (polytetrafluoroethylene, Teflon) is the polymer with the lowest k and DF properties.

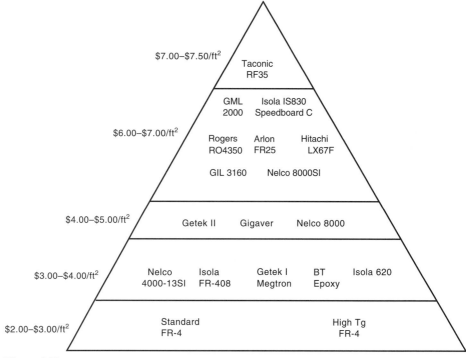

Figure 4.18 Laminate material price pyramid. (Courtesy of Jack Fisher IPC, 2002.)

Thickness tolerance and coefficient of thermal expansion (CTE) are also factors. However, PTFE is a thermoplastic, dimensionally unstable, very slippery material and nothing sticks to it without an aggressive surface treatment. Combining PTFE with other materials, such as glass reinforcement, improves the mechanical properties with minimal affect on the dielectric characteristics. Over the years, materials engineers have compounded many combinations and blends of polymers and fillers to produce a broad variety of desirable characteristics. Figure 4.18 illustrates the general cost structure of materials. The bottom of the pyramid is composed of the conventional epoxy laminates and moving upward through various blends or polymers to the PTFE-based materials at the top. Figure 4.19 follows the same logic and it is noteworthy that performance and price track each other.

Table 4.4 summarizes the critical properties of materials available today. There is a constant evolution improving material characteristics. The best way to stay abreast of these changes is to keep in contact with the material suppliers, either through personal contact or through regular visits to their websites. Many of the North American suppliers are listed in Table 4.5. Keep in mind that this is also constantly changing.

4.4.2 Dielectric Material Constructions and Process Considerations

4.4.2.1 Homogeneous and Hybrid Constructions

Homogeneous or full-body MLB constructions are the traditional or classic MLB that is made with the same materials throughout. Most of the materials identified in Table 4.4 can be used for homogeneous MLB constructions. Exceptions are those that

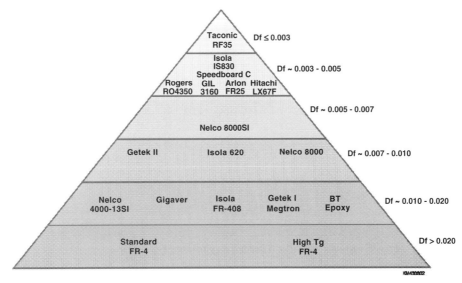

Figure 4.19 Laminate material performance pyramid. (Courtesy of BPA Ltd., 2002.)

do not have bond plies or prepregs of the same composition. These include the thermosetting hydrocarbons, PTFE-based laminates, and hybrid polyesters. Hybrid constructions are those using a mixture of dielectric materials to achieve the desired electrical characteristics and product costs. A common hybrid construction is an MLB with several epoxy-glass-based layers for power distribution and one or two layers of a high-frequency material for RF or high-speed digital signals [8]. Figure 4.20 illustrates a hybrid structure providing high-speed structure for microwave circuits in combination with a standard FR-4 digital MLB structure. The possibilities are virtually unlimited when combining materials into hybrid structures. In addition to performance issues, material costs and process affects must be considered. Even though the possibilities seem unlimited, there are practical limitations, largely driven by producibility.

Hybrid constructions may also be used to control the X- and Y-axis expansions in order to minimize stress on solder joints when designing with large-size and small-pitch array packages. Epoxy and polyimide laminates, utilizing aramid reinforcement, result in an MLB composite with a low CTE in X and Y. These are cited in Table 4.4 as "High Technology, Focus on Controlled Thermal Expansion (X-Y CTE)." The aramid reinforcements enables an MLB composite to have an X-Y CTE in the range of 6 ppm/°C below the value at T_g, compared to 15 ppm/°C for conventional laminates. Another method is to incorporate copper–invar–copper (CIC) foil into the board construction. When used as plane layers, one layer below the external top and bottom layers for symmetry, the extremely low CTE of the invar (1.3 ppm/°C) coupled with its high tensile strength, controls the overall CTE of the composite to the range of 4 to 8 ppm/°C. The disadvantages of this material are: it is very expensive, it is typically 150-μm thick therefore a special prepreg may be needed, nonstandard etchants are required, an inert metal such as nickel must be plated onto the invar in the holes prior to electroless copper processing, and PTH connections to the CIC plane must be made using conventional mechanical drilling, the invar is extremely hard to drill, and the typical hit rate for carbide bits may be less than 10. All of this adds considerably to the manufacturing cost of the PCB.

Table 4.4 Summary of Critical Properties of Dielectric Composite Materials

Resin	Reinforcement or filler	T_g	k	DF	CTE (Z) (ppm/°C)	Relative cost[a]
Standard Technology Multilayers *Laminates and Prepregs*						
Multifunctional epoxy	Woven glass	140 DSC	4.3 @ 1 MHz	0.027 @ 1 MHz	74	1.00
Multifunctional epoxy	Woven glass	170 DSC	4.3 @ 1 MHz	0.027 @ 1 MHz	61	1.03
High Technology, Focus on Temperature Resistance *Laminates and Prepregs*						
Polyimide	Woven glass	260 DSC	3.5–3.8 @ 1–10 GHz	0.011 @ 10 GHz	35	2.50
High Technology, Focus on Controlled Thermal Expansion (X-Y CTE) *Laminates and Prepregs*						
Epoxy	Woven aramid fiber (Kevlar)	175 TMA	3.9 @ 1 MHz	0.016 @ 1 MHz	80 X/Y 6/6	10.00
Epoxy	Nonwoven aramid fiber (Thermount)	175 TMA	3.45–3.85 @ 1 MHz	0.015–0.022 @ 1 MHz	110–120 X/Y 6/6	8.00
Polyimide	Nonwoven aramid fiber (Thermount)	240 TMA	3.5–3.8 @ 1 MHz	0.013–0.015 @ 1 MHz	80–90 X/Y 6/6	15.00
Polyimide	Woven quartz	>250 TMA	3.5 @ 1 MHz	0.009 @ 1 MHz	50 X/Y 9/9	
High Technology, Focus on Thermal Conductivity *Prepregs*						
Multifunctional epoxy	Woven glass thermally conductive fillers[b]	105	4.1 @ 1 MHz	0.029 @ 1 MHz	37	7.50
High Technology, Focus on Lower Dk and Loss Tangent *Laminates and Prepregs*						
Polyphenylene oxide	Woven glass	180 DMA	4.0 @ 1 MHz	0.010–0.015 @ 1 MHz	61	1.70
Enhanced multifinctional epoxy	Woven glass	240 DMA 210 DSC	3.9 @ 1 MHz	0.009 @ 1 MHz	36	1.60
Bismaleimide/ triazine & epoxy (BT)	Woven glass	180 TMA	4.1 @ 1 MHz	0.015 @ 1 MHz	61	1.50

Table 4.4 (*Continued*)

Resin	Reinforcement or Filler	T_g	k	DF	CTE (Z) (ppm/°C)	Relative cost[a]
High Speed/Low Dk and Low Loss High Speed, and Microwave and RF Applications Laminates						
Hydrocarbon, thermosetting	Woven glass, ceramic filled	>280	3.48 ± .05 @ 10 GHz	0.004 @ 10 GHz	50	5.00
Hydrocarbon, thermosetting	Woven glass, ceramic filled	>280	3.38 ± .05 @ 10 GHz	0.0027 @ 10 GHz	46	5.00
PTFE	Random glass microfiber	N/A	2.20 ± .02 @ 10 GHz	0.0009 @ 10 GHz	237	20.00
Cyanate ester	Woven glass	250 TMA	3.4–3.8 @ 1 MHz	0.006–0.009 @ 1 MHz	40	10.00
Polyester hybrid laminate	Woven glass	N/A	3.4 @ 10 GHz	0.0017 @ 10 GHz	60	N/A
Liquid crystal polymer[c] (LCP)	None	N/A	2.9 @ 1–10 GHz	0.002 @ 1–10 GHz	17	N/A
Bond Plies						
Cyanate Ester	PTFE matrix	220 TMA	2.6–2.7 @ 1 MHz–10 GHz	0.004 @ 1 MHz 10 GHz	N/A	18.00
Multifunctional Epoxy	PTFE matrix	140 TMA	3.0 @ 1 MHz	0.02 @ 1 MHz	N/A	15.00
Polychloro-trifluoroethylene, PCTFE	None		2.35 @ 1 MHz	0.0025 @ 1 MHz	N/A	N/A
Hydrocarbon, thermosetting	Woven glass, ceramic filled	N/A	3.54 @ 10 GHz	0.004 @ 10 GHz	N/A	N/A
Bismaleimide/triazine & epoxy coated on PTFE	Woven glass	N/A	3.19 @ 1 MHz	0.002 @ 1 MHz	N/A	N/A
Liquid Crystal Polymer[a] (LCP) (lower melt temperature for bonding)	None	N/A	2.9 @ 1–10 GHz	0.002 @ 1–10 GHz	17	N/A

[a] Based on conventional epoxy laminate being 1.0.
[b] Thermal conductivity in W/m K is 3 compared to 0.3 for conventional epoxy laminate and 398 for copper.
[c] Very new materials in development, limited commercial application as of this writing.

4.4.2.2 Process Affects of Combining Electronic Materials

Adhesion, flatness, and registration must be considered when combining dielectric materials.

1. Adhesion Thermosetting materials, conventional epoxy, polyimide, bismaleinide triazines (BTs), and so forth, use a chemical curing reaction called cross-linking. The essence

Table 4.5 Some North American PCB Laminate Material Supplier Contacts

Material supplier	WebSite Link
Arlon	http://www.arlon-med.com/
Dupont	http://www.dupont.com/teflon/films/
Gore	http://www.goreelectronics.com/
Honeywell	http://www.aclar.com/
Isola	http://www.isola-usa.com/
Nelco	http://parknelco.com/
Polyclad	http://www.polyclad.com/
Rogers	http://rogers-corp.com/
Taconic	http://www.taconic-add.com/
Thermagon	http://www.thermagon.com/

of this reaction is that when the material is cured it becomes permanently solid and has unique physical characteristics. These materials do not soften or flow after they are cured. Conversely, thermoplastic materials do not cure, they can go back and forth through liquid and solid forms. As liquids they can adhere to a solid surface and when they solidify, they bond tenaciously. Common household applications of thermoplastics are "hot melt" adhesives.

Products such as polychloro-trifluoroethylene (PCTFE) are tough films with excellent physical and electrical properties (see Table 4.4). They can bond MLB details together under pressure by reaching the melting temperature. There are two major challenges to using PCTFE to bond layers together. First, the process window is very narrow. The material begins to soften (not flow) at 190°C and it decomposes at 246°C. Allowing for safety margins, the bonding window is only from 204 to 221°C. This is very tight for most commercial laminating presses. However, consider that the bonding process can be adequately controlled to produce a good MLB bond. One standard performance test in the MLB process is solder float or thermal stress as discussed earlier in Section 4.3.6. Solder float is conducted at 288°C, 42°C above the materials decomposition temperature. Or, for example, if the MLB structure has soldered connections and PTHs, the soldering temperature may approach or exceed the bond plys melting temperature. Any of these conditions result in delamination and scrap of the MLB structure. Similarly, FEP (fluorinated ethylenepropolyene) is a bond ply that may be used to laminate PTFE materials. FEP melts at 288°C and, like the PCTFE material, has a very narrow process window. This process temperature exceeds the laminating capability of most PCB manufacturers.

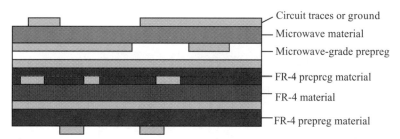

Figure 4.20 Hybrid design with stripline on top of FR-4 multilayer using high-frequency prepreg. (Courtesy Taconic Inc.)

Generally, PTFE materials do not adhere well, nor do things bond well to their surfaces. This includes plating, soldermask, legend, and bond plies. Surface treatment is required to render a bondable condition. Some chemical treatments can be used, but these are very unstable and hazardous. Plasma etch is the preferred surface treatment.

2. Flatness Unsymmetrical MLB structures can warp, which may cause them to not meet flatness requirement, and the subsequent component assembly process will be impacted to the extent that either components cannot be reliably attached or solder joints will be stressed and possibly fail later under thermal or mechanical stress. For this reason, symmetrical stacks are preferred to unbalanced or unsymmetrical structure, and this means having the same kind and order of dielectric materials, metal materials, and layer thicknesses starting from the middle and moving out to the top and the bottom of the board.

3. Registration Many of the high-performance materials do not have fiber-glass reinforcement and, therefore, are dimensionally unstable during the mechanical and thermal stresses of processing. Layer-to-layer registration in MLBs is critical and becomes tighter and tighter as circuit features become smaller. Conventional scaling routines used for woven-glass-reinforced materials do not work consistently on un-reinforced laminates. Manufacturing processes must be characterized to determine the pad-to-hole relationships for hybrid structures.

4.5 SURFACE FINISHES

Finishes in general were discussed in Section 4.3.2.8. In addition to that information, the topic of *lead free* must be mentioned. The lead-free initiative for electronics is driven by the European Community (EC) through two directives. RoHS Directive 2002/95/EC (Restriction of Hazardous Substances in Electrical and Electronic Equipment) and WEEE directive 2002/96/EC (Waste Electrical and Electronic Equipment). If North American original equipment manufactures (OEMs) want to sell products to the European Union (EU), they must comply and the compliance date is approaching rapidly as of this writing. This discussion is not going to argue the issues involved, only point out what is ahead for the PCB manufacturing community. The scope is very broad in terms of the definitions of the types of electronics affected. It is virtually everything from light bulbs to complex IT/telecommunications systems.

Some of the major factors to the manufacturer include:

- Lead-free assemblies will have a major effect on acceptable choices for laminates.
- Most existing materials cannot be used in lead-free assemblies without modification.
- New material types will be needed, particularly those that have high T_g.
- 140 T_g epoxy laminate are compatible with lead-free assembly process.

Lead-free solder joints will be formed at temperatures around 245°C, which is 35 to 40°C higher than the conventional SnPb eutectic solder. The decomposition temperature of the laminate is a bigger factor than T_g. Epoxy laminates with $140T_g$ have higher decomposition temperatures than $170T_g$ laminates and therefore may be more resistant to the abuses of lead-free soldering.

From the perspective of a PCB manufacturer, until the proverbial dust settles, which will take several years, there will be a lot of "finger pointing" between the OEMs, electronic manufacturing service sector, component manufacturers, laminate suppliers, and PCB manufacturers over how this problem should be handled. As a board manufacturer, I fear that issues may unjustly point to the PCB and its manufacturer rather than the component, assembly process, and/or laminate material. We, as an industry, must be proactive to assure that all parties understand the future risks.

4.6 ADVANCED PCB STRUCTURES

4.6.1 High-Density Interconnection (HDI) or Microvia

High-density interconnection is defined as circuit boards that have vias smaller than 150 μm, which qualifies them to be called microvias, and copper trace features smaller than 100 μm [9]. Since microvias are an enabling technology for HDI, we tend to use the terms HDI and microvias interchangeably, which is not entirely correct, but is probably close enough. Like hybrid PCBs, HDI may also be a combination of very small and fine features with conventional MLB features. Anyone using a cellular phone or personal data assistant (PDA) is using a product containing HDI. Some time ago I was curious about the technology used in a wrist watch–camera offered by Casio. At the expense of one camera, I learned that HDI was present in several forms. Figures 4.21 through 4.24 show traces with 50-μm width and spaces, 125-μm-diameter mechanically drilled vias (note that these vias do not have capture pads, they interconnect directly to a copper trace), 75-μm-diameter laser formed blind microvias through 50-μm-thick resin dielectric coated on copper foil (RCC), and buried and blind vias in an epoxy nonwoven aramid laminate (Thermount), laser formed and interconnected using a conductive silver-filled paste. This is a patented process known as ALIVH for any layer interstitial via hole.

The drivers for HDI and microvias have been functional density and cost. Via formation using lasers has proven to be a very cost-effective manufacturing method due to the extremely high speed with which vias can be formed, 500 to 1000 times faster than conventional drilling. Couple this with the fact that the vias can be very small, smaller than practical to drill mechanically, and no drill bits are involved. Since the microvias are blind, that is, they do not go through the structure, routing channels in the subcomposite

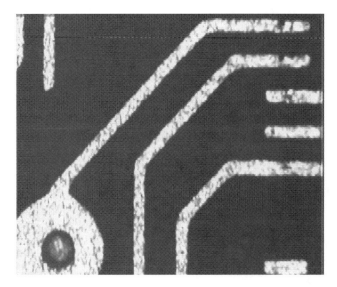

Figure 4.21 50-μm lines in Casio wrist watch–camera.

4.6 Advanced PCB Structures **135**

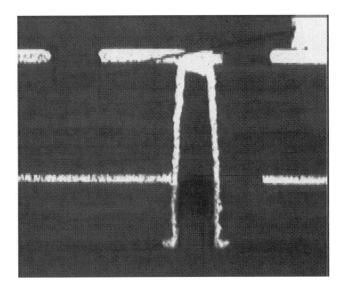

Figure 4.22 125-µm mechanically drilled HDI blind via in Casio wrist watch–camera.

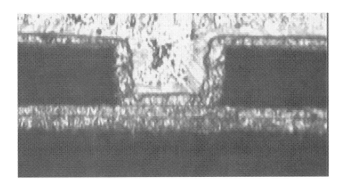

Figure 4.23 75-µm laser-ablated HDI microvia, through 50-µm resin-coated copper with 4-µm copper foil on top and 25-µm copper plating in the via.

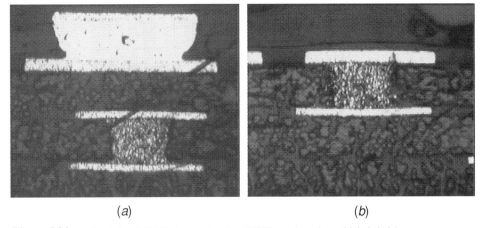

(a) (b)

Figure 4.24 (*a*) Buried and (*b*) blind vias using the ALIVH (any layer interstitial via hole) process.

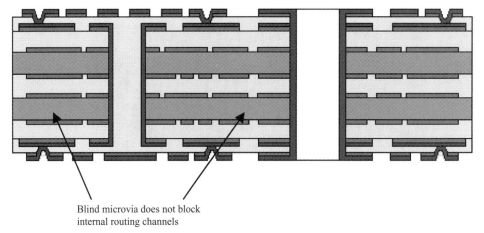

Figure 4.25 Eight-layer microvia construction L1-2, L7-8 on L2-6 buried via subcomposite with through vias L1-8. Note that internal routing channels are clear.

structure are not blocked (Fig. 4.25), therefore higher routing densities can be achieved. Improvements in the image and etch processes have enabled the production of etched feature sizes in the 100-μm range to become commonplace. Hence HDI has satisfied the need. Figure 4.26 shows a plated microvia. Note the robustness of the PTH compared to a hole going through a thick stack of material. Figure 4.27 illustrates simple to complex common microvia structures.

Five categories of performance classification for microvia structures are given in Reference 10 including: chip carrier, hand held (cell phone, pagers), high performance (avionics, military, medical), harsh environment (automotive, space), and portable (laptops, PDAs). Design and construction classification is given in Reference 11 and includes six types. Classification is given in Table 4.6 and is based on construction. Types I through VI cover the known materials and constructions at the time of the writing of the guideline. Given time there probably will be additional interconnection methods.

Figure 4.28 shows the type I construction, the basic and most common structure. Vias are either blind or go all the way through the structure. There are no buried vias. Blind vias only connect the outermost layer with the first inner layer and may be on one or both sides of the structure. The core may be multilayered or double sided. No sequential lamination steps are used.

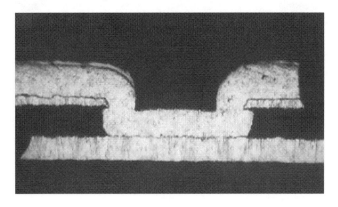

Figure 4.26 Microvia connecting two layers. Note the very low aspect ratio of the via and the relative thickness of the copper plating to the foil and dielectric.

4.6 Advanced PCB Structures

Table 4.6 IPC 2315 Construction Classification of Microvia Structures [11]

Type	Construction	Figure
I	1C0 or 1C1 with through vias	4.28
II	1C0 or 1C1 vias in core, may have through vias	4.29
III	=>2C =>0 vias in core, may have through vias	4.30, 4.31, 4.32
IV	=>1P =>0	4.33
V	Coreless construction with layer pairs	4.34
VI	Alternate constructions	4.35

where
$\quad\quad$ C = an active double-sided or MLB core
$\quad\quad$ P = a passive core
$\quad$ 1, 2, or 0 = number of microvia layers either side of the core

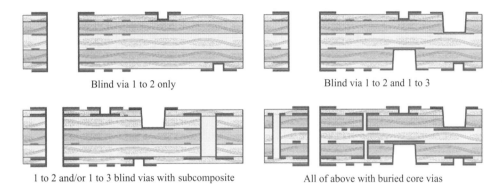

Blind via 1 to 2 only $\quad\quad\quad\quad$ Blind via 1 to 2 and 1 to 3

1 to 2 and/or 1 to 3 blind vias with subcomposite $\quad\quad$ All of above with buried core vias

Figure 4.27 Common HDI microvia structures, simple to complex.

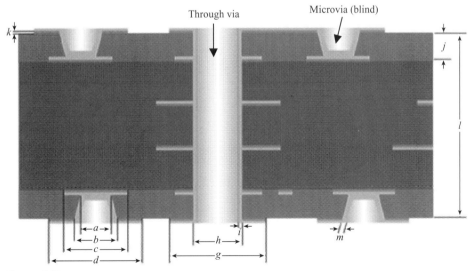

Figure 4.28 Type 1 HDI construction, through vias from surface to surface and microvias may be on one or both sides of the structure [11].

138 Chapter 4 Organic Printed Circuit Board Materials and Processes

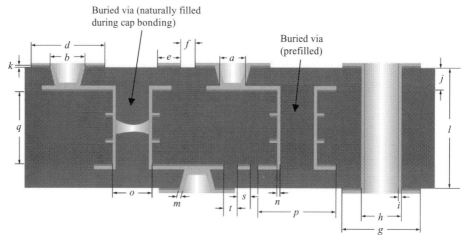

Figure 4.29 Type II HDI construction. This construction is type I with the addition of buried vias [11].

Figure 4.29 illustrates the type II construction. This construction adds the option of buried vias. It may or may not have surface-to-surface vias, and microvias only connect the outermost layer with the first inner layer and may be on one or both sides of the structure. The core may be multilayered or double sided. The core is fabricated as a separate detail through drilling and imaging of its outermost layers. Vias, which will subsequently become buried, may be filled with a via fill material and planarized. Via fill materials may or may not be electrically or thermally conductive (all types are commercially available). If the vias are not prefilled, they will fill to some extent during the bonding of the microvia caps, which become the outermost layers of the finished PCB. This additional bonding or lamination process is commonly known as sequential lamination. The degree of via fill in this process depends on the bonding materials used and thickness of the buried via subassembly as well as the blind via size.

Type III structures are illustrated in Figures 4.30 and 4.31. Type III structures are characterized by having at least two microvia layers on either one or both sides of the board. These vias may be staggered as illustrated in Figure 4.30 or stacked (Fig. 4.31).

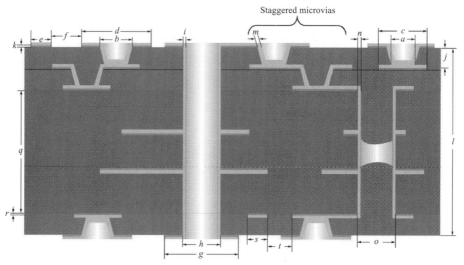

Figure 4.30 Type IIIa HDI construction [11].

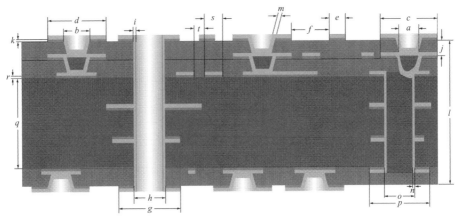

Figure 4.31 Type IIIb HDI construction [11].

Structures are sequentially laminated, and buried vias may be prefilled or naturally filled by the process as with type II. A more complex version of type III is illustrated in Figure 4.32 in which many staggered microvia layers are built up, and vias may be filled or created with a conductive post material.

Figure 4.33 illustrates the type IV passive core structure. The microvia structure is built through sequential lamination on an existing predrilled passive substrate. The passive core in the substrate does not perform an electrical function, but it may be used for thermal or CTE management.

Type V (Fig. 4.34) construction describes an HDI structure in which there are both plated microvias and conductive paste interconnections through a colamination process. There is essentially no core to this type of construction since all layer pairs have the same characteristics. Type V construction uses the fabrication of an even number of layers that are laminated together at the same time the interconnections are made between the odd and even layers. This is neither built up or sequential; it is a single lamination process. The layer pairs are prepared using conventional processes. They are joined together using prepreg resin systems or some other form of dielectric into which conductive adhesive has been placed.

Type VI is called the alternative construction because it does not use plated holes/vias for interconnection. As illustrated in Figure 4.35, in this construction, electrical interconnection and mechanical structure are formed simultaneously. The layers may be formed sequentially or colaminated and the conductive interconnection may be formed by means other than electroplating such as anisotropic films/pastes, conductive paste, dielectric piercing posts, and so forth. The figure illustrates piercing posts. The posts, which are made up of a conductive element, are attached to an unreinforced layer of copper, and the bonding process and interconnections are made by adding prepreg and laminating the PWB together in one step.

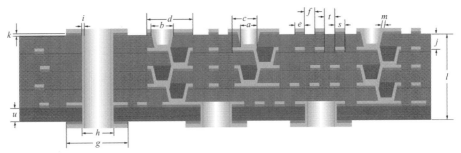

Figure 4.32 Type IIIc HDI construction [11].

140 Chapter 4 Organic Printed Circuit Board Materials and Processes

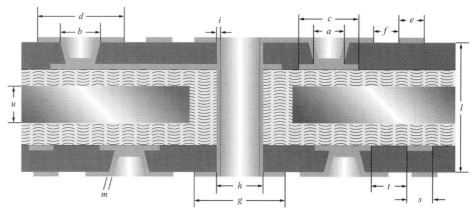

Figure 4.33 Type IV HDI construction, passive core [11].

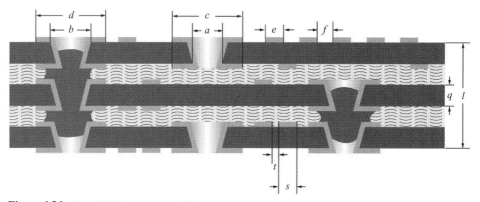

Figure 4.34 Type V HDI construction [11].

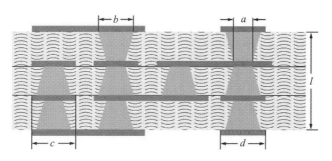

Figure 4.35 Type VI HDI construction [11].

4.7 SPECIFICATIONS AND STANDARDS

4.7.1 The IPC, a Brief History

The IPC, originally known as the Institute for Printed Circuits, was founded as a nonprofit trade association in 1957 during the early stages of the printed circuit board industry. Its membership began to include the emerging electronics manufacturing industry, earlier known as *board stuffers*, that evolved through the CMs or *contract manufacturers* to the giant electronics manufacturing segment known as the EMS, or *electronic manufacturing service* providers. At that time its name was changed to the Institute of Interconnecting and Packaging Electronic Circuits and maintained the acronym of IPC. In 1999, IPC changed its name from Institute of Interconnecting and Packaging Electronic Circuits to IPC. IPC now is not an acronym, but a name always associated with the clever tag line of "Association Connecting the Electronics Industries."

The IPC is still guided by its mission statement adopted in 1991 [10]:

IPC is a United States–based trade association dedicated to furthering the competitive excellence and financial success of its members worldwide, who are participants in the electronic interconnect industry. In pursuit of these objectives, IPC will devote resources to management improvement and technology enhancement programs, the creation of relevant standards, protection of the environment, and pertinent government relations. IPC encourages the active participation of all its members in these activities and commits to full cooperation with all related national and international organizations.

4.7.2 Relevant Standards to Organic Printed Circuit Boards

IPC is our standards organization. Figure 4.36 shows the total scope of the standards program. Developing standards is an enormous task, complicated by the facts that technology is constantly changing and advancing and the work is done by volunteers. It is recommended that the reader visit IPC's website (ipc.org) and review the current standards as well as the ongoing standards activity to develop a sense of the magnitude of this initiative.

The following discussion is intended only to provide the reader an overview of the organic printed circuit board standards and standards development including reference to sources for virtually unlimited details.

On the Design Track side of Figure 4.36 in the block titled PCB you will find IPC-2221 [7]. This is the standard for generic design of organic printed circuit boards. Under it falls a family of standards that include the design details for rigid boards, flex circuits, Personal Computer Memory Card International Association (PCMCIA) cards, multichip module—laminate (MCM-L) (organic MCMs), HDI (high-density interconnecting), and discrete wiring. In the center of the tree, under the Assembly Track, in the block titled PCB/Acceptance you will find IPC-6011 [11]. This is the standard for generic performance of printed circuit boards. Under it falls a family of standards that include the specific performance requirements for rigid boards, flex circuits, PCMCIAs, MCM-L (organic MCMs), HDI (high-density interconnections), and high-frequency circuits. At the bottom of the tree, falling under PCB/Acceptance are three very important blocks covering the standards for materials including foils, laminates (rigid, flexible, and HDI), and reinforcements.

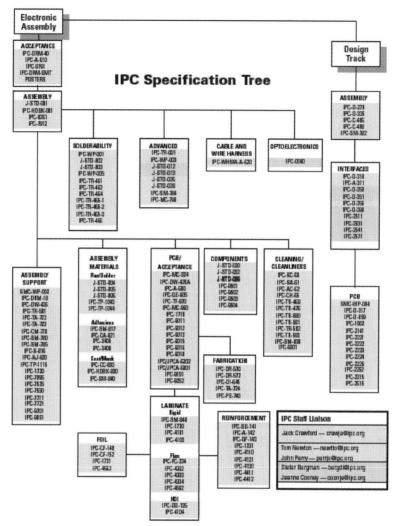

Figure 4.36 IPC standards program, specification tree.

4.8 KEY TERMS

IPC-J-50 [12] is a very complete listing of virtually all terms and definitions for interconnecting and packaging circuits. Following is an extraction of the key terms from that resource.

- *Annular Ring* That portion of conductive material completely surrounding a hole.
- *Aspect Ratio (Hole)* Ratio of the length or depth of a hole to its preplated diameter.
- *B-Stage* Intermediate stage in the reaction of a thermosetting resin in which the material softens when heated and swells, but does not entirely fuse or dissolve, when it is in contact with certain liquids.
- *Bare Board* Unassembled (unpopulated) printed board.
- *Base Material* Insulating material upon which a conductive pattern may be formed. (The base material may be rigid or flexible or both. It may be a dielectric or insulated metal sheet.)
- *Blind Via* Via extending only to one surface of a printed board.

- *Board Thickness*   Overall thickness of the base material and all conductive materials deposited thereon.
- *Bow, Warp, Twist (Sheet, Panel, or Printed Board)*   Deviation from flatness of a board characterized by a roughly cylindrical or spherical curvature such that, if the product is rectangular, its four corners are in the same plane (*See* Warp, Twist).
- *Breakout (Hole Breakout)*   Condition in which the hole is not completely surrounded by the land.
- *Buried Via*   Via that does not extend to the surface of a printed board.
- *C-Staged Resin*   Resin in its final state of cure.
- *Cap Lamination*   Process for making multilayer printed boards with surface layers of metal-clad laminate bonded in a single operation.
- *Card*   Printed board.
- *Circuit Card*   Printed board.
- *Circuitry Layer*   Layer of a printed board containing conductors, including ground and voltage planes.
- *Clearance Hole*   Hole in a conductive pattern that is larger than, and coaxial with, a hole in the base material of a printed board.
- *Component Hole*   Hole that is used for the attachment and/or electrical connection of component terminations, including pins and wires, to a printed board.
- *Component Side*   Primary side.
- *Conductive Pattern*   Configuration or design of the conductive material on a base material.
- *Conductor*   Single conductive path in a conductive pattern, for example, line, path, trace, track.
- *Conductor Spacing*   Observable distance between adjacent edges (not center-to-center spacing) of isolated conductive patterns in a conductor layer.
- *Conductor Thickness*   Thickness of a conductor including additional metallic coatings but excluding nonconductive coatings.
- *Conductor Width*   Observable width of a conductor at any point chosen at random on a printed board as viewed from directly above unless otherwise specified.
- *Coupon*   Test coupon.
- *Desmear*   Removal of friction-melted resin and drilling debris from a hole wall.
- *Dielectric*   Material with a high resistance to the flow of direct current and that is capable of being polarized by an electrical field.
- *E-Glass*   Low-alkali lime alumina borosilicate glass with good electrical properties.
- *Electrodeposition*   Deposition of a conductive material from a plating solution by the application of electrical current, for example, electroplating.
- *Electroless Deposition*   Deposition of conductive material from an autocatalytic plating solution without the application of electrical current, for example, electroless plating.
- *Etch Factor*   Ratio of the depth of etch to the amount of lateral etch, that is, the ratio of conductor thickness to the amount of undercut.
- *Etchant*   Solution used to remove the unwanted portion of material from a printed board by a chemical reaction.
- *Etchback*   Controlled removal by a chemical process, to a specific depth, of nonmetallic materials from the side walls of holes in order to remove resin smear and to expose additional internal conductor surfaces.
- *Etching*   Chemical, or chemical and electrolytic, removal of unwanted portions of conductive or resistive material.
- *Fingers*   Edge board contacts.
- *Foil Lamination*   Process for making multilayer printed boards with surface layer(s) of metal foil bonded in a single operation.
- *Fused Coating*   Metallic coating, usually a tin or solder alloy, that has been melted and solidified to form a metallurgical bond to a basis metal.
- *Fusing*   Melting of a metallic coating (usually electrodeposited tin or tin–lead) on a conductive pattern, followed by solidification.
- *Hole Pattern*   Arrangement of all the holes in a printed board or production board.
- *Hole Void*   Void in the metallic deposit of a plated through hole that exposes the base material.
- *Immersion Plating*   Chemical deposition of a thin metallic coating over certain basis metals that is achieved by a partial displacement of the basis metal.
- *Inner-Layer Connection*   Conductor that connects conductive patterns on internal layers of a multilayer printed board, for example, a plated through hole.
- *Laminate*   Product made by bonding together two or more layers of material.
- *Laminate Void*   Absence of resin or adhesive in an area that normally contains them.
- *Land, Pad, Terminal Area*   Portion of a conductive pattern usually, but not exclusively, used for the connection and/or attachment of components.
- *Landless Hole*   Plated through hole without land(s).

- *Layer-to-Layer Registration* Degree of conformity of a conductive pattern, or portion thereof, to that of any other conductor layer of a printed board.
- *Layer-to-Layer Spacing* Thickness of dielectric material between adjacent layers of conductive patterns in a printed board.
- *Legend* Format of letters, numbers, symbols, and patterns that are used primarily to identify component locations and orientations for convenience of assembly and maintenance operations.
- *Lifted Land* Land that has fully or partially separated (lifted) from the base material, whether or not any resin is lifted with the land.
- *Microsectioning* Preparation of a specimen of a material, or materials, that is to be used in a metallographic examination. (This usually consists of cutting out a cross section, followed by encapsulation, polishing, etching, staining, etc.)
- *Minimum Annular Ring* Minimum width of metal(s) at the narrowest point between the edge of a hole and outer edge of a circumscribing land.
- *Misregistration* Imperfect registration.
- *Multilayer Printed Board* General term for a printed board that consists of rigid or flexible insulation materials and three or more alternate printed wiring and/or printed circuit layers that have been bonded together and electrically interconnected.
- *Panel, Manufacturing Panel, Production Panel* Rectangular sheet of base material or metal-clad material of predetermined size that is used for the processing of one or more printed boards and, when required, one or more test coupons.
- *Pattern Plating* Selective plating of a conductive pattern.
- *Photoprint* Process of forming a circuit pattern image by hardening a photosenstive polymeric material by passing light through a photographic film.
- *Photoresist* Material that is sensitive to portions of the light spectrum and that, when properly exposed, can mask portions of a base metal with a high degree of integrity.
- *Photoresist Image* Exposed and developed image in a coating on a base material.
- *Photo Tool, Artwork, Artwork Master, Production Master, Working Master* Photographic product that is used to produce a pattern on a material.
- *Plated Through Hole (PTH)* Hole with plating on its walls that makes an electrical connection between conductive patterns on internal layers, external layers, or both, of a printed board.
- *Prepreg* Sheet of material that has been impregnated with a resin cured to an intermediate stage, that is, B-staged resin.
- *Printed Board* General term for completely processed printed circuit and printed wiring configurations. (This includes single-sided, double-sided, multilayer boards with rigid, flexible, and rigid-flex base materials.)
- *Registration* Degree of conformity of the position of a pattern (or portion thereof), a hole, or other feature to its intended position on a product.
- *Resin Recession* Presence of voids between the plating of a plated through hole and the wall of the hole as seen in microsections of plated through holes that have been exposed to high temperatures.
- *Resin Smear* Base material resin that covers the exposed edge of conductive material in the wall of a drilled hole. (This resin transfer is usually caused by the drilling operation.)
- *Resin-Rich Area* Location in a printed board of a significant thickness of unreinforced surface-layer resin that is of the same composition as the resin within the base material.
- *Resin-Starved Area* Location in a printed board that does not have a sufficient amount of resin to completely wet out the reinforcing material. (Evidence of this condition is often in the form of low-gloss dry spots or exposed fibers.)
- *Resist* Coating material that is used to mask or protect select areas of a pattern during manufacturing or testing from the action of an etchant, plating, solder, and the like.
- *Screen Printing* Transferring of an image to a surface by forcing a suitable media with a squeegee through an imaged-screen mesh.
- *Solder Coat* Layer of solder that is applied directly from a molten solder bath to a conductive pattern.
- *Solder Leveling* Solder coating process that causes redistribution and/or partial removal of excess molten solder from a printed board by applying sufficient heat and mechanical force.
- *Solder Mask* Solder resist.
- *Solder Plug* Core of solder in a plated through hole.
- *Splay* Tendency of a rotating drill bit to make off-center, out-of-round holes that are not perpendicular to the drilling surface.

- *Supported Hole* Hole in a printed board that has its inside surfaces plated or otherwise reinforced.
- *Tenting* Covering of holes in a printed board and the surrounding conductive pattern with a resist.
- *Test Coupon, Coupon* Portion of quality conformance test circuitry that is used for a specific test, or group of related tests, in order to determine the acceptability of a product.
- *Unsupported Hole* Hole in a printed board that does not contain plating or other type of conductive reinforcement.
- *Via* Plated through hole that is used as an interlayer connection but in which there is no intention to insert a component lead or other reinforcing material.
- *Warp, Twist* See Bow.

REFERENCES

1. MIL-PRF-55110, *Performance Specification, Printed Wiring Board, Rigid, General Specification for*, Defense Electronics Supply Center, Dayton, OH, May 1997.
2. MIL-PRF-31032, *Performance Specification, Printed Circuit Board/Printed Wiring Board, General Specification For*, Defense Electronics Supply Center, Dayton, OH, November 1995.
3. IPC-4101A, *Specification for Base Materials for Rigid and Multilayer Printed Boards*, Bannockburn IL: IPC, December 2001.
4. IPC-6011, *Generic Performance Specification for Printed Boards*, Bannockburn IL: IPC, July 1996.
5. IPC-6012A, *Qualification and Performance Specification for Rigid Printed Boards,* IPC, Bannockburn, IL, July 2000
6. IPC-A-600, *Acceptability of Printed Boards*, Bannockburn IL: IPC, November 1999.
7. IPC-2221, *Generic Standard on Printed Board Design*, Bannockburn IL: IPC, February 1998.
8. Weis Vince, *Combining Dielectrics in Multilayer Microwave Boards*, Arlon Materials for Electronics, Rancho Cucamonga CA, October 1997.
9. IPC 6016 Qualification and Performance Specification for High Density Interconnect (HDI) Layers or Boards, Bannockburn IL: IPC, May 1999.
10. IPC Web site www.ipc.org.
11. IPC-6011, Generic Performance Specification for Printed Boards, Bannockburn IL: IPC, July 1996.
12. IPC/JPCA-2315 Design Guide for High Density Interconnects (HDI) and Microvias Bannockburn IL: IPC, June 2000.

EXERCISES

4.1. What are the levels of interconnection and which one encompasses printed circuit boards (PCBs)?

4.2. Approximately, what is the age in years of the PCB technology?

4.3. In terms of conductor width and space dimensions, expressed in mm, how have these features changed over the years?

4.4. What does the PCB manufacturer receive from his customer to enable him to produce PCBs?

4.5. From industries' primary performance specifications describe the three product classifications and their significance.

4.6. Who on the PCB manufacturer's technical staff determines how the product is going to be manufactured?

4.7. In terms of layers of circuitry, what are the three basic types of PCBs?

4.8. What manufacturing processes are common to each of these types?

4.9. What are photo tools? How are they used? How are they originated? How are they made?

4.10. What is scaling and why is it important?

4.11. What is a manufacturing panel?

4.12. How many different PCB raw material choices are there?

4.13. What are the basic raw materials for PCB construction?

146 Chapter 4 Organic Printed Circuit Board Materials and Processes

4.14. Why is foil construction more cost effective than cap construction?
4.15. Why are PCB laminate raw materials dimensionally unstable?
4.16. What does the PCB designer do to accommodate the known dimensional instabilities of laminate materials?
4.17. What does the PCB manufacturer do to accommodate the known dimensional instabilities of laminate materials?
4.18. Are virtually all PCBs manufactured with similar processes?
4.19. What are inner layers? What are outer layers?
4.20. What is photoresist? Is it the same as "dry film"? How are they applied? What is their photographic polarity?
4.21. How are photo tools registered to each other during the manufacture of a pair of inner layers?
4.22. How is the image transferred to the inner layer?
4.23. What is develop-etch-strip (DES)?
4.24. What is artwork? What is artwork compensation?
4.25. Why is postetch punching the preferred method of putting tooling holes in inner layers?
4.26. Why is AOI used in the inner-layer process?
4.27. Why is it necessary to treat the copper on the inner layer prior to laminating into an MLB structure?
4.28. Chemically and physically, what happens when an MLB structure goes through the lamination process?
4.29. How small a via can be drilled mechanically? How small with a laser?
4.30. Why is desmear or etchback necessary after hole formation prior to plating?
4.31. Why is it necessary to make hole surfaces conductive if they are going to be eventually plated through?
4.32. How does inner-layer image and exposure differ from outer-layer image and exposure?
4.33. What is the difference between an etch resist and a plating resist?
4.34. What type of etch resist is used for outer layers?
4.35. What is the typical copper plating thickness in a PTH?
4.36. What is key about the PTH?
4.37. What is soldermask?
4.38. What is legend?
4.39. Why are different surface finishes used on PCBs?
4.40. What are the primary surface finishes used?
4.41. How is the final profile of the PCB made?
4.42. Is electrically testing bare PCBs necessary? Why?
4.43. How does golden board testing compare to netlist testing?
4.44. What is a flying probe tester?
4.45. Why does the PCB manufacturer run an analysis called net compare?
4.46. What are microsections? Why are they important? What information is derived from microsection analysis?
4.47. What is contract review? From what is the term derived? What is a more common expression meaning the same thing?
4.48. What are the two dielectric properties most critical to PCB performance?
4.49. How does cost relate to these dielectric properties?

4.50. How does performance relate to these dielectric properties?

4.51. What is the difference between a homogeneous or full-body MLB construction and a hybrid construction?

4.52. What thermal and mechanical properties of dielectrics are critical to PCB performance?

4.53. What are the disadvantages of thermoplastic materials used as bond plies?

4.54. What causes MLBs to be bowed or warped, that is, not flat?

4.55. Why is flatness important? What can be done to assure flatness?

4.56. What is RoHS?

4.57. What is WEEE?

4.58. What are the four major misconceptions related to the lead-free solder initiatives for PCBs?

4.59. What is HDI?

4.60. How is HDI enabling to the development of electronic products?

4.61. What U.S. trade organization provides development and maintenance of PCB and related electronic packaging standards?

4.62. Does it provide key terms and definitions?

4.63. Why is it important to have standards?

Chapter 5

Ceramic Substrates

AICHA A. R. ELSHABINI AND FRED D. BARLOW III

5.1 CERAMICS IN ELECTRONIC PACKAGING

Ceramics have a number of positive attributes that make them attractive for electronic applications. This chapter provides the reader with an overview of the processes and materials that are used to fabricate electronic components and circuits using ceramic substrates.

5.1.1 Introduction and Background

Modern ceramic substrates and electronic packages are sophisticated combinations of glasses, ceramics, and metals that form compact, cost-effective solutions for a variety of applications. Due to the unique sintering process utilized to fabricate these materials, a wide range of conductors, dielectrics, resistive materials, and even magnetic materials can easily be incorporated into a given ceramic product. In addition, many of the ceramic technologies utilized today are inherently multilayer in structure and provide tremendous design flexibility for high-density electronic circuitry. Consequently, ceramics are widely used as thick-film substrates, thin-film substrates, insulators, and/or structures that are capable of withstanding temperatures up to 1000°C.

5.1.2 Functions of Ceramic Substrates

The function of an electronic substrate is to provide the base onto which thin- and thick-film circuits are fabricated by the deposition of a variety of materials. The substrate must provide the necessary mechanical support and rigidity to produce a reliable, functional circuit, it must have adequate thermal management ability to ensure proper temperature operation, and it must possess proper electrical insulation to withstand circuit voltages without conduction (dielectric breakdown). One can think of the substrate as the foundation on which circuit traces and components are mounted and supported. Since ceramic materials have high thermal conductivity, good chemical stability, and are resistant to thermal and mechanical shock, they are often used to package electronic circuits [1].

Advanced Electronic Packaging, Second Edition, Edited by Richard K. Ulrich and William D. Brown
Copyright © 2006 the Institute of Electrical and Electronics Engineers, Inc.

5.1.3 Ceramic Advantages

Ceramic substrates are essentially metal oxides, nitrides, and/or glasses fired at elevated temperatures. The high-temperature firing results in a hard and brittle structure that possesses many properties that are very desirable for electronic applications. Specifically, ceramic materials possess high mechanical strength and low thermal expansion, allowing them to withstand variations in operating conditions and high electrical resistivity over a wide temperature range. They also possess adequate dielectric strength to withstand applied voltages without dielectric breakdown, high chemical inertness to most chemicals and etchants, a relatively low dielectric constant, and low dissipation factor that minimize capacitive effects and electrical losses. Additionally, these materials generally exhibit high thermal conductivity and a high tolerance to temperature extremes, which allows for effective thermal management. When used as a substrate for electronic circuits, material properties of interest include mechanical characteristics (compression strength, tensile strength, modulus of elasticity, dimensional stability, flexural strength, and thermal coefficient of expansion), physical characteristics parameters (camber, surface finish, specific gravity, and water absorption), chemical parameters (materials compatibility and chemical reactivity), electrical parameters (dielectric constant, dissipation factor, resistivity, and dielectric strength), and thermal characteristics (thermal conductivity, thermal coefficient of expansion, and heat capacity) [2, 3]. Compared to many other electronic packaging materials, ceramics have a high modulus of elasticity, are rigid materials, which ensures minimum distortion under high loading and high-temperature conditions. They also have a higher compressive strength than alloy steel and a higher tensile strength than porcelain. In addition, they offer higher strength than glass, extremely high dimensional stability, low differential magnitudes of thermal expansion, and high chemical inertness relative to various processing and operating conditions.

5.1.4 Ceramic Compositions

Ceramic materials commonly used for electronic substrates are exposed to processing temperatures in the range from 850 to 1900°C during their fabrication, thus becoming naturally immune to lower temperature processing. Basic ceramic compositions include electrical porcelain [50% clay, $Al_2Si_2O_5(OH)_4$, and 25% each of flint, SiO_2, and feldspar, $KAlSi_3O_8$], steatite [commercial steatite compositions are based on 90% talc, $Mg_3Si_4O_{10}(OH)_2$, plus 10% clay], cordierite ($Mg_2Al_4Si_5O_{18}$, which is useful for high-temperature applications), forsterite (Mg_2SiO_4), alumina or aluminum oxide (Al_2O_3), beryllia or beryllium oxide (BeO), magnesia (MgO), zirconia (ZrO_2), and aluminum nitride (AlN). High-alumina porcelains have a great tolerance for compositional variations, and, since the dielectric constant is nearly constant through a wide range of temperatures, they are of most interest in electronic device applications.

Steatite porcelains (with electrical properties that vary at low frequencies) are low-loss materials commonly used as components for variable capacitors, coil forms, and general structure insulation. Cordierites possess a low thermal expansion coefficient and, consequently, high thermal shock resistance [4]. Forsterites have a higher resistivity and a lower electric loss with increasing temperature due to the absence of alkali ions in the vitreous phase. Beryllia or beryllium oxide is stable in air, vacuum, hydrogen, carbon monoxide, argon, and nitrogen at temperatures up to 1700°C. Magnesia is suitable for insulating thermocouple leads and for heating core elements. Aluminum oxide (alumina, Al_2O_3), beryllium oxide (beryllia or BeO), aluminum nitride (AlN), and silicon carbide (SiC) are common ceramics used as electronic substrates [5–11].

5.1.5 Ceramic Substrate Manufacturing

Ceramic substrate manufacturing is initiated by processing high-purity powders through a ball mill to create particles of the proper size. Organic binders, solvents, plasticizers, and other additives are mixed with the powder to provide stability, packing density, and grain uniformity. These additives are needed to achieve a specific rheology so that the material flows freely during processing and maintains cohesiveness and uniform characteristics. Then, the mixture is either pressed from a powder into a final shape and subjected to a heat treatment to sinter the material or a sheet of uniform thickness is cast from a slurry, dried, and punched to the proper shape, followed by a heat treatment to sinter the material. Substrates can also be extruded by forcing the material in fluid form to pass through a die.

The heat treatment that follows the formation of a substrate material often starts with prefiring at 300 to 900°C to remove most of the organic additives (about 99.9%). Then, the substrates are sintered by firing at higher temperatures to remove the plasticizers, any remaining organic binders, and other additives. This high-temperature firing results in material shrinkage and densification. Recrystallization of the material also occurs during the sintering process, resulting in a strong substrate with a smooth surface. The steps involved in the various methods used to manufacture ceramic substrate are shown in Figure 5.1 [12–15].

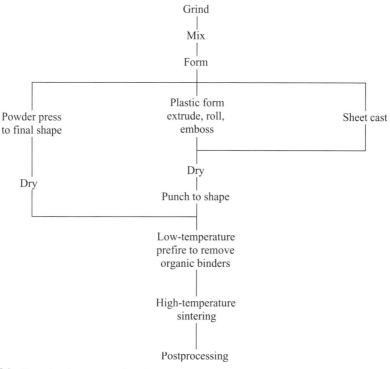

Figure 5.1 Ceramic substrate manufacturing.

5.2 ELECTRICAL PROPERTIES OF CERAMIC SUBSTRATES

Electrical parameters of interest for ceramic substrates include volume resistivity, dielectric constant, dissipation factor, and dielectric strength. These parameters may be critical under certain operating conditions, such as high frequency and/or high voltage. The electrical resistivity, ρ, in Ω-cm, is a measure of the resistance that a material offers to current flow under an applied electric field. Usually, the resistivity is high in value, exceeding 10^{14} Ω-cm for most ceramic substrates at room temperature. As the temperature increases, the electrical resistivity of these ceramics decreases to around 10^6 Ω-cm at 1000°C [16–18]. Electrical property values for most of the ceramics used in electronic packaging are given in Table 2.2.

The dielectric constant, k, of a material is a measure of the ability of a material to store electric charge relative to vacuum and is a dimensionless quantity. In general, the dielectric constant value of ceramics at room temperature varies from about 4 to 10, depending on the type of ceramic (i.e., composition), the temperature, the frequency of operation, the particle size, and the purity of the material, and increases with increasing temperature for most ceramics. Dielectric constant values are normally provided at a measurement frequency of 1 MHz, although, in general, k may depend on the frequency of operation. For some electronic packaging applications, a low dielectric constant is desired in order to minimize capacitive effects. Typical values of k for a number of ceramic materials are given in Table 2.2.

The dissipation factor (DF) [also referred to as the dielectric loss or loss tangent (tan δ)] of a dielectric material is a measure of the real or resistive component of capacitance and is a relative measure of the energy loss in a capacitor. DF is of the form

$$\mathrm{DF} = \tan \delta = R_s \, \omega \, C_s = \varepsilon'' / \varepsilon' \tag{5.1}$$

where

$\omega = 2\pi f$

f = frequency, Hz

R_s = series resistance, Ω

ωC_s = reciprocal of capacitive reactance, Ω^{-1}

ε'', ε' = imaginary and the real components of the complex permittivity ε^*, that is,

$\varepsilon^* = \varepsilon' - i\,\varepsilon''$

Since DF is a relative measure of the losses in a dielectric due to heating, a low DF is desired in order to avoid excessive electrical losses. Values of DF for a variety of ceramic materials are given in Table 2.2. Table 2.3 gives values of DF for several high-alumina-content ceramics. In general, however, tan δ is of the order of 0.0001, 0.0012, 0.005, and 0.05 for Al_2O_3, BeO, AlN, and SiC, respectively, and varies from approximately 0.0014 for 75% aluminum oxide to 0.0001 for 99% aluminum oxide [19, 20].

The dielectric strength of ceramics in volts/millimeter (in general, voltage per unit length) varies considerably as a function of temperature, frequency, and a material's physical properties (such as density, porosity, purity, and physical dimensions of a sample), decreasing sharply as the frequency and/or temperature of operation increases. Adequate dielectric strength is needed so that a capacitor dielectric does not break down under required applied voltages. A dielectric strength of ≥ 10 KV/mm has been observed for most ceramics (26 to 24 for Al_2O_3, 9.5 for BeO, and 10 to 14 for AlN).

5.3 MECHANICAL PROPERTIES OF CERAMIC SUBSTRATES

In general, ceramics are brittle materials, but they exhibit sufficiently good mechanical strength and thermal expansion properties to allow them to endure processing, handling, storage, and operation as an electronic substrate. When they do fail, the failure is usually along the perimeter of the substrate in the form of microcracks as a result of internal stresses, which can often be relieved by an annealing process. Such mechanical failures can be aggravated by a rise in temperature. Due to the ceramic's brittle nature, specific mechanical testing, such as Young's modulus, compression strength, tensile strength, thermal coefficient of expansion, dimensional stability, and thermal shock failure mechanisms, is recommended. Table 5.1 provides mechanical and physical properties of some ceramics at room temperature [21–25].

Table 5.1 Mechanical and Physical Properties of Some Ceramics at Room Temperature[a]

Parameter	Alumina (Al_2O_3)	Beryllia (BeO)	Aluminum Nitride (AlN)	Silicon Carbide (SiC)
Young's modulus (GPa)	360	350	340	400
Compression strength(ksi)	290–380			
Tensile strength	17–35 ksi or 22,000–38,000 lb/in.^{2}			
Flexural strength	43,000–55,000 (lb/in.^{2}) or 250–400 N/mm^2	>250 MPa or 170–240 N/mm^2	>300 MPa or 280–320 N/mm^2	—
Thermal coefficient of expansion[b] (25–200°C), TCE (ppm/°C)	5.3–6.7	6.9	2.7–4.1 (silicon is 2.6 GaAs is 5.7)	3.7
Coefficient of thermal endurance F (representing thermal shock resistance)	3.7 (0.9 for glass and 13.0 for silica)	3.0	4.6	3.7
Thermal conductivity (W/m-K)	20–35[c]	250	160–190	270
Density (kg/m^3) or (g/cm^3)	3.9–3.8	2.9–3.0	3.28–3.3	3.1–3.2
Surface finish	~25 μin.	As low as 5 μin.	<0.6 μm <25 μin.	
Dielectric strength (kV/mm)	18	>26	>15	>100
Dielectric strain (V/mil), 25 mils thick	240–210			
Camber	0.002 in./in. for 10 mils, 0.003 in./in. for 25 mils, and 0.004 in./in. for 40 mils	<0.003 in./in.	<0.0025 in./in.	—

[a] As the purity of material increases (and/or increase in alumina content), the value increases.
[b] The thermal coefficient of expansion of a given material is the slope of the linear thermal expansion versus temperature. The differential magnitudes of thermal expansion between two materials are considered key design parameters and should be kept to a minimum.
[c] Increases with wt% of aluminum oxide, with 96% alumina being considered as standard.

5.4 PHYSICAL PROPERTIES OF CERAMIC SUBSTRATES

The surface finish of a ceramic substrate is a function of the microgranular structure and the density of the ceramic–glass composite. Small grains in as-fired ceramic structures form a smooth surface and are used primarily for thin-film or fine-line thick-film applications. The centerline average (CLA) of a ceramic substrate is a measure of the surface smoothness. The CLA requirement for thick-film ceramic substrates is typically 0.381 to 1 μm (15 to 40 μin.) and, for thin-film substrates, is typically 0.127 μm (5 μin.) or less. Camber (μm/mm) is defined as the overall deviation in thickness of a substrate from one side to the other side, measured along a diagonal. Substrates 635μm (25 mil) thick with 1 to 3 μm/mm (1 to 3 mil/in.) camber are available as standard products. Aluminum nitride substrates may possess a camber less than 3 μm/mm.

5.5 DESIGN RULES

Specific design rules used for a substrate fabrication vary from one fabrication facility to another due to variations in the process equipment used and the material set employed. However, some general guidelines are presented here in order to provide a context for the capabilities of the technology. In general, key geometries must be limited by the ranges of what can be achieved in mass production. These geometries include the conductor width, the conductor spacing, the via size, and the via pitch, to name a few. Vias are simply metal-filled holes used to connect one circuit layer to the next. Table 5.2 summarizes some of the key design rules and gives representative examples of the ranges one would normally find for production houses. The values are broken into two sets, labeled "standard" and "custom." The standard design guidelines fall within the capabilities of virtually all manufacturers, while the custom guidelines are only available from a small number of manufacturers, respectively.

Table 5.2 Some Key Design Rule Range of Values for standard and custom Manufacturers of Ceramic Substrates

Parameters	Standard	Custom
Maximum substrate size (fired)	6 in. × 6 in.	6 in. × 6 in., or 8 in. × 8 in.
Number of layers	<15	>15
Layer thickness	0.010 in.–0.025 in.	0.005 in. 0.010 in.
Tolerances		
Length and width	±0.8%, or ±2 mils	±0.5%, or ± 1 mil
Thickness	±10%, or ±2 mils	±5%, or ± 1 mil
Chamber	3 mils/in.	1 mil /in.
Conductor width	0.007 in. minimum	0.004 in. minimum
Conductor spacing	0.007 in. minimum	0.004 in. minimum
Via diameter	0.007 in. nominal	0.005 in. nominal
Via cover pad	0.014 in.	0.006 in.
Via center to center	0.024 in.	0.010 in.
Via center to center (with one conductor)	0.034 in.	0.020 in.
Sheet resistivity	0.012 Ω/□	0.008 Ω/□
Via hole resistance	0.010 Ω/□	0.005 Ω/□

5.6 THIN FILM ON CERAMICS

One popular method of circuit and component fabrication is the combination of thin films and ceramic substrates. This section provides an overview of thin-film processes and their application to ceramic substrates.

5.6.1 Introduction and Background

Thin-film circuits are composed of films with thicknesses that vary from a few nanometers to a few micrometers ($\sim$0.2 to $\sim$100 μin.) and formed by deposition techniques such as evaporation, sputtering, anodization, and chemical vapor deposition. The selection of materials to process electrical components in thin-film form is determined by the desired electrical properties and the deposition techniques. Since thin-film technology relies on a process involving atoms or molecules, the deposition conditions are critical in determining the properties of the resulting film. Commonly used substrate materials for thin-film deposition include alumina, glass, beryllia, aluminum nitride, and silicon. Ceramic materials are often used for thin-film applications since ceramics possess high thermal conductivity and good chemical stability and are resistant to thermal and mechanical shock.

5.6.2 Deposition Techniques

Vacuum evaporation is a common thin-film deposition technique. It consists of evaporation or sublimation of a solid or liquid material by heating it to a sufficiently high temperature under high-vacuum conditions, that is, on the order 10^{-5} to 10^{-9} Torr, where 1 Torr is equal to a pressure of 1 mmHg. Metals that can be evaporated possess a relatively high vapor pressure at low temperature, thus their evaporation temperature does not significantly exceed their melting temperature. In this process, a gas or vapor of the material is produced and it condenses onto a cool substrate surface to form a film. The deposited films can vary in thickness—the film can be as thin as a monolayer or as thick as a few micrometers. Electron beam evaporation, thermal evaporation, and flash evaporation utilize identical vacuum chambers, substrate holders, and deposition monitoring equipment but may vary in how the source material is heated and evaporated.

Complex materials or alloys are difficult to evaporate using electron beam or thermal evaporation since their components have different vapor pressures. In flash evaporation, a small quantity of the source material is deposited onto a preheated surface, the material is heated, and it evaporates right away, preserving the desired composition. With the advances in control systems and microprocessors, flash evaporation has been largely replaced with multisource evaporation in which each component of a final desired film is evaporated simultaneously from a different source. By controlling each sources rate independently, the final film composition can be controlled.

Sputtering is a cold process of vapor deposition involving the removal of material from a solid cathode (a metal or insulator) through positive ion bombardment. As a result, high-melting-point materials can be deposited onto virtually any substrate material, including ceramic. The source atoms form a thin uniform layer of the sputtered material on a substrate surface. Sputtering can be achieved at temperatures between 400 and 500°C for low-vapor-pressure and high-melting-point materials [31–34].

Electroplating and anodization are both electrochemical processes based on the same principle. However, anodization is used to oxidize a material, such as a pure metal, while electroplating is used to deposit metals. The reader is refereed to Chapter 3 entitled Processing Technologies for a more detailed discussion of these processes.

5.6.3 Thin-Film Substrate Properties

Some important properties of substrates used to grow defect-free films for thin-film circuits include good surface smoothness, a proper thermal expansion coefficient to minimize residual stresses in the films, good mechanical strength and thermal shock resistance to enable the substrate to withstand the rigors of processing and normal usage, high thermal conductivity to remove the heat to allow the realization of high component densities, chemical stability and inertness to withstand etchants during thin-film processing, low porosity to prevent the entrapment of gases, low cost, high electrical resistance, and good uniformity of material composition.

5.7 THICK FILMS ON CERAMICS

Ceramic substrates provide the base on which thick-film circuits are fabricated. Ceramic materials are used in substrate applications primarily due to their mechanical strength, high electrical resistivity over a broad temperature range, and chemical inertness relative to a variety of processing conditions. Since ceramic substrates can withstand temperatures in excess of 1000°C, thick-film materials are often fired at temperatures in the 1000 to 1200°C range.

5.7.1 Introduction and Background

Thick-film technology consists of applying specially formulated pastes (or inks) onto a ceramic substrate in a definite pattern and sequence using screen printing and high-temperature firing to produce electrical components, interconnections, and/or a complete circuit. These pastes contain an organic binder, which makes them thixotropic in nature with a dual viscosity—viscous at rest and flowing when a shear force is applied [35]. The pastes also have a functional component that gives the film the desired electrical properties. Depositing successive layers of pastes results in multilayer interconnection structures containing integrated passive components. Discrete components, such as diodes, transistors, and integrated circuits, can be added to the substrate using conventional attachment techniques. (See Fig. 5.2.)

The thickness of a film is typically on the order of 12 to 25 μm (0.5 to 1 mil). After the film is printed on a substrate, it is then dried to remove the volatiles, while the binder remains. The film is then fired in a multizone belt furnace. The organic binders and solvents are burned out in the first zone. The metallic elements may be either oxidized or reduced to develop resistors with the required characteristics of resistivity, temperature, and voltage coefficients, or they may be left in a metallic state to form conductive interconnects. Then, sintering of the glass materials is performed to adhere the film to the substrate and protect the metallic elements. A uniform airflow is maintained through the furnace in a direction opposite to that of the advancing substrates in order for the volatile organic materials to be removed [36–39].

In conventional thick-film technology, the interconnect structure and integrated passive devices are formed by successive printing and firing of conductive and insulating layers.

5.7 Thick Films on Ceramics 157

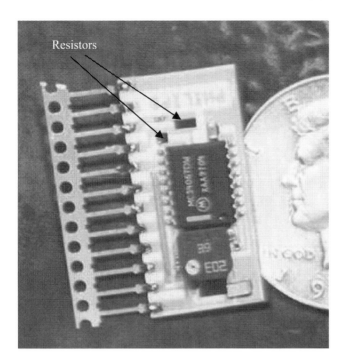

Figure 5.2 Thick-film circuit on alumina. The silver traces are conductors and the black squares are resistors. Both conductors and resistors are screen printed onto the substrate.

The pastes are applied by the conventional screen-printing process, followed by drying at 100 to 150°C to remove solvents. Resistors, capacitors, inductors, and conductors can all be patterned. As many as four dielectric printing and firing steps, with accompanying via fill printing and firing, may be required for each layer. Firing takes place at temperatures from 500 to 1100°C. This essentially sequential process becomes less cost effective as the number of conductive layers increases.

In the past, screen printing in a high-volume production operation has been limited to 200- to 250-μm vias and 150-μm lines and spaces. Today, a number of manufacturers produce circuits with 100-μm vias and 75-μm space and trace design rules. The trends are toward smaller and smaller vias on tighter pitches with finer lines and spaces. Design rules that include ∼30- to 50-μm vias are under evaluation.

To obtain finer lines and smaller vias, one can use a photoimagable thick-film process for patterning. This process involves the use of a photoactive paste printed on a substrate and exposed through artwork (or a mask) to define circuit characteristics, lines, and vias. The materials are developed in an aqueous solution and then fired using conventional thick-film techniques. Copper, silver, and gold metallizations are used, and layer counts of up to 10 circuit layers are possible.

5.7.2 Screen Preparation and Inspection

A screen mesh is made by weaving stainless steel wires to form a long clothlike sheet. Stainless steel is the material of choice for printing screens because it provides a high degree of control and precision in comparison to other materials, and it is resistant to wear and stretching. Common meshes are woven in a plain weave pattern, with each wire simply going over and under only one wire at a time. The mesh count designates the number of

158 Chapter 5 Ceramic Substrates

wires per unit length and can vary from 80 wires per inch used to print solder paste for coarse screening to 400 wires per inch for fine-line printing. The screen mesh is stretched to a prescribed tension over a screen frame that is made from cast aluminum. The bottom of the frame is machined to be parallel to and at a fixed distance from the top surface of the screen because it is important, when screen printing a circuit component, that the screen be equidistance from the substrate over the entire substrate. The screen parameters are an important factor in determining the thickness of the deposition that is printed during the screen-printing process.

Screen printing is the process by which a thick-film paste is applied to a substrate in a desired pattern. The pattern that is screen printed onto a substrate is determined by the pattern on the printing screen. The pattern on the printing screen is formed by exposing a stainless steel mesh screen, which has been coated with a photosensitive material (referred as an emulsion), to ultraviolet (UV) light through a film mask containing the desired pattern, which consists of black areas (which block UV light) and clear areas (which transmit UV light). The thickness of the emulsion on the printing screen must be controlled very accurately and care must be taken to align the film positive (or pattern) to the wire mesh of the screen before the exposure step. The emulsion is hardened by exposure to UV light, so the emulsion on the printing screen under the clear areas of the film mask is hardened. The protected portion of the emulsion on the printing screen, which is not exposed to the UV light, can be removed from the printing screen by simply washing the screen in water at 30 to 40°C (90 to 110°F) for 30 to 60 seconds, leaving openings in the emulsion corresponding to the dark areas in the film mask. A light-pressure water spray is used to wash the emulsion from the pattern area. Then, the screen is allowed to dry thoroughly. If heat is used to accelerate the drying process, it should not exceed 50°C (~120°F). Screen tension must be controlled if fine pattern definition and thickness control are to be maintained (a screen tension gage is used). The screen is then mounted on the screen printer and proper adjustments are made to secure it.

5.7.3 Screen-Printing Process

The screen printer, shown in Figure 5.3, is one of the main pieces of equipment used in thick-film production. The mounting plate on the printer screen must allow easy access to the screen frame for changes and to the bottom of the screen for easy cleaning. The screen mount must hold the screen frame rigidly in place throughout the travel of a squeegee, which is used to force paste (or ink) down through the screen. The substrate holding fixture must

Figure 5.3 A typical low-volume screen printer (*left*) and a mounted screen (*right*).

be capable of accurately locating and registering the substrate, provide a method of rigidly fixing or holding the substrate in a position (vacuum), must wear well, must be easy to replace (to accommodate various substrate sizes), and provide a flat surface. Standard terms associated with the screen-printing process include downstop, squeegee pressure, squeegee speed, attack angle, hydroplaning, and breakaway or snap-off distance. Downstop is a printer feature that limits the maximum up and down travel of the squeegee during the printing cycle. Screen damage may result from an improperly set downstop since this feature prevents the squeegee from putting too much pressure on the screen during the print step. The squeegee pressure simply controls the force applied by the squeegee on the screen. The angle of attack is the angle that the squeegee forms with the plane of the screen. Hydroplaning refers to a condition where the squeegee floats on a thin layer of ink on the screen's surface instead of directly on the screen. As the squeegee moves across the screen, it locally deforms the screen such that the snap off, or distance between the screen and substrate, is reduced to zero. As the squeegee proceeds to the end of its travel, the screen "snaps" back into normal free position.

The squeegee material must be compatible with the pastes. Therefore, neoprene and polyurethane are the most commonly used materials. The hardness of the material should be between 50 and 90 durometer (soft and hard limits). The squeegee angle of attack is normally between 45° to 60°. The exact pressure applied to the squeegee should be determined by experiment (pressure affects line definition and uniformity of the print).

5.7.4 Substrate Cleaning and Process Environment

Ultrasonic cleaning of prefired substrates with an appropriate solvent is advisable to remove grease, oil, or deposited particulate material. Rinsing is done very thoroughly with deionized water. After inspection and cleaning, the substrates should be stored in a clean, dry, and dust-free area. The facility should have a controlled atmosphere area for material preparation and screen printing to maintain cleanliness. Aqueous cleaning of tape materials is seldom performed since these materials are provided in a clean state and susceptible to damage caused by moisture absorption. Tacky adhesive rollers are normally preferred to remove any particulate debris that may arise during processing.

The furnace is usually located outside the screen-printing area to simplify management of the atmosphere control system. The furnace is vented to the outside to eliminate the organic by-products of the burn-off cycle of the firing process. The thick-film facility should be air conditioned, humidity controlled, and dust controlled. For narrow linewidths, dust control in the screening area is an absolute necessity. Temperature control is important because it has an effect on the viscosity of pastes. Humidity control has an effect on the rate of solvent evaporation, as well as the emulsion life of the screens. The room temperature should be maintained at approximately $21°C$ ($\sim 70°F$). Humidity must be controlled at a low level to prevent absorption of moisture by tape ceramic substrate materials. The temperature should be held to $\pm 1°C$ and humidity should be stable to within $\pm 5\%$.

5.7.5 Thick-Film Formulations

Inks or pastes printed and fired on a substrate have some basic features. They contain particles of metals and/or metal oxides, glass (metal oxides mixture), a binder, and a solvent to make the paste fluid in nature. Metallic conductive components are formed from pastes containing one or more precious metals in fine powder form (1 to 10 μm in diameter).

Structural shape and particle morphology are critical parameters that affect the desired electrical characteristics, and controlling these parameters ensures uniformity of the fired film properties.

Films that use a glass (or frit) have a relatively low melting point (on the order 500 to 600°C), but, in general, they pose a problem for subsequent component assembly processes due to the presence of glass on the surface. Chemically, the molten glass may react with any glasses in the substrate. In addition, the glass flows into and between the features on the substrate surface. Both mechanisms result in adhesion of the film to the substrate surface. The glass creates a matrix for the active particles, holding them in contact with each other to promote sintering and to provide electrical passage of current. Common thick-film glasses are based on Ba_2O_3/SiO_2 with modifiers such as PbO, Al_2O_3, Bi_2O_3, ZnO, and BaO, which are used to change the physical characteristics of the film.

In the case of metal-oxide-bonded films, a pure metal is mixed with the paste, and it reacts with oxygen atoms on the surface of the substrate to form an oxide. The conductor adheres to the oxide by sintering during the firing stage. During firing, the oxides react with broken oxygen bonds on the surface of the substrate, resulting in molecular bonded materials with good adhesion. This class of materials usually needs to be fired at 900 to 1000°C. In addition, some materials, referred to as mixed bonded systems, utilize both reactive oxides and glasses, resulting in a material with good properties and a moderate firing temperature.

The organic binder, a nonvolatile organic, serves the purpose of holding the active elements and the adhesion elements in suspension until firing of the film takes place. It also provides a paste with the desired fluid characteristics needed for screen printing. The organic binder, such as ethyl cellulose and various acrylics, starts to burn off at a temperature of about 350°C and has to oxidize completely during firing in order to avoid film contamination.

The solvent, or thinner, which provides the proper flowing paste viscosity, is volatile in nature and may evaporate at about 100°C. Terpineol, butyl carbitol, and complex alcohols are used as solvents.

In addition, some plasticizers may be added to the solvent to modify the thixotropic nature of the paste to promote the printing process. The paste must have a minimum pressure required to produce a flow. The paste must also possess a nonlinear shear rate/shear stress ratio. This means that, as the shear rate increases (due to squeegee pressure and motion), the paste becomes thinner.

All of these ingredients are mixed together in proper proportions and milled for a sufficient amount of time to ensure thorough mixing, distribution, and dispersion within the paste, and to complete the formulation process. The percent solids parameter, 85 to 92% by weight, measures the ratio of the weight of the active and adhesion elements to the total weight of the paste. This parameter must be controlled precisely to produce a good flow of the paste with well-defined lines in the fired film.

5.7.6 Heat Treatment Processes for Pastes

Thick-film components include passive components, such as conductors, resistors, dielectrics (both capacitors and insulators), varistors, filters, couplers, transmission lines, and other components. To achieve these electrical elements with desired electrical properties, thick-film materials are formulated into pastes or inks. These pastes or inks differ in density, viscosity, and solids content, depending on the function they must provide in an electronic circuit. As stated previously, the printing process requires a screen-printing machine and the proper screens for the function(s) being created. Printing a paste onto a substrate is accomplished using a squeegee to press the paste through the openings in the screen. The paste at the bottom of the screen contacts and wets the substrate due to the surface tension of the paste. The substrate pulls the paste through the screen openings when the screen snaps back from the substrate after the squeegee operation.

After depositing the film by screen printing, sufficient time should be allowed for the paste to settle and for the mesh impression in the printed paste to disappear. The settling time varies from 5 to 20 min, depending on the viscosity of the paste. The film is then placed in an oven with circulating air for 15 min at 100 to 150°C to remove the organic solvent. The organic solvent should be removed slowly since rapid evaporation of the solvent may cause voids and blisters to form.

After drying, the particles in the film are bound together with a plasticlike material (ethylcellulose or similar), which is used to bond the film to the substrate and to hold the particles together until the substrate is fired. The film is then fired in a belt or box furnace following a particular firing profile, usually achieved through four to eight heating zones. In the preheat or binder burn-out portion of the furnace, heat decomposes the ethylcellulose, which combines with oxygen in the air to form carbon dioxide and water vapor (both are properly ducted away). The organic binders and solvents are burned out in this first phase of the firing process. The film then is subjected to a temperature of the order of 480 to 650°C to initiate melting of the glass particles in the film. The glass bonds the film to the substrate. Also, the metal particles in the conductor films begin to sinter or join together. At a peak firing temperatures of the order of 850°C for 10 min, the film achieves its desired electrical attributes. Cool down at a specific rate is necessary to avoid stresses in and oxidation of the fired film.

5.7.7 Thick-Film Metallizations

The basic constituents of a thick-film conductor paste are four main ingredients. The first is a conductive metallic phase that typically consists of fine noble-metal powders or alloys. Examples include gold (Au) for high reliability circuits, silver (Ag), silver–palladium (Ag–Pd) in ratios that vary from 1 : 2 to 1 : 12 by weight, palladium–gold (Pd–Au) in a ratio of 1 : 2.5 by weight, platinum–gold (Pt-Au) in a ratio of 1 : 3.5 by weight, and copper (Cu). Aluminum (Al) and nickel (Ni) are also used. The particle sizes range from submicrometer to a few micrometers. The selection of the particle size, distribution, surface chemistry, and shape depends on the interfacing materials, as well as the application. The sheet resistivity, based on a fired thickness of 0.5 mil, varies from 3 to 5 mΩ/square for Au, 12 to 16 mΩ/square for a ratio of 6 Ag to 1 Pd, and 2 to 4 mΩ/square for Cu to 40 to 70 mΩ/square for Ni.

The second ingredient is a binder phase or bonding agent, inorganic in nature, that typically consists of a mixture of glass powders. The third component is an organic, which serves as the carrier agent for the inorganic constituents that provide the proper paste rheology for screen printing. The final ingredient is an organic suspension medium. While noble metals are fired in an air atmosphere, copper firing requires a nitrogen environment.

Thick-film conductor pastes are used to create interconnections, solderable lead and device attachment pads, thick-film resistor terminations, crossover connections, capacitor electrodes, low value resistors, and the like. The factors considered in selecting a conductor composition include ultimate solderability, resistance to leaching, adhesion of the fired films, suitability for wire and chip bonding, compatibility with resistor and dielectric films, and attainable line definition. Resistance to aging effects caused by multiple firing processes is also important.

5.7.8 Thick-Film Dielectrics

Functions of the dielectric pastes include crossover dielectrics and high-k capacitors. A crossover dielectric is a low dielectric constant insulator capable of separating two conductor patterns through several firing steps with good isolation and minimum stray capacitance. The requirements of a crossover dielectric include control of resoftening during the top conductor firing, a low dielectric constant to prevent capacitive coupling between insulated circuit layers, a low electric loss to avoid dielectric heating, a minimum tendency to form pinholes, a high resistance to thermal shock, and a low sensitivity to water vapor. High dielectric constant capacitors are often based on barium titanate. The recommended procedure for high dielectric constant dielectrics is to first print and fire the bottom electrode. A dielectric paste is then screen printed on top of this lower metal electrode, followed by a firing process. The dielectric should overlap the electrode pattern by at lease 10 mils on all sides to eliminate surface leakage. Often, the dielectric is double printed to reduce the probability of pinhole formation. Then, a top electrode is screen printed on top of the dielectric layer, followed by a firing cycle. The top electrode should fall completely within the bottom electrode area. The dielectric constant depends on the dielectric composition, electrode materials, and the firing profile.

Dielectric compositions for insulator fabrication include ceramic-filled glasses (glass and refractory oxides) and crystallizing dielectrics (crystallizable glass and crystallization agents). The raw materials most commonly used in high-k dielectric pastes for capacitor fabrication are barium titanate ($BaTiO_2$), lead titanate ($PbTiO_3$), and lead zirconate titanate (formulated from its oxide constituents). Barium titanate often contains additives of strontium, calcium, or lead. After the initial mixing, the dielectric formulations are heated in powder form to initiate a thermochemical reaction among the constituents for compound formation, to burn-off volatile impurities, and to eliminate both carbon dioxide and water vapor. This is achieved at about 1000°C for about 2 h. Postcalcination grinding is followed by ball milling to small sizes, 1 to 10 µm for screen printing in a carrying vehicle such as deionized water or alcohol, or tetrachloroethylene, and then dried, reducing all compositional inhomogeneity. Raw material purity (99.7% and higher), stoichiometry (excess major cations can affect the temperature and time of film sintering, modifying the microstructure and dielectric properties), and particle size (larger particle sizes can result in low-density fired films possessing large intergranular voids) are important factors in determining the dielectric electrical properties. Commercially available dielectric materials for capacitor fabrication typically have dielectric constant values that vary between 20 and ~12,000. The firing cycle greatly affects the capacitance density and loss factor of the dielectric material and the film adhesion. During the firing process, the nonvolatile portion of the organic vehicle decomposes. The firing process should proceed slowly in order to allow the organic material to burn out or completely decompose prior to the melting of the glass.

5.7.9 Thick-Film Resistors

Resistor pastes contain an insulating glass frit, an electrically conducting powder, an organic agent, and additives to enhance electrical attributes. Resistor materials are classified into organometallic, or resinate, systems and cermet systems, which consist of metal particles in a glass frit. Precious metal resinates are solutions of metal chlorides in organic solvents or organometallic compounds in which the metal atom is attached to an oxygen atom linked to a carbon atom. Cermets are materials resulting from a fused structure of conductive or resistive material in a vitreous nonconductive binder. These resistive elements can include oxides or oxide compounds of indium, thallium, ruthenium, palladium, tungsten, and other noble elements. Resistor properties are microstructurally and stoichiometrically dependent, affecting the sheet resistivity, the temperature coefficient of resistance, the electrical noise, and the stability of values with time under load [40, 41].

Resistor compositions are typically comprised of a resistive component, a glassy phase, an organic suspension medium, and an organic diluent. Compositions are available that yield postfired sheet resistivities from 1 Ω/square to 5 MΩ/square, with the resistive component being palladium–palladium oxide–silver compositions, mixtures of precious metals other than silver and palladium, ruthenium oxide, thallium oxide, indium oxide, tin oxide, tungsten–tungsten carbides, or tantalum–tantalum nitride. Some of the important electrical properties of thick-film resistors are the temperature coefficient of resistance (TCR), the voltage coefficient of resistance (VCR), the long-term drift under temperature and load, and electrical noise.

The conduction process in thick-film resistors is extremely complex. The active materials behave like metals to a certain extent. A metal typically has a positive TCR. The glass behaves basically like an insulator. In an insulator, TCR is slightly positive. However, in very thin layers, when some of the active material has been dissolved in the glass during the firing process, the glass behaves like a semiconductor with a negative TCR. To a very large extend, controlling the amount of active material controls the resistance of a paste. The more active material, the lower the value of sheet resistance. Pastes with lower sheet resistivity tend to take on metallic characteristics and exhibit a positive TCR. Also, thin resistors tend to have a more negative TCR than thick resistors. Short and narrow resistors have a more positive TCR than long, wide resistors due to diffusion of the conductor material.

5.8 LOW-TEMPERATURE COFIRED CERAMICS (LTCC)

While many applications exist for traditional thick films and thin films on ceramics substrates, some applications benefit from a multilayer implementation. This section describes one approach to the fabrication of multilayer ceramic substrates and components.

5.8.1 LTCC Technology

In the late 1970s and early 1980s, a number of applications began to demand very high interconnect densities in high-reliability substrates. Conventional thick film on prefired substrates was hard pressed to meet these requirements. As a result, a new technology, low-temperature cofired ceramic, was developed that offered many of the advantages of thick film with the capability to produce much higher layer counts and, therefore, higher circuit densities. With the standard print and fire thick film or multilayer thin film, the yield of a substrate drops sharply as the layer count increases. This is due to the fact that each layer is added on top of previous ones so that an error on the third layer destroys any progress made on the initial two layers, resulting in a scrapped substrate [42–44]. In contrast, LTCC, like high-temperature cofired ceramics (HTCC), allows each layer in a substrate to be fabricated separately and inspected so that the total substrate yield is not as heavily affected by the layer count. The key advancement in LTCC is the lower firing temperatures that allow it to utilize noble conductors that have melting points far too low for inclusion in HTCC substrates. These conductors are far higher in electrical conductivity, and therefore, the electrical performance of LTCC packages is generally superior to HTCC products.

Some of the advantages of this technology include:

- Conductor thicknesses of 10 μm (about three times the skin depth at 1 GHz) can be achieved with high conductivity metals.
- Inductors and resonators with high-quality factors can be realized for radio frequency (RF) applications.
- Very fine structures can be realized (line resolution), and numerous ceramic layers can be stacked. Thus, higher interconnect and packaging densities are possible [45, 46].
- The ability to postprocess the substrates using thick-film technology, or thin-film technology on the top and/or bottom surface for greater interconnect density.
- Parallel processing allows careful inspection of individual layers, therefore, the yield is higher and the process is more cost effect.
- The process is hermetic by nature.
- Embedded (or buried) passives are possible.

Low-temperature cofired ceramic is a technology that possesses advantages of both conventional thick-film technology and high-temperature, cofired ceramic systems, as summarized in Table 5.3. These include high conductivity metal usage (noble metals), low-k dielectrics, low processing temperatures (around 850°C), low capital investment, high print resolution of conductors, single firing, good dielectric thickness control, low surface roughness, and a virtually unlimited number of ceramic layers.

Two basic types of dielectric materials are used in LTCC fabrication, alumina glass composites and crystallizable glasses. In alumina and glass ceramics, such as DuPont Green Tape™ 951, the glass softens and wets the alumina powder during firing, providing a dense hermetic structure. Most of the critical parameters of the fired composition, such as firing temperature, coefficient of thermal expansion (CTE), mechanical strength, and thermal

Table 5.3 Comparison of Thick-Film Technology and High-Temperature Cofired Ceramic Technology

	Thick film	High-temperature cofired ceramic
Dielectric	Paste composition glass + fillers	Tape 90–96% Al_2O_3
Dielectric constant	7–10	9.5
Conductor metallurgy	Ag, Au, Pd/Ag Cu	W, Mo, Mo–Mn
Sheet resistance mΩ/sq @ 12.5 μm thickness	2–4	16–30
Firing	800–950°C 45–60 min Air (precious metals) N_2 (Cu)	>1500°C >24 h H_2 atmosphere
Vias	Screen printing 0.010 in. diameter 0.025 in. center	Punching 0.005 in. diameter 0.010 in. center
Line resolution Line/spaces	0.006/0.006 in.	0.004/0.004 in.
Multilayer processing	Sequential, many steps	Cofired, fewer steps
Substrate required	Yes	No
Lead attachment	Direct	Plating required
Resistor compatibility	Yes	No
Capital investment	Low	High

conductivity, are dictated by the nature and properties of the glass. Typically, the fired properties are designed to match the CTE of standard alumina ceramics or silicon devices and to have dielectric constant from 6 to 9. When the glass softens during firing, it allows the dielectric to conform to the setter, the structure on which it is fired, producing a flat finished part.

One example of a glass–ceramic or crystallizable material is magnesium aluminum silicate, such as Ferro A6. These materials are generally less sensitive to the effects of multiple firings, such as postfiring, but the initial firing process is more critical than for alumina glass composites. Also, the ceramic does not soften as much, when fired, as alumina-filled glass parts. The most significant feature of the glass–ceramic system is its stability when fired in nitrogen for use with copper conductors.

The process used to fabricate an LTCC substrate involves thick films printed onto unfired ceramic sheets that are made flexible through the use of polymeric additives. These unfired, or green sheets, are flexible in nature much like thick wax paper and are analogous to the unfired clay used in pottery that becomes the final clay pot. The process consists of blanking and registration, via formation, via filling, conductor printing, inspection, collating and registration, lamination, cutting, burnout and firing, and postfired processing, as shown in Figure. 5.4.

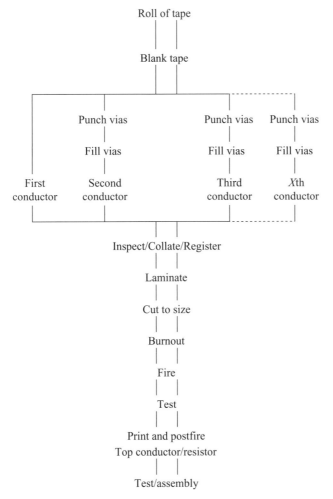

Figure 5.4 Cofired ceramic process flow.

5.8.2 Tape Handling and Cleanroom Environment

Since each tape layer in a given substrate is processed separately, it is critical that the individual layers be kept as clean as possible. Normally, this involves a series of measures including operator protection and the use of a clean room environment for all process operations. The clean rooms used for this technology normally are rated to class 1000 or class 10,000. These designations indicate the maximum number of particles larger than 0.5 μm that are allowed per cubic foot of air. These levels are not as stringent as the class 10 and class 1 cleanrooms used in semiconductor fabrication facilities since the feature sizes used in ceramic substrates are larger.

Unlike prefired ceramic substrates, tape substrates are flexible and must be supported during process operations, such as punching, filling, and printing. This is normally done in one of two ways, either using frames or plastic sheet backing. Framing has been the traditional approach to the problem and involves the use of thin stainless steel sheets, which are formed in the shape of a window frame. An individual tape layer is then attached to the frame with adhesive, as shown in Figure 5.5. This approach works well for tape layers with thicknesses greater than 75 μm (3 mils). However, for thin tape layers and/or large panel sizes, the layers are fragile and tend to tear. As a result, many operations are moving toward the use of plastic sheets as a backing material. Some of the newer tape systems offer plastic sheet backings, which can be easily punched and serve as a support for the tape layer throughout the process sequence.

Many of the processes used in fabricating a substrate, such as the punching or cutting of the tape, tend to generate particulate matter. As a result, great care must be taken to ensure that debris does not become incorporated into a final substrate. Two methods are generally used to accomplish this, removal of the waste tape at the source and the use of adhesive rollers to clean the tape layers after each process step. For example, most punching systems

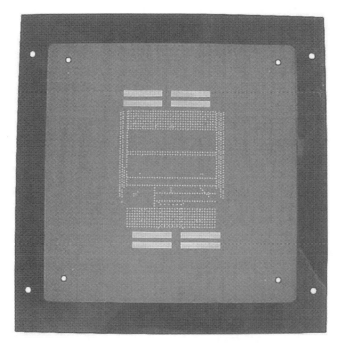

Figure 5.5 Layer of green LTCC mounted to a stainless steel frame.

incorporate vacuum systems that extract the waste tape particles, such as via slugs, to a sealed storage container. These vacuum systems are effective at removing the large particles and waste tape material but may leave small particulate materials behind. A subsequent pass with an adhesive roller will remove the fine particles and leave a clean surface. These rollers are tacky enough to adhere particles to their surface without damaging the tape or forming a permanent bond to the tape layer.

5.8.3 Via Formation

A key aspect of LTCC substrate fabrication is via formation. The vias serve as electrical and/or thermal interconnections between the different layers of a substrate. Since a large number of layers may be required for any given design, and thousands of vias are typically used for each layer, this process must be highly automated. For low-volume production and prototyping, computer numerically controlled (CNC) punching or laser systems are normally used. For higher volume applications, custom die and/or gang punch configurations may be used in order to speed up the process. The two basic mechanisms used for this process are direct punching with a punch and die or laser cutting. Punching has replaced laser cutting as the primary cutting technique for vias, while lasers still find applications in cutting of complex cavity structures that are time consuming to punch.

5.8.3.1 Laser-Cut Vias

Laser cutting systems usually employ a laser head that is moved into position above a fixed blank tape of LTCC. Computer-driven stepper motors to aim the laser point at a given location on the tape surface control the laser. A shutter also allows the computer system to turn the laser system on and off in order to cut lines, vias, or other complex shapes. These systems are normally based on powerful CO_2 or Nd:YAG lasers. The flexibility of this technique has significant merit in that not only can large numbers of small vias be created but complex cavities can be formed in a fraction of the time required by a punch system. These systems also offer significant capabilities for premachining a substrate's outer dimensions prior to firing because the laser can effectively trace out complex shapes, including circular substrate geometries that are difficult to achieve with postmachining techniques.

5.8.3.2 Punched Vias

The most popular technique for via formation in tape ceramics is CNC punching. Key advantages are extremely smooth and clean vias, high speed, and the ability to use several punches simultaneously to increase the overall process speed.

Via punching systems are simple in concept and operate much like a paper hole punch on a much finer scale. As illustrated in Figure 5.6, a metal punch and die are aligned to each other on opposite sides of a tape layer being punched. In most cases, the punch and die are held in fixed positions and the tape layer is moved in the x and y axes in order to position the punch and die at desired coordinates. The punch is then forced through the tape and into the die. The result is a slug of tape material exiting the hollow die and a via being formed in the tape layer. This whole process normally occurs in approximately one-tenth of a second for commonly available commercial punch systems. The tape layer is then moved to the next desired via location and the process repeats. The slugs are small pieces of the LTCC material that are removed to create the via, not unlike the small round pieces of paper

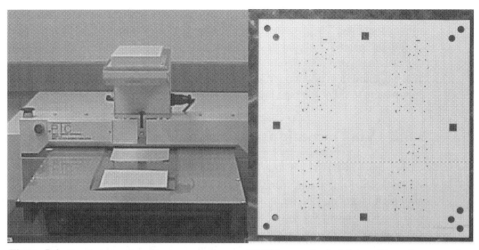

Figure 5.6 Punch system that aligns a punch above an LTCC layer and a die below and is used to create via holes in LTCC. On the left is a 3-in.-square LTCC layer that has been punched.

created by a paper hole punch. Just as with a paper punch, the slug material is removed and collected by the system as waste. This process allows for very large numbers of vias to be created in a short time frame. In addition, cavities and slots can be created by spacing multiple punches adjacent to one another in order to punch the outline of the desired slot or cavity opening.

Punches are usually made from carbide steel since the tape material is very soft and easy to fracture, but the individual particles of which the tape is composed are quite abrasive. Both round and square geometries are available in sizes ranging from 50 μm to several millimeters. The die is normally sized slightly larger than the punch by ∼10 to 25 μm for 150-μm via diameters. This clearance created between the two allows for slight misalignment of the two and/or any punch deflection that may occur. The alignment between the punch and the die is absolutely critical to maintain the close tolerance and avoid damage to the die and/or punch. While the carbide steel is quite hard, it is also rather brittle, and the most common failure mode is snapping of the punch, followed distantly by damage to the die opening. The mere handling and loading of very small punches can be quite challenging to avoid undue damage or strain. Figure 5.7 illustrates a typical punch and mounting holder. The holder supports the punch and the ball bearing surfaces allow easy extension for punching the vias in the desired locations.

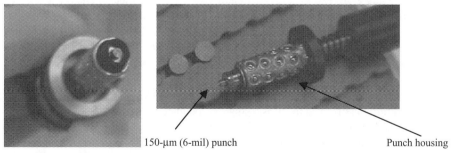

150-μm (6-mil) punch   Punch housing

Figure 5.7 Typical punch and mounting holder. Large 2.36-mm (93-mil) alignment punch (*left*) is used to create registration holes in tape layers, while much smaller punches (*right*) are used to create vias.

Figure 5.8 Typical die and holder. Insert shows picture of a 150- and 100-μm dies. Note the 150-μm punch visible in the bottom left corner of the insert.

A typical die and holder are illustrated in Figure 5.8. Note that the die and holder are hollow and serve as the conduit for extracting the undesired slug material from the system. Most punching systems allow for the simultaneous loading of several punches into the system at a given time. While it may only use one punch size or configuration at a time, the machine can instantly switch between the punches on the fly within a fraction of a second. A common approach is to use several small via punches in combination with a square punch and a larger registration punch. A typical system running a set of two small via punches may use a 100-μm (4 mils) punch and a 150-μm (6 mils) punch, or a combination of 150-μm (6 mils) and 200-μm (8 mils) punches. This allows the designer more versatility in the design process since two sizes of signal and ground vias are available. The square punch is commonly used to create cavities and slots, which may be used to recess other components in the final fired substrate body. Finally, registration punches are used to create registration marks that may be used in subsequent process operations. These registration punches are normally much larger in size, for example, 2.3 mm (93 mils) is a common size.

5.8.4 Via Fill

Once the vias are formed, they are merely small openings in the individual tape layers. To be useful as electrical interconnects, they must be filled with metal to provide a conductive path between the two layers. This is accomplished using stencil printing or a bladder fill operation, which effectively injects conductive ink into the via openings. The ink is then dried to remove the solvents. The result is a solid metal-filled via. An image of a via prior to and after via fill is shown in Figure 5.9. All of the vias in a given layer are filled in one rapid operation. It is important to note that the quality of the filling process is important to the overall signal integrity of the final substrate. In some cases, vias may be overfilled or underfilled. Both of these conditions are undesirable since underfilled vias may result in open connections and overfilled vias may lead to electrical shorts or excessive topography in the final fired substrate.

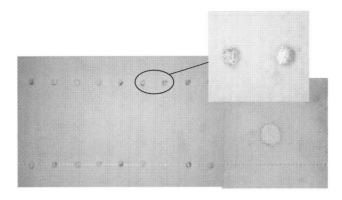

Figure 5.9 Image of a via prior to (*left and top*) and after via fill (*bottom right*).

5.8.4.1 Stencil-Filled Vias

Perhaps the most popular way to fill vias is by using a stencil printing operation. This approach uses a metal foil suspended on a metal frame, as shown in Figure 5.10. The foil is etched or cut to produce an area of small openings at the desired via locations. This foil is commonly referred to as a stencil and is used in a screen-printing operation to force ink through the stencil openings and into the vias in a given tape layer. Just as in the case of punch files, a different stencil is required for each of the unique layers in a given design. The advantages of this approach are that it can be highly automated and requires only a small amount of paste at a given time.

5.8.4.2 Bladder-Filled Vias

An alternative via fill approach is a bladder fill operation. This equipment also uses a stainless steel stencil but without the frame. The metal foil is identical in configuration except that it also includes a number of tooling and registration features that are used to align it to the bladder fill machine and the tape layer to be filled. A relatively large quantity of the desired ink is then placed on top of the stencil and spread out to cover the entire design area. A rubber bladder is placed on top of the ink, and this whole assembly is then loaded into the equipment. In use, the operator places a tape layer on a vacuum stage, and the machine then aligns the stencil bladder assembly with this tape layer such that the holes

Figure 5.10 Framed stainless steel stencil used for filling vias in LTCC with a screen-printing system.

in the stencil are directly above and in contact with the open vias. Air pressure is then applied to the back of the rubber bladder, forcing the ink through the stencil openings and into the vias. By adjusting the amount of air pressure and the duration of its application, the amount of material injected into the vias can be controlled. Once the operation is complete, the tape layer is removed and another identical tape layer can be filled in the same way. This process is repeated until the ink between the stencil and bladder is consumed. As in the case of stencil printing, a different stencil is required for each unique layer of a design, and the stencils must be individually loaded into the system one at a time. In practice, a large number of layers can be filled with one application of ink and loading of the stencil bladder assembly, assuming that they are all the same layer in a given substrate design. The system is then reloaded with the appropriate stencil for the next layer, and a large number of pieces of tape with that given design are filled. This process of reloading the system with a unique stencil would normally be repeated six times for a six-layer substrate design in order to produce a set of substrates.

The benefit of this approach is that it offers significant control over the injection process. However, it requires the application of a relatively large amount of paste/ink to the stencil for even a small number of filling operations.

5.8.5 Screen-Printing Considerations for Tape Materials

Significant adhesive forces exist between a tape layer and the bottom of a screen during the printing process. This is due to the sticky nature of the inks used in thick-film printing. In addition, the tape layer must be fixed in a rigid position to prevent warping of the material or shifting of its position. For prefired ceramics, a vacuum chuck is the normal solution to these problems. However, tape layers are deformed by the small number of vacuum openings in most chucks. Therefore, most systems designed for printing tape layers use porous stone vacuum chucks to support and hold the tape layers flat during the printing process. Figure 5.11 illustrates a typical printer and chuck designed for a low-volume LTCC fabrication facility.

Figure 5.11 Typical printer fixture designed for a low-volume LTCC fabrication facility.

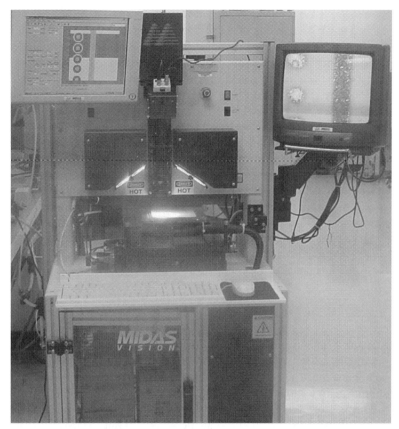

Figure 5.12 Photograph of an automatic inspection system used for multilayer ceramic production.

5.8.6 Inspection

A positive attribute of the LTCC process is the fact that it is fundamentally parallel. Each layer in a given substrate is fabricated separately and can be inspected prior to committing a given tape layer to a final substrate. Usually, an inspection process follows all of the major process steps to ensure that minimal effort is expended on defective layers and that all of the layers in a given substrate meet the design specifications. To achieve this objective, automated inspection systems are used to verify the tape features versus the design. Normally, these systems store the design information in the form of a drill fill or a Gerber file and compare each layer in the design for open or missing vias, as well as printing errors. The importance of this inspection process cannot be overstressed since each layer in a design may contain thousands of vias and a similar number of printed interconnects. Figure 5.12 illustrates an automatic inspection system designed to optically test vias in a unfired LTCC layer prior to subsequent processing.

5.8.7 Tape Layer Collation

All of the above processes used to fabricate an LTCC substrate are performed in parallel, as noted previously. At some point, the layers must be sorted and stacked to form the required internal substrate interconnections. This process is normally referred to as layer collation. The key issues associated with this process are the correct ordering of the layers, as well as the alignment of the layers to each other. The ordering requirement is obvious, but the complexity of the alignment problem may not be. For example, for designs based on 200-mm tape blanks, a typical worst-case layer-to-layer alignment of 50μm is required to create the desired electrical interconnects. This corresponds to 0.025% of the blanks overall dimension. In practice, most systems achieve 25μm or better alignment by aligning each layer to the immediately adjacent layer and tacking it in place. A complete substrate sequence would normally be alignment of layers 1 and 2 and then tacking them together, alignment of layer 3 to the stacked combination of layers 1 and 2 and then tacking them together again, and so forth. In this way, each layer is carefully aligned to the next one and held in place for subsequent processing with this "stack-and-tack" process.

The alignment process is normally done with machine vision. These vision systems align features on the tape layers, punched or printed registration marks on each layer, to the next tape layer and automatically rotate or translate a layer to produce the required alignment. For low-volume production, machines exist that allow for the manual alignment of layers to each other with the assistance of video cameras, which magnify and show the relative position of the alignment marks.

The tacking process is normally produced with heat and pressure, which locally bonds the layers to each other. This process simply holds the registration of the layers until the substrate is laminated. Some equipment manufactures utilize heated posts to effectively spot weld the layers together in a number of locations around the perimeter of the substrate. Others utilize heated bars to seam "weld" the edges of the substrate to provide a continuous seal around the substrate, which ensures an effective bond and, therefore, the alignment of the layers. Typically, temperatures in the range of 50 to 90°C and pressures of 100 to 200 psi are used for this process, but the parameters may vary considerably depending on the composition of the tape. All of these processes only bond the substrate layers at the edges, since this bonding process is destructive due to local distortion of the tape, and the bonded areas are normally cut away from the finished product.

Solvent welding can also be employed to tack the layers together. This approach is popular for laboratory-scale operations due to its simplicity. In this case, small drops of a suitable solvent are placed around the edges of layers, and, when pressed together, the solvent effectively melts the layers together, thereby providing the required bond.

An alternative method for collating the layers is the use of a lamination fixture, which incorporates mechanical pins to hold the layers in place. Registration holes are provided in each layer during the punching or blanking stages, and each tape layer is aligned to this set of pins with these registration holes. Since all the holes are located in the same place and the pins are fixed in position and precisely ground to size, adequate layer-to-layer alignment can be established for many applications. This process is currently falling out of favor in many facilities due to its limitations. Key issues are the tearing of tape layers caused by stress concentration at the registration hole sites, as well as the limitations in tape stretching, which, in combination, effectively limits the layer-to-layer alignment to the 50- to 100-μm range. In addition, the increasing use of thinner tape materials exacerbates these problems.

5.8.8 Lamination

Once the layers required for a given substrate are collated and tacked together, a more intimate bond is required to ensure substrate integrity after the final firing process. To perform this step, heat and pressure are used to bond the polymers in each LTCC layer to one another. Two primary approaches are isostatic lamination and uniaxial lamination.

Figure 5.13 Low-volume system designed to collate and align LTCC substrates prior to lamination and firing.

Uniaxial lamination is based on a simple hot-press concept. Two heated platens are used to apply the required force to the collated substrate layers. The process conditions may vary significantly depending on the nature of the substrate and the tape materials used, but typical values are 70°C and 3000 psi. In many cases, a fixture is used to support the collated substrate layers during this process. This fixture may be the same structure used for a mechanical alignment process during collation or a dedicated lamination fixture. This process has the advantage of being easily integrated into a high-volume assembly line since layers can be collated, tacked, and laminated in one rapid operation. The disadvantage is the difficulty in achieving completely uniform pressure across large panel sizes. Any nonplanarity in the platens creates unequal pressure across the part, which may impact the uniformity of the substrate shrinkage since shrinkage is, in part, determined by the lamination pressure. Figure 5.13 illustrates a low-volume system designed for collating and tacking LTCC substrates.

Isostatic lamination is an alternative to uniaxial presses for lamination of LTCC substrates. In this case, the part to be laminated is vacuum bagged to create a watertight seal. The entire substrate is then immersed in water, which is heated to the desired processing temperature in a sealed pressure vessel. Pressure is then applied to the substrate by pressurizing the vessel to the desired value, for example, 1000 to 3000 psi. This process is considerably more complicated than unaxial lamination but has several key advantages, including superior temperature and pressure uniformity. However, a disadvantage is the complexity of this process in a high-volume production facility. The vacuum seal used to protect the substrate must be very high quality, otherwise leaks may occur that destroy the substrate material.

5.8.9 Firing

While the next step in the production of an LTCC substrate is often referred to as firing, the process actually includes several phases, which achieve a number of important transformations, all within a fixed temperature and time profile. It is important to consider the substrate properties prior to and after firing to understand the key issues of this process. Prior to firing, the substrate is a combination of layers, each of which is composed of metal traces and vias, as well as unfired LTCC material. At this point, the materials are actually a combination of polymeric "carriers" and the inorganic materials, which make up the final product. The polymers give the tape material its flexibility and give the inks the viscosity desired for printing. However, these materials are not desired in the final substrate and must be removed. This process is often referred to as burnout and involves heating the laminated substrate to a temperature at which the polymers and solvents in the substrate decompose and evolve out of the substrate as gases. This process normally occurs at 300 to 400°C. What is left behind is the inorganic materials that compose the final substrate. By continuing to raise the temperature to 800 to 900°C, the ceramic and/or glasses within the substrate sinter and form a dense ceramic body that is the final ceramic substrate.

Figure 5.14 illustrates both a box and a belt furnace used to fire LTCC substrates. Belt furnaces are the preferred approach for high-volume production simply because of their ability to allow the continuous throughput of substrates. A typical furnace consists of a long tube through which the parts being fired move on a conveyor belt. This belt forms a continuous loop and carries the components through the furnace, discharges them at the output of the furnace, and then loops back around to pickup a new set of substrates. The belt is normally a mesh belt fabricated in a continuous loop from a refractory metal capable of withstanding the ~1000°C furnace temperatures. The heat portion of the furnace normally contains a number of zones, each of which is held at a fixed temperature. By moving the parts being fired through the different zones at some controlled rate, a firing profile, or temperature versus time profile, may be precisely controlled. The furnace tube normally contains several zones of ventilation to remove volatile materials, which evolve from the fired materials during different periods of the process. In addition, careful design of the airflow allows a more uniform temperature distribution within the furnace. In general, a key challenge with a belt furnace is temperature uniformity across the belt, particularly for large belt widths, which are often 12 in. or larger. A lack of uniformity of this temperature profile can lead to portions of a large panel shrinking differentially from side to side.

In contrast, box furnaces are not generally optimized for high-volume production since they are fundamentally batch-processing systems. In general, a box furnace profile may require a full 8-h shift for a set of parts to complete the firing cycle. The positive aspect of box furnaces is their far superior temperature uniformity and, in many cases, improved temperature and ventilation control. For this reason, box furnaces are preferred for low-volume production, research activities, or complex material sets that are sensitive to small changes in the firing profile.

176 Chapter 5 Ceramic Substrates

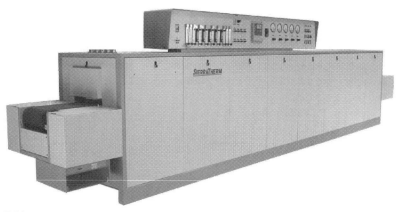

Figure 5.14 (*Top*) Box furnace used for low-volume LTCC production. (*Bottom*) Nitrogen atmosphere belt furnace used for an 850°C copper firing process. (Courtesy Robert A. Station, Sierratherm Production Furnaces.)

A major difference between prefired ceramics and tape ceramics is the fact that tape ceramic materials conform, to some degree, to the surface on which they are sintered. Therefore, the surface roughness and the composition of the surface is an important aspect of properly firing a multilayer tape ceramic substrate. During the firing process, a setter is used to support the tape substrate material while the sintering process takes place. Setters are simple flat plates made from refractory materials. Key properties of these setters are the surface roughness, a lack of adhesion or reaction with the tape and metal system being used, and, in some cases, the thermal conductivity. As described above, the surface roughness is important since irregularities in the setter are picked up, to some degree, by the ceramic substrate. As a result, in general, a smooth defect-free surface is desired. Since different tape systems undergo different chemical reactions during firing, care must be taken to ensure that the setter material is compatible with the tape system. For an inappropriate setter, materials may react with the fired substrate, and, in severe cases, form a bond between the substrate and the setter, resulting in the loss of both the setter and the substrates(s). The manufacturer of the tape system in use can normally provide recommendations as to the optimum setter material. Finally, in many cases, the thermal conductivity of the setter may play a role in the uniformity of the firing profile and the effective heating and cooling rates that the substrate experiences. In general, this profile is more critical for crystallizing tape systems than for glass alumina systems.

5.8.10 Postprocessing

In many cases, a fired LTCC panel does not constitute a finished substrate product and may require several process steps prior to final testing and assembly. The two primary postprocess considerations are machining of the substrate and postfired thick-film formulations that may be applied.

5.8.10.1 Postfired Materials

Postfired thick films allow for the inclusion of materials in a substrate that are not compatible with the initial sintering process required to form the substrate. As described below, these inks offer additional advantages for a number of applications.

For a metallization to be fired on or in an LTCC system during the initial firing process, it must be compatible with the sintering process that occurs within the tape body itself. In many cases, this places restrictions on the nature of the metallization, eliminating or reducing key attributes, such as the adhesion of wire bonds to the metal, as well as solder wetting to the metal surface. For this reason, a number of manufacturers offer postfired metallization, which can be applied to the top and the bottom surfaces of a fired LTCC substrate and then fired. The most common compositions are metals designed for wire bonding with either fine gold wire or heavy-gage aluminum wire, as well as compositions designed for solder attachment. These compositions are either gold, silver–palladium alloys, or platinum alloys, and are tailored to the specific attachment or bonding need. The materials are screen printed on the top and/or bottom of the fired LTCC panel, and, then, the panel is fired at the optimum profile for the metallization system in question. Via connections to the internal circuitry are created by printing the postfired metallization on top of the via metallization protruding from the surface of the panel. The firing process creates an electrical connection between the postfired metallization traces and the vias.

Historically, silver has not been used for military and aerospace applications due to concerns about electromigration at high temperatures and/or voltages. However, silver is far less expensive than gold or platinum. Therefore, it is highly desirable for many applications. While applications may be able to utilize silver throughout the substrate construction, an alternative is to use the so-called mixed-metal systems that allow incorporation of silver within the ceramic body and gold metallization on the exposed surfaces. Since silver is contained within the ceramic, concerns regarding electromigration and/or tarnishing of the metal are largely eliminated. These mixed-metal systems are normally composed of inner and outer ink compositions, as well as interface inks that are used between the silver and gold metallizations.

5.8.10.2 Substrate Machining

A positive attribute of LTCC manufacturing is that the material can be machined to final shape while still in a green state. Premachining is often performed with a hot knife that utilizes a heated blade and a substrate heater to easily cut through the polymer matrix of the green unfired material. While the ease of this operation is attractive, the accuracy is unacceptable in some cases because of uncertainty in the shrinkage of the substrate. In many cases, the error in the substrate shrinkage may be on the order of ±0.2% of the overall substrate dimensions. For a 100-mm premachined part, this error would result in the possibility of as much as 0.2-mm tolerance in the overall parts dimensions. However, for a 10-mm part that was premachined, the error would only be 20 µm and may be negligible. For this reason, some applications may require postmachining processes that allow for much tighter tolerance than would be expected from premachined components. Two primary methods are used to postmachine LTCC substrates: laser cutting systems and diamond saws.

Diamond saws are in common use for a variety of cutting tasks, from dicing silicon wafers to cutting prefired ceramics. These machines are equipped with a diamond impregnated blade and a cooling system that floods the substrate being cut, as well as the blade, with coolant during the cutting process. These machines are normally equipped with vision alignment systems and perform very accurate cuts with uncertainties in position on the order of 10 µm or less, in some cases. The limitations of this technique are that only straight lines can be cut and the process is often time consuming. A diamond blade is required because the ceramics that make up common LTCC systems are very hard and abrasive and quickly destroy a blade composed of a softer material.

An alternative to diamond saws for postmachining of substrates is a laser system. This approach offers the advantage of cutting small holes, as well as complex circular and/or irregular geometries that are difficult to achieve with a conventional saw. A variety of systems exist, but the most common ones are based on CO_2 lasers. Some systems utilize a programmable stage to move the workpiece with a fixed laser position, while other systems hold the substrate in a fixed system and use mirrors to move the laser to the desired location. By moving the laser or substrate and selectively turning the on and off, very complex cuts can be made rapidly.

5.8.11 Design Considerations

To maximize the overall yield of a product and facilitate the production process, a number of design-related issues should be considered during the planning stages. These considerations include design rules, which are normally provided by the fabrication facility, that include limitations on the via size, pitch, and conductor geometries. In addition, care should be taken to consider the process technology and avoid potential problems by balancing metal loading within the part and correcting for shrinkage. These issues are described in detail in the following sections.

5.8.11.1 Metal Loading

A key issue in the design of LTCC substrates is the effect of metal loading on the shrinkage of the substrate. Metal loading refers to the density of the metal interconnects, as well as any ground planes within the substrate structure. In general, metal loading decreases the shrinkage, and, if not balanced throughout the part, may result in different values of shrinkage in various areas of the substrate. The impact of this differential shrinkage usually manifests itself as warping of the substrate and/or a nonuniform final shape. For example, substrates that are intended to have a final square shape may appear slightly trapezoidal if the metal loading is not balanced across the part. This problem can easily be avoided by

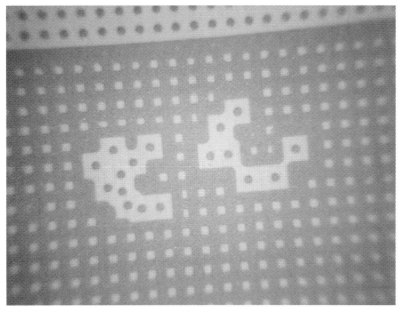

Figure 5.15 Meshed ground plane printed on an LTCC layer.

simply balancing the design across the surface of the substrate, as well as vertically among the substrate layers. For example, if a design requires a ground plane on the bottom surface, an equivalent amount of metal should be applied on one of the top layers of the substrate. Meshed ground planes may also be used to decrease the density of metal loading throughout a substrate, as illustrated in Figure 5.15.

5.8.12 Shrinkage Prediction and Control

Since the starting material used to create LTCC substrates is a combination of polymeric materials and glass ceramic powders, the final dimensions of the substrate are not those of the laminated part. This is due to the fact that the polymers are removed during firing and the glass–ceramic powders sinter together to increase the overall density of the final substrate. The net effect is that the initial panels of LTCC start out at a given size and shrink in the x and y directions, as well as in thickness (z direction). The amount of shrinkage is on the order of 10% but varies with the tape manufacturer since it is a strong function of the tape composition and process conditions used to create the substrate. A number of process factors, including firing profiles and lamination conditions, can have a significant impact on the overall shrinkage of the final part. However, the substrate design also affects shrinkage since the amount of metal loading in the substrate decreases or increases the shrinkage slightly.

In general, this issue is not normally a major problem since it can be corrected at the artwork stage by simply scaling up the artwork used to create the substrates by the shrinkage factor. The key then becomes one of controlling the shrinkage to a precise degree, normally 0.2 to 0.3%. This control strategy often includes the use of an initial shrinkage test part that establishes the shrinkage of a design for a given material set and process conditions. This design can then be produced in volume by controlling the material properties and process conditions.

5.9 HTCC FABRICATION PROCESS

Another technology that has been used for many years to create electronic components and packages is high-temperature cofired ceramics (HTCC). This section outlines the materials and processes used in this technology.

5.9.1 HTCC Process

The HTCC process has much in common with the LTCC process, and, in fact, LTCC is a refinement of the initial HTCC technology. The principle difference, as the name implies, is the firing temperature, where low refers to the 850°C firing that is typical of LTCC, and high means temperatures as high as $\sim$1600°C for HTCC. This difference may appear as semantics; however, the higher temperatures used in HTCC have a number of key ramifications, which include the need for refractory metals, as well as the ability to create relatively pure ceramic substrates without the need for large percentages of sintering agents. Refractory metals are required since they must not melt during the firing process. Common metals used in these technologies include tungsten, molybdenum, and moly-manganese. This restriction results in higher electrical losses in HTCC substrates due to the poor electrical conductivity of these metals. This weakness in HTCC technology is also a strength in the sense that a relatively pure multilayer alumina (92 to 94%) or aluminum nitride substrate can be produced with this technology since the high firing temperatures are ideal for sintering these materials. This purity results in superior strength and thermal conductivity when compared to LTCC materials.

The process for HTCC is virtually identical to LTCC except that the materials used are different and the firing profile and conditions are different. All the other processes, such as printing, punching, and lamination, are still required. In general, HTCC materials are fired in a hydrogen atmosphere to prevent oxidation of the metal conductors, which can easily occur at these high temperatures. Special measures are required for these firing ovens since hydrogen is explosive.

5.9.2 Multilayer AlN

While HTCC substrates are superior to LTCC substrates in terms of thermal conductivity, a number of applications stretch the limitations of both technologies due to high levels of power dissipation. This trend has been growing in the past few years, and a need exists for multilayer substrates with the ability to remove very large amounts of heat. As a result, multilayer AlN technology has been developed that allows the creation of HTCC-like substrates with a higher thermal conductivity. Since HTCC alumina normally exhibits a thermal conductivity of $\sim$20 W/m-k, and AlN is almost 10 times that at $\sim$170 W/m-K, far superior high-power packages can be fabricated with this technology.

5.10 HIGH-CURRENT SUBSTRATES

For power electronics and high-power RF systems, it is not at all unusual for very high currents and voltages to exist in a small substrate area. This creates two primary challenges for the packaging engineer: minimizing the electrical losses and thermal management. Large currents demand low-resistivity materials since very large power losses—and therefore heat—are generated in lossy conductors. Likewise, high-power systems are accompanied by high-power dissipation and high operating temperatures in most cases. As a result, ceramics with thick high-conductivity metallizations are the norm for these types of applications. Ceramics make the most sense for these applications due to their high thermal conductivity and stability at high operating temperatures. The following sections outline the common methods used to meet these challenges.

5.10.1 Direct Bonded Copper Process

For many power systems, hundreds of amperes of current may flow through a substrate in a relatively small area. To prevent severe losses in the conductors, the metallization must be very thick and low in resistivity. One approach to this problem is direct bonded copper (DBC), which was developed by General Electric in the mid-1970s. Unlike thick-film or thin-film conductors, DBC can be purchased with metal thicknesses up to 0.65 mm (25 mils). Combined with the low resistivity of copper, 0.12 mΩ/□ and a high thermal conductivity substrate, such as AlN, this approach creates nearly the ideal substrate for this type of application.

The DBC process begins with either an alumina, beryllia, or AlN substrate upon which a Cu foil of the desired thickness is placed. This assembly is then heated to $\sim$1065°C in the presence of a controlled amount of oxygen. Under these conditions, a bond is formed between the copper foil and the substrate, which is very strong and is stable up to 850 to 900°C. In the case of alumina, a bond is formed between the copper and the aluminum oxide as a eutectic alloy at a 1.6 atomic percentage of O_2 at 1065°C via the reaction,

$$CuO + Al_2O_3 = CuAl_2O_4 \tag{5.2}$$

A copper oxide coating that normally is present on the metal's surface provides the required oxygen. Notice that this reaction takes place just below the melting point of copper, which is 1083°C. For aluminum nitride, the substrate is first oxidized to form a thin coating of alumina on the surface of the part via the reaction,

$$4AlN + 6O_2 = 2Al_2O_3 + N_2 \tag{5.3}$$

which is usually performed at a temperature of 1200°C in the presence of oxygen. After the oxidation step, the process proceeds as if the substrate was composed of alumina rather than aluminum nitride. The strength of the bond between the AlN and alumina is similar due to the common bonding mechanism, with peel strengths greater than 50 N/cm.

Once the DBC metal is attached to the substrate, the copper metallization can be etched with processes similar to those used in the printed circuit board (PCB) industry. The key difference is that the metal is much thicker than what is used in PCBs and requires special etching considerations. The principal method used to etch these conductors is spray etching with a caustic solution. Spray etching is necessary since the amount of material that must be removed is significant and a standard dunk etch process is ineffective due to saturation of the solution at the copper interface. In contrast, the spray process presents a fresh solution to the copper surface and flushes away the reacted solution. While etching DBC, care must be taken to consider the conductor widths and spaces that can be achieved. It is unrealistic to expect to etch a 0.65-mm (25-mil) thick conductor to a resolution of less than 0.65 mm (25 mils). The general guidelines are traces and spaces no smaller than 0.25 mm (10 mils) for even relatively thin DBC metals. Thicker metals may require even larger conductor geometries to ensure high substrate yields.

Applications that require precise etching of DBC conductors must consider the fact that the etching process is lateral, as well as vertical. The result is that the etchant has a tendency to produce finer traces than the photomask used due to this lateral etching. An etch-back factor, which is often determined experimentally for a given etching system, is normally used to expand the artwork to account for this overetch.

The DBC process is simple and offers a number of capabilities beyond plain metallized substrates, including metallizations that extend out beyond the edge of the substrate to form leads, as well as multilayer substrates, and via connections. Leaded DBC substrates are formed by using copper foil that is larger than the substrate desired. In most cases, this foil is stamped to size and may include a variety of prefabricated connections. Multilayer DBC is formed by simply stacking layers of metal foil and substrates, followed by the bonding process described previously. Via connections between these layers are created through holes drilled in the ceramic and the inclusion of rectangular or spherical copper inserts, or the welding of the adjacent copper metallizations through the via hole.

5.10.2 Active Metal Brazing (AMB)

While DBC offers an ideal solution for a number of applications, some applications are best suited to a metal thickness between thick-film printed conductors and DBC. For these cases, active metal braze substrates are an option. In this process, a copper foil is bonded to a ceramic substrate using a braze alloy. This process is analogous to soldering, however, braze alloys are typically classified as alloys with melting points higher than 500°C. The majority of these alloys utilize titanium as a key component, which initiates a reaction between the substrate surface and the braze alloy. In this way, the braze material wets and forms a strong bond to both the foil and the substrate. The copper foil is often grooved on the side that is bonded to the substrate to facilitate the process. Common braze alloys include Ti–Cu–Ag (e.g., 68.8%–26.7%–4.5%), as well as Ti–Al–Cu–N compositions.

The process flow for active metal braze substrates usually involves coating the braze alloy on the ceramic substrate in a paste form or as a metal foil. The copper foil is then placed on top of the braze alloy and the whole assembly is heated in an inert atmosphere. The braze alloy melts and forms a strong bond with the copper and substrate. In many cases, the braze alloy and copper are patterned prior to bonding to eliminate the need for etching of the braze alloy, which can be difficult to etch. This approach also avoids the need to remove unwanted activated ceramic areas, which are conductive and can be difficult to etch.

Alternatively, a complete coating of braze alloy and copper foil can be placed on the ceramic. In this case, the copper is then etched much like a DBC substrate. However, an etching or abrasion process must be used to remove the unwanted braze alloy and activated substrate layer.

5.11 SUMMARY

Ceramics offer a number of attractive features for electronic applications including high thermal conductivity, a low CTE, which closely matches most semiconductor materials, and a high breakdown voltage. The most popular, and therefore, widely used ceramics include Al_2O_3, BeO, AlN, LTCC, and HTCC. These substrates can be fabricated into electronic components or circuits using thin-film, conventional additive thick-film, or multilayer ceramic technologies. The selection of the technology depends on the application requirements. Thin-films on ceramics offer very fine-line resolution capability and find most application in high-density digital or high-frequency circuits since these applications require fine precise lines and spaces. Thick film, on the other hand, does not offer the same level of line resolution, but it is lower in cost and can often handle higher power applications due to the thicker metal. The multilayer technologies of HTCC and LTCC excel in applications that require hermetic packages and/or very high circuit densities. LTCC has been growing in popularity for microwave and millimeter wave applications due to its design flexibility and low loss at high frequencies.

Ceramics also find applications in high-power circuits such as motor drives and power conversion equipment. In fact, power modules built from DBC on alumina or aluminum nitride substrates are the industry standard for these types of applications. These materials excel in such applications due to the very high thermal conductivity and high current-carrying capability that the thick DBC conductor and ceramic substrates provide.

REFERENCES

1. U. Chowdhry and A. W. Sleight, Ceramic Substrates for Microelectronic Packaging, *Ann. Rev. Mat. Sci.*, Vol. 17, pp. 323–340, August 1987.
2. S. Sarraute, Sorensen, O. Toft; Hansen, E. Rubaek, Fabrication Processes of Magnetic and Dielectric Multilayer Ceramics, *Key Engr. Mat.*, Vols. 132–136, Pt 2, pp. 1151–1154, 1997.
3. W. D. Kingery, Ceramic Materials Science in Society, *Ann. Rev. Mat. Sci.*, Vol. 19, pp. 1–21, August 1989.
4. R. Roy, D. K. Agrawal, H. A. McKinstry, Very Low Thermal Expansion Coefficient Materials, *Ann. Rev. Mat. Sci.*, Vol. 19, pp. 59–81, August 1989.
5. S.-H. Lee, G. L. Messing, E. R. Twiname, A. Mohanram, C. A. Randall, and D. J. Green, Co-sintering of Multilayer Ceramics, *Key Engr. Mat.*, Vols. 206–213, No. I, pp. 257–260, 2001.
6. R. M. German, *Sintering Theory and Practice*, New York: Wiley, 1996.
7. W. D. Kingery, H. K. Bowden, and D. R. Uhlmann, *Introduction to Ceramics*, 2nd ed., *Wiley Series on the Science and Technology of Materials*, New York: Wiley, 1976.
8. S. Ahne, W. Rossner, and P. Greil, Study of a Multilayer-Based Technique with Sintering Additives for Microstructuring of Single-Phase Ceramics, *Key Engr. Mat.*, Vols. 132–136, Pt 2, pp. 960–963, 1997.
9. Jong-Bong Lim, Jin-Ok Son, Sahn Nahm, Woo-Sung Lee, Myong-Jae Yoo, Nam-Gi Gang, Hwack-Joo Lee, and Young-Sik Kim, Low-Temperature Sintering of B_2O_3-Added $Ba(Mg_{1/3}Nb_{2/3})O_3$ Ceramics, *Jpn. J. Appl. Phys., Part 1: Regular Papers and Short Notes and Review Papers*, Vol. 43, No. 8 A, pp. 5388–5391, August 2004.
10. J. D. Katz, Microwave Sintering of Ceramics, *Ann. Rev. Mat. Sci.*, Vol. 22, pp. 153–170, August 1992.
11. Peter Greil, Advanced Engineering Ceramics, *Adv. Mat.*, Vol. 14, No. 10, pp. 709–716, May 17, 2002.
12. Jennifer A. Lewis, Binder Removal from Ceramics, *Ann. Rev. Mat. Sci.*, Vol. 27, pp. 147–173, August 1997.
13. R. N. Master, L. W. Herron, and R. R. Tummala, Cofiring Process for Glass-Ceramic/Copper Multilayer Ceramic Substrate, Proceedings—Electronic Components Conference, 1991, pp. 5–9.
14. W. B. Hillig, Strength and Toughness of Ceramic Matrix Composites, *Ann. Rev. Mat. Sci.*, Vol. 17, pp. 341–383, August 1987.
15. K. T. Faber, Ceramic Composite Interfaces: Properties and Design, *Ann. Rev. Mat. Sci.*, Vol. 27, pp. 499–524, August 1997.
16. William Main, Sami Tantawi, and Joy Hamilton, Measurements of the Dielectric Constant of Lossy Ceramics, *Intl. J. Electr.*, Vol. 72, No. 3, pp. 499–512, March 1992.
17. Housaku Sato, Kazutoshi Ayusawa, Minoru Saito, Kazutami Kawamura, Izumi Kawakami, and Koshi Nihei, High Dielectric Constant Ceramics Applied to Microwave Resonators, Proceedings of the Third IEEE/CHMT International Electronic Manufacturing Technology Symposium, Anaheim, CA, 1987, pp. 149–153.
18. E. A. Nenasheva and N. F. Kartenko, High Dielectric Constant Microwave Ceramics, *J. Eur. Cer. Soc.*, Vol. 21, No. 15, pp. 2697–2701, 2001.
19. J. Mazierska, M. V. Jacob, A. Harring, J. Krupka, P. Barnwell, and T. Sims, Measurements of Loss Tangent and Relative Permittivity of LTCC Ceramics at Varying Temperatures and Frequencies, *J. Eur. Cer. Soc.*, Vol. 23, No. 14, pp. 2611–2615, 2003.
20. L. M. Atlas, H. Nagao, and H. H. Nakamura, Control of Dielectric Constant and Loss in Alumina Ceramics, *J. Am. Cer. Soc.*, Vol. 45, No. 10, pp. 464–471, Oct. 1962.
21. Y. Ohya, Z. Nakagawa, and K. Hamano, Grain-Boundary Microcracking Due to Thermal Expansion Anisotropy in Aluminum Titanate Ceramics, *J. Am. Cer. Soc.*, Vol. 70, No. 8, pp. c184–c186, Aug. 1987.
22. S. X. Mao and C. Q. Ru, Effect of Microcracking on Electric-Field-Induced Stress Intensity Factors in Dielectric Ceramics, *Phil. Mag.*, Vol. 83, No. 2, pp. 277–294, Jan. 11, 2003.
23. B. C. Shin and H.-G. Kim, Grain-Size Dependence of Electrically Induced Microcracking in Ferroelectric Ceramics, *J. Am. Cer. Soc.*, Vol. 72, No. 2, pp. 327–329, Feb. 1989.
24. N. Laws and J. C. Lee, Microcracking in Polycrystalline Ceramics Elastic Isotropy and Thermal

Anisotropy, *J. Mech. Phys. Solids*, Vol. 37, No. 5, pp. 603–618, 1989.
25. T. DEN, M. J. JAAP, C. W. RADEMAKER AND C. L. HU, Residual Stresses in Multilayer Ceramic Capacitors: Measurement and Computation, *J. Electr. Pack. Trans. ASME*, Vol. 125, No. 4, pp. 506–511, Dec., 2003.
26. H. M. CHAN, Layered Ceramics: Processing and Mechanical Behavior, *Ann. Rev. Mat. Sci.*, Vol. 27, pp. 249–282, Aug. 1997.
27. J. G. PEPIN, High Fire Multilayer Ceramic Capacitor Electrode Technology, Proceedings—Electronic Components Conference, 1991, pp. 328–334.
28. Y.-Q. WU, Y.-F. ZHANG, G. PEZZOTTI AND J.-K. GUO, Effect of Glass Additives on the Strength and Toughness of Polycrystalline Alumina, *J. Eur. Cer. Soc.*, Vol. 22, No. 2, pp. 159–164, Feb. 2002.
29. G.-D. ZHAN AND A. K. MUKHERJEE, Carbon Nanotube Reinforced Alumina-Based Ceramics with Novel Mechanical, Electrical, and Thermal Properties, *Intl. J. Appl. Cer. Technol.*, Vol. 1, No. 3, pp. 161–171, 2004.
30. N. DANEU, A. RECNIK, S. BERNIK AND D. KOLAR, Microstructural Development in SnO_2-doped ZnO-Bi_2O_3 Ceramics, *J. Am. Cer. Soc.*, Vol. 83, No. 12, pp. 3165–3171, Dec. 2000.
31. H. SANKUR, J. T. CHEUNG, AND W. GUNNING, Laser-Assisted Evaporation: A Thin Film Deposition Technique, *Doga Turk Fizik Astrofizik Dergisi*, Vol. 14, No. 1, Suppl., pp. 40–54, 1990.
32. T. TAKAGI, Ionized Cluster Beam Technique for Thin Film Deposition, *Zeitschrift Physik D (Atoms, Mol. Clusters)*, Vol. 3, Nos. 2–3, pp. 271–278, 1986.
33. J. KRISHNASWAMY, A. RENGAN, J. NARAYAN, K. VEDAM, AND C. J. MCHARGUE, Thin-Film Deposition by a New Laser Ablation and Plasma Hybrid Technique, *Appl. Phys. Lett.*, Vol. 54, No. 24, pp. 2455–2457, June 12, 1989.
34. A. R. ZOMORRODIAN, A. MESSARWI, AND N. J. WU, AES and XPS Study of PZT Thin Film Deposition by the Laser Ablation Technique, *Cer. Intl.*, Vol. 25, No. 2, pp. 137–140, 1999.
35. H. RANGCHI, B. HUNER, AND P. K. AJMERA, A Model for Deposition of Thick Films by the Screen Printing Technique, Proceedings of the 1986 International Symposium on Microelectronics, 1986, pp. 604–609.
36. B. WALTON, Principles of Thick Film Materials Formulation, *Radio Electr. Engr.*, Vol. 45, No. 3, pp. 139–143, March 1975.
37. H. W. MARKSTEIN, Progress in Thick Film Pastes, *Electr. Packaging Production*, Vol. 22, No. 9, pp. 41–48, Sept. 1982.
38. B. WALTON, Thick Film Pastes and Substrates, *Hybrid Microelectr. Technol.*, pp. 41–52, 1984.
39. H. W. MARKSTEIN, Firing Thick Film Hybrids, *Electr. Packaging Production*, Vol. 26, No. 9, pp. 30–32, Sept. 1986.
40. C. C. Y. KUO, Thick Film Nichrome Resistors, 7th European Passive Components Symposium, CARTS – EUROPE '93, 1993, pp. 111–117.
41. M. PRUDENZIATI, F. ZANARDI, B. MORTEN, AND A. F. GUALTIERI, Lead-Free Thick Film Resistors: An Explorative Investigation, *J. Mat. Sci. Mat. Electr.*, Vol. 13, No. 1, pp. 31–37, Jan. 2002.
42. C. Q. SCRANTOM AND J. C. LAWSON, LTCC Technology: Where We Are and Where We're Going—II, IEEE MTT-S International Topical Symposium on Technologies for Wireless Applications, Proceedings, 1999, pp. 193–200.
43. J. HIROTA, T. FUNAKI, AND A. MIWA, LTCC Module Using Flip-Chip Technology for Mobile Equipment, *Proc. SPIE Intl. Soc. Opt. Engr.*, Vol. 4587, pp. 283–286, 2001.
44. G. PASSIOPOULOS, LTCC Technology for High Performance Low Power Base Station Applications, *Proc. SPIE Intl. Soc. Opt. Engr.*, Vol. 5231, pp. 85–90, 2003.
45. D. I. AMEY, M. Y. KEATING, M. A. SMITH, S. J. HOROWITZ, P. C. DONAHUE, AND C. R. NEEDES, Low Loss Tape Materials System for 10 to 40 GHz Application, *Proceedings 2000 International Symposium on Microelectronics*, SPIE, Vol. 4339, pp. 654–658, 2000.
46. R. R. DRAUDT, M. A. SKURSKI, M. A. SMITH, D. I. AMEY, S. J. HOROWITZ, M. J. CHAMP, AND E. POLZER, Photoimageable Silver Cofireable Conductor Compatible with 951 Green Tape™, 12th European Microelectronics and Packaging Conference. Proceedings, 1999, pp. 219–226.

EXERCISES

5.1. What are the advantages of low-temperature cofired ceramic tape? What are the disadvantages of this material?

5.2. State the main factors that govern: (a) a minimum size for resistor design and (b) a maximum size for resistor design.

5.3. State the properties of substrates used for thick-film circuits. What are the candidate materials to possess such properties?

5.4. For a 12 in. × 12 in. stainless steel screen with a wire mesh count of 325, a wire size of 1.1 mil, and a weave thickness of 0.00023 in., determine the wet, dry, and fired thickness of a thick-film conductor for a hybrid/indirect emulsion of 15 μm.

5.5. Choose the proper wire mesh count to achieve a fired thick-film resistor of 1.1769 mil in thickness using stainless steel wire with a weave thickness of 0.004 in. and a hybrid/indirect emulsion of 20 μm.

5.6. Design a 1000-Ω ±1% 0.5-W resistor using rectangular design. The available sheet resistivity of the resistor paste is 500-Ω/square. The resistor paste is rated for 50 W/in.² for the operating range of temperature.

5.7. A 5000-Ω ±0.5% 1/3-W resistor is to be designed using the top hat resistor design. A 500-Ω/square thick-film resistor paste is available, and the paste is rated for 60 W/in.². Determine the resistor width W, the resistor length L, the hat length L_{hat}, and the hat width W_{hat}. $R_u = 50\%$ $R_{desired}$.

Chapter 6

Electrical Considerations, Modeling, and Simulation

S. S. ANG AND L. W. SCHAPER

6.1 INTRODUCTION

This chapter deals with the electrical considerations important in electronic packaging and with the modeling and simulation of electrical interconnects to give realistic predictions of the electrical performance of an electronic assembly. To begin, it is essential to recognize how and why the characteristics of interconnects can affect the performance of an electronic system.

6.1.1 When Is a Wire Not a Wire?

In an ideal world, a wire is a zero-resistance electrical connection between two circuit elements; it adds or takes away nothing from circuit performance. At low frequencies, and with wires of sufficient diameter, this is indeed the case. As frequencies rise and dimensions shrink, however, wires, pins, circuit board traces, power distribution planes, and other interconnect elements exhibit resistive, capacitive, and inductive effects that can severely impact system performance. These "parasitic" circuit elements can be extracted from the geometry of the interconnect path by analytical means or by computer programs that solve the electric and magnetic field distributions around conductor geometries. Once these parasitics are known, or modeled as circuit elements, they can be added to the overall system schematic and included in a circuit's performance simulation. So the answer to the question is that a wire can never be considered only a wire!

6.1.2 Packaging Electrical Functions

Chapter 1 listed the functions of electronic packaging; that list included two electrical functions: power distribution and signal interconnect. Although these two functions are performed with (primarily) copper metallic connections, the requirements on those connections are very different. Most obvious is the difference in number; an electronic system may need to distribute power at only a few voltage levels to many integrated circuits (ICs) or other

---

Advanced Electronic Packaging, Second Edition, Edited by Richard K. Ulrich and William D. Brown
Copyright © 2006 the Institute of Electrical and Electronics Engineers, Inc.

circuit elements, whereas there may be thousands of signal interconnects, each connecting only a few circuit elements. Likewise, the electrical requirements for these two types of interconnects are very different. Power distribution requires very low impedance conductors that can carry high currents; even very small parasitic inductances can generate noise that disturbs IC operation. Signal conductors need to carry high-frequency signals without distortion or attenuation. Their capacitive and inductive parasitics make them vulnerable to interfering fields from other signal conductors.

Thus, the major electrical design objectives in electronic packaging are to maintain signal fidelity in signal paths and to minimize noise generation in electrical power conductors while minimizing the cost. To maintain signal fidelity, it is desirable to minimize unwanted or parasitic effects, such as delta-I noise, in the power distribution system and crosstalk between interconnects. Delta-I noise is caused by the parasitic inductance of a power distribution system and its inability to change the supply current at the same rate as the switching circuits. Crosstalk is caused by the interconnect mutual capacitances and inductances. These effects are caused by parasitics associated with practical electrical conductors or interconnects, due to their proximity to each other in electronic packages. For example, a rectangular conductor has a resistance associated with current flow through it, a charge storage effect as a result of its parasitic capacitance, crosstalk due to its parasitic capacitance and inductance, and signal delay due to its parasitic capacitance and resistance. Consequently, a good portion of the power dissipation in high-performance complementary metal–oxide–silicon (CMOS) microprocessors occurs in their off-chip drivers in order to accomplish signal transmission. In dynamic random-access memories (DRAMs), much of the power is consumed in driving and sensing long column and row lines and in off-chip drivers. Hence, chip-to-chip global interconnects dictate the overall system or package performance and consume a major portion of the required power.

The capacitance of chip-to-chip interconnects is at least an order of magnitude larger than that of on-chip interconnects, which makes package-level communication even more challenging than chip-level signal transmission. To improve overall system performance, the size of devices that drive long on-chip interconnects and chip-to-chip interconnects, such as in a multichip module or circuit board, are increased. For high-speed package design, wide-bandwidth interconnects are required. Thus, both time-domain and frequency-domain analyses of the interconnects are necessary.

Perhaps the most difficult aspect of electrical design is understanding the generation of unwanted signals (or noise) and their impact on circuits. An electronic package designer should understand various parasitics that affect the electrical performance of an electronic package because the problems caused by many of these parasitics, if known beforehand, can be minimized through proper or "creative" electrical design. Some of these parasitics, such as electrical resistance, can be reduced by using wider and thicker interconnects or eliminated using superconductor interconnects. In this chapter, electrical design considerations for electronic packages and interconnect substrates are discussed.

6.2 FUNDAMENTAL CONSIDERATIONS

A fundamental electrical design consideration for electronic packaging is the minimization of parasitic resistance, capacitance, and inductance of the interconnects. In the following sections, the electrical design considerations for digital systems are discussed. Electrical design begins with the selection of the appropriate package(s) and partitioning of functions between chips and packages to meet cost and performance specifications. In interconnect substrate or printed wiring board design, the electrical design branches off into a signal distribution system and a power distribution system. The power distribution system is first considered. Next, the designer generates the logic and timing designs using simulation tools. Lastly, the designer generates the physical design, including the layout of the chips and their interconnect routings. This is usually accomplished using available layout tools such as those from Mentor Graphics and Cadence.

In digital design, a net refers to the network of interconnects between drivers and receivers. A critical net is first identified whose delay limits the maximum possible speed of the system. Appropriate net topology is then chosen to satisfy the requirements for this critical net.

6.2.1 Resistance

Resistance is a basic phenomenon related to current flow in a conventional or "normal" conductor. The term *normal* is used to distinguish a regular metal conductor from a *superconductor*. The Ohm's law description of resistance is given as

$$R = \frac{V}{I} \, \Omega \tag{6.1}$$

where V is the voltage across and I is the current flowing in a conductor. The unit of resistance is the ohm (Ω). A major effect of resistance, or ohmic loss, is a direct current (DC) voltage drop in the power distributing conductors. Consequently, it is desirable to minimize the interconnect resistance since it results in power loss and signal delay problems. Power loss, or dissipation, P, in power distribution conductors is proportional to the resistance of the conductor, and is given by $P = I^2 R$. Similarly, signal delay is, to a first order, directly proportional to the interconnect resistance, or RC product, where C is the parasitic capacitance.

Figure 6.1 shows the pinout for a single-chip package or an electronic packaging substrate. To reduce the DC power losses and parasitic inductance in the power distribution

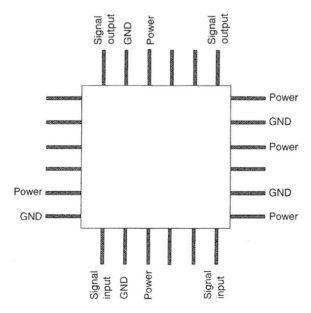

Figure 6.1 Signal and power leads going into a package.

system, the power and ground connections are usually made to a multiple number of pins. It is not unusual to have about 20% of the pins in a 256-pin quad flatpack (QFP) or pin grid array (PGA) package dedicated only to power and ground connections.

6.2.1.1 Conventional Metals

Current flow in conventional metals is a result of the movement of electrons under the influence of an applied electric field. However, the electrons do not travel through the metal unimpeded because of resistance, which results from scattering of the electrons. In general, electron scattering is a combination of intrinsic and extrinsic scattering. One source of intrinsic scattering is other electrons in the metal. Another source is lattice vibrations (or phonons). Thus, electron scattering is intrinsic to metals and gives rise to electrical resistance. Furthermore, resistance can also be caused by defects in the metal, such as impurities, grain boundaries, and twin domains, which are a function of the quality of the metal. Scattering from such defects results in extrinsic resistance.

The material property associated with the resistance of a normal metal is resistivity, ρ, and, for thin-film metallic conductors used in electronic packages, is a function of temperature, crystallinity, defect structure, purity, and applied electric field. Matthiessen's rule states that the total resistivity of a sample is the sum of all the individual contributions, that is,

$$\rho = \rho_{\text{temp}} + \sum \rho_i \tag{6.2}$$

where ρ_{temp} is the temperature-dependent term and ρ_i is the contribution due to defects as noted previously. The temperature dependence of resistivity (it increases with temperature) arises due to interactions of charge carriers with each other and lattice vibrations. Impurities also change the resistivity of a metal because impurity atoms (a) may have a different valence than the host metal atom, (b) may be a different size than the host atom, and (c) may screen charge carriers.

A metal with a lower value of resistivity has a lower resistance to DC flow. Table 6.1 gives the bulk resistivity for several common metals. Silver has the lowest resistivity and, thus, is the best current conductor. However, because of cost considerations and their reasonably low resistivity, aluminum and copper are the most commonly used metals in integrated circuits, and both aluminum and copper are used in electronic packaging substrates. Copper, when used in packaging, is usually plated with nickel. Gold, a noble metal, also finds application in electronic packaging substrates. It is important to note that the resistivity of thin-film metal is a function of the technique used to deposit it. In general, the resistivity of a metal deposited by sputtering, evaporation, or plating is higher than its bulk form value.

Table 6.1 Bulk Resistivity of Metal Conductors at 300 K [1]

Metal	Resistivity (Ω-cm)
Aluminum	2.73×10^{-6}
Copper	1.73×10^{-6}
Gold	2.27×10^{-6}
Silver	1.63×10^{-6}

Interconnect currents in electronic packages must not exceed the current density at which an effect, referred to as electromigration, is initiated. Electromigration is the mass transport of the interconnect metal in the direction of electron flow, which can lead to void and extrusion formation. It is a unique and serious mode of reliability failure in interconnects. For aluminum interconnects, the critical current density is about 10^5 A/cm^2. Thus, the current flowing in a typical 1-μm-wide × 0.5-μm-thick aluminum interconnect should be restricted to no more than 0.5 mA. Hence, the width and thickness of an interconnect are important parameters in electronic package design.

Typical conductors in thin-film MCM substrates are 1 to 5 μm in thickness and 5 to 20 μm wide. A typical 2-μm-thick, 10-μm-wide aluminum conductor with a resistivity of 3.6×10^{-6} Ω-cm has a resistance per unit length of 18 Ω/cm. In circuit boards, copper signal conductors are thicker (12 to 25 μm) and wider (50 to 150 μm), so they have a far lower resistance. It is important to note that, in addition to the DC resistance discussed previously, a package designer must also deal with alternating current (AC) resistance (or the skin effect) of interconnects, especially for high-speed circuits.

6.2.1.2 Electrical Resistance of Rectangular Conductors

Electrical conductors used in electronic packaging usually are rectangular in shape. An electronic packaging substrate may have many electrical conductors of varying linewidths, depending on their current-carrying requirements but of the same thickness since they are formed by a single metal deposition process. Thus, it is convenient to introduce a concept known as sheet resistance for the calculation of their DC resistances.

Consider a rectangular conductor of linewidth w, length L, and thickness t, as shown in Figure 6.2. The resistance of the conductor is

$$R = \frac{\rho L}{A} = \frac{\rho L}{wt} = \rho_s \frac{L}{w} \tag{6.3}$$

where $\rho_s = \rho/t$ is known as the sheet resistance in ohms/square and L/w is the number of squares. The sheet resistance is a constant for a process since the thickness of the conductor is fixed. Thus, the resistance of rectangular conductors of varying widths can be determined if their lengths and widths are known.

An electrical interconnect in an electronic packaging substrate is seldom a straight line. Often, there are right-angle bends along its path, and a right-angle bend contributes only 0.56 of a square to the resistance of an interconnect as compared to "straight" squares [1]. A meander resistor, shown in Figure 6.3, is often used in high-frequency circuits since it has a low self-inductance. However, right-angle bends introduce signal discontinuities. This effect is usually characterized using time-domain reflectometry (TDR).

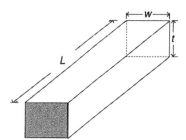

Figure 6.2 Rectangular conductor with a linewidth w, a length L, and a thickness t.

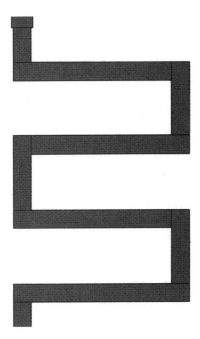

Figure 6.3 Meander resistor.

A designer must be concerned with more than the bulk resistance of an electrical conductor. Often, the contact resistance between a bond wire and a bond pad or between interlayer interconnects (i.e., vias) is more significant than the bulk resistance of the interconnect. This is especially true in an MCM and in some printed wiring boards, where there may be many vias between several different interconnect layers.

Example 6.1

Calculate the DC resistance of the aluminum conductor as shown in Figure 6.4. The thickness of the evaporated aluminum is 2 μm with an average resistivity of 3.6×10^{-6} Ω-cm.

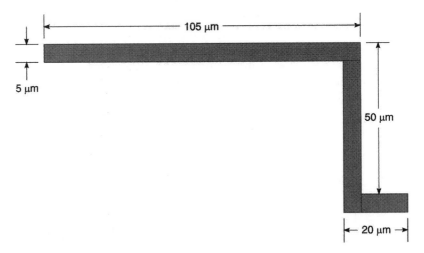

Figure 6.4 Aluminum conductor for Example 6.1.

SOLUTION The sheet resistance of the evaporated aluminum conductor is

$$\rho_s = \frac{\rho}{t} = \frac{3.6 \times 10^{-6}\,\Omega - \text{cm}}{2 \times 10^{-4}\,\text{cm}} = 1.8 \times 10^{-2}\,\Omega/\text{square}.$$

The number of *straight* squares is

$$N_s = \frac{100}{5} + \frac{45}{5} + \frac{15}{5} = 32$$

The number of *corner* squares is

$$N_c = 2$$

Thus, the DC electrical resistance of the aluminum conductor is

$$R_{\text{DC}} = [32 + 2(0.56)]\rho_s = 0.596\,\Omega.$$

Often, in an interconnect path between chips, conductor geometry and material can vary, and the path must be modeled as a string of resistors.

Example 6.2

Find the resistance of the interconnect path from a chip wire bond pad in a ceramic pin grid array (PGA) to the copper trace in the printed wiring board when the PGA is soldered in place.

The interconnect path is made up of several segments, as shown in Figure 6.5:

where

A = aluminum wire bond, 1 mil in diameter and 100 mils long
B = tungsten trace on the PGA bonding layer, 2 mils thick, 10 mils wide, and 4 mm long
C = tungsten via between ceramic layers, 5 mils in diameter and 20 mils long
D = long tungsten trace, 2 mils thick, 6 mils wide, and 1.5 cm long
E = another via like C
F = gold-plated Kovar pin, 20 mils in diameter by 0.2 inch long

(The use of mixed English and metric units in the statement of the problem reflects the situation in the industry, where a transition from English to metric units is in various stages.)

SOLUTION

$$R = \rho(L/A) \quad 1\,\text{mil} = 25\,\mu\text{m} \quad 1\,\text{in.} = 2.54\,\text{cm}$$
$$\rho_{\text{Al}} = 2.6 \times 10^{-6}\,\Omega\text{-cm}$$
$$\rho_w = 5.5 \times 10^{-6}\,\Omega\text{-cm}$$
$$\rho_{\text{Kovar}} = 47 \times 10^{-6}\,\Omega\text{-cm} \text{ (assume gold is thin enough to neglect)}$$

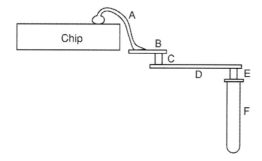

Figure 6.5 Illustration of an interconnect path.

A $L = 100\,\text{mil} = 2.5 \times 10^{-1}\,\text{cm}\quad d = 2.5 \times 10^{-3}\,\text{cm}\quad A = \pi d^2/4 = 4.909 \times 10^{-6}\,\text{cm}^2$
$R_A = (2.6 \times 10^{-6}\,\Omega\text{-cm})(2.5 \times 10^{-1}\,\text{cm})/(4.91 \times 10^{-6}\,\text{cm}^2) = 0.1324\,\Omega$

B $L = 0.4\,\text{cm}\quad t = 5 \times 10^{-3}\,\text{cm}\quad W = 2.5 \times 10^{-2}\,\text{cm}$
$R_B = (5.5 \times 10^{-6}\,\Omega\text{-cm})(0.4\,\text{cm})/(5 \times 10^{-3}\,\text{cm})(2.5 \times 10^{-2}\,\text{cm}) = 0.0176\,\Omega$

C $L = 5 \times 10^{-2}\,\text{cm}\quad d = 1.25 \times 10^{-2}\,\text{cm}\quad A = \pi d^2/4 = 1.227 \times 10^{-4}\,\text{cm}^2$
$R_C = (5.5 \times 10^{-6}\,\Omega\text{-cm})(5 \times 10^{-2}\,\text{cm})/(1.227 \times 10^{-4}\,\text{cm}^2) = 0.00224\,\Omega$

D $L = 1.5\,\text{cm}\quad t = 5 \times 10^{-3}\,\text{cm}\quad W = 1.5 \times 10^{-2}\,\text{cm}$
$R_D = (5.5 \times 10^{-6}\,\Omega\text{-cm})(1.5\,\text{cm})/(5 \times 10^{-3}\,\text{cm})(1.5 \times 10^{-2}\,\text{cm}) = 0.110\,\Omega$

E $E = C R_E = 0.00224\,\Omega$

F $L = 0.508\,\text{cm}\quad d = 5 \times 10^{-2}\,\text{cm}\quad A = \Pi d^2/4 = 1.964 \times 10^{-3}\,\text{cm}^2$
$R_F = (47 \times 10^{-6}\,\Omega\text{-cm})(0.508\,\text{cm})/(1.964 \times 10^{-3}\,\text{cm}^2) = 0.0122\,\Omega$

$R_{\text{TOTAL}} = R_A + R_B + R_C + R_D + R_E + R_F = 0.2744\,\Omega$

Note that almost all of this resistance comes from the wire bond and the tungsten path. The resistance of the pin and the vias could be ignored. Although 0.27 Ω is not a very large resistance, a high-power chip in this package could draw 5 A of current. If this pin were the only power or ground pin, the resulting voltage drop would be 1.35 V, far too much. This is why high-power chips in PGAs will typically have multiple power or ground pins; the voltage drop must be very low. ■

6.2.2 Self and Mutual Inductance

Inductance is the circuit parameter used to describe an inductor, an energy storage device associated with magnetic fields. Inductance is defined as the ratio of the voltage across the inductor to the rate of change of current through it, that is,

$$L = \frac{v}{\frac{di}{dt}} \text{ henrys} \tag{6.4}$$

where v is in volts, i is in amperes, and t is in seconds. Note that if the current is constant, the voltage across an ideal inductor is zero. Thus, an ideal inductor appears as a short circuit to a constant, or DC, current. It should also be noted that current flowing in an inductor cannot change instantaneously. This is why an arc is observed when a switch in an inductive circuit is opened.

Inductance can be classified as either self-inductance or mutual inductance. The self-inductance, L_s, of the current-carrying conductor shown in Figure 6.6 is defined by

$$L_s = \frac{N\Phi}{I} = \text{total number of flux linkages per ampere} \tag{6.5}$$

where N is the number of turns of the conductor, Φ is the flux density, and I is the magnitude of the current flowing in the conductor. Ampere's circuital law states that the line integral of a static magnetic flux density around any given closed path must equal the product of the permeability of free space, μ_0, and the total current enclosed by that path. From this law, the magnetic intensity, H_i, internal to the conductor, assuming that the current density, $I/\pi\rho^2$, is uniform in the conductor, is given by

$$\int \overline{H}\,\overline{ds} = I\left(\frac{r}{\rho}\right)^2 \quad 0 \leq r \leq \rho \tag{6.6}$$

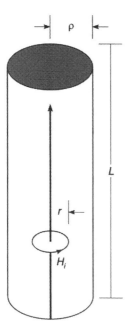

Figure 6.6 Current-carrying conductor for the calculation of self-inductance.

where ρ is the radius of the circular conductor. Thus,

$$H_i = |\overline{H}| = \frac{I}{2\pi r}\left(\frac{r}{\rho}\right)^2 \tag{6.7}$$

The flux density, B_i, internal to the conductor is then

$$B_i = \mu H_i = \frac{\mu I}{2\pi r}\left(\frac{r}{\rho}\right)^2 \qquad 0 \le r \le \rho \tag{6.8}$$

The self-inductance, L_s, resulting from the internal magnetic field is

$$L_s = \frac{l}{I}\int_0^\rho B_i \left(\frac{r}{\rho}\right)^2 dr = \frac{\mu l}{8\pi} \tag{6.9}$$

Note that the internal self-inductance of a current-carrying conductor is directly proportional to its length and is independent of the current flowing in it.

Besides the internal inductance within the wire, there is another contribution to self-inductance caused by the magnetic field around the wire. It can be shown that the total self-inductance of a straight wire is

$$L = 0.002l \left[\ln\frac{2l}{\rho} - \frac{3}{4}\right] \tag{6.10}$$

where L is the self-inductance in microhenrys, and l is the wire length in centimeters.

Magnetic fields emanating from one conductor can intersect another conductor, resulting in signal coupling between the two. This coupling is referred to as mutual inductance. The mutual inductance of the two parallel conductors shown in Figure 6.7 can be found by applying the Biot–Savart law [2], which states that, due to a differential current element

$I\,dy$, the magnitude of the magnetic flux density at a point P is directly proportional to the product of the current I, the length of the current element dy, and the sine of the angle between the current element and the line PQ joining the point P to the current element. Thus, the magnetic flux density, $B(x, y)$, of a current-carrying conductor with a magnitude of I is given by

$$B(x, y) = \frac{\mu I}{4\pi} \int_{-l/2}^{+l/2} \frac{\sin\theta}{r^2} dy_0 = \frac{\mu I}{4\pi} \int_{-l/2}^{+l/2} \frac{x}{r^3} dy_0$$

$$= \frac{\mu I}{4\pi x} \left[\frac{y + \frac{l}{2}}{\sqrt{\left(y + \frac{l}{2}\right)^2 + x^2}} - \frac{y - \frac{l}{2}}{\sqrt{\left(y - \frac{l}{2}\right)^2 + x^2}} \right] \qquad (6.11)$$

where μ is the permeability of the medium. The total flux, Λ, linking the conductor is then given by

$$\Lambda = \int_d^\infty dx \int_{-l/2}^{+l/2} B(x, y)\, dy \qquad (6.12)$$

If the medium is vacuum or air, then $\mu = \mu_0 = 4\pi \times 10^{-7}$ H/m, and the mutual inductance M is given by

$$M = \frac{\Lambda}{I} = \frac{\mu_0 l}{2\pi} \left[\ln\left(\frac{l}{d} + \sqrt{1 + \left(\frac{l}{d}\right)^2}\right) - \sqrt{1 + \left(\frac{d}{l}\right)^2} + \frac{d}{l} \right] \qquad (6.13)$$

$$= \frac{\mu I l}{2\pi} \left[\ln\left(\frac{l}{d} + \sqrt{1 + \left(\frac{l}{d}\right)^2}\right) - \sqrt{1 + \left(\frac{d}{l}\right)^2} + \frac{d}{l} \right]$$

Note that the mutual inductance of the two parallel conductors depends on their spacing and length and is independent of the magnitude of the current flowing in them.

Interconnects associated with an electronic package, such as bond wires, power and signal conductors, and the like, have a small amount of inductance associated with them. It is this parasitic inductance with which a package designer must contend. Furthermore, the power and ground planes of electronic packaging substrates have parasitic inductance associated with them that gives rise to delta-I noise. Consequently, parasitic inductance is generally classified as planar or signal-path inductance depending on its origin.

Example 6.3

Determine the mutual inductance for the two parallel conductors shown in Figure 6.7. The length of the conductors is 100 μm, and the separation between the conductors is 200 μm.

SOLUTION From Eq. (6.13),

$$M = 200 \times 10^{-13} \left[\ln\left(\frac{100}{200} + \sqrt{1 + \left(\frac{100}{200}\right)^2}\right) - \sqrt{1 + \left(\frac{200}{100}\right)^2} + \frac{200}{100} \right]$$

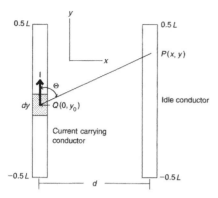

Figure 6.7 Coordinate system for the derivation of mutual inductance.

or

$$M = 4.9 \times 10^{-12} \text{ H} = 4.9 \text{ pH}.$$

Notice that the mutual inductance is independent of the magnitude of the current flowing in the conductors. ∎

Mutual inductance is particularly important in the case of a return circuit for power distribution, where the mutual inductance cancels out most of the conductor self-inductance. The return circuit problem is shown in Figure 6.8. When the switch is closed, current I flows in the loop.

The voltage induced in leg 1 by the mutual inductance M_{12} is opposed to the voltage drop caused by L_1 since the current in leg 2 is equal and opposite to the current in leg 1. Likewise, the voltage induced in leg 2 is opposed to the voltage developed across L_2. The whole loop equation is

$$\begin{aligned} V &= L_1 \frac{dI}{dt} - M_{12} \frac{dI}{dt} + IR + L_2 \frac{dI}{dt} - M_{12} \frac{dI}{dt} \\ &= (L_1 + L_2 - 2M_{12}) \frac{dI}{dt} + IR \end{aligned} \quad (6.14)$$

The term $(L_1 + L_2 - 2M_{12})$ represents the equivalent inductance of the return circuit. Since the mutual inductance between two equal conductors is always less than the self-inductance, there is never a complete cancellation of the inductive effect, but the higher the mutual inductance (wires close together or planes with small spacing), the smaller the voltage fluctuations will be in the power distribution system. This is why power is generally distributed on planes (power and ground plane pairs) in circuit boards and multichip modules.

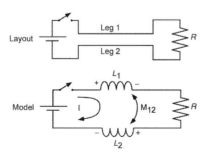

Figure 6.8 Circuit illustration of mutual inductance.

Example 6.4

A return circuit is made up of two conductors, 1 mm in diameter and 10 cm long. What is the loop inductance if the conductor spacing is 1 mm? What if the spacing is 10 mm?

SOLUTION The inductance of one wire is

$$L(\mu H) = 0.002(10)\left[\ln\frac{2 \times 10}{0.05} - \frac{3}{4}\right]$$
$$= 0.02\,[5.99 - 0.75] = 0.105\,\mu H$$

The mutual inductance is

$$L = 0.002\ell\left[\ln\left(\frac{\ell}{d} + \sqrt{1 + \left(\frac{\ell}{d}\right)^2}\right) - \sqrt{1 + \left(\frac{d}{\ell}\right)^2} + \frac{d}{\ell}\right]$$

First, the case of two wires 10 mm apart:

$$L = 0.002 \times 10\left[\ln\left(\frac{10}{1} + \sqrt{1 + 10^2}\right) - \sqrt{1 + \left(\frac{1}{10}\right)^2} + \frac{1}{10}\right]$$
$$= 0.02\,[\ln(10 + 10.05) - 1.01 + 0.1]$$
$$= 0.02\,[2.998 - 1.01 + 0.1] = 0.042\,\mu H$$

Next, the two wires 1 mm apart:

$$L = 0.002 \times 10\left[\ln\left(\frac{10}{0.1} + \sqrt{1 + \left(\frac{10}{0.1}\right)^2}\right) - \sqrt{1 + \left(\frac{0.1}{10}\right)^2} + \frac{0.1}{10}\right]$$
$$= 0.02\,[5.2983 - 1 + 0.01] = 0.086\,\mu H$$

Thus, the total inductance of the return circuit when the wires are 10 mm apart is

$$L = L_1 + L_2 - 2M_{12} = 2 \times 0.105 - 2 \times 0.042 = 0.126\,\mu H$$

And, with the wires 1 mm apart, the inductance is

$$2 \times 0.105 - 2 \times 0.086 = 0.038\,\mu H$$

Clearly, close spacing is key to reducing inductance in a return circuit. ∎

6.2.2.1 Planar Inductance

Power and ground planes, such as those found in an interconnect substrate, contribute to parasitic inductance as noted in the last section. Assume that the power and ground planes, shown in Figure 6.9, carry currents of equal magnitude, but in opposite directions, just as the return circuit above. Then, the magnetic field between them is approximately uniform. Furthermore, outside the two planes, the magnetic field is negligible due to field cancellation. From Ampere's circuital law, the magnetic intensity is equal to I/W where I is the magnitude of the current and W is the width of the planes, and the magnetic flux density, B, is given by $\mu_0(I/W)$. Thus, the inductance of the system is

$$L = \frac{\mu_0 lh}{W}. \tag{6.15}$$

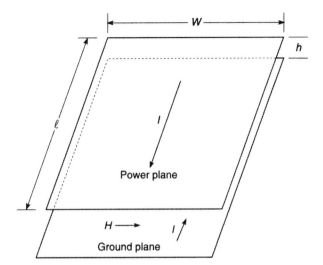

Figure 6.9 Power and ground planes for an MCM.

where *l* is the length of the plane and *h* is the separation between planes. One-half of L is associated with the ground plane and the other half is associated with the power plane so that the inductance of the ground plane, L_g, and the inductance of the power plane, L_p, are given by

$$L_g = L_p = \frac{\mu_0 l h}{2W} \tag{6.16}$$

In electronic packaging substrates, power and ground planes are usually connected via multiple leads to external power supplies to minimize parasitic inductive and resistive effects since the inductance of a circuit element is reduced if the return current path is brought closer to the circuit element.

6.2.2.2 Signal-Path Inductance

For a packaged chip, signals travel from the device to external package pins via interconnect conductors and wire bonds. Consequently, packaged chips are affected by bond wire and package pin parasitic inductances. In electronic packaging substrates, chips are connected to the substrate interconnects using either wire bonds or flip-chip bumps. Typical bond wire inductance is from 1 to 5 nH, depending on the distance between the chip bond pad and the substrate bond pad. However, this inductance does not increase linearly with pad-to-pad distance, especially for short bond wires, because of the required bond wire loop [3]. For a flip-chip electronic packaging substrate, a solder bump typically has a parasitic inductance of about 0.01 nH. Thus, the major advantage of the flip-chip electronic packaging substrate over the wire bond electronic packaging substrate is a reduction in parasitic inductance as a result of shorter interconnections between the chip and the electronic packaging substrate.

As noted previously, pins also contribute to the package parasitics. For example, surface-mount (SMT) package pins contribute about 1 to 12 nH to the parasitic inductance, while those of PGA packages contribute about 2 nH. Hence, the signal-path inductance of a packaged chip is reduced if the length of the package pins is shorter.

6.2.3 Capacitance

Capacitance is a measure of the charge-storing capacity of two parallel plates when a voltage is applied between them. A unit of capacitance, the farad (F), is the ability to store 1 coulomb (C) of charge at a potential of 1 V. Thus,

$$C = \frac{Q}{V} \text{ farad} \tag{6.17}$$

where Q is the stored charge in coulombs and V is the potential between the two plates.

It is also desirable to minimize the parasitic capacitance of electronic packages because it is another source of signal delay. Consequently, a very low relative dielectric constant insulating material should be used whenever possible since the relative dielectric constant is a measure of a material's total polarizability and determines its charge storage capacity with respect to a vacuum. Thus, the relative dielectric constant of a material is expressed as a dimensionless quantity, ε_r, that, when multiplied by the permittivity of free space (vacuum), ε_0, yields a value for the dielectric constant in MKS (meter, kilogram, second) units.

The interconnect capacitance of an electronic system, such as a chip or an electronic packaging substrate, is determined by the dimensions of the interconnect, the dielectric constant, and thickness of the dielectric between the interconnect and the reference plane (i.e., ground plane). In equation form, it is given by the first-order relationship

$$C_i = \frac{\varepsilon_0 \varepsilon_r W l}{d} \tag{6.18}$$

where W is the width of the interconnect, d is the thickness of the dielectric between the interconnect and the ground plane, and l is the interconnect length. Equation (6.18) is only approximate for high frequencies and narrow conductors due to increased fringing fields and dielectric losses of the material. For example, if the linear dimensions of each plate are large compared to the dielectric thickness, the electric field lines look like those shown in Figure 6.10. As the dimensions of the plates become small, the "fringing" effects become significant as can be seen in Figure 6.11.

For many cases involving interconnects, we must look at the capacitance of a wire above a ground plane. In this case, the fringing capacitance becomes extremely significant, as in the case of the "microstrip" transmission line: a flat conductor above a ground plane, as will be described. The capacitance per centimeter of length is shown in Figure 6.12.

For practical conductors in thin-film substrates or printed wiring boards, where the conductor width to dielectric thickness ratio cannot be made $<< 1$ because of manufacturing

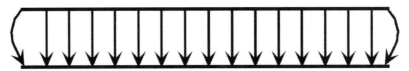

Figure 6.10 Illustration of electric field lines in a parallel-plate capacitor.

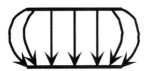

Figure 6.11 Illustration of electric field lines fringing in a parallel-plate capacitor.

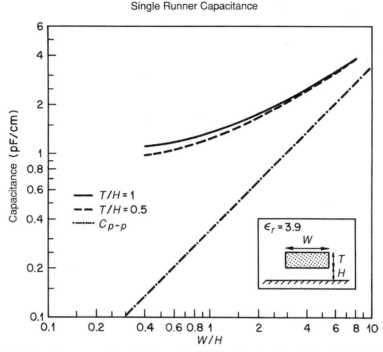

Figure 6.12 Capacitance per unit length of a conductor versus the ratio of conductor width to distance above the ground plane.

considerations, it becomes difficult to achieve a capacitance of less than 1 pF/cm, a value far higher than the parallel-plate formula would indicate. This is the reason for using low dielectric constant materials ($E_r < 3$) in interconnect substrates. Relative dielectric constants of the dielectrics commonly used in electronic packaging substrates, such as silicon dioxide and polyimide, are 3.9 and 3.5, respectively. The benzocyclobutene (BCB) dielectric from Dow Chemical has an E_r of 2.6; conventional FR-4 printed wiring board material has an E_r of 4.7.

In electronic packaging substrate applications, a large decoupling capacitance is often desired to minimize power supply problems. Thus, a high relative dielectric constant material is often used between the power and ground planes. Otherwise, external decoupling chip capacitors are needed.

6.2.3.1 Stray Capacitance

With increasing interconnect density, the parasitic, or stray, capacitance of an electronic package, multichip module, or printed wiring board becomes the predominant speed-limiting factor. Stray capacitance not only contributes to signal delay but also increases power consumption. One way to reduce both the power dissipation and signal delay at high-capacitance nodes is to limit the voltage swing at these nodes. However, as minimum feature size is scaled down and chip size increases, interconnect capacitance becomes more important than device capacitance because metal–oxide–semiconductor (MOS) gate capacitance decreases as feature sizes are scaled down. However, interconnect capacitance increases because the interconnect capacitance per unit length remains approximately constant and global wire length increases as chips become larger. Historically, as a device is scaled down, designers tend to increase the circuit density on a chip, with a concomitant increase in interconnect requirements. Wire bonds and package leads typically have a parasitic capacitance of about 1 pF.

6.2.3.2 Decoupling Capacitance

The primary purpose of decoupling capacitors in electronic packaging substrates is to provide a reservoir of stored charge to maintain a constant voltage as the current demand varies. Also, they are often placed at strategic points in electronic packaging substrates to reduce switching noise, as discussed in Section 6.6.4. Because of what they are intended to do, decoupling capacitors must have a low parasitic inductance. Unfortunately, all capacitors have some parasitic inductance associated with them, which limits their ability to deliver charge rapidly. In fact, even very small (0.1 µF) capacitors can become self-resonant at low frequencies (10 to 20 MHz). However, decoupling capacitors with low parasitic inductance are commercially available [4].

6.2.4 Parameter Extraction Programs

As has been shown in the previous sections, wires are not just wires. Every section of a conductor has associated resistance, inductance, and capacitance. The inductance and capacitance are functions, not only of the geometry of the individual conductor, but also of the relationship of that conductor to all other surrounding conductors. This is why accurate modeling of these parasitic effects is so challenging.

The task has been made easier by the availability of computer analysis tools that solve Maxwell's equations around arbitrary conductor geometries. In many cases, the conductor geometry can be imported into the analysis program from a drawing program such as Autocad. Then various algorithms can be applied to determine conductor resistances, and the capacitance and inductance matrices among all conductors. The result is SPICE-equivalent circuit model values for the parasitics introduced by the conductors. These can be inserted in the circuit schematic to determine the impact of parasitics on circuit performance.

Many software packages are available for this kind of analysis, and there is a whole field of research into what models provide the most accurate values for the least computer run time. Some programs are best for two-dimensional (2D) problems, and some can be used for full three-dimensional (3D) analysis. One popular set of programs comes from Ansoft Corporation (Four Station Square, Suite 660, Pittsburgh, PA 15219; 412–261–3200).

Potential users of any electromagnetic modeling software need to research the capabilities of the various packages available at any point in time. A discussion of the details of these computer aids is well beyond the scope of this book.

6.3 SIGNAL INTEGRITY AND MODELING

Having examined the methods for computing the parasitics associated with interconnects, we turn to using those parasitics to determine the signal integrity of complete interconnect structures, including the drivers and receivers. In this section, we look at modeling using lumped R, L, and C elements.

6.3.1 Digital Signal Representation and Spectrum

As stated previously, a major objective of the electrical design of an electronic package is to maintain signal fidelity. We can envision the typical digital signal as a square wave comprised of a series of sine waves whose amplitudes drop off with increasing frequency. Figure 6.13 shows a typical digital signal along with its frequency spectrum. Its bandwidth is given by [5]

$$BW = \frac{0.35}{t_r} \quad (6.19)$$

where t_r is the rise time of the signal, or the time it takes the signal to transition between 10 and 90% of the voltage swing. Conversely, the fall time, t_f, of the signal is defined as the time it takes the signal to transition between 90 and 10% of the voltage swing as shown in Figure 6.13. From Eq. (6.19), it is obvious that a digital signal with a rise time of 1 ns will have a bandwidth of 350 MHz. Thus, the highest sine wave component is 350 MHz.

Figure 6.14a shows a digital signal having a 50% duty cycle or, in digital terminology, a 101010⋯ digital signal stream. This digital signal stream can be expressed in terms of a Fourier series as

$$V_d(t) = 0.5\, V_a + \sum_{0}^{\infty} \frac{V_a[1 - \cos(n\pi)] \sin(2\pi n f_s) t}{n\pi} \quad (6.20)$$

where V_a is the amplitude of the digital signal and f_s is the frequency. The cosine terms are zero since the signal waveform is an odd function. The first term in Eq. (6.20) is the average, or DC, value of the digital signal. The fundamental component, f_s (or $1/T_s$), is $0.637 V_a$, or 63.7% of the amplitude of the digital signal. Similarly, the magnitude of the third harmonic of the digital signal, $3 f_s$, is $0.21 V_a$, or 21% of V_a. The digital signal stream of Figure 6.14a can be constructed by using various frequency components, as shown in Figure 6.14b.

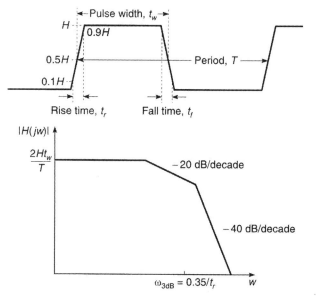

Figure 6.13 Typical digital signal and its frequency spectrum.

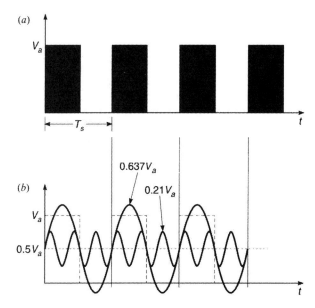

Figure 6.14 Digital signal stream and its frequency components.

An interconnect does not transmit all the sine wave components of a digital signal, resulting in signal distortion. The frequency at which the transmitted power drops to 50% of the incident power is defined as the 3-dB point, or bandwidth, of the interconnect. The corresponding wavelength determines if the interconnect should be treated as a lumped circuit modeled by discrete resistors, inductors, and capacitors or as a transmission line. In general, the interconnect should be treated as a transmission line if the length of the interconnect is longer than the wavelength of the signal.

The pulse design wavelength is a more precise method of determining whether an interconnect should be treated as a lumped-element circuit or as a transmission line. It is defined as [5]

$$\lambda = \frac{\frac{c}{\sqrt{\varepsilon_r}}}{N(BW)} \tag{6.21}$$

where $c = 3 \times 10^8$ cm/s is the speed of light, ε_r is the relative dielectric constant, and N is an empirical integer used to describe the quality of the signal. An interconnect should be treated as a transmission line if its length exceeds 12.5% of its pulse design wavelength, otherwise it is modeled using lumped-element circuits.

Example 6.5

A digital signal with a rise time of 1 ns must travel the length of a 2-cm aluminum interconnect, which is embedded in silicon dioxide ($\varepsilon_r = 3.5$). Determine its bandwidth and pulse design wavelength if the quality of the signal is 2.

SOLUTION The bandwidth of the signal is

$$BW = \frac{0.35}{1 \times 10^{-9}} = 350 \, \text{MHz}$$

The pulse design wavelength is

$$\lambda = \frac{\frac{3 \times 10^8}{\sqrt{3.5}}}{2(350 \times 10^6)} = 0.23 \text{ m}$$

Since $\lambda/8 = 2.86$ cm is greater than the length of the aluminum interconnect, the interconnect should be treated as a lumped circuit. ∎

6.3.2 Driver and Receiver Models

Thus far we have concentrated on determining the parasitics associated with interconnection wiring. We turn next to a few simple models of components that will be needed in simulating electrical performance.

6.3.2.1 Drivers

For the purpose of simulating the electrical performance of packaging, a simple model of the chip output driver is required. CMOS and bipolar CMOS (BiCMOS) drivers will be considered.

The off-chip CMOS driver generally consists of a series of gates, each one containing larger transistors capable of driving the next, until the final stage is sized to drive the 50 to 100 pF off-chip load. Electrically, this looks like the diagram in Figure 6.15.

For convenience, the packaging engineer is only concerned with the final stage and the effective output impedance. An appropriate model is shown in Figure 6.16. Sometimes

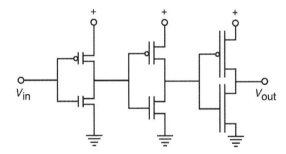

Figure 6.15 Illustration of typical construction of an off-chip CMOS driver circuit.

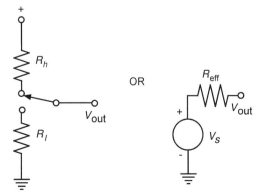

Figure 6.16 Simple model of the final or output stage of a CMOS driver circuit.

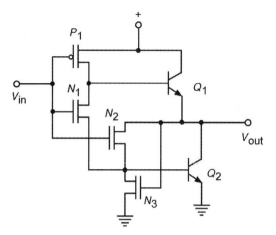

Figure 6.17 Generic BiCMOS driver circuit.

the switch model is useful for conceptual purposes, but it is not easy to use in computer simulations.

In a simulation using the voltage source, the rise and fall times of the output pulse can be controlled to reflect transistor characteristics, and R_{eff} can be sized to represent the effective dynamic resistance. For typical CMOS drivers, 10 Ω would produce a fast driver, and 100 Ω would yield a slow one.

Increasingly, BiCMOS drivers are being used for off-chip drive because of their higher current drive capability. A generic circuit is given in Figure 6.17. This circuit can be quite fast since paths are provided for grounding the bases of Q_1 and Q_2 when they are off. The output voltage swing is less than V_{dd}, going from V_{BEon} to ($V_{dd} - V_{\text{BEon}}$). There is no static power dissipation. The same simulation model can be used as for the CMOS driver, with V_s and R_{eff} adjusted to account for the lower voltage swing and lower R_{eff}.

6.3.2.2 Receivers

A receiver, or chip logic input, is merely a gate with very low capacitance. However, to protect these sensitive transistors from static discharge during assembly, a pair of diodes is usually connected so that any voltage present at V_{in} greater than the breakdown voltage is shunted to the power supply buses as shown in Figure 6.18. These diodes, when not in the breakdown region, have capacitance associated with them. The wire bond pad itself also has capacitance to V_{ss}. Thus, a realistic receiver model is simply a capacitor. A capacitance value of 2 to 3 pF is a good rule of thumb.

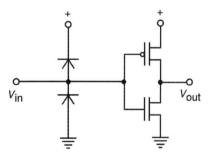

Figure 6.18 Realistic receiver model.

6.3.2.3 Capacitors

We have mentioned that discrete capacitors are used in digital systems to "decouple" the power supply and to reduce digital noise in the power distribution system. This function is discussed later.

The capacitor itself must be examined. Just as wires have parasitics, so too a capacitor is not purely a capacitor. Because of the materials used as conductors, there is some finite resistance. Additionally, because of the geometry of leads, terminals, and plates, there is always some residual inductance. An appropriate model for a capacitor is shown in Figure 6.19. This gives rise to the frequency response shown in Figure 6.20.

For conventional chip capacitors of a given size, such as the 0805 (80 mils by 50 mils) or 1206 (120 mils by 60 mils), the inductance is about 0.5 to 1.0 nH no matter what the value of C, so different capacitors have different resonant frequencies. A typical 0.1-µF chip has a resonant frequency around 20 to 25 MHz. Special low-inductance capacitors have an inductance as low as 60 pH, so they have much higher resonant frequencies.

Having discussed how to model the essential circuit elements, we can proceed to examining the behavior of actual circuits incorporating their parasitic elements.

6.3.3 *RC* Delay

The propagation velocity of an electrical signal on two parallel conductors suspended in air is 30 cm/ns, which is the speed of light. In a practical electronic packaging substrate structure, however, the signal conductors are typically embedded in FR-4 (dielectric constant = 4.7) or polyimide (typical dielectric constant = 3.5), and the signal propagation velocity drops to 13.8 and 16 cm/ns, respectively. In other words, the velocity at which a signal is propagated is inversely proportional to the square root of the dielectric constant of the material in which the conductors are embedded.

Signal delay is caused by interconnect parasitic resistance and capacitance. With increasing interconnect density, the parasitic capacitance becomes an important limiting factor in signal transmission integrity. The effect is known simply as *RC* delay, or propagation delay. Propagation delay is defined as the elapsed time between the two 50% amplitude points of the transmitted and received signals, as shown in Figure 6.21.

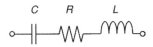

Figure 6.19 Model for a practical capacitor.

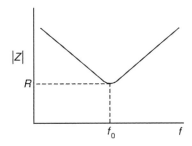

Figure 6.20 Frequency response of a capacitor.

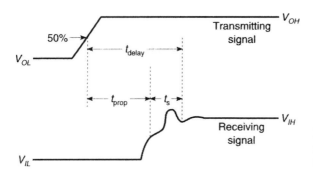

Figure 6.21 Definitions for propagation delay (t_{prop}), settling delay (t_s), and interconnect delay (t_{delay}).

Consider the equivalent circuit for an electronic packaging substrate interconnect structure consisting of a CMOS off-chip driver driving an input or a receiving gate, as shown in Figure 6.22. The driver is modeled by its Thevenin equivalent circuit; a voltage source, V_s, and an equivalent output resistance, R_s. The input gate is modeled by its equivalent capacitance, C_{in}. The transmission line is modeled by an inductance, L_l, a series DC resistance, R_l, and a capacitance, C_l. The output lead, or pad, is modeled by its parasitic inductance, L_x, and capacitance, C_x, and the input pad is modeled by its parasitic inductance, L_y, and capacitance, C_y. The propagation delay, or RC delay, is determined by the resistances of the driver and interconnect, and the capacitances of the interconnect and the receiving gate. The propagation delay, t_{prop}, is given by

$$t_{prop} = R_s(C_x + C_l + C_y + C_{in}) + \tfrac{1}{2} R_l C_l + R_l(C_y + C_{in}). \tag{6.22}$$

Equation (6.22) does not contain the effect of the parasitic inductances. However, it should be noted that the parasitic inductances (L_x, L_y, L_l) increase the propagation delay by only 10 to 30%. In practice, the time it takes for any electrical noise on the signal to settle, t_s, defined in Figure 6.21, must be taken into account in determining the interconnect delay. Thus, the interconnect delay is

$$t_{delay} = t_{prop} + t_s \tag{6.23}$$

Example 6.6

Consider the circuit shown in Figure 6.22 with the following parameters: $R_s = 50\,\Omega$, $C_x = 2$ pF, $C_l = 3$ pF, $C_y = 2$ pF, $C_{in} = 5$ pF, and $R_l = 0.05\,\Omega$. Determine the maximum operating frequency for this circuit.

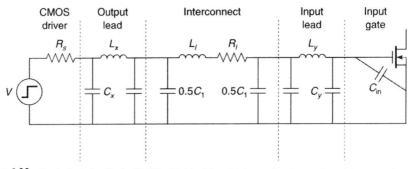

Figure 6.22 Equivalent circuit of a CMOS off-chip driver driving an input gate via an interconnect.

SOLUTION The propagation delay of the circuit can be found by using Eq. (6.22):

$$t_{\text{prop}} = 50\,\Omega(12\,\text{pF}) + 0.5(0.05\,\Omega)(3\,\text{pF}) + 0.05\,\Omega(7\,\text{pF}) = 6 \times 10^{-10}\,\text{s}$$

It should be noted that all but the first term in the equation can be neglected. The maximum operation frequency of the circuit is the inverse of the propagation delay. Thus,

$$f_{\max} = \frac{1}{t_{\text{prop}}} = 1.66 \times 10^9\,\text{Hz} = 1.66\,\text{GHz}$$

■

Example 6.6 shows what can be done with so-called lumped-circuit analysis. That is, the capacitance and inductance of a signal interconnect, which we know are distributed along the signal line, are modeled as discrete circuit elements. There is a lot of "know-how" that determines when this kind of analysis will do the job, or when the more involved transmission line (distributed parameter) analysis described in the next section must be invoked. The well-known rule of thumb is that a transmission line model must be used if $t_r < 5t_f$, where t_r is the rise time and t_f is the time of flight along the line. For polyimide and silicon dioxide dielectrics, the propagation velocity is 15 cm/ns; with epoxy–glass, it is 13 cm/ns. Thus, the time of flight along the line is easily determined from layout geometry. The rise time is determined by the chip technology (typically 1 to 3 ns for CMOS and 0.5 to 1 ns for BiCMOS packaged ICs), and by the ratio of source resistance, R_s, to line impedance, Z_0. If $R_s \gg Z_0$, transmission line effects can be neglected for any length line since the line charging rate will be dominated by RC delay.

The above rule of thumb simply tells us when the inductance of a line will be important and when it will not. For short lines, an RC model such as that shown in Figure 6.23 is appropriate. For longer lines, the inductance, L, limits the ability of the source to drive current fast enough to charge C, giving rise to the transmission line model shown in Figure 6.24.

Any line can be modeled as a set of cascaded sections of this type. Since line lengths of interconnects, particularly in some interconnect structures, often are just a little too long to allow RC models, it is useful to model the line as two or three sections of lumped RLC rather than invoking a transmission line analysis program, many of which do not handle the lossy lines typical of thin-film interconnects. Particularly for the bused interconnections

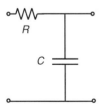

Figure 6.23 RC model for a short transmission line.

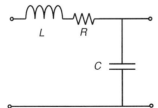

Figure 6.24 Model for a long transmission line.

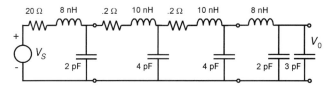

Figure 6.25 Model of transmission line using several sections of lumped *RLC*.

typical of many electronic packaging substrates, the total line length may be long, but the bus may consist of many sections that can be modeled with the *RLC* model.

Example 6.7

A CMOS chip mounted in a 68-lead chip carrier drives a 4-cm length of line with $C = 2$ pF/cm, $L = 5$ nH/cm, and $R = 0.1$ Ω/cm. The receiver is also in a 68-lead chip carrier. Model the circuit and determine the signal delay if the driver has a 1-ns rise time and a 20-Ω output impedance.

SOLUTION The model is as shown in Figure 6.25. The time of flight is 4/15 ns, or 0.266 ns. Five times this is 1.33 ns, which is greater than t_r, so transmission line analysis must be used. For simplicity, the line is modeled as two 2-cm long sections in series. The 8-nH/2-pF loading at each end is the average lead loading of a 68-lead surface-mount device (SMD).

The circuit is easily modeled using pSPICE. The input and output waveforms are shown in Figure 6.26. ∎

The concept of crosstalk, the influence of one signal line on another because of parasitic inductive and capacitive coupling, has been alluded to and is discussed in the transmission line case in the next section. Modeling crosstalk between a driven signal line and a quiet

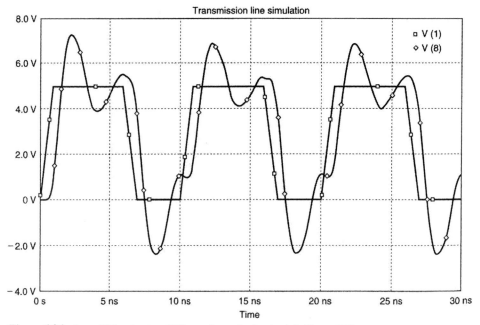

Figure 6.26 Input (V1) and output (V8) waveforms for the circuit in Figure 6.25.

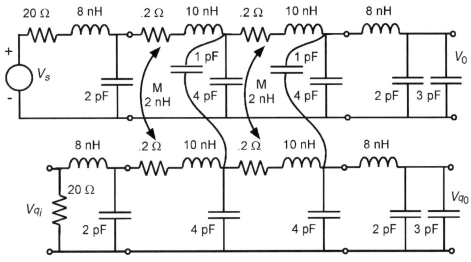

Figure 6.27 Circuit of figure 6.25 with coupling capacitance and mutual inductance added.

signal line is straightforward in many applications. The line geometries and coupled length determine the mutual inductance and coupling capacitance.

Consider the line in the previous example with an adjacent 4-cm line coupled to the driven line by 0.5 pF/cm of coupling capacitance and 1 nH/cm of mutual inductance. The resulting circuit is shown in Figure 6.27. The voltages induced on the quiet line at V_{qi} and V_{qo} are shown in Figure 6.28 for the same driving pulse as the previous example.

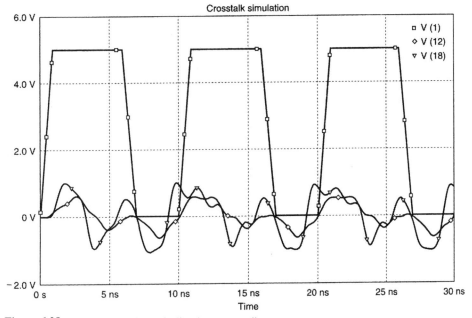

Figure 6.28 Voltage induced on quiet line due to crosstalk.

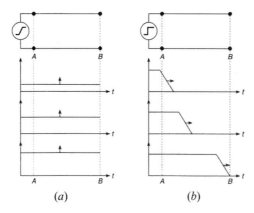

Figure 6.29 Signal transmission in a pair of interconnects with a digital signal of (*a*) a slow rise time and (*b*) a fast rise time.

6.4 TRANSMISSION LINES

In high-speed devices, interconnects must operate in an acceptable manner from DC to high frequencies. However, it is difficult to design a properly terminated, wide-bandwidth, microelectronic transmission system since discontinuities, such as vias and bends in signal conductors, adversely affect the signal waveform. Thus, the proper design of a system requires a thorough time-domain analysis of all interconnects, including the effects of components such as bond wires to IC chips and the chips themselves.

Figures 6.29*a* and 6.29*b* compare the signal transmission environments in a pair of interconnects for digital signals with two different rise times. The signal with a slower rise time should be modeled by a lumped-element circuit, as described in the previous section, since the signal in the interconnect is equal in magnitude and phase along its entire length. On the other hand, when the rise time of the signal is fast, it can be seen from Figure 6.29*b* that the voltage signal varies in magnitude and phase over the length of the interconnect. Thus, a transmission line is a distributed-parameter network where voltages and currents can vary in both magnitude and phase over its length. Consequently, a transmission line is often represented schematically as a two-wire line, as shown in Figure 6.30, and can be modeled as a lumped-element circuit, as shown in Figure 6.31, where L represents the total self-inductance per unit length of the two conductors, C represents the shunt capacitance per unit length due to the close proximity of the two conductors, R represents the resistance

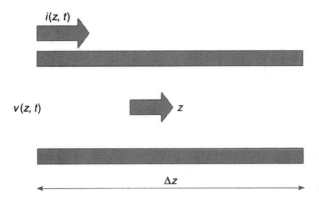

Figure 6.30 Voltage and current definitions for an incremental length of a transmission length.

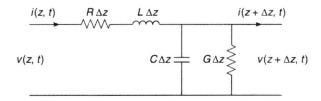

Figure 6.31 Lumped circuit model for a transmission line.

per unit length due to the finite conductivity of the conductors, and G represents the shunt conductance per unit length due to the dielectric loss in the material.

Applying Kirchhoff's voltage law to the lumped-element equivalent circuit gives

$$v(z,t) - R\,\Delta z\, i(z,t) - L\,\Delta z\frac{di(z,t)}{dt} - v(z+\Delta z,t) = 0 \qquad (6.24)$$

and applying Kirchhoff's current law yields

$$i(z,t) - G\,\Delta z\, v(z+\Delta z,t) - C\,\Delta z\frac{dv(z+\Delta z,t)}{dt} - i(z+\Delta z,t) = 0 \qquad (6.25)$$

Taking the limit as $\Delta z \to 0$ yields the time-domain form of the transmission line equations

$$\frac{dv(z,t)}{dz} = -Ri(z,t) - L\frac{di(z,t)}{dt} \qquad (6.26)$$

and

$$\frac{di(z,t)}{dz} = -Gv(z,t) - C\frac{dv(z,t)}{dt} \qquad (6.27)$$

For the sinusoidal steady-state condition, with cosine-based phasors or cosine-based time-harmonic excitation, the above equations can be simplified to

$$\frac{dV(z)}{dz} = -(R + j\omega L)I(z) \qquad (6.28)$$

and

$$\frac{dI(z)}{dz} = -(G + j\omega C)V(z) \qquad (6.29)$$

where ω is the angular frequency. These two equations can be solved simultaneously to yield wave equations for $V(z)$ and $I(z)$ as

$$\frac{d^2 V(z)}{dz^2} - \gamma^2 V(z) = 0 \qquad (6.30)$$

and

$$\frac{d^2 I(z)}{dz^2} - \gamma^2 I(z) = 0 \qquad (6.31)$$

where

$$\gamma = \alpha + j\beta = \sqrt{(R + j\omega L)(G + j\omega C)} \qquad (6.32)$$

is the complex propagation constant, β is the phase constant or wave number, and α is the attenuation constant. The solutions to the wave equations are

$$V(z) = V_0^+ e^{-\gamma z} + V_0^- e^{\gamma z} \qquad (6.33)$$

and

$$I(z) = I_0^+ e^{-\gamma z} + I_0^- e^{\gamma z} \tag{6.34}$$

where the $e^{-\gamma z}$ term represents wave propagation in the +z direction, and the $e^{\gamma z}$ term represents wave propagation in the $-z$ direction. Thus, the voltage wave at any point, z, along the transmission line is the sum of the forward-traveling voltage wave and the backward-traveling voltage wave. The current on the transmission line can be obtained by substituting Eq. (6.33) into Eq. (6.28) to yield

$$I(z) = \frac{\gamma}{R + j\omega L} \left(V_0^+ e^{-\gamma z} - V_0^- e^{\gamma z} \right) \tag{6.35}$$

The characteristic impedance of the transmission line is defined as

$$Z_0 = \frac{R + j\omega L}{\gamma} = \sqrt{\frac{R + j\omega L}{G + j\omega C}} \tag{6.36}$$

and the voltages and currents on the transmission line are related by

$$\frac{V_0^+}{I_0^+} = Z_0 = -\frac{V_0^-}{I_0^-} \tag{6.37}$$

Thus, the characteristic impedance of the transmission line in ohms is the ratio of the voltage and current traveling in the same direction. The current on the transmission line can also be written as

$$I(z) = \frac{V_0^+}{Z_0} e^{-\gamma z} - \frac{V_0^-}{Z_0} e^{\gamma z} \tag{6.38}$$

The time-domain voltage expression is

$$v(z, t) = |v_0^+| \cos(\omega t - \beta z + \phi^+) e^{-\alpha z} + |v_0^-| \cos(\omega t + \beta z + \phi^-) e^{\alpha z} \tag{6.39}$$

where ϕ^{+-} is the phase angle of the complex voltage v_0^{+-}. The wavelength on the transmission line is

$$\lambda = \frac{2\pi}{\beta} \tag{6.40}$$

and the phase velocity is

$$v_p = \frac{\omega}{\beta} = \lambda f \tag{6.41}$$

The phase velocity is defined as the velocity at which a fixed phase point on the wave travels.

In many practical transmission lines, the loss of the line is very small so, with $R = G = 0$, the propagation constant simplifies to

$$\gamma = \alpha + j\beta = j\omega\sqrt{LC} \tag{6.42}$$

and $\alpha = 0$.

The characteristic impedance of a lossless transmission line is

$$Z_0 = \sqrt{\frac{L}{C}} \tag{6.43}$$

which is constant so long as a constant geometrical relationship is maintained between the signal conductor and the reference plane. However, if the line narrows, its capacitance decreases and its characteristic impedance increases.

The general solutions for voltage and current on a lossless transmission line are

$$v(z) = v_0^+ e^{-j\beta z} + v_0^- e^{j\beta z} \tag{6.44}$$

and

$$i(z) = \frac{v_0^+}{Z_0} e^{-j\beta z} - \frac{v_0^-}{Z_0} e^{j\beta z} \tag{6.45}$$

The wavelength is

$$\lambda = \frac{2\pi}{\beta} = \frac{2\pi}{\omega\sqrt{LC}} \tag{6.46}$$

and the phase velocity is

$$v_p = \frac{\omega}{\beta} = \frac{1}{\sqrt{LC}} \tag{6.47}$$

Example 6.8

Consider a lossless two-wire transmission line in air, as shown in Figure 6.30. The series inductance per unit length and shunt capacitance per unit length are 10 nH/m and 10 pF/m, respectively. Determine the characteristic impedance for this transmission line.

SOLUTION   From Eq. (6.42), the characteristic impedance is

$$Z_0 = \sqrt{\frac{10 \times 10^{-9}}{10 \times 10^{-12}}} = 31.6\,\Omega.$$

∎

6.4.1 Microstrip Transmission Lines

The most common type of transmission line for interconnects is the microstrip line depicted in Figure 6.32. The fact that the microstrip is not completely embedded in the dielectric complicates its analysis because it has some of its field lines in the dielectric region, concentrated

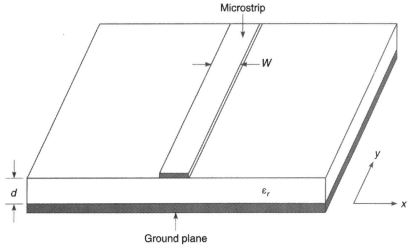

Figure 6.32   Microstrip transmission line.

between the conductor and the ground plane, and some fraction in the air region above and to the sides of the conductor. Since the speed of signal propagation in the microstrip is related to the dielectric material by $(3 \times 10^8)/\sqrt{\varepsilon_r}$ m/s, it is expected that two different velocities exist in the dielectric region and in the air $(3 \times 10^8$ m/s) such that the microstrip cannot support a pure transverse-electromagnetic (TEM) wave. A TEM wave has neither E_z nor H_z field components. Therefore, an exact analysis of the microstrip is complicated. An approximate analysis yields the effective dielectric constant, which can be interpreted as the dielectric constant of a homogeneous medium that replaces the air and dielectric regions of the microstrip. The effective dielectric constant of a microstrip is given approximately by [6]:

$$\varepsilon_e = \frac{\varepsilon_r + 1}{2} + \frac{\varepsilon_r - 1}{2\sqrt{1 + \frac{12d}{W}}} \tag{6.48}$$

where ε_r is the relative dielectric constant of the actual dielectric region of the structure. Given the dimensions of the microstrip, the characteristic impedance can be calculated as

$$Z_0 = \frac{60}{\sqrt{\varepsilon_e}} \ln\left(\frac{8d}{W} + \frac{W}{4d}\right) \tag{6.49}$$

for $W/d \leq 1$ and

$$Z_0 = \frac{120\pi}{\sqrt{\varepsilon_e}\left[\frac{W}{d} + 1.393 + 0.667\ln\left(\frac{W}{d} + 1.444\right)\right]} \tag{6.50}$$

for $W/d \geq 1$. As noted previously, the characteristic impedance of a microstrip can be increased by decreasing the width of the signal line. However, the presence of adjacent lines also reduces the characteristic impedance of a microstrip.

In electronic packaging substrates, buried microstrip lines and offset strip lines, as shown in Figures 6.33 and 6.34, respectively, are more commonly encountered than the surface microstrip. The characteristic impedance for a buried microstrip is

$$Z_0 = \frac{60}{\sqrt{\varepsilon_r + 1.41}} \ln\left(\frac{5.98h}{0.8w + t}\right) \tag{6.51}$$

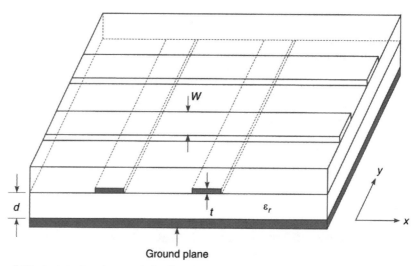

Figure 6.33 Buried microstip transmission lines.

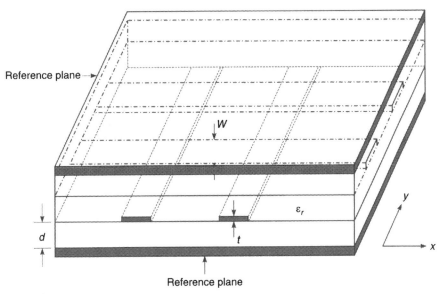

Figure 6.34 Two layers of offset strip lines.

For the same h and ε_r, a buried microstrip line must be narrower or thinner than a surface microstrip to achieve a similar characteristic impedance.

There is no simple empirical relationship to describe the characteristic impedance for an offset strip line, but it has a lower characteristic impedance than the same size microstrip line with a similar dielectric constant since it has two reference planes. In electronic packaging substrates, the reference planes (either power or ground) tend to be meshed to promote adhesion between layers, especially if a polyimide dielectric is used. This allows trapped solvent to escape from the polyimide during subsequent cures. The meshing not only increases the DC resistance of the power and ground planes but also affects the characteristic impedance and propagation delay of the signal lines.

6.4.2 Termination Reflections

When a transmission line is terminated by an arbitrary load (i.e., Z_L is not equal to Z_0), the voltage and current on the line are the sum of the incident and reflected waves, and the total voltage and current at the load are related by the load impedance. So, at $z = 0$,

$$Z_L = \frac{V(0)}{I(0)} = \frac{V_0^+ + V_0^-}{V_0^+ - V_0^-} Z_0. \tag{6.52}$$

Solving for V_0^-,

$$V_0^- = \frac{Z_L - Z_0}{Z_L + Z_0} V_0^+ \tag{6.53}$$

The voltage reflection coefficient, Γ, is defined as the amplitude of the reflected voltage wave normalized to the amplitude of the incident voltage wave, or

$$\Gamma = \frac{V_0^-}{V_0^+} = \frac{Z_L - Z_0}{Z_L + Z_0} \tag{6.54}$$

The total voltage and current waves on the line can then be written as

$$V(z) = V_0^+ \left(e^{-j\beta z} + \Gamma e^{j\beta z}\right) \quad (6.55)$$

and

$$I(z) = \frac{V_0^+}{Z_0} \left(e^{-j\beta z} - \Gamma e^{j\beta z}\right) \quad (6.56)$$

Thus, the voltage and current on the line consist of the superposition of the incident and reflected waves. Such waves are called standing waves. When the load impedance is equal to the characteristic impedance, $\Gamma = 0$, and there is no reflected wave. Such a load is said to be matched to the line.

When a transmission line is terminated in a short circuit, $Z_L = 0$, $\Gamma = -1$, and the reflected wave is inverted with respect to the incident wave. The voltage and current on the transmission line are given by

$$V(z) = V_0^+ \left(e^{-j\beta z} - e^{j\beta z}\right) = -2j V_0^+ \sin \beta z \quad (6.57)$$

and

$$I(z) = \frac{V_0^+}{Z_0} \left(e^{-j\beta z} + e^{j\beta z}\right) = \frac{2 V_0^+}{Z_0} \cos \beta z \quad (6.58)$$

From these equations, it can be seen that the voltage is zero and the current is maximum at the load.

The input impedance is

$$Z_{\text{in}} = j Z_0 \tan \beta l \quad (6.59)$$

Note that the input impedance is purely imaginary for any length l, so it can only have values between $j\infty$ and $-j\infty$. When $l = 0$, $Z_{\text{in}} = 0$. However, when $l = \lambda/4$, $Z_{\text{in}} = \infty$ (i.e., an open circuit). Note that the input impedance is periodic in l, repeating for multiples of $\lambda/2$.

When the load is an open circuit, such as the input of a gate (i.e., $Z_L = \infty$), the voltage and current on the transmission line are

$$V(z) = V_0^+ \left(e^{-j\beta z} + e^{j\beta z}\right) = 2 V_0^+ \cos \beta z \quad (6.60)$$

and

$$I(z) = \frac{V_0^+}{Z_0} \left(e^{-j\beta z} - e^{j\beta z}\right) = \frac{-2j V_0^+}{Z_0} \sin \beta z \quad (6.61)$$

In this case, the current is zero, the voltage is maximum at the load, and the input impedance is

$$Z_{\text{in}} = -j Z_0 \cot \beta l \quad (6.62)$$

which is purely imaginary for any length l.

Now consider terminated transmission lines of some particular lengths. If $l = \lambda/2$, then

$$Z_{\text{in}} = Z_L \quad (6.63)$$

Thus, a half-wavelength line (or any multiple of $\lambda/2$) does not alter or transform the load impedance, regardless of the characteristic impedance. If the line is a quarter-wavelength long, or, more generally, $l = \lambda/4 + n\lambda/2$, for $n = 1, 2, 3, \ldots$, the input impedance is given

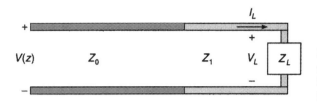

Figure 6.35 Transmission line of characteristic impedance Z_0 feeding a line of different characteristic impedance Z_1.

by

$$Z_{in} = \frac{Z_0^2}{Z_L} \quad (6.64)$$

Such a line is known as a quarter-wave transformer since it has the effect of transforming the load impedance in an inverse manner, depending on the characteristic impedance of the line.

Consider a transmission line of characteristic impedance Z_0 feeding a line of a different characteristic impedance, Z_1, as shown in Figure 6.35. If the line is infinitely long, or if it is terminated in its own characteristic impedance so that there are no reflections from its ends, then the input impedance seen by the feed line is Z_1, with a reflection coefficient, Γ, of

$$\Gamma = \frac{Z_1 - Z_0}{Z_1 + Z_0} \quad (6.65)$$

Note that not all of the incident wave is reflected. Some of it is transmitted onto the second line with a voltage amplitude given by Γ. The voltage for $z < 0$ is

$$V(z) = V_0^+ (e^{-j\beta z} + \Gamma e^{j\beta z}) \quad (6.66)$$

where V_0^+ is the amplitude of the incident voltage wave on the feed line. The voltage wave for $z > 0$, in the absence of reflection, is outgoing only and is given by

$$V(z) = V_0^+ T e^{-j\beta z} \quad (6.67)$$

The voltage must be continuous at $z = 0$ and yields for the transmission coefficient, T,

$$T = 1 + \Gamma = 1 + \frac{Z_1 - Z_0}{Z_1 + Z_0} = \frac{2 Z_1}{Z_1 + Z_0} \quad (6.68)$$

The insertion loss (IL) given by

$$\text{IL} = -20 \log |T| \, dB \quad (6.69)$$

is defined as the transmission coefficient between two points in a circuit.

Now consider a load resistance, R_L, driven by a gate driver through a transmission line, as shown in Figure 6.36. The successive reflections of the wave from each end of the

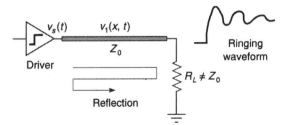

Figure 6.36 Driver driving a load resistance, R_L, through a transmission line with Z_0.

220 Chapter 6 Electrical Considerations, Modeling, and Simulation

transmission line create a damped ringing signal. The expression for the voltage traveling down the line as a function of distance and time can be written as

$$V_I(x,t) = V_s(t)u(t - xt_d) \qquad \text{for } t < t_d \qquad (6.70)$$

where t_d is the propagation delay of the line, $u(t)$ is the unit step function occurring at $t = 0$, and $V_s(t)$ is the source voltage at the sending end of the line. When the incident voltage, $V_I(x,t)$, reaches the end of the long line, a reflected voltage, v_r, will occur if R_L is not equal to the characteristic impedance of the line. The reflection coefficient at the load, Γ_L, can be obtained using Ohm's law. The voltage at the load is $v_I + v_r$, which must be equal to $(I_I + I_r) R_L$ by Ohm's law, where $I_I = v_I/Z_0$ and $I_r = -v_r/Z_0$ (the minus sign is due to the fact that v_r is traveling toward the source). Therefore,

$$v_I + v_r = \left(\frac{v_I}{Z_0} - \frac{v_r}{Z_0}\right) R_L \qquad (6.71)$$

Hence, the reflection coefficient is given by

$$\Gamma_L = \frac{R_L - Z_0}{R_L + Z_0} \qquad (6.72)$$

and the reflection coefficient at the source is

$$\Gamma_s = \frac{R_s - Z_0}{R_s + Z_0} \qquad (6.73)$$

A general expression for the total line voltage at any point on the line, x, as a function of time, t, can be found by summing the incident voltage from Eq. (6.70) with similar voltage contributions from the various orders of reflection due to Γ_L and Γ_s. Thus

$$v(x,t) = v_s(t)[u(t - t_d x) + \Gamma_L u[t - t_d(2l - x)] + \Gamma_L \Gamma_s u[t - t_d(2l + x)] \\ + \Gamma_L^2 \Gamma_s u[t - t_d(4l - x)] + \Gamma_L^2 \Gamma_s^2 u[t - t_d(4l + x)] + \cdots + V_{dc} \qquad (6.74)$$

From this equation, it is evident that higher order reflection coefficient terms are present in $v(x,t)$ as time progresses. Thus, ringing is a potential problem if the time it takes for the wave to travel down the line is longer than one-quarter of its rise time.

A space–time or reflection diagram can be used to illustrate the concept embedded in the above equation. In the reflection diagram, distance along the transmission line is plotted along one axis and transit time along the other, so that a wave traveling at a constant velocity is represented by a line of constant slope [7]. Consider the simple transmission line circuit shown in Figure 6.37a with a constant voltage source, V_s, and a source impedance of Z_0. The switch closes at $t = 0$ to simulate a step function voltage applied to the input end of the transmission line at point A. Initially, the voltage at A is $+V_s/2$ since the source impedance is equal to the characteristic impedance of the transmission line. The voltage reflection coefficient is $+1$ at the open circuit end (C), and it is zero at the source since the source impedance is equal to the characteristic impedance of the transmission line, Z_0. The waveform at any point along the transmission line can be obtained by summing the voltage waveforms shown in Figure 6.37b. At the midpoint of the transmission line, a wave of magnitude $+V_s/2$ arrives at $T/2$, where T is the transit time or the time for a wave to propagate once along the transmission line. A similar voltage wave arrives at the midpoint of the transmission line at $3T/2$. Thereafter, the voltage at B remains at $+V_s$. The waveform at A corresponds to the initial wave $+V_s/2$, which appears at $t = 0$, followed by the reflected wave $+V_s/2$ at $t = 2T$. At C, the waveform is found by considering the waveform a short

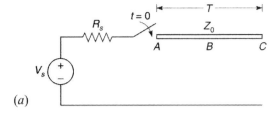

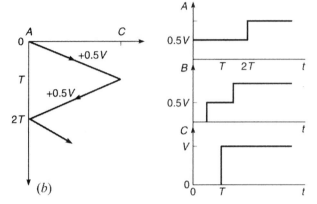

Figure 6.37 (a) simple transmission line circuit with a constant-voltage source and a source impedance of Z_0 and (b) reflection diagram and voltage waveforms.

distance Δx from the discontinuity and then allowing Δx to tend to zero. It can be seen from the figure that the two waves appear to converge on $t = T$ at C, and so the voltage there is zero until $t = T$, when it rises to $+V_s$. Note that, although the voltage waves have the same polarity, the currents corresponding to the forward- and backward-traveling waves are opposed to one another. Thus, at any point on the line, the current rises to $+V_s/2Z_0$ upon arrival of the initial voltage wave and falls to zero when the second voltage wave arrives to charge that point on the line to $+V_s$. At the input, the current is $+V_s/2Z_0$ from $t = 0$ to $t = 2T$.

The four most common interconnect termination schemes in electronic packaging substrates are shown in Figures 6.38 to 6.41. However, the Thevenin equivalent parallel termination schemes shown in Figures 6.38 and 6.39 are not generally used for CMOS since the DC voltage at the common terminal of the two resistors usually falls between the high and low input voltage when no signal is present on the line. The AC termination scheme shown in Figure 6.40 prevents a DC current from flowing through the load termination, resulting in a reduction in power consumption [8]. The series termination scheme of Figure 6.41 does not allow the reflected signal to be re-reflected, thus reducing ringing noise on the line.

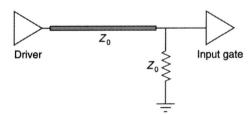

Figure 6.38 Matched parallel termination scheme.

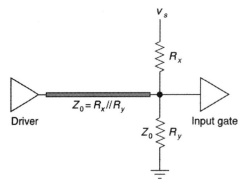

Figure 6.39 Matched parallel termination scheme.

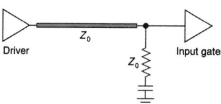

Figure 6.40 AC termination scheme.

The smallest propagation delay on a transmission line is obtained if the first signal arriving at the end of the transmission line attains sufficient magnitude to switch the receiving gate. This is known as *first incident switching*. The reflection noise must be small for the settling delay to be close to zero. It should be noted that the propagation delay may be as much as five times the time-of-flight delay if first incident switching is not achieved.

Example 6.9

Consider the interconnection system shown in Figure 6.42. The source impedance is $3Z_0$, while the load impedance is $Z_0/4$, where Z_0 is the characteristic impedance of the transmission line. Draw the reflection diagram and the voltage waveforms at the input and output ends of this system.

SOLUTION The reflection coefficient at the source is

$$\Gamma_s = \frac{3Z_0 - Z_0}{3Z_0 + Z_0} = \frac{1}{2}$$

and the reflection coefficient at the load is

$$\Gamma_L = \frac{\frac{Z_0}{4} - Z_0}{\frac{Z_0}{4} + Z_0} = -\frac{3}{5}$$

Figure 6.43 shows the reflection diagram and voltage waveforms at A and B for the interconnection system. As can be seen, the voltage at A is initially at 1 V for $2T$.

From $2T$ to $4T$, the voltage at A can be calculated using Eq. (6.74) as

$$V'_A = V_s + \Gamma_L(V_s) + \Gamma_L\Gamma_s(V_s) = 1 + \left(-\tfrac{3}{5}\right) + \left(\tfrac{1}{2}\right)\left(\tfrac{3}{5}\right)(1) = \tfrac{1}{10} V$$

From $4T$ to $6T$, the voltage at A is

$$V''_A = V'_B + \Gamma_L{}^2\Gamma_s{}^2(V_s) = \tfrac{14}{50} + \left(-\tfrac{3}{5}\right)^2\left(\tfrac{1}{2}\right)^2(1) = \tfrac{37}{100} V$$

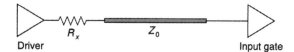
Figure 6.41 Series termination scheme.

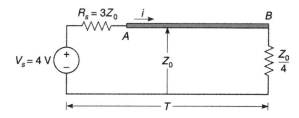
Figure 6.42 Interconnect with a source impedance of $3Z_0$ and a load impedance of $Z_0/4$.

The voltage at the load is initially 0 V from 0 to T, then it is 2/5 V from T to $3T$, and 14/50 V from $3T$ to $5T$. ∎

6.4.3 Signal Line Losses and Skin Effect

The predominant signal interconnect losses are resistive losses. Dielectric losses are important only at very high frequencies. The effect of the resistive loss in a transmission line is to attenuate the signal voltage and increase its rise time, as shown in Figure 6.44. The transmitted voltage is given by

$$V_t = V_s \, e^{-(RL/2 Z_0)} \tag{6.75}$$

where R is the resistance per unit length and L is the length of transmission line.

At signal frequencies, current concentrates in the skin of a conductor with the current density decreasing with distance from the outer surface. The skin depth, δ_s, is defined as the

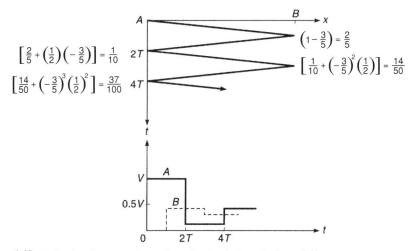
Figure 6.43 Reflection diagram and waveforms for circuit shown in figure 6.42.

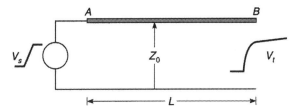

Figure 6.44 Effect of resistive loss of the transmission line on the signal.

depth at which the current density is 37% of the current density at the conductor surface. Accordingly,

$$\delta_s = \sqrt{\frac{\rho}{\pi \mu f}} \qquad (6.76)$$

where ρ is the resistivity of the conductor, μ is the magnetic permeability, and f is the signal frequency. The current density decreases exponentially to about 5% of its surface value at three times the skin depth. If the signal line thickness exceeds $6\delta_s$, the series resistance is approximately proportional to the skin depth, that is, linearly proportional to the square root of the operating frequency. The skin effect causes a significant increase in resistance at frequencies above those for which the skin depth becomes half the thickness of the conductor. Thus, skin depth plays a significant role in signal loss for digital systems operating at clock frequencies above several hundred megahertz. For a rectangular signal line over a flat reference plane, the skin-effect resistance per unit length, R_s, is given by [9]

$$R_s = \frac{1}{\sigma} \frac{\sqrt{\pi f \sigma \mu}}{2(W+T)} \qquad (6.77)$$

where W is the width of the signal line, σ is the conductivity of the conductor, and T is the thickness of the dielectric layer. Table 6.2 lists the skin depth at 10 GHz for the metals commonly used in electronic packages. It shows that most of the current flow in the conductor occurs in an extremely thin region near the surface of the conductor.

Example 6.10

Determine the frequency at which the skin effect results in a 50% loss in effective thickness for a 2-µm-thick and 10-µm-wide aluminum conductor.

SOLUTION For a 50% loss in thickness, the skin depth, δ_s, is 1 µm. Therefore,

$$f = \frac{\rho}{\pi \mu_0 \delta_s^2} = 7092 \, \text{MHz}$$

■

6.4.4 Net Topology

A net is defined as the network of interconnects in an electronic packaging substrate that join a set of digital drivers and receivers. The word *topology* refers to the "science of place," so the topological properties of a circuit are those that are invariant with stretching, squeezing, bending, or twisting of the circuit graph. Thus, two networks are topologically equivalent if they differ only in the circuit elements that make up their branches.

Table 6.2 Skin Depths at 10 GHz for Several Common Metals

Metal	Skin Depth (μm)
Aluminum	0.81
Copper	0.66
Gold	0.79
Silver	0.64

The maximum speed at which an electronic packaging substrate can operate is determined by the propagation delay of its "critical" net. Thus, the selection of a net topology is an important design criterion. Figure 6.45 shows the net topologies used to minimize reflection noise in an electronic packaging substrate, and some particular points should be made about these particular topologies. For example, a ring topology does not require a matching termination at the expense of increased interconnect length. The most common point to the multipoint topology is the daisy chain, and the stub length is selected by the amount of acceptable ring noise and the rise time requirement. The faster the rise time of the signal, the shorter the stub length must be.

For a high-speed electronic system, the point-to-point net or the far-end terminated net may be used to yield only one time-of-flight across the signal line length. The point-to-point net requires a large number of drivers and signal input/output (I/O) on the package. The far-end terminated net requires a small output impedance for the drivers and a small DC resistance for the signal line. To reduce DC power consumption, the far end of a signal line may be left open-circuited to eliminate the DC current drain on the driver circuit after

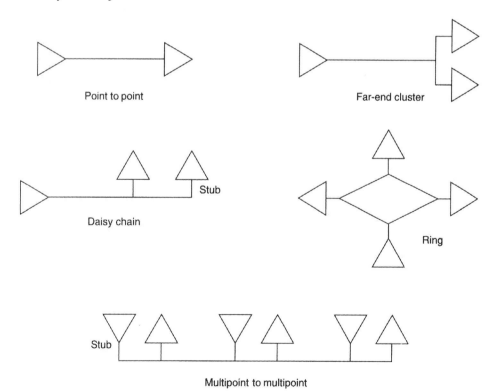

Figure 6.45 Net topologies used to minimize reflection noise in an MCM.

the receiver circuits have completed the transition. To minimize the instantaneous power during switching of the driver circuits, their output impedance should be increased. For a net with several receiver circuits connected at various points along the signal line, and with the far end open circuited, it takes two time-of-flights or longer to bring the input voltage of all receiver circuits above their threshold voltage [10].

6.5 COUPLED NOISE OR CROSSTALK

Coupled noise, or crosstalk, is the electrical noise caused by mutual inductance and capacitance between signal conductors due to their close proximity to each other. In other words, a signal can appear on one interconnect because of coupling from another interconnect. Thus, if not controlled through proper design, crosstalk can cause digital system failure due to false signals appearing on an interconnect. Figure 6.46 shows a possible electric field distribution for a signal conductor in an electronic packaging substrate environment. As can be seen, the conductor on the right has an electric field coupled to it. Thus, the two signal conductors are coupled by the electric field, resulting in crosstalk. Since the electric field coupling between the center signal conductor and the left conductor is very weak, the crosstalk between the two is also very small. Thus, the degree of crosstalk between two adjacent signal conductors depends on their proximity to each other. Consequently, the maximum crosstalk allowed for a given electronic circuit determines the minimum spacing between conductors, and thus, determines the minimum allowable signal line pitch. Typical signal conductor pitch in electronic packaging substrates is about 100 to 250 µm in order to limit the amount of crosstalk to an acceptable level. As the spacing increases, the mutual inductance and capacitance decrease, as does the crosstalk noise level.

Figure 6.47a shows two identical signal transmission lines terminated by the characteristic impedance of the line. A voltage wave is assumed to propagate along line 1 from the source at A to the matched load at B. Since the source and load impedances are equal to the characteristic impedance of the line, the voltage wave is completely absorbed. Induced waves on line 2 are also absorbed by the matched loads so that multiple reflections are eliminated. Figure 6.47b shows the equivalent circuit for the system of Figure 6.47a. The transmission line is modeled by its inductance L, capacitance C, mutual inductance L_m, and mutual capacitance C_m. The capacitive coupling, K_c, and inductive coupling, L_c, are defined as

$$K_c = \frac{C_m}{C} \qquad (6.78)$$

and

$$L_c = \frac{L_m}{L} \qquad (6.79)$$

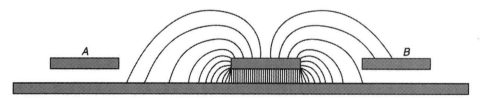

Figure 6.46 Electric field distribution for signal conductors in an MCM environment.

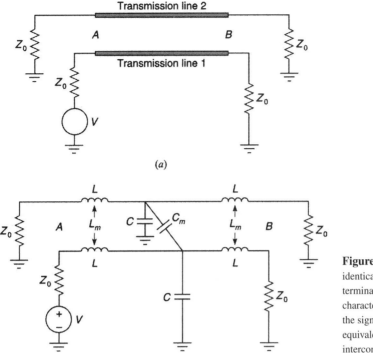

Figure 6.47 (a) two identical signal lines terminated by the characteristic impedance of the signal line and (b) equivalent circuit for the interconnect system shown in figure 6.47a.

where C_m is the mutual capacitance between the lines in per unit length, C is the self-capacitance per unit length of each line, L_m is the mutual inductance per unit length, and L is the self-inductance per unit length of the line. Consider the case for weak coupling where the effects on line 1 of any voltages and currents induced in line 2 can be neglected. Also, assume that the characteristic impedance of the two lines is not affected by the capacitive and inductive couplings. Then, the magnitude of the voltage waves induced on line 2 should not exceed 10 to 20% of the magnitude of the voltage wave on line 1. Figure 6.48 shows the induced voltages due to inductive coupling between the lines. Assuming that the rate of change of current on line 1 is di/dt, the voltage induced in a small section dx of line 2 is

$$dv_1 = (L_m\,dx)\frac{di}{dt} \tag{6.80}$$

In terms of the characteristic impedance, the above equation can be expressed as

$$dv_1 = \left(\frac{L_m\,dx}{Z_0}\right)\left(\frac{dv}{dt}\right) \tag{6.81}$$

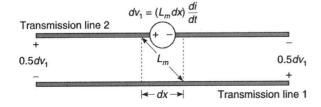

Figure 6.48 Induced voltage on transmission line 2 due to inductive coupling.

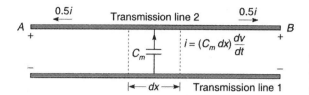

Figure 6.49 Induced voltage on transmission line 2 due to capacitive coupling.

The polarity of the induced voltage is such that it opposes the change in flux linkage between the two lines, according to Lenz's law. From Kirchhoff's voltage law, the induced voltage produces a voltage wave $+0.5(dv_1)$ traveling toward point A and a voltage wave $-0.5(dv_1)$ traveling toward B, as indicated in Figure 6.48. At the same section, the mutual capacitance experiences a rate of change of voltage dv/dt, thus injecting a current $(C_m\,dx)dv/dt$ into line 2. From Kirchhoff's current law, this current divides as shown in Figure 6.49, and results in voltage waves traveling toward A and B given by

$$dv_c = 0.5(C_m\,Z_0\,dx)\left(\frac{dv}{dt}\right) \tag{6.82}$$

The induced incremental voltage wave on line 2 is the result of both inductive and capacitive coupling, and it consists of forward- and backward-traveling components. The forward-traveling wave, propagating toward B, is given by the expression

$$dv^+ = 0.5\left(\frac{dv}{dt}\right)\left(C_m\,Z_0 - \frac{L_m}{Z_0}\right)dx \tag{6.83}$$

and the backward-traveling wave, propagating toward A, is given by

$$dv^- = 0.5\left(\frac{dv}{dt}\right)\left(C_m\,Z_0 + \frac{L_m}{Z_0}\right)dx \tag{6.84}$$

The formation of the forward-traveling wave on the coupled line is shown in Figure 6.50. At B, the induced voltage wave can be found by integrating Eq. (6.83) over the entire length of the line according to

$$v^+ = \int_0^l 0.5\left(\frac{dv}{dt}\right)\left(C_m\,Z_0 - \frac{L_m}{Z_0}\right)dx \tag{6.85}$$

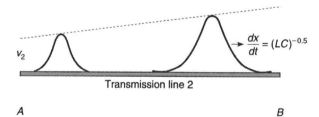

Figure 6.50 Formation of forward-traveling wave on transmission line 2 (not drawn to scale, $v_1 \gg v_2$).

Since the induced voltage wave travels at the same velocity as the voltage wave on line 1, the rate of change of the voltage is constant for each point on the wave. Thus,

$$v^+ = 0.5 \left(\frac{dv}{dt}\right)\left(C_m Z_0 - \frac{L_m}{Z_0}\right) l \tag{6.86}$$

which can also be expressed in terms of the transit time and propagation velocity of the transmission line. Substituting $Z_0 = \sqrt{L/C}$ gives

$$v^+ = 0.5 \left(\frac{dv}{dt}\right)\left(\frac{C_m}{C} - \frac{L_m}{L}\right) T \tag{6.87a}$$

or

$$v^+ = 0.5 \left(\frac{dv}{dt}\right)(K_c - L_c) T \tag{6.87b}$$

Thus, the waveform of the forward-traveling wave on line 2 is the time differential of the voltage waveform on line 1. The amplitude of the induced voltage wave is proportional to the transit time along the line and is also controlled by the line geometry. When K_c is equal to L_c, there is no forward-traveling voltage wave on line 2. For practical cases, the dominant effect is due to L_m, so that a positive-going wave on line 1 produces a negative voltage wave on line 2. Note that the induced voltage wave arrives at B in time phase with the voltage wave on line 1.

The formation of a backward-traveling voltage wave on the coupled line is shown in Figure 6.51. The incremental induced voltage wave generated at any point along line 2 begins as soon as dv/dt of the voltage wave in line 1 becomes significant at that point, and it travels in the opposite direction to the incident voltage wave, as shown in Figure 6.51. Also notice in Figure 6.51 that the incremental voltage wave generated by the passage of the wavefront moves off to position dv_1. As the wavefront progresses toward B, more incremental voltage waves are generated so that the resulting time waveform for the induced voltage at x-x is the sum of time-displaced versions of the incremental voltage wave. However, by the time incremental wave dv_2 is generated, dv_1 moves off to dv_1', so the time for which the waves overlap is effectively half that for the wavefront itself. Therefore, the expression for the backward-traveling voltage wave is [7]

$$v^- = -\frac{1}{4}\left(C_m Z_0 + \frac{L_m}{Z_0}\right) \int_0^l \frac{dv}{dt} dx \tag{6.88}$$

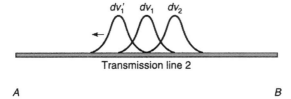

Figure 6.51 Formation of backward-traveling wave on transmission line 2 (not drawn to scale, $v_1 \gg v_2$).

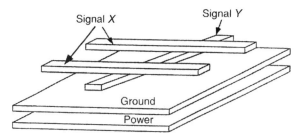

Figure 6.52 Conventional MCM design.

After integration,

$$v^- = \frac{1}{4}\left(\frac{C_m}{C} + \frac{L_m}{L}\right)v \qquad (6.89)$$

Note that the backward-traveling voltage wave has a form similar to the input signal, and it always exists since there is no possibility of cancellation, as is the case for the forward-traveling wave.

6.6 POWER AND GROUND

The power distribution design for an electronic packaging substrate is an extremely important consideration, and it is often done before the signal distribution. The reason for this is that it ultimately affects the signal distribution because it is involved in determining the DC and AC voltage drops. It is important to keep in mind that ohmic loss determines the DC voltage drops, whereas inductive and capacitive reactances determine the AC voltage drops.

A conventional MCM topology consists of the power distribution system (separate power and ground planes) in addition to two signal (X signal and Y signal) planes, as shown in Figure 6.52. This topology, a natural extension of printed wiring board construction, is not an efficient design for the manufacturing of (MCM-D's), because, for most MCM-Ds, the cost of a metal layer does not depend on the interconnect density, due to the costs associated with the deposition, photolithography, and etching techniques employed. In fact, every metal layer costs approximately the same, no matter if it is a solid plane or an interconnect layer with greater than a 300-cm/cm^2 wiring density. For these reasons, attempts have been made to maximize the density of each layer and eliminate the need for a four-layer structure. The Interconnected Mesh Power System (IMPS) is a low-cost MCM topology [11] with a low impedance power distribution structure, signal lines with controlled characteristic impedance, and low crosstalk levels using only two layers of metal. It employs a meshed power and ground distribution, as shown in Figure 6.53, where power

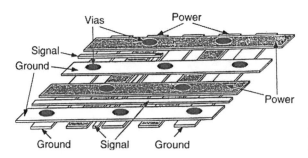

Figure 6.53 The IMPS MCM design.

and ground lines are inserted in an alternating pattern into the space between the signal lines on both metal layers. Thus, the signal lines of IMPS have adjacent power and/or ground lines that resemble a coplanar structure, as well as the intervening orthogonal lines on the lower layer, which basically form a loaded coplanar structure that can be tailored to achieve the desired characteristic impedance.

6.6.1 Dynamic Power Distribution

The purpose of a power distribution system (PDS) is to supply a constant, noise-free voltage to the integrated circuits in a system, board, electronic packaging substrate, or individual chip package. In order to do this, the PDS must exhibit a very low characteristic impedance at all frequencies where noise voltages may exist. Noise voltages are generated by the intermittent current demands of loads, especially in CMOS circuits. Internal gates in CMOS integrated circuits produce current swings as they operate, but particularly important are the off-chip drivers that may drive wide buses (64 bits or so). When many of these drivers switch from low to high or high to low at the same time, very large transient currents must be delivered by the PDS. Even a small inductance in the PDS will generate a noise voltage, which could cause false triggering of other gates. This is known as the simultaneous switching driver problem.

Example 6.11

A microprocessor has a 32-bit-wide address bus. Each driver is designed to switch a 50-pF load in under 2 ns. The output voltage is 10 V. Assuming no resistance in the interconnects, what is the driver dynamic impedance and the resulting current spikes?

SOLUTION The voltage developed across the capacitive load is given by

$$v(t) = v_0 \, e^{(-t/R_0 C)}$$

The time constant $R_0 C$ must be 2 ns or less. Solving for R, we have

$$R_0 = \frac{2 \times 10^{-9}}{50 \times 10^{-12}} = 40 \, \Omega$$

The induced current spike from each driver due to charging of the capacitor is

$$i = C \left(\frac{dv}{dt} \right) = C \, v_0 \left(-\frac{1}{R_0 C} \right) = -\frac{v_0}{R_0} = -0.25 \, \text{A}$$

For a 32-bit output, the magnitude of the current spike is then

$$i_{\text{total}} = 32i = 8 \, \text{A}$$

∎

6.6.2 Power System Impedance

Although it is generally desirable to reduce the DC resistance of power and ground planes, low impedance power and ground planes are extremely important for electronic packaging substrates. For a good design, the frequency-dependent driving point impedance of the power distribution at the circuit terminals must be kept very small compared to the impedance of the circuit load to avoid large potential drops in the distribution system. A typical impedance plot

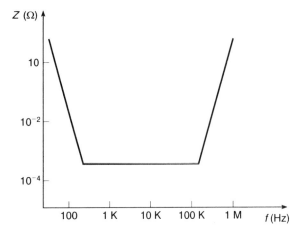

Figure 6.54 Typical impedance plot for a power distribution system as seen from the IC.

for a power distribution system, looking back from the circuit load, is shown in Figure 6.54. At very low frequencies, the power distribution network appears to be capacitive, while, at very high frequencies, it appears to be inductive. In the midrange, the capacitance compensates the inductance to yield a very small impedance for the power distribution system. The goal is to design the power distribution system so that its impedance is flat and resistive throughout the frequency range required by the operating speed of the electronic system.

6.6.3 Resonance of Decoupling Capacitance

Resonance occurs in a circuit when the capacitive reactance is equal to the inductive reactance. Since the two reactances are out of phase with each other, that is, their phase differs by 180°, they cancel each other. Consequently, the impedance of the circuit is purely resistive at resonance. The resonant frequency is defined as

$$\omega = \frac{1}{\sqrt{LC}} \tag{6.90}$$

where L is the parasitic inductance and C is the sum of the parasitic decoupling capacitances of the power distribution system. At resonance, the impedance of the power system is minimum. Usually, capacitive effects dominate the PDS impedance below resonance and inductive effects dominate the PDS impedance above resonance. Of course, different decoupling capacitors will have different resonant frequencies. The effects of all of the decoupling capacitors must be considered in PDS design.

6.6.4 Power Distribution Modeling

Since the actual DC supply, or converter, may be located some distance away from the load, a large inductive return circuit could be present. Decoupling capacitors act as localized charge reservoirs to negate the effect of the inductive loop. (Remember that the inductance would limit the ability of the PDS to deliver a constant voltage with varying current demand.) The closer to the load, the less inductance in the loop, and the higher the frequency accommodated by the decoupling capacitor. The PDS impedance can thus be tailored to be below a certain value from very low to very high frequency. It is limited on the low end by

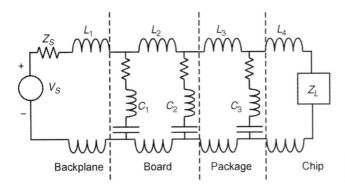

Figure 6.55 Model of a power distribution system.

the power supply and on the high end by the inductances within a chip package, as is shown in Figure 6.54. The PDS model might be as shown in Figure 6.55.

From an equivalent circuit point of view, the inductors in each loop could be combined, but they are kept separate to indicate that there is no true ground in the system. Since all ground connections have inductance, ground potential will vary depending on how well the PDS has been designed. In this model, L_1 represents the total inductance between a circuit board and the power supply, perhaps 1 μH. C_1 is a large electrolytic capacitor on the board. To keep the PDS impedance around 0.1 Ω at frequencies as low as 10 kHz, this capacitor could be 100 μF. But since tantalum electrolytic capacitors are ineffective above 100 kHz, an additional 10 μF in polyester foil capacitance must be paralleled with the electrolytic. This 10 μF could consist of several capacitors in parallel. It could also be partly made up by the many 0.1-μF ceramic chip capacitors represented by C_2 since the inductance of power and ground planes within the board, L_2, is very low. Usually, there is one 0.1-μF chip for every 2 to 4 ICs, yet these capacitors are themselves self-resonant around 20 MHz, and they are still electrically distant from the chip with package and lead inductances L_3 and L_4 in between.

The key to achieving stable voltage at the chip in today's high-performance pin grid array (PGA) or ball grid array (BGA) ceramic packages is to include power and ground planes within the package, separated by thin layers of high ε_r ceramic dielectric—a distributed capacitor. A capacitance of ~50 nF with an equivalent inductance <50 pH is possible, giving good decoupling up to ~100 MHz. Alternatively, several discrete low-inductance capacitors could be included within a ceramic or printed wiring board-based package, with direct connections to the package power and ground planes.

There still remain the bonding wires between the package and the chip. These long (~100 mils), skinny (1 mil) wires have ~1 nH of inductance, so many of them must be paralleled to achieve a low composite inductance. It is not unusual to have 20 to 40 bonds each on V_{dd} and V_{ss} on a high-performance processor. (This is a good reason for flip-chip attachment; the inductance of a solder bump is ~0.05 nH.) It is left as an exercise for the reader to determine $|Z|$ vs. f for the PDS just described.

6.6.5 Switching Noise

Switching noise (or delta-I noise) is caused by the switching of logic circuits in the packaging structure of a digital system. It consists of voltage spikes that appear at the power supply terminals of the chip in response to switching activity of the logic circuits. These spikes propagate through other drivers and appear at the inputs of the corresponding receiver circuits. In order to avoid having these spikes cause intermittent logic errors during operation of the digital circuitry, the magnitude of these spikes must be below the noise tolerance of the logic family used [12].

Delta-I noise is caused by the inductance of the power supply distribution system of the package by preventing the power supply from changing the supply current at the same rate as the circuit switches. The noise voltage generated by simultaneous switching of N output drivers is given approximately as

$$\Delta V = N L_{\text{eff}} \frac{\Delta i}{\Delta t} \tag{6.91}$$

where L_{eff} is the effective inductance of the power and ground connections, $\Delta i/\Delta t$ is the peak rate of change of current, Δi is the current required by each driver during the switching event, and Δt is the rise or fall time of the signal. Thus, the magnitude of the delta-I noise is proportional to the total power supply current slew rate of the driver and the effective inductance of the power supply path. Since reducing the current slew rate of the driver ultimately degrades the switching speed of a digital system, a reduction in the effective inductance of the packaging structure is the usual approach to reducing delta-I noise. Low-impedance decoupling capacitors have been shown to reduce delta-I noise [2, 13].

Switching noise in CMOS circuits has been shown to exhibit a sublinear behavior due to the negative feedback effect, which reduces the switching current as the switching noise increases [14]. For CMOS output drivers, the worst-case switching current is the saturation current. However, in practice, the switching current is smaller than the saturation current since CMOS transistors are not fully saturated during switching. Thus, the maximum current, $I_{d,\text{max}}$, sourced to ground by each driver is

$$I_{d,\text{max}} = \frac{K}{2}[V_{in} - V_t - V_n]^2 \tag{6.92}$$

where V_n is the ground noise induced by n output drivers switching simultaneously, V_t is the threshold voltage, V_{in} is the supply voltage, and K is the transconductance parameter defined as

$$K = \frac{\mu_n C_{ox} W}{L} \tag{6.93}$$

for an n-channel device. The maximum total current sourced to ground by n identical output drivers switching simultaneously is

$$I_{t,\text{max}} = n \frac{K}{2}(V_{in} - V_t - V_n)^2 \tag{6.94}$$

The noise appearing on the ground bus connected to the source of the N-channel MOS (NMOS) switching transistor is

$$V_n = L_{g,\text{eff}} \frac{dI_{t,\text{max}}}{dt} \tag{6.95}$$

where $L_{g,\text{eff}}$ is the total effective lumped inductance of the ground bus. Assume that the total current flowing through the ground bond pad has a triangular waveform. This is true if the transient switching current is all due to CMOS driver "through current" [14]. Then, the noise voltage, V_n, is

$$V_n \approx L_{g,\text{eff}} \frac{I_{t,\text{max}}}{T} \tag{6.96}$$

where T is the time required for the switching current spike to increase from zero to its maximum peak value $I_{t,\max}$. Note that T depends on whether the switching current is controlled by the through current (overlap current) or discharging current. Studies have shown that T remains almost constant as a function of the number of outputs switching simultaneously for commonly used package parasitics and loads [15]. From Eqs. (6.94) and (6.96), the ground noise voltage as a function of the number of simultaneously switching signals is given by [14]

$$V_n^2 - 2V_n \left(V_{in} - V_t + \frac{T}{L_{g,\text{eff}} nK} \right) + (V_{in} - V_t)^2 = 0 \tag{6.97}$$

From this equation, it is clear that the ground noise is not a linear function of the number of simultaneously switching CMOS outputs because negative feedback reduces the switching current when the ground noise increases. From Eq. (6.97), the ground noise can be found as [14]:

$$V_n = V_{in} - V_t + \frac{T}{L_{g,\text{eff}} nK} \left[1 - \sqrt{1 + 2(V_{in} - V_t) \frac{nKL_{g,\text{eff}}}{T}} \right] \tag{6.98}$$

Note that as n approaches infinity, the noise voltage approaches $V_{in} - V_t$. The conventional linear assumption for the ground noise of four identical output drivers yields a 50% or higher ground noise value compared to that obtained from Eq. (6.98). The sublinear model yields a smaller ground noise as the number of output drivers increases compared to that obtained using a linear model or Eq. (6.91).

Consider the equivalent circuit for two CMOS drivers driving a load through a transmission line, as shown in Figure 6.56. Each of the two driver circuits draws current from the V_{ss} supply and delivers it to another supply, V_{dd}, through a transmission line and a load resistor, R_L. One end of the load resistor is connected to the signal line and the other is connected to the power supply V_{dd} through parasitic inductance L_{via}, due to vias connecting the signal plane to the plane above, and L_{gnd}, due to the ground plane. L_{ss} and L_{dd} are the parasitic inductances associated with the power and ground conductors. L_{pwr} is the parasitic inductance of the power plane. If a sufficiently large decoupling capacitance C_d, which acts

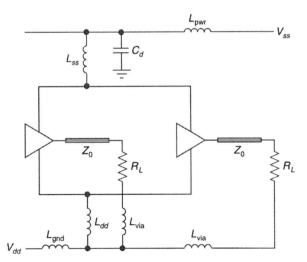

Figure 6.56 Equivalent circuit of two CMOS drivers driving a load through a transmission line.

as a charge reservoir, is connected between V_{ss} and ground, then the return current can be provided instantaneously by the decoupling capacitor.

In practice, decoupling capacitors are commonly placed at strategic points in a circuit near switching devices to reduce delta-I noise. However, a practical decoupling capacitor always has a parasitic inductance associated with it, as does the current path on the power plane from the power vias under the chip to the vias of the decoupling capacitor. Consequently, the total loop inductance of a decoupling capacitor is the most important parameter limiting its effectiveness in reducing delta-I noise [13]. Since this inductance always has a finite value, the delta-I noise is never eliminated, but it can be reduced significantly.

Example 6.12

An electronic packaging substrate consists of 20 drivers that switch simultaneously. The characteristic impedance of the transmission lines in the module is 50 Ω with a 50 Ω matched load. The supply is 5 V with an effective parasitic inductance of 0.5 nH, while the effective parasitic inductance of the ground is 1 nH. For a rise time of 2 ns, determine the magnitudes of the delta-I noise in the power and ground planes.

SOLUTION The current for each driver is

$$i = \frac{V_{cc}}{R_L} = \frac{5\text{ V}}{50\text{ }\Omega} = 100\text{ mA}$$

The rate of change of the current is

$$\frac{di}{dt} = \frac{100\text{ mA}}{2\text{ ns}} = 50 \times 10^6 \text{ A/s}$$

The noise voltage induced in the supply plane is

$$V_{ss} = NL'_{\text{pwr}}\frac{di}{dt} = 20(0.5 \times 10^{-9})(50 \times 10^6) = 0.05\text{ V}$$

while the noise voltage induced in the ground plane is

$$V_{\text{gnd}} = NL'_{\text{gnd}}\frac{di}{dt} = 0.1\text{ V}$$

■

6.7 OVERALL PACKAGED IC MODELS AND SIMULATION

An accurate model of a chip package depends on individual package parameters, which vary widely. This model looks at the problem conceptually and introduces the interaction between power distribution and signal transmission.

The generic model is shown in Figure 6.57. This model assumes that "perfect" ground is the ground plane in the printed wiring board and that V_S, coming from the board is noise free. L_{P1} and L_{G1} represent the inductances between board power and ground and package power and ground, respectively. No matter how many pins or leads are in parallel, some residual inductance will be present. C_{PKG} is whatever internal decoupling capacitance is within the package. L_{P2} and L_{G2} are the paralleled wire bond inductances between package and chip. Note that the chip power supply is V_{CHIP} and that it will vary depending on the currents being drawn.

The package loading on the signal lead is L_S and C_S, but C_S returns to package ground. The transmission line between packages is conventionally modeled, but it is referenced to

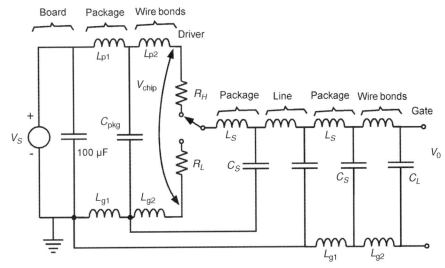

Figure 6.57 Generic model of a chip package.

board ground. The receiver package mirrors the driver package, and the actual signal voltage appears on the receiver chip as V_0.

If all that was going on was one signal switching from low to high, this model would not be very significant. But what happens if many drivers are switching at the same time, as in the case of a 64-bit bus? There will be a huge current surge, sometimes in the tens of amperes/nanoseconds, flowing through L_{P1} and L_{P2} as all of the output lines charge. (The worst case of this "simultaneous switching" problem assumes that all outputs except one undergo the same transition.) Any output remaining high will see a large voltage dip, as will the internal logic circuitry. Similarly, the switching of many outputs from high to low will cause all capacitance connected to those outputs to discharge to ground through L_{G2} and L_{G1}, causing a voltage rise in any low-state outputs. A large enough "power bounce" or "ground bounce" can cause false switching and logic errors. It is, therefore, critical to keep power and ground package inductance low by paralleling many pins or leads, and by incorporating power and ground planes into the package.

6.7.1 Simulation

The emphasis in this chapter has been on modeling, because good models are essential for achieving accurate simulation results. For the packaging engineer, simulation, or determining the performance of the model, is a computational exercise using packaged software. There is another side of simulation that deals with algorithms for the solution of field and circuit problems, and how these algorithms can run faster, but this is a specialty unto itself and beyond the scope of this book.

In electrical performance, there are two types of simulation: circuit and timing. A circuit simulator, such as pSPICE, solves for currents and voltages in an entire multinode, multimesh circuit. It is the type of simulation used on the circuit models that have been discussed. Circuit simulations are always used on analog circuits, but for large digital systems, a complete circuit simulation would be too time consuming to be cost effective. Usually, particular critical paths are examined, or ground bounce is determined by combining the effects of many drivers into one, or other simplifying assumptions are made. Unfortunately, there is a great deal of art to this; no hard and fast rules exist. Trial and error will provide the experience needed to build good judgment about which elements are critical to the model and which can be omitted.

For large digital systems, timing simulations are often done. Once particular paths have been examined with circuit simulations and net delay parameters become clearer, these delays can be introduced into a large timing simulation of the entire system. This will determine critical nets, bus and clock skew, race conditions, and the like and, thus, the overall circuit performance.

Timing simulators, because they do not solve circuits, but simply manipulate logic signals and delays, are much faster than circuit simulators, although less accurate as to the actual shape of waveforms. Both must be used in the analysis of a complex digital system.

Any simulation is only as good as the model data provided. It is essential that the packaging engineer be able to extract good parasitic models for interconnect structures and correlate them with measurements. Thus, we finish off this chapter with a discussion of the most useful time-domain measurement technique.

6.8 TIME-DOMAIN REFLECTOMETRY

Time-domain reflectometry (TDR) is an echo measurement technique in which a step voltage is propagated down an interconnect being characterized, and then the incident and reflected voltage waves along the line are monitored [16]. From this information, the characteristic impedance and the nature (resistive, inductive, or capacitive) of any discontinuity along the line are determined. As such, the TDR waveform can also be used to determine the nature of defects, such as open circuit, short circuit, or increased impedance along the path. Thus, by knowing the length of the interconnect, the propagation delay of a signal traveling along the interconnect can be measured. TDR techniques can also be applied to the characterization and modeling of electronic packages [17] and printed circuit board (PCB) interconnect measurements [18]. In practice, the determination of signal discontinuities in these electronic packages or substrates can be hampered by the rise time of the TDR signal as well as the waveform capture instrument.

Figure 6.58 shows the setup of a TDR system. A positive-going incident voltage wave, generated by a fast step generator, is applied to the interconnect lines under test. The voltage is measured on an oscilloscope connected at point A. The voltage wave travels at the velocity of propagation down the line. No wave is reflected if the load impedance, Z_L, is equal to the characteristic impedance of the line, Z_0. However, part of the incident wave will be reflected if a mismatch exists at the load, as shown in Figure 6.59.

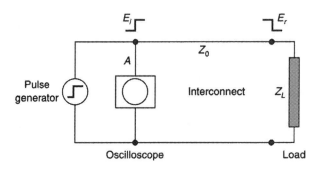

Figure 6.58 Setup for a time-domain reflectometer (TDR).

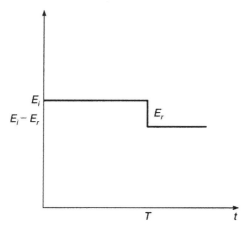

Figure 6.59 Waveform at point A when Z_L is not equal to Z_0.

The shape of the reflected wave is a function of both the nature and magnitude of a mismatch as revealed in the reflection coefficient, Γ. From Eq. (6.53), the load impedance is related to the characteristic impedance by

$$Z_L = Z_0 \frac{1+\Gamma}{1-\Gamma} \qquad (6.99)$$

Thus, knowledge of the incident voltage wave, E_i, and the reflected voltage wave, E_r, allows Z_L to be determined in terms of Z_0, or vice versa. Figure 6.60 shows the waveform corresponding to an open-circuit termination with $\Gamma = +1$ and E_r equal to E_i. Figure 6.61 shows the waveform corresponding to a short-circuit termination with $\Gamma = -1$ and E_r equal to E_i. When the line is terminated in $Z_L = 2Z_0$, the resulting waveform is shown in Figure 6.62 with $E_r = E_i/3$. When the line is terminated in $Z_L = Z_0/2$, the resulting waveform is shown in Figure 6.63 with $E_r = -E_i/3$.

Figure 6.64 shows a practical TDR waveform for a microstrip transmission line with two different characteristic impedances terminated with a load. The waveform was acquired using a 50 Ω coaxial cable from the TDR equipment to a soldered joint on the microstrip. The peak at point X in the waveform results from the signal discontinuity at the solder joint. As can be seen, the characteristic impedance of the first section of the microstrip is higher

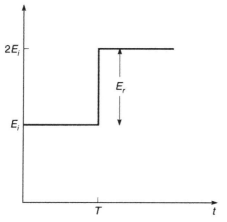

Figure 6.60 Waveform at point A for an open-circuit termination.

240 Chapter 6 Electrical Considerations, Modeling, and Simulation

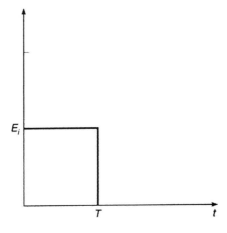

Figure 6.61 Waveform at point A for a short-circuit termination.

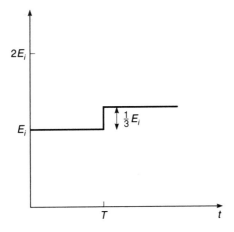

Figure 6.62 Waveform at point A when $Z_L = 2Z_0$.

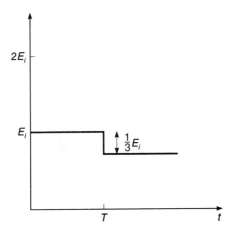

Figure 6.63 Waveform at point A when $Z_L = 1/2Z_0$.

6.8 Time-Domain Reflectometry

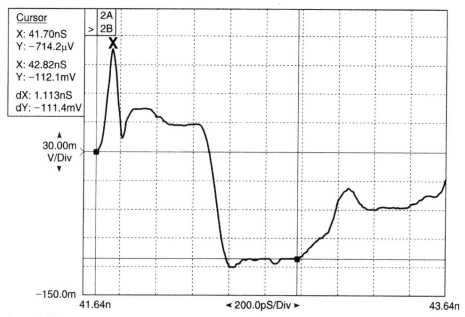

Figure 6.64 Time-domain reflectometry waveform of a practical microstrip transmission line terminated with a load.

than 50 Ω, while it is less than 50 Ω for the second section. Furthermore, the load impedance is seen to be less than 50 Ω.

Example 6.13

Briefly describe the possible structure of a transmission line with the time-domain reflectometry waveform shown in Figure 6.65.

SOLUTION The transmission line consists of a sequence of three sections with $Z_3 < Z_1 < Z_2$, and the load is an open circuit.

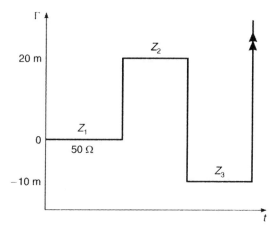

Figure 6.65 Time domain reflectometry waveform of a transmission line structure.

Since the characteristic impedance for section 1, Z_1, is given as 50 Ω, Z_2 and Z_3 can be found. The characteristic impedance for section 2, Z_2, is

$$Z_2 = Z_1 \frac{1+\Gamma}{1-\Gamma} = 50 \frac{1+20\,\text{m}}{1-20\,\text{m}} = 52\,\Omega$$

while the characteristics impedance for section 3, Z_3, is

$$Z_3 = Z_1 \frac{1+\Gamma}{1-\Gamma} = 49\,\Omega$$

∎

6.9 SUMMARY

This chapter discusses the electrical design considerations for electronic packages, in particular, electronic packaging substrates. It emphasizes the importance of parasitic resistances, inductances, and capacitances of the interconnects, which must be properly accounted for during the design process. The parasitic resistance in a signal conductor decreases the amplitude and increases the rise time of a transmitted signal. The parasitic inductance and capacitance introduce crosstalk in electronic packages and electronic packaging substrates. For high-speed devices, interconnects must be considered as distributed-parameter networks or transmission lines. Furthermore, proper load termination is required in high-speed circuits to mitigate signal reflection that degrades the signal and causes ringing in the transmitted signal along the interconnect. Crosstalk in the range of 2 to 4% is generally considered acceptable for electronic packaging substrates. Switching noise or delta-I noise arises from the inability of a power supply distribution system to change the current it supplies at the rate required by the switching circuit due to parasitic inductance in the supply and ground buses. Time-domain reflectometry is a useful technique for characterizing the electrical performance of an interconnect.

REFERENCES

1. A. B. GLASER AND G. E. SUBAK-SHARPER, *Integrated Circuit Engineering—Design, Fabrication and Applications*, Reading, MA: Addison-Wesley, 1977.
2. C. T. A. JOHNK, *Engineering Electromagnetic Fields and Waves*, New York: Wiley, 1975, p. 286.
3. C-T TSAI, Package Inductance Characterization at High Frequencies, *IEEE Trans. Components, Packaging and Manufact. Technol.*, Vol. 17, No. 2, pp. 175–181, 1994.
4. *Low Inductance Capacitor Arrays*, Myrtle Beach, SC: AVX Corporation.
5. D. A. DOANE AND P. D. FRANZON, eds., *Multichip Module Technologies and Alternatives: The Basics*, New York: Van Nostrand Reinhold, 1993, p. 532.
6. D. M. POZAR, *Microwave Engineering*, Reading, MA: Addison-Wesley, 1990, pp. 67–83.
7. C. W. DAVIDSON, *Transmission Lines for Communications*, London; MacMillian, 1978.
8. P. D. FRANZON, et al., Tools to Aid in Wiring Rule Generation for High-Speed Interconnects, Proc. Design Automation Conf., IEEE and ACM, Anaheim CA, pp. 461–471, June 1992.
9. C. S. CHANG, Electrical Design of Signal Lines for Multilayer Printed Circuit Boards, *IBM J. Res. Develop.* Vol. 32, No. 5, 1988.
10. C. S. CHANG, Resistive Signal Line Wiring Net Designs in Multichip Modules, *IEEE Trans. Components, Hybrids, and Manufact. Technol.*, Vol. 16, No. 8, pp. 909–918, 1993.
11. L. W. SCHAPER, S. S. ANG, Y. L. LOW, AND D. R. OLDHAM, Electrical Characterization of Interconnected Mesh Power System (IMPS) MCM Topology, *IEEE Trans. Components, Packaging, Manufact. Technol.*, Part B, Vol. 18, No. 1, February 1995.
12. G. A. KATOPIS, Delta-I Specification for a High Performance Computing Machine, *Proc. IEEE*, Vol. 73, No. 9, pp. 1403–1415, Sept. 1985.
13. R. DOWNING, P. GEBLER, AND G. KATOPIS, Decoupling Capacitor Effects on Switching Noise, *IEEE Trans. Components, Hybrids, Manufact. Technol.*, Vol. 16, No. 5, pp. 484–489, August 1993.
14. R. SENTHINATHAN AND J. L. PRINCE, Simultaneously Switching Ground Noise Calculation for Packaged CMOS Devices, *IEEE J. Solid-State Circuits*, Vol. SC-26, No. 11, p. 1724, November 1991.
15. R. SENTHINATHAN, J. L. PRINCE, AND S. NIMMAGADDA, Effects of Skewing CMOS Output Driver Switching on the Simultaneously Switching Noise, Proc. of the 1991 IEEE/CHMT International Electronics Manufacturing Technology Symposium, p. 342, September 1991.

16. TDR Fundamentals: For Use with HP 54120T Digitizing Oscilloscope and TDR, Hewlett-Packard Application Note 62, 1988.
17. TDR Techniques for Characterization and Modeling of Electronic Packaging, TDA Application Note, www.tdasystems.com/support.htm.
18. PCB Interconnect Characterization from TDR Measurements, TDA Application Note, www.tdasystems.com/support.htm.
19. H. B., BAKOGLU, *Circuits, Interconnections and Packaging for VLSI*, Reading, MA: Addison-Wesley, 1990, Chapters 4–7.
20. R. SENTHINATHAN AND J. PRINCE, *Simultaneous Switching Noise of CMOS Devices and Systems*, Kulwer Academic, 1994.
21. R. TUMMALA AND E. RYMASZEWSKI, *Microelectronics Packaging Handbook*, New York: Van Nostrand Reinhold, 1989, Chapter 3.

Exercises

6.1. What are the major electrical design objectives in electronic packaging?

6.2. What is desirable in order to maintain signal fidelity?

6.3. What is the cause of delta-I noise?

6.4. What is the cause of crosstalk?

6.5. What is the most difficult aspect of electrical design for an MCM?

6.6. What are the fundamental electrical design considerations in electronic packaging?

6.7. Define *interconnects* in electronic packaging?

6.8. What are the general procedures in the electrical design of a digital MCM?

6.9. Define *net* and *critical net* in a digital design?

6.10. What is *resistance*?

6.11. What are the major effects of resistance in electronic packages?

6.12. What is the physical cause of resistance in a normal metal?

6.13. What are the major factors to be considered when designing an interconnect in an MCM?

6.14. What is sheet resistance?

6.15. What are the effects of a right-angle bend in an MCM electrical conductor?

6.16. Determine the DC resistance of a copper conductor as shown in Figure 6.3. The thickness of the plated copper conductor is 10 µm with an average resistivity of 2×10^{-6} Ω-cm.

6.17. As an electronic packaging engineer, you are required to design a thin-film aluminum resistor on a high-frequency packaging structure with the following specifications:
Total resistance = 1 Ω
Sheet resistance of aluminum conductor = 2×10^{-2} Ω/square
Width of aluminum conductor = 100 µm
Draw the geometry of the aluminum resistor that you designed based on the above specifications.

6.18. Define *inductance*.

6.19. Define *Ampere's circuital law*.

6.20. What is the *self-inductance* of a conductor?

6.21. Determine the self-inductance of a conductor of length 1 m carrying a current of 1 An.

6.22. Define *mutual inductance*.

6.23. Define *Biot–Savart law*.

6.24. Determine the mutual inductance for the two parallel conductors shown in Figure 6.5. The length of the conductor is 200 µm and the separation between the conductors is 200 µm.

6.25. How do you classify the parasitic inductances in an MCM?

6.26. Define *capacitance*.

6.27. An interconnect of width 10 μm and length 100 μm is on top of a ground plane. The dielectric separating the ground plane and the interconnect has a thickness of 3 μm. Determine the capacitance for the interconnect if the dielectric is silicon dioxide ($\varepsilon_r = 3.9$).

6.28. What are the major effects of stray capacitance in an electronic package?

6.29. What can be done to reduce both the power dissipation and signal delay at high-impedance nodes?

6.30. What is the major purpose of a decoupling capacitor in an MCM?

6.31. Define *rise time* and *fall time* for a digital signal.

6.32. A digital signal with a rise time of 2 ns is to traverse an aluminum interconnect of 1 cm length embedded in polyimide ($\varepsilon_r = 3.9$). Determine its bandwidth and pulse design wavelength if the quality of the signal is 5. Should this interconnect be treated as a transmission line? Justify your answer.

6.33. Define *propagation delay*.

6.34. Draw a lumped equivalent circuit for an interconnect structure consisting of a CMOS off-chip driver driving two receiving gates.

6.35. Consider the circuit shown in Figure 6.10 with the following parameters: $R_s = 50\,\Omega$, $C_x = 1$ pF, $C_l = 5$ pF, $C_y = 5$ pF, $C_{in} = 5$ pF, and $R_l = 0.1\,\Omega$. Determine the propagation delay and the maximum operating speed for this circuit.

6.36. Draw a lumped equivalent circuit for an interconnect structure consisting of a CMOS off-chip driver driving three equivalent receiving gates and express the propagation delay in terms of their parasitic elements.

6.37. A 2-μm-thick aluminum conductor forms the signal path between two unpackaged CMOS chips on an MCM substrate. The signal conductor is 8 μm above a ground plane, with polyimide dielectric ($\varepsilon_r = 3.6$) in between. The distance between the CMOS driver on one chip and the receiver on the other is 5 cm. What is the optimum width of the signal conductor for minimum rise time? Assume the driver has a 50-Ω output impedance and the receiver can be modeled as a 4-pF capacitor.

6.38. What is the major challenge presented to the interconnect in a high-speed electronic system?

6.39. What is a transmission line?

6.40. What is the criterion for an interconnect to be treated as a transmission line?

6.41. Draw a general circuit model for a transmission line. Label all the circuit elements.

6.42. Define the *characteristic impedance* for a transmission line.

6.43. Define the *phase velocity* in a transmission line.

6.44. Define the *characteristic impedance* for a lossless transmission line.

6.45. Consider the lossless two-wire transmission line in air shown in Figure 6.30. The series inductance per unit length and shunt capacitance per unit length are 15 nH/m and 10 pF/m, respectively. Determine (a) the characteristic impedance and (b) the phase velocity for this transmission line.

6.46. Draw a simple microstrip structure with its electric field lines.

6.47. Can a simple microstrip support a TEM wave? Why?

6.48. What are the two most commonly encountered microstrip lines in electronic packaging susbtrates?

6.49. A 10-cm-long MCM substrate transmission line, 20 μm wide, is located 10 μm above a ground plane, separated by a layer of polyimide. The line connects the CMOS driver and receiver of Exercise 37. Compare the relative performance of lines fabricated in 2-μm-thick aluminum and 5-μm-thick copper. Model in pSPICE using lumped circuit approximations. The driver rise time is 1 ns.

6.50. What is the relationship between the voltage and current on a transmission line when it is terminated in an arbitrary load with Z_L not equal to Z_0?

6.51. Define *voltage reflection coefficient*.

6.52. Define *standing wave* on a transmission line.

6.53. What is the relationship between the voltage and current at the load when a transmission line is terminated with a shorted load end?

6.54. What is the relationship between the voltage and current at the load in a transmission line when the load end is open?

6.55. What is the relationship between the input and load impedances in a half-wavelength transmission line?

6.56. The load and characteristic impedances of a quarter-wavelength transmission line are $100\,\Omega$ and $50\,\Omega$, respectively. Determine its input impedance.

6.57. Consider the interconnection system shown in Figure 6.42. The source impedance is $2Z_0$ while the load impedance is $Z_0/2$, where Z_0 is the characteristic impedance of the transmission line. Determine the source and load reflection coefficients. Draw the reflection diagram and the voltage waveforms at the input and load ends for this system.

6.58. Consider the interconnection system shown in Figure 6.42 with no source impedance. The characteristic impedance of the interconnect is $50\,\Omega$ with a load impedance of $100\,\Omega$. Determine (a) the voltage reflection coefficient at the source and (b) the voltage reflection coefficient at the load. Draw the reflection diagram for a propagation time of $4T$. Label all the quantities at each time interval.

6.59. Draw the four common termination schemes used in electronic packaging substrates.

6.60. Under what condition does a transmission line yield the smallest propagation delay?

6.61. What are the predominant losses in signal lines in an MCM?

6.62. Define the *skin effect* in a current-carrying conductor.

6.63. The skin depths for silver, gold, and aluminum conductors at $10\,\text{GHz}$ are 0.64, 0.79, and 0.81 µm, respectively. Based on the given data, which conductor material would you choose for an application that required the transmission of the 10-GHz current signal. Justify your answer.

6.64. Name and draw the five common topologies used to minimize reflection noise in an MCM.

6.65. What is coupled noise or crosstalk?

6.66. Assuming two interconnects have a similar characteristic impedance, source impedance, and load termination impedance, briefly describe how crosstalk is induced into one interconnect line when a pulse is introduced at the other interconnect line.

6.67. For the two interconnects described in Exercise 66, under what condition(s) will there be no induced forward-traveling voltage wave.

6.68. Two identical signal lines having a characteristic impedance of $50\,\Omega$ are terminated by load impedances of $50\,\Omega$ as shown in Figure 6.30a. The length of the signal line is $2\,\text{cm}$. The rate of change of the source is $10^9\,\text{V/s}$, the mutual inductance between the two signal lines is 2 nH, and the mutual capacitance is 2 pF. Determine the magnitude of the crosstalk induced on transmission line 2.

6.69. A digital signal processor has a 32-bit wide output bus. Each output driver is designed to switch a 100-pF-capacitive load in under 1 ns with an output voltage of 5 V. Neglecting the resistance in the interconnect, determine the driver dynamic impedance and the resulting output spike.

6.70. What is the design goal for a power distribution system?

6.71. Over the frequency range from $1\,\text{MHz}$ to $1\,\text{GHz}$, determine the power distribution impedance versus frequency for a printed wiring board 5 in. by 8 in. The power distribution system consists of solid 1-mil copper power and ground planes spaced 10 mils apart, with FR-4 material

246 Chapter 6 Electrical Considerations, Modeling, and Simulation

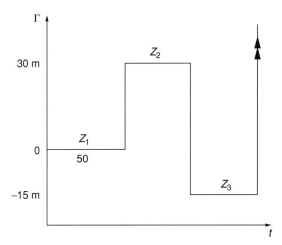

Figure 6.66 Time-domain reflectometry waveform for exercise 78.

($\varepsilon_r = 4.7$) in between. Four 0.1-µF and four 10-nF chip capacitors are distributed around the board. Assume the impedance of the power supply to board connection is essentially infinite. (Remember you are finding the impedance from a chip's eye view.) Assume the 0.1-µF decoupling capacitor has parasitics of 500 pH and 400 mΩ, and the 10-nF capacitor has parasitics of 400 pH and 300 mΩ. Model using pSPICE.

6.72. What is *delta-I* noise?

6.73. What is commonly used to reduce *delta-I* noise?

6.74. An MCM consists of 10 drivers switching simultaneously. The characteristic impedance of the transmission line is 50 Ω with a 50-Ω matched load. The power supply is 5 V with an effective inductance of 0.2 nH, while the effective inductance of the ground plane is 1 nH. For a rise time of 1 ns, determine the magnitude of the delta-I noise in the power and ground planes.

6.75. Compare the relative performance of flip-chip and wire bond attachment on the on-chip power distribution voltage fluctuation of a CMOS chip during a current surge of 10 A over 2 ns. The chip has 328 total I/O, with 30 power and 40 ground connections. V_{DD} is 5 V. Assume the chip is on an MCM substrate with very low impedance power distribution at all frequencies of interest.

6.76. Design an MCM power distribution system capable of delivering 5 V± 0.1 V to the chips contained on it. Assume that the chips are wire bonded and the substrate has no integral decoupling capacitor. The largest chip produces substantial current spikes of 10 A/ns over 2 ns. Other chips operate asynchronously and draw 13 An. How many wire bonds are needed for power and ground on the large chip? How many and what size decoupling capacitors are needed? Assume the current spike in question is supplied entirely by on-MCM substrate decoupling.

6.77. What is time-domain reflectometry?

6.78. For the time-domain reflectometry waveform shown in Figure 6.66, determine the characteristic impedances Z_2 and Z_3 if the reference impedance in section 1 is 50 Ω. What is the load termination in this structure?

Chapter 7

Thermal Considerations

RICK J. COUVILLION

7.1 INTRODUCTION

Thermal management of electronic components and systems has driven the development of many advanced heat transfer techniques for the past 60 years. The benefits of better reliability, increased power, and miniaturization would not have been possible without development of improved thermal analysis and design methods.

The expanding use of electronics after World War II made obvious the need for thermal management and spurred the growth of thermal control technology. Cooling of vacuum tubes required evolutionary development of enhanced air cooling. The development of integral liquid coolant passages was required in some applications. Transistors greatly reduced dissipated power per component but created new problems due to greatly increased component density and the reduction in component reliability with increased temperature.

Through the 1960s and 1970s, much work was done consolidating previous work in enhanced air cooling and liquid cooling. Over the same period, investigations of more novel methods were begun, including immersion cooling, augmented boiling, heat pipes, thermoelectric coolers, and microchannel cooling. These remained in the prototype stage until the early 1980s, but increased packaging density, as a result of very large scale integration (VLSI) and multichip module (MCM) technologies, created the need for commercialization of these methods [1]. This chapter explores most current methods.

7.1.1 Heat Sources

Electronic systems generate heat in many places. The I^2R losses in wiring generate some heat but not enough to be a primary consideration. A significant source of heat is the power supply used in many systems where alternating current (AC) voltage is converted into the various direct current (DC) voltage levels needed. But the primary concern here is the heat generation at each chip and collection of chips, whether they are mounted on printed wiring boards (PWBs), hybrid packages, or MCMs.

One or more monolithic chips consisting of active and passive elements are bonded to a substrate, which is, in turn, packaged in a number of different ways described earlier in this text. The energy generation within each chip die generally occurs within 1 mil of the top surface and, depending on function and sophistication, the time-averaged heat generation

---

Advanced Electronic Packaging, Second Edition, Edited by Richard K. Ulrich and William D. Brown
Copyright © 2006 the Institute of Electrical and Electronics Engineers, Inc.

Table 7.1 Packaging Thermal Parameters 2003 (2012)

	Handheld	Cost–performance	High performance	Auto	Memory
Power dissipation (W)	2 (3.2)	75 (109)	129 (174)	14 (14)	1.5 (3)
Chip size (mm^2)	59 (77)	430 (750)	430 (750)	59 (77)	560 (1580)
Junction temperature (°C)	115 (115)	100 (100)	100 (100)	180 (180)	100 (100)
Ambient temperature (°C)	55 (55)	45 (45)	45 (45)	170 (170)	45 (45)
Chip heat flux (W/cm^2)	3.4 (4.2)	17.4 (14.5)	30.0 (23.2)	23.7 (18.2)	0.27 (0.19)
Overall[a] HT Coeff (W/cm^2 °C) ×100	5.7 (7.0)	31.6 (26.4)	54.5 (42.2)	237 (182)	0.49 (0.35)

[a] Chip heat flux/($T_{junction} - T_{ambient}$).

ranges up to about 130 W in high-performance packages. The resulting heat flow per unit chip surface area (heat flux, q'') averaged over time varies from less than 1 to 30 W/cm^2. As references, the reader should note that the solar heat flux at the equator is approximately 0.1 W/cm^2, and the heat flux at the surface of a 100-W lightbulb is approximately 1 W/cm^2.

The objective in thermal management is to remove heat generated at rate q in Watts at the chip dies using the temperature difference available. The available temperature difference is the chip die temperature, often called $T_{junction}$, minus the temperature of the environment, $T_{ambient}$, where the heat is eventually dumped. The ratio $(T_{junction} - T_{ambient})/q$ is the *overall thermal resistance* $R_{thermal}$ (°C/W). The ratio of available temperature difference to heat flux is often called the *specific thermal resistance*, m^2 °C/W. Its inverse, the ratio of heat flux to available temperature difference is often called the *overall heat transfer coefficient*, U (W/m^2 °C).

Present and projected future cooling needs for chip-level electronic components are shown in Table 7.1, which summarizes roadmaps provided by the National Electronics Manufacturing Initiative (NEMI) [2] and the Semiconductor Industry Association [3]. These are projections made in 1996 and 1997 through the year 2012, but the data shown in Table 7.1 is for 2003 and 2012. Examples of equipment in the categories shown in the table are:

- Hand held—personal digital assistants (PDAs) and cell phones
- Memory—dynamic random-access memory (DRAM) and static random-access memory (SRAM)
- Automotive—under hood environment, mining equipment, some military
- Cost/performance—desktop and notebook personal computers (PCs)
- High performance—workstations and servers

The automotive category presents a real challenge. The heat fluxes are similar to the cost–performance category, except they must be achieved with high junction temperatures and a small temperature difference due to the high ambient temperature under the hood of an automobile. Another interesting trend is the reduction in heat flux expected in a number of categories between 2003 and 2012. The expected increase in packaging density is expected to be more than offset by a drop in transistor switching energy.

Heat generation on a PWB is primarily from the packages mounted on it plus minor amounts from discrete and printed resistors. These can range from 5 to 30 W. The resulting

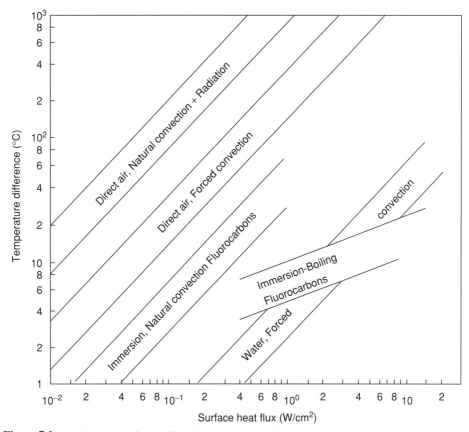

Figure 7.1 Achievable heat fluxes with various cooling methods.

heat flux on a board surface is usually 0.2 W/cm² or less. In some cases, primary cooling requires conduction in the plane of the PWB to cooled connectors on the edge of the PWB. In this situation, the board can provide significant resistance and may require measures to lower the resistance.

7.1.2 Approaches to Heat Removal

Various cooling methods are able to provide a range of heat fluxes for a range of temperature differences. Figure 7.1 shows the range of heat fluxes and temperature differences that can be achieved with various cooling methods. As the packaging level goes from the chip die to its packaging (level 1), to a board (level 2), to a motherboard (level 3), to a cabinet (level 4), the area over which the heat is distributed increases so the heat flux decreases correspondingly. This "heat spreading" effect allows heat removal at one or more packaging levels, where the heat flux and temperature difference are consistent with the available cooling method. However, each increase in packaging level creates additional resistance that raises the temperature at the silicon die. The challenge is often moving the heat from its source to another location, along a path of low thermal resistance, where an available cooling method can be applied.

The thermal resistance at level 1 is associated with conduction of heat from the die to its substrate to the surface of the package. There is often significant resistance in the attachment of the die to the substrate (contact or interface resistance). From the package surface, heat can be removed by a coolant and/or conducted to a board, level 2. At the board level, heat can be removed by a coolant and/or conducted through the board to its edge where the heat is removed or moved to a higher packaging level.

The simplest and cheapest temperature control method is simply cooling with air using air movement created by the presence of the heat source, known as natural or free convection. As shown in Figure 7.1, for a chip-to-air temperature difference of 100°C, natural convection and radiation combined can remove up to 0.1 W/cm^2, while adding a fan to provide airflow (forced convection) can achieve up to 1 W/cm^2 at the same 100°C temperature difference. Other more expensive and complex cooling systems and their approximate heat removal capability include:

- Immersion cooling in refrigerants with natural convection, 1 W/cm^2 at $\Delta T = 30°C$
- Immersion cooling in a boiling refrigerant, 10 W/cm^2 at 10°C
- Forced convection with water, 10 W/cm^2 at 40°C

All thermal management systems must be designed along with the electrical system. The electrical system uses electrical insulating materials that are usually also poor thermal conductors. The mismatches in the coefficient of thermal expansion (CTE) of materials bonded together create stresses as a package warms and cools when turned on and off. These conflicts make thermal design quite a challenge.

7.1.3 Failure Modes

Much of the description of temperature-induced failure modes in this section relies heavily on the work of Kraus and Bar-Cohen [1]. The primary objective of thermal management is the prevention of catastrophic thermal failure, defined as immediate, thermally induced, total loss of electronic function of a specified component. This kind of failure is usually a result of thermal fracture of a mechanical support element (case or substrate) or separation of the leads from the external electrical connections. Catastrophic failure can also result from semiconductor material failure due to overheating.

The secondary objective of thermal management is to achieve the desired reliability and lifetime of the electronic system. Control of operating temperature achieves this objective. The failure rate, the inverse of reliability, of each chip increases almost exponentially with operating temperature. For silicon transistors, the failure rate increases by a factor of 5 to 7 as the operating temperature is increased from 25 to 130°C. The lifetime of most electronic components are shortened by extended use at higher temperatures, a result of creep in the chip and bonding materials and other causes. Given the very large numbers of components in an electronic system, the junction temperatures of critical components must be minimized to increase reliability and lifetime.

Temperature cycling in excess of ±15°C around an average operating temperature also reduces reliability. Thermal stress fatigue is induced when two adjacent materials with different coefficients of expansion are subjected to a cycling temperature level. This can occur between two layers of a PWB, between a chip and a substrate, between a substrate and a bonding material, and between a surface-mounted chip and the PWB to which it is soldered.

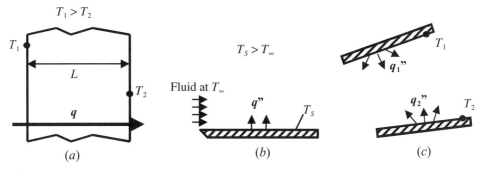

Figure 7.2 Heat transfer modes: (*a*) conduction through a solid, (*b*) convection from surface to moving fluid, and (*c*) net radiation between two surfaces.

7.2 HEAT TRANSFER FUNDAMENTALS

Heat transfer is the movement of energy due to a temperature difference. Whenever there is a temperature difference between regions within a medium, for example, within a solid, or between different media, for example, between a hot surface and cooler air flowing over the surface, heat transfer occurs. Our intuition and the second law of thermodynamics tells us that the transfer must be from the warmer medium or region to the cooler medium or region.

The mechanisms of heat transfer depend on the media involved and are usually called *modes* of heat transfer. The three modes are *conduction, convection, and radiation* and are illustrated in Figure 7.2. When a temperature gradient exists within a continuous, nonmoving medium, solid or liquid, heat is transferred through the medium via the conduction mode. Convection heat transfer occurs when a surface is in contact with a fluid, liquid or gas, at a different temperature. Heat is exchanged between the fluid and the surface. Radiation heat transfer occurs when two surfaces at different temperatures exchange energy in the form of electromagnetic energy emitted by the surfaces. Radiation can occur in a vacuum since no medium between the two surfaces is required as in conduction and convection.

7.2.1 Heat Transfer Rate Equations

In all three modes of heat transfer, the rate of heat transfer increases as the magnitude of the driving temperature difference increases. The relationship between heat transfer rate and temperature difference is shown in the following sets of rate equations.

7.2.1.1 Conduction

Heat within electronic components must be transported from within the component to its surface via conduction. At the surface, the energy can be removed via convection or radiation. Fourier's law for one-dimensional heat transfer via conduction in a solid is

$$q = -kA_c \frac{dT}{dx} \quad (7.1)$$

or

$$q'' = \frac{q}{A_c} = -k\frac{dT}{dx} \quad (7.2)$$

where

q = rate of heat transfer (W)
A_c = cross-sectional area of the solid in the direction of heat flow (m²)
q'' = heat flux (W/m²)
k = *thermal conductivity* of the solid in the direction of heat flow (W/m K)
dT/dx = thermal gradient within the solid (°C/m)

The thermal conductivity is an intrinsic property of the material. Values for various solids at room temperature are given in Table 7A.1 in the Appendix and in Chapter 2. Note that the conductivity can vary over several orders of magnitude in materials used in electronic applications. Thermal conductivity varies to different degrees with temperature and can also be anisotropic, that is, dependent on direction. For example, a PWB can have a significantly different effective conductivity in the direction of board thickness than in the direction of the length and width due to layered construction.

Assuming constant thermal conductivity, and applying Eq. (7.1) to Figure 7.2a gives

$$dT/dx = (T_2 - T_1)/L$$

and

$$q = kA_c(T_1 - T_2)/L \tag{7.3}$$

Note that this equation can be written as

$$q = \frac{T_1 - T_2}{L/kA_c} \tag{7.4}$$

Comparing this to Ohm's law written for electrical current I resulting from a potential difference E across a resistance R, that is,

$$I = E/R$$

gives the following analogies:

q analogous to I

$T_1 - T_2$ analogous to E

$R_{\text{cond}} = L/kA_c$ analogous to R

The *conduction thermal resistance*, R_{cond}, has units of °C/W and represents the temperature difference $T_1 - T_2$ required to drive 1 W through the wall via conduction.

7.2.1.2 Convection

Convection occurs between a surface and a fluid moving over it when they are at different temperatures. Convection may be classified by how the fluid motion is created. *Forced convection* occurs when the fluid motion is created by forces far removed from the surface, for example, the wind, a pump, or a fan. Consider a PWB mounted in a personal computer. The board is cooled by airflow created by a fan mounted in the case. The surfaces of the chips mounted on the PWB are being cooled via forced convection with the air motion being created by a fan, that is, a force far removed from the surfaces being cooled. Consider the same PWB being cooled without a fan. In this case, convection exists, but the airflow results from buoyant forces created by the temperature difference between the chip surfaces and the ambient air. This is called *natural or free convection*. Usually, forced convection results in higher rates of heat transfer for the same temperature difference. However, if the forced airflow is relatively small and a large temperature difference creates large buoyant forces, the forced and natural convection components may be comparable and result in *mixed convection*.

There are also convection cooling processes involving a liquid-to-vapor phase change, that is, boiling. In boiling, heat transfer fluid motion is effected by the creation of vapor bubbles at the surface that detach from the surface and rise through the surrounding liquid.

Convection processes are described by *Newton's law of cooling*:

$$q = hA_s(T_s - T_\infty) \tag{7.5}$$

or

$$q'' = \frac{q}{A_s} = h(T_s - T_\infty) \tag{7.6}$$

where

q = heat transfer rate (W)
A_s = surface area (m^2)
q'' = heat flux (W/m^2)
T_s = surface temperature (°C)
T_x = fluid temperature (°C)
h = convection heat transfer coefficient (W/m^2°C)

The value of h is also called the film conductance or the film coefficient. It is a function of surface geometry, the nature of the fluid motion, and a number of fluid properties. The determination of h, given all of this information, is where most effort in the study of convection is directed. Until that material is covered later in this chapter, a value of h will be given so that problem solution can proceed. Typical values of h are given in Table 7.2. Ranges of h for various fluids are shown in Figure 7.3.

Equation (7.6) can be written as

$$q = \frac{T_s - T_\infty}{1/hA_s} \tag{7.7}$$

Comparing this equation to $I = E/R$ again gives an electrical analogy with $T_s - T_\infty$ as the driving potential analogous to potential difference E and a *convection thermal resistance* $R_{\text{conv}} = 1/hA_s$ analogous to electrical resistance R.

Table 7.2 Typical Values of the Convection Heat Transfer Coefficient

Process	h (W/m^2 K)
Free convection	
Gases	2–25
Liquids	50–1,000
Forced convection	
Gases	25–250
Liquids	50–20,000
Convection with phase change	
Boiling or condensation	2500–100,000

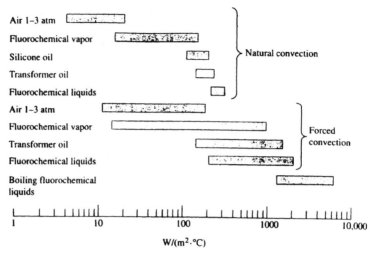

Figure 7.3 Heat transfer coefficients for various fluids.

7.2.1.3 Radiation

Matter emits thermal radiation at its surface when its temperature is above absolute zero. This radiation is in the form of photons of varying frequency, and they need no medium to transport them. The maximum possible heat emission from an ideal radiating surface is given by the Stefan–Boltzmann law:

$$q = \sigma A_s T_s^4 \tag{7.8}$$

or

$$q'' = \sigma T_s^4 \tag{7.9}$$

where T_s is the *absolute* temperature of the surface (K) and σ is the Stefan–Boltzmann constant (5.67×10^{-8} W/m² K⁴. An ideal radiating surface is called a *blackbody* or a black surface. The emission from a nonideal surface is

$$q'' = \varepsilon \sigma T_s^4 \tag{7.10}$$

where ε is a property of the surface called *emissivity*. The emissivity varies between 0 and 1 where $\varepsilon = 1$ for a black surface. The quantity σT_s^4 is called the *blackbody emissive power* E_b. A black surface is also a perfect absorber, that is, it absorbs all incident radiation. Nonblack surfaces reflect some of the incident radiation and, therefore, absorb less. The absorbed radiation is

$$q_{abs} = \alpha q_{inc} \tag{7.11}$$

where α, the *absorptivity*, is the fraction of incident radiation absorbed, and q_{inc} is the rate at which radiant energy is incident on the surface. For a black surface, $\alpha = 1$. The emissivity and absorptivity of a surface are often both functions of the wavelength distribution of photons emitted and absorbed, respectively, by the surface. However, in many cases, it is reasonable to assume that both σ and α are independent of wavelength. This is called a *gray surface*. For a gray surface, $\sigma = \alpha$.

Two surfaces at different temperatures that can "see" each other can exchange energy via radiation. The net exchange rate between the two depends on the relative size of the surfaces, their relative orientation and shape, and their emissivity and absorptivity. However, when all of the radiation emitted by one gray surface of area A_1 is incident on another much larger surface of area A_2, the net heat exchange rate between the two is

$$q = \varepsilon_1 A_1 (E_{b1} - E_{b2}) = \varepsilon_1 A_1 \sigma (T_{s1}^4 - T_{s2}^4) \tag{7.12}$$

This situation is common when all the radiation emitted by A_1 is incident on a large enclosure, A_2. Equation (7.12) can be written as

$$q = \frac{E_{b1} - E_{b2}}{1/\varepsilon_1 A_1} \tag{7.13}$$

Comparison again to $I = E/R$ shows that $E_{b1} - E_{b2}$ is analogous to the driving potential in an electric circuit, and $(\varepsilon_1 A_1)^{-1}$ is analogous to electrical resistance. This is a convenient analogy when only radiation is being considered, but if convection and radiation are both occurring at a surface, the convection is described by a driving potential based on the difference in the first power of the temperatures while the radiation is described by the difference in the fourth power of the temperatures. In cases like this, it is often useful to express the radiation given by Eq. (7.12) as

$$\begin{aligned} q_{\text{rad}} &= \varepsilon_1 A_1 \sigma (T_{s1} - T_{s2})(T_{s1} + T_{s2})(T_{s2}^2 + T_{s1}^2) \\ &= A_1 h_{\text{rad}} (T_{s1} - T_{s2}) \end{aligned} \tag{7.14}$$

where

$$h_{\text{rad}} = \varepsilon \sigma (T_{s1} + T_{s2})(T_{s2}^2 + T_{s1}^2) \approx 4\varepsilon \sigma (T_{s1} T_{s2})^{3/2} \tag{7.15}$$

is often called a *radiation heat transfer coefficient*. The disadvantage of this form is that h_{rad} depends on the temperature of the surface, which is often the desired result of the calculation. Note that a *radiation resistance* corresponding to the temperature difference $T_{s1} - T_{s2}$ can be defined as

$$R_{\text{rad}} = \frac{1}{h_{\text{rad}} A_1} \tag{7.16}$$

The total heat transfer from the surface by convection and radiation, where $T_\infty = T_{s2}$, is then

$$q = q_{\text{rad}} + q_{\text{conv}} = (T_{s1} - T_\infty)(A_1 h_{\text{rad}} + A_1 h)$$

The temperature $T_{s1} - T_\infty$ is a temperature difference in which either degrees Kelvin or Celsius can be used; however, absolute temperatures must be used to calculate h_{rad}.

7.2.2 Transient Thermal Response of Components

When an electronic component is turned on, it starts generating heat internally. As the temperature of the component rises with time, its ability to dump heat to its environment also increases. Eventually, the temperature of the component rises to a level that allows it to dump heat at the same rate it is being generated in the component. At that point, steady state is reached, and the temperature of the component stays constant unless the rate of generation, the temperature of the environment, or the thermal resistance between the component and its environment changes. For example, if the component is being cooled by convection, increasing the velocity of airflow would increase the heat transfer coefficient, and the component would begin changing to a lower steady-state operating temperature. An energy balance for the component can be written as

$$\dot{E}_{stored} = \dot{E}_{in} - \dot{E}_{out} + \dot{E}_{gen} \quad (7.17)$$

where $\dot{E}_{in}$ and $\dot{E}_{out}$ are at the component boundaries, and $\dot{E}_{gen}$ is the rate of heat generation inside the component. Let us assume that the component can be treated as a lumped mass at temperature T, that is, there is little temperature difference between the interior of the component and its surface. Then, we can write $\dot{E}_{stored} = mC_p(dT/dt)$, where m is the component mass, C_p is the mass-averaged specific heat of all the masses that make up the component, and dT/dt is the rate of component temperature change. For a component cooled by convection, $\dot{E}_{in} = 0$, $\dot{E}_{out} = hA_s(T - T_\infty)$, and the energy balance becomes a simple first-order differential equation:

$$\frac{dT}{dt} = \frac{\dot{E}_{gen}}{mC_p} - \frac{hA_s}{mC_p}(T - T_\infty) \quad (7.18)$$

If the component is initially at T_∞, the solution is

$$T - T_\infty = \frac{\dot{E}_{gen}}{hA_s}(1 - e^{-t/t_c}) \quad (7.19)$$

where $t_c = mC_p/hA_s$ is the system time constant. After $t = 3t_c$ to $4t_c$, the system has essentially reached steady state, and the eventual steady-state temperature is $T_{ss} = T_\infty + \dot{E}_{gen}/hA_s$. Note that the approximate time to reach steady state and the eventual steady-state temperature can both be determined by examining the differential equation without solving it.

Example 7.1

A square chip with heat-generating circuits on the bottom is mounted in a substrate so that there is no heat transfer from the sides and bottom. The chip can be treated as a lumped mass. The chip is at $T_\infty = 20°C$ at $t = 0$ when it is energized and starts generating 0.9 W/cm². For $w = 5$ mm $= 0.005$ m, $\delta = 1$ mm $= 0.001$ m, $k = 150$ W/m K, $\rho = 2000$ kg/m³, $C_p = 750$ J/kg K, and $h = 300$ W/m² K, what is the eventual steady-state temperature? Approximately how long will it take to reach steady state. (See Fig. E7.1)

SOLUTION

$$\dot{E}_{gen} = q''w^2 = 0.225 \text{ W}$$
$$T_{ss} = T_\infty + \dot{E}_{gen}/hA_s = 20°C + 30°C = 50°C$$
$$m = \rho w^2 \delta = 5 \times 10^{-5} \text{ kg}$$
$$t_c = mC_p/hw^2 = 5s \rightarrow t_{ss} \approx 15 \text{ to } 20 \text{ s}$$

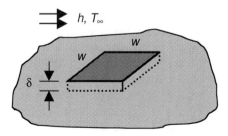

Figure E7.1

There is a criterion for determining if the lumped-mass assumption is valid. Heat leaving the bottom of the chip at T_c in the example above is conducted through the chip, leaves the top surface by convection, and is dumped to the air at T_∞. The temperature drop across the chip, $\Delta T_{internal}$, is $qR_{cond} = q\delta/kA_c$, and the temperature drop from the top surface to the air, $\Delta T_{external}$, is $qR_{conv} = q(1/hA_s)$. The ratio $\Delta T_{internal}/\Delta T_{external}$ is called the *Biot number*, Bi:

$$\text{Bi} = h(V/A_s)/k = h\delta/k$$

where V is the component volume, and A_s is the convection surface area. If Bi < 0.1, the lumped-mass assumption is valid. In the example above, Bi $= 0.002$. ∎

7.2.3 Conduction in Various Shapes

The concepts of heat transfer rate, temperature gradient, thermal conductivity, and thermal resistance were introduced for heat transfer through a plane wall in Section 7.2.1.1. Those same concepts will now be extended to other geometries.

7.2.3.1 Conduction in a Cylinder

Consider a long hollow cylinder of inside radius r_1, outside radius r_2, and length Z shown in Figure 7.4. The inside of the cylinder is at temperature T_1, and the outside is at temperature T_2. The radial heat transfer rate through any cross section at arbitrary radius r is

$$q = -kA_c \frac{dT}{dr} = -k(2\pi r Z)\frac{dT}{dr} \tag{7.20}$$

where $A_c = 2\pi r Z$ is the cross-sectional area at radius r.

If steady state has been reached and there is no energy generation in the hollow cylinder, the first law of thermodynamics requires that the heat transfer rate, q, be the same at every radial location. Thus, Eq. (7.20) can be rearranged as

$$-dT = \frac{q}{2\pi Z k}\frac{dr}{r}$$

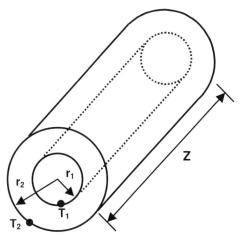

Figure 7.4 Radial conduction through a hollow cylinder.

Integrating between limits T_1, T_2, and r_1, r_2 gives

$$T_1 - T_2 = \frac{q}{2\pi Z k} \ln \frac{r_2}{r_1}$$

or

$$q = \frac{T_1 - T_2}{\ln r_2/r_1 / 2\pi Z k} \tag{7.21}$$

Again note the electrical analogy if the conduction resistance of a hollow cylinder is defined as

$$R_{\text{cond,cyl}} = \frac{\ln(r_2/r_1)}{2\pi Z k} \tag{7.22}$$

These results are useful when heat generated by an electronic package must be conducted radially through a cylinder.

7.2.3.2 Layered Composites in Series

Consider a composite slab made of two layers of different thickness and conductivity as shown in Figure 7.5. The two layers are well bonded at the interface. The left side of layer 1 is maintained at T_L, and the right side of layer 2 is maintained at T_R. Clearly, there will be heat transfer from left to right through the composite slab if $T_L > T_R$. If the slab is at steady state and no heat is stored in either layer, the heat transfer rate must be the same in both slabs. The two layers are then thermal resistances in series, and the temperature differences in the two slabs $(T_L - T_{12})$ and $(T_{12} - T_R)$ add together to give the overall temperature difference, $T_L - T_R$. The heat transfer through the composite slab can be written as

$$q = \frac{T_L - T_R}{L_1/k_1 A + L_2/k_2 A} \tag{7.23}$$

where $L_1/k_1 A$ and $L_2/k_2 A$ are the resistances of layers 1 and 2, respectively.

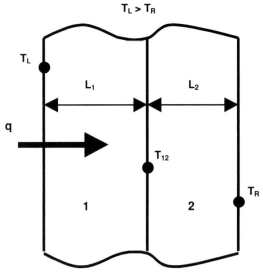

Figure 7.5 Slab resistances in series.

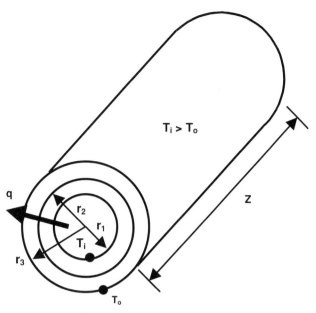

Figure 7.6 Cylindrical resistances in series.

Layered cylinders also form resistances in series. Consider the well-bonded long cylindrical layers shown in Figure 7.6. The radial heat transfer through the layers is

$$q = \frac{T_i - T_0}{\ln(r_2/r_1)/2\pi k_A Z + \ln(r_3/r_2)/2\pi k_B Z} \tag{7.24}$$

7.2.3.3 Layered Composites in Parallel

Now consider the composite solid shown in Figure 7.7. The left-hand face of the composite is again maintained at T_L, while the right-hand face is maintained at T_R. The top and bottom are both well insulated. In this case, heat flows from left to right with part of the heat transfer occurring in the top layer and the remainder in the bottom, but the temperature difference across both layers will be $T_L - T_R$. Thus, the two layers form parallel thermal resistances.

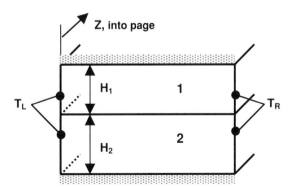

Figure 7.7 Slab resistances in parallel.

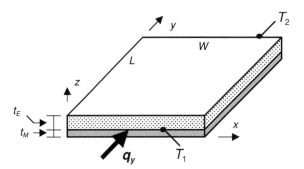

Figure 7.8 Equivalent conductivity with ground layer.

The heat transfer rate through the parallel layers is

$$q = (T_L - T_R)\left[\frac{1}{R_1} + \frac{1}{R_2}\right]$$

where

$$R_1 = \frac{L}{k_1 Z H_1}$$
$$R_2 = \frac{L}{k_2 Z H_2}$$

are the thermal resistances of the two layers with Z equal to the depth of the layers into the page. Note that the equivalent resistance of the parallel combination is

$$R_{\text{parallel}} = \left[\frac{1}{R_1} + \frac{1}{R_2}\right]^{-1} \tag{7.25}$$

7.2.3.4 Effective Conductivity

Consider a glass–epoxy board with a metal ground layer as shown in Figure 7.8. The heat transfer in the y direction will be in parallel paths through the glass–epoxy and metal layers and can be written as

$$q_y = \frac{k_M t_M W(T_1 - T_2)}{L} + \frac{k_E t_E W(T_1 - T_2)}{L} = \frac{k_{\text{eq},y}(t_M + t_E)W(T_1 - T_2)}{L}$$

where subscripts M and E refer to metal and glass–epoxy, respectively. The equivalent conductivity in the y direction, $k_{\text{eq},y}$ can be solved for as

$$k_{\text{eq},y} = \frac{k_M t_M}{t_M + t_E} + \frac{k_E t_E}{t_M + t_E} = f_M k_M + (1 - f_M) k_E \tag{7.26}$$

where f_M is the fraction of the $W \times (t_M + t_E)$ cross section that is metal. The same approach can be used to determine $k_{\text{eq},z}$ with metal vias through a board as shown in Figure 7.9. In

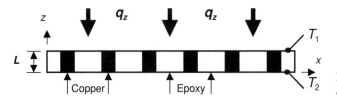

Figure 7.9 Equivalent conductivity with vias.

this case, f_M is the fraction of the board area, looking in the z direction, which is metal vias instead of epoxy–glass.

7.2.3.5 Contact Resistance

The composite slabs in series, discussed previously, were described as well-bonded at the interface. This implied that the two layers were in perfect thermal contact and that the temperature of both layers was the same at the interface. Often, when two surfaces are brought into contact, surface roughness prevents good thermal contact and creates voids between the two surfaces that inhibit heat flow. This inhibition of heat flow is characterized as a *contact resistance*:

$$R_{\text{cont}} = \frac{1}{h_{\text{cont}} A_c} = \frac{R''_{\text{cont}}}{A_c} \qquad (7.27)$$

where

h_{cont} = contact conductance (W/m² °C)
$R''_{\text{cont}} = 1/h_{\text{cont}}$ = contact resistance coefficient (m² °C/W)
A_c = cross-sectional area of layer at interface (m²)

The contact resistance causes a "jump" in the temperature at the interface and is usually modeled as a function of surface roughness, conductivity of the two layers, the contact pressure, and the conductivity of any fluid that may be introduced at the interface to fill in the gaps.

Example 7.2

The heat flux created by a 12.7-mm × 12.7-mm thin integrated circuit chip is 3 W/cm². The top of the chip is cooled by a coolant at $T_\infty = 20°C$ creating a convection coefficient of $h_t = 1000$ W/m² °C. The bottom of the chip is attached to a substrate such that the contact conductance $h_c = 10^4$ W/m² K. The substrate is $t = 5$ mm thick and has a conductivity k_b of 1 W/m K. The bottom of the substrate is cooled by 20°C air with convection coefficient $h_b = 40$ W/m². Draw the thermal circuit for this arrangement and predict the steady-state chip temperature. (See Fig. E7.2a.)

SOLUTION First calculate the various resistances and other values the circuits shown in Figures E7.2b and E7.2c, either of which can be used:

$A_c = A_s = (12.7 \text{ mm})^2 = 161 \text{ mm}^2 = 1.61 \times 10^{-4} \text{ m}^2$
$q_{\text{chip}} = q''_{\text{chip}} A_c = 3 \times 10^4 \text{ (W/m}^2) \times 1.61 \times 10^{-4} \text{ m}^2 = 4.84 \text{ W}$
R_{cvt} = convective resistance on top = $[1000 \text{ W/m}^2 \text{°C} \times 1.61 \times 10^{-4} \text{ m}^2]^{-1} = 6.21 \text{°C/W}$
R_{cont} = contact resistance = $[10^4 \text{ W/m}^2 \text{°C} \times 1.61 \times 10^{-4} \text{ m}^2]^{-1} = 0.621 \text{°C/W}$
R_{bd} = substrate conduction resistance = $0.005 \text{m}/(1 \text{ W/m K} \times 1.61 \times 10^{-4} \text{ m}^2) = 31.1 \text{°C/W}$
R_{cvb} = convective resistance on bottom = $[40 \text{ W/m}^2 \text{°C} \times 1.61 \times 10^{-4} \text{ m}^2]^{-1} = 155 \text{°C/W}$

Perform a heat balance on the T_{chip} node.

$$q_{\text{chip}} = q_{\text{up}} + q_{\text{down}} = \frac{T_{\text{chip}} - T_\infty}{R_{\text{cvt}}} + \frac{T_{\text{chip}} - T_\infty}{R_{\text{cont}} + R_{\text{bd}} + R_{\text{cvb}}}$$

Solving gives $T_{\text{chip}} = 49°C$. Using this value allows calculation of $q_{\text{up}} = 4.68$ W and $q_{\text{down}} = 0.16$ W. ∎

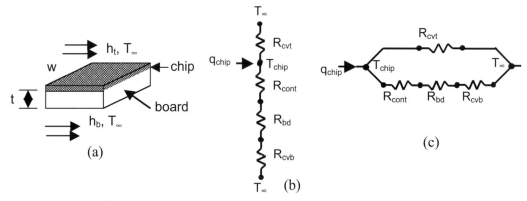

Figures E7.2a–E7.2c

7.2.3.6 Two- and Three-Dimensional Conduction Resistance

The thermal resistance of more complex shapes can be expressed in terms of a conduction *shape factor, S*

$$R_{\text{cond}} = \frac{1}{kS} \qquad (7.28)$$

where S (meters) is tabulated in heat transfer texts for a number of common two- and three-dimensional (2D and 3D) geometries, and k is the conductivity of the material. As an example, consider a package of chips mounted on the outside of a $w \times w \times Z$ square heat sink as shown in Figure 7.10 and cooled by a liquid flowing in the circular channel of diameter d at temperature T_{fluid} with convection coefficient h. The thermal circuit for this situation is also shown, and the operating temperature of the chips can be solved for as

$$T_{\text{chips}} = T_{\text{fluid}} + q\left[\frac{1}{Sk} + \frac{1}{\pi dZh}\right]$$

where q is the rate of heat transfer from the chips to the fluid, and S is the shape factor for heat transfer between a cylinder in the center of a square solid and the surface of the solid.

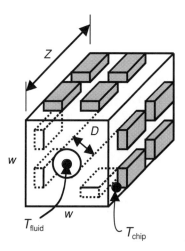

Figure 7.10 Multidimensional conduction.

For this geometry, the shape factor is known to be

$$S = \frac{2\pi Z}{\ln(1.08\, w/D)}$$

The nature of the shape factor can be understood by recalling the conduction resistance of a slab

$$R_{\text{cond}} = \frac{L}{kA_c} = \frac{1}{k(A_c/L)}$$

Comparison shows that $S = A_c/L$ for a slab. For more complex shapes, S can be thought of as the ratio of a mean cross-sectional area to a mean heat transfer path length.

7.2.3.7 Small Heat Sources on Much Larger Solids

The heat source in many applications is often a small heat source maintained on the surface of a large conducting body, for example, a small chip at one temperature mounted on the surface of a relatively large substrate at a different temperature. The heat generated by the chip "spreads out" across the area of the substrate while also being conducted through the substrate. Finding the resulting chip temperature is a fairly complex problem involving a two- or three-dimensional analysis. Thinking of the heat from the chip as first spreading out across the area of the substrate and then being conducted through the substrate provides the basis for a simpler analysis [4], that is,

$$q = \frac{\Delta T}{R_{\text{spr}} + R_{\text{cond}}} \tag{7.29}$$

where

R_{spr} = spreading resistance
R_{cond} = conduction resistance of the body on which the heat source is mounted
ΔT = temperature difference between the heat source and boundary temperature of the body

Consider the small circular and square heat sources shown in Figure 7.11 as the surfaces of very large conductors. The shape factors S for a disk of diameter D and an $L \times L$ square on a large flat surface are $D/0.54$ and $L/0.55$, respectively. The resulting spreading resistances are

$$R_{\text{spr}} = \frac{0.54}{Dk} \tag{7.30}$$

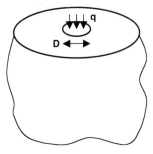
(a) Circular source

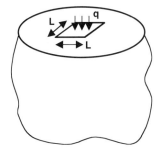
(b) Square source

Figure 7.11 Small heat sources on large areas.

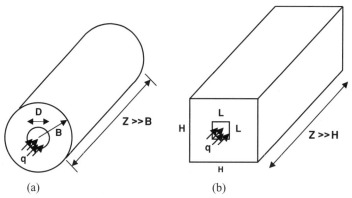

Figure 7.12 Small heat sources on finite areas.

for the circular source in Figure 7.11a and

$$R_{\text{spr}} = \frac{0.55}{Lk} \tag{7.31}$$

for the square source in Figure 7.11b. For a circular source on the finite conducting body shown in Figure 7.12a,

$$R_{\text{spr}} = \frac{0.54}{Dk}\left[1 - \frac{D}{B}\right]^{3/2} \tag{7.32}$$

The conductivities in Eqs. (7.30) to (7.32) are those of the body on which the source is mounted.

For the square source on a square finite body as shown in Figure 7.12b, using Eq. (7.32) with $0.55/L$ instead of $0.54/D$ and $D/B = L/H$ gives reasonable results. The conduction resistance of the conducting bodies in Figure 7.12a and 7.12b are

$$R_{\text{cond}} = \frac{4Z}{k\pi D^2}$$

for the cylinder and

$$R_{\text{cond}} = \frac{Z}{kH^2}$$

for the square.

7.2.4 Overall Resistance

Consider conduction through the arrangement shown in Figure 7.13. The left-hand face is maintained at T_L, while the right-hand face is exposed to a fluid at T_∞ with a convection heat transfer coefficient h. There is a contact conductance, h_c, at the interface of the two layers. The heat transfer through the layers is to be determined.

In the absence of heat generation within the layers, the heat transfer through both layers, across the interface and across the convection resistance must be the same at steady state. Therefore, the resistances of each layer, the contact resistance at the interface, and

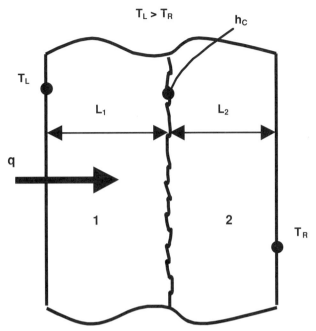

Figure 7.13 Multilayer wall with contact and convection resistances.

the convection resistance at the right-hand surface are all in series. The overall driving temperature difference is $T_L - T_\infty$, and the overall resistance is

$$R_{\text{overall}} = \frac{L_1}{k_1 A} + \frac{1}{h_c A} + \frac{L_2}{k_2 A} + \frac{1}{h A}$$

where A is the cross-sectional area of the wall in the direction of heat flow and the surface area for convection.

Now reconsider Figure 7.7. Instead of the right-hand face being maintained at T_R, this face is now exposed to an environment at T_∞ with convective heat transfer coefficient h. This is now a series-parallel circuit with resistances shown in Figure 7.14. The heat transfer

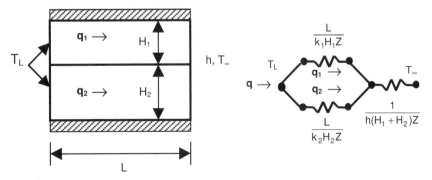

Figure 7.14 Parallel–series resistances.

through this combination is

$$q = \frac{T_L - T_\infty}{\left[\frac{k_1 Z H_1}{L} + \frac{k_2 Z H_2}{L}\right]^{-1} + \frac{1}{hZ(H_1 + H_2)}}$$

where the denominator is the overall resistance, and $T_L - T_x$ is the driving temperature difference.

Two examples of overall resistance have been given in this section. One could easily envision other parallel and/or series combinations of thermal resistances that could be reduced to an overall thermal resistance via use of the electrical resistance analogy.

7.2.4.1 Manufacturer Thermal Resistance Definitions

The thermal resistance of a commercial chip carrier involves conduction along complex series and parallel paths from the integrated circuit (IC) junction to the surface of the carrier and out through the leads. The equivalent resistance of this complex combination is usually called the *junction-to-case resistance* θ_{jc} (°C/W). This resistance is relatively constant except for the relatively small variation of the chip and carrier materials with temperature. A value of this resistance is usually supplied at a recommended operating temperature by the manufacturer.

If the chip is being air cooled, a second resistance represents the convection resistance between the carrier surface and leads and the cooling air. It is commonly called the *case-to-air resistance* θ_{ca} (°C/W). Due to the variation in heat transfer coefficient on the outside of the case, this resistance decreases as the velocity of the cooling air increases. The overall thermal resistance between the IC junction and the cooling air is the sum of θ_{jc} and θ_{ca}.

$$\theta_{ja} = \theta_{jc} + \theta_{ca} \tag{7.33}$$

This resistance is often shown as a function of air velocity by the manufacturer for a particular chip.

7.2.4.2 Heat Sinks

The flip-chip module shown in Figure 7.15 is cooled primarily by convection from its top surface, and it is necessary to increase the heat transfer while keeping the chip operating at the same temperature. One alternative is to reduce the convective resistance at the top surface with the addition of additional surface area by attaching a finned heat sink as shown.

The heat transfer from the heat sink (HS) is

$$q_{HS} = h \eta_0 A_{HS} (T_0 - T_\infty) \tag{7.34}$$

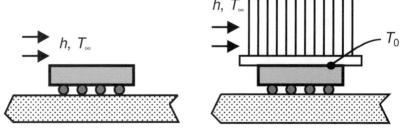

Figure 7.15 Addition of a heat sink.

and the *heat sink resistance* is

$$R_{HS} = \frac{1}{h\eta_0 A_{HS}} \qquad (7.35)$$

where A_{HS} is the total surface area of the heat sink, and η_0 is the *area effectiveness*. The area effectiveness would be 1 if all of A_{HS} were at T_0; however, the surface temperature of the heat sink clearly decreases with distance from the base of the heat sink. The area effectiveness accounts for this.

Other parameters are often used to characterize heat sinks. In cases where the footprint area of the heat sink is important, the *array heat transfer coefficient* is of interest:

$$h_A = \frac{q}{LW(T_0 - T_\infty)} \qquad (7.36)$$

where L and W are the length and width of the base of the heat sink. In cases where the space occupied by the heat sink is at a premium, the volumetric heat transfer coefficient is important:

$$h_V = \frac{q}{V_{HS}(T_0 - T_\infty)} \qquad (7.37)$$

where V_{HS} is the volume occupied by the heat sink.

These heat sink performance parameters depend on a number of things, including the heat transfer coefficient, the shape and size of the heat sink and its fins, and fin material thermal conductivity. In some cases, these parameters can be determined by analysis, but often the geometry is so complex that they must be measured as a function of air velocity. In heat sinks used with natural convection, the air velocity depends on $T_0 - T_\infty$. Heat sink vendors usually supply this kind of performance information. Some heat sinks have their own small fan attached, making the air velocity and resulting heat sink resistance nearly constant.

Example 7.3

We want to double the power output from the chip in Example 7.2 while keeping the chip temperature at 49°C. It is decided to purchase and attach a heat sink to the top of the chip. Assuming the heat sink can be attached with no contact resistance, find the heat sink resistance required. (See Fig. E7.3a.)

SOLUTION All the resistances remain the same, except that the convection resistance at the top, R_{cvt}, is replaced by the resistance of the heat sink, R_{HS}. The q_{chip} is doubled to 9.68 W. (See Figs. E7.3b and E7.3c.)

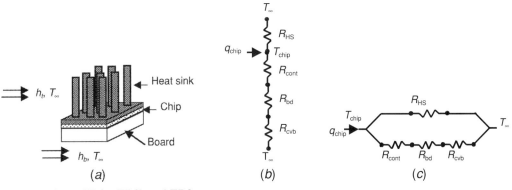

Figure E7.3a, E7.3b and E7.3c

The heat balance on the T_{chip} node yields

$$q_{chip} = \frac{T_{chip} - T_\infty}{R_{HS}} + \frac{T_{chip} - T_\infty}{R_{cont} + R_{bd} + R_{cvb}}$$

where only R_{HS} is unknown. Solving yields $R_{HS} = 2.0°C/W$. As noted above, this replaces the convection resistance at the top, $R_{cvt} = 6.2°C/W$. ■

7.2.5 Forced Convection Heat Transfer

Thus far, the convection heat transfer coefficient, h, has been assumed to be known. The heat transfer coefficient is actually a function of a number of things including:

- Fluid properties
- Nature of the fluid flow
- Surface geometry

The relationship among heat transfer coefficient and these things will now be presented.

Most forced convection heat transfer coefficients are correlated in terms of dimensionless groups. One of these is the *Nusselt number*, a measure of the ratio of heat transfer by convection to the heat transfer when the fluid is stationary and only conduction occurs:

$$\text{Nu} = \frac{h L_{char}}{k_{fluid}} \tag{7.38}$$

where

$\quad$ Nu = Nusselt number (dimensionless)
$\quad L_{char}$ = characteristic length for the geometry being considered (m)
$\quad k_{fluid}$ = thermal conductivity of the fluid (W/m K)

Reynolds number is a measure of the inertia of the moving fluid compared to the shear stresses within the fluid:

$$\text{Re} = \frac{U_{char} L_{char} \rho}{\mu} \tag{7.39}$$

where

$\quad U_{char}$ = characteristic velocity of the fluid flow (m/s)
$\quad \rho$ = fluid density (kg/m^3)
$\quad \mu$ = fluid viscosity (kg/m s)

The viscosity μ is a thermophysical property that relates the shear stress within the moving fluid to the velocity gradient within the fluid.

Reynolds number is the most important dimensionless group in forced convection. It is used to characterize the flow as laminar or turbulent. When the flow is laminar, the fluid moves over the surface in ordered layers. As the Reynolds number increases, minor disturbances in the flow field begin to be amplified instead of dissipated, and small pockets of swirling motion called eddies begin to appear in the flow field. At very high Reynolds numbers, the size and number of eddies increases and the ordered layers disappear. This condition is called turbulent flow. As the flow changes from laminar to turbulent, the analytical results and empirical relations used to calculate the heat transfer coefficients change.

Another dimensionless group appearing in heat transfer correlations is the *Prandtl number* of the fluid:

$$\text{Pr} = \nu/\alpha = C_p \mu / k \tag{7.40}$$

where

ν = fluid kinematic viscosity = μ/ρ (m²/s)
α = fluid thermal diffusivity = $k/\rho C_p$ (m²/s)
C = fluid specific heat (J/kg K)

Note that the Prandtl number is a dimensionless thermophysical property of the fluid since it is the ratio of two properties with the same dimensions.

Most forced convection heat transfer data and/or analytical results yield correlations of the form

$$\text{Nu} = B\,\text{Re}^p\,\text{Pr}^n \qquad (7.41)$$

where B, p, and n are dimensionless correlation constants that depend on the surface geometry and the nature of the flow field, primarily whether it is laminar or turbulent. Usually $n \approx 1/3$. Correlations relating Nu to Re and Pr for a number of different geometrics can now be presented.

7.2.5.1 Laminar Flow over Flat Plates

Consider a fluid flowing over a stationary flat plate at velocity U as shown in Figure 7.16. The plate temperature is maintained at T_s. The fluid velocity at the plate surface is zero, and the fluid temperature at the plate surface is T_s. The fluid velocity increases as the distance from the plate increases until the fluid velocity essentially is the free-stream velocity U. The region close to the wall, where the fluid velocity is between zero and U, is called the velocity boundary layer, and its thickness is denoted as δ. Likewise, the fluid temperature changes from T_s at the plate to T_x as the distance from the plate increases. The region where the fluid temperature is between T_s and T_∞ is called the thermal boundary layer, and its thickness is denoted as δ_t. Both δ and δ_t increase as the distance x from the leading edge of the plate is increased. If the Prandtl number of the fluid is greater than 1, the velocity boundary layer is thicker than the thermal boundary layer at any location x, and the opposite is true for a fluid Prandtl number less than 1. The increasing thickness of the boundary layers causes the heat transfer coefficient to decrease as the downstream distance x increases.

The heat flux at any location x is

$$q_x'' = h_x(T_s - T_\infty) \qquad (7.42)$$

where the x subscript indicates that the heat flux and heat transfer coefficient are *local* values, that is, dependent on location x. The *local heat transfer coefficient* can be determined analytically as [5]

$$\text{Nu}_x = \frac{h_x X}{k} = 0.332\,\text{Re}_x^{1/2}\,\text{Pr}^{1/3} \qquad (7.43)$$

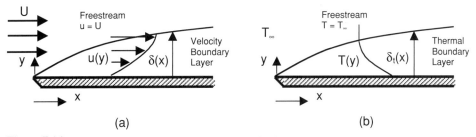

Figure 7.16 (a) Velocity and (b) thermal boundary layer development.

where the x subscript on Re and Nu indicates that x should be used as the characteristic length. The fluid properties are evaluated at the film temperature

$$T_f = \frac{T_s + T_\infty}{2}$$

The heat transfer from the plate between the leading edge, $x = 0$, and location x is

$$q_{0 \to x} = \bar{h}_x Z x (T_s - T_\infty) \qquad (7.44)$$

where

$\bar{h}_x$ = the *average heat transfer coefficient* between 0 and x
Z = depth of the plate

The average heat transfer coefficient is determined by calculating the average Nusselt number:

$$\overline{\mathrm{Nu}}_x = \frac{\bar{h}_x x}{k} = 0.664 \, \mathrm{Re}_x^{1/2} \, \mathrm{Pr}^{1/3} \qquad (7.45)$$

7.2.5.2 Turbulent Flow over a Flat Plate

At some distance downstream from the leading edge, the flow becomes unstable and becomes turbulent. The distance at which the transition from laminar to turbulent flow begins can be expressed in terms of a critical Reynolds number

$$\mathrm{Re}_{x_c} = \frac{U x_c}{\nu} = 5 \times 10^5 \qquad (7.46)$$

where x_c is the critical distance where the transition begins. This transition is illustrated in Figure 7.17.

For turbulent flow, the local Nusselt number is [5]

$$\mathrm{Nu}_x = \frac{h_x x}{k} = 0.0296 \, \mathrm{Re}_x^{4/5} \, \mathrm{Pr}^{1/3} \qquad (7.47)$$

and the average Nusselt number for 0 to x is

$$\overline{\mathrm{Nu}}_x = \frac{\bar{h}_x x}{k} = 0.037 \, \mathrm{Re}_x^{4/5} \, \mathrm{Pr}^{1/3} \qquad (7.48)$$

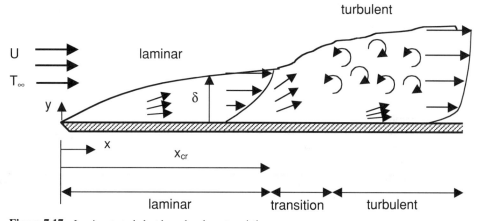

Figure 7.17 Laminar to turbulent boundary layer transistion.

When the length of the plate is such that the flow over the entire plate is laminar, Eq. (7.45) can be used to calculate an average heat transfer coefficient. Often, some projection from the surface at the leading edge "trips" the flow causing it to be turbulent beginning at $x = 0$. In this case, Eq. (7.47) or (7.48) should be used. However, when the flow is not "tripped" and the length of the plate is long enough that a laminar-to-turbulent transition occurs upstream of the trailing edge of the plate, both laminar and turbulent boundary layers are present on the plate, and both contribute to the average heat transfer coefficient. The average heat transfer coefficient in this case is

$$\overline{\mathrm{Nu}}_x = (0.037 \, \mathrm{Re}_x^{4/5} - 871)\mathrm{Pr}^{1/3} \tag{7.49}$$

The second term in parentheses represents the difference between the turbulent and laminar Nusselt numbers from 0 to x_c, while the first term is the average Nusselt number for tripped flow.

7.2.5.3 Unheated Starting Length on a Flat Plate

All previous expressions for Nu have assumed that the plate temperature is T_s starting at $x = 0$. In many applications, an *unheated starting length* ξ exists, that is, the plate temperature is T_∞ for $x < \xi$ and T_s for $x > \xi$. For example, a PWB being treated as a flat plate for heat transfer purposes often does not have active thermal elements placed on the leading edge of the board. In this case, the local Nusselt number for laminar flow is [6]

$$\mathrm{Nu}_x = \frac{\mathrm{Nu}_{x,\xi=0}}{\left[1 - (\xi/x)^{0.75}\right]^{0.33}} \tag{7.50}$$

and for turbulent tripped flow is

$$\mathrm{Nu}_x = \frac{\mathrm{Nu}_{x,\xi=0}}{\left[1 - (\xi/x)^{0.9}\right]^{0.11}} \tag{7.51}$$

7.2.5.4 Constant Heat Flux on a Flat Plate

In many problems, a constant heat flux is applied to the plate instead of the plate being maintained at constant temperature T_s as assumed thus far. For constant heat flux with laminar flow [6],

$$\mathrm{Nu}_x = 0.453 \, \mathrm{Re}_x^{1/2} \mathrm{Pr}^{1/3} \tag{7.52}$$

For turbulent flow with constant heat flux,

$$\mathrm{Nu}_x = 0.0308 \, \mathrm{Re}_x^{4/5} \mathrm{Pr}^{1/3} \tag{7.53}$$

In constant flux cases, the heat transfer is simply the product of heat flux and plate area. However, the variation in heat transfer coefficient with x requires that the plate temperature vary with x as

$$T_s(x) = T_\infty \pm \frac{q''}{h_x} \tag{7.54}$$

where the plus sign is for heat input to the fluid, and the minus sign is for heat removal.

Example 7.4

A PWB is covered with 8-mm × 8-mm chips, each dissipating 160 mW. The board is cooled by blowing 27°C air across it at 10 m/s. The chips trip the flow and cause it to be turbulent throughout.

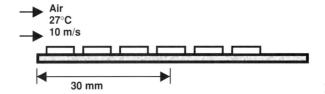

Figure E7.4

We need to determine the temperature of a chip whose center is located 30 mm downstream from the leading edge. (See Fig. E7.4.)

SOLUTION First look up properties at the film temperature $T_f = (T_s + T_\infty)/2$. Since finding T_s is the objective, we need to make a reasonable guess for T_s so we can calculate the T_f at which we look up properties. Assume $T_s = 53°C$; then $T_f = 40°C$. Properties at $40°C$ (313 K) are $k = 0.0265$ W/m K, $\alpha = 17.6 \times 10^{-6}$ m²/s, Pr = 0.71, $\rho = 1.025$ kg/m³.

$$\text{Re}_x = Vx/\nu = 17{,}045 \quad \text{at} \quad x = 0.03 \text{ m}$$

Note the flow would be laminar if the flow were not tripped since $\text{Re}_x < 500{,}000$. Use Eq. (7.53) since the flow is turbulent, and the chips produce a constant heat flux at the surface.

$$\text{Nu}_x = 0.0308 \, \text{Re}_x^{4/5} \, \text{Pr}^{1/3} = 66.7 \qquad h_x = \text{Nu}_x k/x = 58.9 \text{ W/m}^2 \text{ K}$$

We can now calculate the chip temperature. $T_s = T_\infty + q_s''/h_x = 69.4°C$, where $q_s'' = 0.160 \text{ W}/(0.008 \text{ m})^2 = 2500 \text{ W/m}^2$. The guess at T_s was reasonable and did not significantly affect the result. As noted above, if the flow were not tripped, the boundary layer at $x = 30$ mm would be laminar, and Eq. (7.52) would have been used. The results would be $\text{Nu}_x = 52.6$, $h_x = 46.5 \text{ W/m}^2 \text{ K}$, $T_s = 80.8°C$. ■

7.2.5.5 Cylinders and Spheres in Crossflow

The average heat transfer coefficient for crossflow over a cylinder of diameter D can be calculated using the Zhukauskas correlation [7]:

$$\overline{\text{Nu}_D} = \frac{\overline{h}D}{k} = B \, \text{Re}_D^p \, \text{Pr}^n (\text{Pr}/\text{Pr}_s)^{0.25} \tag{7.55}$$

where all properties are evaluated at T_∞ except Pr_s, which is evaluated at T_s. The constants B and p are given in Table 7.3. The constant $n = 0.37$ for Pr < 10, and $n = 0.36$ for Pr > 10.

For crossflow over a sphere of diameter D, Whitaker recommends [8]:

$$\overline{\text{Nu}_D} = \frac{\overline{h}D}{k} = 2 + (0.4 \, \text{Re}_D^{1/2} + 0.06 \, \text{Re}_D^{2/3}) \text{Pr}^{0.4} (\mu/\mu_s)^{0.25} \tag{7.56}$$

Table 7.3 Constants in Zhukauskas Correlation for Cylinders in Crossflow

Re_D	B	p
1–40	0.75	0.4
40–1,000	0.51	0.5
1,000–200,000	0.26	0.6
200,000–1,000,000	0.076	0.7

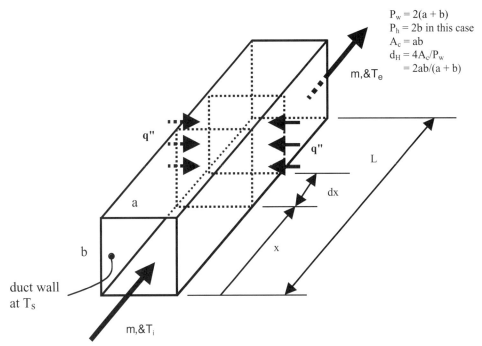

Figure 7.18 Flow in a duct.

where all properties are evaluated at T_∞ except μ_s, which is evaluated at T_s.

7.2.5.6 Forced Convection in Ducts

Previous forced convection heat transfer coefficients were for flow over the exterior of a body. We will now consider heat transfer to a fluid flowing within a closed duct due to the interior surface of the duct being different from the fluid flowing in the duct. Closed ducts occur often in electronic cooling applications. For example, a closed rectangular duct is formed by the space between two boards in a PC case, and air flowing in this passage cools the boards.

Before the calculation of heat transfer coefficients between the interior surface of a duct and the fluid flowing in it can be addressed, the fluid temperature variation along the duct length as a result of heating must be examined. Consider the duct cross section shown in Figure 7.18. Applying the first law of thermodynamics to the section dx wide gives

$$q'' P_h \, dx = \dot{m} C_p \, dT \tag{7.57}$$

where

q'' = heat flux at the duct wall (W/m²)
P_h = heated perimeter of the duct (m)
$\dot{m}$ = mass flow rate of the fluid (kg/s)
C_p = specific heat of the fluid (J/kg °C)
dT = temperature rise of the fluid from x to $x = dx$ (°C)

Constant Heat Flux For a constant heat flux q'', Eq. (7.57) can be integrated to yield the fluid temperature at location x as

$$T(x) - T_i = \frac{q'' P_h x}{\dot{m} C_p} \tag{7.58}$$

or

$$T_e - T_i = \frac{q'' P_h L}{\dot{m} C_p} \tag{7.59}$$

where x is the downstream distance from the inlet to the duct, L is the length of the duct, T_i is the fluid temperature at the duct inlet, and T_e is the fluid temperature at the duct exit. The surface temperature of the duct wall, that is, the chips on the board, can then be found as

$$T_s(x) - T(x) = \frac{q''}{h} \tag{7.60}$$

where h is the heat transfer coefficient on the inside of the duct. Finding h will be addressed later.

The mass flow rate is

$$\dot{m} = \rho A_c V = \rho Q \tag{7.61}$$

where

ρ = fluid density (kg/m³)
A_c = cross-sectional area of duct (m²)
V = average fluid velocity (m/s)
Q = volumetric flow rate (m³/s)

Constant Wall Temperature Now consider a duct where the duct surface temperature is constant at T_s. In this case, the heat flux in Eq. (7.57) is written as $q'' = h(T_s - T)$, and Eq. (7.57) can be rearranged to give

$$\frac{d(T - T_s)}{(T - T_s)} = \frac{h P_h}{\dot{m} C_p} dx$$

This can be integrated to give

$$\frac{T(x) - T_s}{T_i - T_s} = \exp\left[\frac{-h P_h x}{\dot{m} C_p}\right] \tag{7.62}$$

or

$$\frac{T_e - T_s}{T_i - T_s} = \exp\left[\frac{-h P_h L}{\dot{m} C_p}\right] \tag{7.63}$$

The heat transfer to the fluid can then be calculated as

$$q = \dot{m} C_p (T_e - T_i) \tag{7.64}$$

Now we can consider the calculation of the heat transfer coefficients. Like external flow, the most important dimensionless group is the Reynolds number. For flow in a duct, the Reynolds number is

$$\text{Re} = \frac{V d_H \rho}{\mu} = \frac{V d_H}{\nu} \qquad (7.65)$$

where d_H is the hydraulic diameter

$$d_H = \frac{4 A_c}{P_W} \qquad (7.66)$$

and P_W is the wetted perimeter of the duct. Using Eq. (7.61) and the definition of d_H allows Re to be written in a form that is often more convenient:

$$\text{Re} = \frac{4 \dot{m}}{\mu P_W} = \frac{4 Q}{\nu P_W} \qquad (7.67)$$

For Re $>$ 10,000, the flow is usually turbulent. The flow is laminar for Re $<$ 2300. Reynolds numbers between 2300 and 10,000 are usually flows in transition.

The heat transfer coefficients to be presented are for *fully developed flow*. The flow in a duct becomes fully developed some distance downstream from the duct inlet. When fluid enters the duct, boundary layers form at the walls of the duct and become thicker as the flow moves downstream. At some distance downstream, the boundary layers growing on the walls merge, and the flow is then said to be fully developed. Until that point, the heat transfer coefficient decreases as the boundary layers thicken. After that point, the heat transfer coefficient remains constant. Addressing heat transfer coefficients in the developing flow region is beyond the scope of this book.

7.2.5.7 Laminar Flow in Ducts

The fully developed Nusselt number for rectangular ducts of dimension $a \times b$ depends on the aspect ratio a/b as shown in Table 7.4. The characteristic length used is $d_H = 2ab/(a+b)$. Nusselt numbers for circular ducts are also given with diameter d as characteristic length.

7.2.5.8 Turbulent Flow in Ducts

The Dittus–Boelter equation is often used for both circular and noncircular ducts and either constant heat flux or constant duct temperature cases [9]. The hydraulic diameter is used as

Table 7.4 Fully Developed Laminar Nu for Rectangular and Round Ducts

a/b	$\text{Nu}_{q''=\text{const.}}$	$\text{Nu}_{T_s=\text{const.}}$
1	3.61	2.98
2	4.12	3.39
3	4.79	3.96
4	5.33	4.44
6	6.05	5.14
8	6.49	5.60
heated both sides	8.24	7.54
insulated one side	5.39	4.86
Round	4.36	3.66

the characteristic length for noncircular ducts, and the diameter is used for circular ducts. Properties are evaluated at $(T_i + T_e)/2$:

$$\text{Nu} = 0.023 \, \text{Re}^{4/5} \, \text{Pr}^n \tag{7.68}$$

where $n = 0.4$ for heating the fluid ($T_s > T$), and $n = 0.3$ for cooling the fluid ($T_s < T$).

Example 7.5

Two 30-cm × 10-cm PWBs are mounted 1.25 cm apart, forming a 10-cm × 1.25-cm rectangular duct into which 1.9×10^{-3} kg/s of 27°C air flow and cool the PWBs. Chips on the PWBs create a heat flux of 0.1 W/cm² inside the duct. What is the air temperature at the exit of the duct, and what is the highest chip surface temperature? Note that only the 10-cm sides of the duct are heated, not the 1.25-cm sides.

SOLUTION

Properties at the average of the inlet and outlet temperatures are needed. Assume $T_e = 73°\text{C}$ and look up properties at 50°C (average of 27 and 73°C). At 50°C, $k = 0.0271$ W/mK, $\mu = 18.6 \times 10^{-6}$ kg m/s², Pr = 0.71, $\rho = 1.059$ kg/m³, C = 1016 J/kg K.

For this arrangement $L = 30$ cm, $A_c = (1.25 \text{ cm} \times 10 \text{ cm}) = 12.5 \text{ cm}^2 = 1.25 \times 10^{-3}$ m², $P_W = 2(10 + 1.25)$ cm $= 22.5$ cm, $d_H = 4A_c/P_W = 2.22$ cm, $P_h = 2 \times 10$ cm $= 20$ cm, $q_s'' = 1000$ W/m².

Calculate the outlet temperature of the air. $T_e = T_i + q_s'' P_h L/(\dot{m}C) = 58.1°\text{C}$ (Be careful with units.)

Now that we have the air temperature at the exit, we need the heat transfer coefficient since $T_{se} = T_e + q_s''/h$. Start with the Reynolds number Re $= 4\dot{m}/\mu P_W = 1816$. The flow is laminar, with constant heat flux, and the duct aspect ratio is 10/1.25 = 8. From Table 7.4, Nu = 6.49, and $h = \text{Nu} \, k/D_h = 7.9$ W/m² K. Now we can calculate T_{se}:

$$T_{se} = T_e + q_s''/h = 58.1 + 126.6 = 184.7°\text{C}$$

This is probably too high. Let us try to remedy the situation by increasing the airflow by a factor of 10 to 1.9×10^{-2} kg/s. First, recalculate the exit air temperature. $T_e = T_i + q_s'' P_h L/(\dot{m}C) = 30.1°\text{C}$. Then, using the same properties gives Re $= 18,160$. Now the flow is turbulent, and Eq. (7.68) is appropriate:

$$\text{Nu} = 0.023 \, \text{Re}^{4/5} \, \text{Pr}^{0.4} = 51.2 h = 62.4 \text{ W/m}^2 \text{ K} \quad T_{se} = T_e + q_s''/h = 30.1 + 16.0 = 46.1°\text{C}$$

The increased mass flow helped in two ways; it lowered the exit temperature of the air, and the increased h lowered the temperature difference between the chips and the air. This improved cooling comes with a cost; the fan power required to create this additional airflow will increase by a factor of 500 or so. ∎

7.2.6 Natural or Free Convection Heat Transfer

Consider a PWB mounted vertically with active heat generating chips distributed across the board. No fluid movement is provided by external means, so the fluid velocity far removed from the board is zero. Close to the board, the fluid is warmed above T_∞, the temperature of the fluid far removed from the board. Warming the fluid close to the board above T_∞ creates buoyancy forces that cause the fluid to rise. As shown in Figure 7.19, the velocity boundary layer in this case is that region where the velocity of the fluid is nonzero. The thermal boundary layer is the region where the temperature is above T_∞.

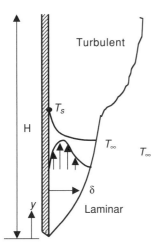

Figure 7.19 Natural convection boundary layer.

The *Grashof number* is a measure of the ratio of buoyancy forces generated by the temperature difference between the board and T_∞, and the shear stresses within the fluid. It is given by

$$\text{Gr} = \frac{g\beta(T_s - T_\infty)L_{\text{char}}^3}{\nu^2} \tag{7.69}$$

where β is the volumetric coefficient of thermal expansion, a thermophysical property of the fluid that measures how the density of the fluid changes with temperature, and g is the gravitational acceleration. The characteristic length, L_{char}, depends on the geometry. For a flat vertical plate, for example, the characteristic length can be the vertical position on the board in calculating local heat transfer coefficients and the height of the board when calculating an average heat transfer coefficient for the whole board.

The *Rayleigh number* is the product of Grashof and Prandtl numbers, and is given by

$$\text{Ra} = \text{Gr}\,\text{Pr} \tag{7.70}$$

This product often appears in correlations for Nusselt numbers.

Most correlations for the Nusselt number are of the form

$$\text{Nu} = B\,\text{Gr}^p\,Pr^n$$

or just

$$\text{Nu} = B\,\text{Ra}^p \tag{7.71}$$

where B, p, and n are constants depending on geometry and flow conditions.

7.2.6.1 Laminar Natural Convection on a Vertical Plate

Laminar-free convection on a vertical flat plate with constant plate temperature T_s will be considered first. The flow remains laminar until the height y is reached where $\text{Ra}_y = 10^9$, and the laminar to turbulent transition begins. The local Nusselt number in the laminar zone is correlated as [5]

$$\text{Nu}_y = 0.443\,\text{Ra}_y^{1/4} \tag{7.72}$$

The average Nusselt number from 0 to height y is

$$\overline{\mathrm{Nu}}_y = 0.59\, \mathrm{Ra}_y^{1/4} \tag{7.73}$$

The y subscript on Nu and Ra indicate that y is the characteristic length. All properties are evaluated at the film temperature $T_f = (T_s + T_\infty)/2$.

7.2.6.2 Turbulent Natural Convection on a Flat Plate

For tripped flow, the Nusselt number for turbulent flow is [5]

$$\overline{\mathrm{Nu}}_y = \mathrm{Nu}_y = 0.10\, \mathrm{Ra}_y^{1/3} \tag{7.74}$$

7.2.6.3 Constant Heat Flux on a Vertical Flat Plate

For a constant heat flux input condition, as opposed to constant wall temperature, the temperature difference $[T_s(y) - T_\infty]$ is zero at the bottom of the plate where $y = 0$, and increases with y. Equations (7.73) and (7.74) may still be used if the temperature difference used in Ra_H is $T_s(H/2) - T_\infty$, where H is the plate height. Then, for a heat flux input,

$$q'' = \overline{h}\,[T_s(H/2) - T_\infty]$$

If the flow is laminar, Eq. (7.73) can be used for $\overline{h}$ giving

$$\frac{q''}{[T_s(H/2) - T_\infty]} = \overline{h} = 0.59 \frac{k}{H} \left[\frac{g\beta H^3}{\nu^2}\mathrm{Pr}\right]^{0.25} [T_s(H/2) - T_\infty]^{0.25} \tag{7.75}$$

which can be solved for $[T_s(H/2) - T_\infty]$. The temperature difference at other locations is then [10]

$$T_s(y) - T_\infty = 1.15(y/H)^{0.2}[T_s(H/2) - T_\infty] \tag{7.76}$$

For tripped turbulent flow, the temperature difference varies little with y.

7.2.6.4 Natural Convection from Inclined and Horizontal Plates

For an inclined plate tilted at angle θ from the vertical, replacing g in the Grashof number with $g\cos\theta$ gives satisfactory results for the bottom surface of a heated plate or top surface of a cooled plate when θ is less than $60°$. For the top surface of a heated plate or bottom surface of a cooled plate, the flow situation is too complex for a simple correction to a vertical plate correlation. For horizontal plates, the characteristic length used in the Grashof number is $L_c = A_s/P$ where A_s is the surface area and P is the perimeter of the plate. For hot plates facing up or cold plates facing down, use [11]

$$\overline{\mathrm{Nu}}_{L_c} = 0.54\, \mathrm{Ra}_{L_c}^{1/4} \quad 10^4 < \mathrm{Ra}_{L_c} < 10^7 \tag{7.77a}$$

$$\overline{\mathrm{Nu}}_{L_c} = 0.15\, \mathrm{Ra}_{L_c}^{1/3} \quad 10^7 < \mathrm{Ra}_{L_c} < 10^{11} \tag{7.77b}$$

For hot plates facing down or cold plates facing up, use

$$\overline{\mathrm{Nu}}_{L_c} = 0.27\, \mathrm{Ra}_{L_c}^{1/4} \quad 10^5 < \mathrm{Ra}_{L_c} < 10^{10} \tag{7.78}$$

Figure E7.6

Example 7.6

Components are mounted on the bottom of a large horizontal square board 60 cm on a side (Fig. E7.6). Assume all power dissipated must be removed at the top surface of the board where there is quiescent air at 27°C. If the top surface of the board cannot exceed 68°C, what is the maximum power dissipation if (a) only natural convection is considered and (b) if radiation is also considered, the board has an emissivity of 0.8, and the board is in a large enclosure at 27°C.

SOLUTION

(a) First look up air properties at the film temperature $T_f = (27 + 68)/2 \approx 48°C$. At 48°C, properties are $k = 0.0271$ W/mK, $\nu = 17.5 \times 10^{-6}$ m^2/s, $\alpha = 25.7 \times 10^{-6}$ m^2/s, $\beta = 3.12 \times 10^{-3}$ K^{-1}. Then calculate the Rayleigh number:

$$\text{Ra} = (g\beta \, \Delta T L_c^3)/(\alpha \nu) = 9.41 \times 10^6$$

where

$$\Delta T = 41°C \quad \text{and} \quad L_c = A_s/P = (60 \text{ cm})^2/(4 \times 60 \text{ cm}) = 15 \text{ cm} = 0.15 \text{ m}$$

Equation (7.77a) for a hot plate facing up is appropriate. Nu = 0.54 Ra$^{1/4}$ = 29.9; $h = \text{Nu} \, k/L_c = 5.4$ W/m^2 °C.

$$q_{\text{conv}} = hA_s \, \Delta T = 79.7 \text{ W}$$

(b) Equation (7.12) may be used in this case to calculate the net radiation loss:

$$q_{\text{rad}} = \varepsilon \sigma A_s (T_s^4 - T_\infty^4) = 88.5 \text{ W} \qquad q_{\text{tot}} = q_{\text{conv}} + q_{\text{rad}} = 168.2 \text{ W}$$

As is often the case with natural convection, the radiation loss is comparable to the convection loss. ∎

7.2.6.5 Natural Convection from Cylinders and Spheres

For a vertical cylinder of height H, a modified version of correlations for a vertical plate is used [12]:

$$\overline{\text{Nu}}_{H,\text{cyl}} = \overline{\text{Nu}}_{H,\text{plate}} \lfloor 1 + L_*^{0.9} \rfloor \tag{7.79}$$

where $L_* = (H/D)/\text{Gr}_H^{1/4}$.

For a horizontal cylinder of diameter D, use [13]

$$\overline{\text{Nu}}_D = \left[0.60 + 0.322 \, \text{Ra}_D^{1/6} \right] \quad 10^{-5} < \text{Ra}_D < 10^{12} \tag{7.80}$$

For a sphere, use [10]

$$\overline{\text{Nu}}_D = \frac{0.589 \, \text{Ra}_D^{1/4}}{\left[1 + (0.469/\text{Pr})^{9/16} \right]^{4/9}} \quad \text{Ra}_D < 10^{11} \tag{7.81}$$

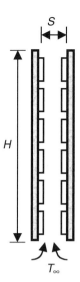

Figure 7.20 Vertical channel.

7.2.6.6 Natural Convection in Vertical Parallel-Plate Channels

Consider heated plates of height H, separated by distance S, as shown in Figure 7.20. Bar-Cohen and Rohsenow presented correlations of the form [14]

$$\overline{\mathrm{Nu}}_s = \left[\frac{C_1}{(\mathrm{Ra}_s S/H)^2} + \frac{C_2}{(\mathrm{Ra}_s S/H)^{1/2}} \right]^{-1/2} \text{ constant } T_s \quad (7.82a)$$

$$\mathrm{Nu}_{S,H} = \left[\frac{C_1}{\mathrm{Ra}_s^* S/H} + \frac{C_2}{(\mathrm{Ra}_s^* S/H)^{2/5}} \right]^{-1/2} \text{ constant } q_s'' \quad (7.82b)$$

for all values of S/H, where

$$\mathrm{Ra}_s = g\beta \Delta T S^3 / \nu \alpha \quad (7.83a)$$

$$\mathrm{Ra}_s^* = g\beta S^4 q'' / k\nu\alpha \quad (7.83b)$$

Equation (7.82a) is for constant-temperature plates, and the result $\overline{\mathrm{Nu}}_s$ is an average Nu for the whole plate. Equation (7.82b) is for plates with constant heat flux, and the result $\mathrm{Nu}_{S,H}$ is the local value at the top of the plate where the highest surface (chip) temperature occurs. The constants C_1 and C_2 are given in Table 7.5. Properties are evaluated at $T_f = (T_s + T_\infty)/2$ in (7.82a) and at $(T_{s,H} + T_\infty)/2$ in (7.82b).

Table 7.5 also contains optimum plate spacings S_{opt} and S_{max} determined by Bar-Cohen and Rohsenow [14]. The value S_{opt} maximizes heat transfer from multiple isothermal plates, while S_{max} maximizes the heat transfer from each plate. The values of S_{opt} and S_{max} include the variation of heat transfer coefficient with S, and also the variation of heat transfer area with S. For plates with constant heat flux, S_{opt} maximizes the heat transfer per unit value for a specified surface temperature $T_s(H)$, while S_{max} minimizes $T_s(H)$ for a specified heat flux. The values of S_{opt} and S_{max} in Table 7.5 assume the plate thickness is very small compared to S.

Table 7.5 Constants C_1 and C_2 in Bar-Cohen and Rosenhow Correlation for Vertical Channels

Surface Condition	C_1	C_2	S_{opt}	S_{max}/S_{opt}
Symmetric isothermal plates ($T_{s,1} = T_{s,2}$)	576	2.87	$2.71(Ra_s/S^3 H)^{-1/4}$	1.71
Symmetric isoflux plates ($q_{s,1} = q_{s,2}$)	48	2.51	$2.12(Ra_s^*/S^4 H)^{-1/5}$	4.77
Isothermal/adiabatic plates ($T_{s,1}, q_{s,2=0}$)	144	2.87	$2.15(Ra_s/S^3 H)^{-1/4}$	1.71
Isoflux/adiabatic plates ($q_{s,1}, q_{s,2-0}$)	24	2.51	$1.69(Ra_s^*/S^4 H)^{-1/5}$	4.77

Example 7.7

Reconsider the two PWBs in Example 7.5. It is proposed that they be cooled by natural convection. They are arranged with the 1.25 cm top and bottom removed so that the two boards form a channel $S = 1.25$ cm wide and $H = 10$ cm tall. For the same heat flux of 0.1 W/cm² on each board, what will be the temperature of the hottest chip?

SOLUTION Again using air properties at 50°C; $k = 0.0271$ W/m K, $\nu = 17.5 \times 10^{-6}$ m²/s, $\alpha = 25.7 \times 10^{-6}$ m²/s, $\beta = 3.12 \times 10^{-3}$ K^{-1}. Equation (7.82b) is appropriate with $C_1 = 48$, $C_2 = 2.51$, $S = 1.25$ cm, $H = 10$ cm, and the Rayleigh number modified for the constant flux case.

$$Ra_s^* = (g\beta q'' S^4)/(k\alpha \nu) = 61,246 \quad Nu_{S,H} = \text{Eq. (7.82b)} = 3.62$$
$$h_H = Nu_{S,H} k/S = 7.84 \text{ W/m}^2 \text{ K} \quad T_{s,H} = T_\infty + q''/h_H = 154.5°C$$

This is still hot, but better than the lower mass flow case in Example 7.5. ∎

7.2.6.7 Mixed Convection

The relative effects of natural and forced convection are determined by examining the dimensionless ratio Gr/Re². For Gr/Re² << 1, natural convection may be ignored, and for Gr/Re² >> 1, forced convection may be neglected. Mixed convection occurs when Gr/Re² is on the order of 1. Mixed convection is commonly expressed by

$$Nu_u^n = Nu_F^n \pm Nu_N^n \tag{7.84}$$

where the subscripts F and N refer to purely forced and purely natural convection, respectively. The exponent $n = 3$ is most commonly used. The plus sign is used when natural convection assists, and the minus sign is used when natural convection opposes.

The reader should note that the Nusselt numbers in 7.84 must be based on the same length parameter. As an example, consider mixed convection between vertical parallel plates. The natural convection Nusselt number, Nu_N, is based on plate spacing S, while the forced convection Nusselt number, Nu_F, is usually based on the hydraulic diameter $2S$. Either Nu_F or Nu_N must be adjusted so that they are both based on the same length.

7.3 AIR COOLING

Air cooling is still the most common means of thermal management because it is inexpensive and convenient. Much empirical data and analyses are published, and design criteria are well developed. However, the capability is limited to relatively low chip heat fluxes of roughly 0.05 W/cm² at 100°C difference for natural convection and 1 W/cm² at 100°C difference for forced convection. Advanced forced convection methods involving finned heat sinks and other augmentation can achieve up to 1.5 W/cm² at 100°C difference.

Chip junction temperatures, T_{chip}, can be written as [15]

$$T_{chip} - T_{amb} = \Delta T_{amb\text{-}in} + \Delta T_{air} + P\theta_{ja} \tag{7.85}$$

where

T_{amb} = ambient air temperature
ΔT_{amb-in} = temperature rise from ambient to air channel inlet
ΔT_{air} = air temperature rise in channel
P = chip power (W)
θ_{ja} = junction-to-air resistance (°C/W)

The temperature rise, $\Delta T_{amb\text{-}in}$, is usually due to fan motor power dissipation. The rise ΔT_{air} can be calculated using methods presented earlier, for example, Section 7.2.3.6. The power P for a chip is usually specified, while θ_{ja} is given by a manufacturer or calculated. The high-performance and cost–performance categories described in Section 7.1 are characterized by the use of heat-sink-assisted air cooling in an attempt to reduce the case-to-air, θ_{ca}, component of the θ_{ja} resistance above. Heat sinks with self-contained fans are now common in both desktop and notebook computers. Heat removal rates as high as 70 W with chip-level heat fluxes of 25 W/cm² are common.

7.4 LIQUID COOLING

Beyond air cooling, the next cooling level is usually some form of liquid cooling. Liquid cooling can either be single phase or two phase. Single-phase cooling can occur via both natural and forced convection, and two-phase cooling is accomplished via evaporation or boiling.

7.4.1 Single-Phase Liquid Cooling

Single-phase liquid cooling can be direct, where the liquid is in direct contact with the circuitry. The principle advantage of direct liquid cooling is a large reduction in the case-to-air θ_{ca} component of θ_{ja} in (7.85). However, direct liquid cooling is expensive and often unreliable. The choice of liquids is limited. Indirect liquid cooling is not as effective as direct liquid cooling due to the introduction of conduction and convection resistances associated with separating the fluid from the circuitry. However, many of the concerns in direct liquid cooling are addressed.

7.4.2 Two-Phase Liquid Cooling

Two-phase liquid cooling can be used in a number of ways. The heat removal rate depends on the application, the liquid, and how the liquid is delivered to the heat sources. The method of delivery can be one of the following:

- Pool boiling—immersion of the circuitry into a pool of evaporating/boiling liquid, usually a refrigerant
- Forced convection boiling—forced flow of the liquid over the heat producing surfaces
- Jet impingement—spraying of the liquid directly onto the heat source

Currently, pool boiling (or immersion cooling) is most commonly used.

7.4.2.1 Immersion Cooling

Immersion cooling is accomplished by simply submerging an electronic module in a liquid, usually a refrigerant. The resulting evaporation or boiling results in very high heat transfer coefficients. A number of modules designed to take advantage of this effect are shown in Figure 7.21.

Boiling and evaporation both involve a change from liquid to vapor. Boiling, compared to evaporation, involves creation of vapor bubbles on the surfaces of the heat sources. These

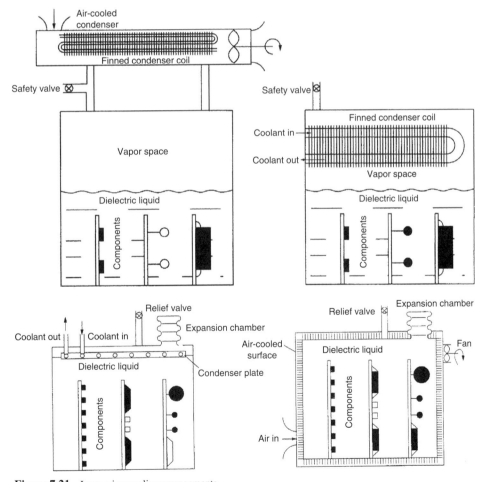

Figure 7.21 Immersion cooling arrangements.

284 Chapter 7 Thermal Considerations

bubbles break away from the heat source, rise to the surface, and the vapor is released. Boiling begins when the heat source surface is 3 to 10°C above the saturation temperature of the liquid. Bubbles then begin to grow on very small cavities in the surface. Eventually the bubbles become so large that buoyant forces overcome the surface tension forces holding the bubbles in place. An excellent description of the boiling process is given by Kraus and Bar-Cohen [1] and is repeated here.

> *In the analysis and correlation of boiling phenomena, it is convenient to distinguish between* **pool boiling**, *referring to boiling in an initially quiescent liquid, and* **flow boiling**, *referring to boiling in the presence of a strong velocity field, as may occur, for example, in pipe flow. Furthermore, since the boiling process depends primarily on the temperature of the heated surface, both* **subcooled boiling**, *during which the bulk fluid is below the saturation temperature, and* **saturated boiling**, *during which the bulk fluid is uniformly at the saturation temperature, may be observed.*
>
> *In boiling, the variation of heat flux along a heated element or surface results in a characteristic temperature response at the surface, reflecting a progression through particular regimes of ebullient heat transfer. This behavior is generally represented by a log–log plot of the heat flux q'' versus the surface superheat, $T_s - T_{sat}$. A typical "boiling curve" for the saturated pool boiling of Refrigerant 113 (R-113) at atmospheric pressure is shown in Figure 7.22.*
>
> *The initial, preboiling part of the curve, labeled I in Figure (7.22), represents thermal transport by natural convection, resulting from the temperature difference between the heated surface and the liquid. At point a, sufficient superheat is available to initiate the growth of vapor bubbles at nucleation sites on the surface, and boiling commences. In region II, modest increases in surface superheat result in the activation of many more nucleation sites and a rapid increase in the frequency of bubble departure at each site. The violent disturbance of the hot boundary layer along the heated surface and gross circulation induced by the motion of the vapor bubbles leads to a very steep rise in the heat flux from the heater in this nucleate boiling regime. At point b, bubble departure frequency at each nucleation site is so high that the trailing bubbles merge with the leading bubble and vapor columns rooted at particular nucleation sites appear. Beyond this inflection point, bubble interference constrains the incremental increase in heat flux resulting from higher wall superheat, and, at point c, bubble interference is so severe that the flow of liquid to the surface is halted and* dryout *occurs. This peak or critical heat flux represents*

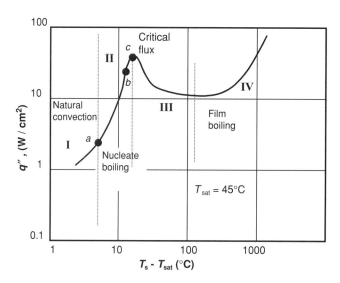

Figure 7.22 Boiling curve for R-113 at 1 atm.

a local maximum in the boiling curve, and higher heat fluxes can be obtained only in the film boiling regime, labeled IV, at much higher wall superheats than are encountered in nucleate boiling. In region IV, a vapor layer blankets the heater, and relatively ineffective thermal conduction through the vapor (augmented by radiation) must be relied on to transfer the heat from the heated surface to the liquid. Region III on the boiling curve corresponds to unstable film boiling and is characterized by rapid, local oscillations between nucleate and film boiling.

Because of the complex interactions of the distinct heat transfer mechanisms active in ebullient thermal transport, it is not yet possible to represent the entire boiling curve by a single analytical function. Rather, it is common practice to provide separate analytical or semiempirical relations for each inflection point and boiling regime.

7.4.2.2 Nucleate Boiling

Most current applications operate in regions I and II in Figure 7.22. Region I is natural convection and has been discussed in earlier sections. Region II is the *nucleate boiling regime*. Rohsenow has presented the most useful correlation for nucleate pool boiling as [16]

$$q_s'' = \mu_l h_{fg} \left[\frac{g(\rho_l - \rho_v)}{\sigma} \right]^{1/2} \left(\frac{c_{p,l} \Delta T_e}{C_{s,f} h_{fg} \mathrm{Pr}_l^n} \right)^3 \tag{7.86}$$

where

μ = viscosity
h_{fg} = heat of vaporization
σ = surface tension
c_p = specific heat
ρ = density
g = gravitational acceleration
$\Delta T_e = T_{\text{surface}} - T_{\text{saturation}}$

The subscripts l and v refer to liquid and vapor. The coefficient $C_{s,f}$ and exponent n depend on the surface–liquid combination and are available in the literature [17,18]. The exponent n is usually taken to be 1.0 for water and 1.7 for other fluids. The value $C_{s,f} = 0.013$ correlates a wide range of data fairly well if a value is not known.

The critical heat flux (point C in Figure 7.22) is the highest heat flux at which nucleate boiling occurs. Heat fluxes in excess of this value result in values of ΔT_e two orders of magnitude higher than in nucleate boiling. Zuber [19] obtained the following expression for the maximum heat flux:

$$q''_{\max} = 0.149 h_{fg} \rho_v \left[\frac{\sigma g(\rho_l - \rho_v)}{\rho_v^2} \right]^{1/4} \tag{7.87}$$

The experimental constant 0.149 replaces Zuber's value of $\pi/24$ obtained analytically [19]. Strictly speaking, this expression only applies to horizontal surfaces, but it seems to work well for other geometries, including cylinders, spheres, and ribbons.

Example 7.8

Chips on a board are to be cooled by immersion in FC-72 at atmospheric pressure. The chips produce a heat flux at their surface of 5 W/cm². (a) What will the chip surface temperature be? (b) How high can the heat flux be raised, and what would the chip surface temperature be at that heat flux? Use $C_{s,f} = 0.005$.

SOLUTION (a) Use $q'' = 5$ W/cm^2 and values from Table 7A.4 in the Appendix using Eq. (7.86), being very careful with units, and solving for ΔT_e yields 15.9°C. Then, the chip surface temperature is $T_s = T_{\text{sat}} + \Delta T_e = 56.6 + 15.9 = 72.5°$C.

(b) Using values from Table 7A.4 and Eq. (7.87), again being very careful with units, gives the critical heat flux $q''_{\text{max}} = 15.4$ W/cm^2. This heat flux in Eq. (7.86) gives $\Delta T_e = 23.2°$C and $T_s = 79.8°$C. ∎

7.5 ADVANCED COOLING METHODS

Currently, the methods discussed previously herein are in most common use. However, in special applications, other methods are available but not commonly used. They are discussed in the following sections.

7.5.1 Heat Pipes

A heat pipe is simply a tube filled with a two-phase mixture used as a heat transfer medium. The tube also contains a capillary material, usually called a wick, running down its length. The wick occupies only a fraction of the cross-sectional area of the tube. Heat pipes can transfer heat at a high rate over large distances with little temperature difference and no moving parts.

Figure 7.23 shows a heat pipe that removes heat from an electronic package and delivers it to a finned surface at the other end of the heat pipe, where it is dumped to air. Liquid is vaporized in the evaporator section on the warmer end. The vapor created migrates to the other end of the tube, driven by the vapor pressure difference, where it condenses. The condensate travels back to the warm end of the tube via capillary action in the wick, driven by the difference in the saturation level of the wick. The effective conductivity of a heat pipe using water, based on its total cross-sectional area, can be 250 to 1000 times that of copper. It is also much lighter since it is essentially a hollow tube. However, one must remember that the heat pipe is just one of a number of resistances between the chip junctions and the eventual heat sink.

The type of fluid chosen depends on the operating pressure and temperature. For a pure fluid, the pressure inside a heat pipe is the saturation pressure at the average heat pipe temperature. For example, the pressure inside a water heat pipe operating at an average temperature of 100°C would be 1 atm. Working fluids vary from lithium that can operate in the 850 to 1600°C range to helium that can operate from 2 to 5 K.

The purpose of the wick is to move the liquid from the condenser section to the evaporator section at a sufficient rate. The performance of the wick depends on its cross-

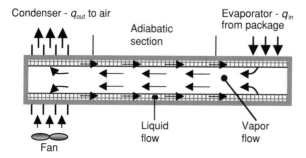

Figure 7.23 Typical heat pipe arrangement.

sectional area, porosity, and the nature of the pore structure. Heat pipes are not restricted to cylindrical cross sections, and they do not have to be straight. In some applications, packages can be mounted on the walls of heat pipes of rectangular cross section. The pipe can have bends and turns if necessary. In some cases, chip modules are cooled by evaporating fluid in direct contact with the chips, condensing the vapor on the walls of the package and returning liquid to the chips with a wick built into the walls. The performance of a heat pipe often degrades with time, usually caused by contamination of the fluid when the ends are sealed and by contamination of the wick during installation. Leakage of air and other noncondensable gases also degrades performance.

7.5.2 Thermoelectric Cooling

Most engineers are familiar with thermocouples. See Figure 7.24. As shown in (Figure 7.24a), two dissimilar p–n materials connected at two junctions at different temperatures generate a voltage $V = \alpha_S(T_{hot} - T_{cold})$, where α_S is the Seebeck coefficient for the two materials. If a load were connected across this voltage as shown in (Figure 7.24b), a current would flow. Maintaining this current would require heat addition at the hot junction and heat removal from the cold junction. In thermodynamics, a device where heat is input at high temperature, part of the heat input is converted to work, and the remainder, is rejected at low temperature, is called a heat engine. Conversely, if a current is forced in the opposite direction as shown in (Figure 7.24c), heat absorbed by the cold junction and the input power is rejected by the hot junction, that is, the thermocouple pair acts as a *thermoelectric cooler* (or refrigerator).

Designing a thermoelectric refrigerator requires trade-offs among Seebeck coefficient, losses proportional to the electrical resistivity, ρ_{TE}, of the p–n materials, and heat transfer from the hot to cold side (the wrong direction) proportional to the thermal conductivity of the materials, k_{TE}. A figure of merit Z_{TE} for a p–n combination has been defined as

$$Z_{TE} = \frac{\alpha_S^2}{k_{TE}\rho_{TE}} \qquad (7.88)$$

As Z_{TE} increases, the possible capacity and efficiency increases. For current available materials, Z_{TE} ranges from 0.002 to 0.005 K^{-1}. Ongoing research may result in increases of a factor of 10 over the next decade.

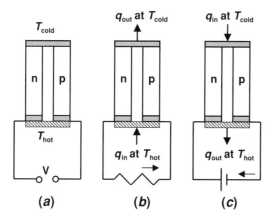

Figure 7.24 Thermocouple arrangements: (a) thermocouple, (b) heat engine, and (c) refrigerator.

288 Chapter 7 Thermal Considerations

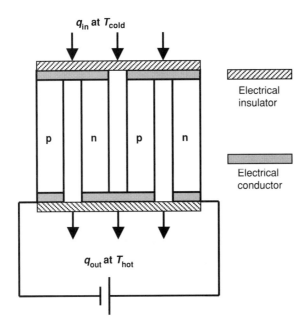

Figure 7.25 Two p–n thermocouples connected in series electrically, in parallel thermally.

A combination of two couples connected electrically in series and thermally in parallel is shown in Figure 7.25. If chip(s) at T_{cold} need to reject heat to an environment at T_∞, the thermoelectric cooler can be used to "amplify" the temperature difference available to drive the heat transfer from $T_{cold} - T_\infty$ to $T_{hot} - T_\infty$. However, more heat must be transferred from the surface at T_{hot} since the power input to the cooler must be rejected in addition to the heat generated by the package. For example, a cooler with $Z_{TE} = 0.005$ K^{-1} removing 1 W at $T_{cold} = 350$ K would require 1.8 W power input and reject 2.8 W at $T_{hot} = 450$ K. Air-cooled heat sinks are often used for the hot surface. Commercially available thermoelectric coolers are essentially parallel and/or series combinations of thermocouples arranged to absorb heat at one surface and reject it at another. Manufacturers supply information on refrigeration capacity and power input requirements at varying operating temperatures. As with regular refrigerators, the capacity decreases and the power input increases as the temperature difference $T_{hot} - T_{cold}$ increases.

Peltier discovered the effects described above in 1834. However, the poor thermoelectric properties of available materials made practical use of thermoelectric coolers difficult until semiconductor materials became available in the 1950s.

7.5.3 Microchannel Cooling

Consider an integrated circuit on the surface of a substrate. Small parallel channels can be cut through the substrate under the circuitry, and a liquid can be circulated through the channels. Such an arrangement provides cooling on the order of 600 W/cm^2 with a 60°C temperature difference.

The channels in the substrate can be fabricated with directional etching or with precise mechanical sawing. The resulting channels are open on one side and covered with a cover plate. A schematic of a possible microchannel arrangement is shown in Figure 7.26. Note that the analysis of such an arrangement is essentially the same as the earlier analysis of air cooling in a duct.

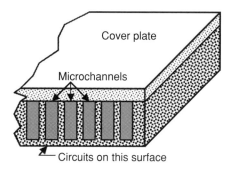

Figure 7.26 Microchannel cooling schematic.

7.6 COMPUTER-AIDED MODELING

Computer-aided modeling for thermal analysis typically falls into two categories—solids modeling (SM) and computational fluid dynamics (CFD). Solids modeling programs calculate the temperature distribution within the solid parts of a package. Computational fluid dynamics programs calculate the velocity, pressure, and temperature distributions within a fluid flowing over the boundaries of the package. Both divide the solid or fluid domain into small subregions called elements and generate an approximate numerical solution for the desired quantities.

7.6.1 Solids Modeling

Solids modeling programs typically generate an approximate solution by:

1. Dividing the solid into elements, that is, discretizing the domain. The corners of each element are nodes.
2. Writing interpolation polynomials for the temperature within each element as a function of location within the element. Often, the functions are simply linear. For example, a linear function of x and y within a 2D triangular element (3 nodes) would be just $T_J(x, y) = a_J + b_J x + c_J y$, where T_J is the temperature within element J, and $a_J + b_J x + c_J y$ are constants to be determined for element J. Higher order interpolation polynomials can be used by placing additional nodes in the element boundaries.
3. The unknown constants within each element are solved for by minimizing the error between the approximate solution and true solution subject to satisfying the governing partial differential equation and boundary conditions and maintaining continuity of temperature at the interfaces between elements. Note that the error can be minimized without actually being determined.
4. With the unknown constants for each element in hand, the approximate temperature at any location can be calculated and plotted if desired.
5. In most SM models, package boundaries where convection is present require the user to supply the heat transfer coefficient h everywhere along the boundary. Unless the boundary geometry can be approximated as one for which good correlations are available, the user estimate of h is questionable. CFD can be used to determine h.

7.6.2 Computational Fluid Dynamics

Computational Fluid Dynamics programs have many similarities to SM programs but are much more complex. The fluid flow region is discretized as in SM, but a number of difficult partial differential equations (PDEs) must be solved simultaneously for each element along with appropriate boundary conditions. A PDE is written for each of the following conserved quantities: mass, momentum in each direction being considered, and energy. For a 2D (x, y) domain, momentum must be conserved in the x and y directions; in a 3D domain, a PDE conserving momentum in the z direction must also be written.

To simplify the solution process, the mass and momentum conservation equations are often solved separately from the energy equation, that is, they are decoupled. This can be done if the fluid properties are assumed independent of the fluid temperature. The resulting velocity distribution is input to the energy equation, which can then be solved for the temperature distribution. If the fluid properties vary strongly with temperature, the mass, momentum, and energy equations are coupled and must be solved simultaneously.

Once the temperature distribution in the flow stream is known, the heat transfer coefficient at the solid boundary can be determined by differentiating the result, that is,

$$q_s'' = h(T_s - T_{\text{fluid}}) = -k \left(\frac{\partial T}{\partial n} \right)_{\text{boundary}}$$

and solving for h, where n is the direction normal to the boundary. The heat transfer coefficient can be calculated everywhere along the solid boundary of the fluid and input to the solids model if needed. This decouples the CFD model from the SM model.

7.6.3 Levels of Decoupling

Decoupling allows complex problems to be broken up into a manageable number of parts. If the mass and momentum conservation equations in a CFD package can be decoupled from the energy equation, the problem is simplified by not accounting for property variation of the fluid with temperature. Decoupling the CFD solution for the temperature distribution in the fluid from the SM solution for temperature distribution in the solid package results in tremendous savings. The number of elements and nodes required to solve these two problems simultaneously requires massive amounts of memory and processing time. Decoupling usually gives good results, but there is some sacrifice in accuracy.

7.6.4 Typical Results

A schematic of a computer case analyzed at the University of Minnesota [20] is shown in Figure 7.27. The temperature distribution in the solids is also shown. The velocity distribution inside the case resulting from a CFD package is shown in Figure 7.28.

7.7 SUMMARY

The purposes of this chapter were to

1. Describe the thermal management issues and failure mechanisms.
2. Give an overview of common thermal management methods and their limits of applicability.

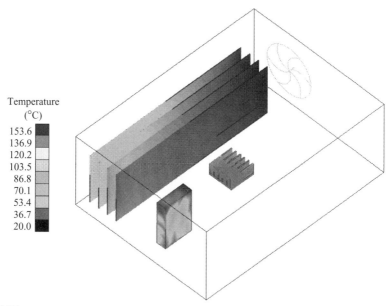

Figure 7.27 Case schematic and temperatures.

3. Provide an introduction to the heat transfer and thermodynamics fundamentals required to understand thermal management issues.
4. Introduce enough heat transfer methods to analyze common thermal management systems.
5. Briefly describe some novel management methods.

The thermal/electric resistance analogy is a very useful concept that allows easier understanding of thermal management methods and allows people with little heat transfer and thermodynamics background to analyze fairly complex cooling methods. The approach in this chapter was to describe

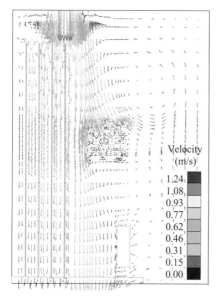

Figure 7.28 Case velocity distribution.

the various types of thermal resistances and show how they could be combined in parallel/series combinations to model thermal management systems. Basic heat transfer methods required to calculate the thermal resistances were then introduced.

Later the chapter, an overview of more novel management methods was given. A brief description of each method was given, along with its capabilities and examples of its relatively specialized applications.

REFERENCES

1. A. D. KRAUS AND A. BAR-COHEN, *Thermal Analysis and Control of Electronic Equipment*, New York: McGraw-Hill, 1983.
2. National Electronics Manufacturing Technology Roadmaps, Herndon, VA: NEMI Inc., 1996.
3. National Technology Roadmap for Semiconductors: Technology Needs, Washington DC: Semiconductor Industry Association, 1997.
4. B. B. MIKIC, Course Notes, M.I.T. Special Summer Session on Thermal Control of Modern Electronic Components, 1978–1979.
5. F. P. INCROPERA AND D. P. DEWITT, *Fundamentals of Heat and Mass Transfer*, 3rd ed., New York: Wiley, 1990.
6. W. M. KAYS AND M. E. CRAWFORD, *Convective Heat and Mass Transfer*, New York: McGraw-Hill, 1980.
7. A. ZHUKAUSKAS, Heat Transfer from Tubes in Cross Flow, in J. P. HARTNETT and T. F. IRVINE, JR., eds., *Advances in Heat Transfer*, Vol. 8, New York: Academic, 1972.
8. S. WHITAKER, *AIChE J.*, Vol. 18, p. 361, 1972.
9. F. W. DITTUS AND L. M. K. Boelter, *University of California Publications on Engineering*, Vol. 2, Berkeley, 1930, p. 443.
10. S. W. CHURCHILL, Free Convection Around Immersed Bodies, in E. U. Schlünder, Ed.-in-Chief, *Heat Exchanger Design Handbook*, Section 2.7.7, New York: Hemisphere, 1983.
11. R. J. GOLDSTEIN, E. M. SPARROW, AND D. C. JONES, Natural Convection Mass Transfer Adjacent to Horizontal Plates, *Int. J. Heat Mass Transfer*, Vol. 16, pp. 1025–1053, 1973.
12. W. J. MINKOWYCZ AND E. M. SPARROW, Local Nonsimilar Solutions for Natural Convection on a Vertical Cylinder, *J. Heat Transfer*, Vol. 96, pp. 178–183, 1974.
13. S. W. CHURCHILL AND H. H. S. CHU, Correlating Equations for Laminar and Turbulent Free Convection from a Horizontal Cylinder, *Int. J. Heat Mass Transfer*, Vol. 18, pp. 1049–1053, 1977.
14. A. BAR-COHEN AND W. M. ROHSENOW, Thermally Optimum Spacing of Vertical Natural Convection Cooled Paralleled Plates, *ASME J. Heat Transfer*, Vol. 106, p. 116, 1984.
15. D. P. SERAPHIM, R. C. LASKY, AND C.-Y. LI, *Principles of Electronic Packaging*, New York: McGraw-Hill, 1989.
16. W. M. ROHSENOW, A Method of Correlating Heat Transfer Data for Surface Boiling Liquids, *Trans. ASME*, Vol. 74, p. 969, 1952.
17. R. I. VACHON, G. H. NIX AND G. E. TANGER, Evaluation of Constants for the Rohsenow Pool-Boiling Correlation, *J. Heat Transfer*, Vol. 90, pp. 239, 1968.
18. W. M. ROHSENOW, Boiling, in W. M. ROHSENOW AND J. P. HARTNETT, eds., *Handbook of Heat Transfer*, Chapter 13, New York: McGraw-Hill, 1973.
19. N. ZUBER, On the Stability of Boiling Heat Transfer, *Trans. ASME*, Vol. 80, p. 711, 1958.
20. R. TUMMALA, *Fundamentals of Microsystems Packaging*, New York: McGraw-Hill, 2001.
21. U. P. HWANG, Boiling Heat Transfer of Silicon Integrated Circuits Mounted on a Substrate, *ASME Heat Transfer Division*, Vol. 20, pp. 53–59, 1981.
22. University of Minnesota, Minneapolis, ME 5348—Heat Transfer in Electronic Equipment, Spring 2000.

APPENDIX: THERMOPHYSICAL PROPERTIES FOR HEAT TRANSFER CALCULATIONS

Table 7A.1 Thermophysical Properties of Select Materials

Material	$k(\text{W/m }°\text{C})$
Aluminum	
Pure	136
2024 T4	121
6061 T6	156
7075 T6	121

Table 7A.1 (*Continued*)

Material	k(W/m °C)
Beryllium copper	82.7
Brass 70 Cu–30 Zn	100
Copper	
Pure	381
Drawn wire	287
Gold	296
Iron wrought	58.8
Kovar	15.6
Lead	32.7
Magnesium	157
Silicon	153
Steel 1020	55.4
Tin	62.3
Titanium	15.6
Zinc	102
Alumina	
95% pure	29.4
90% pure	12.1
Beryllia	
99.5% pure	242
95% pure	156
Diamond	2000
Glass	
Soft	0.98
Pyrex	1.26
Mica	0.59
Epoxy	
Unfilled	0.21
Filled	2.16
Fiberglass	0.26
FR-4	0.30
Mylar	0.19
Nylon	0.24
Phenolic paper	0.28
Plexiglass	0.19
Polyimide	0.33
Polyvinyl chloride	0.16
Rubber	
Butyl	0.26
Silicone	0.19
Silicone grease	0.21
Teflon	0.19
Thermal grease	1.10
Air	0.026
Water	0.658

Source: James W. Dalley, *Packaging of Electronic Systems*, New York: McGraw-Hill, 1990.

Table 7A.2 Properties of Dry Air at Atmospheric Pressure

Temperature		Density	Vol. coef. thermal expansion	Specific heat	Thermal conductivity	Thermal diffusivity	Absolute viscosity	Kinematic viscosity	Prandtl number	$g\beta/\nu^2$
T		ρ	$\beta \times 10^4$	C_P	k	$\alpha \times 10^6$	$\mu \times 10^6$	$\nu \times 10^6$		$\times 10^{-8}$
K	°C	(kg/m³)	(1/K)	(J/kg K)	(W/m K)	(m²/s)	(N s/m²)	(m²/s)	Pr	(1/K m³)
273	0	1.252	3.66	1011	0.0237	19.2	17.456	13.9	0.71	1.85
293	20	1.164	3.41	1012	0.0251	22.0	18.240	15.7	0.71	1.36
313	40	1.092	3.19	1014	0.0265	24.8	19.123	17.6	0.71	1.01
333	60	1.025	3.00	1017	0.0279	27.6	19.907	19.4	0.71	0.782
353	80	0.968	2.83	1019	0.0293	30.6	20.790	21.5	0.71	0.6
373	100	0.916	2.68	1022	0.0307	33.6	21.673	23.6	0.71	0.472
473	200	0.723	2.11	1035	0.0370	49.7	25.693	35.5	0.71	0.164
573	300	0.596	1.75	1047	0.0429	68.9	39.322	49.2	0.71	0.0709
673	400	0.508	1.49	1059	0.0485	89.4	32.754	64.6	0.72	0.035
773	500	0.442	1.29	1076	0.0540	113.2	35.794	81.0	0.72	0.0193
1273	1000	0.268	0.79	1139	0.0762	240	48.445	181	0.74	0.00236

Source: K. Raznjevic, *Handbook of Thermodynamic Tables and Charts*, New York: McGraw-Hill, 1976.

Table 7A.3 Properties of Dry Saturated Water

Temperature		Density	Vol. coef. thermal expansion	Specific heat	Thermal conductivity	Thermal diffusivity	Absolute viscosity	Kinematic viscosity	Prandtl number	$g\beta/\nu^2$
T		ρ	$\beta \times 10^4$	C_P	k	$\alpha \times 10^6$	$\mu \times 10^6$	$\nu \times 10^6$		$\times 10^{-9}$
K	°C	(kg/m^3)	(1/K)	(J/kg K)	(W/m K)	(m^2/s)	(N s/m^2)	(m^2/s)	Pr	(1/K m^3)
273	0	999.3	−0.7	4226	0.558	0.131	1794	1.789	13.7	
293	20	998.2	2.1	4182	0.597	0.143	993	1.006	7.00	2.035
313	40	992.2	3.9	4175	0.633	0.151	658	0.658	4.30	8.833
333	60	983.2	5.3	4181	0.658	0.159	472	0.478	3.00	22.75
353	80	971.8	6.3	4194	0.673	0.165	352	0.364	2.25	46.68
373	100	958.4	7.5	4211	0.682	0.169	278	0.294	1.75	85.09
473	200	862.8	13.5	4501	0.665	0.170	139	0.160	0.95	517.2
573	300	712.5	29.5	5694	0.564	0.132	92.3	0.128	0.98	1766

Source: K. Raznjevic, *Handbook of Thermodynamic Tables and Charts*, New York: McGraw-Hill, 1976.

Table 7A.4 Properties of Immersion Cooling Fluids at 1 atm

	FC-40	FC-87	FC-72	FC-84	FC-77	FC-43	R-113	Water
T_{sat} (°C)	156.6	30	56	83	100	172	48	100
r_l (kg/m³)	1,870	1,633	1,592	1,575	1,590	1,545	1,511	958
r_v (kg/m³)	25	11.58	12.68	13.28	14.31	18.35	7.4	0.59
m (kg/m s)	2.47×10^{-3}	4.20×10^{-4}	4.50×10^{-4}	4.20×10^{-4}	4.50×10^{-4}	3.90×10^{-4}	5.03×10^{-4}	2.70×10^{-4}
c_p (J/kg K)	1,252.8	1,088	1,088	1,130	1,172	1,255	979	4184
h_{fg} (J/kg)	71,162	87,927	87,927	79,553	83,740	71,179	146,824	2,257,044
k (W/m K)	6.38×10^{-2}	5.51×10^{-2}	5.45×10^{-2}	5.35×10^{-2}	5.70×10^{-2}	6.50×10^{-2}	7.02×10^{-2}	6.83×10^{-1}
Pr	48.5	8.29	8.98	8.87	9.25	7.53	7.01	1.65
s (N/m)	1.60×10^{-2}	8.90×10^{-3}	8.50×10^{-3}	7.70×10^{-3}	8.00×10^{-3}	4.50×10^{-3}	1.47×10^{-2}	5.89×10^{-2}
b (1/K)	1.12×10^{-3}	1.50×10^{-3}	1.50×10^{-3}	1.40×10^{-3}	1.20×10^{-3}	1.70×10^{-3}	2.00×10^{-4}	7.50×10^{-4}
Dielectric constant	1.83	1.71	1.72	1.71	1.75	1.68	2.4	78
MW (kg/kmol)	650	288	338	388	438	670	187	18
q''_{max} (W/m²)	21.9	14.8	14.9	13.5	15.0	11.7	23.0	135.7

EXERCISES

7.1. A 6-mm × 6-mm chip is 4 mm thick and mounted horizontally on a substrate. The heat-producing circuits are on the bottom of the chip. Assume all heat generated by the chip circuits is removed at the top by air at 20°C with a heat transfer coefficient of 150 W/m² K. Use $k_{chip} = 150$ W/m K. If the maximum chip temperature cannot exceed 80°C, what is the maximum chip power output? What is the temperature at the chip surface?

7.2. The chip in Exercise 7.1 is placed horizontally on a small 25-mm × 25-mm board that is 5 mm thick and has a thermal conductivity of 1 W/m K. There is a contact conductance of 10^4 W/m² K between the chip and board. The top of the chip is cooled by air at 20°C with $h = 500$ W/m² K. The bottom is also cooled by air at 20°C, but the heat transfer coefficient is only 200 W/m² K. If the chip operates at a heat flux of 20 W/cm², what will the temperature of the chip circuits be if (a) the circuits are on the top and (b) on the bottom?

7.3. The chip in Exercise 7.2 is to be operated at 75°C with a heat flux of 30 W/cm². A heat sink that increases the effective surface area of the top of the chip is to be attached. What must the effective area of the heat sink be?

7.4. Electronic packages 5 mm × 5 mm are mounted on a printed circuit board that is cooled by air at 25°C flowing at 11 m/s parallel to the board. Each package produces 50 mW. Assume no heat is transferred to the board. What is the package surface temperature at a location 2.3 cm downsteam from the edge of the board if (a) the flow is tripped at the edge and (b) the flow is not tripped. If the junction-to-surface resistance of the chip is 30°C/W, what is the junction temperature in both cases?

7.5. Two 10-cm × 40-cm boards are mounted 2.5 cm apart as shown in Figure 7.29. Air at 20°C enters the passage formed by the boards. The chips on the boards produce a heat flux of 0.5 W/cm² from each board. For an air mass flow rate of 0.04 kg/s, find the highest chip surface temperature.

7.6. A vertical printed circuit board is 12 cm tall × 25 cm long. Chips on the board produce a uniform heat flux of 0.03 W/cm². The board is cooled via natural convection by air at 25°C. What is the highest surface temperature?

7.7. The two boards in Exercise 7.5 are mounted vertically, that is, the vertical dimension is 40 cm. Natural convection causes air at 20°C to provide cooling by flowing vertically through the passage formed by the boards. The chips on the boards produce a heat flux of 0.5 W/cm² from each board. What is the surface temperature of the chips at the top of the boards? What is the temperature if a fan is added that creates an upward velocity of 1 m/s?

7.8. A 5-mm × 5-mm integrated circuit package is immersed in R-113 at atmospheric pressure. If the package surface temperature cannot exceed 70°C, what is the maximum allowable heat output? What is the surface temperature if the package operates at 70% of the maximum.

7.9. Consider the silicon chip in Figure 7.30 to be cooled with water circulated through etched microchannels. A 10-mm × 10-mm chip has 50 parallel microchannels of width $W = 50$ μm and

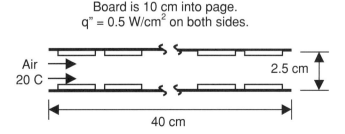

Figure 7.29

298 Chapter 7 Thermal Considerations

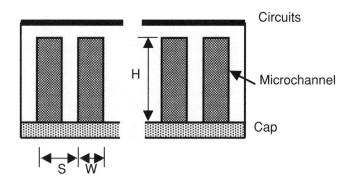

Figure 7.30

height $H = 200\,\mu\text{m}$. The heat flux produced by the chip maintains the surface of the microchannels at 350 K. All heat generated by the circuits is transferred to the water. What are the water exit temperature and heat flux?

Chapter 8

Mechanical Design Considerations

WILLIAM F. SCHMIDT

8.1 INTRODUCTION

Mechanical design plays an important role in electronic packaging, primarily by imposing limitations. To increase circuit density of a package, large chips with more input/output (I/O) are used, but this causes increased strain on the chip-to-substrate interconnection that may result in a premature failure. Mechanical failures can produce a significant loss of components during fabrication, testing, or in the normal operating life of the system. Examples of problems are plentiful; surface-mount solder joints can break if a card is flexed during assembly, vibration, or shock can cause connection failure, copper circuit lines can fail through flexing or thermal expansion, chips crack or debond from substrates when temperature changes occur, and the like. As a result, mechanical failure is one of the major concerns of engineers and designers, especially since the high degree of integration, speed, and power level and the high cost of production have imposed more stringent reliability requirements on individual components.

This chapter is, therefore, devoted to an introduction and application of solid mechanics, and a subset known as strength of materials, as they apply to electronic packaging. The basic concepts of strain, stress, and material behavior are presented in a somewhat general form. This is followed by some general approximations, which produce a more tractable theory. Various failure theories, including fatigue and fracture, are outlined, followed by a few analytical solutions, which provide insight into the behavior of layered assemblies. The final topic covered is a brief overview of the finite-element method and its application to component design.

There is a vast amount of literature related to the topics covered in this chapter. Here, only the fundamentals are presented, and the reader is encouraged to make reference to other work to gain a more complete understanding.

8.2 DEFORMATION AND STRAIN

All material experiences some form of deformation or distortion when acted on by an external force. For example, a rubber band stretches when forces act on the ends or a balloon changes its shape as it is inflated. In these examples, the deformations are rather large and easily seen by an observer. A metal or ceramic bar will also stretch when acted

---

Advanced Electronic Packaging, Second Edition, Edited by Richard K. Ulrich and William D. Brown
Copyright © 2006 the Institute of Electrical and Electronics Engineers, Inc.

Chapter 8 Mechanical Design Considerations

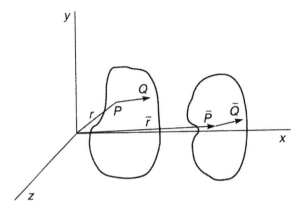

Figure 8.1 Reference and deformed configurations of a body.

on by forces on the ends, but the amount is usually so small that it is not seen directly by observation.

The intent of this section is to develop the appropriate mathematical representation of the deformation process. We begin by defining an initial reference configuration occupied by the material, as shown in Figure 8.1, and consider this to be the undeformed state.

A point P is located in the configuration by a position vector $\mathbf{r} = (x\mathbf{i} + y\mathbf{j} + z\mathbf{k})$, and a nearby point Q is located by the vector $d\mathbf{r}$ with the differential line segment between the two points having a length ds given by $ds^2 = d\mathbf{r} \cdot d\mathbf{r}$ where $\bullet$ is the dot product of the two vectors. Due to some deformation of the configuration, it displaces to a new position referred to as the deformed configuration. The two points, P and Q, now occupy a new position and are labeled $\bar{P}$ and $\bar{Q}$ with the position vector to $\bar{P}$ given by $\bar{r} = (\bar{x}i + \bar{y}j + \bar{z}k)$. The displacement vector from the point P in the undeformed state to the point $\bar{P}$ in the deformed state is designated by $\mathbf{U} = u\mathbf{i} + v\mathbf{j} + w\mathbf{k} = \bar{\mathbf{r}} - \mathbf{r}$. The line segment between $\bar{P}$ and $\bar{Q}$ in the deformed configuration is given by $\overline{d\mathbf{r}}$ with a length $\overline{ds^2} = \overline{d\mathbf{r}} \cdot \overline{d\mathbf{r}}$. The strain measure is now defined by the equation

$$\overline{ds^2} - ds^2 = \overline{d\mathbf{r}} \cdot \overline{d\mathbf{r}} - d\mathbf{r} \cdot d\mathbf{r}$$
$$= (d\mathbf{r} + d\mathbf{U}) \cdot (d\mathbf{r} + d\mathbf{U}) - d\mathbf{r} \cdot d\mathbf{r} \qquad (8.1)$$
$$= 2 d\mathbf{r} \cdot d\mathbf{U} + d\mathbf{U} \cdot d\mathbf{U}$$

where

$$d\mathbf{r} = dx\,\mathbf{i} + dy\,\mathbf{j} + dz\,\mathbf{k}$$

and

$$d\mathbf{U} = du\,\mathbf{i} + dv\,\mathbf{j} + dw\,\mathbf{k}$$

Assuming that the components of the displacements u, v, and w are continuous functions, the chain rule of calculus gives

$$du = \frac{\partial u}{\partial x}dx + \frac{\partial u}{\partial y}dy + \frac{\partial u}{\partial z}dz \qquad (8.2)$$

with similar expressions for v and w. To determine the resulting expression for $\overline{ds^2} - ds^2$, it is convenient to express dU and dr in matrix form as

$$d\mathbf{U} = \lfloor dx \quad dy \quad dz \rfloor A \begin{Bmatrix} \mathbf{i} \\ \mathbf{j} \\ \mathbf{k} \end{Bmatrix}$$

where

$$A = \begin{bmatrix} \dfrac{\partial u}{\partial x} & \dfrac{\partial v}{\partial x} & \dfrac{\partial w}{\partial x} \\ \dfrac{\partial u}{\partial y} & \dfrac{\partial v}{\partial y} & \dfrac{\partial w}{\partial y} \\ \dfrac{\partial u}{\partial z} & \dfrac{\partial v}{\partial z} & \dfrac{\partial w}{\partial z} \end{bmatrix}$$

and

$$d\mathbf{r} = \lfloor dx \quad dy \quad dz \rfloor \begin{Bmatrix} \mathbf{i} \\ \mathbf{j} \\ \mathbf{k} \end{Bmatrix}$$

Then,

$$\overline{ds^2} - ds^2 = \lfloor dx \quad dy \quad dz \rfloor [2A^T + AA^T] \begin{Bmatrix} dx \\ dy \\ dz \end{Bmatrix} \qquad (8.3)$$

where A^T is the transpose of the matrix A, and the term $2A^T$ in Eq. (8.3) provides the usual linear strain measures, while the term AA^T, contains terms involving products and squares of the first partial derivatives of u, v, and w. In the study of electronic components presented in this chapter, the investigation is restricted to problems where it can be assumed that u, v, and w are small compared with the characteristic dimensions of the body, and the various partial derivatives of u, v, and w are small compared to unity. This last assumption permits the squares of the partial derivatives to be neglected compared to the derivatives, or AA^T can be neglected compared to A^T and Eq. (8.3) can be simplified to

$$\overline{ds^2} - ds^2 = \lfloor dx \quad dy \quad dz \rfloor [2A^T] \begin{Bmatrix} dx \\ dy \\ dz \end{Bmatrix}$$

Performing the indicated matrix operations gives

$$\overline{ds^2} - ds^2 = 2 \left\{ \frac{\partial u}{\partial x} dx^2 + \left(\frac{\partial u}{\partial y} + \frac{\partial v}{\partial x} \right) dx\, dy + \left(\frac{\partial u}{\partial z} + \frac{\partial w}{\partial x} \right) dx\, dy \right.$$
$$\left. + \frac{\partial v}{\partial y} dy^2 + \left(\frac{\partial v}{\partial z} + \frac{\partial w}{\partial y} \right) dy\, dz + \frac{\partial w}{\partial z} dz^2 \right\} \qquad (8.4)$$

The deformed state of the material is described by the nondimensional multipliers of $dx^2, dy^2, dz^2, dx\, dy, dx\, dz$, and $dy\, dz$, which are called the linear strain measures or simply linear strains. The customary notation is to write Eq. (8.4) as

$$\overline{ds^2} - ds^2 = 2[\varepsilon_x\, dx^2 + \varepsilon_{xy}\, dx\, dy + \varepsilon_{xz}\, dx\, dz + \varepsilon_y\, dy^2 + \varepsilon_{yz}\, dy\, dz + \varepsilon_z\, dz^2] \qquad (8.5)$$

with

$$\varepsilon_x = \frac{\partial u}{\partial x} \qquad \varepsilon_y = \frac{\partial v}{\partial y} \qquad \varepsilon_z = \frac{\partial w}{\partial z}$$

$$\varepsilon_{xy} = \frac{\partial u}{\partial y} + \frac{\partial v}{\partial x} \qquad \varepsilon_{xz} = \frac{\partial u}{\partial z} + \frac{\partial w}{\partial x} \qquad \varepsilon_{yz} = \frac{\partial v}{\partial z} + \frac{\partial w}{\partial y}$$

where $\varepsilon_x, \varepsilon_y, \varepsilon_z, \varepsilon_{xy}, \varepsilon_{yz}, \varepsilon_{xz}$ are called the components of strain, and the equations defining them are the strain–displacement relations.

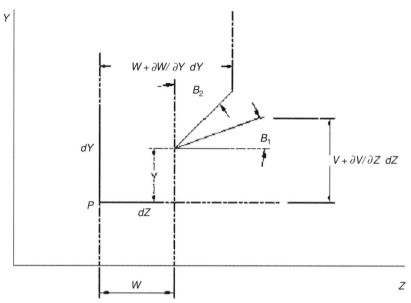

Figure 8.2 Shear strain for two material fibers.

The physical interpretation of the components of strain for the linear case is easy to provide. For example, consider a line segment in the undeformed configuration defined by the points P and Q and oriented parallel to the x axis. Assume that a force is applied to the body, which results in this line segment being elongated so that in the deformed configuration the line remains parallel to the x axis. In this case, ε_x is simply the change in length $(\overline{ds^2} - ds^2)$ of the line segment divided by the original length dx^2. The same is true for ε_y and ε_z and, as a result, these components are referred to as normal extensional or longitudinal strains. These strains are easily measured in a one-dimensional loading situation using simple measurement techniques. For example, if a wire is marked in two locations separated by a known distance and a force is applied to the wire and the distance between the two marks is again measured, with the force acting, then the difference between the separation distance before the force was applied and after the force was applied is the amount the wire stretched over the selected distance. This divided by the original distance between the marks is equal to the strain in the wire. That is, the normal strain is given by the change in length divided by the original length.

The other three strain components, ε_{xy}, ε_{xz}, and ε_{yz}, relate the change in angular orientation between two material lines. In particular, we can consider two material lines dy and dz in the undeformed configuration shown in Figure 8.2.

Assume that after the deformation, the point P is at position $\bar{P}$ and the lines have rotated some amount defined by the angles B_1 and B_2. The sines of the angles are given by

$$\sin B_1 = \left(\frac{\partial v}{\partial z} dz\right) \bigg/ dz = \frac{\partial v}{\partial z}$$

and

$$\sin B_2 = \left(\frac{\partial w}{\partial y} dy\right) \bigg/ dy = \frac{\partial w}{\partial y}$$

Based on the earlier assumption that $\partial v/\partial z$ and $\partial w/\partial y$ deformations are small, the sine of the angle can be replaced by the angle. Therefore, the change in the angular orientation between the two lines is $B_1 + B_2 = \partial v/\partial z + \partial w/\partial y = \varepsilon_{yz}$. Deformations that result in changes of angular orientation of material lines are called shear, and the components ε_{xy}, ε_{yz}, and ε_{xz} are called shear strains.

The discussion to this point has been concerned with what are commonly called mechanical strains. That is, there is a deformation of the body due to some system of external forces acting on the body. Another type of strain, which is important in the design of electronic components, is thermal strain. Thermal strains occur when the temperature of the body is changed. In general, if the temperature of a block of material is increased, the material expands, while, if the temperature is reduced, the material contracts. This expansion/contraction behavior is the reason there are expansion joints in highways and on bridges. The reason for the expansion of the material with an increase in temperature is that, in any material, the atoms are oscillating in their lattice sites with small amplitudes. When the temperature of the material is increased, the amplitudes increase, which results in an increase in the interatomic distance. Macroscopically, this means that the material will expand. In addition, for a uniform temperature change, the deformation is such that the overall size changes, but the shape remains the same. That is, a sphere becomes a larger sphere, a cube will have larger edge lengths, or a cylinder will have an increased length and diameter for a uniform increase in temperature. In terms of the strain, this means that there are equal normal strains in all directions, but no shearing strains for a material that is subject to a uniform temperature change without any constraints placed on its motion. The thermal strain, which results from unrestrained expansion due to a temperature change, ΔT, is given by $\varepsilon_x = \varepsilon_y = \varepsilon_z = \alpha(T)\Delta T$, where $\alpha(T)$ is the coefficient of thermal expansion, T is the temperature and ΔT is the temperature change. For many materials, α is not a strong function of the temperature and can be treated as a constant over a reasonably large variation in temperature.

8.3 STRESS

In this section, a brief introduction to various force systems and the concept of stress in a body are presented. There are three classifications of force distributions, which are of most interest in the design of electronic components: (1) point forces, (2) surface–force distributions, and (3) body–force distributions.

The point force is defined as a force acting at a point on the surface of a body. This is a mathematical abstraction because it is not possible to actually apply a force at a point. However, in many applications, the actual force acting on a body can be treated as a point force. The surface–force distribution acts on the surface of the body and is specified as force per unit area. In some texts, this distribution is referred to as a surface traction. The body–force distribution is a force field that acts throughout a body. Two common body–force distributions are the force of gravity and inertial forces. These distributions are expressed per unit mass or per unit volume of the body elements they influence.

A body responds to the application of forces by deforming and developing internal forces. In our study, we will be considering the special case of Newton's laws of motion for bodies at rest. Namely, the sum of forces in any three mutually perpendicular directions and the sum of moments about any three mutually perpendicular lines must be zero for equilibrium to exist. In equation form, the conditions are

$$\sum F_x = 0 \quad \sum M_x = 0$$
$$\sum F_y = 0 \quad \sum M_y = 0 \quad (8.6)$$
$$\sum F_z = 0 \quad \sum M_z = 0$$

304 Chapter 8 Mechanical Design Considerations

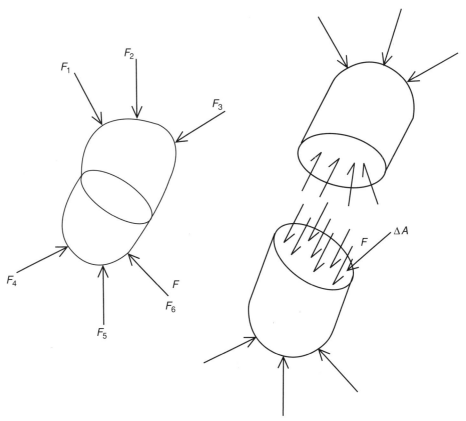

Figure 8.3 Forces on a body in equilibrium.

where

M_x = moment about the x axis

F_x = force in the direction of the x axis

The other terms have similar definitions.

Stress is a measure of the internal force per unit area within a body. Consider the body in equilibrium shown in Figure 8.3a, which has some system of forces acting due to external forces and support forces. The body is cut into two parts with a free-body diagram for each part, as shown in Figure 8.3b. Since both parts must be in equilibrium, there must be forces acting on the areas exposed by the cut, which balance the external forces on the respective pieces. In addition, by Newton's third law, these internal forces must be equal in magnitude and opposite in direction. In effect, these forces are such that when we join the body together, they cancel each other. The nature of the force F on an internal section depends on which section is examined. Consider a small area element, ΔA, on some part of one exposed surface, as shown in Figure 8.4. On this element, there is a force acting, ΔF, which can be resolved into three mutually perpendicular components; one normal, ΔF_n, and two tangential, ΔF_{t1} and ΔF_{t2}, to ΔA. The normal stress at a point in the direction of ΔF_n is then defined as

$$\sigma_n = \lim_{\Delta A \to 0} \frac{\Delta F_n}{\Delta A} \tag{8.7}$$

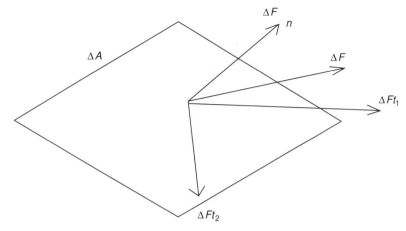

Figure 8.4 Differential area element.

and the shear stresses at the same point in the directions ΔF_{t1} and ΔF_{t2} are defined as

$$\tau_{t1} = \lim_{\Delta A \to 0} \frac{\Delta F_{t1}}{\Delta A}$$
$$\tau_{t2} = \lim_{\Delta A \to 0} \frac{\Delta F_{t2}}{\Delta A} \quad (8.8)$$

It is convenient in most cases to expose internal surfaces, which are parallel to the coordinate axes. In the rectangular Cartesian coordinate case, we can isolate an infinitesimal volume element from the interior of the body and show the stresses at a point, as in Figure 8.5. There are nine components of stress. The notation for normal stress is that the subscript tells the direction in which it acts. The shear stress has two subscripts. The first indicates the normal to the plane on which the stress acts and the second gives the direction in which it acts. For example, τ_{yz} is a shear stress on a plane normal to the y axis acting in the z direction.

The body under consideration is acted on by some system of forces and is in equilibrium. Therefore, any portion of the body must be in equilibrium. From this, a relationship between various components of stress can be determined by considering the stress at a point P having coordinates (x, y, z) and the stress at a point Q having coordinates $(x + \Delta x, y + \Delta y, z + \Delta z)$ as depicted in Figure 8.6.

Stress is converted to force by multiplying the stress on a face by the area of the face. This gives a force diagram and allows Eq. (8.6) to be used. Rather than writing out all six equations, here only the force resultant in the y direction and the resultant moment about a line parallel to the y axis passing through the center of the element are considered. The resultant force equation is

$$\Sigma F_y = -\sigma_y \Delta x \Delta z + \left(\sigma_y + \frac{\partial \sigma_y}{\partial y}\Delta y\right)\Delta x \Delta z + \left(\tau_{zy} + \frac{\partial \tau_{zy}}{\partial z}\Delta z\right)\Delta x \Delta y - \tau_{zy} \Delta x \Delta y$$
$$+ \left(\tau_{xy} + \frac{\partial \tau_{xy}}{\partial x}\Delta x\right)\Delta y \Delta z - \tau_{xy} \Delta y \Delta z = 0$$

which reduces to the following when terms are combined

$$\left(\frac{\partial \sigma_y}{\partial y} + \frac{\partial \tau_{zy}}{\partial z} + \frac{\partial \tau_{xy}}{\partial x}\right)\Delta x \Delta y \Delta z = 0 \quad (8.9)$$

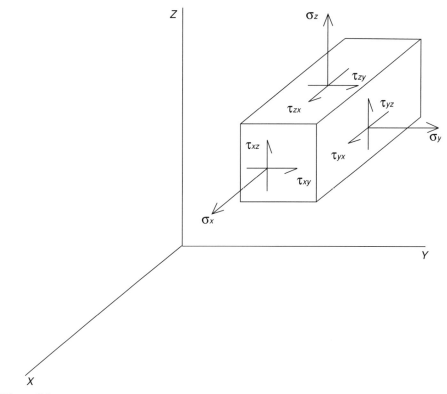

Figure 8.5 Stress components.

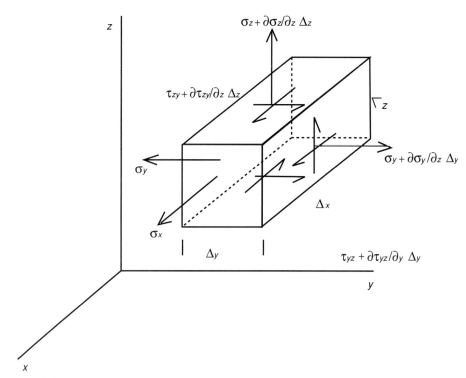

Figure 8.6 Change in stress over differential section of a body.

306

The resultant moment equation is

$$\Sigma M_y = \left(\tau_{zx} + \frac{\partial \tau_{zx}}{\partial z}\Delta z\right) \Delta x \, \Delta y \frac{\Delta z}{2} - \left(\tau_{xz} + \frac{\partial \tau_{xz}}{\partial x}\Delta x\right) \Delta y \, \Delta z \frac{\Delta x}{2}$$

$$+ \tau_{zx} \Delta x \, \Delta y \frac{\Delta z}{2} - \tau_{xz} \Delta y \, \Delta z \frac{\Delta x}{2} = 0$$

and gives

$$(\tau_{zx} - \tau_{xz}) + \frac{1}{2}\left(\frac{\partial \tau_{zx}}{\partial z}\Delta z + \frac{\partial \tau_{xz}}{\partial x}\Delta x\right)\Delta x \, \Delta y \, \Delta z = 0 \tag{8.10}$$

Now, divide Eqs. (8.9) and (8.10) by $(\Delta_x \Delta_y \Delta_z)$ and take the limit as Δ_x, Δ_y, and Δ_z go to zero. Repeating this process for the other coordinate axes gives the equilibrium equations in terms of the stress components:

$$\frac{\partial \sigma_x}{\partial x} + \frac{\partial \tau_{yx}}{\partial y} + \frac{\partial \tau_{zx}}{\partial z} = 0$$

$$\frac{\partial \tau_{xy}}{\partial x} + \frac{\partial \sigma_y}{\partial y} + \frac{\partial \tau_{zy}}{\partial z} = 0$$

$$\frac{\partial \tau_{xz}}{\partial x} + \frac{\partial \tau_{yz}}{\partial y} + \frac{\partial \sigma_z}{\partial z} = 0$$

and

$$\tau_{xy} = \tau_{yx} \qquad \tau_{xz} = \tau_{zx} \qquad \tau_{yz} = \tau_{zy} \tag{8.11}$$

The last three equations, involving only the shear stress, show that the state of stress at a point is defined by only six independent components. These six components of stress are related to the six components of strain by the properties of the material. The equations that relate the stress to the strain are called constitutive relations and depend on the type of material being considered.

8.4 CONSTITUTIVE RELATIONS

In the two previous sections, the concepts of stress and strain were developed. The equations from these two sections are applicable to any type of continuous body undergoing small displacements. They are basic laws because they in no way depend on the type of material under consideration. To predict the components of stress and strain, based on external forces or temperature changes, it is necessary to relate the components of stress and strain. The particular equations, which relate the stress to the strain in a continuous solid, are called constitutive laws or relations since they depend on the structure of the material. It is through the constitutive relations that the material properties of the body enter the problem. In addition, since the primary concerns are with stresses and deformations caused by applied forces and temperature fields, the constitutive relations of interest are based on the material's mechanical and thermal behavior.

8.4.1 Elastic Material

The simplest type of material response for a solid falls under the general heading of elastic material obeying Hooke's law. The general form states that the six independent components of stress may be expressed as linear functions of the six independent components of strain.

Since there are six stress components as linear functions of six strain components, there are, in general, a total of 36 material coefficients. For a homogeneous material, the properties are the same at all points, and, therefore, the coefficients are constants. This large number of constants is, at first, overwhelming, but, fortunately, for most materials the mechanical properties are not dependent on any particular direction. Such a material is said to be isotropic, and, for this case these 36 constants reduce to only two independent elastic constants. These two constants must be determined experimentally.

One of these constants, usually designated as E and called Young's modulus, or the elastic modulus, is determined by a uniaxial tensile test. This test uses a slender uniform cross-section length of the material subjected to axial forces at its ends. The nature of the loading is such that only one component of stress, σ_x, is nonzero and is related to the strain by

$$\sigma_x = E\varepsilon_x \tag{8.12}$$

The stress in Eq. (8.12) can be determined based on the force developed by the test machine and the cross-sectional area of the test specimen. Extensive experimentation has shown that the force in the test specimen is uniformly distributed across the section at approximately three times the cross-sectional dimension away from the ends. Therefore, the normal stress in this region of the test specimen is given by the applied force divided by the cross-sectional area.

In a similar manner, using a torsion test, that is, the twisting of a circular bar, the relation

$$\tau_{xy} = G\gamma_{xy} \tag{8.13}$$

can be obtained with G being the modulus of rigidity or shear modulus for the material.

The situation becomes more involved when more than one component of stress is nonzero. This is the result of the lateral contraction occurring when a specimen of material is stretched, which is called the Poisson effect, first observed in 1830 by S. D. Poisson (1781–1890). Assuming that the material is stretched along the x axis, then the normal strain ε_x is accompanied by lateral contractions $-\nu\varepsilon_y$ and $-\nu\varepsilon_z$, where ν is called Poisson's ratio, a constant for the material. The three material constants introduced are related by

$$E = 2G(1+\nu) \tag{8.14}$$

In the general case, when all components of stress and strain are present in a homogeneous isotropic elastic body, they are related as

$$\begin{aligned} \varepsilon_x &= \frac{1}{E}[\sigma_x - \nu(\sigma_y + \sigma_z)] & \varepsilon_{xy} &= \frac{1}{G}\tau_{xy} \\ \varepsilon_y &= \frac{1}{E}[\sigma_y - \nu(\sigma_z + \sigma_x)] & \varepsilon_{yz} &= \frac{1}{G}\tau_{yz} \\ \varepsilon_z &= \frac{1}{E}[\sigma_z - \nu(\sigma_x + \sigma_y)] & \varepsilon_{zx} &= \frac{1}{G}\tau_{zx} \end{aligned} \tag{8.15}$$

or, if Eqs. (8.15) are solved for the stress components, there results

$$\begin{aligned} \sigma_x &= \frac{E}{1+\nu}\varepsilon_x + \frac{\nu}{1-2\nu}\varepsilon & \tau_{xy} &= G\varepsilon_{xy} \\ \sigma_y &= \frac{E}{1+\nu}\varepsilon_y + \frac{\nu}{1-2\nu}\varepsilon & \tau_{yz} &= G\varepsilon_{yz} \\ \sigma_z &= \frac{E}{1+\nu}\varepsilon_z + \frac{\nu}{1-2\nu}\varepsilon & \tau_{zx} &= G\varepsilon_{zx} \end{aligned} \tag{8.16}$$

where $\varepsilon = \varepsilon_x + \varepsilon_y + \varepsilon_z$.

Example 8.1

A copper wire of length P and diameter d is mounted on a printed circuit board (PCB) with each end fixed by a solder connection. Determine the stress and the strain in the wire if the temperature of the board and wire is increased by 100°C. The properties of the PCB are $\alpha_B = 15$ ppm/°C, $E = 14$ GPa, and for the copper wire, $\alpha_C = 15$ ppm/°C, $E = 117$ GPa.

The temperature increase results in both the board and the wire expanding in size. Since the coefficient of thermal expansion of the board is greater than that of the wire, the board will expand more than the wire. Assume the board is so stiff compared to the wire that it stretches the wire to the final position. Now, locate a coordinate system with the origin at one solder connection and the x axis along the length of the wire. To fix measurements, assume the origin of the coordinate system is the one point that does not change position when the temperature is increased. The displacement of any point along the length of the wire is then given by

$$u = (\alpha_B - \alpha_C)\Delta T\, x$$

The strain can be determined using Eq. (8.5) as

$$\varepsilon_x = \frac{\partial u}{\partial x} = (\alpha_B - \alpha_c)\Delta T$$

Notice that the strain is the same at every point along the wire. The stress can be found using Eq. (8.12). The only nonzero strain component is ε_x; therefore

$$\sigma_x = E_C \varepsilon_x = E_c(\alpha_B - \alpha_C)\Delta T$$

The actual force, P, on the wire due to the thermal expansion can be found by using

$$P = \sigma A = \pi d^2 E(\alpha_B - \alpha_C)\Delta T/4$$

where $A = \pi d^2/4$ is the cross-sectional area of the wire.

8.4.2 Plastic Material

There are situations of importance in electronic packaging where the elastic constitutive equations are not adequate for describing the material response. If a sample of material, such as solder, is considered and a tensile test is conducted, the resulting stress–strain curve would resemble the curve shown in Figure 8.7.

The curve has a region, which is approximately linear, and it is in this region that the slope is measured to give the value of the modulus E. For stress levels up to the yield stress, the point Y on the curve corresponding to a stress σ_{yp}, the material behaves in an essentially linear manner. For stress levels below the yield stress, the specimen will return to a zero state of strain if the load is removed. If a force is applied such that the stress level is greater than the yield stress, say to point A on the curve, and then the force is removed, the specimen response will follow the line AB. In this case, there will be a residual strain ε_p in the specimen. If a continuously increasing load is again applied, the response will closely follow the line from B to A and then go along the solid line until it ultimately fails at some maximum stress level. The behavior of the material beyond the yield point Y is nonelastic. That is, upon removal of the load, the body does not return to its original configuration. The deformation and associated strain in this region is called plastic.

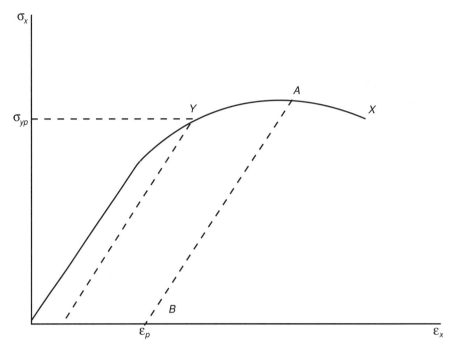

Figure 8.7 Uniaxial stress–strain diagram.

If, in loading the specimen from point B, the stress level obtained is less than the stress at point A and the load is removed, the residual strain will be the strain at point B. In effect, the yield point has been increased to the value at point A. This increase in yield stress is referred to as strain hardening. The true point of initial yielding is difficult to locate experimentally and is typically defined as the value of stress resulting in a small specified residual strain (usually 0.002) when the specimen is unloaded.

The analysis of materials that exhibit elastic–plastic behavior is very complex. There are one-dimensional models, which can be used for attachment materials for die to substrate connection. Goldman presented a one-dimensional model of the form [1]

$$\begin{aligned} \tau &= G\gamma \quad \text{for} \quad \tau \leq \tau_{yp} \\ \tau &= \tau_{yp}(\gamma - \gamma_{yp})^n \quad \text{for} \quad \tau > \tau_{yp} \end{aligned} \quad (8.17)$$

where γ_{yp} and τ_{yp} are the shear strain and shear stress at yield, respectively. The value of n is determined from an experiment, is generally less than unity, and varies from 0.3 to 0.5. Equation (8.17) gives rise to nonlinear differential equations when used in conjunction with the equations from the previous sections. As a result, the solutions for elastic–plastic behavior are normally found using a numerical approach.

8.4.3 Creep

The mechanical properties of a material have, to this point, been discussed only for a static or slowly applied load at a constant temperature and for materials that behave in an elastic manner for some portion of the loading. In some cases, electronic components must be used in an environment for which either loadings exist over a long period of time, or temperatures are high or fluctuate, or a combination of both. These effects are not considered in detail in this chapter, although the means for determining the strength in these situations is discussed. In the following section, the behavior under a constant load at various conditions is discussed and, in a later section of the chapter, the fluctuating loads and temperature situations are considered. When a material has to support a constant load for a long period of time, it may continue to deform or relax until it either does not function properly or it fails by rupturing. This time-dependent behavior is known as creep. As an example of this behavior, consider a connector, which uses a pair of spring tabs to clamp a card forced in between them to make an electrical connection. These springs are usually designed to wipe across the gold-plated surface of the contact pads. To prevent corrosion from degrading the contact over time, the springs must maintain some minimum contact force for the product's lifetime. Creep in the solder or in the spring could cause the contact force to decrease over time to less than the minimum requirement. The amount of creep and resulting force relaxation is dependent on the operating temperatures of the device. Increasing the temperature increases the creep rate. This means that the deformation is not only dependent on the applied load, but also depends on the duration of the loading and the temperature during the loading.

In the general case, both stress and temperature play a significant role in the rate of creep. A mechanical property that is very important is the creep strength. This value represents the largest initial stress a particular material can withstand for a specified time without exceeding a given creep strain. The creep strength is, therefore, a function of temperature, time of loading, and allowable creep strain.

8.5 SIMPLIFIED FORMS

In the previous few sections, the general relations for stress and strain and the relationship between them, that is, the constitutive relations, were developed. These general equations give rise to a problem having a number of unknown variables and associated differential equations, which can be solved for the unknown variables of interest. Unfortunately, in all but a few special situations, solutions cannot be determined in a simple manner or in terms of elementary functions. Various methods and techniques exist to reduce the general form to a more tractable set of equations. The intent of this section is to reduce the general equations for a few situations, which have direct application in electronic packaging.

8.5.1 Plane Stress and Plane Strain

The three-dimensional problem can be reduced to one of two dimensions provided certain constraints are imposed. Consider a thin chip loaded by forces applied along the boundary edges parallel to the plane of the chip. If the x-y plane is parallel to the plane of the chip, then the stress components σ_z, τ_{xz}, and τ_{yz} are zero on both plane faces and are assumed to be zero within the chip. The state of stress is then given by σ_x, σ_y, and τ_{xy} and is called plane stress.

A second, more useful, situation for packaging is plane strain. Consider a simple package of a chip bonded to a substrate as shown in the Figure 8.8. Later in this chapter, consideration will be given to the stresses that develop in the bond between the chip and substrate when there is a temperature change applied to the assembly. For now, the assumptions are that any external forces acting do not vary with z and are not in the direction of z. Under these conditions, the strains ε_z, γ_{yz}, and γ_{xz} on a section near the middle of the

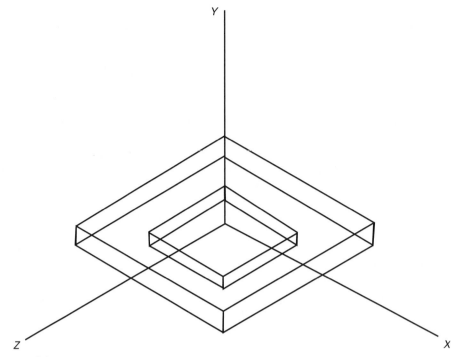

Figure 8.8 Die–substrate configuration.

assembly are zero. In this case, Hooke's law [Eq. (8.15)] gives

$$\sigma_z = -\nu(\sigma_x + \sigma_y) - E\alpha \Delta T \qquad (8.18)$$

for σ_z, where the thermal strain has been included. Substituting Eq. (8.18) into the first two equations of Eq. (8.15) gives

$$\varepsilon_x = \frac{1}{E}[\sigma_x(1-\nu^2) - \nu(1+\nu)\sigma_y] + (1+\nu)\alpha \Delta T$$
$$\varepsilon_y = \frac{1}{E}[\sigma_y(1-\nu^2) - \nu(1+\nu)\sigma_x] + (1+\nu)\alpha \Delta T$$

and, again, the problem is reduced to two dimensions.

8.5.2 Beams

One method for analyzing packages, such as shown in Figure 8.8, is to consider a unit width strip cut parallel to the xy plane and then analyze the stresses in the strip. The problem is simplified even more by replacing the stresses on the cross section by equivalent resultant forces and moments. The general class of problems, which are addressed in this manner, is called beams. To simplify the discussion, only a simple prismatic bar of length P will be considered, with an axial force $F(x)$, transverse distributed load $p(x)$, and shear τ_0 as external forces acting on the section, as shown in Figure 8.9.

The bar, of course, is a three-dimensional body, but, because of relative dimensions, it is usually possible to predict the behavior using an idealization of the bar. Usually, the length of the bar must be significantly larger than the depth, three or more times the depth.

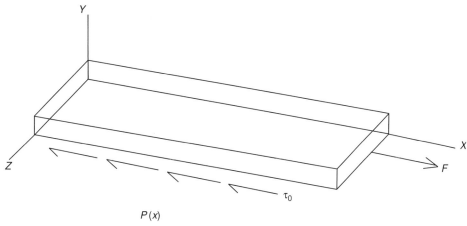

Figure 8.9 Unit width element cut from die.

The forces in the bar are defined in terms of stresses on the cross-sectional planes, and equilibrium conditions are expressed in terms of the stress resultants. Stress resultants are internal forces and moments, which result when various components of stress are integrated over the cross-sectional area on which they act. These give a resultant normal force F_x, a shear force V_y, and a moment M_z as follows

$$F_x = \int \sigma_x \, dA$$
$$V_y = \int \tau_{xy} \, dA \qquad (8.19)$$
$$M_z = \int y \bullet \sigma_x \, dA$$

In Eq. (8.19), y is measured from the centroid of the cross section. The equilibrium equations can be determined in terms of these resultants and the external forces by cutting a section of the bar having a length Δx and isolating this section in a free-body diagram, as shown in Figure 8.10.

Applying the equations of statics to the force system shown in Figure 8.10 gives the following relations:

$$\Sigma F_{Rx} = -F_x + (F_x + \Delta F_x) - \tau_0 \Delta x = 0$$
$$\Sigma F_{Ry} = V - (V + \Delta V) - P \bullet \Delta x = 0$$
$$\Sigma M_{Rz} = -M_z - V\Delta x - \frac{t}{2} \bullet \tau_0 \Delta x + (M_z + \Delta M_z) + P\Delta x \bullet \frac{\Delta x}{2} = 0$$

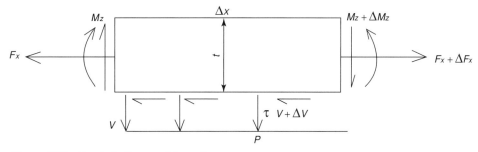

Figure 8.10 Free-body diagram of die section.

Dividing each equation by Δx and then taking the limit as $\Delta x \to 0$ gives the following equilibrium relations for the beam:

$$\frac{dF_x}{dx} = \tau_0$$
$$\frac{dV}{dx} = -P \qquad (8.20)$$
$$\frac{dM}{dx} = V + \frac{t}{2}\tau_0$$

The other relation needed involves the strain, which occurs in the bar. There are several texts related to mechanics of materials, which give detailed accounts of the relationship between strain and the various other terms in Eq. (8.20). Here, only the results necessary for analysis later in the chapter will be listed.

For the beam, the stresses σ_y and σ_z are zero and, from Hooke's law,

$$\varepsilon_x = \frac{\sigma_x}{E} + \alpha\,\Delta T$$

In terms of the resultants, the normal stress is given by

$$\sigma_x = \frac{F_x}{A} + \frac{My}{I}$$

and the axial strain becomes

$$\varepsilon_x = \frac{F_x}{AE} + \frac{My}{EI} + \alpha\,\Delta T \qquad (8.21)$$

or

$$\varepsilon_x = \frac{F_x}{AE} + \frac{y}{R} + \alpha\,\Delta T$$

where A is the cross-sectional area, $1/R = M/EI$ is the curvature of the deformed beam, y is the distance from the centroid of the cross section to any point on the cross section, and I is the area moment of inertia about the z axis at the centroid of the cross section. Notice that the strain due to bending (when M is not zero) is zero at the centroid of the section.

When a strip cut from a wide section is under consideration, then a form of plane strain conditions exist and, for the strip, $\sigma_y = 0$ and $\sigma_z = \nu\sigma_x$, so that the strain becomes

$$\varepsilon_x = \frac{(1-\nu)F}{AE} + \frac{y}{R} + (1+\nu)\alpha\,\Delta T \qquad (8.22)$$

with

$$\frac{1}{R} = \frac{(1-\nu^2)M}{EI}$$

The first term on the right side of the equality in Eqs. (8.21) and (8.22) represents the strain due to the axial force in the beam. The second term on the right in these equations gives the strain in the beam when bending occurs. Notice that, since the coordinate origin is on the centroid of the cross section, there are compressive strains on part of the cross section and tensile strains on another portion due to the bending. At the $y = 0$ level, the strain due to the bending is zero. The third term, obviously, corresponds to the thermal strain. Equations (8.21) and (8.22) are utilized later in the chapter to determine the shear stress at the chip–substrate interface.

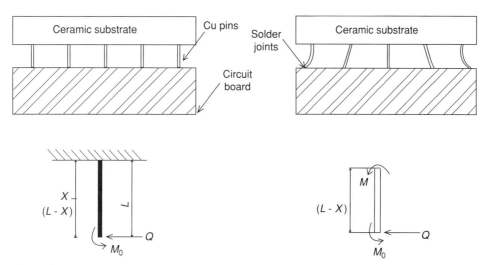

Figure 8.11 (*a*) Surface-mount package at room temperature, (*b*) the same package after reaching operating temperature, (*c*) free-body diagram of a package pin, and (*d*) the pin of part (*c*) cut at location *x*.

Next, an example of how the stress in the pins of a surface-mount package may be determined is examined. Figure 8.11*a* illustrates a surface-mount package at room temperature. In Figure 8.11*b*, the same package is shown after reaching operating temperature. It is desired to determine the maximum stress in the copper pins in terms of the pin diameter d, length P, and spacing a, for a temperature increase ΔT from room temperature. By inspection of the figure, it is clear that the maximum stress occurs in the pins located at the greatest distance from the center of the package. This is due to the difference in thermal expansion coefficients and the displacement due to temperature being dependent on the length of the part.

The pin under consideration is removed and represented in a free-body diagram in Figure 8.11*c*. In this figure, Q is the force the PCB exerts on the end of the pin and M_0 is a moment, which may or may not exist, depending on the end condition. That is, if M_0 is not zero, then physically it represents the moment necessary to keep the end of the pin at $x = P$ from bending. This would correspond to a solder connection on the top of the PCB. If $M = 0$, then this represents the case when the pin passes through a hole in the PCB and is soldered on the bottom side. In this case, the pin would be free to bend at $x = P$. The analysis here will consider the case when $M < 0$. This is the worst case in that it will produce the largest stress. Why?

To determine the stress, $\sigma_x = E\varepsilon_x$ can be used with σ_x given by Eq. (8.21). For this problem, Eq. (8.21) reduces to $\varepsilon_x = My/(EI)$. The force F_x is zero because there is no restraint on vertical motion (the weight of the ceramic device is negligible) and the thermal strain ($\alpha \Delta T$) does not contribute to the normal stress in this problem. The problem now reduces to finding the value of M. The moment M can be determined by considering a section of the pin cut at a location x, as shown in Figure 8.11*d*. Summing moments about the point x gives

$$\Sigma M = M + M_0 - Q(\ell - x) = 0$$

or

$$M = Q(\ell - x) - M_0$$

Unfortunately, the values of Q and M_0 are not known. What is known is the amount one end of the pin moves relative to the other, defined as δ, due to the temperature change, and is given by

$$\delta = (\alpha_p - \alpha_c)\Delta T \quad (2a)$$

where α_p and α_c represent the coefficients of thermal expansion for the PCB and ceramic, respectively.

The value of δ represents the amount the end of the beam deflects and can be related to the bending moment M in the beam using the relationship between the curvature $(1/R)$ and the moment given following Eq. (8.21), that is, $1/R = M/EI$. If v represents the displacement of the beam at any point along its length, then, from elementary calculus, it is known that the curvature is

$$\frac{1}{R} = \frac{d^2v}{dx^2} \bigg/ \left[1 + \left(\frac{dv}{dx}\right)^2\right]^{3/2}$$

For problems such as this one, the term dv/dx, which is the slope of the pin at the location x, is very small and considerably less than unity. As a result, the squared term can be neglected compared to unity and the curvature becomes

$$\frac{1}{R} = \frac{d^2v}{dx^2}$$

Also, $1/R = M/EI$, so that

$$\frac{M}{EI} = \frac{d^2v}{dx^2}$$

The moment M is known in terms of Q and M_0, and the above equation can be written as

$$EI\frac{d^2v}{dx^2} = Q(\ell - x) - M_0$$

This equation can easily be integrated to obtain

$$EI\frac{dv}{dx} = Q(\ell x - x^2/2) - M_0 x + C_1$$

and, integrating a second time gives

$$EIv = Q\left(\frac{\ell x^2}{2} - \frac{x^3}{6}\right) - \frac{M_0 x^2}{2} + C_1 x + C_2$$

where C_1 and C_2 are constants of integration.

The constants of integration, C_1 and C_2, can be found using known conditions at the ends of the pin. These conditions are referred to as the boundary conditions. In this particular problem, the pin is held at $x = 0$ so that there is no displacement and the end is not able to rotate. Therefore, at $x = 0$, $v = 0$, and $dv/dx = 0$. The first condition gives $C_2 = 0$ and the second gives $C_1 = 0$. The condition that $M_0 < 0$ means that, at $x = P$, the slope of the pin is zero due to the solder preventing rotation of the end. Therefore, at $x = P$, dv/dx must be zero, which gives

$$0 = \frac{Q\ell^2}{2} - M_0 \ell$$

or
$$M_0 = \frac{Ql}{2}$$

Also, at $x = P$, the displacement v must be equal to δ, which allows the determination of Q in terms of δ when the above equation is used. In equation form, at $x = P$,

$$\begin{aligned} EIv &= Q\left(\frac{\ell^3}{2} - \frac{\ell^3}{6}\right) - \frac{(Q\ell/2)\ell^2}{2} \\ &= \frac{Q\ell^3}{12} \end{aligned}$$

and $v = \delta$ gives

$$\delta = \frac{Q\ell^3}{12EI}$$

or

$$(\alpha_p - \alpha_c)\,\Delta T(2a) = \frac{Q\ell^3}{12EI}$$

which may be solved for Q as

$$Q = 24EI(\alpha_p - \alpha_c)\,\Delta T\,\frac{a}{\ell^3}$$

The above equation permits the determination, of the force on the end of the pin, in terms of parameters of interest to the packaging engineer. The moment M can now also be determined in terms of these same parameters. Since the largest stress is the one of interest and the stress is directly proportional to the bending moment M, the largest value of M is desired. By inspection of the equation for M, it is clear that the moment will have its largest value at $x = 0$ and is equal to

$$\begin{aligned} M_{\max} &= Q\ell - M_0 = Q\ell - \frac{Q\ell}{2} = \frac{Q\ell}{2} \\ &= 12EI(\alpha_p - \alpha_c)\,\Delta T\,\frac{a}{\ell^2} \end{aligned}$$

The stress is given by $\sigma = My/I$ and, for the largest value of σ, the largest values of M and y should be used. The maximum value of y is

$$y_{\max} = d/2$$

which yields

$$\begin{aligned} \sigma_{\max} &= \frac{M_{\max}\,y_{\max}}{I} \\ &= 6E(\alpha_p - \alpha_c)\,\Delta T\,\frac{ad}{\ell^2} \end{aligned}$$

which is the desired result. In this equation, E is the modulus of the copper wire.

8.6 FAILURE THEORIES

Failure in the mechanical sense is generally defined as any change in the shape, material properties, or integrity of a part, which makes it incapable of performing its intended function. Mechanical failure of an electronic device can be the result of one or any combination of responses of the device to loads and the environment. Failures may occur at the lowest system level, but the effects are often seen at the system level. These external influences can occur during manufacturing, testing, handling, or operation. As a result, it is necessary to be able to predict failure conditions in order to avoid them. Failure theories seek to correlate a calculated level of stress in the device with limits for the material obtained from tests of the material for each component. In this section, some of the more common theories for failure are presented.

8.6.1 Static Failure

The major static failure modes of interest in the design of microelectronic components are ductile rupture and brittle rupture. Ductile rupture occurs when the plastic deformation in a material results in a progressive local reduction in cross-sectional area. This type of failure can occur in solder joints. Brittle failure occurs when the deformation results in a breaking apart of the interatomic bonds. Preexisting flaws usually serve as stress concentration sites, which initiate rapidly growing fracture surfaces. Silicon and ceramics often fail by brittle fracture as a result of the thermal expansion mismatch between these materials and other bonded layers. In this section, the criteria, in terms of stress, for these failure theories are outlined.

The theory for ductile failure, which is best suited to the behavior of a material such as solder, is the distortional energy theory, or Von Mises theory. The theory states that a given material failure by yielding will occur at a critical level of distortional energy. Therefore, the distortional energy for a general loading must be determined and then equated to the distortional energy for a known situation, usually a tension test.

An expression for the distortional energy is required to utilize the theory. The distortional energy can be found using the total strain energy density. The general expression for the strain energy density (work/unit volume of material) of an elastic body is

$$\omega = \tfrac{1}{2}\left(\sigma_x \varepsilon_x + \sigma_y \varepsilon_y + \sigma_z \varepsilon_z\right) + \tfrac{1}{2}\left(\tau_{xy}\gamma_{xy} + \tau_{xz}\gamma_{xz} + \tau_{yz}\gamma_{yz}\right) \quad (8.23)$$

Equation (8.23) can be expressed in terms of either the stress components only or the strain components by using the constitutive relations for an elastic material, Eqs. (8.15) and (8.16). Replacing the strain components using Eq. (8.15) results in

$$\omega = \frac{1}{2E}(\sigma_x^2 + \sigma_y^2 + \sigma_z^2) \bullet \frac{\nu}{E}(\sigma_x \sigma_y + \sigma_y \sigma_z + \sigma_z \sigma_x) \\ + \frac{1}{2G}(\tau_{xy}^2 + \tau_{yz}^2 + \tau_{xz}^2) \quad (8.24)$$

It is convenient to consider changing the orientation of the coordinate system to coincide with the principal directions of stress, which are the eigenvectors or characteristic vectors associated with the matrix of the stress components. The stresses acting along the principal directions are called principal stresses and are the eigenvalues or characteristic values of the same matrix, which is explicitly given as

$$\begin{bmatrix} \sigma_x & \tau_{xy} & \tau_{xz} \\ \tau_{yx} & \sigma_y & \tau_{yz} \\ \tau_{zx} & \tau_{zy} & \sigma_z \end{bmatrix} \quad (8.25)$$

This is a symmetric matrix for which the eigenvalues $\sigma_1, \sigma_2, \sigma_3$ can be found by setting the determinate of the following matrix to zero:

$$\begin{vmatrix} \sigma - \sigma_x & \tau_{xy} & \tau_{xz} \\ \tau_{yx} & \sigma - \sigma_y & \tau_{yz} \\ \tau_{zx} & \tau_{zy} & \sigma - \sigma_z \end{vmatrix}$$

The result is a cubic equation in σ of the form

$$\sigma^3 + S_1\sigma^2 + S_2\sigma + S_3 = 0 \qquad (8.26)$$

where

$$S_1 = -(\sigma_x + \sigma_y + \sigma_z)$$
$$S_2 = \sigma_x\sigma_y + \sigma_x\sigma_z + \sigma_y\sigma_z - \tau_{xy}^2 - \tau_{xz}^2 - \tau_{yz}^2$$
$$S_3 = -\left(\sigma_x\sigma_y\sigma_z + 2\tau_{xy}\tau_{yz}\tau_{xz} - \sigma_x\tau_{yz}^2 - \sigma_y\tau_{xz}^2 - \sigma_z\tau_{xy}^2\right)$$

The roots of Eq. (8.26), which are all real, are the principal stresses and can be found from

$$\sigma_1 = A\cos\left(\frac{\theta}{3}\right) - B$$
$$\sigma_2 = A\cos\left(\frac{\theta}{3} + \frac{2\pi}{3}\right) - B \qquad (8.27)$$
$$\sigma_3 = A\cos\left(\frac{\theta}{3} + \frac{4\pi}{3}\right) - B$$

with

$$A = 2\sqrt{-Q} \qquad Q = \frac{3S_2 - S_1^2}{9} \qquad B = \frac{S_1}{3}$$
$$\theta = \cos^{-1}\left(\frac{R}{\sqrt{-Q^3}}\right) \qquad R = \frac{9S_1S_2 - 27S_3 - 2S_1^3}{54}$$

The advantage of the principal stresses and associated planes, called principal planes, is that there are no shear stresses acting on the planes and the planes are orthogonal. In addition, it can be shown that one of the principal stresses is the maximum normal stress acting at the point, and another is the minimum in the algebraic sense. This will be useful when brittle failure is considered.

Consider Eq. (8.22) for the strain energy density and assume that, for convenience, the *xyz* coordinate system is rotated so that it coincides with the principal directions. The strain energy density is then given in terms of the principal stress components as

$$\omega = \frac{1}{2E}(\sigma_1^2 + \sigma_2^2 + \sigma_3^2) - \frac{\nu}{E}(\sigma_1\sigma_2 + \sigma_2\sigma_3 + \sigma_3\sigma_1) \qquad (8.28)$$

To determine the distortional energy, the stress must be resolved into two parts: a distortional portion and a hydrostatic portion. Note that a hydrostatic stress, uniform pressure on all surfaces, does not cause a body of material to distort. The distortional strain energy is obtained by subtracting the strain energy for a hydrostatic state of stress from the total strain energy. Using $\sigma_1 = \sigma_2 = \sigma_3 = P$ for the hydrostatic state of stress and substituting into Eq. (8.28) gives ω_H, the strain energy, for uniform pressure P, or the hydrostatic strain energy as

$$\omega_H = \frac{1}{2E}(3P^2) - \frac{\nu}{E}(3P^2)$$

or

$$\omega_H = \frac{1-2\nu}{2E}(3P^2) = \frac{1-2\nu}{6E}(\sigma_1 + \sigma_2 + \sigma_3)^2 \tag{8.29}$$

The distortional energy, ω_d, is then given as

$$\omega_d = \omega - \omega_H \tag{8.30}$$

$$\omega_d = \frac{1+\nu}{6E}[(\sigma_1 - \sigma_2)^2 + (\sigma_2 - \sigma_3)^2 + (\sigma_1 - \sigma_3)^2]$$

To obtain a failure criterion, the general expression for the distortional energy, Eq. (8.30), must be equated to a known test situation, usually the tension test at yield. The tension test is a relatively simple experiment to conduct and provides the information required for the failure theory for many materials. In a tension test, all stress components are zero except one, for example, σ_x, which equals σ_{yp} at the point of the material yielding. Also, since all components are zero but the one, the principal stresses reduce to $\sigma_1 = \sigma_x, \sigma_2 = \sigma_3 = 0$, and therefore, the distortional energy for the tension test, ω_T, is, from Eq. (8.30):

$$\omega_T = \frac{1+\nu}{6E}(2\sigma_{yp}^2) \tag{8.31}$$

Equating Eqs. (8.30) and (8.31) gives the condition for failure and the criterion for no failure can be written as

$$S = \frac{\sqrt{2}}{2}[(\sigma_1 - \sigma_2)^2 + (\sigma_2 - \sigma_3)^2 + (\sigma_3 - \sigma_1)^2]^{1/2} \leq \sigma_{yp} \tag{8.32}$$

In Eq. (8.32), the symbol S is used to designate the effective or Von Mises stress. The process used when checking a design is to compute the stresses in the assembly and determine the locations where failure may be expected. Then, the stresses at this location are used to find principal stresses, which are substituted into Eq. (8.32). If the value of S is less than σ_{yp}, then there is no yielding and the design can be considered adequate, otherwise, the design must be modified.

Brittle failure occurs when the material shows no yielding and the stress–strain curve continues smoothly to fracture at a stress designated as σ_u, the ultimate stress. Brittle failures occur rapidly with little or no indication of the impeding failure. One failure theory applicable to brittle materials, such as ceramics, is the maximum normal stress theory. Failure is said to occur when the largest principal stress reaches the limit σ_u, found from a tension test. If we order the principal stresses so that $\sigma_1 > \sigma_2 > \sigma_3$, then, in equation form, the theory is expressed as

$$|\sigma_1| < \sigma_u \quad \text{or} \quad |\sigma_3| < \sigma_u \tag{8.33}$$

8.6.2 Fracture Mechanics

Unless extreme care is exercised in the forming of devices, small flaws generally occur in the material. These flaws may be manufacturing defects, small edge cracks due to die forming, or even cracks that form under service conditions. Fracture mechanics is the science of predicting under what circumstances an existing imperfection, usually a crack, will propagate and eventually result in total failure. Usually the designer determines the axial stress present in the component and then linear elastic fracture mechanics can be used to determine the maximum allowable flaw size above which fracture will occur. The use of fracture mechanics in the design of electronic components has been limited but is, at present, expanding. In this section, a very brief outline is given of the basis for the technique.

The basis for fracture mechanics is the recognition that, by combining the analysis of the gross elastic changes in a component that occur as a sharp crack grows with measurements of the energy required to produce a new fracture surface, it is possible to calculate the average stress (without a crack) that will cause crack growth in a component. The method uses a stress intensity factor, K, which is a measure of the stress near the crack and is usually of the form $K = A\sqrt{a\pi\sigma}$ where A depends on the geometry of the component, a is the crack length, and σ is the stress in the near vicinity of the crack front. The value of K is compared to the critical stress intensity factor for the material K_c. If $K < K_c$, then the crack will not propagate. The critical stress intensity factor, or fracture toughness, has been determined for a number of materials and is a material property, which is tabulated in handbooks. The fracture toughness is found using a standard test where a tension specimen with a crack on its edge is loaded. When a point is reached such that a small increase in crack size causes the release of more potential energy than is required to form a new surface, the situation becomes unstable and the crack grows at an accelerating rate. This point determines the critical stress intensity factor.

8.6.3 Fatigue

The designer of devices, which may be subjected to a cyclic type of loading, thermal or mechanical, must know the strain and stress limitations of the materials under these conditions. The following is a short overview of the history and material responses, which occur for cyclic loadings. Failure under the conditions of cyclic loadings is termed fatigue failure and always occurs at stress levels less than would be predicted by the static failure theories discussed previously. As an illustration of fatigue, consider a simple experiment with a standard paper clip. You are probably not able to pull on the wire of the paper clip to the point of ductile failure, but you can bend the wire. If the wire is bent 90° in one direction and then 90° in the opposite direction, then, after a relatively small number of cycles, it will break. This is a fatigue failure. In effect, the material tires out when subjected to repeated loads above a certain level.

The early studies of fatigue began with iron chain used in Germany in 1829, which was failing at load levels below those obtained under dead weight tests. Later, around 1850, railroad axles were breaking at loads well below their static capacity and, because of the importance of this rapidly expanding industry, a great deal of attention was focused on the problem. During the next 20 years, a large amount of effort was devoted to the problem, not only because of the axles but also because some bridges were experiencing premature failures. The effects of fatigue were established in these early years, but the complete explanation of the phenomenon is still not available today. The work over the years has indicated that plastic deformation is the initiator of fatigue failure. The plastic deformation occurs because there are microcracks present in a material that give rise to local areas of stress concentration. The stress concentration causes large stress levels and more plastic straining that, in turn, causes the crack to grow in size and join with other microcracks. The joining of the microcracks eventually results in a crack that propagates and grows to the point where the cross section cannot support the external forces. It is at this stage that fatigue rupture occurs. The crack growth depends, primarily, on the material properties, the magnitude of the stress, the rate of application of the external forces, the temperature, the surface condition, and the past history.

Two types of tests are performed to determine the behavior of various materials when subjected to a fatigue loading. One type imposes an external force on the system, which produces a certain tensile stress level. The force is then applied in the opposite sense to produce the same compressive stress level. This cycling continues until the test specimen ruptures. Such testing is called stress controlled because, at each cycle, the same stress level is achieved. Each such test produces one data point, the number of cycles to failure for the given stress level. For sufficiently low stress levels, some tests run without rupture occurring with the test usually being stopped at 10^8 cycles. The results of the tests are plotted on what is referred to as an S-N diagram with stress on the vertical axis and N or, more commonly, log N, plotted on the horizontal axis. For ferrous materials and titanium, if the stress scale is also logarithmic, the results essentially form a straight-line segment with a change to horizontal in the vicinity of a million cycles. A stress level below the horizontal part of the curve implies that the specimen would have an infinite life. The stress at the horizontal portion defines what is called the endurance limit, but has been, more recently, referred to as the fatigue limit, which is preferred. The strength for a finite life is called the fatigue strength at N number of cycles. Nonferrous materials do not exhibit a break to the horizontal in the S-N curve, so there is not a well-defined fatigue limit. One final note regarding the stress-controlled testing is related to the reproduceability of results. There is always scatter, and the life N, at a given stress level, may vary by a factor of 10. Usually, S-N curves are drawn at the center of the scatter and, as a result, S-N curves represent a 50% survival probability.

The second type of test used to determine the behavior of materials for fatigue is strain controlled. Here, the maximum strain is maintained constant, rather than the stress level. The maximum stress generally will change during these tests. Usually, materials, which are initially soft, solder, for example, become harder, that is, strain hardening. In this case, the stress corresponding to the imposed strain increases in the initial strain cycling. Materials that are initially hard generally show cyclic softening in the initial cycling. Tests show that, for low strains, the life is governed by the elastic strain, which means the life is stress controlled. For high strains, and a corresponding short life, the plastic strain is dominant and the strain controls the results. This high strain usually occurs when a stress concentration exists.

In practice, the main parts of any electronic package are never intended to be stressed beyond the yield point. However, a point of stress concentration, the root of a notch, or an inclusion may experience stresses above the yield stress, and thus, become a critical stress point. Most of the material around this point behaves elastically and, as a result, only the root of the notch is subjected to a strain-controlled situation. These situations require the use of data from strain-controlled tests for proper design. Since a portion of the material exhibits elastic behavior, and another portion, the region near the flaw, exhibits plastic behavior, the overall behavior is elastic–plastic.

The behavior of material under elastic–plastic fatigue is generally represented by the equation giving the total strain amplitude ε_a as $\varepsilon_a = \varepsilon_{ae} + \varepsilon_{ap} = (\sigma_f/E)(2N)^b + \varepsilon_f(2N)^c$. Here, σ_f represents the fatigue strength coefficient and is the failure stress for one strain cycle of the material. The exponent b is the slope of the logarithmic S-N line and varies from -0.5 to -0.25, and ε_{ae} is the elastic strain amplitude for the tests. The plastic strain for failure at one cycle is denoted as $\varepsilon_f N$ and the fatigue ductility exponent ,c, is the slope of the log ε_{ap} – log $2N$ curve.

In summary, there are two approaches to treating problems in fatigue. The first approach is based on the S-N curves, a stress-controlled test, which is applicable to situations when the yield point of the material has not been exceeded. Generally, in this approach, the number of cycles is very large and the fatigue situation is referred to as high-cycle fatigue. The second approach is based on strain-controlled fatigue tests and is generally applicable to the situation termed low-cycle fatigue. This approach is used when the yield point is exceeded. Low-cycle fatigue is frequently a concern in electronic packaging. Processing, assembly, and shipping may all cause failures, but the most common cause of fatigue is thermal cycling, the most common being power on and power off. Fatigue damage of multimaterial assemblies due to thermal cycling occurs as a result of one of the following: repeated thermal expansion and contraction, thermal stress due to nonuniform temperature distributions, or repeated reversals of stress due to a mismatch of thermal expansion coefficients in the layers. Thermal fatigue is much more complicated than mechanical fatigue due to the effect that the temperature has on materials. One of these complications is high-temperature creep, which can occur during each cycle. This creep can make a determination of the actual stress state by analytical means nearly impossible.

The currently used model for low-cycle fatigue in electronic components is based on the Coffin–Mason model given by

$$\frac{\Delta \varepsilon_p}{2} = \varepsilon_f \left(2N_f\right)^c$$

where N_f is the number of cycles, $\Delta \varepsilon_p$ is the plastic strain accumulated over a cycle, and c and ε_f are material properties. The status of fatigue in electronic packaging today is in its infancy and the majority of studies in this area, at present, are experimental. Thus, investigators must rely on the results of thermal cycling tests to determine the thermal fatigue limits of materials and structures.

8.7 ANALYTICAL DETERMINATION OF STRESS

Solutions to a few assembly problems, which can be obtained using analytical procedures, are given in this section. The original work in this area was done by Timoshenko in 1925 when the problem of a bimetal thermostat was solved under some very limiting assumptions [2]. Since that time, a number of solutions have been presented. A few of these solutions were found using the general equations of elasticity. These solutions are in the form of complicated series and are not suitable for easy application. Other solutions have been obtained using concepts from beam theory and give a more reliable result. The solutions developed here are based on beam theory.

8.7.1 Bi-Material Assembly–Axial Effects

The first situation considered is the bond between a chip and a substrate. The problem is simply two materials bonded together at their interface. Denote the top material layer by 1 and the lower material layer by 2, with the thickness, moduli, and coefficients of linear thermal expansion designated by t_1, t_2, E_1, E_2, and α_1 and α_2, respectively. The behavior of this system under several different assumptions when there is a uniform temperature change, ΔT, in both materials is investigated.

The first case assumes that the properties of the bond between the two slabs can be neglected. This, in effect, implies that the bond is so small in thickness that it does not influence the behavior of the system. In addition, assume that the strip is very wide, and examine a unit width section in the middle. This section will be considered a beam, that is,

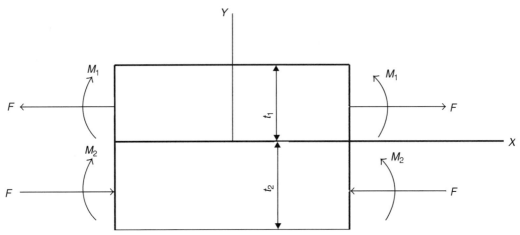

Figure 8.12 Section of bi-material assembly.

its length is much greater than the thickness and plane sections prior to deformation remain plane after the deformation. The free-body diagram for the strip is shown in Figure 8.12.

The resultant normal force F on each strip must be equal because the external force on the system is zero. The radius of curvature after the system experiences a temperature change can be related to the force F, using equilibrium, that is, the sum of the moments must be zero, and the moment curvature relations. For the system under consideration, the sum of the moments about the z axis at $y = 0$ is

$$M_R = M_1 + M_2 - \frac{t_1 + t_2}{2} F = 0 \tag{8.34}$$

For a beam of unit width cut from the assembly, the moment curvature relation gives

$$M_i = \frac{E_i I_i}{(1 - v_i^2) R} = \frac{E_i t_i^3}{12(1 - v_i^2) R} = \frac{D_i}{R} \quad i = 1, 2 \tag{8.35}$$

where the radius of curvature, R, for each member is assumed to be equal. This is valid if the strips are thin. Substituting Eq. (8.35) into Eq. (8.34) provides a solution for the radius of curvature in terms of the material properties, thicknesses, and force F as

$$\frac{1}{R} = \frac{t_1 + t_2}{2(D_1 + D_2)} F \tag{8.36}$$

Equation (8.36) contains two unknowns, R and F. Another equation relating these quantities can be obtained using the strain compatibility condition at the interface. The strain at the lower surface of the upper member must equal the strain at the upper surface of the lower member when the system deforms. The strains that arise are due to the axial force, the bending moment, and the temperature change. In equation form, the compatibility condition is

$$\frac{F}{E_1 t_1} + \frac{t_1/2}{R} + \alpha_1 \Delta T = -\frac{F}{E_2 t_2} - \frac{t_2/2}{R} + \alpha_2 \Delta T \tag{8.37}$$

Combining Eqs. (8.36) and (8.37) allows a direct solution for the force F as

$$F \left\{ \frac{1}{E_1 t_1} + \frac{1}{E_2 t_2} + \frac{t_1 + t_2}{2(D_1 + D_2)} \left(\frac{t_1 + t_2}{2} \right) \right\} = (\alpha_2 - \alpha_1) \Delta T \tag{8.38}$$

or

$$F = \frac{(\alpha_2 - \alpha_1)\Delta T}{\lambda_1 + \lambda_2 + \lambda_{12}} = \frac{(\alpha_2 - \alpha_1)\Delta T}{\lambda}$$

where

$$\lambda = \lambda_1 + \lambda_2 + \lambda_{12}, \lambda_i = \frac{1-\nu}{E_i t_i} \quad \text{and} \quad \lambda_{12} = \frac{(t_1 + t_2)^2}{4(D_1 + D_2)}$$

The force given by Eq. (8.38) enables one to compute the moments M_1 and M_2 using Eqs. (8.35) and (8.36). The axial stress in the strips can be computed once the force is determined. The solution given by Eq. (8.38) is the simplest form of solution for modeling electronic circuits. This solution is only valid away from the edges of the strip because the force F was considered constant along the length. As the ends of the strip are approached, the force F must go to zero. Therefore, in the next model, the case when F varies along the length of the section is considered. The analysis proceeds in essentially the same manner.

In the second example, a model is constructed for a chip bonded to a substrate under the assumption that there is no significant bending, but the axial force does vary along the length, and, hence, there is shear in the bond material. The system is again assumed to be subjected to a uniform change in temperature and there are no external forces acting. A strip of unit width is cut from the assembly and a section of length Δx is isolated, as shown in Figure 8.13.

Applying the force equilibrium requirement to each section and taking the limit as Δx goes to zero gives the following

$$\frac{\partial F_1}{\partial x} = \tau(x) \quad \text{and} \quad \frac{\partial F_2}{\partial x} = -\tau(x) \tag{8.39}$$

Since shear in the bond layer is being considered, the compatibility condition relates the displacements U_1 and U_2 on the top and bottom surfaces of the bond section to the shear

$$U_1 - U_2 = K\tau \tag{8.40}$$

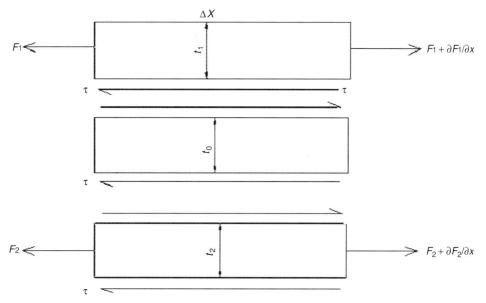

Figure 8.13 Free-body diagram of components.

326 Chapter 8 Mechanical Design Considerations

where $K = t_0/G$. Note that Suhir defines K as $2t_0/3G$ based on an interfacial compliance. This compliance is the inverse of the shear stiffness of the bond layer [3].

The displacements U_1 and U_2 can be determined from the expressions for the strain. The strains at the bottom surface of the top layer and the top surface of the bottom layer, when bending is neglected, are

$$\frac{\partial U_1}{\partial x} = \lambda_1 F_1 + \alpha_1 \Delta T \quad \text{and} \quad \frac{\partial U_2}{\partial x} = \lambda_2 F_2 + \alpha_2 \Delta T \tag{8.41}$$

respectively, with λ_i defined as before. Equations (8.39), (8.40), and (8.41) can be combined into a differential equation by differentiating Eq. (8.40) and using Eq. (8.39) to yield

$$K \frac{\partial \tau}{\partial x} = \lambda_1 F_1 - \lambda_2 F_2 + (\alpha_1 - \alpha_2) \Delta T \tag{8.42}$$

Differentiating Eq. (8.42) and using Eq. (8.39) to eliminate F_1 and F_2 gives an equation involving only the shear stress as the unknown:

$$\frac{\partial^2 \tau}{\partial x^2} - \frac{1}{K}(\lambda_1 + \lambda_2)\tau = 0 \tag{8.43}$$

Let $k^2 = 1/K(\lambda_1 + \lambda_2)$, then the solution of Eq. (8.43) is

$$\tau = A \sinh kx + B \cosh kx \tag{8.44}$$

The coordinate system has its origin at the center of the assembly and the assembly is symmetric about the origin. The symmetry condition for the shear stress requires that, at $x = 0$, the shear must be zero, and therefore, $B = 0$. At the end, $x = l$, $F_1 = F_2 = 0$ and, from Eq. (8.42), the condition for the shear stress is

$$K \frac{\partial \tau}{\partial x} = (\alpha_1 - \alpha_2) \Delta T \quad \text{at} \quad x = \ell \tag{8.45}$$

Differentiating Eq. (8.44) and using Eq. (8.45) gives

$$KkA \cosh k\ell = (\alpha_1 - \alpha_2) \Delta T$$

which may be solved for A. Knowing A and B provides the solution for the shear stress at any x as

$$\tau = \frac{(\alpha_1 - \alpha_2) \Delta T}{Kk \cosh k\ell} \sinh kx \tag{8.46}$$

The solution for the shear stress τ, given by Eq. (8.46), increases as x increases, and, therefore, the maximum shear stress occurs at $x = l$ and is given as

$$\tau_{\max} = \frac{(\alpha_1 - \alpha_2) \Delta T}{Kk} \tanh k\ell \tag{8.47}$$

In many situations, there is not a continuous bond between the chip and substrate, but only an edge bond, such as a solder bump. Equation (8.43) can be applied to this situation by simply restricting the solution to the region of connection. In particular, consider the package shown in Figure 8.14, where the bond region begins at $x = c$.

The solution to Eq. (8.43) in the region $c \leq x \leq \ell$ is then

$$\tau = A \sinh k(x - c) + B \cosh k(x - c) \tag{8.48}$$

The previous boundary condition at $x = 0$ cannot be used in this case, but a condition at $x = c$ using Eqs. (8.40), (8.41), and (8.42) can be applied. Evaluate Eq. (8.42) at $x = c$ and

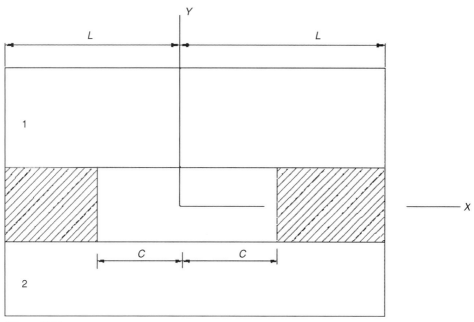

Figure 8.14 Partial bond assembly.

integrate Eq. (8.41) to find

$$U_1 = (\lambda_1 + \alpha_1 \Delta T)c \quad \text{and} \quad U_2 = (\lambda_2 + \alpha_2 \Delta T)c \quad (8.49)$$

Then, substitute Eq. (8.49) into Eq. (8.40) to obtain

$$cK\frac{\partial \tau}{\partial x} = [\lambda_1 F_1 - \lambda_2 F_2 + (\alpha_1 - \alpha_2) \Delta T] \bullet c = K\tau$$

or

$$\frac{\partial \tau}{\partial x} = \frac{\tau}{c} \quad (8.50)$$

where all quantities are evaluated at $x = c$. Equation (8.50), together with the condition at $x = l$ expressed by Eq. (8.45), determines the constants A and B in Eq. (8.48). Equation (8.50) gives

$$(AKc - B) = 0 \quad \text{therefore} \quad B = Akc$$

and, from Eq. (8.45),

$$A = \frac{(\alpha_1 - \alpha_2) \Delta T}{kK\{kc[\sinh k(\ell - c)] + \cosh k(\ell - c)\}}$$

The solution can now be written as

$$\tau = \frac{(\alpha_1 - \alpha_2) \Delta T [\sinh k(x - c) + kc \cosh k(x - c)]}{kK[kc \sinh k(\ell - c) + \cosh k(\ell - c)]} \quad (8.51)$$

Notice that if $c = 0$ and we evaluate at $x = l$, then Eq. (8.51) gives the same result as expressed by Eq. (8.47).

8.7.2 Bi-Material Assembly–Bending Effects

The next case to consider is similar to the previous case except that bending is included. This affects both the equilibrium equations, as well as the strain at the interface. The free-body diagram is obtained by cutting the assembled system at an arbitrary x location as shown in Figure 8.15.

Force equilibrium requires that $F_1 = -F_2$ and $V_1 = -V_2$ where V_1 and V_2 are the shear forces on the section. The equilibrium requirement for the moment about the z axis at $y = 0$ and x gives the following:

$$M_1 + M_2 - t_1 \frac{F_1}{2} + t_2 \frac{F_2}{2} = 0 \quad (8.52)$$

which is the same as Eq. (8.34) and, if the force equilibrium condition $F_1 = -F_2$ is used, then

$$\frac{1}{R} = \frac{t_1 F_1 - t_2 F_2}{2D} \quad (8.53)$$

where $D = D_1 + D_2$. The displacements are related by Eq. (8.40), and again, the strains at the interface must be computed. The strains are given by

$$\frac{\partial U_1}{\partial x} = \alpha_1 \Delta T + \lambda_1 F_1 - \frac{t_1}{2R} - K_1 \frac{\partial \tau}{\partial x}$$

$$\frac{\partial U_2}{\partial x} = \alpha_2 \Delta T + \lambda_2 F_2 + \frac{t_2}{2R} + K_2 \frac{\partial \tau}{\partial x} \quad (8.54)$$

where the first two terms on the right are the same as in the previous case. The third term represents the strain due to bending, evaluated at the interface for each material, that is, the bottom surface of material 1 and the top surface of material 2. The fourth term is a correction factor introduced by Suhir to account for the actual nonuniform distribution of the force F on the cross section [4]. The term is included here to illustrate the effect it has on the overall solution. Suhir defined the constants K_1 and K_2 as

$$K_i = \frac{2(1 + \nu_i)t_i}{3E_i} = \frac{t_i}{3G_i} \quad i = 1, 2 \quad (8.55)$$

and calls them coefficients of interfacial compliance. Physically, they represent the inverse of the stiffness of the bond layer and provide a measure of the bond's resistance to deformation.

The compatibility condition at the interface is again given by Eq. (8.40). Differentiating Eq. (8.40) and using Eq. (8.54) gives an equation involving the various unknown quantities

$$(\alpha_1 - \alpha_2) \Delta T + \lambda_1 F_1 - \lambda_2 F_2 + \frac{t_1}{2R} + \frac{t_2}{2R} - (K_1 + K_2 + K)\frac{\partial \tau}{\partial x} = 0 \quad (8.56)$$

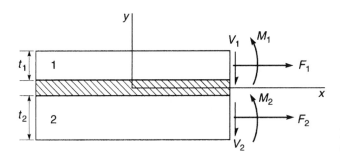

Figure 8.15 Free-body diagram of section of assembly.

Equation (8.53) can be used to eliminate R. Equation (8.39) is differentiated with respect to x and Eq. (8.40) is used to eliminate F_1 and F_2. The result is a second-order differential equation for the shear stress on the interface:

$$(K_1 + K_2 + K)\frac{\partial^2 \tau}{\partial x^2} - \left[\lambda_1 + \lambda_2 + \frac{(t_1 + t_2)^2}{4D}\right]\tau = 0 \tag{8.57}$$

or

$$\frac{\partial^2 \tau}{\partial x^2} - k^2 \tau = 0$$

where

$$k^2 = \frac{\lambda_1 + \lambda_2 + \lambda_{12}}{K_1 + K_2 + K}$$

Equation (8.57) is identical to Eq. (8.43) except that k^2 is redefined to include the bending terms. The solution to Eq. (8.57) is found using the same boundary conditions as before and is given by

$$\tau(x) = \frac{k(\alpha_2 - \alpha_1)\Delta T \, \sinh(kx)}{\lambda \, \cosh(k\ell)}$$

for the complete bond. A similar result may be found for the edge bond using the boundary conditions used to develop Eq. (8.51) for edge bonding. Once the shear stress is known, it is relatively straightforward to determine the normal stress in either of the two layers. The normal stress in the layer is composed of two parts, one due to the normal force F in the layer and one due to the bending moment. The normal force F is given by

$$F = \int_{-\ell}^{x} \tau(\xi)\,d\xi = \frac{(\alpha_2 - \alpha_1)\Delta T}{\lambda}\left[1 - \frac{\cosh(kx)}{\cosh(k\ell)}\right]$$

and the bending moment is given by

$$M_i = \frac{D_i}{R} \quad \text{with} \quad \frac{1}{R} = \frac{t_1 + t_2}{2(D_1 + D_2)}F$$

For design purposes one is normally interested in the maximum normal stress in a layer that, in layer 1, is given by

$$\sigma_{1\,\max} = \frac{F}{A} + \frac{M(t_1/2)}{I} = \frac{F}{t_1} + \frac{6M_1}{t_1^2}$$

for the strip of unit width. Replacing M_1 and F gives

$$\sigma_{1\,\max} = \left[\frac{1}{t_1} + \frac{3D_1(t_1 + t_2)}{t_1^2(D_1 + D_2)}\right]\left\{\frac{(\alpha_2 - \alpha_1)\Delta T}{\lambda}\left[1 - \frac{\cosh(kx)}{\cosh(k\ell)}\right]\right\}$$

8.7.3 Peeling Stress

The various analyses considered to this point have not considered the stress normal to the interface, which is commonly referred to as the peeling stress. This stress is important in terms of the design of the bond and can be estimated now that the case with bending has been considered. If a section of the chip is isolated in a free-body diagram, as shown in

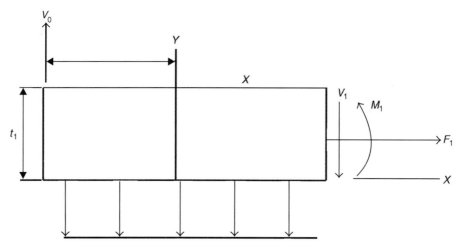

Figure 8.16 Die interface force distribution.

Figure 8.16, the peeling stress, $p(x)$, can be related to the other resultants using moment equilibrium about the z axis at x and $y = 0$ to obtain

$$M_1 - V_0(x + \ell) + \int_{-\ell}^{x} \int_{-\ell}^{\xi} P(\xi)\,dq - \frac{t_1}{2}F_1 = 0$$

or

$$V_0(x + \ell) - \int_{-\ell}^{x} \int_{-\ell}^{\xi} P(\xi)\,d\xi = M_1 - \frac{t_1}{2}F \tag{8.58}$$

The term on the right side of Eq. (8.58) can be expressed in terms of $F = F_1 = -F_2$ using Eqs. (8.52) and (8.53), and the equilibrium condition, $F_1 = F_2$, as

$$M_1 - \frac{t_1}{2}F = -M_2 + \frac{t_2}{2}F = \frac{D_1 t_2 - D_2 t_1}{2D}F = aF \tag{8.59}$$

where

$$a = \frac{D_1 t_2 - D_2 t_1}{2D}$$

In Eq. (8.58) V_0 is a concentrated force used to replace the distributed peeling stress at the edge, which acts upward over a small area. This simplification was introduced by Suhir and allows for an easier computation of the peeling stress [5, 6]. The peeling stress can be found by differentiating Eq. (8.58). Taking the first derivative gives

$$V_0 - \int_{-l}^{x} P(\xi)\,d\xi = a\frac{\partial F}{\partial x} = a\tau \tag{8.60}$$

and differentiating again produces

$$P(x) = -a\frac{d\tau}{dx} \tag{8.61}$$

The shear stress has been determined in previous examples, and therefore, the peeling stress can easily be computed using Eq. (8.61). For example, if Eq. (8.61) is used, then

$$P(x) = \frac{-a(\alpha_1 - \alpha_2)\,\Delta T}{K}\frac{\cosh kx}{\cosh k\ell} \tag{8.62}$$

Equation (8.60) can be used to determine the concentrated force V_0 by using the condition that, at either end, the shear force is zero. Evaluating Eq. (8.60) at $x = l$ gives

$$V_0 = \alpha \tau_{\max} \tag{8.63}$$

8.7.4 Tri-Material Assembly

The previous examples all reflected the effect of the adhesive layer on the general stresses developed during thermal cycling, but do not actually consider stress within the adhesive. The only stress was the shear stress on the contact surface. In this section, the results of the previous section are extended to include three layers, a substrate, an adhesive or solder, and a chip. The system is again assumed to have been fabricated at some elevated temperature, and the stresses that arise when the system is cooled are to be determined.

Figure 8.17 shows the model with the three layers, designated as 1, 2, and 3, and a free-body diagram for a cut at some arbitrary x location. The approach is essentially the same as in the earlier examples. Namely, equilibrium is used to relate the radius of curvature to the axial forces, and then, the compatibility conditions at the interfaces are used to eliminate all unknowns except for the shear stress. This procedure is general and can easily be applied to any number of layers.

The moment resultant about the z axis at x and $y = 0$ gives

$$M_1 + M_2 + M_3 - \tfrac{1}{2}(t_1 + t_2)F_1 - \tfrac{1}{2}(t_2 + t_3)F_2 = 0 \tag{8.64}$$

Since $M_i = D_i/R$, the above equation becomes

$$\frac{1}{R} = \frac{t_1 + t_2}{2D} F_1 + \frac{t_2 + t_3}{2D} F_2 \tag{8.65}$$

with $D = D_1 + D_2 + D_3$.

Equation (8.65) contains three unknowns; R, F_1, and F_2. Two additional equations are obtained from the compatibility of the strains at the interface between layers. The strain dU_1^B/dx at the bottom of layer 1 must equal the strain dU_2^T/dx at the top of layer 2, and similar conditions for layers 2 and 3. In equation form,

$$\frac{\partial U_1^B}{\partial x} = \frac{\partial U_2^t}{\partial x} \quad \text{and} \quad \frac{\partial U_2^B}{\partial x} = \frac{\partial U_3^t}{\partial x} \tag{8.66}$$

The various strains are given as in Eq. (8.54) with appropriate numbers used for subscripts, depending on the material under consideration. This results in the following:

$$\begin{aligned}
\frac{\partial U_1^B}{\partial x} &= \alpha_1 \Delta T - \lambda_1 F_1 + \frac{t_1}{2R} + K_1 \frac{\partial \tau_1}{\partial x} \\
\frac{\partial U_2^B}{\partial x} &= \alpha_2 \Delta T - \lambda_2(F_2 - F_1) + \frac{t_2}{2R} + K_2 \frac{\partial \tau_2}{\partial x} \\
\frac{\partial U_2^T}{\partial x} &= \alpha_2 \Delta T - \lambda_2(F_2 - F_1) - \frac{t_2}{2R} - K_2 \frac{\partial \tau_1}{\partial x} \\
\frac{\partial U_3^T}{\partial x} &= \alpha_3 \Delta T + \lambda_3 F_2 - \frac{t_3}{2R} - K_3 \frac{\partial \tau_2}{\partial x}
\end{aligned} \tag{8.67}$$

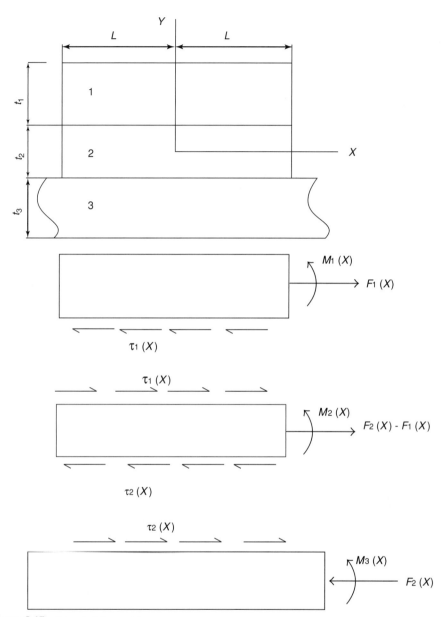

Figure 8.17 Tri-material assembly.

Substituting Eq. (8.67) into Eq. (8.66), and using Eq. (8.65), gives the following two equations:

$$K_{23}\frac{\partial \tau_2}{\partial x} + \lambda_{20} F_1 - \lambda_{23} F_2 = (\alpha_3 - \alpha_2)\Delta T$$
$$K_{12}\frac{\partial \tau_1}{\partial x} - \lambda_{12} F_1 + \lambda_{20} F_2 = (\alpha_2 - \alpha_1)\Delta T \tag{8.68}$$

where

$$K_{ij} = K_i + K_j, \qquad \lambda_{20} = \lambda_2 - \frac{(t_1+t_2)(t_2+t_3)}{4}D \quad \text{and} \quad \lambda_{ij} = \lambda_i + \lambda_j + \frac{(t_i+t_j)^2}{4D}$$

Equation (8.68) can be converted to a set of two second-order differential equations by differentiating each with respect to x and using the relation between the axial force and shear force given by

$$F_i = \int_{-\ell}^{x} \tau_i(\eta)\, d\eta \tag{8.69}$$

An alternate approach used by Suhir is to integrate each equation and obtain a set of integral equations that can be solved in a relatively straightforward manner. To create the integral equations, Eq. (8.65) is multiplied by $d\eta$ and integrated from 0 to x to obtain

$$K_{12} - \lambda_{12} \int_0^x F_1\, d\eta - \lambda_{20} \int_0^x F_2\, d\eta = (\alpha_2 - \alpha_1) \Delta T\, x$$
$$K_{23} - \lambda_{20} \int_0^x F_1\, d\eta - \lambda_{23} \int_0^x F_2\, d\eta = (\alpha_3 - \alpha_2) \Delta T\, x \tag{8.70}$$

The solution is assumed to be of the form $\tau_i = A_i \sinh(kx)$ where A_i and k are to be determined. Substituting the assumed solutions into Eqs. (8.70) and noting that

$$F_i = \int_{-\ell}^{x} \tau_i\, dn = \frac{A_i}{k}(\cosh kx - \cosh k\ell) \tag{8.71}$$

gives the following two algebraic equations:

$$\left\{ K_{12} A_1 - \lambda_{12} \frac{A_1}{k^2} + \lambda_{20} \frac{A_2}{k^2} \right\} \sinh(kx) + \left\{ \lambda_{12} \frac{A_1}{k} - \lambda_{20} \frac{A_2}{k} \right\} \cosh(kl)\, x = (\alpha_2 - \alpha_1) \Delta T\, x$$
$$\left\{ K_{23} A_2 + \lambda_{20} \frac{A_1}{k^2} - \lambda_{23} \frac{A_2}{k^2} \right\} \sinh(kx) + \left\{ -\lambda_{20} \frac{A_1}{k} + \lambda_{23} \frac{A_2}{k} \right\} \cosh(kl)\, x = (\alpha_3 - \alpha_2) \Delta T\, x$$
$$\tag{8.72}$$

Equation (8.72) must be valid for all x, which imposes the requirement that the coefficients of x and $\sinh kx$ must be zero. Employing these conditions gives the following two sets of algebraic equations:

$$(k^2 K_{12} - \lambda_{12}) A_1 + \lambda_{20} A_2 = 0$$
$$(k^2 K_{23} - \lambda_{23}) A_2 + \lambda_{20} A_1 = 0 \tag{8.73}$$

and

$$\lambda_{12} A_1 - \lambda_{20} A_2 = \frac{k(\alpha_2 - \alpha_1) \Delta T}{\cosh(k\ell)}$$
$$-\lambda_{20} A_1 + \lambda_{23} A_2 = \frac{k(\alpha_3 - \alpha_2) \Delta T}{\cosh(k\ell)} \tag{8.74}$$

For a solution other than $A_1 = A_2 = 0$, the determinate of the coefficients of A_1 and A_2 in Eq. (8.73) must be zero. The determinate is

$$(k^2 K_{12} - \lambda_{12})(k^2 K_{23} - \lambda_{23}) - \lambda_{20}^2 = 0$$

which, when solved for k^2, gives

$$k^2 = \frac{r \pm (r^2 - 4 K_{12} K_{23} s)^{1/2}}{2 K_{12} K_{23}} \tag{8.75}$$

where $r = \lambda_{12} K_{23} + \lambda_{23} + K_{12}$ and $s = \lambda_{12}\lambda_{23} - \lambda_{20}^2$.

Solving Eq. (8.74) gives the values for A_1 and A_2, which completes the solution. The shear stresses on the two interfaces are now given as

$$\tau_1 = k\,\Delta T[(\alpha_2 - \alpha_1)\beta_1 + (\alpha_3 - \alpha_2)\beta_2]\frac{\sinh(kx)}{\cosh(k\ell)}$$
$$\tau_2 = k\,\Delta T[(\alpha_2 - \alpha_1)\beta_2 + (\alpha_3 - \alpha_2)\beta_1]\frac{\sinh(kx)}{\cosh(k\ell)}$$
(8.76)

where k is obtained using Eq. (8.75),

$$\beta_1 = \frac{\lambda_{23}}{\lambda_{12}\lambda_{23} - \lambda_{20}} \quad \beta_2 = \frac{\lambda_{20}}{\lambda_{12}\lambda_{23} - \lambda_{20}} \quad \beta_3 = \frac{\lambda_{12}}{\lambda_{12}\lambda_{23} - \lambda_{20}}$$

Notice that the solution is the same functional form as all of the previous solutions. The only difference in the various solutions is a result of the basic assumptions made in formulation that affect the constants in the solution, but not the form.

Example 8.2

The various solutions obtained in the previous sections are used to obtain the shear stress distribution in the solder bond between a gallium arsenide chip and a diamond substrate. A unit width strip is cut from the assembly. Since the chip is 5 mm by 5 mm, the length, l, used in all of the equations, is 2.5 mm. The gallium arsenide has the following properties; $\alpha = 5.58$ ppm/°C, $E = 0.123 \times 10^6$ N/mm^2, $\nu = 0.3$, and $t = 0.635$ mm. The diamond properties are $\alpha = 2.0$ ppm/°C, $E = 1.18 \times 10^6$ N/mm^2, $\nu = 0.148$, and $t = 0.635$ mm. The gold–tin solder bond is over the entire surface area of the chip, that is, a continuous bond. The properties of the solder are $\alpha = 16.8$ ppm/°C, $E = 0.705 \times 10^5$ N/mm^2, $\nu = 0.41$, and $t = t_0 = 0.0508$ mm. Figure 8.18 shows the results obtained using the stated values. The curves labeled 1a or 1b use Eq. (8.46) with $K = t_0/G$ or $2t_0/3G$, respectively. The curves labeled 2a or 2b have the same corresponding K values but use Eq. (8.57). In case 2a, the compliances K_1 and K_2 are zero, while in case 2b, their values are computed and used. The curve labeled 3 represents Eq. (8.76). In all cases, the shear stress increases as the end is approached, as

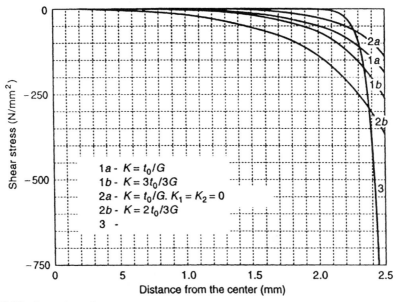

Figure 8.18 Comparison of analytical solutions.

is expected. The differences are in the rate of increase and the maximum values obtained. It is not possible to state which is the best. These results provide trends and insight into the behavior of the system. To obtain more details of the solution, numerical methods or experimental tests must be used. The next section discusses the numerical method known as the finite-element method.

8.8 NUMERICAL FORMULATIONS

The analytical approaches discussed previously are useful to the designer for assessing trends in the behavior of electronic components and provide useful insight into an assembly's behavior. Once a preliminary design is developed for an assembly, it is necessary to do extensive testing to be sure the device has the required reliability. Testing for mechanical reliability can be accomplished by fabricating the device, then subjecting it to service conditions. This approach is usually very expensive and time consuming. An alternative method of testing is to construct a numerical model and perform computer simulation. This approach has the advantage of allowing changes in design without the fabrication expense. The cost of this approach is usually much less than the cost for fabrication and mechanical testing. The success of the numerical modeling depends on the ability of the designer to construct a model that truly simulates the assembly, and the accuracy of the simulation procedure. A method that is used and has been successful in the simulation of electronic assemblies is the finite-element method.

8.8.1 Finite-Element Method

The finite-element method (FEM), or finite-element analysis (FEA), is a numerical procedure for solving the differential equations of physics and engineering. Despite its very visible success in analyzing difficult engineering problems in structural mechanics, heat transfer, electronic packaging, electromagnetics, and other areas, it has not existed very long in its current form. The basis of the method is to divide the region of interest into a finite number of subregions called elements. In each element, the unknown function of interest is represented by a specific function containing the value of the unknown function at specified points. The equations describing the problem under consideration are then employed to arrive at the properties of the elements in terms of these functions. The partial differential equations of the entire region can then be replaced by a system of algebraic or ordinary differential equations.

The idea of dividing a region into subregions is certainly not new. This was the method used by oriental mathematicians to determine π using a unit circle by dividing it into a large, but finite, number of rectangles. Much later, Courant, in 1943, developed an approximate solution using triangular subdomains and piecewise continuous approximations [7]. His approach involved all the basic concepts of the procedure now known as the finite-element method. Therefore, we may say the inception of the method occurred about 50 years ago. Little was done with the ideas presented by Courant, however, so there was no real development that occurred until about the mid-1950s. In 1956, a paper was published by Turner et al. where the finite-element approach was used in structural and solid mechanics with application to airframe design [8]. This paper provided a basis for a number of other researchers, primarily in the aerospace and defense industry, to begin to seriously consider the approach. Digital computing also began to emerge at this time, and both have experienced significant developments. The term finite elements came into the literature for the first time in a paper by Clough in 1960 [9]. The use of the method increased considerably in the period from 1956 to 1970, although almost all applications were in structural and solid mechanics. During this period, nearly a thousand publications appeared on the subject. The method is strongly dependent on computer resources and, as the digital computer evolved, more and more applications followed. Today, there are numerous textbooks devoted to the subject, as well as a number of companies devoted to developing and selling finite-element software. Applications in the area of electronic packaging are numerous. The proceedings of a conference on packaging will list a number of papers in which the method is used.

The purpose of this section is to provide information concerning the method and to define terms in common use in any of the many commercial software products. The general conceptual aspects will be presented to define various terms used in conjunction with the method. A simple example will follow to illustrate the conceptual aspects. Finally, an example of a packaging problem will be given with solutions determined using the commercial package ANSYS.

The fundamental concept of the method is that any continuous function can be approximated by a discrete model composed of a set of piecewise continuous functions defined over a finite number of subdomains. The basic aspects of the finite-element method, as applied to a field problem over some region, is as follows. Here, the function to be determined over the region of interest is assumed to be f.

1. A discrete model of the region is constructed by dividing the region into a number of subregions called elements. If the problem is two dimensional, the elements may be triangular, quadrilateral, or more general polygons. Typically, triangle and quadrilaterals are used. In three dimensions, brick elements, or tetrahedrons, are used.

2. A number of points on the boundary or within the element are identified. These are called nodal points or nodes. For a triangle, three nodes, one at each vertex, or four, each vertex and the centroid, or some other combination, may be used.

3. A specified function is used within each element that approximates the unknown solution. This function is called the interpolation polynomial and, in many cases, a linear polynomial is used.

4. The interpolation polynomial is expressed in terms of the nodal values for the element to generate shape functions. The shape functions are the multipliers of the various nodal values in the element.

5. The element relations are assembled to give a piecewise continuous function for the total region. This function is used in the field equations and, by means of a weighted residual approach or functional minimization, a set of algebraic or differential equations are generated with the nodal values as unknowns.

6. The boundary conditions are applied and the equation system is solved.

7. The element resultants, for example, stress, are computed.

As an illustration of the concepts outlined, a simple example problem is considered. Suppose the solution to the differential equation

$$\frac{d^2 f}{dx^2} + b = 0 \qquad (8.77)$$

in the region $0 \le x \le L$ is desired with boundary conditions $f(O) = 0$ and $f(L) = 0$. The region is simply a line of length L and, when divided into subregions, simply results in

Figure 8.19 Finite-element model.

line segments. Using three elements for the finite-element model gives the three segments as shown in Figure 8.19.

The second concept requires that nodes be selected. In this case, the ends of the line segments will be used and each element then has two nodes. The interpolation polynomial selected is a linear function that gives

$$f^e(x) = a_1 + a_2 x \tag{8.78}$$

as the approximation to f in an element e. Three nodes could have been selected, the ends of the line segment and its center or four, or more. In the case with three nodes, the interpolation polynomial would be quadratic and contain three constants. To compute the shape functions, an arbitrary element e, shown in Figure 8.20, with nodes i and j is selected.

The constants a_1 and a_2 in the interpolation polynomial are found in terms of the nodal values f_i and f_j using $f_e(x_i) = f_i = a_1 + a_2 x_i$ and $f_e(x_j) = f_j = a_1 + a_2 x_j$. Solving these two equations for a_1 and a_2 and substituting into Eq. (8.78) gives

$$f^e(x) = N_i f_i + N_j f_j \tag{8.79}$$

with $N_i = (x_j - x)/\ell$ and $N_j = (x - x_i)/\ell$ where $\ell = x_j - x_i$. Here, N_i and N_j are the shape functions. Equation (8.79) applies to each element with the node coordinates x_i and x_j changing from element to element. The algebraic system of equations for each element is obtained by using Eq. (8.79) in Eq. (8.77). Since Eq. (8.79) does not exactly satisfy the differential equation, there is an error, $\varepsilon(x)$, that results when the substitution is made and the equation is written as

$$\frac{d^2 f^e}{dx^2} + b = \varepsilon(x) \tag{8.80}$$

The error, $\varepsilon(x)$, is made zero in the sense that it is made orthogonal to the shape functions N_i and N_j over the region of the element. In equation form, the orthogonality condition gives

$$\int_{x_i}^{x_j} \varepsilon(x) N_i \, dx = \int_{x_i}^{x_j} \left[\frac{d^2 f^e}{dx^2} + b \right] N_i \, dx = 0$$
$$\int_{x_i}^{x_j} \varepsilon(x) N_j \, dx = \int \left[\frac{d^2 f^e}{dx^2} + b \right] N_j \, dx = 0 \tag{8.81}$$

Integrating the term $(d^2 f^e/dx^2) N_s$ by parts in Eq. (8.81a) and the analogous term in Eq. (8.81b), then using Eq. (8.79) to replace f^e, and completing the integrations gives the algebraic system for the element in the form

$$\frac{K}{\ell} \begin{bmatrix} 1 & -1 \\ -1 & 1 \end{bmatrix} \begin{Bmatrix} f_i \\ f_j \end{Bmatrix} = K \begin{Bmatrix} q_i \\ -q_j \end{Bmatrix} + \frac{b\ell}{2} \begin{Bmatrix} 1 \\ 1 \end{Bmatrix} \tag{8.82}$$

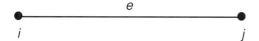

Figure 8.20 Typical element.

338 Chapter 8 Mechanical Design Considerations

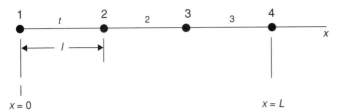

Figure 8.21 Assembled model.

where $q_i = df_i/dx$ at node i with an analogous definition for q_j. A similar equation can be written for each element. The three-element model selected can be assembled into the total system by joining the three elements together to obtain the assembled system shown in Figure 8.21.

Notice that in the assembled model, the node on the right of element e and the node on the left of element $e + 1$ coincide. Since there are no discontinuities in the solution at this point, the element relations are combined to form the total system of equations by imposing continuity of the function f at the nodes. This gives the following system of equations for the unknown function values f_k, for $k = 1, 2, 3$, and 4.

$$\frac{K}{\ell} \begin{bmatrix} 1 & -1 & 0 & 0 \\ -1 & 2 & -1 & 0 \\ 0 & -1 & 2 & -1 \\ 0 & 0 & -1 & 1 \end{bmatrix} \begin{Bmatrix} f_1 \\ f_2 \\ f_3 \\ f_4 \end{Bmatrix} = K \begin{Bmatrix} -q_1 \\ 0 \\ 0 \\ -q_4 \end{Bmatrix} + \frac{b\ell}{2} \begin{Bmatrix} 1 \\ 2 \\ 2 \\ 1 \end{Bmatrix} \quad (8.83)$$

Equation (8.83) represents the total system of equations for the model selected with $l = L/3$. The boundary conditions for the problem give $f_1 = f_4 = 0$ and the system of equations, Eq. (8.83) reduces to two equations for f_2 and f_3. The solution is $f_2 = f_3 = b\ell^2/K = bL^2/9K$. This sample problem can be solved exactly to give $f = (b/2K)(Lx - x^2)$. At node 2, $x = L/3$, and at node 3, $x = 2L/3$. The exact solution gives $f(L/3) = f(2L/3) = bL^2/9K$. In this simple example, the finite-element solution gives the exact values at the nodes. Because the values for f at the nodes are known, Eq. (8.82) can be used to find the values of q at the nodes. Here, the q's would be considered the resultants, step 7.

The example just completed is simple, yet the steps to obtain the solution are exactly the ones used for a more complicated equation. If the problem is two dimensional, the integrations are over the element area. For three-dimensional problems, the volume is used.

8.8.2 Commercial Codes

There are a number of commercial finite-element software packages available. Many of them are for structural analysis/solid mechanics and apply to mechanical design in electronic packaging. Some programs are limited to structural behavior (e.g., beams, plates, shells) exclusively and do not meet the needs of the packaging engineer. Others, such as ABAQUS, ADINA, ANSYS, MSC/NASTRAN, and MARC, are well suited [10]. In this section, ANSYS, a software package produced by Swanson Analysis Systems, Inc. is used to analyze a current packaging problem. The case considered is the gallium arsenide chip bonded by gold–tin solder to a diamond substrate, which was used for the analytical solution example. The properties for the various materials are given in Table 8.1. Most of the properties are dependent on temperature and the finite-element code does allow temperature-dependent

Table 8.1 Material Properties for a Current Packaging Problem

Material	Thermal expansion (ppm/°C)	Elastic modulus (N/mm² × 10⁻⁶)	Poisson's ratio
GaAs	5.58	0.123	0.3
Solder Au–Sn	16.8	0.0705	0.41
Diamond	2.0	1.18	0.148

properties. However, for simplicity, the values used in this case were for ambient temperature and considered constant.

The gallium–arsenide chips are 5 mm² and 0.635 mm thick. The solder covers the entire surface and is 0.0508 mm thick. The diamond layer is 10 mm² with a thickness of 0.635 mm. In this problem, the stress that results during the manufacture of the assembly is of interest. The bond is made by heating the three layers to 300°C and pressing the assembly together at this temperature. The assembly is then cooled to room temperature. The system is symmetric and one quarter of the chip surface will be analyzed. The finite-element model is shown in Figure 8.22. There are 1200 elements for the GaAs, 800 elements for the solder, and 1600 elements for the diamond. A uniform temperature change of 272°C is specified for every element. Figure 8.23 shows the deformed shape of the assembly after it cools. The resulting stress distribution along the symmetry line is shown in Figure 8.24. The stresses are as expected with the maximum occurring near the edge. The tensile stress in the middle of the chip is due to thermal mismatch and could lead to failure of the chip. Also plotted is the Von Mises stress for comparison to the ultimate strength for the gallium arsenide material.

Figure 8.24 also shows a comparison of the shear stress found by the finite-element method to some of the analytical solutions presented in a previous section. The solution 2a compares most favorably with the finite-element solution. Unfortunately, there is no fundamental reason to expect any one of the solutions to give the best results. The design engineer must make the judgment based on the information available.

Number of elements/die =1200
Number of elements/solder = 800
Number of elements/substrate =1600

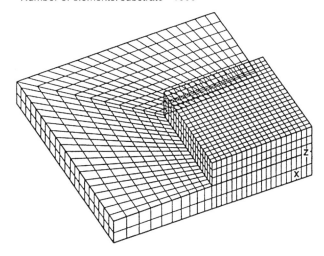

Figure 8.22 Finite-element model of assembly.

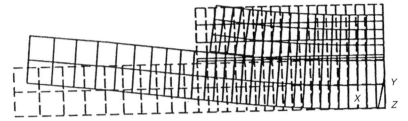

Figure 8.23 Deformed configuration due to temperature change.

The results from the finite-element analysis provide a tremendous amount of information. It is easy to examine the various parameters of interest using computer graphics. For example, the deformed shape of the assembly may provide insight for design changes or stress contours at level in the system can also be obtained.

8.8.3 Limitations and Hazards

The finite-element method is a powerful tool for the design engineer. The software packages available enable the designer to solve extremely complex problems. Unfortunately, there is no way to be sure the results are accurate. An error in an input code to the program can go undetected and the program will provide results. The results, however, may be for a situation somewhat different from what the user intended. Another possible hazard is in constructing the model. In complex geometries with various materials, it is easy to lose track of elements and their corresponding properties. The software will run, but again, the results will be in error. Improvements in graphical display capabilities in recent years have helped to reduce the occurrence of this problem. With multicolored displays of the model, it

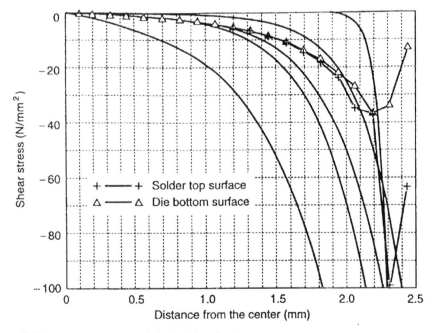

Figure 8.24 Finite-element and analytical solutions for shear stress.

is not as easy to overlook incorrect material types or interfaces as it is with a monochromatic display. Regardless, the designer must use caution and always compare the computer results to estimates of the solution.

In conclusion, the finite-element method can be of assistance in the design, but as many checks as possible of the numerical solution should be made. Comparison to analytical results or experimental results is the best check. Since this is not always possible, a good designer should examine limiting cases where the trend in the FEM solution should match the engineer's intuition.

8.9 SUMMARY

The chapter provided an introduction to stress and strain and the relationship between them as applied to electronic packaging. Concepts of creep, fatigue, and thermal loading of packages were presented, along with the detailed development of equations for the stresses in layered packages. The shear stress and peeling stress relations were derived for dice–substrate bonds subject to thermal loading for any set of elastic material properties. A brief introduction to the finite-element method was also presented, along with some comparisons of the analytical solutions to the FEM solutions.

REFERENCES

1. L. S. GOLDMAN, Geometric Optimization of Controlled Collapse Interconnections, *IBM J. Res. Develop*, Vol. 13, pp. 251–265, 1969.
2. S. TIMOSHENKO, Analysis of Bi-Metal Thermostats, *J. Optical Soc. Am.*, Vol. 11, pp. 233–255, 1925.
3. E. SUHIR, Calculated Thermally Induced Stresses in Adhesively Bonded and Soldered Assemblies, Proc. ISHM International Symp. on Microelectronics, Atlanta GA, Oct. 1986, pp. 383–392.
4. E. SUHIR, Stresses in Bi-Metal Thermostats, Applied Mechanics Division of the American Society of Mechanical Engineers, Winter Annual Meeting, Anaheim, CA, December 7–12, 1986.
5. E. SUHIR, Thermally Induced Interfacial Stresses in Elongated Bimaterial Plates, *Appl. Mech. Rev.*, Vol. 42, No. 11, Part 2, Nov. 1989.
6. E. SUHIR, Calculated Interfacial Stresses in Elongated Bimaterial Plates Subjected to Bending, *J. Electronic Packaging*, Vol. 111, p. 277, December 1989.
7. R. COURANT, Variational Methods for the Solution of Problems of Equilibrium and Vibrations, *Bull. Am. Math. Soc.*, Vol. 49, pp. 1–23, 1943.
8. M. J. TURNER, E. H. DILL, H. C. MARTIN, AND R. J. MELOSH, Large Deflections of Structures Subjected to Heating and External Loads, *J. Aerospace Sci.*, Vol. 27, pp. 97–102, 1960.
9. R. W. CLOUGH, The Finite Element Method in Plane Stress Analysis, *J. Struct. Div.*, ASCE, Proc. 2d Conf. Electronic Computation, pp. 345–378, 1960.
10. (a) ABAQUS: Hibbit, Karlsonn & Sorenson, 1080 Main St., Pawtucket, RI 02860-4847, 401-727-4200. (b) ADINA: Adina R & D Inc., 71 Elton Ave., Watertown, MA 02172, 617-926-5189. (c) MSC/NASTRAN: MSC, 815 Colorado Blvd., Los Angeles, CA 90041-9777, 213-258-9111. (d) MARC: MARC Analysis Research Corp., 24 Frank Lloyd Wright Dr., Ann Arbor, MI 48106, 313-998-0540. (e) ANSYS: DRD Corporation, 5506 South Lewis Ave., Tulsa, OK 74105, 918-743-3013.

BIBLIOGRAPHY

CHEN, W. T., Thermal Stress in Bonded Joints, *IBM J. Res. Develop.*, Vol. 23, No. 2, March 1979.

CONNALLY, J. A., Micromechanical Fatigue Testing, *Experimental Mechanics*, pp. 953–956, June 1993.

DASGUPTA, ABHIJIT, Thermomechanical Analysis and Design, in *Handbook of Electronic Packaging Design*, Michael Pecht, ed., New York: Marcel Dekker, 1991.

EISCHEN, J. W., Realistic Modeling of Edge Effect Stresses in Bimaterial Elements, *J. Electronic Packaging*, Vol. 112, pp. 16–23, March 1990.

GOLAND, M., The Stresses in Cemented Joints, ASME Annual Meeting, New York, Nov. 29–Dec. 3, 1943.

GRIMADO, P. B., Interlaminar Thermoelastic Stresses in Layered Beams, *J. Thermal Stresses*, Vol. 1, pp. 75–86, 1978.

HONG, BOR ZEN, Time-Dependent Inelastic Deformation of Thin Film Polimide: A nonlinear Viscoelasticity Theory Based on Overstress and Experiment, *Ad. Electronic Packaging*, EEP-Vol. 4-1, pp. 277–286, ASME 1993.

HSU, T. R., A Continuum Damage Mechanics Model Approach for Cyclic Creep Fracture Analysis of Solder Joints,

Adv. in Electronic Packaging, EEP-Vol. 1, pp. 127–137, ASME, 1993.

HU, JUN MING, Temperature Dependence of the Mechanical Properties of GaAs Wafers, *J. Electronic Packaging*, Vol. 113, p. 331, December 1991.

LAU, J. H., A Note on the Calculation of Thermal Stresses in Electronic Packaging by Finite Element Methods, *J. Electronic Packaging*, Vol. 111, p. 313, December 1989.

LEE, CHIN C., Highly Reliable Die Attachment on Polished GaAs Surfaces Using Gold-Tin Eutectic Alloy, *IEEE Trans. Component, Hybrids, Manufact. Technol.*, Vol. 12, pp. 406–409, 1989.

LEE, CHIN C., A Low Temperature Bonding Process Using Deposited Gold-Tin Composites, *Thin Solid Films*, Vol. 208, pp. 202–209, 1992.

LING, SHARON X., A Design Model for Through-Hole Components Based on Lead-Fatigue Considerations, *Adv. Electronic Packaging*, EEP-Vol. 1, pp. 217–225, ASME, 1993.

MIRMAN, B., Interlaminar Stresses in Layered Beams, *J. Electronic Packaging*, Vol. 114, p. 389, December 1992.

MIRMAN, ILYA B., Effects of Peeling Stresses in Bimaterial Assembly, *J. Electronic Packaging*, Vol. 113, p. 431, December 1991.

MORGAN, H. S., Thermal Stresses in Layered Electrical Assemblies Bonded with Solder, *Trans. ASME*, Vol. 113, pp. 350–354, December 1991.

NAKANO, YUICHI, Thermal Stress in Bonded Joints Subjected to Uniform Temperate Change, *Adv. Electronic Packaging*, EEP-Vol. 1, pp. 11–16, ASME, 1993.

NISHIGUCHI, MASANORI, Highly Reliable Au-Sn Eutectic Bonding with Back-Ground GaAs LSI Chips, *IEEE Trans Compendium Hybrides Manuf. Tech* (USA) Vol. 14, pp. 523–528, Sept. 1991.

OLSEN, DENNIS R., Properties of Die Bond Alloys Relating to Thermal Fatigue, *IEEE Trans. Components, Hybrids, and Manufact. Technol.*, Vol. CHMT-2, No. 2, June 1979.

PAN, TSUNG-YU, Deformation in Multilayer Stacked Assemblies, *J. Electronic Packaging*, Vol. 112, March 1990.

ROYCE, BARRIE S. H., Differential Thermal Expansion in Microelectronic Systems, InterSociety Conference on Thermal Phenomena in The Fabrication and Operation of Electronic Components (I-THERM) '88, Los Angeles, CA, May 11–13, 1988.

SHEPHARD, M. S., Global/Local Analyses of Multichip Modules: Automated 3-D Model Construction and Adaptive Finite Element Analysis, *Adv. Electronic Packaging*, EEP-Vol. 4-1, ASME, 1993.

SHUKLA, R. K., A Critical Review of VLSI Die-Attachment in High Reliability Applications, *Solid State Technol.*, Vol. 28, pp. 67–74, July 1985.

SOLOMON, H. D., The Solder Joint Fatigue Life Acceleration Factor, *Trans. ASME*, Vol. 113, pp. 186–190, June 1991.

SUHIR, E., Die Attachment Design and Its Influence on Thermal Stresses in the Die and the Attachment, 1987 Proceedings: 37th Electronic Components Conference; May 11–13, 1987, the Boston Park Plaza Hotel & Towers, Boston, pp. 508–517 MA.

SUHIR, E., Approximate Analysis of the Interfacial Shearing Stress in Cylindrical Double Lap Shear Joints, With Application to Dual-Coated Optical Fiber Specimens Subjected to Tension, *Adv. Electronic Packaging*, EEP-Vol. 4, pp. 1–10, ASME, 1993.

SUHIR, E., An Approximate Analysis of Stresses in Multilayered Elastic Thin Films, Applied Mechanics Division of the American Society of Mechanical Engineers, Winter Annual Meeting, Chicago, IL, November 28 to December 2, 1988.

SUHIR, E., Interfacial Stresses in Bimetal Thermostats, Applied Mechanics Division of the American Society of Mechanical Engineers, Winter Annual Meeting, Dallas, TX, November 25–30, 1990.

TIERSTEN, H. F., A Global-Local Procedure for the Thermoelastic Analysis of Multichip Modules, *Advances in Electronic Packaging*, EEP-Vol. 1, pp. 103–118, ASME, 1993.

VAN KESSEL, C. G. M., The Quality of Die-Attachment and Its Relationship to Stresses and Vertical Die-Cracking, 33rd Electronic Components Conference, Orlando, FL, May 16–18, 1983.

WANG, CHEN Y., A Eutectic Bonding Technology at a Temperature below the Eutectic Point, *IEEE Trans. Components, Hybrids, Manufact. Technol.*, Vol. 15, pp. 502–507, 1992.

WILLIAMS, H. E., Asymptotic Analysis of the Thermal Stresses in a Two-Layer Composite with an Adhesive Layer, *J. Thermal Stresses*, Vol. 8, pp. 183–203, 1985.

YAMADA, S. E., A Bonded Joint Analysis for Surface Mount Components, *J. Electronic Packaging*, Vol. 114/1, March 1992.

EXERCISES

8.1. Describe how the elastic modulus and the coefficient of thermal expansion affect the reliability performance of solder joints in electronic assemblies subjected to thermal cycling.

8.2. A 20-μm-diameter copper wire having an initial length of 6 cm is tested in tension. The engineering strain is measured and found to be 0.03. What is the final length of the wire?

8.3. If the elastic modulus of the copper wire in Exercise 8.2 is $E = 117$ GPa, then what is the force exerted on the wire in the test?

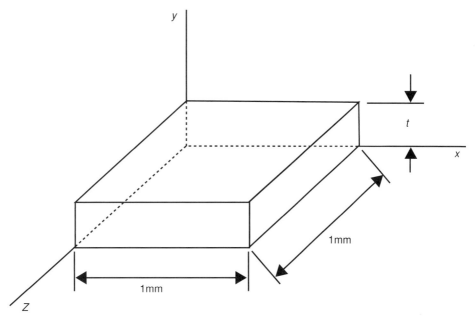

Figure 8.25 Surface-mount chip.

8.4. Determine the stress in the wire of Exercises 8.3 and 8.4. Is this a normal or shear stress? Does the stress depend on position along the length of the wire?

8.5. The displacement of a chip ($E = 138$ GPa, $\nu = 0.3$) during a processing operation was measured and is given by (see Figure 8.25)

$$u = 0.003x + 0.002y$$
$$v = -0.001x + 0.0005z$$
$$w = 0.006x + 0.003y - 0.003z$$

a. What are the components of strain?

b. What is the stress in the chip?

c. Consider the edge defined by the two points ($x = 0, y = 0, z = 0$) and ($x = 1, y = 0, z = 0$) prior to the processing. What is the location of this edge after the deformation? Is the edge straight or curved?

d. Is this deformation the result of a uniform temperature change? Why?

8.6. For the surface-mount package shown in Figure 8.26, determine the following:

a. The maximum stress in the copper pins in terms of the pin diameter d and length a, for a temperature increase ΔT.

b. If the stress in the pins is to be reduced would it be best to:

  (1) increase the length a
  (2) increase the diameter d
  (3) decrease the length a
  (4) decrease the diameter d
  (5) a combination of 1 to 4

Justify your results.

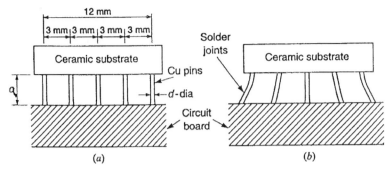

Figure 8.26 Surface-mount package.

8.7. A 10-μm-diameter copper wire used on a circuit board is observed using an optical microscope. When the board is at room temperature, the length of the wire is measured as 100 μm and when the devices on the board are energized the temperature of the board increases and the length of the wire increases to 120 μm. What is the normal strain in the wire?

8.8. If the elastic modulus of the copper wire in Exercise 8.7 is $E = 117$ GPa, then what is the stress in the wire when the board is energized?

8.9. List three of the major thermal stress-induced failure modes that may occur in electronic packages.

8.10. Assume that the temperature increases for each of the four cases listed below. For each case indicate if the stress that results is axial, $\sigma = F/A$, bending, $\sigma = My/I = Ey/R$, neither, or both. In cases where both are present, explain why both are present.

  a. Smooth silicon wafer on a smooth flat surface.

  b. Bar between two rigid supports that do not move.

  c. Two thin slices of two materials with different coefficients of thermal expansion bonded together.

  d. Two materials with different coefficients of thermal expansion bonded together, having their ends placed between two rigid supports that do not move. The top and bottom surfaces are free.

8.11. During wave-soldering operations, the temperature difference between the top and bottom surface of a PCB may reach up to 100°C. Assume the board is 0.062 in. thick.

  a. Compute the curvature of the board, assuming a linear temperature gradient through the thickness and constant board properties with temperature.

  b. What stress would develop on the top and bottom surfaces of the board if the board is bent back to the flat configurations?

  c. What harmful effects would you anticipate due to the curvature?

  Properties:

PCB	Ceramic	Copper
$\alpha = 19$ ppm/°C	$\alpha = 6$ ppm/°C	$\alpha = 16$ ppm/°C
$E = 14$ GPa	$E = 138$ GPa	$E = 117$ GPa
$\nu = 0.16$	$\nu = 0.3$	$\nu = 0.28$

8.12. Determine the radius of curvature as a function of die thickness for a diamond substrate having a unit depth and a thickness of 0.625 mm for a temperature change of 150°C when the die thickness varies from 0.025 to 0.4 mm. For diamond $\alpha_s = 2.0$ ppm/°C, $E = 1.18 \times 10^{-6}$ N/mm², and, for the die, $\alpha_d = 5.6$ ppm/°C, $E = 0.123 \times 10^{-6}$ N/mm².

8.13. Repeat Exercise 8.12 for the case when there is a solder layer 0.05 mm thick between the die and substrate. For the solder, use $\alpha = 16.8$ ppm/°C, $E = 0.071 \times 10^{-6}$ N/mm², and $\nu = 0.41$.

8.14. A two-layer tape-automated bonding (TAB) film is made from polyamide attached to a copper film. Derive

  a. An expression for the direct tensile force in the copper (and direct compressive force in the polyamide), due to a temperature rise of T degrees. Thickness, moduli of elasticity, and coefficient of thermal expansion should be denoted by t, E, and α, respectively, with subscripts p for polyamide and c for copper.

  b. the radius of curvature due to a temperature rise of T degrees.

8.15. Having found forces and moments at the ends of interfaces between polyamide and copper, calculate expressions for

  a. The maximum tensile stress in the copper

  b. The maximum compressive stress in the polyamide

8.16. A ceramic chip carrier (see Figure 8.27) is to be designed to operate in a temperature range from -55 to $125°$C. The properties of the substrate, solder, and component are given in Table 8.2. The thickness and length of the solder may be varied to improve the performance of the package. Determine the value of the maximum shear stress for $h_s = 0.0254, 0.0508$, and 0.0762, and $P_s = 0.26, 0.39, 0.52$, and 3.15 (all dimensions are mm). Use Eq. (8.57) with $K_1 = K_2 = 0$ and $K = h_s/G_s$. Which combination of h_s and P_s would you select?

8.17. Determine the maximum stress in the lead for the gullwing leaded surface-mount package mounted on a FR-4/epoxy board as shown in Figure 8.28 for the two levels given in Table 8.2.

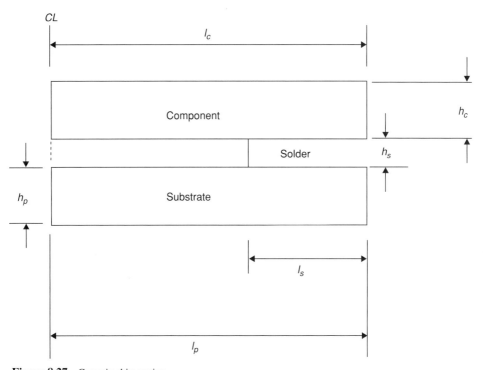

Figure 8.27 Ceramic chip carrier.

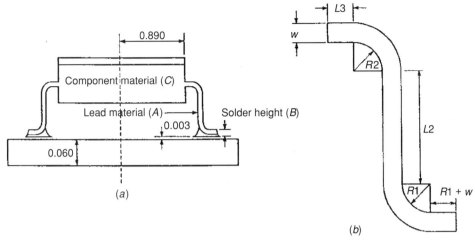

Figure 8.28 Gullwing leaded surface-mount package: (*a*) dimension variable of the complete structure and (*b*) lead dimension variables.

Table 8.2 Substrate, Solder, and Component Properties for Problem Exercise 8.16

	Substrate	Solder 60Sn–40 Pb	Component Al$_2$O$_3$
E (MPa)	17,200	30,800	303,460
CTE (ppm/°C)	16	24.7	5.26
Poisson's ratio	0.28	0.4	0.22

Table 8.3 Values for Dimensions of Some Component Parts and the Type of Material for Exercise 8.17

ID	Variable name	Levels	Level 1	Level 2
A	Lead material	2	Kovar	Copper
B	Solder height	2	2 mils	6 mils
C	Package material	2	Ceramic	Plastic
	Maximum temperature (T_{max})	2	30°C	125°C
	Minimum temperature (T_{min})	2	−55°C	20°C
G	Lead stiffness (K)	2	4 lb/in.	300 lb/in.
G1	Lead thickness (t_{lead})	2	5 mils	10 mils
G2	Height of stem (h_{lead})	2	50 mils	8 mils
G3	Lead radius (R)	2	20 mils	6 mils

Table 8.4 Material Properties for All Components in the Package of Exercise 8.17

No.	Material	Used for	E (Mpsi)	Poisson's ratio	CTE (ppm/°C)	Shear modulus (Mpsi)
1	Eutectic solder	Solder	3.62	0.40	21	
2	Kovar	Lead	20.02	0.30	5.87	
3	Copalloy 194	Lead	17.11	0.31	16.3	
4	Copper	Bondpad	18.7	0.345	16.7	
5	Alumina	Component	43.51	0.22	7.1	
6	Epoxy Glass	Component	3.05	0.30	30.6	
7	FR4	PWB	2.18	0.241	20.6 x	0.305 xy

The lead foot is soldered to the PWB bond pad with eutectic solder. The values for dimensions of some component parts and the type of material are given in Table 8.3. Table 8.4 gives the material properties for all components in the package. The definition of some symbols and values for fixed dimensions are given below:

$w = $ lead thickness $= G1$ (Table 8.3)
$L_2 = $ height of stem $= G2$ (Table 8.3)
$R_1 = $ bottom lead radius $= G3$ (Table 8.3)
$R_2 = $ upper lead radius $= 0.010$ in.
$L_3 = $ lead shoulder length $= 0.025$ in.
Lead width (not shown in Fig. 8.28) $= 0.010$ in.

Chapter 9

Discrete and Embedded Passive Devices

RICHARD ULRICH

9.1 INTRODUCTION

Resistors, capacitors and inductors make up what are called *passive devices*. They provide impedance, current-to-voltage phase angle, and energy storage that are used to condition all sorts of analog and digital signals and are present in just about every electronic system ever made. They are generally two-port devices so that the signal passes through. The impedance of an ideal passive is not a function of signal strength (current or voltage) and is not a function of signal frequency for an ideal resistor but falls with increasing frequency for an ideal capacitor and rises with frequency for an ideal inductor. These sort of behaviors are very handy for making filters, as well as in energy storage, voltage modification, line termination, current control, and many, many other applications. They are nonswitching components, unlike active components such as transistors that have a third port used to command an impedance change in the device. Also, unlike actives, passives have no gain; they cannot amplify, although they can use their impedance to decrease the current or voltage of signals.

When the first edition of this book was written in 1997, there was no chapter dedicated to passive devices. At the time, passives were just an assembly issue as far as packaging was concerned. You would purchase a manufactured resistor, capacitor, or inductor that came in its own little enclosure, and you mounted it on the printed wiring board (PWB) and that was all that related to packaging. However, passives are currently starting on a major change in the way they can be included in microelectronic systems; they can be integrated or embedded (the two terms are synonymous) directly into the circuit boards [1–3] just as active devices were integrated into silicon 40 years earlier. This is a profound difference in the way passives are built that results in the printed circuit board (PCB) shop having to get into the business of manufacturing not only interconnect substrates but also the passives that go into them. This means that the packaging community requires a deeper level of understanding in the fundamentals of the material science, fabrication techniques, and electrical performance of passive devices. Obviously, the industry would not go to all this trouble if it was not worth it, and the potential rewards are considerable. This chapter will

---

Advanced Electronic Packaging, Second Edition, Edited by Richard K. Ulrich and William D. Brown
Copyright © 2006 the Institute of Electrical and Electronics Engineers, Inc.

concentrate on what is inside discrete or surface-mount passives as well as the developing field of embedded passives and is organized as follows:

- How passives appear in systems today
- Film-based fabrication
- Resistors, capacitors, and inductors
- Electrical characteristics
- Embedded passives
- Example application: decoupling

The discussion will be limited to organic substrates only since they comprise the vast majority of PWB in all applications today and in the foreseeable future.

9.2 PASSIVES IN MODERN ELECTRONIC SYSTEMS

Tremendous progress has been made in the past four decades in miniaturizing and integrating transistors for logic applications onto silicon. By comparison, passive components at the circuit board level have made only incremental advances in size and density. Consequently, passive components occupy an increasingly larger area and mass fraction on PCBs and are a major hurdle to the miniaturization of many electronic systems. This is particularly true for analog and mixed-signal applications that use a larger number of passives than typical digital systems. Almost no through-hole passives are used anymore; they have been replaced with smaller rectangular surface-mount components with solder joints at both ends. The size of these modern discretes is described by a number such as 0603, which indicates a size of 60 × 30 mils (1.5 × 0.75 mm). The 0402 (1.0 × 0.5 mm) size is one of the most popular, and the smallest discrete passives available today are 0201 (0.50 × 0.25 mm), which represents a considerable challenge in handling, attachment, and inspection [4]. The 0201 may represent the smallest size that will be practical for surface mount, although 01005 components have been prototyped. Passives tend to move down a case size every 4 years while, following Moore's law, integrated circuits (ICs) double their transistors/cm^2 about every 1.5 years. Figure 9.1 shows these sizes relative to a dime, which is 18 mm in diameter.

Figure 9.1 Sizes of surface-mount passives.

9.2 Passives in Modern Electronic Systems

Figure 9.2 Cell phone RF section utilizing surface-mount passives.

Figure 9.2 shows a cell phone radio frequency (RF) section that utilizes 0402 and 0603 resistors and capacitors surrounding a 6 × 6 mm packaged integrated circuit. About a trillion passive devices were placed in electronic systems in 2000, with the vast majority utilizing surface-mount technology. How these are actually placed and affixed to the board is the topic of the next chapter (Electronic Package Assembly) so we will not go into anymore detail about surface-mount technology here.

Today, each mounted passive costs about half a cent to purchase, and about 1.3 cents for conversion (assembly, testing, inspection, and rework) for a total installed cost of around 1.8 cents. This is an average cost only; some passives, such as low-inductance decoupling capacitors, can cost up to $2 each just to purchase. The present total market for passive devices is around $20 billion annually. In terms of numbers of components, there are more passive devices than active devices in just about any application. An Ericsson CF388 PCS 1900 cell phone has 380 components, including 322 passives and 15 ICs, for a passive-to-active ratio of 21 : 1. Digital systems, such as desktop and laptop computers, weigh in at somewhat lower ratios: between 5 and 15 passives for every IC. See Table 9.1.

In terms of area and lead count, an individual surface-mount passive is almost always smaller than any packaged IC and usually has only two connections, but, because they are more numerous, the relative total footprints and total number of device-to-board connections are closer to equal to those of ICs. Figure 9.3 shows part of a board from a Nokia 6161 cell phone with the location and footprint of surface-mount discrete passives marked in white. As you can see, a considerable fraction of the board space is taken up by passives. In a typical cell phone, Global Positioning System (GPS) receiver, computer, or just about any electronic system, the board space taken up by ICs includes about a billion components in the form of logic transistors and capacitors, while the other half of the board space contains only a few hundred passives.

A breakdown of the 405 individual passive components by number and value for this same phone is shown in Table 9.2 [6]. Additionally, there were 15 ICs and 40 miscellaneous surface-mount devices such as power transistors and diodes for electrostatic discharge (ESD) protection, all mounted onto 6.2 in.^{2} of board area for an average passive density of 85/in.^{2}.

The number of discrete passives in a model series of desktop computers over the years is given in Table 9.3 [7]. Some trends are clear: a rapid increase in the total number of passives utilized, a total switch from through-hole or "leaded" to surface-mount technology (SMT) components, and the initiation of the use of passive arrays—multiple passives in one surface-mount package. Mobile wireless, including cell phones, will account for the largest share of the increase in passive usage in coming years, but other significant new markets

Table 9.1 Passive and IC Count for Consumer Products [5]

System	Total passives	Total ICs	Ratio
Cellular Phones			
Ericsson DH338 Digital	359	25	14 : 1
Ericsson E237 Analog	243	14	17 : 1
Philips PR93 Analog	283	11	25 : 1
Nokia 2110 Digital	432	21	20 : 1
Motorola Md 1.8 GHz	389	27	14 : 1
Casio PH-250	373	29	13 : 1
Motorola StarTAC	993	45	22 : 1
Matsushita NTT DOCOMO I	492	30	16 : 1
Consumer Portable			
Motorola Tango Pager	437	15	29 : 1
Casio QV1O Digital Camera	489	17	29 : 1
1990 Sony Camcorder	1226	14	33 : 1
Sony Handy Cam DCR-PC7	1329	43	31 : 1
Other Communication			
Motorola Pen Pager	142	3	47 : 1
Infotac Radio Modem	585	24	24 : 1
Data Race Fax-Modem	101	8	13 : 1
PDA			
Sony Magic Link	538	74	7 : 1
Computers			
Apple Laptop Logic Board	184	24	8 : 1
Apple G4	457	42	11 : 1

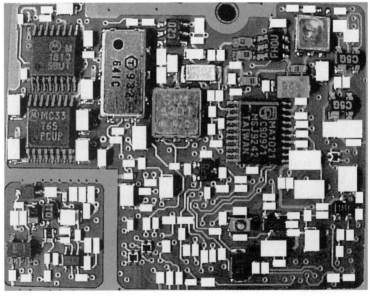

Figure 9.3 Cell phone board showing the footprints of surface-mount passive components marked in white.

Table 9.2 Distribution of Sizes and Values for Surface-Mount Passive Components in a Nokia 6161 Cell Phone

Size	Values	Quantity
Capacitors		
0402	< 100 pF	100
0402	1 nF	37
0402	15 nF	20
0603	30 nF	22
0603	100 nF	29
0805	250 nF	2
0805	1 µF	12
1206	2 µF	2
1310	9 µF	2
Electrolytic	10 µF	6
		Total: 232
Resistors		
0402	1–10 MΩ	109
0402 Dual array	1–20 MΩ	18
0603	1–22 MΩ	16
0805	1–22 MΩ	4
1206	1–22 MΩ	2
		Total: 149
Inductors		
0603	1–100 nH	16
0805	2–500 nH	3
1206	< 220 µH	5
		Total: 24
	Total Passive Components: 405	

include Bluetooth and automotive applications. The 2000 National Electronic Manufacturing Initiative (NEMI) roadmap predicts that cell phone sales will reach 1 billion units annually by 2004, which will require replacing half the cell phones in use today, and there should be 2 billion Bluetooth devices operating by 2005. Telecommunications has replaced computers as the top user of passives.

Table 9.3 Number and Type of Passive Components in Personal Computers

	486	Pentium 120	Pentium 200	Pentium II	Pentium III
Leaded capacitors	73	1	0	4	0
SMT capacitors	0	158	225	348	695
Capacitor arrays (>1 per package)	0	0	32	140	200
Leaded resistors	92	0	0	0	0
SMT resistors	0	146	188	635	1000
Resistor arrays (>1 per package)	0	64	148	346	300
Total passives	**165**	**369**	**593**	**1473**	**2195**

354   Chapter 9   Discrete and Embedded Passive Devices

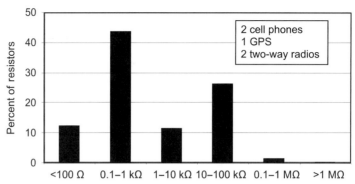

Figure 9.4   Distribution of resistor values in portable consumer equipment.

An analysis of two cell phones, one GPS receiver, and two two-way radios produced the resistor and capacitor distributions shown in Figures 9.4 and 9.5 [8]. The required values extend over many orders of magnitude for resistors and capacitors. Inductors range in value from about 1 to 50 nH, but there are usually far fewer inductors than capacitors and resistors in most consumer microelectronic products. It has been observed that 40% of capacitors in a cell phone are under 1 nF, and 80% of inductors in handheld products are less than 200 nH.

In summary, all types of electronic systems are becoming more complex while simultaneously being under pressure to be smaller and lighter. The numbers of passives are steadily increasing, and the required range of values is very wide. Since passives are not shrinking in size as fast as transistors on ICs, passives are becoming a limiting factor in system size, especially for mixed-signal applications. Manufacture and placement of 0201 discretes may represent a limiting size and density for SMT devices.

9.3   DEFINITIONS AND CONFIGURATIONS OF PASSIVES

Discrete Passive Component   This is a single passive element in its own leaded or surface-mount technology package. An example would be a single resistor, capacitor, or inductor in an 0603 package as shown in Figures 9.1 and 9.2. This will typically have two contacts to be soldered to the board. Presently, the vast majority of passives are utilized in this manner.

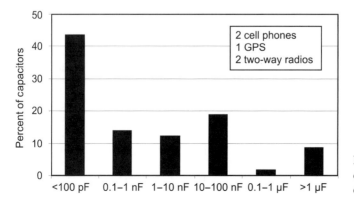

Figure 9.5   Distribution of capacitor values in portable consumer equipment.

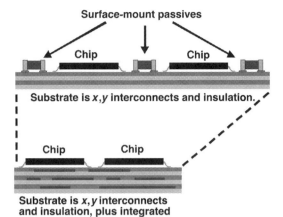

Figure 9.6 Embedded passive components formed within the layers of the primary interconnect substrate.

Embedded or Integrated Passive Component This is a general term for multiple passive components that share a common substrate and packaging (see Fig. 9.6). They may be housed inside the layers of the primary interconnect substrate, or they may be on the surface of a separate substrate that is then placed in an enclosure and surface mounted on the primary interconnect substrate, in which case they would be called a *passive array* or *passive network*.

Passive Array Multiple passive components of like function (all resistors or all capacitors) formed on the surface of a separate substrate and packaged in a single SMT case (Fig. 9.7). This case is then mounted on the primary interconnect substrate of the system. The number of leads will typically be twice the number of internal components in the array, but more leads may be provided to reduce the total inductance in capacitor arrays or fewer leads may be present if some of the components are connected internally, such as for voltage dividers. Inductors are not normally arrayed since their separate electromagnetic fields would interfere with one another in close proximity. The passive array does not always reduce the number of leads that must be attached but does increase the efficiency of their attachment since more connections are made with one alignment and mounting. This is the lowest level of passive integration and involves many of the same manufacturing techniques used for discretes.

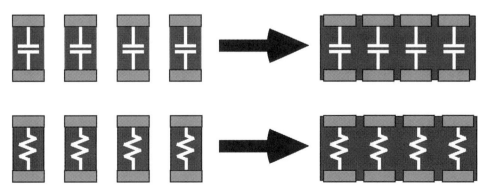

Figure 9.7 Integrated passive arrays.

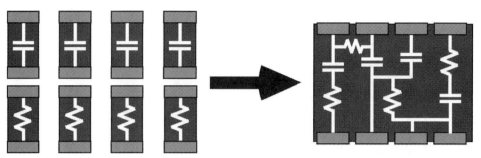

Figure 9.8 Integrated passive network.

Passive Networks Multiple passive components of more than one function are formed on the surface of a separate substrate and packaged in a single SMT case (Fig. 9.8). This case is then mounted on the primary interconnect substrate of the system. These typically have some internal connections to form simple functions such as filters or terminators. The number of leads can vary with functionality and the number of internal elements. This approach generally does reduce the number of leads to be connected since some passive-to-passive connections are made within the package.

Both passive arrays and networks of various types are available from several manufacturers and are in common use in all types of electronic systems in the form of surface-mount components as small as 0402. They are particularly useful in digital systems where parallel data buses require RC termination or pull-up/pull-down resistors for many lines in a small footprint. Their commercial penetration is probably less than 5% at this time but is expected to increase. Embedded passives have very little commercial penetration at this time but is also expected to rise in the coming years.

9.4 FILM-BASED PASSIVES

The same film-based processes that are used to make surface-mount discretes are often found as embeddable materials. The latter is an emerging technology, and the preferred materials and strategies for fabrication directly into organic boards are still being determined. This section is a broad overview of the various film-based methodologies used in both incarnations of passives, and subsequent sections will describe the specific materials and processes for resistors, capacitors, and inductors. At the time of this writing, there are five basic technologies on the market for embedded resistors and four for embedded capacitors, most of which became available in just the past 2 years. Due to the planar nature of embedded passives, the formation and patterning of films are central to their fabrication, but have long been used in discrete components as well. For instance, anodized Ta has been a staple of surface-mount capacitors for decades and is under active development as an embeddable dielectric. The main difference is that, in an 0402 package, the film is folded up in three dimensions instead of being completely flat. The emphasis on film-based passives in this chapter is due to their widespread use in discrete components, their universal use in embedded passives, and the planar emphasis of this book.

There are three broad classes of films required: conductive, resistive, and dielectric. Conductive films are those that are needed for carrying current with a minimum of voltage loss, such as the top and bottom plates of capacitors and the spiral windings of a planar inductor. There is no advantage to having parasitic resistance in this type of film; that would only degrade the performance of the capacitor or inductor. Therefore, these would normally be metals or else very conductive metal-filled polymer thick films with resistances less than about 0.1 Ω/square. Narrow tolerance and repeatability is not a major issue as long as the overall resistance is low enough for the application. Of course, it might be desirable for embedded passives to utilize the same metal used as interconnect on the substrate. Resistive films would be used in making embedded resistors only and would be specified on the basis of providing predictable, reproducible values of resistance in a sufficiently small footprint. A wide variety of materials could be used for this, ranging from resistive alloys (NiCr, CrSi, TaN_x, NiP, TiN_xO_y) to ceramic–metal nanocomposites (cermets) to carbon-filled polymers. Resistivities of 100 to 10,000 Ω/square are required to efficiently cover the entire range of resistor values commonly found in electronic systems. Dielectric films would be used to form embedded capacitors, and a vast array of materials with a wide range of dielectric constants are feasible from simple unfilled polymers ($k = 2$–5) for small-valued capacitors to amorphous metal oxides ($k = 9$–50) to highly ordered mixed oxides for the highest possible dielectric constants ($k > 1000$). There are essentially no fabrication issues for inductors; they are simply shaped conductors made from the interconnect metallization already present in the boards.

These three types of films can be formed subtractively or additively by sputtering, chemical vapor deposition (CVD), evaporation, anodization, dry oxidation, sol–gel, spin-on, doctor blade coating, chemical conversion, and many others. Etching options include liquids and a variety of reactive and unreactive, directed or nondirected plasmas. And, they can be modified by annealing, exposure to chemicals, or an array of surface treatments. A classification scheme of films based on manufacturing methods is shown in Table 9.4.

Embedded passives are not a new idea; they have been used for decades in ceramic substrates. Thick-film pastes of conductors and dielectrics are used to form resistors, capacitor dielectrics, and spiral inductors that are fired simultaneously with the green tape of the insulating layers. However, the firing requirements mean that this technology is not transferable

Table 9.4 Comparison of Electronic Film Technologies

Thin film	Polymer thick film	Ceramic thick film
Sputtering, CVD, evaporation similar to IC technology; requires vacuum	Spin-on, cast, screen print, or stencil polymer materials with functional filler; then cure polymer	Cast, screen print, or stencil glass paste, functional filler, and organic binder; then fire at to remove binder
100–250°C	100–250°C	600–1000°C
Subtractive processing	Additive processing	Additive processing
Thickness <few microns	Thickness = few microns to mils	Thickness = few microns to mils
Metals, oxides, but rarely polymers	Filled polymers	Filled glasses, oxides
Higher capacitance since thin	Lower capacitance since thick	Medium capacitance since thick, but has high k
Capable of smaller width and spaces down to microns	Widths and spaces down to mils	Widths and spaces down to mils
Better dimensional tolerance	Less controllable tolerance	Less controllable tolerance
Most have stable values with time and humidity	Reliability and stability a limiting concern for many, and a major area of R&D on these materials	Stable values with time and humidity
Most expensive	Least expensive	In between

to organic substrates. Glass has been used for embedded passive substrates by Intarsia, and silicon had a brief period of use as an multichip module (MCM) substrate; Bell Labs and nChip utilized an embedded decoupling capacitor as part of the buildup over a silicon substrate for their MCM—deposited (MCM-D) designs. Organic substrates make up the vast majority of interconnect boards due to their low cost, and it is here that embedded passive efforts are most important. In addition to not being conducive to firing, vacuum processing is not available in many board shops. Whatever methods and materials are chosen must be compatible with the board's conductors and insulation layers already in place, embedded passives already in place, and any fabrication steps that follow so the idea of embedding passives still has a considerable amount of development ahead of it.

9.5 RESISTORS

Surface-mount resistors can be had from a fraction of an ohm to many megaohms in packages as small as 0402 and even 0201. You can even buy a "zero-ohm resistor" if you just need to short a couple of pads. Embeddable resistor materials and processes make up a rapidly growing product segment.

9.5.1 Design Equations

Many discrete and all embedded resistors are fabricated either by creating, through additive or subtractive processing, a layer of resistive material in series with an interconnect line (Fig. 9.9). In keeping with the concept of a planar, stacked assembly, the resistor will be a film of material, probably between a few hundred angstroms and a few microns thick.

Assuming that all of the resistance is in the resistor material and not in the interconnects (usually Cu) or in the connections between the interconnect and the resistor, the resistance of the structure is

$$R = \frac{\rho L}{Wt}$$

where

R = resistance, Ω

ρ = resistivity of the material, Ω-cm

L = length of the strip, cm

W = width of the strip, cm

t = thickness of the strip, cm

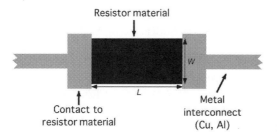

Figure 9.9 Layout for a simple embedded resistor.

The resistivity of a material is an intrinsic property and is a function of composition and microstructure. For thin films, the resistivity is generally higher than that of bulk materials. The reciprocal of resistivity is conductivity in $(\Omega\text{-cm})^{-1}$, sometimes referred to as "siemens/cm."

Sheet resistance is defined by

$$R = \left(\frac{\rho}{t}\right)\left(\frac{L}{W}\right) = R_s N_s$$

where

$R_s = \rho/t$ = sheet resistance, Ω/square

$N_s = L/W$ = number of squares

The sheet resistance is the resistance of a square of material ($L = W$) when the electrical contacts cover opposite edges completely, as in the second picture in Figure 9.10. The size of the square is irrelevant as long as length equals width and the contacts completely cover two opposing sides. When long, narrow materials are used, the resistor is thought of as squares in series:

Resistors consisting of many squares are usually formed in a serpentine pattern to fit into an allocated substrate area with corner squares counted as 0.556 square, since the current does not have to traverse the entire side-to-opposite-side distance. Thus, a material with a resistivity of 1 mΩ-cm that is 1 μm thick would have a sheet resistance of 10 Ω/square, and this number, multiplied by the number of squares, would give the value of the resistor in ohms. The benefit of expressing films of resistor materials in Ω/square is that it does not matter what the resistivity or the thickness is as long as it gives the desired sheet resistance. For example, in sputtering TaN$_x$ resistors, changing the sputtering conditions such as vacuum level, temperature, power, gas composition and so forth will almost always affect both the material's film thickness and resistivity simultaneously. It may be impossible to change conditions so that only resistivity or film thickness is adjusted to desired values due to the inevitable cross-dependencies of processing conditions. However, if the sheet resistance

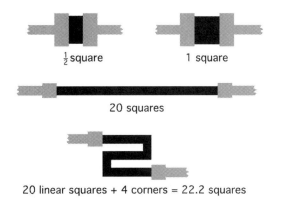

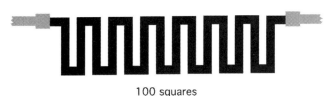

Figure 9.10 Resistor geometries expressed as numbers of squares.

is the target variable, processing conditions can be optimized to give the desired value of Ω/square without having to measure or specify both the thickness and resistivity. Also, the sheet resistance is a quantity easily measured with standard four-point probes or from simple test structures, again without having to measure either of its two constituent parts.

The temperature coefficient of resistivity (TCR), is defined as the temperature derivative of dimensionless resistivity and is usually expressed in ppm/°C:

$$\text{TCR} = \frac{1}{R}\frac{\partial R}{\partial T} = \frac{1}{R_{T_1}}\frac{R_{T_2} - R_{T_1}}{T_2 - T_1}$$

For resistor materials instead of the resistors themselves, the values of R may be replaced by the material's resistivity in Ω-cm. For most applications, the ideal values of TCR would be zero so that the value of the resistor is constant under any operating temperature, but less than 200 ppm/°C is usually considered low. See Table 9.5.

9.5.2 Sizing Embedded Resistors

Required resistor values in common electronic systems span a large range: from less than 10 Ω to well over 1 MΩ. It is practical to build a serpentine embedded resistor with between

Table 9.5 Resistance Properties of Common Electronic Materials [1]

Material	Resistivity range (μΩ-cm)	Film thickness	Sheet resistance (Ω/square)	TCR (ppm/°C)
Ag	1.6	2 μm	0.0080	4100
Cu	1.7	2 μm	0.0085	4330
Au	2.4	2 μm	0.0120	4000
Al	2.7	2 μm	0.0135	
Ni	6.9	2 μm	0.0345	6750
Ta	bcc: 13 beta: 180	500 Å	2.6	3800
Cr	13	500 Å	2.6	3000
Ti	42	500 Å	8.4	
TaN$_x$, CrSi, NiCr, TiN$_x$, NiP	100–500	500 Å	20–100	±50 with process optimization
NiP (Ohmega-Ply)	~2000	1000–4000 Å	Up to 250; 1000 in development	0–100
NiP (MacDermid)			Up to 100 higher in development	
TiN$_x$O$_y$	Up to 7000	500 Å	up to 1400	±100 with process optimization
LaB$_6$ (DuPont)	10^7	10 μm	10,000	±200
PTF (several vendors)	Very wide, depending on filler	1–2 mil	10–10^7	~200
Cermets	10^4–10^{10} depending on metal/glass ratio	1 μm	100–10^8	Close to zero or slightly negative

about 0.1 and perhaps a hundred squares; resistor patterns with more squares may require excessive footprints, are more prone to yield and tolerance problems, and tend to show greater parasitic capacitance at high frequencies due to coupling between the meandering strips. Taking these as the boundaries of the design envelope, a 10-Ω resistor fabricated from 0.1 square would require a material with 100 Ω/square, while a 1-MΩ resistor with 100 squares would require 10,000 Ω/square. It would be possible to cover this range with embedded resistors if they could be made with two different materials:

1. Resistor material with about 100 Ω/square for values from 10 to 10,000 Ω.
2. Resistor material with about 10,000 Ω/square for values from 10,000 to 1 MΩ.

If these two materials are utilized, then resistances from 10 Ω to 1 MΩ can be covered with 0.1 to 100 squares. For resistances outside of this range, embedded resistors might not be practical. In fact, depending on the application and the number of extreme-value resistors required, it may only make sense to use one material to cover a certain range and to use SMT discretes for the rest. It should also be noted that it is possible to make very low valued resistors from the interconnect metallization. For example, 2 μm of Cu gives about 10 mΩ/square so 100 squares would give only 1 Ω. Subohm values are often used for current sensing.

Once a resistor material and process is selected, it is a simple matter to determine the length/width ratio to give any required value of resistance from the Ω/square of the material. However, to separate these two measures out of the ratio and establish the actual footprint of the resistor, the following factors must be considered.

Heat Dissipation Embedded resistors must be designed so that the temperature rise during use will not heat them to the point that their value drifts significantly or that failure mechanisms are accelerated to the point of affecting their reliability. Large-area resistors are favored regardless of their number of squares. For resistors integrated into circuit boards, three-dimensional thermal simulations are usually necessary to predict operating temperatures at all parts of the components and assembly. The thermal operating limits of some passives are given as maximum power, but it should be remembered that heat causes temperature, and it is temperature that causes failure. The maximum continuous operating temperature for FR-4 is usually around 130°C.

Tolerance Tolerance and precision improves for larger areas. A ±10% value precision is considered acceptable for many resistors, but many applications require tolerances down to 1%.

Parasitic Capacitance Long serpentine resistors will exhibit characteristics of capacitors at high frequencies due to coupling between adjacent strips resulting in a drop in resistor impedance. Small numbers of squares and large strip spacing are favored to minimize this regardless of the total area.

9.5.3 Materials for Resistors

Single-Component Metals Because the density of charge carriers is high, on the order of the atomic density ($\sim 10^{22}$ cm^{-3}), and because electrons are very mobile, metal conductivities are the highest of any materials at room temperature, ranging from about 2 to 50 μΩ-cm with Cu at a bulk value of 1.7 and Al at 2.7 μΩ-cm. The high-end is represented by Ti at 43, Mn at 140 μΩ-cm and β-Ta at 180 μΩ-cm. All of these are too conductive to form resistors with values higher than a few ohms, even when formed into films as thin as 1000 Å, about the lower limit for reproducible thicknesses on common substrates. Also, their TCR values tend to be very high, up to thousands. For these reasons, single-component metals are not very useful as integrated resistor materials.

Metal Alloys Mixtures of metals exhibit higher resistivities than single-component materials, a few as high as 160 μΩ-cm, and they also typically have lower TCR values. But most metal–metal alloys, such as elemental metals, have resistivities too low for use as embedded resistors. Some binary metal-containing compounds have been investigated that could serve at the low end of the required resistance range, and three seem the most practical: NiCr, TaN_x, and CrSi. TaN_x may be the best of the group due to its ease of processing, low TCR and stability against value change with time, but all three have been demonstrated as embedded resistors for organic substrates. They are typically sputtered to form submicrons films. However, sheet resistances above about 200 Ω/square have not been achieved with this class of materials to date, making them suitable only for the lower half of resistor values.

Nickel Phosphide While most thin-film compounds demonstrated for embedded resistor applications are sputtered, it is also possible to electroplate NiP, a process that is more familiar to most board shops. These materials can be electrolytically or electrolessly plated to give up to 170 μΩ-cm with a TCR of about 100 ppm/°C for 12 to 14% P in Ni. Electroplated NiP is also used as the resistive material for Ohmega-Ply, one of the few mature commercialized integrated passive systems, providing up to 250 Ω/square, and for MacDermid's M-Pass to give up to 100 Ω/square. Ohmega-Ply is a subtractive process where the NiP is electrolytically plated and M-Pass is an additive process that uses selective electroless plating over areas pretreated with a catalyst.

Semiconductors Undoped semiconductor materials, on the average, contribute only one charge carrier (electron or hole) per billions or trillions of atoms. As a result, resistivities are much higher than for metals; undoped Si is 250,000 Ω-cm, almost a trillion times that of Cu. However, the concentration of charge carriers is a strong function of temperature and impurity levels, which makes these materials too variable for use as resistors. Since increasing temperature liberates charge carriers in exponential amounts, the TCR is strongly negative, about −73,000 ppm/°C for pure Si at room temperature. Furthermore, their resistivity is a strong function of trace amounts of impurities, which is a key feature in making integrated circuits possible. Polycrystalline Si can be sputtered onto organic substrates, but its extreme TCR behavior would make it useful for thermistor applications only.

Cermets Most of the resistor materials described above are useful at only the lower end of the range required for common systems. High-value resistors, above about 100 kΩ, made from reasonably small numbers of squares, require films with sheet resistances in excess of 1000 Ω/square; this is not achievable with any known metal–metal alloy. Stable high-resistivity materials may be obtained by combining a metal and a ceramic insulator into a two-phase structure known as a *cermet*. These are nanostructured compounds, with metal being the distributed phase and ceramic the continuous phase. They are commonly used on ceramic substrates where they are made by firing metal–glass pastes, although it may be possible to form them at lower temperatures. The most commonly used of these is Cr plus SiO, which can provide useful resistors with high values, low TCR, and good stability. Also 70% Cr in SiO gives 1100 μΩ-cm and TCR = 0, while 55% approaches 10,000 μΩ-cm [9].

Polymer Thick Film (PTF) PTF materials are a very promising technology for embedded resistors due to their low cost of materials and processing, wide range of resulting resistance, and low-enough curing temperature for organic substrates, but there are significant problems with value drift and reliability to overcome before they can be widely used. Most PTF systems under development are epoxy-based polymers with micron-scale carbon or graphite fillers, and the conduction mechanism is contact bridging between filler particles. They are provided as viscous liquids that may be screen-printed or stenciled, then cured at temperatures generally below 200°C. One available product cures in 45 min at 165°C or may be snap cured with infrared (IR) in only 5 min. The result is a film of several microns thickness with a practical width no smaller than about 3 to 5 mils due to limitations in screen printing or stenciling, although 1 mm is a more common lower limit due to heat removal considerations. Not only is the processing simple and suitable for FR-4 and flex, but also the achievable sheet resistances cover a very wide range, from 1 to 10^7 Ω/square. There is rarely a need to use more than 10 squares due to the wide range of resistivities supplied by the manufacturers, so resistor footprints can be quite compact. As printed, their tolerance is no better than about 5 to 10%, but they can be laser trimmed. Solder will not wet them, due to the epoxy polymer binder, so they can be printed and cured before other surface-mount components are added without the need for solder masking. PTF is additive so there is no patterning and etching required. The films are generally 1 to 2 mils thick wet and about half that dry, and a single gram may cover 200 to 400 cm^2 with little waste. Several board shops offer PTF as an option, although generally on noncritical applications where some drift is not harmful, such as in rheostats.

The problem with these materials is that the resistivity tends to increase with time both in the bulk of the material and at the metal–polymer interfaces, especially in humid environments. There are several reasons for this, including oxidation at the Cu–polymer interface, delamination from the contacts, swelling of the polymer material with moisture, cracking of the polymer material due to coefficient of thermal expansion (CTE) mismatch, and possibly others. Better contact stability may be obtained by using larger contact areas, Au- or Ag-plated contacts, Au- or Ag-filled epoxy materials over the Cu termination as a transition material, or by applying proprietary oxidation inhibitors, all at additional cost. Considerable research is being performed on this class of materials to solve these limitations. Several companies offer PTF products and services, but value stability should be evaluated under the conditions of the intended use and not extrapolated from the company's test conditions. As issues related to long-term stability and reliability can be resolved, PTF will become a major player in embedded resistors.

9.6 CAPACITORS

A capacitor, discrete or embedded, is simply a thin dielectric material sandwiched between two metal plates. In surface-mount form, the maximum capacitance available in, say, an 0402 is rising constantly and is presently a few microfarads, so it is not necessary to go to larger case sizes to cover the range of almost any electronic system. If embedded, the entire assembly must be very thin to be integrated into the board, less than a mil at most and, if put into a surface-mount enclosure, the very same area is required, but is folded up in three dimensions. A very wide range of capacitor dielectric materials are potentially useful, some of the most important are shown in Table 9.6.

Ideally, their dielectric constant should be flat with regard to frequency, temperature, voltage, and time, and the material should be capable of being bent and stretched to a

Table 9.6 Dielectric Constants for Common Dielectrics [1]

Composition	Dielectric constant	Dissipation factor (%)
Teflon	2.0	0.02
Polyethylene	2.3	0.02
BCB	2.7	0.1
Parylene	2.7	0.01–0.1
Low e BT resin	2.7	0.2
BPA cyanate	3.1	0.4
Polycarbonate	3.1	0.1
Mylar	3.2	0.4
SiO_2	3.7	0.03
Polyimide	3–4	0.2–1.0
Epoxies	3–6	0.4–0.7
Epoxy resin for FR-4	3.9	1.2
FR-4	3–5	0.5–1.5
E glass	5	0.09
SiO	5.1	0.01
Si_3N_4	7–9	
BeO	7–9	<0.1
ZnO	8	
AlN	9	
Al_2O_3	9	0.4–1
Si_3N_4	9.4	<1
MgO	9.5	
YO_x	12–17	
$BaTiO_3$ (amorphous)	17	
NbO_x	20	
Ta_2O_5 (amorphous)	24	0.2–1
SnO_2	25	14
PbO	26	
SiC	40	
HfO	23, 40	~1
WO_2	42	0.6
Ta_2O_5 (polycrystalline)	50	
TiO_2	31 (anatase)	2–5
	78 (brookite)	
	117 (polycrst)	
	~40–60 (film)	
$BaTiO_3$ (tetragonal)	Up to thousands	5
$BaSrTiO_3$	Up to thousands	
$PbZr_xTi_{1-x}O_3$	Up to thousands	
$Ba_{0.8}Pb_{0.2}(Zr_{0.12}Ti_{0.88})O_3$	Up to thousands	

reasonable degree so that it will not suffer from the effects of CTE mismatch with other board materials during normal temperature excursions, and so it can be used in applications with little packaging such as smart cards. It should be amendable to mass production at an economic cost using common materials and patterning techniques that do not harm other parts of the board or components already in place. Certainly, any material and process technology will compromise on some of these issues, which is why no one perfect capacitor dielectric has yet been identified from the hundreds of journal and proceedings articles to date. However, several useful materials are now being commercialized.

9.6.1 Paraelectrics and Ferroelectrics

There are two broad classes of dielectrics: paraelectrics and ferroelectrics [10]. The important difference from the point of view of embedded passives is that ferroelectrics generally have much greater dielectric constants than do paraelectrics, sometimes by as much as three orders of magnitude, because of a mobile ionic charge that can move within the crystal lattice structure. For example, the common ferroelectric barium titanate is a perovskite crystal with a Ti cation at the center that can shift back and forth within the confines of this crystal a distance larger than the ions can move in a typical paraelectric such as tantalum oxide. This motion creates a considerably larger dipole arm than is possible in Ta_2O_5, resulting in a much larger dielectric constant for $BaTiO_3$. However, to achieve this very high dielectric constant, ferroelectrics must be in an oriented, single-crystal form, which usually requires a curing temperature of at least 600°C in oxygen, far in excess of what can be tolerated by organic boards. Adapting ferroelectric dielectrics to organic substrates is a major goal of capacitor integration development.

The dielectric constant of ferroelectric materials are a strong function of temperature, frequency, film thickness, and voltage. In those regards, it is not a constant at all! A temperature coefficient of capacitance (TCC) of under about 200 ppm is considered low, but the TCC of ferroelectrics can not only be much higher than this but can also change sign at certain temperatures. Figure 9.11 shows the effects of temperature on two ferroelectrics and one paraelectric, demonstrating the widely varied types of temperature-driven behavior

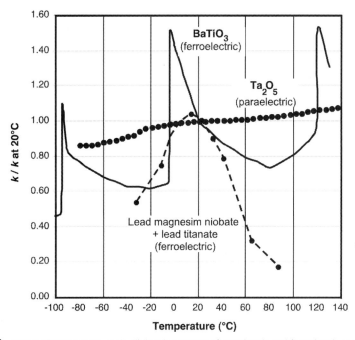

Figure 9.11 Effect of temperature on the dielectric constant of paraelectric and ferroelectric materials.

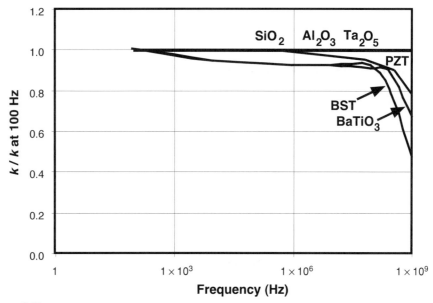

Figure 9.12 Effect of frequency on the dielectric constant of paraelectric and ferroelectric materials.

ferroelectrics can exhibit [11–13]. The peaks in the $BaTiO_3$ data are due to crystal transitions such as the tetragonal to cubic conversion at 120°C.

Frequency, film thickness, and bias can also affect the dielectric constant of both paraelectric and ferroelectric dielectrics by acting through the relevant polarization mechanisms. For instance, for a material to exhibit a constant k value with frequency, the dipole must reverse direction at the same rate for the polarization to remain in synchronization with the field. As the frequency increases, it may outrun the ability of the particular dipole to keep up with the reversals, resulting in the dipole arm being effectively shortened, resulting in a decrease in dielectric constant with frequency. Of the charge storage mechanisms described above, only the ionic motion in ferroelectrics, such as the Ti^{4+} in the BaO_3^{4-} cage, is affected at frequencies below the infrared range. Figure 9.12 shows the ratio of the dielectric constant measured at various frequencies to the value at low frequencies for three ferroelectrics and three paraelectrics [14–16]. While the paraelectrics show no significant decrease and the ferroelectrics show a sharp dropoff, it should be remembered that ferroelectrics may start with such a high k that, even at gigahertz frequencies, they may still have much higher dielectric constants. This fact is important in matching dielectric materials to integrated capacitor applications. The actual degree of this dropoff, where it begins in frequency and, for that matter, the dielectric constant of the ferroelectrics are highly dependent upon crystal structure and orientation. Amorphous $BaTiO_3$ gives a k of only 17, similar to paraelectrics, and also has flat frequency response.

Table 9.7 is a summary of the relative dielectric properties of paraelectrics and ferroelectrics. Ferroelectrics such as $BaTiO_3$, $Pb_xZr_{1-x}TiO_3$, and $Ba_xSr_{1-x}TiO_3$ can exhibit dielectric constants up to three orders of magnitude higher than those of paraelectric materials such as SiO_2, Al_2O_3, Ta_2O_5, and BCB. However, the dielectric properties of ferroelectrics are typically a stronger function of temperature, frequency, film thickness, and bias resulting in significant nonlinearities in their performance. Also, the dielectric constant of some ferroelectrics degrades with time. All of these factors must be kept in mind when determining what dielectric material is right for a specific application.

Table 9.7 Comparison of Paraelectrics and Ferroelectric Dielectrics

	Paraelectrics	Ferroelectrics
k	2–50	Up to 1000's
k vs. T	Little dependence, <500 ppm/°C	Can be highly dependent due to crystal phase transitions and ion mobility
k vs. frequency	Little dependence	Decreases significantly, typically above a few GHz
k vs. film thickness	No dependence since amorphous	Highly dependent due to effects on crystal structure
k vs. bias	No dependence	Decreases with DC bias
Dielectric fatigue	None	k can decrease significantly with cycles and time
k vs. film structure	Little or no dependence	Film must be crystalline
Cure requirements	None	May require up to 700°C in O_2

9.6.2 Sizing Dielectric Areas

Because capacitors are area ruled, the best way to express their value is as capacitance per unit area or *specific capacitance*. The following equation uses convenient units:

$$\text{Specific capacitance in } \frac{\text{nF}}{\text{cm}^2} = 0.885 \frac{\text{dielectric constant}}{\text{dielectric thickness in } \mu m}$$

Table 9.8 shows some performance data for common dielectrics.

Figure 9.13 shows the length required for the side of a square planar capacitor to provide a given amount of capacitance for various dielectric materials. The x axis is the total capacitance of the structure, not the specific capacitance, and the y axis is the required length of one side of the square plate of dielectric material in mils. Since capacitors are area ruled, the lines have a slope of $\frac{1}{2}$ on log–log coordinates. A few representative dielectrics are shown for some practical thicknesses. Other dielectric materials of known dielectric constant and thickness may be interpolated into the diagram to show their required sizes. The four horizontal dashed lines represent the areas of common surface-mount components along with a 10-mil keep-away distance. For comparison purposes, these surface-mount components were converted to square areas so that the "square plate width" for these units is the average

Table 9.8 Specific Capacitance Achievable from Various Dielectrics

Dielectric	Dielectric constant	Thickness (μm)	Specific capacitance (nF/cm²)
BCB	2.7	2.0	1.2
SiO_2	3.7	0.2	16
Ferroelectric powder in epoxy matrix	45	5.0	8
SiO	6	0.2	27
Al_2O_3	9	0.2	40
Ta_2O_5	24	0.2	110
TiO_2	40	0.2	180
Barium titanate	∼2000	1.0	1800

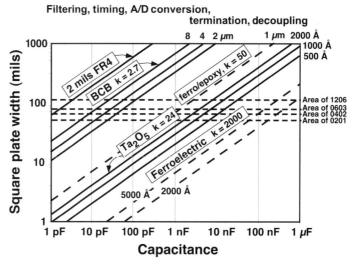

Figure 9.13 Square plate sizes required for various embedded capacitor technologies.

side length required. It is clear from this diagram that embedded capacitors do not necessarily have a smaller footprint than their surface-mount counterparts. For instance, using 2000 Å of a paraelectric such as Al_2O_3 or Ta_2O_5 would result in a smaller footprint only for capacitor values below a few nanofarads. For micron thicknesses of benzocyclobutene (BCB), polyimide, SiO_2, or SiN, the crossover is around 10 to 100 pF. However, embedded passives are placed below the surface of the interconnect substrate so, no matter their footprint, they take up no surface area, enabling the system to either be smaller or to have more ICs.

It can also be seen from Figure 9.13 that embedded capacitors do not necessarily have a smaller footprint than their surface-mount counterparts. For instance, using 2000 Å of a paraelectric such as Al_2O_3 or Ta_2O_5 would result in a smaller footprint only for capacitor values below a few nanofarads. For micron thicknesses of BCB, polyimide, SiO_2, or SiN,

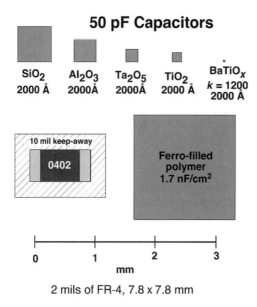

Figure 9.14 Relative sizes of embedded and surface-mount capacitors to give 50 pF.

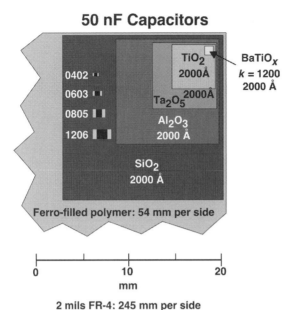

Figure 9.15 Relative sizes of embedded and surface-mount capacitors to give 50 nF.

the crossover is around 10 to 100 pF. Figures 9.14 and 9.15 show the relative sizes of planar embedded capacitors and surface-mount discretes for values of 50 pF and 50 nF, respectively. The surface mounts are shown with a 10-mil keep-away distance around them. The median size of a capacitor in a cell phone is about 1 to 10 nF.

9.6.3 Dielectric Materials Used in Capacitors

The dielectrics described above can be formed by sputtering, various forms of CVD, anodization, sol–gel, hydrothermal techniques, and many others. All of the commercialized dielectrics, to date, can be categorized into four classifications:

Thick-Film Paraelectrics These are layers of polymer typically 5 to 50 µm thick that can be used in a single layer within a PCB or stacked as parallel plates in a surface-mount enclosure. Dielectric constants are relatively low, from about 2 to 6. These modest k factors, combined with relatively thick layers, result in specific capacitances that are the lowest of the four categories, ranging from only 0.07 to 0.3 nF/cm^2. Since polymers are paraelectric, the resulting capacitance is quite stable with regard to temperature, frequency, and the like. Processing is easy and familiar to PWB shops and, of the four groups, probably the cheapest. But, specific capacitance is so low that it will be useful for embedding only the smallest valued capacitors on the board. Using this technology, a 10-nF capacitor would require large areas, between about 33 and 140 cm^2. Specific capacitance will not get much higher for this class of materials.

It is also possible to make polymeric dielectrics through coating and curing liquid resins, resulting in layers that range from 2 to 5 µm. Dielectric properties are similar to thick-film polymers, of course, but with higher specific capacitances, up to around 1.5 nF/cm^2. Spin-on BCB has been used for this purpose on an industrial scale to replace low-valued capacitors with embedded versions. This type of buildup processing is becoming increasingly familiar to board shops although it will never completely replace lamination.

Thin-Film Paraelectrics These are nonpolymer materials such as SiO_2 ($k = 3.7$), SiN_x ($k = 8$), Al_2O_3 ($k = 9$), and Ta_2O_5 ($k = 23$) that are formed by either sputtering or anodizing over an underlying metal to form layers as thin as a few hundred angstroms, much thinner than the polymers just described. These dielectric constants are not vastly higher than those of polymers, but the much thinner films provides considerably higher specific capacitances, from 10 to 300 nF/cm^2. The very thin dielectric also provides the lowest parasitic inductance of the group but may be both electrically and mechanically delicate. Submicron metal oxides may not be able to pass many standard ESD tests or, certainly, the high-pot test and may be prone to cracking and breakage when laminated into materials with different CTEs. Also, the thin films may be very sensitive to underlying roughness with regard to yield, reliability, and value tolerance. Processing requires vacuum equipment that, while familiar to the IC industry, is less utilized in PWB manufacturing. There is still much development that can be done with regard to higher k materials and ease of processing. Ta_2O_5 and Al_2O_3 are commonly used in surface-mount capacitors in the form of very fine anodized powders or anodized metal films. They employ self-healing techniques to remove dielectric defects as they form, enabling the use of very thin submicron films with good breakdown voltage and reliability.

Ferroelectrics The highest specific capacitances by far are achievable with pure ferroelectric dielectrics. However, they require firing temperature in excess of 500°C in pure oxygen to achieve the high k phases, making it impossible to process them directly onto organic boards. This is not a problem in low-temperature cofired ceramic (LTCC) substrates and in ceramic surface-mount packages. Considerable development is underway at various companies and universities on ferroelectric films with specific capacitances exceeding 10,000 nF/cm^2, and this seems achievable. They do show a high dependence on frequency, temperature, and voltage that limits their usefulness in some applications requiring high stability and tolerance such as filters, analog/digital (A/D) converters and timing circuits. They are better suited to pure energy storage where the exact value is less important than having a minimum amount of capacitance.

Polymers Filled with Ferroelectric Particles This is the only type of dielectric developed specifically for embedded capacitors; it is unknown in discrete components. Almost any high k ferroelectric material can be produced in quantity as submicron powders. For example, $BaTiO_3$ can be made by the dry calcination of $BaCO_3$ and TiO2 at $>1200°C$, resulting in particles well under a micron in diameter with dielectric constants in the thousands. These high k particles can be mixed with a polymer resin such as epoxy or polyimide at up to 60 to 80% loading by volume, then screen printed, spun-on, or stenciled onto the substrate, and the polymer phase cured at temperatures quite tolerable to organic boards. Multiple printings can eliminate pinholes.

The mixing rules for a two-phase combination of two materials with different dielectric constants are, unfortunately, such that the dielectric constant of the final composite will be much closer to that of the low k material, which is generally the polymer with a k of about 3 to 5. The overall dielectric constant will end up being around 10 to 50. The result is a film as thin as 8 μm that can be screen printed cheaply, made pinhole free, and delivers up to about 5 nF/cm^2. This method is not suitable for paraelectric powders because the mixture of these much lower k powders with polymer would have about the same dielectric constant of the polymer alone.

The advantage of this approach is that much of the processing, and all of the high-temperature steps necessary to get high k from the ferroelectric phase, can be done in advance of application to the organic substrate. Application is additive and the dielectric is applied only where it is wanted so there is no patterning and little waste. No vacuum equipment is required. Because the films are thicker than sputtered, sol–gel, or CVD, the working voltages are higher, on the order of hundreds of volts. However, screen printing or stenciling is not amenable to tight tolerances, and, once the ceramic loadings approach 85% by volume, which amounts to about 98% by weight, adhesion to the metal electrodes is very poor, resulting in air gaps and lowered capacitance. Its behavior with regard to temperature, frequency, and voltage is typical of ferroelectric materials, showing strong dependencies.

The highest composite dielectric constants reported for commercialized products to date do not exceed 36, and films thinner than a few microns are very difficult to produce with good tolerance, so the maximum capacitance densities are around about 5 nF/cm^2. Future products of this type will probably not exhibit much more capacitance than this, but its combination of ease of application and fair amount of capacitance might make it attractive for a number of applications where low stability with regard to temperature and frequency are acceptable.

9.7 INDUCTORS

The component value for resistors and capacitors depends on both their geometry and the electrical properties of their materials. However, for inductors, the materials are not an issue since they are made only from conductor material, usually Cu. For surface mount this is simply wound wire, and for embedded it is spiral-shaped interconnect metallization. It is the geometry alone that sets the inductor's value and parasitics. Although inductors are the easiest of the passives to fabricate, they are the most difficult to accurately model since their behavior is dictated by complex arrangements of magnetic and electric fields [17–20]. Since inductors are magnetic devices, they pose an integration problem not shared by resistors and capacitors; they perform best when there is a sufficient volume of space to allow their magnetic fields to be unimpeded by other structures. As a result, a keep-away distance is required for inductors, especially when they are embedded within the substrate. This avoidance zone applies not only for the layer they are fabricated on but all layers in order to prevent interference with nearby signal lines and ground planes from the inductor's field. Furthermore, embedded inductors perform poorly over conductive substrates such as doped Si and over dense layers of interconnects due to the magnetic field's interaction with conductive materials. Surface-mount inductors are not as prone to these proximity effects since they are generally some millimeters above the PCB surface and may have ferromagnetic materials built in that concentrates the magnetic fields.

Surface-mount inductors are generally solenoid-style coils, perhaps with a ferromagnetic material within the core to enhance the level of inductance. Often the coiled wire is visible on the component. They are designed semiempirically; the manufacturer simply makes many prototypes with various numbers of windings to get the desired value, which can be as high as microhenries in small packages. Embedded inductors are almost always planar spirals—either circular or square—made out of interconnect metallization. Compared with surface mount, their value range is considerably smaller. Typically, they are something like 1 to 8 turns with a total outer diameter of 0.2 to 2.0 mm, which will provide an inductance of around only 1 to 40 nH and a Q perhaps as high as 50. The area efficiency of these structures is actually quite low compared to surface-mount devices. A size comparison between typical surface-mount sizes and two spiral embedded inductors that were made from 6-mil-wide sputtered-and-etched Cu lines is shown in Figure 9.16.

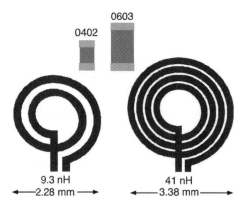

Figure 9.16 Comparison of integrated and surface-mount inductor footprints.

There are four measures that completely specify a simple spiral inductor: linewidth, line spacing, inner diameter, and number of turns. Linewidth and spacing are often set equal to one another. These four numbers set the size and all of the performance factors that can be measured such as inductance, self-resonance frequency, quality factor, parasitic resistance, and parasitic capacitance. There are several empirical and semiempirical equations that can be used to estimate the inductance of simple spirals, most of which use the mutual and self-inductance of concentric rings to approximate a spiral form. These can give results good to within 10 to 20% if there is no interfering metal around and if the substrate is perfectly insulating. These results are usually good enough for rough layout work but, for a more accurate design, full-wave solvers can come to within a few percent while taking into account the presence of nearby metallic structures and the influence of noninsulating substrates such as Si.

Because embedded inductors are fundamentally different from embedded resistors and embedded capacitors, and because they are used in much fewer numbers in typical systems, embedded inductors receive much less attention. However, due to their simplicity, they were one of the first passives embedded. Many small-valued inductors in cell phone filters are now routinely embedded.

There are several strategies for increasing the value of embedded inductors. Ferromagnetic alloys, such as Fe/Ni, can be plated around the windings in order to efficiently guide and concentrate the fields. However, this means increased cost because of considerably more processing and can also result in the addition of parasitic capacitance between the coil and the plated metal. Ideally, the conductors would be completely embedded within the ferromagnetic material, but almost any geometry of core material addition around the conductors is beneficial. Several other topologies, such as solenoids, toroids, and meanders, have also been investigated. Two-layer spirals in series yield approximately 4 times the inductance of only one spiral, twice the Q, but a lower self-resonance frequency, while two-layer spirals in parallel have the same inductance and twice the Q and about the same self-resonance frequency. Both of these can be built with two metal layers, the same as required for just one spiral, with proper via routing.

9.8 ELECTRICAL CHARACTERISTICS OF PASSIVES

A more complete and accurate model of each passive's electrical behavior leads to a better model and subsequent design of the system. For any passive, discrete or embedded, this model should include both the pure value of the ideal component together with its associated parasitics in a proper arrangement. The result will be a simple circuit made up of at least one capacitor, resistor, and inductor in some combination of series and/or parallel to accurately represent the observed frequency-dependent behavior of the component. For a capacitor, that would mean taking into account its resistive and inductive aspects as well. Purchased discrete passives sometimes come with a full set of electrical performance data from their suppliers, enabling designers to create models of systems that will closely match the manufactured article.

In the case of an embedded passive, that particular size, value, and configuration may never have existed before it was designed into a specific product. In the absence of a component spec sheet, it may be necessary to understand what parasitics to expect and how to measure them. For embedded resistors and inductors, the parasitics are not much different from their surface-mount counterparts. But, an embedded capacitor will almost always have much less parasitic inductance than a surface mount. This is a major advantage of embedded capacitors and opens the door for their use in applications such as decoupling where inductance is an important issue. The inductance of the integrated capacitor itself can be so small that it is insignificant compared to that of the leads to it, vias on its leads, and other nearby contributions, while a surface-mount capacitor nearly always has a significant amount within itself. In fact, it can be difficult to even measure the inductance of an embeddable capacitor.

9.8.1 Modeling Ideal Passives

Ideal passives would be characterized by one number: resistance, capacitance, or inductance. The impedance of an ideal resistor is not a function of frequency, while the magnitude of impedance vs. frequency for ideal capacitors and inductors are functions of opposite slope. The electrical models of ideal components are shown in Figure 9.17:

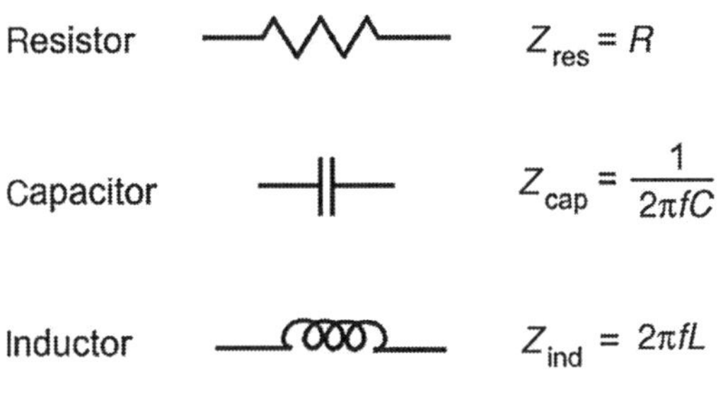

Resistor — $Z_{res} = R$

Capacitor — $Z_{cap} = \dfrac{1}{2\pi f C}$

Inductor — $Z_{ind} = 2\pi f L$

Figure 9.17 Electrical models of ideal components.

where

$Z =$ impedance, Ω
$f =$ frequency, Hz
$R =$ resistance, Ω
$C =$ capacitance, F
$L =$ inductance, H

Plotted together in Figure 9.18, capacitance moves downward and inductance upward with increasing frequency. Capacitors act as zero-order high-pass filters and inductors are low-pass filters. The impedance of an ideal resistor would, of course, be a flat line at an impedance equal to its resistance.

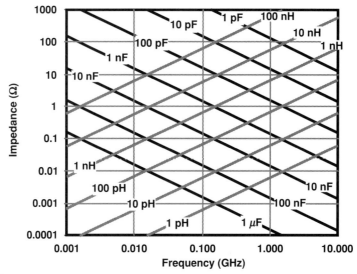

Figure 9.18 Magnitude of impedance for capacitors and inductors.

9.8.2 Modeling Real Capacitors

Figure 9.19 shows a three-component model that is useful for a real capacitor whether it is discrete or embedded [21]:

Figure 9.19 Three-parameter model for a capacitor.

where

C = capacitance, F

R_{AC} = parasitic alternating current (AC) resistance or *equivalent series resistance* (ESR), Ω

L = parasitic inductance or *equivalent series inductance* (ESL), H

The impedance of this arrangement is

$$Z_{cap} = \sqrt{R_{AC}^2 + \left(\frac{1}{2\pi fC} - 2\pi fL\right)^2} \qquad (9.1)$$

This is an effective circuit model to be used in SPICE or other simulations and works well for both discrete and embedded capacitors to match the observed total impedance vs. frequency for a wide range of capacitor sizes, styles, and dielectric materials. The ESR is due to the finite conductance of the top plates, the bottom plates, and the associated leads to the capacitor and represents the resistance seen by a AC signal passing through the component. It is ideally zero. The leakage resistance of the capacitor dielectric is ideally infinite, and the series inductance, ESL, is ideally zero.

The solid line in Figure 9.20 is the impedance vs. frequency for a 100-nF capacitor that exhibits 20 pH of parasitic inductance and 120 mΩ of parasitic resistance. The dotted lines are the behaviors of the three individual components of the model.

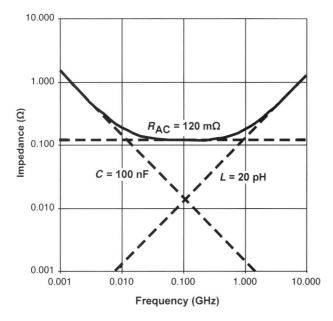

Figure 9.20 Superimposed values of a 100-nF capacitor, a 120-mΩ resistor, and a 20-pH inductor over the measured capacitor impedance.

The point in Figure 9.20 where the capacitive and inductive impedances are equal and the slope changes is the *self-resonant frequency* of the capacitor.

$$f_{\text{srf}} = \frac{1}{2\pi\sqrt{LC}} \tag{9.2}$$

Above this point, the impedance begins to rise with frequency so the overall component is no longer a capacitor at all, it becomes an inductor. For this example, the resonant frequency is 113 MHz. This is also the point at which the phase passes through zero as it moves from $-90°$ toward $+90°$ and is, exactly at $f_{\text{srf}}, 0°$. This frequency is important to almost all applications since it generally marks the upper limit of usefulness as a capacitor.

9.8.3 Differences in Parasitics Between Discrete and Embedded Capacitors

The main difference between the electrical performance of discrete and embedded capacitors is that the embedded capacitors typically exhibit far lower parasitic inductance, which is a major advantage in a wide range of applications [22]. Figure 9.21 shows measured data for a surface-mount and an embedded capacitor, both with equal values of capacitance, 8.5 nF, and equal values of parasitic resistance, 40 mΩ.

The parasitic inductance of the embedded capacitor is almost two orders of magnitude less, giving it about an order of magnitude higher frequency range than the surface-mount component. There are two main reasons that an embedded capacitor almost always exhibits lower inductance than the typical discrete capacitor. The first has to do with the size of the current loop through the component, its interconnects, and its vias. The discrete will have a much larger loop because of the height of the component above to wits contacts to the board, while the entire integrated capacitor is within a mil or two of the level of the metal to which it is connected. The other effect has to do with the reduction of the self-inductance of the capacitor plate structure by mutual inductance. In the conventional surface-mount device (SMD)

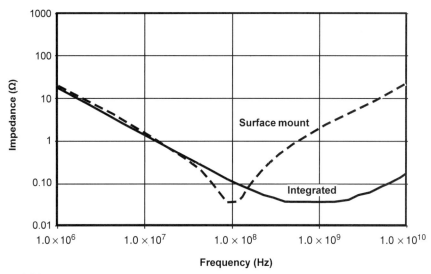

Figure 9.21 Measured electrical performance for a surface-mount and an embedded capacitor.

on the left side of Figure 9.22, current travels from left to right in the plates of both polarities as it travels from the contact on the left to the one on the right. Since the currents are going the same direction, their magnetic fields add, creating parasitic inductance. For the embedded capacitor on the right, the connections to the plates are arranged so that current flows in opposite directions in the plates, so the fields are in opposite polarities, canceling much of the structure's inductance.

For embedded capacitors, it can be shown that the inductance of the structure is directly proportional to the thickness of the dielectric and is best expressed in henries/square just as resistance can be expressed in ohms/square. For a parallel-plate integrated capacitor with the current entering and leaving the same side and the contacts distributed over that entire edge so there is not spreading inductance, the inductance in convenient units is

$$L = 1.26h \tag{9.3}$$

where

$L =$ inductance/square for an integrated capacitor, pH/square

$h =$ plate separation, μm

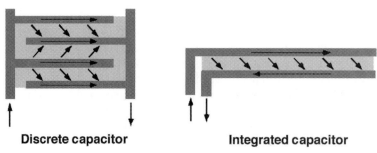

Figure 9.22 Current flow through discrete and embedded capacitors.

For example, an embedded capacitor 0.5 cm wide and 3 cm long made from 0.5 μm of sputtered niobium oxide ($k = 41$) would have a specific capacitance of 73 nF/cm^2, a total capacitance of 110 nF, a specific inductance of 0.63 pH/square, and a total inductance of 3.8 pH. It would have a self-resonant frequency of 250 MHz. The parasitic inductance of a surface-mount part with a capacitance this high would be on the order of 30 to 200 pH, which would result in a correspondingly lower self-resonance frequency. The inductance of the parallel plate part of embedded capacitors can be so low that it is difficult to measure when they are installed in circuit boards because the parasitic inductance of vias and traces that connect to these capacitors may have much more inductance than the capacitor itself.

9.8.4 Modeling Real Inductors

Discrete or embedded inductors follow a similar sort of behavior as capacitors; they act as inductors up to some self-resonance frequency and then, due to capacitive coupling between the windings, lose impedance with frequency [17, 23]. A useful equivalent circuit model is shown in Figure 9.23.

The parasitic resistance is that of the metal windings in either the surface mount or the embedded versions, and must take into account skin effect at high frequencies. The capacitance contains contributions from the windings to any nearby ground planes and between the windings themselves. For embedded inductors, the latter includes the parallel plate capacitance contributions of the metal/metal crossovers and the fringing capacitance contribution between adjacent windings in the spiral. The magnitude of impedance for this model is

$$Z_{\text{ind}} = \frac{1}{\sqrt{\left[\frac{R}{R^2+(2\pi fL)^2}\right]^2 + \left[2\pi fC - \frac{2\pi fL}{R^2+(2\pi fL)^2}\right]^2}} \quad (9.4)$$

The self-resonance frequency, where the impedance turn-around starts, is also given by Eq. (9.2) since they are both derived from the capacitive reactance being equal to the inductive reactance at this frequency. The Q factor is related to the energy efficiency of the inductor and is equal to 2π times the ratio of the energy that comes out of the inductor in one cycle to the energy that went in. Q's under about 20 are considered quite low for most applications. For the model parameters in Figure 9.23:

$$Q_{\text{ind}} = \frac{2\pi fL - 2\pi fC\left[R^2 + (2\pi fL)^2\right]}{R} \quad (9.5)$$

This is typical Q factor behavior, showing a maximum at about 60% of the self-resonance frequency. A curve fit of measured Z vs. f data for a spiral inductor is shown in Figure 9.24 and its Q vs. f is in Figure 9.25. This particular component was a planar spiral with 9 turns of 8 mil conductor for a total outer diameter of 8 mm. The data points are measurements

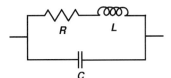

Figure 9.23 Three-parameter model for an inductor.

378 Chapter 9 Discrete and Embedded Passive Devices

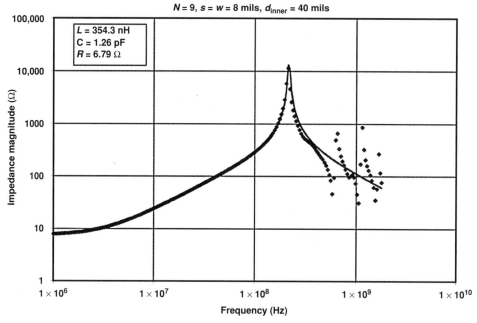

Figure 9.24 Measured inductor performance with curve fit to extract the *LRC* modeling parameters.

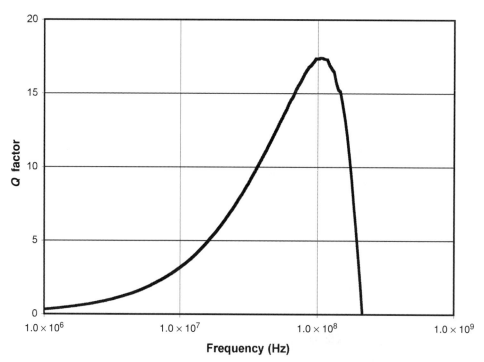

Figure 9.25 Quality factor vs. frequency calculated from the modeling parameters.

from an impedance analyzer, and the solid line is the curve fit with the *LCR* values shown, using Eq. (9.4).

9.8.5 Modeling Real Resistors

The parasitics of discrete or embedded resistors would be capacitance and inductance, both of which are highly dependent on the layout of the resistive material carrying the current, but the phenomenon is too geometry specific to generalize much. The equivalent circuit for an inductor, Figure 9.23, is probably more applicable than that of a capacitor. In geometries where the resistive material winds back on itself as a coil or meandering line, there is the possibility of capacitive coupling between adjacent "windings" that would enable current to jump across from one part of the resistor to the next at high frequency. It would act like a parallel capacitance that would bypass the resistor resulting in a loss of impedance at high frequency. This can be seen in highly meandered embedded resistors or wire-wound discretes. Straightening out the resistive material will remove this problem. Parasitic inductance is not often a problem since most resistors are already fairly high impedance structures to start with.

9.9 ISSUES IN EMBEDDING PASSIVES

The idea of fabricating passive components directly into the PCB dates back to the early 1960s, but industry-wide efforts to make this a standard practice began in the mid-1990s and continues in active development today. Several commercial products for embedding resistors and capacitors are available now, and many others are in development, many of which are described in Sections 9.5 and 9.6. Embedding inductors is not a materials or fabrication issue, so there are no products specifically for that purpose.

9.9.1 Reasons for Embedding Passives

While it is tempting for engineers to look only at the technical aspects, these issues should actually serve as input to business models to determine the ultimate worth of changing technologies. Deciding which engineering option will provide a higher return on each dollar invested cannot be based solely on the goal of lowering the cost to make the same product; the implications are much more nebulous and include concepts such as increased functionality, added product value, higher consumer appeal, and the ability to make products that cannot be realized with the older technology. Some of these issues are difficult to quantify; for instance, what is the added dollar value to a cell phone that is made 20% smaller? There is a real worth to this type of improvement, but it is a function of consumer psychology, which is notoriously hard to enumerate. The merely technical issues often can be quantified fairly accurately. For embedded passives, the reasons in favor of integration can be broken down into the following motivations:

Reduced System Mass, Volume, and Footprint The idea is that individual packages are eliminated, and passives can go underground leaving more room on the surface for ICs. Most embedded resistors and the small-valued end of capacitors are much closer to, or even smaller than, their surface-mount size. However, the real reduction in system form factor with embedded passive technology would come from the fact that embedded passives, regardless of their footprint, can be fabricated under the surface of the board, which frees up top-surface area that was formerly occupied by surface-mount passives. The price paid for this is that additional board layers may be required to accommodate these passives in their embedded form, adding cost and complexity to the manufacturing process, which can offset some of the benefits of freeing up surface area and reducing volume or mass.

Improved Electrical Performance   Due to their simplified structure and lack of leads and contacts, embedded capacitors and resistors tend to have considerably less parasitic inductance than their surface-mount counterparts. This is a major motivation for embedded capacitors especially for decoupling, where multiple surface-mount caps have to be placed in parallel just for the purpose of decreasing their overall inductance.

Increased Design Flexibility   An additional electrical advantage comes about because the value of embedded passives can be specified exactly. If an 18.2-nF capacitor, a 2360-Ω resistor, and a 14.6-nH inductor are needed for a design, the embedded passives can be sized to give those values (within tolerance limits); it is not necessary to choose the next closest value from a catalog of discrete passives or to have to create the values by stringing together discretes in series and/or parallel. Using multiple discretes may take up a large amount of board space, require a large number of lead attach steps, and exacerbates problems associated with parasitics.

Improved Reliability   The use of embedded passives does eliminate two solder joints per passive, which are a major failure point for systems with discrete components. The standard environmental and thermal stress tests will still apply but, while discrete components may be tested on their own to identify failure modes not associated with the solder joints, embedded passives must always be tested in the systems that they will occupy, since the interconnect substrate forms their packaging. Of particular concern is the result of mechanical stresses on large-area thin-film capacitors and resistors brought about by CTE mismatch, layer-to-layer slippage during fabrication, and flexure.

Reduced Unit Cost   Because embedded passives can be formed simultaneously, the incremental cost of producing just one more is nearly zero. This characteristic is attractive for systems requiring dense placement of passives, which is the direction most systems are heading. At present, economic analysis to determine the feasibility of embedded passives is very specific to both the application and the embedded passive technology and is, therefore, very hard to generalize, but there is no doubt that there are considerable savings to be found [24].

9.9.2 Problems with Embedding Passive Devices

Indecision on Materials and Processes   It is not that it cannot be done; scores of research projects have shown that the full required range of R, C, and L values are achievable on just about any substrate, and a couple of dozen products are on the market today that can be used to integrate at least some passives on common FR-4 boards. The problem is that the best materials and processes have not been identified.

Lack of Design Tools   Software is only now emerging that is capable of taking embedded passives into consideration by incorporating them from SPICE-like electrical models, autorouting around them, and optimizing their placement on and among layers. Doing all of this by hand is possible but can be quite tedious, and very few designers are experienced with embedded passives to the point of knowing the pertinent layout issues necessary for taking advantage of their unique electrical and size characteristics. However, progress is being made in this area by design and layout software vendors, and there is no major technological hurdle to enabling these programs to utilize embedded passives effectively and with the same ease of operation as they do for surface-mount boards. The processes must be established first so the software knows the characteristics of the components.

Requires Vertical Integration of Board Shops Companies that previously only made PCBs might now have to get into the business of manufacturing passive components. Not only are the passives integrated into the board, their processing requirements are mixed in as well.

Yield and Tolerance Issues Embedding passives brings a much greater sensitivity to passive yield; one bad component out of many can force the entire board to be scrapped. The problem might not be apparent until the substrate is completed so a considerable amount of fabrication may be wasted if the bad component is formed early on. Rework might be possible but few procedures for this have been developed or reported in the literature. Trimming technology for embedded resistors is just entering the market, but there is no trimming available for embedded capacitors.

Surface-Mount Technology is Improving Meanwhile, surface-mount technology continues to move toward higher board-level densities. On the average, surface-mount passives have gone down by about one case size every 4 years (as opposed to ICs increasing in functional density every 1.5 years). Sizes as small as 0201 (0.50 × 0.25 mm) are in limited use and, although it is hard to imagine them becoming much smaller, there is work in progress for 01005 components. The 0201s may represent the end of cost-effective surface-mount miniaturization due to problems in fabricating, mounting, and inspecting components that small.

Lack of Costing Models The greater the detail necessary to accurately model a system, the less general and more application specific are the results [24, 25]. This is the situation with embedded passives on organic boards. Because the specific processes and materials have not yet been reduced to a small number, it is not currently possible to say under what general circumstances passive integration is and is not economically feasible. No generalized models exist to aid the manufacturer, and all cost evaluations are very application specific. But, application-specific costing can be done, but it requires a knowledge of both the desired architecture of the system, the availability of various passive integration technologies, and the methodology of constructing cost models in general.

9.10 DECOUPLING CAPACITORS

9.10.1 Decoupling Issues

A modern microprocessor might draw an average of a few amps but, if you could inspect the current on a much finer timescale, you would see that it is not nearly constant with time. The IC will draw very large current at the start of each clock cycle, as millions of

382 Chapter 9 Discrete and Embedded Passive Devices

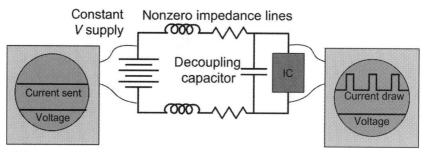

Figure 9.26 Power supply and power distribution system with a decoupling capacitor.

gates change state almost at once, then be almost quiescent for the remainder of the clock cycle. Furthermore, it may demand that this current ramp up with very fast edge rates and, regardless of the current, must always see a nearly constant voltage in order to operate. Most power supplies used in consumer equipment is not capable of this kind of transient performance; they are much happier putting out constant current. And, even if they could provide this sort of power, the inductance and, to a lesser extent, resistance of the PCB's power distribution system would not allow such fast edge rates [26–28].

The solution to this problem is to put capacitors, called decoupling or bypass capacitors, across the power and ground distribution conductors, physically close to the ICs that are demanding the varying current. These capacitors act as short-term low-impedance reservoirs of charge, and supply current that cannot otherwise be supplied by the power supply because of the low-pass filtering action of the parasitic inductances. They are referred to as *decoupling* because they decouple the power distribution system from the current surges required by the IC or "bypass" because they bypass whatever noise is on the power supply conductors to ground. This is shown in basic form in Figure 9.26. Viewed as decoupling capacitors, they act as batteries to run the IC for one clock cycle. In between periods of high current demand, the power supply acts as a battery charger to recharge the capacitor. Viewed as bypass capacitors, they are high-pass filters that short high-frequency noise generated by the IC and prevent it from getting back into the power distribution system.

There are then two tasks for a decoupling capacitor: provide enough charge to run the IC for one clock cycle and be able to provide it as fast as the IC wants it without changing the power/ground voltage by more than a few percent. Using larger capacitance for decoupling helps on the first one, but inductance is the main problem in the second. Resistance can usually be made insignificantly small by adding more metal, but inductance is a geometric phenomena, which is much harder to model and eliminate. While the optimal distribution of capacitors on a PCB is difficult to know with certainty, it has long been understood that the decoupling capacitors placed closer to the ICs should have the lowest inductance. It is for this reason that embedded decoupling capacitors have an important performance advantage over surface mount: embedded caps can have parasitic inductance levels that are almost too low to measure and can be orders of magnitude less than those of discretes.

9.10.2 Decoupling with Discrete Capacitors

A significant problem with decoupling capacitors, no matter where they are located in a power distribution system, is that capacitors are not ideal devices; they have their own

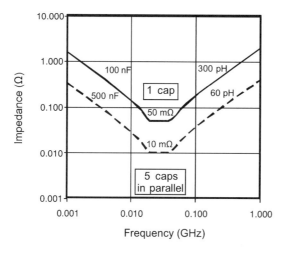

Figure 9.27 Comparison of one decoupling capacitor with five in parallel.

internal parasitic resistance (ESR) and inductance (ESL) as was discussed previously. Because of this, no matter how close the capacitors are to the IC, there is still some inductance and resistance that prevents them from decoupling perfectly. For many discrete chip capacitors of the size normally used for decoupling (10 to 100 nF), the ESR can be hundreds of milliohms and the ESL can be several hundred picohenries. For decoupling, the way to overcome the ESR and ESL limitations is to place multiple capacitors in parallel. All three quantities change favorably; capacitance increases to the sum of the individual components while inductance and resistance decrease. The top curve in Figure 9.27 shows the impedance vs. frequency for a 100-nF capacitor that also has an ESR of 50 mΩ and an ESL of 300 pH, which are representative values for a ceramic chip capacitor. The bottom curve is for five of these identical capacitors in parallel. The effect is to move the entire curve down while maintaining its overall shape. This increases the frequency range that is below a desired impedance, such as 0.10 Ω [29].

It is not unusual, in a complex, multiboard system, to have hundreds of capacitors used for decoupling, both to provide the necessary charge reservoirs as well as to reduce inductive effects. You can see many of these surface-mount caps swarming around a microprocessor, often grouped on the opposite side of the board. All of these discretes can occupy a substantial amount of PWB real estate, contribute to cost, and the solder joints near a hot chip can be a reliability concern.

9.10.3 Decoupling with Embedded Capacitors

Decoupling is an ideal initial application for embedded passives due to their low inductance. It can solve many of the performance problems associated with surface-mount components, frees up board space, and eliminates solder joints. It is difficult to be very specific about how much total capacitance is needed, how the embedded caps should be arranged on the board, and exactly how much inductance is tolerable for each layer of decoupling. That is because these issues are extremely specific to the application and the old rules of thumb for surface mount do not apply.

When decoupling with discretes, much of the capacitance is present only because many capacitors are put in parallel in order to reduce the total inductance. But, due to the inherently low inductance of embedded capacitors, the amount required should be considerably less. The complexity of determining this, particularly with all of the various board, power supply, and IC configurations, probably means that some combination of modeling and experimentation is necessary. Connecting multiple embedded caps in parallel does not further lower their inductance, so a single large-area integrated capacitor can replace the multiple discrete capacitors mounted in parallel, along with all those pick-and-place operations and solder joints. And, surface space near the chip is freed up for other uses.

9.11 FUTURE OF PASSIVES

Of the over approximately one trillion passive devices mounted on organic boards this year, about 95% will be single passive components in a single surface-mount package, a little less than 5% will be surface-mounted passive arrays and passive networks, and very few will be fully integrated into the primary interconnect substrate. The 0402 is the most common surface-mount package at the time of this writing and 0201 is in limited use. The 0201 may become a mainstream size, but implementation problems are proving to be quite difficult and may represent the end of discrete miniaturization. Passive arrays and networks will continue to increase their market share as they replace terminators, filters, and other natural groups of passives. This should top out at some fraction of total passives, probably less than a quarter. The implementation of embedded passives will be an evolution, not a revolution. The use of embedded capacitors for decoupling is a certainty since their parasitic inductance is lower than can be achieved with any surface-mount components. Upcoming high-current and high-speed microprocessors cannot be decoupled any other way. Penetration into commodity boards is an important goal that will be realized gradually as the interrelated issues are resolved. Once cost savings are demonstrated, market share for embedded passives should increase steadily.

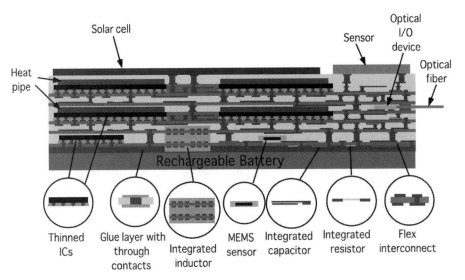

Figure 9.28 Complete integration of an electronic system.

How far can this concept go? Figure 9.28 shows a hypothetical system with passives, chips, and other subsystems integrated together. In this format, all individual component packaging is discarded and the layers of the "board" become the mechanical and environmental protection for each part so that there is very little mass that does not have electrical function. The chips are thinned to fit within one layer of the stack, and the surface is left only for those components that require access to the outside world. The system could hardly be smaller or have shorter interconnects.

REFERENCES

1. R. ULRICH AND L. SCHAPER, *Integrated Passive Component Technology*, Hoboken, NJ: IEEE/Wiley, 2003.
2. R. ULRICH, Moving Embedded Passives from the Lab to the Fab, *CircuiTree*, Vol 16, No. 3, p. 64, March 2003.
3. R. ULRICH AND L. SCHAPER, Putting Passives in Their Place, *IEEE Spectrum*, p. 26 July 2003.
4. *National Electronic Manufacturing Initiative Roadmap*, 2002 edition, Passive Components, p. 1, 2000.
5. R. LADEW AND A. MAKL, Integrating Passive Components, *ISHM '95 Proceedings*, p. 59, 1995.
6. M. LEFTWICH, Between the Layers—A Case Study of PWB Design Options from Current Surface Mount Component Technology to Embedded and Integral Component Technologies, Master's, Univ. of Arkansas, p. 9, 2001.
7. D. LIU, et al., Integrated Thin Film Capacitor Arrays, Proceedings of the International Conf. and Exhibition on High Density Packaging and MCMs, IMAPS, p. 431, 1999.
8. H. KAPADIA, et al., Evaluating the Need for Integrated Passive Substrates, *Advancing Microelectronics*, Vol. 26, No. 1, p. 12, Jan./Feb., 1999.
9. Neugebauer, Resistivity of Cermet Films Containing Oxides of Silicon, *Thin Solid Films*, Vol. 6, p. 443, 1970.
10. R. ULRICH, et al., Comparison of Paraelectric and Ferroelectric Materials for Applications as Dielectrics in Integrated Capacitors, *IMAPS J.*, Vol. 23, No. 2, 2nd Quarter, p. 172, 2000.
11. D. LIU, et al., Integrated Thin Film Capacitor Arrays, Proceedings of the International Conf. and Exhibition on High Density Packaging and MCMs, IMAPS, p. 431, 1999.
12. A. GITELLSON, et al., Physical Properties of (Ba,Sr)TiO$_3$ Ferroelectric Thin Films in Weak Electric Fields, *Soviet Physics Solid State*, Vol. 19, No. 7, p. 1121, July 1997.
13. W. MERZ, The Electric and Optical Behavior of BaTiO$_3$ Single-Domain Crystals, *Phys. Rev.*, Vol. 76, No. 8, p. 1221, 1949.
14. U. SYAMAPRASAD, et al., A Modified Barium Titanate for Capacitors, *J. Am. Ceramic Soc.*, Vol. 70, No. 7, p. C-147, 1987.
15. A. VON HIPPEL, ed., *Dielectric Materials and Applications*, New York: Wiley, p. 300, 1954.
16. H. YOSHINO, et al., Tantalum Oxide Thin Film Capacitors Suitable for Being Incorporated Into an Integrated Circuit Package, IEEE/CHMT '89 Japan IEMT Symposium, p. 156, 1989.
17. E. JONES, Geometry and Layering Effects on the Operating Characteristics of Spiral Inductors, PhD dissertation, Univ. of Arkansas, May 2005.
18. F. W. GROVER, *Inductance Calculations: Working Formulas and Tables*. New York: Van Nostrand, 1946.
19. H. M. GREENHOUSE, Design of Planar Rectangular Microelectronic Inductors, *IEEE Trans. Parts, Hybrids, Packaging*, Vol. PHP-10, No. 2, pp. 101–109, 1974.
20. S. O'REILLY, M. DUFFY, S. MATHUNA, AND S. PAYNE, Routes to Embedded Inductors in MCM-L Technology—Design, CAD and Manufacturing Issues, International Symposium on Microelectronics, 1998.
21. R. ULRICH AND L. SCHAPER, Integrated Passive Component Technology, Hoboken, NJ: IEEE/Wiley, 2003, p. 153.
22. L. SCHAPER AND G. MORCAN, High Frequency Characteristics of MCM Decoupling Capacitors, Proc. 46th ECTC, pp. 358–364, 1996.
23. S. S. MOHAN, M. DEL MAR HERSHENSON, S. P. BOYD, AND T. H. LEE, Simple Accurate Expressions for Planar Spiral Inductances, *IEEE J. Solid-State Circuits,* Vol. 34, No. 10, pp. 1419–1424, 1999.
24. P. SANDBORN, B. ETIENNE, D. BECKER, Analysis of the Cost of Embedded Passives in Printed Circuit Boards, Advanced Embedded Passives Technology website: aept.ncms.org/papers.htm, 2001.
25. M. SCHEFFLER, et al., Assessing the Cost-Effectiveness of Integral Passives, *Microelectronics Intl*. Vol. 17, No. 3, pp. 11, 2000.
26. R. ULRICH AND L. SCHAPER, Decoupling with Surface Mount Capacitors, *CircuiTree*, Vol 16, No. 7, p. 26, June 2003.
27. W. BECKER, et al., Modeling, Simulation and Measurement of Mid-Frequency Simultaneous Switch Noise in Computer Systems, *IEEE Trans. Components, Packaging Manufact. Technol.*, Part B, Vol. 21, No. 2, May 1998.
28. L. SMITH, et al., Power Distribution System Design Methodology and Capacitor Selection for Modern CMOS Technology, *IEEE Trans. Adv. Packaging*, Vol. 22, Issue 3, p. 284, Aug. 1999.
29. R. ULRICH AND L. SCHAPER, Decoupling with Surface Mount Capacitors, *CircuiTree*, Vol 16, No. 6, p. 22, June 2003.

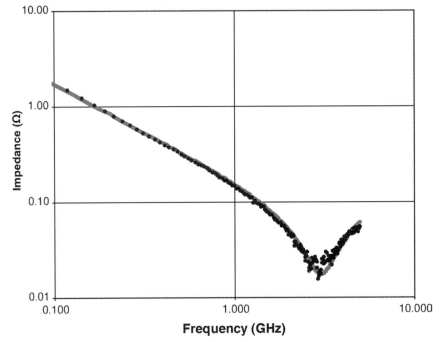

Figure 9.29

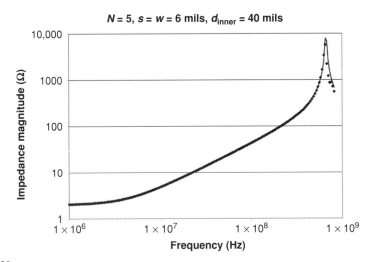

Figure 9.30

EXERCISES

9.1. How big, in millimeters is a surface-mount component with a size code of 0402?

9.2. What is the maximum density in components/cm^2 of 0402 passives you could put on a board with a 10-mil keep-away distance? Keep-away distance is edge to edge of the components.

9.3. What is the resistance in ohms of an integrated resistor when the resistor material is TaN_x with a resistivity of 180 µΩ-cm that is 400 Å thick and is a total of 6.5 mm long, 4 mils wide, and has 6 square 90° bends?

9.4. Design a 5-kΩ meandered resistor made from a material having a resistivity of 420 µΩ-cm using that is 600 Å thick and 6 mils wide. Make the spacing of the meanders equal to the width and make the entire footprint as close to a square as possible.

9.5. What is the percent variation in resistance over the standard military operating range of −55 to +125°C for a resistor that has a TCR of 120 ppm/°C?

9.6. What is the capacitance of 3 cm² of 2000 Å Ta_2O_3?

9.7. What is the highest value of capacitor in nanofarads you could make with the same footprint as an 0603 (without the keep-away distance) using a 2000-Å-thick film of aluminum oxide?

9.8. Typical gate oxides on ICs use silicon dioxide and have a specific capacitance of around 80 nF/cm². What is the dielectric thickness?

9.9. At 1 GHz, what is the impedance of a: (a) 50-nF capacitor and (b) 50-nH inductor?

9.10. A capacitor and an inductor in series show a self-resonance frequency when their impedances are equal. If a 150-pH inductor and a 10-nF capacitor were placed in series, what would their individual impedances be when this happens and at what frequency would this happen?

9.11. For the capacitor in Figure 9.29, what are the *CRL* components of its electrical model?

9.12. For the inductor in Figure 9.30, what are the *CRL* components of its electrical model?

Chapter 10

Electronic Package Assembly

TARAK A. RAILKAR AND ROBERT W. WARREN

10.1 INTRODUCTION

The electronics industry has evolved and come a long way since 1947, when John Bardeen, William Schokley, and Walter Brattain invented the transistor, which was followed by Jack Kilby's device integration to realize the first known integrated circuit. Since then, on-chip, as well as off-chip, technologies have evolved significantly. Packaging technologies needed to make these and similar devices usable have evolved, however, the needs of different types of packages and applications are usually very different. Assembly and manufacturing of such advanced packages, too, have also evolved, especially with consumer pressure to do so in high volumes and at low cost. Topics in this chapter apply to high-volume manufacturing of laminate and ceramic-based packaging and are aimed at offering a glimpse at some of the assembly technologies to give the reader a flavor of some of the challenges and joys of advanced packaging. Figure 10.1 shows a block diagram of a generic assembly process flow to help readers who are new to the field get a fair idea of the overall processes involved.

10.2 FACILITIES

Practically all of advanced electronic assembly—automated or otherwise—needs a controlled environment where the dust and particulates, ambient temperature, and humidity are maintained within a predetermined range. Such control is essential to enable and assure high manufacturing yields at optimal device performance. Automated assembly equipment is used to enable high-volume manufacturing, help reduce undesirable product variability, and reduce the product cost. The following sections address some of these aspects in more detail.

10.2.1 Cleanroom Requirements

The definition of *clean* usually depends on the application and its specific needs. In the context of semiconductors, *cleanroom* refers to a controlled environment where the levels of dust particulates, microbes, and contamination of all kinds are reduced and controlled, along with control of the ambient humidity. Such environmental tailoring is typically realized by

---

Advanced Electronic Packaging, Second Edition, Edited by Richard K. Ulrich and William D. Brown
Copyright © 2006 the Institute of Electrical and Electronics Engineers, Inc.

390 Chapter 10 Electronic Package Assembly

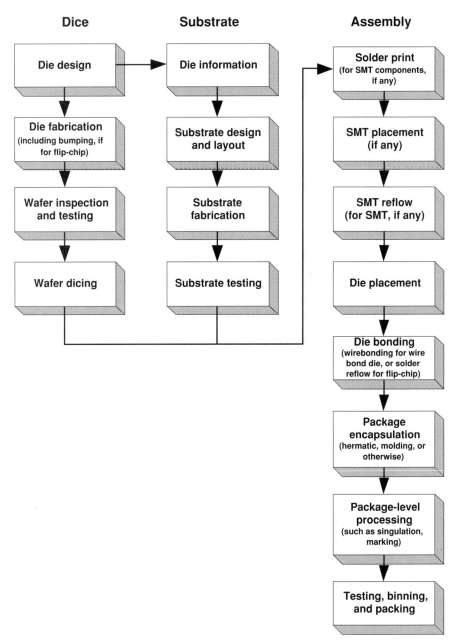

Figure 10.1 Schematic package assembly process flow.

continuously circulating the cleanroom enclosure with forced air circulated through high-efficiency particulate air (HEPA) filters. Such filters can block over 99.99% of particles at a mean particle size of about 0.3 μm from entering into the cleanroom. Two different configurations are possible with unidirectional flow-type cleanrooms: horizontal flow and vertical flow. For horizontal flow, HEPA filters are placed in a wall from where clean air is forced out from one side of the enclosure to the other. In vertical flow, HEPA filters are located in the ceiling and push clean air toward the floor. In either case, air is forced through the filters at a linear rate of about 100 ft per minute, traveling in a uniform, pistonlike fashion known as unidirectional airflow. The airflow is relatively unobstructed, flowing in a straight path and is eventually released to the exterior.

The contamination control industry follows ISO (International Organization for Standardization) guidelines for contamination control [1] to provide a qualified and standardized method for measuring how clean the air is in a cleanroom. Six classes have been established to designate cleanroom cleanliness:

- Class 100000
- Class 10000
- Class 1000
- Class 100
- Class 10
- Class 1

The class number refers to the maximum number of particles, greater than or equal to one-half of a micron, that is allowed in a volume of one cubic foot of cleanroom air. Thus, a class 100 cleanroom, for example, shall not contain more than 100 particles bigger than half a micron in a cubic foot of air. Human users and human activity are some of the most important sources of contamination [2]. Human contributions typically come from skin, hair, dirt, and clothing particulates. The use of cleanroom garments and protocols help in significantly reducing the contamination levels arising from these factors. Cleanroom garments are made of very tightly woven fabrics that are designed to filter human particles and keep them from entering the cleanroom. Hoods for the head help in containing the contamination arising from hair and/or skin flakes. The use of face masks helps in reducing contamination from facial hair, such as moustaches, beards, eyebrows, and eye lashes. Most cleanrooms discourage or disallow the use of cosmetics of any form. The use of vinyl gloves helps in reducing contamination from hand grease. Although body grease does not contribute much to the particulate count in a cleanroom, the use of gloves does improve the overall state of its cleanliness. The use of special cleanroom booties or footwear is mandatory for environments supporting class 10000 or cleaner environments. In cleanrooms with less stringent ratings, it may suffice to wear a hood and an open bottom smock. Further, depending on the type of work being performed inside the cleanroom and any existing cleanroom policies specific to the facilities, additional attire may be needed. Examples include the use of special acid-resistant gloves and a safety apron when handling wet chemicals or the use of protective eyewear when working with lasers or lithography tools. Following appropriate guidelines for the use, storage, and disposal of wet chemicals is critical in ensuring not only the satisfactory utilization and maintenance of the cleanroom environment but also in avoiding preventable accidents.

It is always good practice to avoid using bare hands when handling anything in a cleanroom. Bare hands not only may contain body oils, skin flakes, or other dirt that can contaminate whatever is being handled but may also severely damage or destroy static-sensitive parts. Additional details about cleanroom design, usage etiquette, and operation may be found in the literature [3, 4].

10.2.2 Electrostatic Discharge Requirements

Simply stated, static electricity is the static form of electricity. This form of electricity is generated due to an imbalance of electronic charge on the surface of materials with poor electrical conductivity. Electrostatic discharge (ESD), which is defined as the transfer of charge across two or more bodies that have a different electrostatic potential, is measured in units of volts. While there are not many useful applications for harnessing static electricity, ESD can have severe adverse consequences, particularly in microelectronics where it can affect production yields, reliability, quality, and, hence, the production cost.

Electrostatic charge is most commonly created by the contact and separation of two or more materials. Electrostatic discharge refers to the phenomenon of transfer of charge across objects at different electrostatic potentials. Typical scenarios where electrostatic charge can be generated include the sliding in or out of an electronic device from a bag or magazine or a person walking across the floor [5]. ESD can upset the normal operation of an electronic system, thereby degrading or even destroying it. Also, the presence of ESD in cleanrooms can cause charged surfaces to attract dirt and contaminants, which are difficult to remove. Such contamination can cause a reduction in the yield of package assembly processes.

Electrostatic discharge damage is usually caused by one or more of the following mechanisms. Electrostatic discharge can occur to the device, from the device, or there can be field-induced discharges. The extent of damage caused depends on the device's ability to withstand or dissipate the energy of the discharge. This ability is called the *ESD sensitivity* of the device. Several electronic components are vulnerable to ESD damage at low to moderate potential difference levels. Many disk-drive components have sensitivities of the order of 10 V or less. With trends toward miniaturization of devices, the ESD challenges become even more critical.

10.2.3 Moisture Sensitivity Level (MSL) Requirements

The presence of moisture in electronic packages is usually undesirable for several reasons. Residual moisture trapped inside an electronic package or that ingresses later can cause severe damage to the functionality or reliability of the device. The Joint Electron Device Engineering Council (JEDEC) and the Institute for Interconnecting and Packaging Electronic Circuits (IPC) have developed and maintain standards, that are not just accepted and followed by the industry but were contributed by it. The latest revisions of these standards can be found at the JEDEC website [6].

The current moisture sensitivity standard contains seven levels, as shown in Table 10.1. The purpose of this standard is to classify electronic packages by their sensitivity

Table 10.1 JEDEC Moisture Sensitivity Level Classification[a]

	Floor life		Soak Requirements			
			Standard		Accelerated	
Level	Time	Conditions	Time (h)	Conditions	Time (h)	Conditions
1	Unlimited	<30°C/60% RH	168	85°C/85% RH		
2	1 year	<30°C/60% RH	168	85°C/60% RH		
2a	4 weeks	<30°C/60% RH	6962	30°C/60% RH	120	60°C/60% RH
3	168 h	<30°C/60% RH	1922	30°C/60% RH	40	60°C/60% RH
4	72 h	<30°C/60% RH	962	30°C/60% RH	20	60°C/60% RH
5	48 h	<30°C/60% RH	722	30°C/60% RH	15	60°C/60% RH
5a	24 h	<30°C/60% RH	482	30°C/60% RH	10	60°C/60% RH

[a]www.jedec.org.

to moisture at a fixed reflow temperature, thereby enabling the user to exercise the appropriate precautions for storage and handling during device assembly or rework.

10.2.4 Reflow Temperatures

Packages are usually classified to the lowest level for moisture sensitivity, but to the highest reflow temperatures possible. This enables the assemblers, or the original equipment manufacturers (OEMs), a wide process window during the second-level assembly operations and also helps minimize damage caused by moisture to components and/or devices. Some of the common reflow temperatures are 220, 225, 235, and 240°C. Recent lead-free assembly requirements on packages and an identified need to widen the process window across established, as well as upcoming, assembly infrastructures, have generated the need for reflow at 250°C, as well as at 260°C. With everything else kept constant, an increase in the reflow temperature requirements usually causes a drop in the MSLs achievable by assembled devices. This is due to increased pressure exerted by any residual moisture at higher temperatures and the plasticity or softness of assembly materials, including the solder.

10.3 COMPONENT HANDLING

Depending on their needs and applications, electronic devices and components can be packaged to protect them from moisture and static buildup.

10.3.1 Shipping

Care should be exercised, not only during shipping of parts to protect them from static electricity, moisture, and damage, but also during the loading and unloading of such shipments. Components, particularly those that are static sensitive, should be packed and unpacked at an ESD-safe workstation. Personnel handling the components should wear the proper antistatic attire. The work surface should be grounded appropriately so that the surface is static dissipative. Often, a grounded, static dissipative mat is used to cover a conductive surface. This reduces the possibility of rapid ESD. Similar care needs to be taken during unpacking of components.

Static-sensitive devices are usually shipped in some form of conductive packing, such as an aluminum lined bag (which also are often vacuum sealed to keep out moisture). Other alternatives used for packaging for shipment include the use of carbon-filled polymer foam.

10.3.2 Storage

The storage of electronic products can be broadly classified as short term (typically up to 1 year) and long term (1 to 5 years). Short-term storage may be required in order to stock an adequate amount of inventory. Long-term storage is generally required for products or equipment that are no longer manufactured or are no longer supported by the manufacturer's maintenance organizations. In either case, electronic components are packaged in moisture barrier bags (MBB), together with a desiccant to trap moisture, and a humidity indicator card (HIC) to measure the extent of moisture ingress inside the sealed MBB. Moisture levels are usually kept low—typically in the 40 to 60% relative humidity (RH) range—within the MBB. Sealed bags should be stored in a low-humidity environment, such as nitrogen

Figure 10.2 Humidity indicator card.

cabinets. Products thus packed are generally good for a year, provided the integrity of the MBB is not compromised. Parts intended for long-term storage should be periodically inspected, about once every year, to ensure that the moisture levels inside the MBB are adequately low (this can be verified by looking at the HIC—see Fig. 10.2). The electronic devices should then be sealed in new MBB bags with fresh desiccant material and a new HIC card.

Devices that are designated for assembly after moderate to long-term storage should be baked to remove moisture and may need to be examined for oxidation/corrosion of the metallic interconnects. Intermetallic growth can be an issue for some stored devices, especially if the storage temperature is uncontrolled. The storage area should also be free as possible from vibration since even small, but consistent, vibrations can cause a variety of damage when sustained over extended periods of time.

10.3.3 Handling/Processing

Apart from the need for adequate care while handling all electronic components in view of their small size, some components have to be protected from high-voltage static charges and/or moisture since these can be fatal for the device. Static electricity and charges are easily generated by friction between two or more insulating materials, and the charges are usually retained on the surface due to their low bulk conductivity. Unlike the current that can kill humans, it is the voltage that can be detrimental to electronic devices. Several devices or device components can be damaged by voltages as low as 200 V or lower—see Table 10.2 for some typical breakdown voltages [7]. Consequently, it is important to ensure that the device handler does not carry or transfer a static charge to the device. This can be ensured by controlling the ambient humidity, using an ionizer, or by placing a conductive sheet across the work area.

Reducing the relative humidity directly reduces the dissipation of accumulated charges, thereby increasing the possibility of generating static electricity. Hence, the relative humidity is typically maintained in cleanrooms in the range of about 40 to 60% by using a humidifier, particularly during the winter season when the ambient is generally dry. Circumstances

Table 10.2 Typical Minimum Damaging Voltages for Some Common Electronic Devices [7]

Device type	Minimum damaging voltage (V)
VMOS	30
MOSFET	100
EPROM	100
JFET	140
Operational amplifier	190
CMOS	250
Schottky diodes	300
Film resistors	300
Bipolar transistors	380
SCRs	680
Schottky TTL	1000

that make it difficult to suppress the generation of static electricity often require the use of an ionizer to reduce the impact of static electricity. An ionizer is a device that actively generates suitably, oppositely charged ions to neutralize static electricity in a local area, such as a loading station. Finally, an electrically grounded work surface and a similar wrist strap can provide a conductive path for any accumulated charge to leak away and avoid a charge buildup. It is also good practice to dissipate any charge buildup by momentarily making contact with a grounded metal surface just before handling sensitive devices.

10.4 SURFACE-MOUNT TECHNOLOGY (SMT) ASSEMBLY

System-in-package (SIP) modules are becoming increasingly popular due to their ability to integrate functionality in a very small amount of space and in a very cost-competitive manner. SIP modules take active and passive components and integrate them within the same package. Discrete passive components, in the form factor of surface-mountable devices (SMDs), can be mounted into packages using epoxy attachment or solder attachment assembly processes. Epoxy adhesive or solder paste is screen printed or dispensed onto termination pads on the substrate. The amount of material applied must be sufficient to form a good reliable joint but not enough to short opposing or adjacent SMD terminations together. After the SMD terminations are aligned carefully and placed in position into the adhesive or solder paste, the assembly is thermally cured (epoxy adhesive) or reflowed (solder) to form a solid electrical and mechanical bond between the device and the substrate. The following provides an overview of the processes required for assembly of SMDs into packages using solder paste.

10.4.1 Solder Printing and Related Defects

The solder paste printing process is a critical step in the assembly of a printed circuit board that contains surface-mounted components. Many of the solder defects seen in as-assembled electronic circuits can be attributed to poor design considerations or poorly printed solder paste. This renders significant importance to the accuracy and repeatability of the stencil

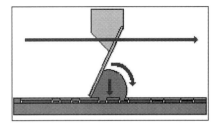

Figure 10.3 Schematic of solder stencil printing.

printing process. In the screen-printing process, solder paste is printed through a metal stencil or mesh screen onto the printed wiring board (PWB) solder pads. There are two main methods used to print solder paste onto PWBs: metal stencil printing and mesh screen printing. For some years now, metal stencil printing has been the dominant method. Since the introduction of laser-cut stainless steel stencils in the beginning of the 1990s, both quality and price have made stencil printing the most commonly used method. Mesh screen printing was used in the very beginning of SMT in the early 1980s and is still the cheapest method, but it cannot be used for fine pitch and small components, such as 0603 size and smaller. During printing, the solder paste on top of the stencil is rolled and pressed into the stencil apertures and onto the PWB solder lands by the movement of an angled squeegee, as seen in Figure 10.3. Modern screen printers are equipped with sophisticated features such as computer control, vision, or laser print control systems, environment control, automatic PWB support setup, and underside stencil cleaning.

Process and material variables that can affect the realization of a reliable solder joint include the stencil thickness, solder print aperture opening (absolute, as well as relative to the pad dimensions on the substrate), spacing between adjacent apertures, size of particles forming the solder paste, and some process variables, such as exposure time to the ambient, and so forth. Solder material particle diameters are typically in the range of 20 to 50 μm, with a varying particle size distribution. The volume fraction, or metal content, can also vary, typically in the range of 86 to 91% metal content [8]. Ingredients of the flux/vehicle system determine the paste print properties, as well as the susceptibility to environmental variations. The flux vehicle system chemistry is generally considered as proprietary information. It is typically made up of resins, solvents, activators, and rheological additives. Specific ingredients and proportions are a function of the nature of the solder type (e.g., eutectic PbSn, high-lead solders, no-lead solders, etc.). Resin-based chemistries form the foundation of most flux systems. It is dissolved in a rosin-soluble solvent whose composition partly determines paste viscosity. Suitable agents that help in reducing the surface tension and enable adequate substrate wetting are added to ensure sufficient dispersion of the solder particles within the paste. The viscosity should be adequately low to allow for consistent and satisfactory solder flow through the stencil aperture holes during the squeegee operation, yet high enough to prevent the paste from slumping or flowing out after it is printed.

Solder wicking, solder splashing, and tombstoning are some of the most commonly seen SMT process defects. Solder wicking is the phenomenon of solder flowing preferentially from one surface to another, either due to poor solderability or due to voids in the surrounding area when the solder is in a molten state and capable of flowing freely. Solder wicking can also occur if there is a significant difference in the reflow temperatures across the two ends of the component. Solder wicking can contribute to a weakening of the solder joints due to solder material migration. In some cases, this can also cause electrical shorts across the components. Solder splashing refers to the boiling and splattering of excessive solder material at a given point. Tombstoning is a phenomenon where a passive component is elevated and removed from the assembled board at one end but remains bonded at the other pad termination. The attached component then essentially "stands" on one of its terminations and causes an electrical open. The quality of solder printing can be as much a design issue as it is a process one. Parameters, such as the ratio of stencil aperture to component pad and the stencil thickness, determine the solder volume dispensed on the substrate pads for subsequent assembly process steps. Misalignment of the deposited solder relative to the substrate pad can significantly affect the quality of component placement and the yield of the SMT process.

10.4.2 Component Placement

In the component placement process, SMD components are placed onto solder paste or adhesive on the PWBs. Three different materials can be used to hold the components after placement: solder paste, staking adhesive, and conductive adhesive. The material used relates to the attachment method chosen for the particular product. However, the placement method is nearly identical in all three cases. Most SMD placement machines move the components from feeders to the placement positions on the PWB using various size vacuum nozzles. Components can be supplied in various forms of feeders, such as tape and reel, stick, tray, or bulk feed.

Components are placed on substrates using automated placement equipment that have placement speeds or CPH (components per hour) typically in the range of 50,000 per hour. Consistent quality levels, high throughput, low placement, and overhead costs are some of the benefits of using automated placement equipment. Placement of components that are either small or those that need to be placed at tight pitches usually make use of vision systems consisting of one or two cameras, along with suitable global or local fiducials, to facilitate accurate component placement. Different types of pickup tools, as well as feeders are needed, depending on the type of component to be placed. Some of the critical factors involved when considering the use of a component placement machine include:

- Placement speed, or the number of components placed per unit time
- Placement accuracy
- Types of components to be placed by the machine
- Types and number of component feeders the machine would support
- Machine footprint
- Other factors, such as compatibility with computer-aided design and computer-aided manufacturing (CAD/CAM) data formats, machine cost, maintenance support, and so forth, could also be important factors

The choice of the appropriate equipment depends not only on the application at hand but also on giving due consideration to future needs. Component placement for smaller devices, such as capacitors, resistors, and inductors, is performed on very fast placement equipment called chip-shooters. Chip-shooters provide extremely fast placement rates with a slight trade-off in placement accuracy over what can be achieved with slower, but more accurate, fine-pitch placement machines. For component recognition and alignment, two main vision system types are available: back lighted and front lighted. On back-lighted systems, components are lighted from behind and the camera only "sees" a shadow of the component. On front-lighted systems, components are lighted from the front so the camera

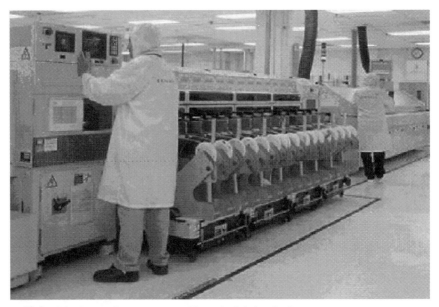

Figure 10.4 Chip-shooter placement system (Skyworks Solution, Inc.).

"sees" actual features on the device. Many SMD placement machines incorporate both types of imaging systems in order to recognize a wide range of component types.

The term *vision on-the-fly* is often used for pick-and-place machine types. In this process, the components are picked up, moved over a camera for recognition and alignment, and then moved to the placement position on the PWB. Some machines use laser or light-emitting diode (LED) systems for alignment of small chips instead of cameras. The components are lighted from one side and a special charge-coupled device (CCD) sensor on the opposite side of the component registers the components size and orientation when rotated on the vacuum nozzle.

The placement accuracy and repeatability of an SMD placement machine is a result of a lot of parameters: accuracy of the moving axis, resolution of vision cameras and laser recognition, vision algorithms, PWB fixation, program coordinates, component data, and the like. SMD placement machines can be fitted and surrounded by a lot of options, including bar-code verification of component part numbers, machine, and feeder performance monitors, co-planarity check systems, and feeder trolleys and automatic nozzle change systems. Figure 10.4 shows a photograph of a chip-shooter in a manufacturing assembly line.

10.4.3 Solder Reflow

Reflow soldering is the process that takes solder from a paste form (a mixture of solder powder, flux, solvent, and activators) to a molten liquidus phase, and then, finally, to a homogeneous solid phase. The reflow process is normally achieved with a conveyorized reflow oven with distinct heating zones and cooling zones. The reflow profile can be characterized by four different phases: preheat, thermal soak, reflow, and cool down. The preheat phase increases the temperature at a controlled rate to prevent thermal damage to the substrate and components. The soak phase brings all of the surfaces of the assembly up to equilibrium temperature and allows the flux to begin reducing metal oxides on the component terminations

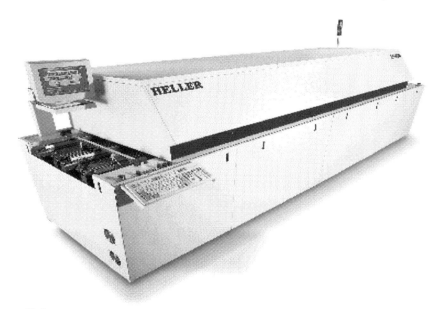

Figure 10.5 Conveyor reflow oven system (Heller, Inc.).

and component lands. The solder reaches its liquidus temperature in the reflow phase, wets the component terminations and lands, and begins forming the intermetallic layers necessary for a good reliable solder joint. The correct cool-down phase allows the solder to solidify with the proper grain structure for good electrical and mechanical properties. Figure 10.5 shows a picture of a conveyor reflow oven system.

10.4.4 Cleaning

Most solder pastes today use either no-clean fluxes that do not require any washing after soldering or water-soluble fluxes that are removed with hot water or hot water with saponifiers. Some forms of cleaning equipment include conveyor-based spray cleaners with wash, rinse, and drying zones and vapor degreasers with immersion tanks and ultrasonic cleaning.

10.5 WAFER PREPARATION

Processed semiconductor wafers, with devices fabricated on them, need to be suitably prepared or postprocessed before they can be used in the assembly process. While the devices can be assembled in a number of different ways, many of the wafer preparation steps described below are common to most assembly configurations. For example, most organizations involved in the assembly business prefer to probe the devices at the wafer level to determine the yields, as well as to identify the known good die (KGD).

10.5.1 Wafer Probing

Wafer probing usually refers to the electrical testing of fabricated devices or chips while they are still in the form of a fabricated semiconductor wafer. During the process of wafer probing,

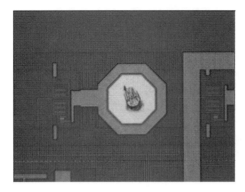

Figure 10.6 Example of acceptable probe damage on bond pads of a wafer for wirebond applications.

a temporary contact is established between the bond pads on the semiconductor device and the electrical contact of the measurement/testing equipment. Usually, the test equipment uses test probes in the form of sharp metallic needles (cantilever based or vertically descending), metallic "pyramids," or flexible and electrically conductive membranes. The specific choice of probe technology to be used usually depends on the pitch and input/output (I/O) count on the device, the frequency range of interest, and any cost/lead time constraints. The test setup enables transmission of electrical signals between the chip and the control unit and analyzes the responses. Dice not meeting the specifications are suitably identified and are not used in subsequent assembly process steps. This way, only KGD are used in assembly. However, as probing can be expensive, it is essential to weigh the cost implications of proceeding with probing to identify KGD versus accepting the assembly fall-outs that are almost certain to result from not probing the die. Probing generally proves cost effective when either the dice and/or other assembly materials are expensive, but the implications have to be carefully weighed when the die cost is not a significant fraction of the total package cost.

Wafer probing can have some undesirable side effects too. During the process of probing wafers, solder bumped or otherwise, the probe tips need to break through any surfacial oxide on the device pads. Since the oxide is often electrically insulating, inadequate pressure on the tips can cause erroneous probe results if the oxide layer is not completely penetrated. On the other hand, applying excessive pressure on the probe tips can either damage the device pads, or the probe tips, or both. Figure 10.6 shows an example of pads on a wirebond device with acceptable damage. Figure 10.7 shows a probed solder bumped device with acceptable damage to the solder bumps.

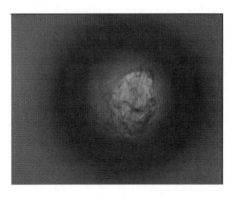

Figure 10.7 Example of acceptable probe damage on solder bumps of a wafer for flip-chip assembly.

Probing flip-chip wafers before solder bumping can cause problems with the wafer bumping process itself. The probe marks on the device pads can interfere with the coverage of the under bump metallurgy (UBM) layer, thereby causing reliability issues. Furthermore, device manufacturers often prefer to probe wafers *after* solder bumping so that yield results can reflect losses due to the wafer fabrication process, as well as that due to the solder bumping process, if any.

10.5.2 Wafer Mounting

Wafer mounting is an important process step in minimizing manual handling of wafers/devices and realizing automation of an assembly line. Wafer mounting enables automated machines to handle wafers safely, while making the process repeatable. Wafer mounting is usually a precursor to process steps such as wafer backgrinding (or thinning), wafer sawing, and the like.

Semiconductor wafers are mounted on suitable frames, which have a stretched, adhesive-coated, polymer tape. Wafers are mounted either using manual, semiautomatic, or completely automatic mounting equipment. In essence, the process involves exposing the adhesive side of the tape by removing the protective polymer layer, attaching the tape onto a ring (usually metallic), and then placing the active surface of the wafer in intimate contact with the adhesive side of the mounting tape. Care must be taken to ensure that the taping process does not allow any contaminants or air bubbles to be trapped between the wafer and the adhesive, as such discontinuities can seriously affect the quality of subsequent processes, such as wafer backgrinding and dicing. The plastic used should be capable of stretching if wafer expansion is to be used. Furthermore, the adhesive layer is required to be robust enough to hold the wafers in place and not allow any of the backgrinding slurry to contaminate the active side of the wafer, yet allow release of the devices during the detaping operation.

10.5.3 Wafer Backgrinding/Thinning

Electronic packages [and other forms of advanced packages, such as optoelectronic, microelectromechanical system (MEMS), etc.] continue to shrink in size, while simultaneously allowing delivery of increasing operating speed, capability, and functionality. Package size reduction has occurred, and continues to occur, not only in the dimensions that determine the package footprint but also in thickness. Reduction in package thickness is a direct contrast to the trend seen in the thickness of the starting semiconductor wafer thickness. For silicon (and silicon–germanium or SiGe), wafer size has seen an increase from sub-6-in. diameter size to the presently fairly common 8-in. diameter, and many wafer fabs are already handling the latest 12-in. diameter Si and SiGe wafers. Increasing the wafer size allows more devices to be fabricated on each wafer, thereby optimizing the process and equipment overhead per device, which reduces the overall cost per device. However, an increase in wafer diameter comes with a corresponding increase in thickness to ensure that the wafer does not warp or buckle due to its own size. Consequently, these wafers have to be suitably thinned to assemble them in thin packages.

The wafer thinning process needs to satisfy at least the following requirements. It must be uniform, that is, the quantity of material removed should be consistent across the entire wafer. The process should not leave stress or damage (scratches, grind marks, etc.). Thinning wafers for flip-chip application poses some additional challenges. In principle, the wafers can be thinned either before or after the solder bumping process. If thinned before the bumping process, the backgrinding process does not have to be any different than that for wirebond wafers. However, most solder bump vendors prefer wafers at their regular thickness since the probability of handling related damage is significantly reduced while working with regular thickness wafers. Also, most of the solder bumping equipment is designed to handle wafers of regular thickness more easily than thinned wafers. Consequently, the assembly logistics have to accommodate a wafer thinning operation *after* the wafers are solder bumped.

Different commercial options are currently available for thinning wafers. One method involves backgrinding the back side of wafers using grinding wheels with a suitable grit size or with a combination of two or more grit sizes. Such grinding machines use a rotating diamond-impregnated wheel in a wet environment to thin, polish, or planarize wafers down to the desired thickness. This method of thinning wafers is well established and works well with wafers for wirebonding. However, for thinning solder-bumped wafers, a special mounting tape is needed that must protect the shape and overall integrity of the solder bumps during the backgrinding process, while simultaneously exerting a uniform pressure on the wafer. This is usually achieved by using a tape whose adhesive thickness exceeds the height of the solder bumps. Since adhesive is sensitive to ultraviolet (UV), background wafers are subjected to UV radiation for a short time (usually 45 to 60 s), which helps in releasing the wafers from the adhesive.

Another way of thinning wafers is to subject them to suitable plasma conditions, which etches the back surface of wafers, thereby thinning them. A commercially available process called atmospheric downstream plasma (ADP) is based on an ion-free chemical etch that thins and removes damage. Essentially, the process is comprised of generating an inert gas thermal plasma at atmospheric pressure, using a direct current (DC) discharge, in the process chamber. A reactant is injected into the plasma stream outside the plasma source. The high temperature of the plasma causes the reactant to decompose, which then uniformly etches the wafers. Using this method, any wafer, with or without solder bumps, with or without grinding damage, can be thinned to yield wafers that are damage free and, hence, mechanically robust. However, as the method tends to be somewhat expensive and slow, the wafers are often mechanically/chemically background to a suitably low thickness first, before subjecting them to ADP.

10.5.4 Wafer Sawing

Once a semiconductor wafer with fabricated devices has been processed (e.g., cleaned, thinned, etc.), the individual devices have to be singulated for assembly. Wafer scribing and wafer sawing are two of the most common techniques used across the semiconductor industry to separate a wafer into individual die. Wafer scribing is more commonly used for singulating brittle materials such as GaAs.

The process of wafer sawing (or dicing, as it is often called) usually begins by washing, rinsing, and drying the wafers. They are then mounted on a dicing tape using suitable automated or (semi) manual equipment. The dicing tape itself is a flexible polyvinyl chloride (PVC) material with synthetic acrylic adhesive bonded to one of its sides. The tape is tough and has high tear strength, as well as a high elongation resistance. Once placed in the dicing machine, the wafer is subjected to cutting wheels that are impregnated with diamond. These wheels cut through the wafers in designated regions called saw-streets (see Figure 10.8). Typically, the saw-street width (also called kerf-width) is in the range of 50 to 200 μm. However, this often depends on several wafer processing factors. Street width is

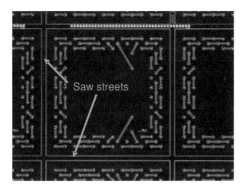

Figure 10.8 Example of saw streets on a wafer with solder bumps.

usually on the lower side of the range for wafer singulation, based on the scribe and break operations. Material type, desired throughput, feed rates, spindle speed, cooling nozzle design, cooling fluid (usually water), type of wafer mounting, blade exposure, diamond particle size, available power, and blade flange design are some of the several factors that affect the sawing operation and its quality. If the wafers are thinned before dicing, the quality of backgrinding, in terms of the defect density on the background surface, can affect the process parameters for the dicing operation. The quality of the dicing process itself is determined by the nature and extent of die chipping and cracking, if any, and the quality of the backgrinding process. In cases where the wafer is likely to suffer chipping, either because of the material brittleness or wafer thinness, a two-step dicing process is used. In this process, the first pass of the dicing blade accomplishes only a partial cut, and uses a relatively thick blade. The subsequent cut uses a thinner blade and cuts through the wafer, as shown schematically in Figure 10.9. Cutting in the crystal plane of the semiconductor is easier than cutting in a direction against a cleavage plane. In a nonmanufacturing environment, it is possible to partially cut the wafer through its thickness and then gently breaking the die apart with little force. However, when cutting against the crystal plane, it is essential to cut all the way through the wafer thickness since the wafer will tend to snap along the crystal plane, when stressed, if it is only partially cut.

10.5.5 Wafer Scribing

The process of scribing and breaking wafers finds its roots in the glass-making industry. Glass workers used these methods to tailor the size and shape of the brittle material. While semiconductor manufacturing has made several refinements in the process of scribing and breaking, the basic principles have remained quite the same, and continue to be the preferred means of singulating devices fabricated on a brittle material such GaAs. The quality of the

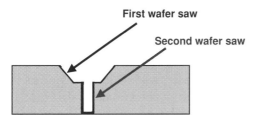

Figure 10.9 Schematic of two-step wafer dicing.

scribing process, as well as that of the subsequent break, depends on the quality of the scribe line generated on the device wafer.

Scribe lines are usually created with a diamond scribe tool. It is essential to minimize debris generation, while keeping the edges on the resultant dice, free of cracks or other nonuniformities. The optimum scribe line leaves a ductile deformation in the surface of the wafer. The deformation leaves a vertical crack in the material. Typically, the deformation is about 1 μm deep and 3 μm wide. The vertical crack extends from the center of the deformation to a depth of about 25 μm. The break operation involves a solid bar making contact and applying pressure to the wafer, or by a roller rolling over the scribed section of the wafer. A stretched adhesive tape supports the wafer to ensure that adjacent die, after the break, do not touch each other. This limits damage due to die chipping or damage caused by the die brushing against each other. Common problems occurring during the scribe and break processes are damage to the die surface caused by scratches on the wafer material from silicon (or any other material that the wafer is made up of) residue. Further, microcracks generated in the street regions during the wafer scribing may propagate into the active area of the die, leading to device functionality or reliability problems. The breaking operation can also cause some yield loss. Small die can suffer from incomplete breaking, and large die are often susceptible to cracks or breakage across the die during the break operation. Yield loss associated with the scribe and break processes is usually higher than for wafer sawing. However, scribe and break offers a higher throughput and is generally more suitable for brittle materials such as GaAs.

10.5.6 Equipment

This section addresses some typical types of equipment used in wafer preparation and handling processes and highlights some of their characteristics.

10.5.6.1 Film Frames and Film-Frame Cassettes

The use of film frames facilitates automation of the wafer sawing process and eliminates the need for manual handling, which can introduce significantly larger variability in the entire process. Film frames are typically metallic. A variety of film frames are available commercially that are compatible with different wafer-saw equipment. Such frames are usually capable of handling wafers having a diameter in the range of 3 to 12 in. The film frames are constructed of hardened magnetic stainless steel and have an electropolished finish. Metal (usually aluminum, but also Teflon) cassettes are available to accommodate about 25 film frames each. Typically, the slot pitch in commercially available, off-the-shelf cassettes is either 0.187 in. (7.5 mils) or 0.250 in. (10 mils).

10.5.6.2 UV Tape Curing System

Ultraviolet-curable tape is often used in the assembly process since it exhibits strong bond strength. This attribute facilitates dicing and reduces device loss incurred due to die flying out during the saw process, which can be a concern, especially for small die. UV-curable tapes, as the name suggests, have to be cured with UV in order to facilitate die release from the tape. As the tape is cured, cross-links are formed in the adhesive material, which help to significantly reduce the adhesive strength for faster die removal. UV-curable tape is particularly useful in backgrinding solder-bumped flip-chip wafers. This tape is fabricated with a thicker adhesive that protects the solder bumps by completely surrounding them during and after the backgrinding process.

Commercially available UV curing systems incorporate a built-in safety interlock, a light-tight enclosure, and an automatic timer. UV power can also be controlled in some systems. Most curing systems allow for the storing of several programs or recipes that can have different exposure settings.

10.5.6.3 Wafer Mounting Stations

Wafer mounting stations can be automatic, semiautomatic, or manual. Most stations handle wafer diameters of up to 8 in. However, some can handle up to 12-in. diameter wafers. Typically, the stations are capable of working with different types of adhesive film frames. The roller assembly is adjustable and a spring-loaded arrangement helps provide for an optimal and consistent wafer–film-frame contact pressure. This helps in realizing an optimum dicing quality. Many equipment manufacturers offer vacuum-assisted chucks to hold the wafer during its mounting. Wafer stages can be programmable and can be heated to temperatures of up to about 80 to 100°C. Fully automatic wafer mounters can handle up to about 25 wafers per process cycle.

10.6 DIE ATTACHMENT

Die-attach materials can perform multiple functions in an electronic package, and die attach is an important step in the assembly process. Apart from the primary function of providing a means of mechanically attaching a die to a substrate, die-attach materials can also provide a thermal path for dissipating the generated heat, act as an intermediate material to partially offset the differences in the thermal expansion coefficients of the die and the substrate, thereby reducing stress at the interface, and provide an electrical reference (usually ground) to the back surface of the die. The importance of reliability of the die attach itself, and of the process, is discussed toward the end of this section, along with some typical characteristics of the die-attach equipment.

Die-attach adhesives can be categorized into two groups, soft and hard. Soft adhesives are comprised primarily of epoxies, polyimides, or organic-based materials filled with thermally conductive fillers. Some soft adhesives are also lead based. Soft adhesives transfer very little thermal mismatch-induced stress to the die and absorb most of the stress themselves. Hard adhesives are highly resistant to fatigue, but they transfer most of the stress to the die, which can cause die cracking. Hard die-attach materials are typically gold-based eutectics or glass-based materials.

10.6.1 Epoxy

The choice of an optimal die-attach adhesive can be critical for the performance of the end product. Epoxy-based die-attach materials are commercially available in different forms, such as snap-cure, fast-cure, and conventional-cure forms. The key element in choosing the most appropriate material for a specific application often depends on requirements, such as the desired throughput, bond performance for reliability, as well as any other thermal, electrical, and mechanical requirements. Process variables, such as the physical properties, application methods, and cure requirements also play a role in determining the best suited material. Most commercial and consumer applications require die-attach materials to meet or exceed the requirements recommended in the industry standards, such as the JEDEC guidelines. For more demanding applications, MIL (military) standards often provide an acceptable baseline.

Important physical properties that need consideration include:

- Material viscosity
- Thixotropic index (this is the ratio of a fluid's viscosity, as measured at two different speeds, with the measurements usually made at speeds that differ by a factor of 10)
- Glass transition temperature
- Moisture absorption
- Shear strength and modulus
- Filler content and type
- Operating temperature ranges
- Pot life

Some of the most commonly used application methods include needle dispensing, stamping or pin transfer, and screen or stencil printing. The application method should be effective, fast, and cheap. Further, the method must not allow air entrapment since voids can cause performance and reliability problems. Dispensing is the most common method since it generally meets these requirements and produces a uniform application of the die-attach material on different substrate types. Stamping or pin transfer allows coverage of large areas with a single application, while printing is preferred when a large number of relatively small areas are to be covered.

10.6.2 Thermoplastics and Thermosets

As noted in Chapter 2, most polymers can generally be grouped into two major categories: thermoplastics and thermosets. Thermoplastics soften or melt when heated and become solid again when cooled. The shape of thermosets, on the other hand, cannot be changed once they are polymerized.

Structurally, thermoplastics are non-cross-linked flexible linear molecules. They can be melted a number of times without inducing any chemical changes. Commercial products with a range of material properties are available. Low-modulus pastes and films help in reducing the thermal mismatch-induced stress at the die-to-substrate interface. Thermosets do not have this reversible phase change behavior. When heated to or beyond the critical temperature, polymerization occurs, causing the thermoset polymers to form chemical bonds across adjacent chains. This forms a three-dimensional network that is significantly more rigid than the thermoplastic structure. These interlinked chains are not free to move, even when thermal energy is available, resulting in the polymer being set into a permanent shape after polymerization.

Thermoplastic adhesives have recently gained wider application in different processes in electronics assembly. These materials now can have higher melting points and excellent dimensional stability. Thermoplastics can operate in practically the same temperature ranges as thermosets. Key advantages of using thermoplastics include their low moisture absorption and low thermal coefficients of expansion.

Another distinct advantage of thermoplastic adhesives is their ability to be used and processed in a dry form. These materials are available commercially in a polymerized form. The bonding process involves softening or melting the polymer while in contact with the adherents, followed by cooling of the assembly. Since there is no chemical transition or change associated with the bonding process, the configuration can be easily reworked by reheating the assemblies. Further, thermoplastics can be manufactured such that they are conductive, thermally and/or electrically, and can be used in the form of a dry film or a viscous paste. The adhesive system can be filled with silver particles when electrical and thermal conductivity is needed. Thus, thermoplastics enable faster processing and assembly with practically unlimited shelf life without the need for unused material refrigeration. However, the additional steps of deposition and drying are required with this approach.

10.6.3 Solder

With customers and applications demanding better MSL (moisture sensitivity level) performances, it is imperative for certain applications to withstand higher reflow temperatures. Molded plastic packages containing devices that dissipate power in the range of about 3 W or more often use soft solder die-attach materials. These configurations are typical for power transistors, thyristors, surface-mount Zener diodes, and other such applications. The dispense step is a sensitive process. The solder volume needs to be optimized by considering the wetting capability of all materials involved. Solder balls can result from an excessive solder volume, which could then short either the component or other components. On the other hand, metal voids can be generated when the solder volume is insufficient. Such voids can increase the contact resistance across the joint or can cause reliability issues during subsequent reflow operations or thermal cycling.

Eutectic tin–lead solder has generally been the solder material of choice for most applications. This solder has good electrical and metallurgical properties, is relatively cheap, and has excellent manufacturability. However, it contains lead and, therefore, is considered to be a high-risk material from an environmental and health perspective. As a result, significant efforts have been and are being directed at developing alternative solder materials that have properties similar to those of eutectic tin–lead. To date, no lead-free material has been developed that is a drop-in substitute for the eutectic tin–lead alloy. Alternatives, such as antimony (Sb) containing solders, have been sought. However, Sb is at least equally as hazardous as lead, if not more. Other, more environmental friendly materials, such as tin–silver–copper (different mix ratios), have their liquidus temperature in the range of about 218°C to about 240°C, with the reflow temperatures being even higher. More information on a variety of material properties of different solder materials may be found at the NIST (National Institute of Standards and Technology) website [9].

10.6.4 Rework

The exact nature and scope of rework depends on the type of component needing rework, as well as the application. This section refers to the rework of second-level subassemblies, namely components or packages that are assembled to a motherboard or a phone board. Such rework proceeds through the following process steps:

- Thermal profiling to melt the solder
- Removal of defective package or circuit component
- Substrate pad repair
- Flux application and solder replenishment

- Placement of good component
- Solder reflow
- Removal of flux residue

During a reflow operation, it is important to avoid any excessive warpage caused by temperature gradients, delamination, discoloration of the surrounding areas, and functional or aesthetic damage.

Semiautomated rework stations usually use a hot-gas-based system. Such a system has two heating mechanisms—one at the top, used for local heating of the component, and one at the bottom that provides heat to the entire assembly/board to minimize the thermal gradient across the board. Most of the rework is performed in an inert ambient, such as a nitrogen, to minimize oxidation of the solder material. Components can be removed and replaced either manually or with the help of an assisted robotic arm. A hot gas nozzle allows for the flow of hot air under the component to be replaced. It is important to optimally heat the components and the board, since an improper profile could cause lifting of the substrate or component pads if the solder has not melted properly before pulling at the defective component. However, excessive heat can damage the adjacent components on the board. Additional information may be found in commercial trade publications and conference proceedings [10, 11].

10.6.5 Die-Attach Equipment

Die-attach equipment can generally be classified by the mode of die attach, such as flip-chip or wirebond. Regardless of the mode of die-attach, most die-attach equipment is capable of handling small and large die, even while dealing with substrate warpage. This equipment has an automatic wafer handling system, which can typically be programmed to pick up die from a magazine for up to 25 different wafer types. Most commercially available equipment is capable of handling die in the range of about 0.15 mm to about 35 mm with a placement accuracy of $\pm 10\,\mu m$ (at ± 3-sigma) at a throughput of up to 4000 units per hour (UPH). Equipment used for wire bonding requires the additional capability for die-attach material dispense. This includes hardware that permits dispensing through nozzles of different shapes and sizes to appropriately handle optimal volumes, as well as software capability that enables the user to define suitable dispense patterns, such as a dot dispense or a cross dispense. Equipment for flip-chip die attach does not need these capabilities. However, they may need a flux substation to suitably apply flux to the bumped die. Further, flip-chip die are picked from the wafers with the bumped side up but have to be placed on the substrates with the bumped side down. Consequently, flip-chip die-attach equipment is equipped with a "flipper" mechanism that flips the die before dipping it in the flux reservoir or placing it on the substrate.

Most die-attach equipment is also equipped with wafer mapping capabilities, which allows programming of the machine to skip bad or defective dice. This can be done either using an ink dot or electronically, where an electronic wafer map is fed into the machine memory that enables the machine to skip defective die or enables it to pick only a certain type of die from a multivariant device wafer. Some machines also allow substrate mapping using a software program that determines the bonding location on the target wafer or substrate for chip-on-wafer applications. Most die-attach equipment has pattern recognition capabilities, which allow better control over die placement or the die-attach dispense process. Different search modes are available, such as edge or feature recognition.

Die ejection from a wafer holder is an important step, particularly for thinned or brittle die since the ejector pins can damage or crack the die during the ejection process. Synchronized die ejection options allow for careful chip handling during the ejection of the die and pick up from the wafer. Also, depending on the application, optional accessories, such as an indexer, and a substrate strip handler may be used for high-volume applications.

10.7 WIREBONDING

Wirebonding is one of the most important and critical manufacturing processes in electronic packaging. The term *wirebonding* refers to the interconnection of the I/O pads of a device with the conductive circuit traces on a circuit board/substrate. Gold or aluminum wires, about 1 mil (25 µm) in diameter, are typically used in the wirebonding process. In the late 1950s, the wirebonding process used thermocompression bonding in which heat and force are used together to produce a solid-phase diffusion bond. Later, the addition of ultrasonic vibration to the wirebonding process made the process significantly more manufacturable and reliable. Wirebonding that incorporates the use of ultrasonics is commonly known as ultrasonic bonding. The combination of ultrasonic vibration (typically 60 kHz) and force scrubs the interface between the wire and the metallized pad, causing a localized temperature rise that promotes the diffusion of molecules across the interfacial boundary and creates a weld. In the case of gold wire, the entire circuit assembly is usually globally heated (100 to 300°C) to further encourage the migration of materials. The addition of globalized heat with ultrasonic energy is commonly known as thermosonic bonding. The vast majority (>95%) of all microelectronic devices are bonded using ultrasonic or thermosonic wirebonding. There are two basic types of wirebonding processes, ball bonding and wedge bonding. The name of each process is derived from the shape of the wire at the point where it is welded to the device.

10.7.1 Thermocompression Wirebonding

Thermocompression bonding and ultrasonic bonding are the two conventional ways of bonding wires to chips. In thermocompression bonding, the die and the wire are heated to a high temperature (around 250°C). The tip of the wire is heated to form a ball. The tool holding it then forces it into contact with the bonding pad on the chip. The wire adheres to the pad due to the combination of heat and pressure. The tool is then lifted up and moved in an arc to the appropriate position on the frame, dispensing wire as required. The process is repeated to bond the wire to the frame, but without forming a ball on the substrate side. This technique cannot be used for bonding devices that cannot withstand high processing temperatures.

10.7.2 Ultrasonic Wirebonding

Ultrasonic wire bonding, along with thermosonic bonding, is one of the conventional ways of wire bonding. This technique relies on pressing together the two bonding interfaces, namely the wire and the bonding surface on the substrate/frame, and subjecting them to an ultrasonic vibration, thereby enabling the formation of the desired bond. Applications where the device is unable to withstand high process temperatures can be interconnected using ultrasonic wirebonding.

10.7.3 Thermosonic Wirebonding

Thermosonic wirebonding partially combines the advantages of both conventional ways of wirebonding, by heating the device (but not as much as in thermocompression bonding), while simultaneously agitating the bonding area ultrasonically. This process is manufacturing friendly, as it produces high yields that are reliable.

10.7.4 Ribbon Bonding

Ribbon bonding is a form of wirebonding that relies on wedge-to-wedge bonding using flattened gold wire that has a rectangular cross section. Ribbon bonding is used in thin- and thick-film circuits as crossovers (connecting a trace over another intersecting trace), and in microwave circuits due to its rectangular shape (e.g., 0.5 mils thick × 10 mils wide) that provides lower inductance and lower skin effect losses. Ribbon wire is typically made of gold but can also be made from silver, copper, or aluminum and is bonded using thermosonic bonding. One of the main advantages of ribbon wire bonding is its higher conductivity and lower impedance. The common areas of application are high-frequency and high-power components, such as microstrip transmission lines and waveguides, along with other active and passive components that operate at frequencies in the range of several gigahertz. Mutual inductance and crosstalk between adjacent ribbon bonds is significantly lower than that of equivalent round wires. Electrical signals at high frequencies are transmitted through the skin effect. For example, the penetration depth of an electrical signal is less than 1 μm at an operating frequency of 10 GHz [12]. Figure 10.10 shows an scanning electron microscopy (SEM) image of a device with ribbon bonds [13].

Because there is relatively little deformation of the ribbon during bonding, there is little scrubbing action to remove oxides and contaminants from the surface. For this reason, it is important to have a very clean surface before bonding to reduce the possibility of low bond strengths.

10.7.5 Ball Bonding

Ball bonding uses high-purity (99.999%) gold wire, typically 0.001 in. diameter. At the beginning of the process, a spherical ball is formed at the end of the wire from a mechanism

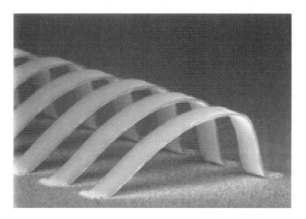

Figure 10.10 SEM image of a device interconnected with a ribbon bond (http://www.kns.com/prodserv/tools-materials/specwire.asp).

10.7 Wirebonding

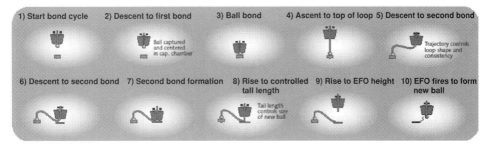

Figure 10.11 Ball bonding process (ESEC).

called an electronic flame-off (EFO) arm. In this process, the EFO arm is rotated into position under the gold wire that protrudes from a ceramic capillary. An electrical arc is formed between the arm and wire that heats the end of the wire, forming a small ball. The capillary descends to the device with the ball captured at its end, and, when contact is made to the bond pad, a combination of ultrasonics and heat is applied, which welds the gold ball to the pad metallization. The first bond is known as the ball bond. The capillary ascends and moves toward and then down to the region of the second bond, forming a tightly controlled wire loop. The capillary makes contact to the second bond pad, welding the wire to the pad and simultaneously shearing the wire off against the side of the capillary. This second bond is known as a wedge bond due to its shape. Normally, ball bonds are placed on the device pads and wedge bonds are placed on the substrate pads, but, in special cases (e.g., to achieve extremely low loop heights), the bonds may be reversed. Figure 10.11 shows a schematic of the ball bonding process.

10.7.6 Wedge Bonding

Most wedge bonding is accomplished with aluminum wire and is accomplished at room temperature. When gold wire is used, heat must be used during bonding. In wedge bonding, the wire is fed under a flat or grooved wedge tool, and the transducer and tool must be oriented in a straight line from the first to second bond before the first bond is initiated. The process of mechanically rotating the transducer or the substrate to line up the first and second bonds for each wire means that wedge bonding speeds are roughly half that of ball bonding. The advantage of wedge bonding is that it can achieve finer pitches and lower bond heights than ball bonding. Another advantage of aluminum wedge bonding is the ability to bond larger diameter (>50 to $500\ \mu m$) aluminum wire for power devices that require more than several amps per wire. Large wirebonds require bonders that are able to provide ultrasonic energies greater than 20 W, versus the 1 W or less used for fine wire. Bonding large wires also requires much higher forces and uses up to 1 kgf versus the 10 to 25 gf required for fine wire. Figure 10.12 shows the wedge bonding process.

10.7.7 Wirebond Testing

Integrity testing of bonded wires is a critical part of a wirebond setup, in-line process control, and final quality and reliability testing. The most common test for measuring wirebond quality is the pull test, which identifies the weakest point in the interconnection.

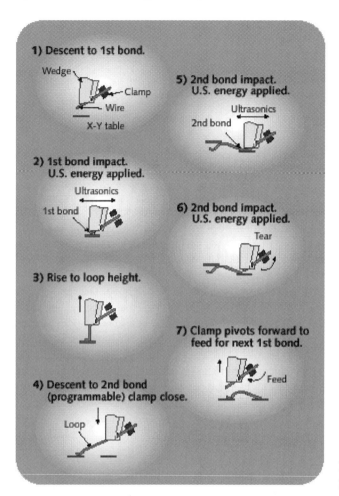

Figure 10.12 Wedge bonding process (ESEC).

Pull testing can be performed in either a destructive or nondestructive mode. In destructive testing, the pull test is performed by inserting a miniature hook in the middle of the bonded wire. The hook is vertically raised at a constant velocity until the wire breaks at its weakest point, at which time the pull tester records the breaking force. The operator then inspects the broken wire and records the failure mode, which could be at either of the bond interfaces, the neck of the bonds, or anywhere along the wire span. The same testing procedure is used in nondestructive testing except that the bond is pulled to some minimum force and the pull is then stopped. This test merely ensures that the bond strength is greater than some minimum force. Figure 10.13 shows a photograph of a wire pull/bond shear tester, while Figure 10.14 shows a schematic of the wire pull setup.

For ball bonds, the weakest point is typically at the neck area just above the ball bond (the heat-affected zone), due to recrystallization of the gold after the electronic flame-off process. A combination of two mechanisms, a large grain size, and bending of the neck above the ball bond, make it the weakest point in the wire. In some cases, wedge bonds can be excessively deformed, causing microcracks at the heel of the bond that will break during pull testing.

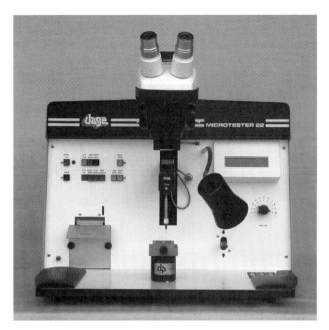

Figure 10.13 Wire pull/bond shear tester (Dage).

Because pull testing does not test the strength of a ball bond effectively due to the large bond areas relative to other regions of the wire, shear testing is often performed in addition to pull testing (see Fig. 10.15). The shear test is used to measure the shear strength of the bond interfaces. Typical failure modes include (mode 2 is preferred):

1. Bond lift—ball separates cleanly from bond pad.
2. Bond shear—ball leaves some gold on the bond pad.
3. Cratering—ball and bond pad lifts, taking portion of underlying dielectric.

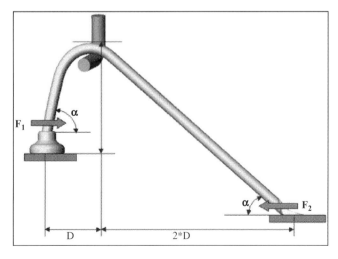

Figure 10.14 Schematic of a wire-pull test setup.

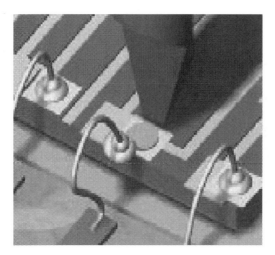

Figure 10.15 Example of ball shear test (Intelliquest).

10.7.8 Tape-Automated Bonding

Tape-automated bonding (TAB) was initially used in ultra-low-cost, very high volume commercial applications in the 1970s. The ability to perform gang bonding of inner bond leads (to the device) and outer bond leads (to the substrate) was an advantage over the much slower serial and manual process of wirebonding. TAB also provided lower pitch capability, lower profile, lower inductance, controlled impedance, and smaller real estate requirements at the PWB level than wirebonding technology. Figure 10.16 shows a schematic of the TAB process.

In the 1980s and 1990s, TAB was adopted because it provided tighter pitch capabilities, a thinner package profile, lower inductance, controlled impedance, and a smaller area at the printed wiring board (PWB) level than wirebonding technology. Since then, though, wirebonding has evolved to a point where TAB no longer holds many of these advantages. However, one advantage that TAB continues to offer over wirebonding is that devices can be pretested and burned-in on the carrier tape to provide only known good devices for final assembly.

TAB technology uses photoimaged and etched conductor patterns on a dielectric carrier tape. The carrier tape is stored on reels similar to a movie film, in widths of 35, 48, and 70 mm. The conductor leads are fanned in from the outer test points, which are typically at 0.050-mm pitch, to the outer lead bonds at 0.15- to 0.5-mm pitch, and then to the inner lead bonds at 0.05- to 0.1-mm pitch. The conductor is held in place by the dielectric tape, with the inner and outer bond leads extending over windows in the tape. The bonding process begins by bonding the inner leads to the chip. The device is maneuvered under the tape until the inner leads are aligned to the chip pads, and then using a thermocompression or thermosonic bonder, the leads are bonded all at once (gang bonded) or single-point bonded one at a time. Once the chips are bonded to the tape, the leads are electrically isolated using a punch tool. The isolated chips are then tested and, if required, burned in. The chips are then excised from the tape frame and, at the same time, the outer leads are formed to their final shape. The chip is then attached to the next level assembly and the outer leads are bonded to the circuit traces. Outer lead bonding is performed using thermocompression or thermosonic bonding or soldering. If soldering is used, a hot thermode is used to make

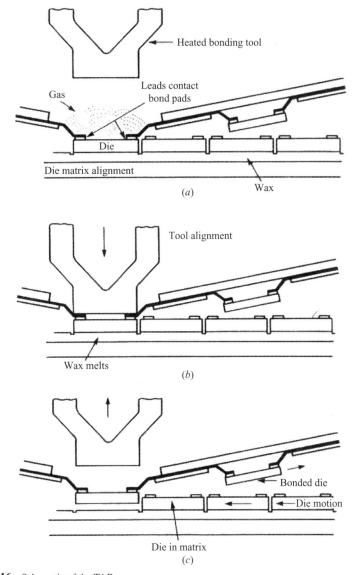

Figure 10.16 Schematic of the TAB process.

contact with one of more rows of outer leads while delivering sufficient heat to reflow solder that is preapplied to the circuit board.

Either the TAB devices or the leads must have gold-plated bumps on them in order to facilitate TAB bonding. Figure 10.17 shows an example of gold bumps on devices used for TAB bonding.

10.7.9 Plasma Surface Treatment

Substrate materials, adhesives, and other advanced materials used in the package assembly process often do not possess the needed physical or chemical properties required for good

Figure 10.17 Gold bumps that enable TAB bonding (MicroFab Technology).

adhesion, and thus, require surface modification [14]. Decreasing package geometries, the use of newer materials, and expectations of satisfactory reliability performance at ever-increasing temperatures are some of the challenges that need to be overcome to facilitate high yields. Plasma surface treatment is one of the popular and effective processes that enable better yields and reliability of electronic packages. Such a treatment results in physical or chemical modification of the outermost molecular layers of the treated surfaces. Plasma surface modification can change the surface through different mechanisms, including the removal of contamination, etching of materials, surface activation, and material cross-linking.

Plasma treatment for surface modification is based on the interaction of plasma-generated species with a surface or an interface. The effectiveness of the plasma process is achieved either by transfer of momentum from neutral atoms (typically argon) to the specimen of interest. Another form of plasma treatment, called reactive ion etching (RIE), utilizes the chemical reactivity of the plasma species. Oxygen, suitably mixed with argon, is the most common composition of RIE plasma systems. However, hydrogen is used in some cases. Since plasma is composed of particles that are positively charged, negatively charged, and electrically neutral, it can interact with a wide variety of materials and contaminants. Contamination usually forms or is deposited from previous semiconductor fabrication or assembly processes, for example, the formation of a metal oxide during thermal exposure or the deposition of epoxy residues from the die-attach process. Assembly processes, such as soldering, wirebonding, and encapsulation can be negatively impacted by the presence of these contaminates. Plasma treatment is used at different stages in package assembly. It may be used before the die-attach process step to improve the adhesion between the die and the ceramic or laminate substrate, before the wirebonding process to improve the "wire-bondability" of the the gold/aluminum wires to the substrate, before the underfill operation in flip-chip assemblies and before the encapsulation/mold process steps to improve bond strength and yields. Plasma cleaning is also found to be effective in removing oxide from exposed metal on the substrate, dice, or solder bumps/balls and returning the system close to its pristine state. Plasma is often used in assembly processes for minimizing, or even eliminating, contamination arising from manual handling of sub assemblies. In the case of metal oxidation, a chemical treatment using hydrogen plasma can reduce copper oxides through the reduction of hydrogen radicals with the copper oxide.

$$CuO + 2H^* \rightarrow Cu + H_2O \tag{10.1}$$

For die-attach epoxy residues, such as resin bleed-out, an oxygen gas plasma reacts with the epoxy and generates gas by-products that volatilize and are then pumped away by the vacuum pump of the plasma system. The addition of argon to the plasma roughens the bond surfaces after the contamination is removed, increasing the surface area of the bond surface and improving bond strength.

Surface activation can occur when certain gases are exposed to a plasma, dissociate, react with the surface and create a new chemical functional group on the surface. This new functional group forms a strong bond with the surface material and to subsequent adhesives, encapsulants, or molding compounds that are applied to the circuit, improving overall adhesion of the materials to each other.

Another form of surface activation is achieved through the use of inert gases to remove atomic species from the surface and generate reactive radicals on the surface. These radicals chemically react with other surface elements, creating a cross-linked surface. This cross-linking improves the adhesion of metal layers or adhesives to the treated surface.

Etching is another form of plasma surface treatment sometimes used to strip materials of coatings and is common in the semiconductor and PWB industries. In optoelectronic assemblies, plasma etching is sometimes used to strip optical fibers of their outer buffer coating prior to processes such as hermetic sealing or pigtailing of laser diodes. Most plasma systems are equipped with a radio frequency (RF) generator that operates at 13.56 MHz. Typical RF power is in the range of 100 to 500 W. The plasma impedance is matched to the 50-Ω output of the generator by an automatic matching unit. Gas flows, pressure, RF power, and duration can be tuned to define specific "recipes" for a given process. The effectiveness of plasma treatments is commonly assessed by measuring the pull strengths of wirebonds performed on processed substrates/dice and comparing the data with their unprocessed counterparts. Such testing enables quantification of the effectiveness of plasma treatments.

Wire pull testing cannot provide details about the underlying mechanism. Contributions to changes in surface energy of the processed parts are measured by performing the *liquid drop test*, which estimates the surface energy by measuring a water drop's contact angle with the surface of interest. Energy dispersive X-ray analysis (EDAX) or X-ray photoelectron spectroscopy (XPS) is used when the chemical nature of plasma-induced surface modifications need to be assessed. Optical microscopy (25 to 1000$\times$) or scanning electron microscopy (SEM) is commonly used for observing the topographical nature of the surface modifications.

10.8 FLIP-CHIP

The flip-chip mode of die attach is another way of realizing an electrical and a mechanical interconnection between a semiconductor die and a carrier substrate. In principle, this involves connecting metallic contacts on the die to a corresponding set of pads on the substrate using an array of solder balls (also called solder bumps, or simply, bumps). The bumps may be distributed either in an area or in a peripheral configuration. The chip is placed face down on a carrier that has a corresponding set of metallized pads. The carrier is typically either laminate-based, ceramic, or a suitable flexible material, such as polyimide. Heat is then applied, causing the solder to reflow onto the substrate pads. Figure 10.18 shows a schematic of the flip-chip mode of die attach.

When properly designed, flip-chip offers the possibility of reducing the die, as well as the package size, enhancing the electrical performance, and lowering assembly costs,

418 Chapter 10 Electronic Package Assembly

Figure 10.18 Schematic of flip-chip mode of die attach.

while simultaneously improving package reliability. Flip-chip assemblies offer lower and more predictable parasitic inductances than their wirebond counterparts. Unlike wirebond assemblies, where the wire-based inductance is directly related to the length of the wire, which has a tendency to increase with increasing die size, packages with flip-chip assemblies offer a constant inductance value for all interconnects on a given die. This allows designers to accurately predict the parasitic components of a circuit, thereby enabling an optimal device design.

Historically, IBM was the first to use what was termed controlled-collapse chip connection (C4) technology, which is now widely known as flip-chip attach (FCA). C4 and flip-chip were revolutionary when they were first introduced because the technology provided high I/O density, uniform chip power distribution, and high reliability. C4 was originally developed for use with ceramic-based chip carriers. The process uses high-lead solder balls (97 wt% Pb with 3 wt% Sn), with a typical solder bump height of about 100 μm. Delco Electronics used its solder bump flip-chip process to build assemblies for automotive applications in the 1980s. Consumer electronic applications, such as mobile phones, usually employ eutectic SnPb solder bumps (63 wt% Sn with 37 wt% Pb). This solder has a eutectic temperature of 183°C, where it melts to form a joint with corresponding pads on the chip carrier. On the other hand, both high-lead and no-lead solders are reflowed at much higher temperatures, with peak temperatures typically around 230°C, or higher. Gold stud bumps have been used by the Japanese watch manufacturers since the 1980s for low-cost applications. Figure 10.19 shows an example of a gold stud bump.

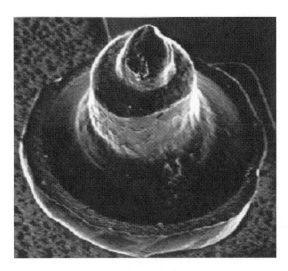

Figure 10.19 Example of a gold stud bump.

10.8.1 Wafer Bumping

Loosely defined, wafer bumping refers to a process for providing a means of electrically and mechanically interconnecting an integrated circuit to a package or interconnect substrate. Wafer bumping can be categorized into two broad types, namely regular bumping and redistribution bumping. Regular bumping refers to a process for wafers (i.e., devices) that is designed for flip-chip applications. A redistribution process, to put it simply, refers to suitably converting an integrated circuit originally designed for wirebonding to a flip-chip configuration. Redistribution essentially repositions contact pads on an integrated circuit (IC) from a perimeter footprint to alternative locations on the chip, thereby allowing for a transition to flip-chip without redesigning the chip. Regular or redistributed bumping can be accomplished using any of several different types of commercially available bumping options. Each bumping technology offers certain advantages but also offers some unique challenges. Some of the commercially available bumping options are discussed in the bumping section of this chapter.

Typically, flip-chip devices, regular or redistributed, have a center-to-center solder bump pad pitch of about 250 μm. The pad geometry, as seen in Figure 10.20, can either be peripheral single row, peripheral multiple rows, or a full area array of bump pads. To convert an originally wirebond device to a redistributed flip-chip device, the wafers are coated with a thin layer of a suitable polymer, typically benzo-cyclobutene (BCB) or polyimide (PI). After applying the polymer coat, it is cured, imaged, and suitably etched to define vias to expose the underlying metal pads. The vias are then filled with metal and metal traces are defined on the BCB layer to redistribute the pad locations to a pitch and a configuration that is more suitable for solder bumping.

Devices that are designed from the beginning for flip-chip application usually do not require redistribution. The flip-chip process involves an under bump metallurgy [UBM, which is also called ball-limiting metallurgy (BLM)], to define the solder-wettable area. The UBM is composed of a wetting layer that enables good adhesion to the top metal surface on the die and a barrier layer that prevents the solder components from diffusing into the device. Then, a UBM wetting layer for the solder is applied. Sometimes, an oxidation barrier is also added to the UBM. The total thickness of all layers forming the UBM is about 1 μm.

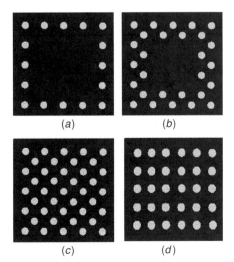

Figure 10.20 Solder bump pad configurations: (a) peripheral single row, (b) peripheral, staggered two rows, (c) staggered full area array, and (d) full area array.

Finally, BCB is used to bump the wafers. The major reason for using BCB for bumping without redistribution is planarization of the wafer topology, allowing for a uniform, planar surface for wafer bumping. This can be a significant advantage to contract bumping vendors, who are likely to receive wafers from numerous fab-houses across the world. Further, BCB is a photo-imageable polymer, which has several properties that make it suitable for electronic packaging applications, such as:

- Low dielectric constant
- Spin-dispensable, making the process more manufacturable
- Low moisture uptake

10.8.1.1 Solder Bumping

Wafer bumping is a process that creates solder bumps on a die surface to provide for electrical connection from a semiconductor chip to a substrate or chip carrier. The solder bumps also provide mechanical support for the die and, in some cases, provide a path for thermal dissipation of the heat generated within the die.

Two of the more commonly used methods for realizing solder bumps on wafers are electroplating of solder and printing of solder paste. Other methods of solder deposition include evaporation and needle dispensing.

Plated Bump Flip-Chip Plated bump technology uses wet-chemical processes to fabricate solder bumps on wafers. The process consists of first removing traces of aluminum oxide, if any, from bond pads on the die. This is followed by plating the UBM. The UBM is approximately a 1-μm thick metallurgical system that provides a mechanical base for the solder bump, as well as a reliable and stable interconnect. The UBM also overlaps the wafer passivation layer to protect the wafer circuitry from corrosion. UBM provides for a low-resistance ohmic contact between the solder bump and the bond pads on the die surface. It also acts as a diffusion barrier, restricting intermetallic formation arising from a potential interaction between the solder bump metals and the bond pad metals. In the absence of such a diffusion barrier, intermetallic formation can quickly result in performance degradation or even reliability failures.

Solder bump plating follows the UBM deposition process. Some of the common solder materials include eutectic tin–lead [63 wt% tin (Sn) and 37 wt% lead (Pb)], high lead (90 wt% Pb and the rest Sn), and lead-free solders [95 wt% Sn and the rest antimony (Sb)]. Plating-based solder bump services are available from several suppliers within and outside of the United States.

Printed Solder Bump Printed solder bumps are created by printing solder paste on the UBM using a stencil or other suitable means, followed by a reflow step to convert the deposited solder paste into a near-spherical solder bump. Printing technology offers better control of the solder bump composition and allows wafers to be bumped with a wider range of solder alloys, such as eutectic tin–lead, lead-free, ternary, and quaternary solders. Printing technology is generally cheaper than electroplating, but electroplating processes are more suitable for solder bump pitches of less than 200 μm. Although solder printing can be used for bumps on a 150-μm pitch, this is accomplished at a significant decrease in bump height, which can have adverse reliability implications after the die is attached to a substrate. One way of realizing higher collapsed heights on substrates after assembly is to use Cu pillar bumped die, a bumping process that is discussed in a later section of this chapter.

Figure 10.21 Gold stud bump, "coined" for better coplanarity.

Adhesive Bump A conductive adhesive, suitably dispensed, provides an electrical interconnect, as well as mechanical support, across a die and a substrate, because the cured adhesive performs the functions of a bump. Adhesive bumping services are available from several licensed suppliers across the United States and elsewhere.

Gold Stud Bump Gold stud bumping is a modified form of the wirebonding process. As in the case of wirebonding, the process begins with the formation of a gold ball by melting the end of a gold wire. The gold ball is then placed on the bond pad of the device. Stud bumping equipment is suitably modified to break the gold wire close to the ball once the bond is made with the device bond pad. Sometimes, an extra process step, called *coining*, is employed that flattens the broken wires into a uniform plane to improve manufacturability in the assembly process. Figure 10.21 shows an example of a stud bump that was subsequently *coined* for better coplanarity. Gold stud bumped devices are placed on their carriers, usually with conductive epoxies, and then cured to make electrical contact. A thermosonic gold-to-gold connection can also be used to establish mechanical and electrical contacts. This bumping technology is common in the electronic displays industry, and also in some consumer applications, such as wrist watches, and the like.

Copper Pillar Bump A relatively new technology, called copper pillar bumps, offers larger standoff heights than most all other bumping technologies, which are generally limited to pitches of 200 μm or less. Also, Cu pillar bumps have a higher height-to-diameter aspect ratio compared with conventional bumps. These bumps use a high melting temperature base, Cu, in this case, which is topped off with either eutectic tin–lead solder or any other suitable solder material. The concept is similar to that employed by IBM in the 1960s, where it used a high-lead base, which was topped off by eutectic tin–lead solder.

Figure 10.22 shows devices bumped with Cu pillar bumps. These bumps allow for smaller pads to be designed on the die, coupled with taller bumps, which can favorably affect manufacturing yields. The underlying copper in the bumps allows for better current-handling capacity per solder joint, while significantly reducing the risk of electromigration and related failure modes. Also, Cu pillar bumps can be an alternative to lead-based, low-α bumps, which are the present standard for memory devices.

Polymer Bump The general process steps for polymer bumps are similar to the bumping process based on solder printing, namely, evaporation or sputtering of the UBM, followed by stencil printing of a conductive silver-filled polymer material with isotropic electrical properties. The polymers are usually thermoset, meaning that they cure with heat, but they can also be thermoplastic, so the bumps soften when heated. Polymer bumps offer the

Figure 10.22 Cu pillar solder bumps.

convenience of a low-temperature process that is compatible with high-volume production and is scalable down to center-to-center bump pitches of about 150 µm. However, a smaller bump pitch is accompanied by smaller bump heights, which may have an effect on device reliability.

10.8.2 Fluxing

The use of flux in the flip-chip assembly of solder bumped chips helps in at least two ways. The use of flux during reflow removes oxide from the solder surface, which ensures a good mechanical and electrical contact across the die and the substrate. Second, the tackiness of the flux helps to physically hold the placed die on the substrate until the reflow is complete. Flux is commonly dispensed using either of the methods discussed in the following subsections of this chapter.

10.8.2.1 Dip

In this method, a flux reservoir forms an integral part of the die pick-and-place equipment. The flux application mechanism is a rotating disk equipped with an adjustable doctor blade. Continuous rotation of this disk, together with controlled clearance of the doctor blade over the disk surface, ensures that the flux film thickness is consistent. The solder bumped die is picked up from the diced wafer, flipped over, dipped in the flux for a preset amount of time (thickness is controlled by doctor blading and the rotating disk), and is then placed on the substrate using a vision system. After populating the entire substrate with die, the strip enters a reflow oven to melt the solder and establish mechanical and electrical contact between the die and the substrate. To facilitate a good solder joint, all solder bumps should be of the same height to enable a uniform coating of flux. This method of dispensing flux is advantageous for a no-clean process, since the flux quantity used, and hence, the post-reflow residue left behind is small.

10.8.2.2 Brush

This process uses flux with low viscosity, which is applied with a brush, not only on the flip-chip pads but also on the die area. The die pick, flip, and place operation is similar to the one described in the previous section, except for the intermediate dipping in the flux. In automatic machines, the brush is wetted by the flux through a tube that has its opening near the base of the brush. The quantity of flux used in this method is higher than that in the dip process, however, this equipment can place die at a faster rate.

10.8.2.3 Spray

The spray fluxing method is similar to the brush method of applying flux, the only difference being that a mist of flux is sprayed over the die footprint area on the substrate instead of brushing it on. The die is placed following the flux spraying and the assembly is then subjected to reflow.

10.8.2.4 Dispense

The process flow is identical to the spray method, except that the flux is dispensed onto the substrates with a nozzle.

10.8.2.5 Die Placement

Once the flux is suitably applied, either to the die or to the substrate, the die is placed on the substrate. Local or global fiducials help the vision system place the die onto the corresponding substrate pads. The surface tension of the molten solder helps align the die on the substrate pads during the reflow step. This property is commonly referred to as *self-alignment* and allows for small die with a 225-μm solder bump pitch to typically be placed with an accuracy of about $\pm 50\,\mu m$ and a $\pm 2°$ rotational tolerance. Dice larger than about 5 mm require better control over the rotational tolerance. Commercially available flip-chip die placement equipment can place in excess of 3000 die per hour with an accuracy of $\pm 10\,\mu m$ or better.

10.8.2.6 Reflow

After placing die on the substrate, the assemblies are subjected to a high-temperature reflow oven where the solder melts to provide a mechanical and an electrical contact between the die and the substrate. Heating in reflow ovens is accomplished by forced-air convection or infrared or both. Reflow temperature profiles are specifically tailored for different applications, considering factors such as thermal mass of the components, their thermal sensitivity, and the temperature requirements of the solder material. A good profile generally provides for a 3- to 4-min exposure of the assemblies inside the heated path and consists of different zones for preheat, flux activation, reflow, and cooling. Care must be taken to ensure an optimal preheat setting since an excessively high preheat ramp rate can cause solder spattering, solder balling, or even microcracking of sensitive components. The activation zone brings all components of the assembly up to a uniform temperature near, but below, the reflow temperature. This also helps evaporate most of the volatile elements within the solder and accounts for about 35 to 50% of the total heat time. This is followed by the reflow zone where the temperature is elevated beyond the solder melting point, thereby establishing a bond between the component and the substrate pads. The cooling cycle, which follows, generally mirrors the ramp side.

10.8.2.7 Cleaning

Since fluxes are usually acidic in nature, cleaning their residue after reflow helps to reduce corrosion of the solder/pad interface. In a high-volume manufacturing environment, flux residues are cleaned by one of two methods: batch cleaning or in-line cleaning. The in-line approach delivers the most consistent cleaning. Batch cleaning is more susceptible to variability. The choice of cleaning chemistry depends on the cleaning challenges. A batch centrifugal cleaner with the appropriate fixtures can clean different types of assembled packages, as well as wafers, from 150 to 200 mm in diameter. Centrifugal cleaners can also be used to clean packages of small geometries. Commercially available centrifugal cleaning systems are compatible with several different cleaning chemistry options, such as saponifier, hydrocarbon, semiaqueous, and aqueous. These cleaners allow parts to be submersed and rotated during the cleaning cycle. The cleaning chemistry can be heated up to about 100°C or so, which is followed by a rinse in deionized water and drying in heated air.

Traditionally, the process of cleaning electronic assemblies uses solvents that can be ozone depleting. Legislation across the world has recommended or, in some cases, such as in the United States has mandated, that electronics manufacturers adopt alternative chemistries as substitutes for established, but environmentally detrimental, cleaning chemistries. No-clean formulations offer another environmentally friendly approach to flux cleaning. Two types of materials are available, one for which the residue is compatible with underfill materials and another that leaves only trace quantities of residue. With either of these materials, it is not necessary to clean the residue after reflow, thereby, not only making the manufacturing process more environmentally friendly but also helping to reduce the manufacturing cost by eliminating the cleaning step from the process flow.

10.8.2.8 Underfill

Underfill encapsulant is a thermosetting polymer material that fills the gap between a die and a substrate and is an important step in the flip-chip assembly process. Underfilling flip-chip assemblies provides the following benefits:

- Underfill helps reduce stress on the bumps caused by a thermal mismatch between the die and substrate.
- Cured underfill keeps the reflowed solder bump in a state of compression, which limits creep flow of the solder joint.
- The presence of an underfill reduces crack initiation based failures originating at the solder bump/substrate interface.

Two types of underfill are commonly used in the electronics industry: flowable and no-flow underfills.

Flowable Underfill Flowable underfill is dispensed around a die after reflow processing. Capillary action drives the underfill into the small gap between the die and the substrate. The dispense pattern for the underfill depends on factors such as die size, the number of solder bumps and their pattern on the die, gap height, and the desired fillet across the die edges. Most underfills require an additional curing step to harden it.

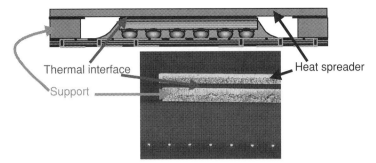

Figure 10.23 Schematic of a thermal spreader attached to a flip-chip assembly.

No-Flow Underfill In the no-flow underfill process, a no-flow encapsulant is dispensed on the substrate before placing the die. After die placement, solder bump reflow and underfill cure are performed simultaneously in a standard reflow oven, thereby offering a higher throughput than that for the flowable underfill. However, the no-flow underfill encapsulant must provide self-fluxing capabilities that leave minimal residue. Also, the size of filler particles, if the encapsulant contains any, needs to be tightly controlled to ensure that the solder bumps make good contact with the substrate bond pads. This requirement poses restrictions on how much the encapsulant can be loaded with particles.

10.8.2.9 Heat Spreader Attach

Power dissipating flip-chip devices often require the use of thermal management materials of high thermal conductivity as a heat spreader. Heat is pulled away from the device before it is dissipated to the environment. An optimal way of managing the heat generated by devices is to minimize the thermal resistance of the heat transfer path, while simultaneously maximizing the contact area for spreading and sinking the heat. This is accomplished by thinning the die as much as possible (typically to the range of 100 to 450 μm) and attaching it to a heat spreader using a suitable thermally-conductive interface, such as a conductive epoxy or tape. Figure 10.23 shows a schematic, as well as a side view, of a package with a thermal spreader attached to it.

10.9 PACKAGE SEALING/ENCAPSULATION/COATING

After the assembly of active and passive components to a circuit board or substrate, it is necessary to protect them from chemical, mechanical, and sometimes electromagnetic hazards, while providing electrical and thermal paths to the outside world. The two most common methods for providing circuit protection are hermetic sealing, and nonhermetic coating. The difference between the two is their susceptibility to penetration by contaminants in the environment. Hermetic sealing provides a relatively impermeable package to the environment and is used where harsh environments or very high reliability is required. Nonhermetic coatings are normally organic coatings, and provide only a modest level of protection from gases and fluids. It must be emphasized that hermetic is a relative term, and that there is no such thing as a perfect hermetic seal. Nonetheless, the penetration of moisture and gas into packages made of metal, glass, or ceramic is orders of magnitude smaller than plastic or organic packages.

10.9.1 Hermetic Package Sealing

A hermetic package consists of a ceramic or metal housing with pins or leads for power, ground, and signals. The device or devices are placed within the housing, and all of the power, ground, and signal lines are connected to the internal package pins or leads with wirebonds. The metal pins or leads exit to the outside of the package through hermetic passages. Once the devices are all wirebonded within the package, a lid is hermetically attached to the top of the housing using either glass or metal, fully encasing the devices and protecting them from moisture and other environmental hazards. While there are some packages that are sealed with polymeric materials, moisture will permeate this seal within hours or days. Only packages made of and sealed with ceramic, metal, or glasses are considered hermetic.

Package hermeticity is defined by the military standard MIL-STD-883 [15] as having a helium leak rate of better than 5×10^{-8} cm^2/s for package volumes less than 0.40 cm^3. One of the greatest advantages of hermetic packaging is its ability to keep moisture away from the devices it contains. Moisture is the principal cause of circuit failures in nonhermetic packaging. Military specification MIL-STD-883 also defines the maximum amount of moisture content allowable to prevent water from condensing onto the devices. For most applications, the maximum moisture allowed is 5000 parts per million by volume (ppmv). Before sealing the packages, devices are baked in a vacuum oven to drive off moisture from die-attach and substrate epoxies, as well as other packaging materials.

Hermetic sealing is accomplished through any of the four principal processes: soldering, brazing, glass sealing, and welding. Solder sealing is performed by melting a metal alloy between the package body and the lid. The most commonly used metal alloy is eutectic 63Pb–37Sn, with a melting point of 183°C. Solder sealing is typically performed in a conveyor oven with zones that provide for preheat, melting, and cool-down.

Brazing uses higher melting temperature alloys than soldering, such as eutectic 80Au–20Sn, with a melting point of 280°C. Brazing with Au–Sn alloy has the advantage of providing a stronger and more corrosion-resistant seal than solder sealing and does not require the use of a flux. Both soldering and brazing require a solderable, metalized seal ring on the package and the lid.

Glass sealing uses vitreous or devitreous glasses that melt in the 400°C range. Due to the relatively high temperatures required to seal packages with glass, the chip attach material must also be glass based or be a metal preform, and a monometallic system (e.g., all aluminum) must be used for wirebonds, bond pads, and internal package leads to avoid reliability problems. Glass sealing is typically performed in a conveyorized oven. Due to the complexity of the glass sealing process and the narrow process window, this type of sealing is not as popular as it once was.

Welding is a popular method for sealing military packages because it provides very high process yields and very high reliability. Welding is most commonly accomplished by applying high-current pulses to the lid with rolling or stationary electrodes, resulting in localized heating of 1000 to 1500°C, which fuses the lid to the package seal ring. Alternative methods of welding utilize an electron beam or a laser. Equipment for these methods is more expensive, but they have the advantage of being able to weld at very high speeds and limit the heat input into the package, which is important for applications with thermally sensitive devices.

10.9.2 Hermetic Package Testing

Hermetic package seal integrity testing is performed by subjecting packages to leakage tests. Gross leak testing measures leaks in the range of 10^{-1} to 10^{-4} atm/cm^3/s, while fine leak testing measures leaks in the range of 10^{-5} to 10^{-12} atm/cm^3/s. Gross leaks are tested by first subjecting the package to a vacuum evacuation of greater than 5 Torr for 1 h, followed by immersion into liquid fluorocarbon (FC-84) for 30 min. The packages are then immersed in another liquid fluorocarbon (FC-40) with a higher boiling point than FC-84. The FC-40 is brought to a temperature higher than the boiling point of FC-84, so that any FC-84 present in a leaky package will expand in volume and leak back out as visible bubbles. An alternative, but less common method than using FC-40, is to use a gas analyzer that detects the FC-84 molecules once the packages are heated to 125°C.

Fine leak testing is performed by "bombing" the packages for 2 h in a pressure chamber filled with helium gas at 60 psig. The packages are then placed in a vacuum chamber attached to a helium mass spectrometer. Any helium escaping from the package is detected by the mass spectrometer.

Another leak test method detects fine and gross leaks at the same time. Packages are first subjected to pressurization in dry nitrogen gas with a mixture of 1% Kr-85, a radioactive tracer gas. After pressurization, a radioactive counter detects any γ emissions resulting from the β decay of Kr-85. If Kr-85 leaks into the package, the γ emissions will penetrate the package walls and the radioactive counter will measure the concentration of Kr-85 that entered it.

10.9.3 Nonhermetic Encapsulation

Hermetic packaging provides optimal protection from the elements, but this comes at a relatively high cost due to the expensive materials and the complex processes required. For most commercial applications, nonhermetic circuit packaging provides sufficient levels of protection, and the materials and processes used are much less expensive and, typically, less complex than for hermetic protection. Encapsulation requires covering the chip surface and wires with a polymer material, thereby sealing it from the environment. Encapsulation can include potting, molding, glob-top, dam and fill, or cavity fill methods. The reliability of encapsulated circuits has improved greatly over the last decade, largely due to improvements in the chemical purity and composition of encapsulation materials. Extractable (mobile) ionic levels of many encapsulant materials (Cl^-, K^+, Na^+), which cause corrosion when moisture is present, are now typically found in the low range of 1 to 5 ppm. These ionic purity levels are almost an order of magnitude better than materials used in the 1980s. The physical properties of polymer materials have also been engineered with improved adhesion, lower residual stress levels, and improved resistance to hot, wet, and biased environments. Circuits encapsulated with many of these new polymer materials can readily survive 10,000 h of biased humidity testing (85%RH/85°C) without failure.

The majority of encapsulant materials are applied either by transfer molding or by dispensing. Transfer molding is by far the most popular method used for plastic encapsulated microcircuits. The transfer molding process works by melting pellets of epoxy contained in a transfer pot and then pressure injecting the molten epoxy through runners and gates into enclosed mold cavities containing the devices. The pressurized and hot epoxy is forced around the devices and wires, and once the mold cavity is completely filled, a higher packing pressure is applied to compact the mold compound to reduce voiding. Tiny vents in the mold chase are large enough to allow air to evacuate during pressurization, but are small enough that mold compound cannot flow through them. After 1 to 3 min at the molding temperature, the mold compound is semicured in the mold chase. The mold then opens, and bottom pins eject the packages from the mold. The packages are then batch cured for several hours to

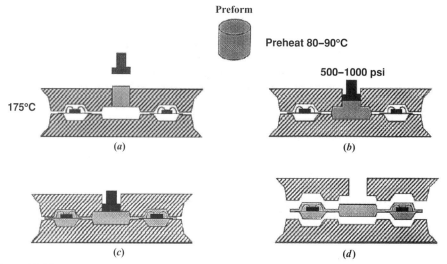

Figure 10.24 Schematic of the molding process: (*a*) charge mold, (*b*) transfer, (*c*) cure, and (*d*) demold.

completely cure the mold compound. Figure 10.24 shows the package molding process schematically.

The dispensing method is used primarily for encapsulating plastic ball grid arrays (PBGAs), by applying an epoxy or silicone material using the glob-top method, the dam-and-fill method, or the cavity-fill method. Figures 10.25, 10.26, and 10.27 show these techniques pictorially.

All encapsulation methods require slight variations of the same manufacturing processes, including cleaning, epoxy application, and curing. Preencapsulation cleaning is critical for removing organic and inorganic contamination and improving adhesion of the encapsulation material to the circuit surfaces for improved reliability. Cleaning processes can include aqueous or solvent-based wet cleaning by spray or immersion techniques, and/or plasma cleaning using reactive or nonreactive gases.

The principle method for applying an encapsulant is by dispensing it onto the devices. The dispensing process involves moving the encapsulation material from a syringe or holding reservoir and through a needle that directs exact amounts of material over and around the device active area and interconnections. Dispensers may be equipped with a number of different pump configurations and complexities (time/pressure, piston pump, auger screw drive), depending on the dispense application and material requirement. Glob-top

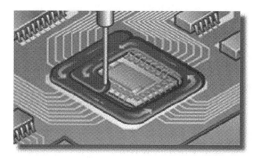

Figure 10.25 Glob-top encapsulation method.

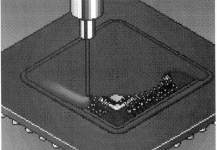

Figure 10.26 Dam-and-fill encapsulation method.

dispensing is the simplest process in which a glob of encapsulant is dispensed over the chip and its interconnects and is allowed to flow over and around the device. The dam-and-fill method is more accurate since it first applies a defining "dam" or wall around the device with a relatively high-viscosity encapsulation material, followed by a "fill" using low-viscosity encapsulation that can flow easily around fine pitch wires or leads. The dam-and-fill method is also used in applications where controlling the maximum encapsulation height is critical. Cavity fill methods are used in multilayer ceramic or plastic BGA packages where the device is mounted inside an open cavity in the package. Once the cavity is filled to the top with encapsulation material, the device and wires are completely covered and protected within the body of the package. Following the dispensing process, the assemblies are thermally cured in a box or belt oven to harden the encapsulation material.

10.10 PACKAGE-LEVEL PROCESSES

Package-level processes can be loosely defined as the final steps in any given package assembly. Specific steps involved depend on the type of packages, their functionality and purpose but may include processes such as singulation of individual packages from their mass production format, marking, or attaching solder balls, in the case of BGA packages.

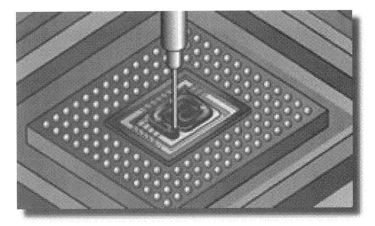

Figure 10.27 Cavity-fill encapsulation method.

10.10.1 Lead Trim, Form, and Singulation

Trim and form is the process by which individual leads of packages are separated from a lead frame strip. The first step is to remove the dambar that electrically isolates the leads. The leads are then trimmed and formed mechanically to specified shapes, such as J-bend or gull-wing for surface-mounted packages. Individual packages are singulated from the lead frame strip by punch tools or sawing, inspected for lead coplanarity, and placed in trays, tubes, or on tape and reel.

10.10.2 Solder Ball Attach and Singulation

Unlike lead frame packages, BGAs use solder balls as the interconnect path from the package to the printed circuit board. Instead of lead forming processes, solder balls are attached to the substrate by applying a flux, placing the balls on the substrate pads, and reflowing the BGA. The reflow process forms a metallurgical joint between the solder ball and the substrate ball pad. Alignment and the reflow profile are key parameters for ball attach to avoid missing ball, solder bridging, and coplanarity problems. Individual packages on laminate strips are singulated by punch tools or sawing, inspected for ball coplanarity, size and position, and placed in trays or on tape and reel.

10.10.3 Marking

Marking is used to place corporate and product identification on a packaged device. Marking allows for product differentiation. Either ink or laser methods are used to mark packages. Laser marking is preferred in many applications because of its higher throughput and better resolution.

10.11 STATE-OF-THE-ART TECHNOLOGIES

Per the Institute for Printed Circuits (IPC) definitions, a technology is termed as *leading edge* when its product is commercially available from some of the world's advanced manufacturers, usually at a premium price. Typically, products in this category represent about 10 to 15% of the world's production and may carry a price premium of the order of 60 to 80% over conventional products. State-of-the-art, on the other hand, is available from a much smaller number of manufacturers and at a significant price premium (up to, and sometimes beyond, 10 times that for a conventional product). Products in this category usually represent 5% or less of the world's production. This section addresses some of the state-of-the-art packaging technologies.

10.11.1 3D and Stacked Die

Regardless of the application (military, consumer, or otherwise), there is always a demand to shrink package size while simultaneously increasing its functionality. Thinning the die and then stacking them on top of each other provides a practical and an attractive option to realize these demands on package performance. Stacked die packages are finding application in cellular phones, avionics, military, "smart-card", and other devices where devices and their controller chips need to coexist in a very small volume of space. Two-die stacks are among the most common in this class of packages, however, 50 or more die of the same

Figure 10.28 Stacked die from Intel, Inc. (www.intel.com/research/silicon/mobilepackaging.htm).

type have been stacked together by Irvine Sensors using its Neo-Stack technology [16]. Intel is developing a technology to stack as many as eight dice, each about 50 μm thick and wirebonded to realize a wirebonded stacked die CSP, as shown in Figure. 10.28 [17].

Die can be stacked over each other in any one of the three configurations or a combination thereof: (i) smaller die placed on top of larger die, (ii) same sized die placed over each other with a suitable spacer placed between the two die, and (iii) larger die placed over a smaller one. In all cases, the dice are thinned typically to a thickness of about 75 μm in order to achieve maximum package heights of about 1 to 2 mm. The method of interconnect across die or between die and substrate can either be wirebond, flip-chip, or sometimes a combination of the two. The major advantage of stacked die is space savings, wherein two or more packages, each assembled with conventional packaging are integrated into a single one.

One of the more common applications of stacked die is in cellular phones and other wireless telecommunication products where memory devices, such as Flash, SRAMS, DRAMS, and dedicated controller chips, are key components. Companies, such as Intel, have developed technologies that enable a flip-chip die and a wirebonded die inside the same package [17]. Also, in some cases, it may be more practical to stack packages rather than die. Such approaches may provide greater flexibility in designing packages with different form factors, but with similar or identical functionality to suit a particular application (see Fig. 10.29).

10.11.2 Radio Frequency (RF) Modules

A radio frequency (RF) module is essentially an electronic package solution that integrates several (or all) RF components needed for a certain application. Typical applications for RF modules include wireless infrastructure, wireless local area network (WLAN), and Bluetooth. The functions of an RF module, depending on the application, may include the following elements: power amplification (PA) and management, a base-band processor, transmit and/or receive modules, harmonic filtering circuitry, and high linearity and low insertion loss RF switches, power amplifier control (PAC) circuitry, and antenna blocks. An RF module is, thus, an integration of several technologies and skills. RF design is generally acknowledged to be as much of an art as a science or a technology, especially recognizing the tolerances associated with various steps in high-volume manufacturing. A few of the distinct technologies that go into the manufacture of an RF module are listed below to give the reader a flavor of the

Figure 10.29 Stacked packages from Intel, Inc.(www.intel.com/research/silicon/mobilepackaging.htm).

432   Chapter 10   Electronic Package Assembly

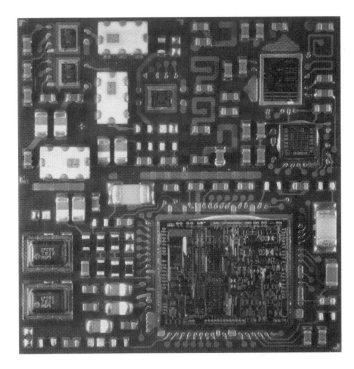

Figure 10.30 Skyworks Solutions Inc.'s Single Package Radio; example of an RF module.

associated challenges. This list is by no means, comprehensive:

- Separate global system for mobile communications (GSMC), data collaboration server (DCS), and personal communication service (PCS) power amplifier blocks
- Internal circuitry for matching to 50 Ω input and output impedances
- Harmonic filtering
- High linearity and low insertion loss p-high electron mobility transistors (PHEMT) RF switches
- Diplexers
- Power amplifier control (PAC)
- Si CMOS integrated circuits
- Heterojunction bipolar transistor (HBT) blocks on gallium arsenide (GaAs)
- Various elements of the assembly technologies

Figure 10.30 shows Skyworks Solutions Inc.'s Single Package Radio, an example of intricacies involved in designing and assembling a complex RF module. From individual components to the entire integrated module, it takes a unique blend of experience and cross-functional technologies to realize an efficient and a cost-effective RF product.

10.11.3 Microelectromechanical Systems (MEMS) and Microoptoelectromechanical Systems (MOEMS)

A microelectromechanical system is the microsized integration of mechanical elements along with controlling electronics that can deliver a complete system on a chip through microfabrication, usually on silicon-based materials. The electronics is usually fabricated using

10.11 State-of-the-Art Technologies **433**

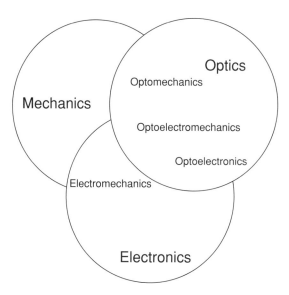

Figure 10.31 MEMS and MOEMS.

integrated circuit processes, such as complementary metal–oxide–semiconductor (CMOS), bipolar, or bi-CMOS, and is integrated with the miniaturized mechanical components that are fabricated from fab-compatible processes. Depending on the application, such processes could be a combination of subtractive processes that selectively etch away the silicon wafer and additive processes that add new layers to form the mechanical and electromechanical devices. One of the most common and widespread applications of this technology is the MEMS accelerometer used in the automotive industry for airbag activation. Other MEMS applications include microfluidics, emission sensing in integrated diagnostics, biochemical sensing using molecular recognition microcantilevers, gas sensors, miniature stepper motors for precision positioning, as well as surgical and scientific microinstruments. Packaging of MEMS devices depends on the MEMS application. However, in sensor applications, a major packaging challenge is protection of the device from the very environment the device attempts to sense and measure. Such devices, especially those with parts that move or rotate, can be susceptible to moisture. Consequently, they may require not only hermetic packaging but also a getter material within the package to extend their useful life.

The MOEMS are MEMS devices with optical functionality. Figure 10.31 schematically shows the areas of MEMS, MOEMS, and other mechanical, electronic, and optical devices.

Modern applications in optics have primarily focused on the generation, manipulation, guidance, and detection of light for information processing. The primary responsibility of MOEMS devices is the manipulation of light in a linear, area, or volume region of space. For clarity, light is defined as electromagnetic radiation, generally with a wavelength in the range of about 0.2 to 20 μm. The defining range for light is important since it sets a lower and an upper limit on the size of the component. The lower bound is a natural consequence arising from the laws of diffraction because the feature sizes of microoptical elements must be about 10 times larger than the wavelength of light intended to interact with the microoptical element to avoid distortion or loss due to diffraction. However, if the device/component performance is based on diffraction, then, of course, these limits do not apply.

When packaging MOEMS devices, one needs to consider the fundamental operating principles of the device or element under consideration. Typically, a MOEMS device or element performs dynamic manipulation of a light beam through movement of microoptical elements. This can involve modulation of either the amplitude or the wavelength of optical signals, introduce temporal delays, cause diffraction, reflection, refraction or just a simple spatial realignment, or complex operations on the light beam through some combination of these individual phenomenon. Packaging such devices cannot impose constraints on their light manipulation ability at any time during the useful life of the devices.

Miniaturization of MEMS devices has already attracted attention and funding and is called nanoelectromechanical systems, or NEMS. Due to space and scope limitations, it would not be possible to elaborate here on the specifics. However, the reader may find some excellent resources in the literature [18–20].

10.11.4 Nanotechnology

Nanotechnology is the art, science, and technology of materials at the nanometer scale and in the realm of atoms and molecules. This exciting field has aroused interest among not only scientists and engineers but also among entrepreneurs, as well as corporate and federal funding agencies. A lot of interest, academic, as well as entrepreneurial, has been generated since the 1991 discovery of high aspect ratio carbon tubes. Sumio Iijima discovered that the tubes were made up of pure carbon and were only about 10 nm in diameter. These entities were subsequently called carbon nanotubes and displayed the same patterned symmetry of graphite [21]. Further, nanotubes demonstrated mechanical and electrical properties that were much different than those of any other known forms of carbon. The tubes had greater resilience, better mechanical strength, higher thermal stability, and larger current-carrying capacity than most of the known materials. Nanotubes are presently being used to provide structural reinforcements in lithium–ion batteries [22]. Field emitters using nanotube components in flat-panel displays are expected to reach the commercial market soon [22]. Nanotubes and several other nanostructures have demonstrated significant promise as enabling factors in applications, such as nanoscale computing devices, fibers stronger but much lighter than steel, biochemical devices and sensors, probe tips, fuel cells, and components that can significantly reduce the weight of objects, such as automobiles and spacecrafts. Characterization and packaging of such devices pose significant challenges. For example, characterizing the electrical properties of a nanotube would involve manipulating a single nanowire from its source into a measurement environment. This is usually achieved by means of microgrippers and/or nanomanipulators (see Figs. 10.32 and 10.33 [23]).

Figure 10.32 Nanomanipulators from Zyvex Corporation.

Figure 10.33 Microgrippers from Zyvex Corporation.

Challenges associated with electrical characterization include realizing a four-point probe station suitable for measuring the conductivity of a nanotube or making an ohmic contact (and ensuring its linear electrical characteristics) across both ends of the nanotube to the measurement pads. Packaging of such a circuit or device element will require significant cross-functional and cross-disciplinary effort.

10.12 SUMMARY

Advanced packaging and low-cost, high-volume assembly of devices involves cross-disciplinary understanding and expertise. It is as much an art as it is engineering and technology. This chapter has attempted to give a flavor of some aspects of this fascinating and mature, yet evolving, field. A comprehensive and more detailed treatment, as the reader would realize, is well beyond the scope of this chapter.

REFERENCES

1. http://www.iest.org; see documents ISO 14644-1 through ISO 14644-7 Standards are published by ISO-International Organisation for Standardization (2004).
2. http://www.clean-air-solutions.com/basic.html.
3. W. WHYTE, *Cleanroom Technology: Fundamentals of Design, Testing and Operation*, London: Wiley Interscience, 2001.
4. M. RAMSTORP, *Introduction to Contamination Control and Cleanroom Technology*, Schauernheim, Germany: VCH, 2000.
5. Electrostatic Discharge Association, see www.esda.org/basics/part1.cfm and related web pages.
6. JEDEC website http://www.jedec.org.
7. The British Amateur Electronics Club web site at http://members.tripod.com/~baec/DEC90/comp.htm.
8. M. J. MINDEL, Solder Paste Rheology as a Function of Temperature, Proceedings of SMI '91, San Jose, CA, pp. 490–495, 1991.
9. Solder material properties at the NIST website: http://www.boulder.nist.gov/div853/lead%20free/props01.html.
10. P. HALLEE; Getting Dressed for BGA Rework, *Surface Mount Tech.* No. 2 p. 72, 1998.
11. J. D. PHILPOTT, T. A. NGUTY, N. N. EKERE, and G. D. JONES, Effect of CSP Rework on Surface Intermetallic Growth, Twenty Fourth IEEE/CPMT International Electronics Manufacturing Technology Symposium, Austin, TX, pp. 141–147, 1999.
12. L. R. LEVINE, Wire Bonding Optoelectronics Packages, *Chip Scale Rev.*, Vol. 5 No. 8, pp. 49–53, 2001.
13. http://www.kns.com/prodserv/tools-materials/specwire.asp.

14. F. D. EGITTO and L. J. MATIENZO, Plasma Modification of Polymer Surfaces for Adhesion Improvement, *IBM J. Res. Devel.*, p. 423, July 1994.
15. http://www.dscc.dla.mil/Downloads/MilSpec/Docs/MIL-STD-883/std883.pdf.
16. http://www.irvine-sensors.com/chip_stack.html.
17. http://www.intel.com/research/silicon/mobilepackaging.htm.
18. P. RAI-CHOUDHURY, ed., *MEMS and MOEMS Technology and Applications* (SPIE PRESS Monograph Vol. PM85); SPIE—The International Society for Optical Engineering, Washington DC, December 2000.
19. M. GAD-EL-HAK, ed., *The MEMS Handbook*, Boca Raton, FL: CRC Press, 2001.
20. N. MALUF, *An Introduction to Microelectromechanical Systems Engineering*, Artech House, 2000.
21. S. IIJIMA, Helical Microtubules of Graphitic Carbon, *Nature*, Vol. 345, pp. 56–58, 1991.
22. J. OUELLETTE, Building the Nanofuture with Carbon Tubes, *Indust. Phys.*, pp.18–21, Dec. 2002/Jan. 2003.
23. http://www.zyvex.com.

EXERCISES

10.1. Explain cleanroom classification nomenclature and with examples state why processes for advanced electronic assemblies need a cleanroom environment.

10.2. State and explain any three design/assembly factors affecting SMT assembly yield. Estimate the impact of number of SMT components in a module to the overall package yield (assume that the SMT yield remains constant and less than 100%).

10.3. Construct a design-FMEA table, listing at least three distinct failure modes involving SMT components in an assembled module. (FMEA is an acronym for "failure modes and effects analysis." Design-FMEA is an exercise aimed at predicting different possible failure modes in a certain process.)

10.4. Describe pros and cons of any two wafer thinning techniques for 8-in. silicon wafers. What extra precautions would have to be considered if the wafer diameter were to increase to 12-in.

10.5. A module with some surface-mountable resistors and two wirebondable dice needs to be assembled. Generate a flowchart for the assembly operations and describe each of the process steps in not more than three sentences each.

10.6. List the salient properties of thermosets and thermoplastics, respectively. Comment and elaborate on which materials would be preferred for packages that may need rework.

10.7. A certain application demands that the parasitic interconnect inductance to be predictable, consistent across all interconnects on the die and as low as possible. State your choice of interconnect (wirebond or flip-chip) with reasons. Also state conditions when the other interconnect will make a better choice based on factors such as the number of interconnects across the die and its substrate, die size, and thickness.

10.8. Estimate the minimum die size (square or otherwise) needed to accommodate 12 solder bumps at the following bump pitches and configurations:
- 1 peripheral row, at 250 μm center-to-center pitch
- 2 peripheral rows, at 250 μm staggered center-to-center pitch
- 1 peripheral row, at 150 μm center-to-center pitch
- 2 peripheral rows, at 150 μm staggered center-to-center pitch

10.9. What is the main function of flux in flip-chip assemblies? When can the use of flux be avoided in a flip-chip assembly process? In cases where use of flux cannot be avoided, when can the extra process step of flux cleaning be eliminated?

10.10. What could be the possible consequences of using a strong flux during assembly and not cleaning the residue?

10.11. What purpose does package encapsulation (molding, glob top, etc.) serve? Under what circumstances would one consider not encapsulating the package, without adversely affecting the package reliability or its performance?

Chapter 11

Design Considerations

J. P. PARKERSON AND L. W. SCHAPER

11.1 INTRODUCTION

This chapter deals with two major topics: the trade-offs that must be made in packaging, starting with the integrated circuit (IC) and system requirements, and the design process for an electronic system, particularly the packaging and interconnect components. The first few sections cover the trade-offs that must be made at all levels of electronic packaging, when the constraints of the various packaging functions come into sharp conflict. These trade-offs must be made at the system conceptual design stage, before any hardware is built or even designed in detail. Yet, the decisions made at this stage will determine the success or failure of a system's implementation. This is why the packaging engineer must be a skilled generalist, for it is he or she, more than anyone else except those who define the system's functions and cost targets, who will determine the success of the design effort. Then, we turn to the design process, describing the steps that the design engineer must go through to turn system concept and block diagram into a complete component, interconnect, and system design.

11.2 PACKAGING AND THE ELECTRONIC SYSTEM

11.2.1 Packaging Functions

Broadly defined, electronic packaging is everything necessary to turn bare IC chips and other individual electronic components into functioning electronic systems. Thus, the "packaging" can account for anywhere from 50 to 80% of the overall system cost and 100% of the blame if the system does not work as intended. It is useful to look at packaging from a "chip's eye view" as well as from a system perspective. To the chip, packaging must provide four functions, which were described in Chapter 1:

1. Signal connectivity, in both quantity and quality, necessary for proper chip functioning
2. Stable and noise-free power at the required voltage levels
3. Protection from corrosion and mechanical damage
4. Removal of heat to keep chip temperature within operational limits

---

Advanced Electronic Packaging, Second Edition, Edited by Richard K. Ulrich and William D. Brown
Copyright © 2006 the Institute of Electrical and Electronics Engineers, Inc.

From the system point of view, packaging must deliver the intent of the system architect and circuit designer with the required size, weight, power consumption, performance, reliability, manufacturability, and cost. Both the chip and the system requirement sets contain conflicts, which must always be resolved by trade-offs. In order to properly make trade-offs, it is essential to have metrics, or performance measures, for all aspects of packaging decisions and to know the relative importance of each metric. For example, the maximum IC operating temperature may be given as 100°C; the chip simply will not operate above this temperature. However, operating at lower temperatures will increase system reliability, which is desirable in many applications. In a fan-cooled desktop box, this may be achieved by blowing more air through the enclosure but may violate a system noise requirement of <60 dBA noise at 1 m in front of the box. Which is the more important requirement, reliability or noise? There is no absolute answer, but the packaging engineer must be able to present such trade-offs to the product design team, and metrics can make the necessary decision process more rational.

11.2.2 System and Packaging Metrics

The goal of advances in electronic packaging, and in electronics in general, is to make electronic functions faster, smaller, and cheaper. Performance metrics can exist for each of these or for various combinations.

Faster   The most often-quoted performance metric for computers is that of speed: MIPS, or millions of instructions per second. Also popular in computer ads is simply processor clock speed, expressed in megahertz or gigahertz, as a 1.8- or 2.4-GHz machine. Although useful for comparing machines with different speed sorts of the same processor, a clock speed metric says nothing about machine throughput when comparing different processors, particularly CISC (complex instruction set computing) and RISC (reduced instruction set computing), because of the varying number of clock cycles required to implement the same instruction. To compare machines better in real computing power, various benchmarks developed to emulate a given mix of computing requirements have been developed. A discussion of benchmarks is beyond the scope of this text, but even the packaging engineer must be aware of the ultimate measures of the art.

Of course, the entire concept of gates switching faster with each generation of IC processing is appropriate only for digital systems. But the same advances in IC technology allow analog and radio frequency (RF) systems to function at multi-gigahertz frequencies, so the concept of *faster* must be understood in the appropriate context. Any single metric may not capture the complexity needed to define "faster" in a given system.

Smaller   "Smallness" metrics generally are not very absolute. Clearly, the cellular phone or laptop computer would not have been possible 30 years ago, but since performance generally increases with each new product generation, it is difficult to make the direct comparison that "function X now occupies $Y\%$ of what it occupied 10 years ago." In computer packaging, however, the metrics of gates per cubic centimeter or bits per cubic centimeter, for memory, are significant, as these reflect the ultimate goal of packing the most electronics in the smallest physical volume.

Another significant packaging metric is that of silicon packaging efficiency, or the ratio of silicon (chip) area to package surface area, be it at the level of individual package, module, or printed circuit wiring board. Various papers over the years have examined this

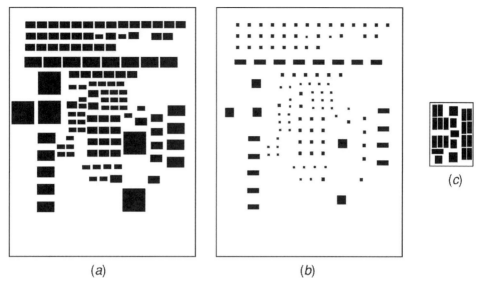

Figure 11.1 (*a*) Printed wiring board (PWB), (*b*) a "phantom" view of the chips within the packages on PWB in part (*a*), and (*c*) the same function as the PWB part (*a*) packaged on an MCM.

metric or its variants [1–3], and it is the significant improvement in this metric that is one of the driving forces for multichip packaging.

For example, consider the board shown in Figure 11.1*a*. This is a fairly representative board with a mix of moderate- and large-scale integration. A "phantom" view of the chips within the packages is shown in Figure 11.1*b*. The ratio of silicon area to board area is low, perhaps the typical 5%. The same function, with glue logic gathered up into two gate arrays, was packaged on the multichip module (MCM) in Figure 11.1*c*. Here, the ratio of silicon to substrate area is approximately 80%, better than an order-of-magnitude improvement. For this application of a space-borne signal processor, described as a "gigaflop in a soup can," such packing density is essential. An enlarged view of this MCM is shown in Figure 11.2 [4].

Cheaper The most startling feature of electronic systems is the steady decline in cost per function. This is sometimes expressed at the machine level as MIPS/$ or $/MIPS, and at the circuit level as cents per gate or cents per bit. Because of the relentless progress in ICs, more and more functionality can be produced for the same cost.

Even on-chip, however, more and more of both area and processing cost are being devoted to wiring, as five, six, and even seven metal layers are applied to ICs. It has virtually become true that "you pay for wire; the transistors are included for free." This is a very different case than in the early days of electronics when expensive components were hooked together with cheap wires. It has been shown that the cost of wire per unit length is roughly equivalent, no matter if that wire is on a chip, an MCM substrate, or a PWB or a backplane connecting several PWBs [1]. Thus, the rationale for integrating more and more functions on a chip for cost reasons is not because wires on a chip are cheaper, but because they are shorter. For the packaging engineer, taking meters of wire out of a system translates directly to cost reduction. The same rationale is driving three-dimensional (3D), or stacked, silicon packaging. Note that this argument pertains to mature technologies. Any technology in its infancy, be it the latest very large scale integration (VLSI) design rule or a

Figure 11.2 Enlarged view of an MCM.

low-volume interconnect substrate, will cost (in ¢/cm of wire) more than the average wire. But over the long term, reducing wire length will reduce system cost.

Reducing wire length also improves performance, of course. When circuit speeds were slow, the delay of interconnection (~1 ns per 6 in. of wire, in PWB-like material) did not matter. But as circuit speeds increased, wiring delays became significant. The delay budget (cycle time) of the mainframe computer roughly became one-third gate delay, one-third wiring delay, and one-third skew and noise [5]. All mainframe makers were the first to adopt multichip packaging to reduce the fraction of system delay due to wiring and gain maximum benefit from faster chip technologies.

As workstations approach the performance of mainframes, and PCs and laptops approach workstations, the need to reduce wiring delay by eliminating individually packaged chips impacts these systems as well.

Composite Metrics As electronic functions approach the goal of creating the "hairy glowing golf ball" [hairy, because of all the interconnects to input/output (I/O) devices, glowing because of the power consumed, and golf ball size so that it can operate synchronously at a clock frequency of several gigahertz], composite metrics can be used to measure overall packaging "goodness." A metric such as gate-Hz/(cm^3-$) relates circuit density, speed, and cost but is quite specialized to CPU-like functions. In general, it will be up to the system design team to develop appropriate metrics for a given product.

11.2.3 System Constraints and Trade-Offs

The packaging engineer must consider all constraints imposed by product design, including cost, size, weight, shape or form factor, human interfaces, power consumption, audible and RF noise generation, thermal requirements, cabling to other devices, and so on. For

11.2 Packaging and the Electronic System

portable equipment, battery life, weight, and power consumption are critical. The above requirements relate to the system as a whole, but each component of the system, from an individual chip to a complex subassembly such as a disk drive, has its own relationship to the rest of the system, just as the system has to the outside world. Thus, each subsystem will have constraints on size, weight, power consumption, I/O connectivity, physical interface, thermal environment, and so on. Often different design teams will be working on various subsystems; on large projects, tens or even hundreds of people may be involved. In such projects, well-defined constraint sets and interfaces are critical to program success.

The requirements placed on an MCM or "subsystem-in-package" design are an excellent example of system constraints. The term *system constraints* means all of the requirements—physical, electrical, and thermal—imposed on the MCM by the system that contains it. Physical constraints include size, weight, and mechanical aspects of connection to the PC board: through hole or surface mount, socketed or soldered. As will be discussed, often the complexity absorbed by the MCM wiring will allow a simple, undemanding interface to the board as well as an easily assembled lead pitch. Except for MCMs with very special requirements, the MCM should look like every other component from the viewpoint of the board assembler.

The electrical interface to the board is usually dictated by the mechanical considerations mentioned above. But the possibility of very short rise time signals leaving the module may impose severe requirements on the parasitics of the module-to-board connection. Special sockets, Z-axis conductive materials (elastomers or adhesives), multilayer flex, shielded or otherwise impedance-controlled connections, and even photonic (fiber-optic) connections are all possibilities for maintaining signal integrity. Once again, the performance of off-module signals must be dealt with very early in the design process, since this may be a major constraint limiting MCM performance in the system.

For many systems, the most severe external constraint on the MCM is heat removal. The thermal path within the MCM and the need for a low-resistance thermal path from chips to heat sink are critical. The external factors include heat sink dimensions, airflow, ambient air temperature, and desired junction (chip) temperature. The problem in MCMs is that system designers, accustomed to dealing with single-chip packages on closely spaced boards with moderate airflow, are reluctant to change cooling strategies to deal with this new component, the MCM. Yet, the MCM can present a higher thermal flux density than any other component. To lessen this thermal concentration by making the MCM bigger would defeat the purpose of using the MCM. So there is often conflict between the electrical performance people, trying to get the MCM as compact as possible, and the thermal analyst, faced with an impossible hot spot. In some instances, a simple computation will demonstrate to system designers that, even with the highest efficiency heat sink, a certain junction temperature is impossible to achieve given the MCM power dissipation, airflow, air temperature, and allowed heat sink volume.

As with electrical and physical system constraints, thermal constraints must be considered very early in the design process, in fact, well before the design of the MCM itself. It is easy to size the MCM. The substrate, if it is separate, is sized to hold the chips it needs to provide the functionality required. The MCM package is usually size constrained by the number of interconnects to the board (depending on whether peripheral or area array attachment is used) or by the substrate size. In either case, a good estimate of package size can be made. This will generally determine heat sink area, and the necessary thermal computation based on chip power can be performed. If nothing appears impossible, the necessary constraints on the MCM can be calculated (e.g., thermal resistance from junction to heat sink or total inductance in power and ground paths) and the comparative performance under these constraints for various MCM technologies can be determined. Trade-offs of cost and performance must be made. All of this work should be done before a technology is selected or a detailed design begun, or else the resulting MCM could be severely out of balance—overconstrained in some areas and very loose in others. The key is matching the technology to the application in every area of performance.

11.2.4 System Partitioning

The term *system partitioning* refers to the need to break up a large electronic function into smaller parts and then package these parts in some logical way. A generalized *hierarchy of partitioning* is shown in Figure 11.3. The largest electronic systems, such as supercomputers and telephone switching systems, occupy many equipment racks and consume huge amounts of power, cooling, and floor space. The initial partitioning is generally into units capable of fitting through a normal size door and that make sense as functional subsystems. Often, the cabling between racks can become cumbersome. Increasingly, this is being done by high-speed fiber-optic links. This decision, too, is a trade-off since generally many electrical

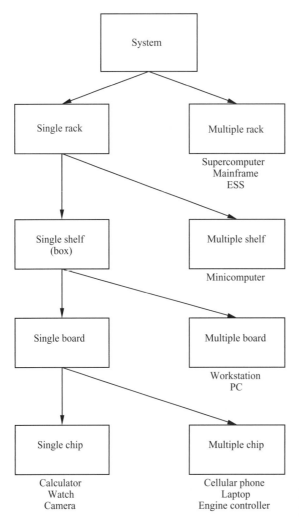

Figure 11.3 Generalized hierarchy of partitioning.

signals must be multiplexed and demultiplexed to take advantage of the fiber bandwidth. Cable and connector cost is reduced, but electronics cost increases. Fortunately, it is only the largest systems that will be packaged in multiple racks, and a great deal of engineering talent will be consumed in making partitioning decisions. The recently graduated engineer will have only a small part of the total system design.

The same is true of the multishelf, single-rack system, such as a large server or minicomputer. Within a single rack, there can be numerous design trade-offs in power distribution, heat removal, and electrical design. The total system function will be partitioned, first into shelves, then into individual cards in each shelf. Particularly important in systems of this kind is modularity, or some limitation on the number of distinct circuit card designs within the overall unit. Often, there will be multiple memory cards, or I/O channel cards, or disk driver cards, and so on. Great economies can be realized from partitioning into duplicated units, even if the units contain some type of programmability so they can serve multiple functions.

Because electronic functions have become so compact, largely due to the functionality of VLSI circuitry, most systems now fall into the "breadbox" or below category. The simple shelf, or box system, is the most complex system on which many design engineers will have an opportunity to work. There are two types: card/backplane and motherboard. The card/backplane system is typical of telephone equipment and some industrial electronics. A card cage houses a passive backplane containing multiple connectors into which cards with active circuitry are plugged. The simple to design (STD) bus card cage shown in Figure 11.4 is an example of this packaging method.

The motherboard is characteristic of "box" electronics, particularly personal computers. Instead of being a passive backplane, the main board of the system holds a great deal of the basic system functionality as well as connectors for "daughterboards" that provide additional functionality, such as memory cards or I/O controllers. In terms of partitioning, the philosophy is to get the minimum system configuration on the motherboard, with expansion slots and connectors used for additional capabilities. However, since many subsystems may be made by subcontractors or because subsystems may be interchangeable among several system designs, it is sometimes more convenient to have an essential subsystem plug into the motherboard, rather than incorporating that functionality on the motherboard itself. Part of this decision process depends on interconnect density. There must be a manageable number of connections between motherboards and daughterboards, so the mating connector does not get too big or expensive.

Even at the single-board-system level, there are important partitioning decisions to be made involving size, cost, and component complexity. Almost any digital system can be made up from small- and medium-scale integration (SSI, MSI) and catalog large-scale integration (LSI) parts such as memories and microprocessors. Yet it is possible to combine many of these functions on a programmable logic device (PLD) or field-programmable gate array (FPGA). How much logic should be incorporated into these parts and how many semicustom and programmable devices should be used in a given system are critical questions.

Similarly, the partitioning question when considering using an MCM or multichip package is equally critical. Determining which of the system's components should be included on the MCM is easy in some cases but very difficult in others. In many microprocessor or digital signal processor systems, the parts requiring very fast data transfer are the processor, the cache, and the memory management unit, and these can easily be incorporated into a moderate-size module. Many similar designs have been implemented, with the goal being to eliminate processor wait states by bringing the cache within one instruction cycle of the processor. This also creates a functional block with processor/cache buses contained entirely on the MCM.

444 Chapter 11 Design Considerations

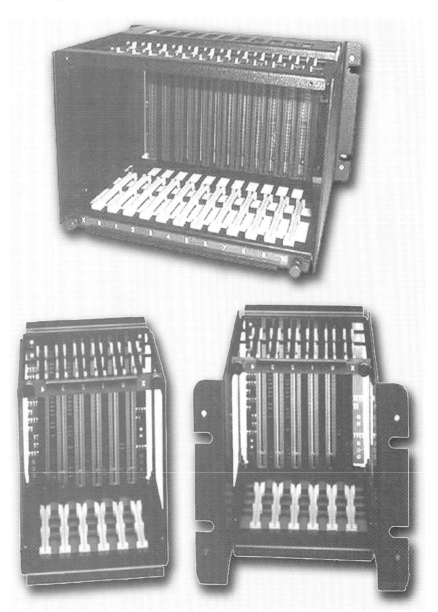

Figure 11.4 STD bus card cage.

Other systems, particularly large military systems, are not so easy to partition. Here, the goal is often size reduction, that is, compressing an entire board onto an MCM. Unfortunately, the resulting MCM may be so large as to be unmanufacturable or only producible at unreasonable cost. Often there is no clear approach to partitioning into a set of MCMs. A useful figure of merit (FOM) for an MCM partition is the complexity absorbed by the MCM, measured as the number of component signal terminals contained on the MCM over the number of MCM-to-board signal interconnections. Table 11.1 lists the relevant data for a few designs.

Table 11.1 Figure of Merit for MCM Interconnect Complexity Reduction

Application	Number chips on MCM	Total chip signal connections	Off-MCM signal connections	FOM
Digital signal processor (DSP) module	9	263	87	3.0
μP module	12	787	203	3.9
Display interface	7	310	90	3.4
Workstation core	10	653	110	5.9

This FOM is important because a significant fringe benefit of MCM use can stem from the fact that the large amount of silicon contained therein requires a large package size, even with very dense chip packaging (small chip pitch). A low number of off-module connections can be accommodated with 40 or 50 mils I/O pitch, versus the 20 or 25 mils pitch needed on high-lead-count, single-chip, VLSI packages, making for a more manufacturable assembly. If an area array package, such as a ball grid array (BGA) is used, a similar larger interconnect pitch is possible. This easier "space transformation" also reduces required local board wiring density, a benefit already mentioned.

The previous discussion deals with partitioning the system into what is on the MCM and what is not. A higher level of the partitioning problem concerns partitioning the entire logic set contained on the MCM into a set of chips. For modules consisting of already manufactured or catalog chips, this is not an issue. But as the performance benefits of MCMs begin to be realized through the use of chips custom designed for the MCM environment, the notion of "optimum partitioning" becomes useful.

Optimum in this case means a global optimization of cost/performance for the particular application. In the single-chip environment, when going off chip involved a large performance penalty because of the large parasitics to be driven and the large multistep driver needed to drive them, as much logic as possible needed to be included on a single chip. These chips are large, low yield, and thus expensive. In the MCM environment, the low off-chip performance penalty may lead to a very different cost performance trade-off: smaller chips and more of them. With much higher chip yield, overall system cost could be reduced while performance is maintained. Given the density of a flip-chip area array chip-to-substrate interconnect, accommodating additional chip-to-chip buses becomes far easier. A more optimum partitioning of wire is also possible if the wiring, both on chip and on module, is designed concurrently. In some cases, it may result in higher performance to run what would be a very long wire on chip, off the chip through a lower resistance module wire, then back onto the chip. Unfortunately, the computer-aided design (CAD) systems necessary to perform this global optimization are not yet available.

11.3 TRADE-OFFS AMONG PACKAGING FUNCTIONS

11.3.1 Signal Wiring

A principal requirement for all levels of electronic packaging is that the medium employed must provide sufficient signal wiring density to interconnect the packaged components. This is true of a multichip substrate, a chip package, a PWB, a connector, a backplane, or a cable harness. It is even true of areas on the chips themselves, since logic blocks, and individual gates, must be interconnected, and the necessary wiring density must be provided by the four, five, or six levels of on-chip metal.

Since the original work by E. F. Rent of IBM in 1960, many analysts have tried to determine an analytical relationship between the number of gates or bits in a logical subunit and the number of external signal connections required by that unit. The unit could be an on-chip logic block, or a chip itself, or an MCM, or a circuit board. Rent's early work was for random logic chips in the CPU of a mainframe computer, and the observed relationship was [6]

$$N_p = 2.5 N_g^{0.6} \tag{11.1}$$

where N_p is the required number of signal pins for N_g gates.

It is obvious, of course, that this kind of curve fit relationship will be extremely dependent on the architecture of the system and the types of chips being considered. Dynamic random-access memory (DRAM) chips pack the most bit density with the fewest number of pins (particularly in an $N \times 1$ configuration), while the byte or word-wide arrangements typical of static RAM (SRAM) require more pins for the same number of bits. Any bus-oriented chips, such as microprocessors, do not need as many I/O as a gate array with the same gate count. This kind of "family" arrangement is shown in Figure 11.5 [7]. At the "high end" of packaging—large MCMs or PWBs—the relationship again changes, since, in general, complete functional subsystems require fewer I/O than the equivalent gate count block in a large CPU.

Thus, Rent's rule, although a convenient fiction, is not really useful for predicting practical wiring requirements, particularly in microprocessor-based systems. It is better to use algorithms that look at pin count and device pitch, as shown in Figure 11.6 [8]. But even this is based on random logic in IBM CPUs in which most nets from one component terminate on one of the nearest-neighbor components. For bus-oriented systems, it is essential to include global buses and their contribution to wiring density requirements. For this reason,

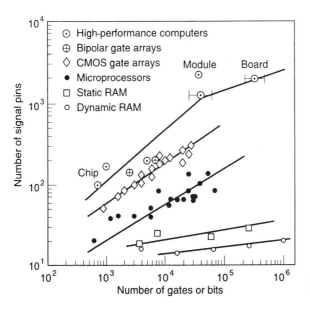

Figure 11.5 Rent's curves according to system and chip types.

11.3 Trade-Offs Among Packaging Functions

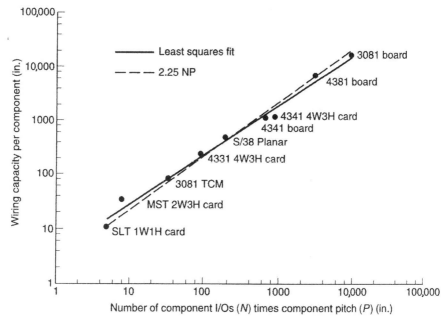

Figure 11.6 Wiring capacity.

a model proposed by Moresco [9] has sometimes been used:

$$W = \frac{A N_{\text{chip}} N_{\text{sc}} F_p}{2} + (1 - A)(N_{\text{chip}} - 1) N_{\text{sc}} F_p + \frac{\sqrt{N_{\text{chip}}} N_{\text{se}} F_p}{V} \quad (11.2)$$

where

- W = total wiring demand, cm
- A = fraction of nets nearest neighbor routed
- $1 - A$ = fraction of nets globally routed
- N_{chip} = number of chips on module
- N_{sc} = number of signal I/O per chip
- N_{se} = number of signal I/O off module
- F_p = chip pitch (wiring area limited footprint), cm
- V = 4 for peripheral I/O off module

On close examination, this formula makes good sense. The first term is the wiring demand from point-to-point nearest-neighbor nets, where the length of wire per net will, on average, equal the chip pitch, and the number of these nets is determined by the other factors (A, N_{chip}, and N_{sc}). The 2 factor derives nets from end points. The second term determines the contribution from globally routed nets as the product of the number of these nets [$(1 - A)N_{\text{sc}}$] times the meandering length of each [$(N_{\text{chip}} - 1)F_p$]. The third term is the contribution from the N_{se} wires off module, which, on average, travel one-fourth of the module side ($\sqrt{N_{\text{chip}}} F_p$).

From W, the necessary wiring density in the interconnect substrate can be determined, assuming the substrate size is known. Note that wiring demand (in centimeters) increases linearly with substrate size, but required wiring density decreases quadratically with size.

448 Chapter 11 Design Considerations

Table 11.2 Theoretical Wiring Capability for Various Interconnection Media

Medium	Wiring pitch (μm)	Number of signal layers	Total cm/cm²	Total in./in.²
Submicrometer VLSI	2–5	2–3	~10,000	~25,000
MCM-D	20–50	2–3	400–1500	1000–3800
MCM-C	200	2–30	100–1500	250–3800
MCM-L	125	<6	<480	<1200

Note: MCM-D, -C, -L—deposited, ceramic, laminate.

Thus, a wire-limited substrate can often be saved by a modest increase in area, rather than by the addition of wiring layers.

In 1984, the tremendous difference between on-chip wiring density and the wiring of the then available interconnection media was referred to as the substrate gap [1]. If anything, the intervening years have exacerbated this gap, save for the introduction of MCMs. Table 11.2 lists typical theoretical wiring densities for various interconnection media.

Note that these are theoretical wiring densities. Various path blockages due to chip or module I/O structures will obviously reduce this density. But not so obviously, in some technologies, signal vias between X and Y signal pairs can severely impact density. Some thin-film implementations require a 50-μm via between thin-film layers, thus cutting wiring density to half of theoretical and requiring more layers than should be needed. This effect is even more pronounced in PWBs where 2 to 3-mil lines must be compromised by 6 to 8-mil or larger drilled holes. This problem is reduced or eliminated by laser drilling, although large catch pads may still be required, because of material instability across large panels, even in advanced microvia PWBs. It is not uncommon to need a 20- or 22-mil catch pad for a 10-mil mechanically drilled hole or a 250-μm catch pad for a 50-μm laser-drilled via. In general, it is much easier to make fine lines than small holes. Figure 11.7 shows the impact of via size on achievable wiring pitch.

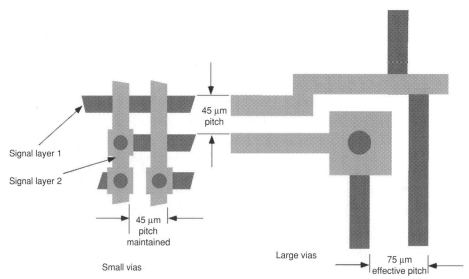

Figure 11.7 Effective wiring density reduction from the impact of large vias and capture pads. Both sets of design rules have 45 μm line pitch [10].

11.3 Trade-Offs Among Packaging Functions

Just as local congestion in a PWB can create a wiring demand requiring more layers than the average wiring demand would indicate, local congestion in MCMs can cause a similar burden. There is rarely a congestion problem in an MCM containing peripherally leaded chips, except in regions between the chips and the module I/O terminals (assuming these are also peripheral) where often large fanouts must occur. In this case, increasing the size of the module slightly is preferable to adding layers.

The most demanding signal wiring density occurs in an area array flip chip, where all of the signals must escape the area under the chip into the rest of the substrate. Often, localized line width reductions, which tax the technology, must be employed to achieve the required wiring density.

A good rule of thumb for MCM substrates is that the wiring density of a substrate should never impact the chip pitch. In other words, the goal of the MCM is to minimize wire length and, thus, keep the chip pitch as tight as the chip mounting technique allows. (Flip chips may require 0.5-mm spacing; wire-bonded chips ~2 mm). If the size of the module is determined by wiring density instead of these attachment limitations, the substrate technology is probably a poor match to the functionality required. It makes as little sense to connect an array of 400 I/O VLSI chips by chip-on-board (COB) as it does to connect 32-lead SRAMs with 20-μm-pitch thin film. In both cases, there is a poor match between wiring demand and substrate wiring capability. The same is true for PWBs. Even advanced microvia PWBs may not have enough local wiring density to cope with the demands of a large flip chip.

A metric for this match is the length of wiring used for a particular design over the theoretical wiring capacity of the substrate technology. Often this metric can be computed from CAD data. Typical MCMs incorporating several memory chips might have 10 to 20% of the wiring used, compared with dense application-specific ICs (ASICs) or switching modules with 60 to 70% signal wiring utilization. Figure 11.8 shows the wiring on the two signal planes in a typical processor/SRAM module. Sixteen percent of the available wiring capacity is used.

As an example of this wiring density and demand analysis, consider the case of determining the necessary wiring density in a MCM. First, of course, some information about the chips to be included on the module must be known. We are looking for an estimate of the number of signal layers that must be provided, so we must consider chips at the maximum size and I/O density.

Chip size has increased gradually over the years. Ten-millimeter chips used to be considered large. Now, a 12-mm square chip is commonplace. A 15-mm square chip is at the upper limit of manufacturability today, and chips of this size are expected to be about as large as can be manufactured due to yield constraints, so we will use a chip of that size for this analysis.

Chip I/O will, of course, be area array to generate the maximum wiring demand on the substrate. Because of considerations of the coefficient of thermal expansion (CTE), there is a practical minimum size of bump interconnect, whether that is by solder or other material. A 125-μm interconnect on 250 μm pitch is reasonable, given the CTE mismatches (or power cycling requirements for silicon on silicon). Thus, on a 15-mm square chip, there is the potential for 3600 I/O, which is far more than most chips with peripheral I/O, where 500 would be extremely large. A good rule of thumb is that half of the bumps would be needed for power and ground and half would be available for signals. Thus, 1800 signal I/O would impose the maximum wiring demand that must be considered.

A reasonable maximum for the number of these chips that would populate an MCM must be determined. Of course, the 1800 signal I/O chip represents the worst-case wiring

Figure 11.8 Typical processor/cache module showing the low density of utilized wiring [10].

demand. Realistic MCMs would include many varieties of chips, particularly memories, which would have far fewer I/O. But to determine the maximum substrate capability, we will use this worst-case chip in a 5 × 5 array. The potential gate count of such an MCM is more than a large mainframe, so this is also an extreme case. With flip chips, we will assume a totally tiled surface (flip chips can be assembled with as little as 0.25 mm between them), so a substrate 8 cm square will be used. This minimal area, with very high chip coverage ratio, again imposes a maximum wiring demand or spreads the total wiring demand over a small area, pushing up the required substrate capability.

Using the Moresco formula, the wiring demand can be computed. Assume 70% of the signal I/O are nearest-neighbor routed and 30% (540) are globally routed (bused). This represents a large number of bus lines. We will also assume 1000 signal I/O off the module. These numbers yield a total wiring demand of 45,000 cm in the module, or 703 cm/cm^2 for our 64-cm^2 module.

What is the wiring capability of a typical MCM substrate? If we assume wires 20 μm wide on 40 μm pitch, a capability regularly produced in MCM-D, a density of 250 cm/cm^2 per layer is available. This, of course, assumes no blocking due to power distribution and microvia capability (vias totally contained within the signal wire). Microvias are normally done, but there would be some blocking from power vias passing through the signal layers to the power distribution planes (one-twelfth of the wiring tracks). However, it appears that four signal layers would accommodate the wiring demand of this extreme example.

In reality, of course, most chips on any module would have far fewer than 1800 signal I/O, so wiring congestion in most areas would be considerably less. In such cases, it is normal to allow somewhat more substrate area for the densest chips, thus making more wire available for those congested regions rather than increasing the number of layers, which increases cost and decreases yield.

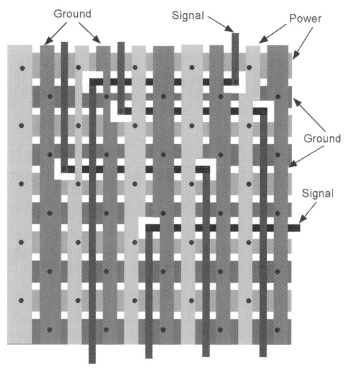

Figure 11.9 IMPS substrate (vias not shown).

For many applications, far less wire is needed than is available on only two signal layers. In these cases, it becomes very costly to provide two signal layers and two more layers for power distribution. The IMPS topology (Interconnected Mesh Power System) allows signals and power to be combined on only two metal layers while retaining the low-impedance characteristics of planar power distribution [11]. A signal line density of 250 cm/cm^2 is easily achieved on MCM-D with two metal layers. Figure 11.9 shows a simple illustration of the IMPS topology.

This discussion has thus far dealt with the quantity of signal interconnect lines, but quality is also important. As in PWBs, signal lines are described by their characteristic impedance, but in many MCM technologies, the lines are also lossy because of their extremely fine cross section. Resistances of 2 to 10 Ω/cm are typical for thin-film lines. Of course, the MCM signal traces are generally far shorter than their PC board equivalents, but the losses must still be included at the design and simulation stage. In complementary metal–oxide–semiconductor (CMOS) systems, often the rise time is greater than five times the line propagation delay, so a lumped circuit equivalent can be used instead of a transmission line analysis. The line resistance simply contributes to the RC delay. For long lines or short rise times, a lossy transmission line analysis must be used [12].

Particularly in bipolar systems, the line resistance can be used as part of a series-terminated impedance-matching scheme, but precise implementations of this scheme requires customizable output drivers and a design system combining chip and substrate design. These topics were covered in detail in Chapter 6.

Electromagnetic field analysis techniques are being used to simulate, not only the characteristics of the lines, but also interconnect structures such as wirebonds [13]. Often, a heavily loaded, on-module bus determines the maximum module operating speed.

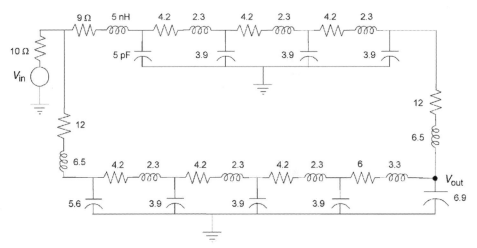

Figure 11.10 Equivalent circuit of a 10-chip module address line with SPICE parameters extracted from physical measurements.

Figure 11.10 shows a lumped circuit model for such a bus line, connecting two VLSI circuits and eight SRAMs. The bus has been structured as a loop in order to minimize delay. Knowing the physical lengths of module interconnects from the chip layout and the per-centimeter wire parasitics, it is easy to compute the circuit parameters. Figure 11.11 shows the signal waveforms at the near and far ends of the bus, with one of the VLSI chips driving. A similar analysis method is also useful for adjusting the clock distribution to minimize skew and reflection problems.

11.3.2 Power Distribution

In electronic packaging, the term power distribution refers to everything necessary to deliver constant, noise-free, direct-current (DC) power at the appropriate voltages to the electronic components that make up the system. This includes everything from the line cord to the

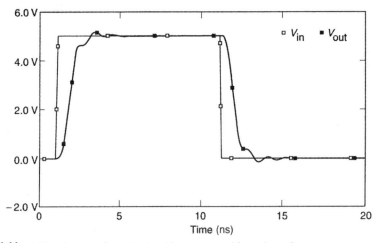

Figure 11.11 Address bus waveform showing driver output and far-end waveform.

power supply [alternating-current (AC) to DC conversion], power distribution wiring, decoupling capacitors, and embedded power and ground planes in circuit boards. On large machines, power systems can be distributed, with one or more bulk rectifiers or off-line switchers supplying an intermediate power distribution voltage, such as 48 V, and on-board or in-shelf DC–DC converters supplying system operating voltages, such as 3.3, 5, and ±12 V. In single-box systems, there is usually a single power supply, often bought as a subsystem by the system assembler, which converts 120 V AC to low-voltage DC. Increasingly, these are off-line switching supplies rather than the less efficient transformer/rectifier/filter supply.

Portable electronic systems are battery powered, often with rechargeable lithium ion or other sophisticated batteries. Chargers can be built in or external, in the line cord or plug-containing transformer box. Even battery-powered systems, however, may have some form of DC–DC conversion to provide necessary voltages, such as for display backlighting in laptop computers.

Power conversion itself, however, is beyond the scope of this book. The noise-free distribution of power, once it has been converted to low-voltage DC, is one of the functions that packaging must provide to the active circuitry. This is usually done by a combination of low-inductance planes within PWBs and chip packages and the provision of decoupling capacitors of different values to serve as local charge reservoirs for the changing current demands of the circuits.

An MCM example will be used to illustrate the concepts involved. As chip power increases and supply voltage drops, it becomes more and more difficult to supply stable, noise-free power to the chip. A 50-W, 3.3-V microprocessor IC draws as much current as can be supplied by 12-gauge copper in normal house wiring. Yet, in MCMs, such currents must be supplied on power and ground planes with a thickness measured in micrometers. Connections from the planes to the chips are through wires, fingers, or solder joints finer than a human hair. Very careful design is required to hold DC voltage drops in this power distribution system to a few percent of supply voltage. Multiple connections from the chips are essential, of course, and in some thin-film MCMs, appliquéd bus bars on the module surface eliminate voltage drops in the power distribution planes by bringing power into the interior of the MCM. Figure 11.12 shows the DC drops associated with such a module, from connector all the way to on-chip power distribution.

Voltage drop problems also occur on chip, as large currents must be brought to the interior circuitry from peripheral interconnects. In very high power chips, this becomes another reason to use area array flip chips, connecting power and ground, as well as signals, where they are needed.

Problems arising from DC voltage drop now only affect the highest power chips in MCMs. A far greater concern for most MCM designs is control of switching noise from simultaneous switching output drivers.

In a module, the switching noise problem becomes easier to manage than in single-chip packages, where package lead inductance is substantial. Typically, in large chips, the power distribution system is broken down into various segments: core logic and perhaps several sets of output drivers. By keeping these separate through the package and onto the board, where adequate energy storage for sharp transients is provided by decoupling capacitors, noise from the outputs is kept out of the core logic, where logic errors could occur. Some ceramic single-chip packages are bringing these attributes of power and ground planes, and even crude decoupling capacitors, into the package. Several independent capacitor sections can be included. However, inductance in package leads and traces can still contribute to noisy power.

454 Chapter 11 Design Considerations

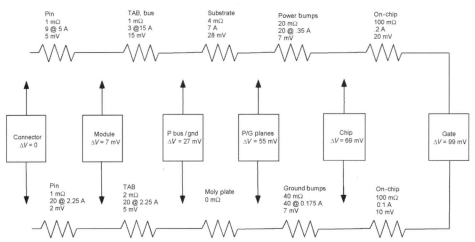

Figure 11.12 Equivalent circuit for DC power distribution on a high-power MCM. Voltage drops are shown at various points in the circuit.

In the MCM, power distribution is generally provided on continuous or perforated power and ground planes. Some thin-film MCMs incorporate a parallel-plate decoupling capacitor with a deposited dielectric. On silicon substrates, a grown SiO_2 layer can provide decoupling if highly doped silicon is used as one capacitor plate. The electrical properties of such large-area capacitors are vastly superior to most multilayer ceramic chip capacitors, which, although they have larger capacitance, also have 300–800 pH of parasitic inductance, making them self-resonant in the 15- to 80-MHz range, where decoupling is essential. Thus, power distribution within the MCM can have far fewer parasitics than single-chip packages, eliminating the need for separation of various power distribution sections in the packages. Figure 11.13 attempts to dramatize this concept with large and small inductor symbols.

However, no amount of MCM decoupling will make up for the separation of the chip from the decoupling charge reservoir by inductive connections. Even short wire bonds or tape-automated bond (TAB) connections have an inductance on the order of 1 nH. With switching currents of several amperes per nanosecond, multiple chip connections for power and ground are still required. Flip-chip solder bumps have inductances ∼20 times lower than wire bonds and are desirable when the ultimate in noise-free performance is necessary.

Although the analysis of both DC and AC power distribution is straightforward, once parasitic values and dI/dt values are known, in many cases, the expected dI/dt is not known, particularly for CMOS systems. Since the MCM environment provides a much lower signal capacitive load than the circuit board environment, drivers do not have to supply the same peak currents, although they may operate faster. It is often necessary to simulate an entire MCM bus structure, with assumptions about worst-case simultaneous switching, in order to determine what dI/dt will be encountered. Output driver models and resistive signal line losses must be included.

In summary, the very different constraint set for MCM power distribution, although offering the opportunity for improved noise control compared with PC boards, requires very

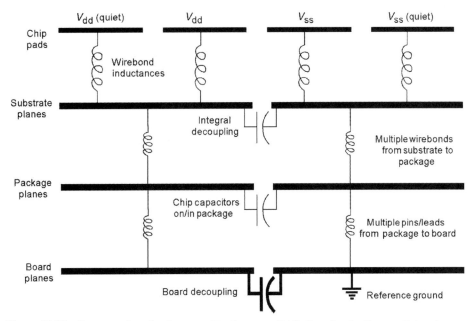

Figure 11.13 Representation of package parasitics for power distribution, showing the use of internal decoupling and power distribution planes.

careful design methods to ensure the desired result. The desired FOM will generally be the achievable noise voltage reduction each design change produces.

11.3.3 Thermal Management

Chapter 7 dealt with the detailed problems of heat removal in electronic equipment, but at this level of system overview, it is appropriate to review the overall thermal management problem, which in some cases can become the principal driver of the form taken by packaging hardware.

Although the shift from vacuum tubes to discrete transistors in the 1950s decreased the thermal management problem, every other electronic advance since then has exacerbated it. Each time circuitry becomes denser, be it from advances in levels of integration on chip or from packaging improvements such as the introduction of surface-mount components, heat removal becomes more difficult. Without the transition from bipolar to CMOS for most applications, and even for high-performance computers, currently possible packaging densities would necessitate elaborate and costly heat removal methods in most systems.

On a system basis, most electronics can be cooled by convection or by the use of small fans. Inside the box, however, localized heat sources, such as MCMs or high-power individual ICs, can produce thermal management problems. Thus, the MCM is a victim of its own success. By allowing very tight chip spacings and increased operating speed, the MCM concentrates a great deal of power dissipation into a very small footprint, thus creating a cooling problem. In fact, in many designs, thermal management emerges as the most important constraint.

The first design issue is the direction of heat flow from the chips, which is intimately tied to chip attachment. Wire-bonded or TAB chips are die bonded to the substrate, so heat flow will be into the substrate, although the ultimate direction will be determined by a "cavity-down" or "cavity-up" configuration. Solder bump or flip TAB chips can use either convergent (through the substrate) or divergent (back-side) heat removal or both.

With chips die bonded to the substrate, heat must flow through the substrate's dielectric layers, which are often polyimide or other materials of low thermal conductivity. It can become necessary to provide thermal vias (metallic conducting paths) through the dielectric to the substrate base itself. However, these vias take up space that may be needed for wiring. In the extreme case, a thermal via the size of the die can be made through the dielectric and the die directly bonded to the substrate base. All wiring is then forced into the "streets" between the die, with severe impact on the MCM wireability.

With flip chip, heat removal through the substrate can be enhanced by providing extra solder bumps not required for electrical connectivity. These can be placed anywhere on the die, even over areas of active circuitry where there are no connection pads. Additional passivation is applied to the die, and solder bumps are created on the surface of this layer. Although hundreds of extra thermal paths can be obtained this way, each bump requires a thermal via in the substrate, creating the trade-off between thermal conductivity and wireability previously described. With both flip chip and flip TAB, the space between the die and the substrate can be filled with some electrically nonconductive material (usually an epoxy) to provide increased heat transfer.

For maximum thermal performance, back-side heat removal, as used in the IBM thermal conduction module and other high-end designs, can be used when the dies are facing down. The advantage is in separating the heat transfer path from the electrical wiring and its associated insulating materials. Many methods of various effectiveness and cost, including pistons, springs, filled elastomers, and conductive pastes, can be used, depending on the requirements.

The principal thermal design issue for the MCM itself (as opposed to the system thermal issues of heat sink design, airflow, etc., which will be discussed later) is the provision of a low-thermal-resistance path from chip to heat sink. Normally, a simple 1D computation using material thicknesses and thermal resistances serves as a first approximation on the effectiveness of the thermal design. But often, a detailed finite-element analysis of all the chips in the MCM and their respective power dissipations and thermal path particulars is required [14]. Figure 11.14 is an example of the kind of analysis that can be done [15].

The severe thermal constraints of MCMs have led to the use of exotic substrate materials, elaborate conduction mechanisms to chip surfaces, immersion of chips in liquids, and other costly schemes which have appeared in the literature and are beyond the scope of this chapter. From the design standpoint, it cannot be stressed too strongly that adequate thermal analysis must be done at the earliest stages of MCM design, as the thermal constraints will often make or break an MCM application.

11.3.4 Interconnect Testing

A complete discussion of the testing of electronic assemblies is well beyond the scope of this chapter, but we mention here some aspects of the testing of interconnection components, such as bare boards, MCM substrates, or chip packages. We also discuss some of the implications of the testing of assembled units on interconnection. Though the notion of testing is not specifically a trade-off, the means for performing tests (e.g., building in test pads) must be considered very early in the design.

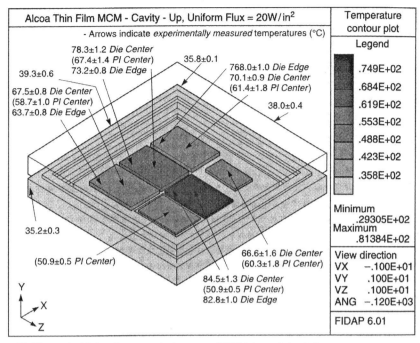

Figure 11.14 Typical finite-element analysis output of MCM thermal simulation.

Many unpopulated packaging elements must be electrically tested for opens in conducting paths and shorts between conductors. For bare circuit boards, this is generally done by a "bed-of-nails" tester in which pins come in contact with the terminals of each net and a computerized switch matrix ensures proper functionality. Often this testing is done on individual layer pairs before the board is assembled. This avoids adding value to "dead" boards. In general, the earlier testing can be done, the better.

Similarly, high-density MCM substrates must be tested before components are mounted. A "known good substrate" must be combined with a "known good die" to achieve the highest probability of successful assembly. Because of the wiring density of MCM substrates, use of a bed-of-nails tester is impractical, since even exotic buckling beam probes cannot achieve the ∼100 μm pitch that may be required to test nets that end at wirebond pads.

Instead, MCMs are generally tested with some form of moving probe tester. More exotic measures, such as electron beam testers that can "see" voltage levels on conductors, have been demonstrated, but they have not been accepted due to cost and complexity (vacuum required). The moving probe, usually two probes, tester is simply a robotic, precise way to "buzz out" substrate wiring by measuring the resistance of nets. Some testers use capacitance to look for short circuits by knowing or learning what capacitance to ground a particular net should exhibit. A net showing greater capacitance must be shorted to another net. As most substrates, and certainly the wafers used in MCM-D, contain thousands of nets, the speed of testing is an issue. Multiple probes, sophisticated control software to prevent collisions, and careful design of moving mass are being used to provide testers capable of hundreds of measurements per second.

Expensive single-chip packages, such as ceramic pin grid arrays, are also tested before chips are die bonded to them. This testing is complicated by the potential for having to

458 Chapter 11 Design Considerations

test on a double-sided part in cavity-up designs (pins or pads on bottom; die cavity and wirebond shelf on top). Robotically controlled, two-probe testers have also been used in this application.

The logic and methods of testing completed functional subassemblies are discussed elsewhere in this book. One method of testing, particularly in MCMs, has a significant impact on interconnection, however. This method assumes the accessibility of all nets in the module, even those that would normally not be brought to an external connection point. By providing test pads for these nets, complete diagnostic probing, sometimes using a robotically controlled logic probe, can be used to debug a module and determine what chip or chips need to be replaced. This test method requires additional wiring and the provision of space for probe pads on the substrate and must be considered at the earliest stages of module design.

In this, then, as in all aspects of the interconnection of ever-denser electronics, a complete "system view" is essential from the very beginning of the design process. The omission of any element can lead to a "fatal flaw" in the overall design and the need to start over.

11.4 TRADE-OFF DESIGN EXAMPLE

To pull all of these concepts of trade-offs and technology selection together, it will be helpful to work through a design example.

The example is a processor core module. There are two VLSI chips and eight static memories on the module. The CPU measures 16 mm square, has 312 I/O on 210 μm pitch, and consumes 15 W. The memory management unit (MMU) is 13 mm square, has 420 I/O on 110 μm pitch, and consumes 12 W. The eight memories are 6.34 mm × 16.13 mm, have 35 I/O, and draw 0.7 W each. Chips are wirebonded. The package for the module has 100 I/O, including a 64-bit bus, miscellaneous control signals, and power/ground. The module internally has a bus architecture with 64-bit data and 36-bit address buses connecting the CPU and MMU. Each SRAM ties into 17 bits of the address bus and 8 bits of the data bus. There is also an 8-bit parity bus, one for each SRAM, connecting CPU, MMU, and SRAM. The module must operate synchronously at 80 MHz. (This is an old example, but the concepts have not changed.) All chips are bipolar CMOS (BiCMOS) with voltage swings from 0.4 to 2.4 V. The total module height must be <1 in. with a heat sink. Assume air cooling with a maximum inlet temperature of 40°C and that a flow of 200 lfpm is available. The maximum junction temperature is 100°C. Appropriate decoupling for the 5-V supply must be provided.

The assignment is to design the module, that is, determine substrate type, layer count, electrical and thermal performance, size, layout, and package type, all for lowest cost.

This kind of module design must include a constraint analysis, floorplanning, and wireability analysis. Constraints could include electrical, wireability, thermal, size, and so on. For this module, chip power is high (15 W + 12 W + 8 × 0.7 W = 33 W), and the module height is constrained, which should raise a red flag about the thermal performance. Any thermal constraint will depend on module size, so the next step would be to floorplan.

Because there are wide buses between the CPU and MMU, many designers would immediately draw the floorplan indicated in Figure 11.15a, where the CPU and MMU are next to each other. But this puts the most power (27 W) at the center of the module, creating a hot spot. Since the module is synchronous, there is no real need for these two chips to be next to each other, assuming enough wiring tracks are available to carry the necessary interconnections. Thus, the floorplan of Figure 11.15b offers a better solution. The hot chips are separated to distribute the heat more evenly, yet nothing is lost in performance.

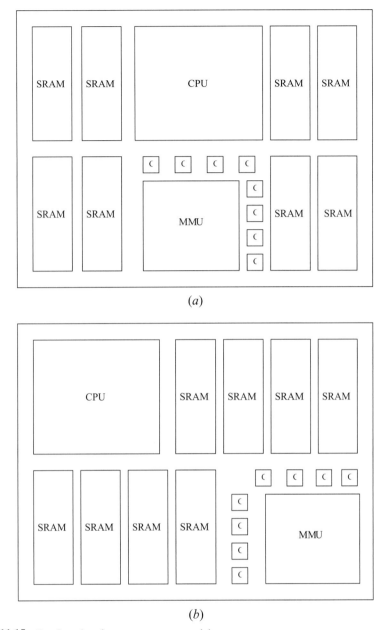

Figure 11.15 Two floorplans for a processor core module.

The exact spacing between the chips is a function of wirebond area and wirebond pad pitch, which depend on substrate type. The 1.5-mm spacing indicated in Figure 11.15b is for an MCM-D structure, in which the 110 μm pad pitch on the MMU can be accommodated by one row of substrate pads. In MCM-C, at least two rows would be needed, and in MCM-L, possibly three, depending on how pads were fanned out to vias. Thus, the substrate area will change depending on technology, just for pads.

Wire density is also critical. The densest area will be around the MMU, where ~350 signal lines will have to pass around two sides of the chip, since the chip is in the corner.

The space is 26 mm, and 350 line-wiring capability implies 74 μm pitch, a good match for two layers of MCM-D. Other technologies would require considerably more signal layers to ensure wireability.

The electrical performance will be determined by the speed of the address bus that connects all 10 chips. The 80-MHz clock implies a new clock pulse every 13 ns, or a clock pulse width of 6.5 ns. Using a rule of thumb, that rise and fall delay should be no more than one-tenth of the clock pulse of $\sim$0.7 ns. Will this be true with the MCM-D substrate of 38×54 mm shown in Figure 11.15b? Figure 11.10 showed the equivalent circuit of an address line on this module, with the bus running in a loop to provide a parallel path and minimize loading. In this case, the lines were modeled as having 5 Ω/cm resistance, 1.2 pF/cm capacitance, and 3 nH/cm inductance, fairly typical values. An additional 3 pF per chip was included for loading. The interconnect parasitics were simply scaled from Figure 11.15b. The resulting bus waveform is shown in Figure 11.11, which shows a rise time of $\sim$0.8 ns, close enough. The assembled substrate would be packaged in a leaded flatpack with generous lead pitch, since only 100 leads are required. A typical ceramic quad flatpack (QFP) would be 7 mm larger on each side than the substrate, leading to a 52×68-mm package.

Thermally, the module heat sink must dissipate 33 W into a 40°C, 200-lfpm airstream with a heat sink of $52 \times 68 \times 23$ mm. (The total of package and heat sink had to be 1 in. or less, and the package could be as little as 0.1 in. high). With a cavity-down configuration, the back of the substrate is bonded to the package, which is in turn bonded to the heat sink with fairly low thermal resistance. The major temperature drop will be between the heat sink and air. A pin fin-type heat sink of this size, with 200 lfpm of air, has a thermal resistance of 1.6 C/W for a temperature rise of 53°C. Given inlet air of 40°C, the heat sink temperature will be 93°C. A computation of the conductive thermal path between chip and heat sink should be done to show the additional conductive thermal resistance, but the temperature rise will be a few degrees. Thus, the thermal constraint is barely satisfied.

This brief exercise demonstrates the trade-off and technology selection methodology used in a particular MCM example, but the same philosophy carries over to all packaging problems. A balance must be achieved among all constraints, and it is often up to the packaging engineer to mediate competing interests to achieve an optimal design. The basic packaging functions of signal interconnection, power distribution, environmental protection, and heat removal must be provided for chips subject to many system constraints. It is essential to provide balance among these constraints.

11.5 PRODUCT DEVELOPMENT CYCLE

This section addresses the design methodology for implementing an electronic system that meets the product development objectives. Due to tremendous international competition in this sector of the economy, an effective design technique must be used if a product is to be successful. Most companies develop a design methodology that addresses issues related to the particular types of products they offer as well as the accepted level of quality and manufacturability. These "in-house" design processes are typically proprietary and consist of a mix of CAD products from various vendors and, often, additional custom software or manual processes available only in that particular company. The best electronic components, interconnect substrates, connectors, and assemblies will not lead to a successful electronic system unless the design of the product effectively applies their characteristics to the design objectives. Many of the terms and processes described are generic while others are specific to products from Mentor Graphics Corporation, a leading supplier of CAD software.

11.5.1 Traditional and Modified Product Cycles

The traditional design development cycle is shown in Figure 11.16. The traditional design cycle is a series sequential process that has evolved over many years. It originated from a design environment that was based on intensive human effort before advanced CAD software existed. The process involved many hours applying engineering experience estimating electrical performance; then engineering prototypes were developed to confirm performance. Often circuit modifications were required and additional prototypes developed. After iterating until acceptable performance was achieved, the final product was developed. Development cycles were frequently long with large and unpredictable variations in development time.

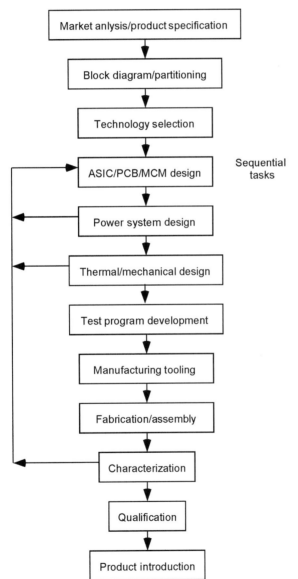

Figure 11.16 Traditional product development cycle.

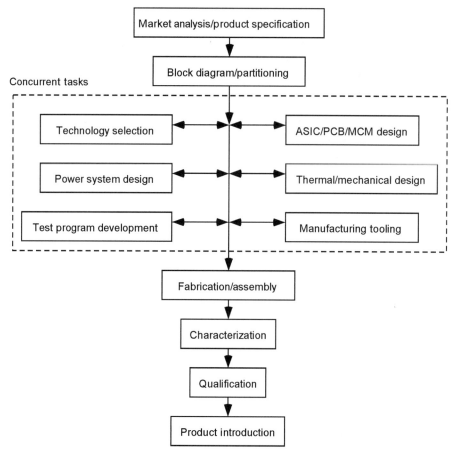

Figure 11.17 Modified product development cycle.

The traditional development cycle (or variations) is still often used. Experience with the type of product being developed, product introduction time frames, product development costs, and expected product market lifetime are factors in determining the suitability of the design approach selected.

The enhanced product development cycle is shown in Figure 11.17. This approach involves using concurrent design techniques to shorten the development cycle. Advanced CAD software allows for more accurate predictions of circuit performance and eliminates the need for engineering prototypes. This is more appropriate for a team-based product design, especially if a reduction in development time is required. The tight coupling of the concurrent steps requires coordination between various designers. As the details of the electronic circuits are determined, the technology selection, power distribution requirements, thermal environment characteristics, and mechanical structures may need to be modified. Concurrent modifications to each of these elements as the product is designed ultimately lead to a design acceptable to all aspects of the product requirements.

Variations of traditional and enhanced development cycles are often the best choice. Experience with the type of product being developed, product introduction time frames, product development costs, design resource availability, and expected product market lifetime are factors in determining the suitability of the design cycle selected. The market success of a product requires that it function as specified, meet cost objectives, and be introduced to the market within the introduction time frame. If the development cycle takes too long and could potentially miss the introduction date, it will risk being canceled by the marketing group. It is an emotional experience for an engineer to have a project canceled for nontechnical reasons like missing a market window.

11.5.2 Market Analysis and Product Specification

Market analysis and product specification involve addressing present or future consumer demand for a product that also meets the profit requirements of the supplier. The electronics industry is very sensitive to economic variations, technology evolution, and competition. A successful product planner must be familiar with future trends in each of these areas. Often errors in these predictions will lead to modifications of a product specification (even during the design cycle) or canceling the product (even if original "time-to-market" objectives are being met).

11.5.2.1 Profit Model

A well-designed and manufactured product can still lead to an unprofitable product and even cause a company to go out of business. A profit model is used to predict the profitability of a product. Profit models used by companies are very sensitive and kept proprietary. A simple profit model is shown in Figure 11.18. Key characteristics of this simple profit model are:

- The sales rate is constant from the time of introduction to peak sales.
- The peak sales occur at a time that is independent of product introduction.
- The sales rate decreases linearly from the time of peak sales to the end of the product life.
- The total sales are the area under the curve.

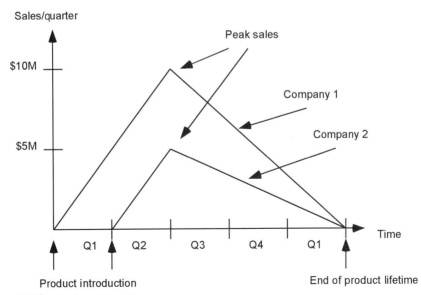

Figure 11.18 Simple profit model.

It should be noted that this is a simplified profit model. Many other factors are not considered, among these are seasonal sales variations (i.e., Christmas sales season), price erosion due to the introduction of competition, or economic changes.

Nonrecurring engineering (NRE) costs are the fixed, up-front costs for developing the product. Recurring costs are the costs of manufacturing a product and the costs of advertising and distributing the product. The profitability of a product is the total sales minus the sum of NRE and recurring costs. As shown in Figure 11.18, a delay in product introduction results in lost sales. These lost sales can result in reductions in the profitability of a product and possibly a product that loses money. This illustrates the sensitivity of a product to meeting the expected introduction time frame.

11.5.2.2 Product Specification

A product specification involves defining the functionality, mechanical/physical interface, and cost guidelines. There are many ways to describe the functionality of an electronic system. It may consist of a "human language" description of the function, truth tables, state diagrams, graphs, register transfer language (RTL) descriptions, or a mixture of these. There are many ways to implement a product from a specification.

Experience and knowledge of applicable technologies are necessary to effectively progress from a specification to the electronic circuits that perform the functions. A simpler design is a better design. If electronic circuits are selected that are more complicated than necessary, it will lead to higher product costs. This will leave a door open to competition. A competitor may implement an identical product specification with lower cost components and take the market (and profits) away from your company.

11.5.3 Block Diagram and Partitioning

Once the market analysis and product specification is complete, the system is described in functional blocks and a block diagram that includes the interaction between blocks is created. This is used to partition the system into smaller parts. The partitioning is done using a "top-down" approach. The system is divided into sections that are implemented and connected together (electrically and mechanically). This may involve dividing the functions into different printed circuit boards (PCBs) (or MCMs) each addressing the specific requirements of that portion of the system. Cabling, thermal requirements, and mechanical construction must be evaluated to properly partition a system. Technology requirements also contribute to partitioning. If a portion of the functionality requires a high-density, impedance-controlled, multilayer PCB, it would not make sense to include large through-hole components in this partition if it can be placed in a different PCB that has similar technology requirements.

11.5.4 Technology Selection

Once a product has been partitioned and functional specifications defined, the technologies necessary for implementation are selected. This includes the selection of semiconductors, passive components, connectors, interconnect substrates, cabling, housing, and thermal components. During the CAD process it may be determined that modifications to the technology selection are necessary. Semiconductor packages may need to be changed to better match interconnect substrate constraints or due to thermal conditions. The number of interconnect layers in substrates may be increased or decreased or the type of interconnect substrate may need to be changed. These modifications may have an impact on cost, assembly, or performance. As a result, the changes must be conveyed to designers of the affected sections to allow effective trade-offs to be determined. The use of more advanced technologies may be attractive to a designer of a particular section, but to optimize the overall product, more advanced technologies may not be justified.

11.5.5 ASIC/PCB/MCM Design

The CAD process used to implement ASICs, PCBs, and MCMs will be described in detail in later sections. This section will give a general description of how they fit with the overall product development flow.

11.5.5.1 ASIC Design

The use of ASICs in a product is a key factor in the technology selection of semiconductor products. Partitioning, performance, cost, and size are greatly affected by the use of ASICs. If commercially available ICs are capable of meeting the product objectives, they are preferable.

One easily overlooked aspect of using ASICs involves protecting the product market. If a product is successful, potential competitors will purchase the product, reverse engineer it, duplicate it, and offer it for sale. They are able to learn how the product was implemented without having to evaluate the trade-offs during the development cycle. If all of the components are commercially available, they can purchase the same components and offer an almost identical product. ASICs prevent competitors from being able to purchase identical components commercially. Determining the functionality contained in a ASIC presents a significant barrier in reverse engineering a product. Even if the functionality of an ASIC is determined, it would still have to be implemented. If the product market lifetime is expected to be relatively short, the impact on profit due to a later introduction may be prohibitive. After a product is introduced and the market established, competition must introduce their version of the product quickly enough to make it cost justifiable.

The ASIC design methodology depends greatly on the type of ASIC selected. The types of ASICs and their characteristics are listed below.

Full-Custom ASIC A full-custom ASIC is an IC designed from scratch for a particular product. The development time is the longest among the types of ASICs. The risk is the highest due to it being designed from scratch. The performance is highest due to the entire IC being optimized for a particular implementation. NRE costs are the highest due to engineering design time and having every IC layer unique (requiring a full set of photomasks). For high-volume products with a long product market lifetime, a full-custom ASIC gives the lowest cost and highest profit.

Cell-Based ASIC (CBIC) A CBIC uses a cell library with proven functionality to construct an IC. There are various types of cell-based ASICs, including channeled, channelless, and structured. The risks are less than for a full-custom ASIC due to using proven cells and design methodology. The development time is longer, performance is better, and NRE costs are higher than for all ASIC types except for full custom. A CBIC is a reasonable alternative to a full-custom ASIC without requiring the expertise needed for a full-custom ASIC.

Masked Gate Array ASIC (MGA) A MGA is an ASIC that uses an established IC with only the metal layers customized. The MGA consists of a "base array" that has a collection of various cells that may be hooked up to implement different functions. Groups of transistors may be connected to create various logic gates; then the logic gates may be connected to implement the intended logic function. MGAs are not limited to digital logic; the same approach may be used to implement analog or mixed-signal ASICs. A MGA is a proper ASIC choice if factors do not justify optimizing to the extent of a full-custom or CBIC ASIC. The performance, design time, and NRE are reduced as compared to a full-custom or CBIC ASIC.

Field-Programmable Gate Array For the purposes of this discussion no distinction will be made between FPGAs and complex programmable logic devices (CPLDs). Under conditions of very short development cycles and low volumes, a FGPA is the best ASIC selection. None of the IC layers are customized for a FPGA. This results in the shortest development time and lowest NRE costs. A FPGA may be customized (programmed) by loading information into the FPGA "in the field." The programming is performed either by customizing the interconnect by "blowing fuses" contained in the IC or by loading RAM, which configures the interconnect. An effective technique for product development is to design a product containing FPGA ASICs, introduce the product quickly (due to the short development time for FPGAs), then after the product volume is confirmed, redesign the FPGAs with other types of ASICs to reduce costs for higher volumes. The initial product based on FGPA ASICs will establish your company as a supplier "capturing the market"; then the cost reduction redesign based on other ASIC types could be introduced, all completely transparent to the consumer.

11.5.5.2 PCB and MCM Design

Design of PCBs and MCMs will be described in detail in later sections. It involves a CAD-intensive process for designing individual PCB and MCM substrates. During the design process, trade-offs involving technology selection and partitioning must be evaluated and design iterations may be needed.

11.5.6 Thermal/Mechanical Design

Thermal and mechanical aspects of a design should not be underestimated. The functional lifetime of a product is often dominated by the thermal and mechanical characteristics. The speed and functional life of a semiconductor are primarily determined by the operating temperature. Shock and vibration experienced during use may lead to damage, resulting in failure. The thermal and mechanical design is tightly coupled to the details of the ASIC/PCB/MCM designs. As the electrical design is developed, conditions are conveyed to the thermal/mechanical designers to assure design objectives are met.

11.5.7 Test Program Development

A testing approach must be planned from the beginning of the block diagramming and partitioning process. In the past, testing was an afterthought to the design process. As products became more complex, a lack of an adequate testing approach often resulted in excessive complexity and costs of testing. Even though each component of the system is tested individually before assembly, testing the fabricated product must be addressed adequately. The objectives are to test the completed product exhaustively enough to guarantee an acceptable level of quality while minimizing the testing complexity and, as a result, the costs of testing.

Built-in self-test (BIST) and board/system-level scan path [joint testability action group (JTAG)] testing are techniques used in a "design-for-testability" methodology. This typically requires additional logic gates and interconnects specifically used for testing. The objectives involve proving correct functionality and the ability to identify a failure location (fault isolation). Isolating a failure location is necessary to allow reworking the product so the defect may be repaired. Large and complex electronic systems rarely function completely after fabrication. Isolating a failure allows for the replacement of failed components. Boards, components, or cables may need to be replaced (among other types of repairs). As a system is fabricated, the cost of replacing a failed component increases by an order of magnitude for each level of completion (component level, board level, subsystem level, system level, and while it is in the field).

11.5.8 Manufacturing Tooling

Manufacturing tooling involves the development of hardware and software needed during the manufacturing process. Depending on the item, there may be long lead times. As a result, in order to minimize the product development time, manufacturing tooling may need to order or begin fabrication as soon as it is defined during the design process. The product casing may require various types of molds, extrusion assemblies, fixtures, or modifications to machinery. Photomasks, screens, stencils, probe cards, and other items may be needed for the fabrication of the ICs, PCBs, and MCMs. Manufacturing tooling for ICs, PCBs, and MCMs will be discussed in a later section.

11.5.9 Fabrication/Assembly

An electronic system usually requires many types of fabrication and assembly processes for a completed product. The lead times for fabricated items and individual components may vary. These must be coordinated effectively so they are available for assembly as required. Long-lead-time items must be ordered as soon as possible to reduce any delays in product introduction. The objective is for each component in a system to arrive at the time it is needed for assembly, also known as "just-in-time" manufacturing.

11.5.10 Characterization

After an electronic system is fabricated, it must be tested over expected manufacturing tolerances and environmental conditions. There will be variations in performance between individual units due to variations within a lot (a group of units manufactured together), lot-to-lot variations, and variations due to component tolerances. There are also performance variations due to temperature, humidity, power supply tolerances, signal strength variations, and other interface or environmental conditions. The results of a product characterization are used to set the final product specifications and expected yields.

11.5.11 Qualification

After a product has been characterized, a trial manufacturing of the product is done. This allows evaluation of the quality level and variation in performance due to a manufacturing environment. The objective is for the manufacturing and assembly factory to operate as it normally would and be able to supply an acceptable product. Often it is found that a product has sensitivities to particular variations in manufacturing processes after a sufficient volume of product has been manufactured. Several lots of product are manufactured and tested. Any failing units go through a failure analysis process to determine the cause of the failure. This information is used to determine if the product should be modified to be more tolerant to these manufacturing steps, giving higher yield and overall level of quality.

The type of qualification required depends on the type of product. Four types of qualifications are commercial, medical, industrial, and military. The acceptable level of defective products sold as good varies greatly among these types of qualifications. Higher levels of quality may add to product costs. Adding to the cost of a product to attain a level of quality that exceeds the requirements of the product use should be avoided. A defective wristwatch may be inconvenient and require the consumer to return or replace the product, but a premature failure in an airline navigation system could have severe consequences. Yet if a wristwatch was produced with the more stringent quality levels of an airline navigation system, the increase in costs may make it significantly more expensive. The quality level of a product must meet the consumer expectations to properly balance the trade-off between quality and cost.

11.5.12 Product Introduction

Once a product has passed qualification, it may be introduced to the market. Pressures due to competition or product introduction windows often cause a product to be introduced before acceptable levels of quality or yield have been proven during qualification. A product may be introduced to capture the market and an enhanced manufacturability redesign started. This is where the marketing/sales department and design engineering may disagree. The design engineers will want their product to be as good as possible before it is sold but the market/sales group will decide if it is good enough to introduce. If the product is good enough to introduce but it is delayed due to further optimization, the market window may close and the product canceled.

11.6 DESIGN CONCEPTS

CAD software is the primary tool used for electronic design. Electronic design automation (EDA) applications are a category of CAD tools used for electronic design. The electronic design may be as simple as a single logic gate or as complex as an entire system. EDA tools consist of various separate applications that interface to each other through an electronic design data model (EDDM). The EDDM is an object-oriented database (or multiple databases) that contains all information necessary to fully describe the design. The EDDM has the following features:

Multiple Descriptions. A design component may have multiple descriptions (models). A digital component may have schematic models, behavioral language models, state machine models, truth tables, hardware description models, Boolean equations, and others. An analog component may have a SPICE model, a Verilog Hardware Description Language mixed-signal (VHDL-MS) model, or another type of model. A component may also have multiple physical descriptions, including bare die (wirebonded), bare die (flip-chipped), plastic leaded chip carrier (PLCC) package, dual in-line package (DIP), and so on. Models are checked and registered with interfaces for various subsequent EDA applications via the EDDM.

Seamless Modifications. Design modifications may be required at various points of the design process. Modifications may be made to the EDDM seamlessly. This reduces the possibility of errors, especially in a concurrent design environment. These modifications may be accomplished without editing the original design description.

Component Libraries. In-house or vendor-supplied libraries may be used in a design. Using libraries that have been previously developed gives a reduction in errors and design time. Library development and maintenance, with revision control, are key elements in an effective design methodology. A concurrent design methodology requires tight control on library development during the design cycle.

Design Views. Multiple design views allow a design to be described at different levels of abstraction. During logic simulation a portion of a design may be represented at the gate level, behavioral level, or others. A design view allows different portions to be described with different models or geometries. A design may be verified with portions of it modeled at a high level; then when these portions are more fully developed at a detailed level, the detailed models may be substituted for verification.

11.6.1 Component Overview

In the Mentor Graphics EDA environment a component is a set of models that describe various details of a distinct element (component). This may include functional, physical, timing, or other technology descriptions. The design viewpoint of a component must be created for EDA applications to use the model for the component. A design view may contain information necessary for analog simulation, digital simulation, mixed-signal simulation, test vector development, PCB layout, and others.

11.6.1.1 Component Symbol Creation

Design Architect is used to create a component symbol (as well as schematics). A symbol may represent a basic element, such as a logic gate, packaged IC, bare die, and power terminal. A symbol may also represent a collection of other components, such as a group of ICs, a partition of a PCB, functional blocks, or an entire PCB.

The component symbol contains a graphic representation of the component to be used for schematic creation. A component symbol may contain a symbol body, pins, and properties. The properties attached to a symbol are used in downstream applications. Figure 11.19 shows the Mentor applications window used for symbol creation. This symbol is for a MC10EP11D IC. It contains the graphic body for the symbol and properties for the reference designator, part number, component name, pin names, and pin numbers. The user interface contains a variety of command interfaces. Function keys, pull-down menus, and command icons are among the command interfaces available in the graphical user interface.

A component symbol is placed on a schematic with other symbols to construct a schematic sheet. Most components represent actual physical devices while others may be used for interconnect descriptions and documentation. An example of this is a port component that defines a signal interface to the schematic.

470 Chapter 11 Design Considerations

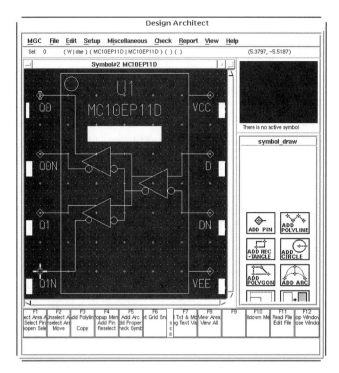

Figure 11.19 Symbol creation using Design Architect.

11.6.1.2 Component Symbol Property Annotation

A component symbol will typically contain many properties. The properties are used by downstream EDA applications and are contained in the design viewpoint(s) for the component. Examples of component properties include (among many others):

Reference Designator. A reference designator allows for a unique reference for a component. For example, R4 may refer to a particular instance of a resistor. The reference designator is commonly included in the silk screen layer for a PCB.

Part Number. A part number is used to define the part number for the component. It may be a supplier part number, distributor part number, or in-house part number. Among other uses, this property is used to create a bill of materials for a design.

Pin Name. A pin property defines the name of a pin on a component. ENABLE may refer to a pin that connects to the enable of a flip-flop.

Pin Number. A pin number defines the pin number of a pin on a physical package. A15 may refer to a pin that corresponds to pin A15 on a PGA package. Pin numbers are used to map pin numbers on a symbol to a physical pin on a component.

TPHL. TPHL defines the delay time for a particular signal when it switches from a logic value of 1 to 0. TPHL refers to the time (T) of propagation (P) high (H) to low (L). This property may be used for symbols that represent a logic function.

Many other properties may be annotated to a symbol. The overall design methodology and appropriate EDA tools determine what properties are needed. Properties have what is known as an "owner." A reference designator property is owned by the symbol body and there is only one occurrence of the reference designator property in a symbol. A pin name property is owned by an individual pin. Each pin on a symbol would have a unique pin name and pin number.

11.6.1.3 Component Modeling

A component may be modeled in various ways. It may be modeled with a schematic (containing component symbols), VHDL code, a behavioral language, a primitive logic gate, and so on. Any and all of these representations may be included for a component model. A design viewpoint is used to select which model(s) are to be used for a particular representation of the component. A component may first be modeled at a higher level, such as VHDL; then lower level models are used as the details of partitioning and modeling are developed. Different models may be used for downstream applications. For example, for digital simulation a VHDL model or a gate-level model may be used; for PCB or MCM layout a physical device model may be used; for thermal analysis a thermal model may be used.

11.6.2 Schematic Overview

Schematic drawing, also known as schematic capture, is the process of creating a drawing that contains components and the interconnection of signals. The Mentor Graphics EDA application Design Architect is used for schematic capture (as well as symbol creation). The schematic contains details of design information that is required by EDA applications used in subsequent design steps. This information includes descriptions of component instances, connections, connectors, packages, timing, test points, engineering notes, and so on. According to Mentor Graphics' design methodology, a schematic may contain multiple sheets. Each sheet is a distinct drawing with the collection of all of the sheets constituting the complete schematic. Electrical connections may be made between sheets. This allows for a schematic to be divided into individual portions (sheets) to make it manageable for printing and reviewing. Having multiple sheets also allows for a concurrent design methodology where different engineers are working at individual sheets simultaneously. An alternative approach is to have a single sheet for each schematic with the concurrently designed schematics interconnected hierarchically (as described next).

11.6.2.1 Schematic Hierarchy

A hierarchical schematic contains a top-level schematic of a system and a hierarchy of schematics and component symbols. The component symbols in the top-level schematic represent either a primitive component (a component with no schematic below) or a schematic (with hierarchy below). The branching downward through component schematics, eventually to physical components, represents the schematic hierarchy.

An electronic system may contain a very large number of components. Creating a single schematic containing all of these components and interconnects would result in a very large, unprintable, and unreadable schematic. A schematic hierarchy allows for the system to be partitioned into functional units with physical components or schematics for each partition of the system.

Figure 11.20 shows an illustration of how a hierarchical schematic is used to represent a large system. This diagram is simplified but shows the basic concept of a schematic hierarchy. The system-level schematic shown contains four component symbols. Each component symbol represents a schematic. Each branch of the hierarchy continues until it terminates on a physical component. Component 1 (at the top level) represents a schematic that contains component 1.1 and component 1.2. Component 1.1 represents a schematic that contains component 1.1.1 and component 1.1.2. All of these components are symbols for physical devices (IC, connector, etc.). Each branch of the schematic hierarchy terminates on symbols for physical components.

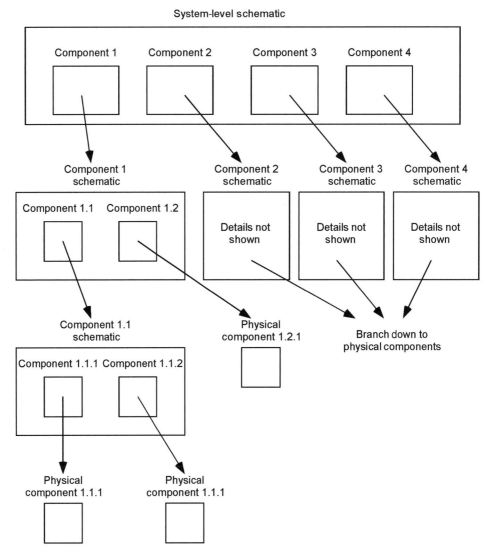

Figure 11.20 Schematic hierarchy.

11.6.2.2 Schematic Creation

Figure 11.21 shows the Design Architect application during schematic creation. This schematic contains a 4-bit arithmetic logic unit (ALU) that is constructed of 1-bit ALUs. Port-in components are used to define the input signals to the schematic (signals CI and buses OP[1 : 0], A[3 : 0], and B[3 : 0]). Port-out components are used to define output signals (signal CO and bus R[3 : 0]). Nets are used to define the connections between components. To connect to a single signal in a bus, a bus ripper component is used. A property on the bus ripper defines which bits on the bus are to be "ripped" from the bus.

The application software contains the same types of user interface commands as for symbol creation (pull-down menus, function keys, etc.). The commands available are those that are appropriate for the particular usage of the application (schematic capture or symbol

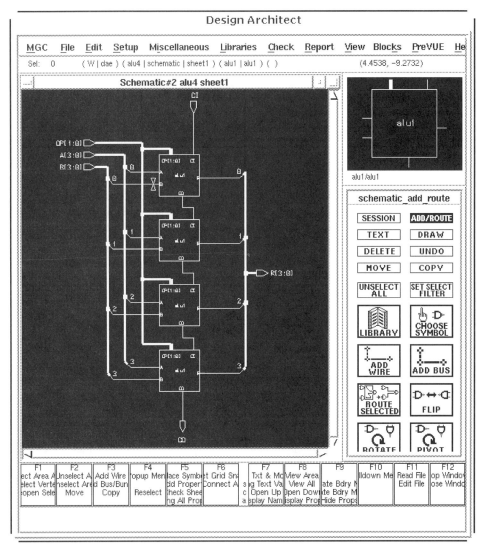

Figure 11.21 Schematic creation using Design Architect.

creation). A large variety of commands are available and allow for diverse styles of schematic creation and high productivity. The schematic editing commands include:

- Add an instance of a component (a symbol placement).
- Modify or create properties.
- Create and modify nets (signals and buses).
- Edit objects (moving, copying, deleting, connecting, etc.).
- Add or edit comment text.
- Add or edit templates for the creation of nets, properties, text, grids, pages, and so on.
- View (zoom in or out).

- Check for rules violations.
- Save and register schematics.
- Online help.

11.6.2.3 Schematic Property Annotation

During schematic creation properties may be added to the schematic and symbol properties may be added or modified (unless they are defined to be "fixed," unchangeable properties). Commonly used properties include:

Net. The net property is attached to the vertex of a net and defines the signal name.

Port. The port property is attached to a port connector (port in or port out) and defines signals that go in or out of the schematic. These signal names correspond to pin names on the symbol for the schematic.

Electrical Class. The electrical class properties are connected to nets and are used by downstream applications. For instance, a particular value of the electrical class property may be used in the layout of a PCB to control the signal router for various routing requirements.

Many other properties may be included in a schematic. The types and values of properties depend on design methodology, component modeling, layout technologies, and others.

11.6.2.4 Schematic Checking

Schematic checking is required to assure that schematic creation rules are not violated. The schematic must be error free before it may be registered for used by subsequent EDA applications. In addition to the schematic rules supplied by Mentor Graphics, user-defined rules may be defined for specific requirements. Among the types of rule checks are:

Unique Reference Designators. A reference designator must be unique in a schematic. The reference designator is used to refer to a particular instance. If a schematic contained two occurrences of R2, one of them must be changed.

Outdated Instances. A schematic may contain a symbol that is outdated due to it being edited after it was placed on a schematic. The rule will give a warning (instead of an error) to alert designers in case they were not aware the symbol had been edited.

Dangling Vertex. If a net does not terminate on a component pin, it represents either an error in the schematic or a poorly drawn schematic.

Unconnected Pin. A warning will be issued for any component pins that are not connected to a net. While this may simply be an unused pin, it is reported as a warning to alert the designer of the possibility of an error.

11.6.3 Design Viewpoint

After design schematics have been created, a design configuration is needed. The design viewpoint contains the configuration information necessary for other applications. The four types of design configuration are primitive, parameter, visible property, and substitute. The design viewpoint object is a database container that contains objects that are dependent on the design viewpoint version and configuration. The purposes of a design viewpoint are:

Property Evaluations. The evaluation of property values and expressions as appropriate for particular applications.

Expansion Level. Defining the lowest level of a design for an application. The design hierarchy is expanded to the level of "primitives." A primitive is the lowest level component in the design. For example, during logic simulation a partition of the design may be modeled with VHDL. The VHDL model is the lowest level (primitive) for that portion of the design. It may later be necessary to model this partition at the gate level. A different design viewpoint would be created that expands the hierarchy down to the gate-level models for this part of the design. It is possible to mix different model types in a design viewpoint for different parts of a design.

Evaluation of Parameters. Some parameters may not be defined with a schematic of symbol properties. Parameter values may be defined and used for the evaluation of properties to control various characteristics of a design.

Property Substitutions. The values of properties may be substituted. These properties will be evaluated for modifications to the design.

Design viewpoints are a powerful tool for product design. The benefits of using design viewpoints include:

Locking the Design. A design viewpoint allows for layout or simulation of a design in a fixed state. Modifications to the design are not reflected in the design viewpoint (unless it is re-created). This is valuable in a concurrent design process.

Checking. The entire hierarchy of a design may be checked for syntax and electrical rules checks.

Selectivity. A design viewpoint allows for selectively excluding portions of a design. This allows for evaluations of a part of a design without including other portions.

Property Modifications. Any properties may be changed, added, or deleted without affecting the source files.

Mixed Simulation. Portions of a design may be simulated using different types of models. A design viewpoint allows mixing different model types.

Model Changes. Models may be changed during simulation without exiting the simulator. A VHDL model may be changed to a schematic model without requiring significant effort.

Model Updates. Component models may be updated to new versions without exiting the simulator.

Model Sharing. Models are contained externally to a design viewpoint. This allows for a single model to be used in multiple designs without duplicating the models.

11.6.4 Back Annotation

Back annotation involves modifying or adding design information contained in property values. Most applications allow for the back annotation of property values. Back annotation of properties are often used for:

Timing Value Estimation. Logic gate delay values may be modified to include design-specific information. Characteristics related to the number and type of loads connected to an output may be evaluated to modify the timing information for simulation. Layout characteristics may be included to model net segmentation to give more accurate simulation results.

Floor Planning. Estimations of timing may be evaluated based on layout floor planning. This allows for increased accuracy during floor planning without needing detailed routing information.

Layout Timing. After routing signals, capacitance may be calculated and evaluated to modify timing parameters for accurate simulation results.

Reference Designators. Reference designator values may be modified during layout and back annotated into the design viewpoint for documentation.

Pin Numbers. Pin numbers may be reassigned during layout to allow for a more optimized design. These changes are back annotated to document the component descriptions needed.

Generic Properties. Additional properties (generic properties) may be added to a design to describe various characteristics such as power dissipation or temperature. These added generic properties may be used in downstream applications.

Model Selection. Annotations of model properties may be used to configure a design and they are saved as part of a back-annotated design viewpoint.

Back annotation results in a design object being created that compliments a design viewpoint. The back-annotation object documents a version of the design that may be used by downstream applications.

11.6.5 Simulation and Evaluation

Many types of simulation are possible. Simulations using Mentor applications include:

Logic Simulation. Digital logic may be modeled and simulated with various types of models (described previously). The primary application for digital simulation is QuickSim II. Other digital simulators are available for specific tasks (PLD, FPGA, etc.).

Analog Simulation. Analog simulations similar to SPICE are available. AccuSim is the Mentor application most often used for analog simulation.

Signal Integrity Analysis. Applications useful for signal integrity analysis are able to model characteristic impedance, transmission line, and cross-talk analysis.

Electromagnetic Analysis. Simulators able to evaluate electromagnetic interference (EMI) and electromagnetic compatibility (EMC) are often necessary for high-speed designs. Some products must meet FCC requirements of EMI and EMC; these tools allow the designer to evaluate a design during simulation.

Thermal Analysis. Thermal simulations evaluate the thermal characteristics of a design. Heat generated by operating electronic circuits must be evaluated to design a proper thermal environment. Heat sinks, heat spreaders, fans, and vents may be necessary to assure a design meets the maximum temperatures of various electronic components.

11.7 PCB/MCM BOARD DESIGN PROCESS

The Mentor applications supply a full range of tools needed for PCB, MCM, and hybrid system design. This section will focus on the PCB tools, but the discussions are applicable to MCM and hybrid design as well. Many of these tools contain an option setting that changes the available commands for PCB, high-speed PCB, MCM, and hybrid design environments. These tools support the following tasks:

- Schematic capture
- Library creation
- Packaging components into physical layout packages
- Interactive and automatic component placement
- Gate and pin swapping
- Interactive and automatic test point generation
- Interactive and automatic signal routing
- Interactive and automatic power routing
- Design rule checking
- Analysis for high-speed signal integrity, reliability prediction, thermal characteristics, and manufacturability
- Cabling tools for creating system cabling and cable harnesses
- Manufacturing data creation

Many other tasks are supported and may be customized for specific design methodologies. A programming language (AMPLE) is available that allows for user-defined commands, batch commands, and other types of customization of the design environment.

11.7.1 PCB Design Flow

Figure 11.22 shows a PCB design flowchart using these applications. Details for these applications will be discussed in the next sections (or have been previously discussed).

11.7.2 Librarian

The librarian application contains a general-purpose graphical editing tool used to create geometries to be used during layout. Materials used for layout are defined along with the layer stackup (layer order). User-defined layers may be defined for documentation and analysis purposes. The three primary tasks of librarian are:

- Creating library geometries
- Creating component mapping and catalog files
- Compiling geometries for use in a specific design

Commercially available libraries may be used in librarian and included in a design. The use of existing libraries increases productivity and reduces errors. These geometry libraries may be modified for use in different manufacturing technologies.

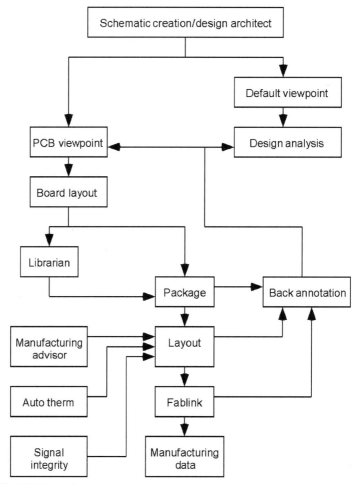

Figure 11.22 PCB design flowchart.

The primary types of geometries created in librarian include:

Artwork Order. The artwork order specifies the layer stackup for the PCB. This includes metal, dielectric, silkscreen, solder, and paste layers.

Board. The board geometry defines the PCB to be used during layout. This includes the ability to define regions for component placement and signal routing.

Breakout. A breakout geometry is used for COB components. The die wirebond pads and corresponding PCB wirebond pads are defined. This allows for moving the PCB wirebond pads during the layout process and automatic generation of wirebond diagrams.

Component. A component geometry represents the graphic composition of a discrete physical component. This may include ICs, passive components, connectors, and testpoints. A component geometry includes pin padstacks (described later), silkscreen shapes, and paste shapes.

Drawing. Drawing components may be created for documentation.

Generic. A generic geometry may be used for documentation, analysis, or other reasons. Alignment markers (fuducials) would be created as a generic geometry.

Manufacturing Panel. After a PCB is designed, it is panelized. This involves replicating the PCB into a pattern appropriate for a manufacturing panel. This may also be done using the fablink application.

Probe. A probe geometry is used for testing access.

Test Fixture. A test fixture geometry is used for the fabrication of a fixture to be used for testing purposes.

Pin Padstack. A pin padstack geometry defines the construction (layer stackup) of layers for contacting to a component pin. This includes blind, through-hole, and surface pins. Rules defining a pin packstack are created. For instance, a pin rule can prevent a wirebond pad from being placed above a buried via for manufacturing reasons.

Via Padstack. A via padstack geometry defines the construction (layer stackup) of layers for creating a via for signal routing. The types of via padstacks are two-layer, buried, blind, and through-hole vias. Rules defining how a via padstack may be used are also created. For instance, a via rule can prevent stacking buried vias for manufacturing reasons (if required).

11.7.3 Package

The package application links the component symbols (from the schematic) to layout geometries created in librarian. A mapping file is created that associates each component symbol to the corresponding physical layout geometry. This mapping function may be done automatically or interactively. The package application also creates documentation files, including a BOM. The design database objects created by package includes the comps, gates, nets, and pins—design objects that are used during later design steps. The design viewpoint used in layout (pcb_design_vpt) is created by the package application. Figure 11.23 shows the package application shell.

11.7.4 Layout

The layout application is used to place components on a PCB, route signal traces, and create the power distribution interconnect. Both component placement and routing may be done automatically or interactively. The three configurations of layout are:

- **General Layout**. General layout is for designing conventional PCBs.
- **High-Speed Layout**. The high-speed layout configuration includes tools applicable for the design and analysis of high-speed PCB designs.
- **MCM/Hybrid Layout**. The MCM/hybrid layout configuration includes tools for the design of MCM or hybrid designs.

11.7.4.1 Component Placement

Although layout provides the ability to automatically place components onto the PCB, it is typically necessary to manually (interactively) adjust the placement of components.

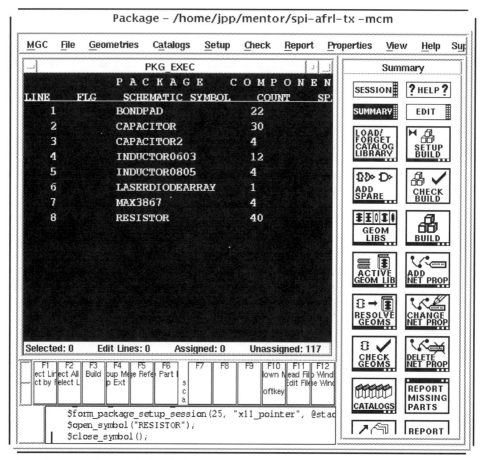

Figure 11.23 Package application shell.

Placement rules guide placement with constraints such as gridding and spacing. Components may be rotated, moved, or swapped to the other side of the PCB (for designs with components on two sides of the PCB). Histograms may be created that convey information about component density and congestion to guide the designer during placement.

11.7.4.2 Signal Routing

Signal routing is guided by the interconnections defined in the schematic. This connectivity-constrained routing approach will guarantee that no errors in routing occur. Required connections that have yet to be routed are shown with a graphic called a guide. A guide connects to all pins on a net. This visual feedback to the designer is useful during component placement as well as during signal routing. A variety of routing rules are defined that are automatically enforced during either automatic routing or interactive routing. These rules include:

Trace Width. Trace widths are defined for each metal layer or for classes of signals (as defined by annotated properties).

Trace Spacing. Trace spacing is defined for each metal layer or for classes of signals.

Snapping. Snapping rules control if signals may be routed vertically, horizontally, diagonally (45° angles), or at any angle.

Topology. It is possible to control weather signals that must be routed in a point-to-point manner (often called a "daisy chain") or if a star-shaped routing structure is allowed. Some high-speed designs require that selected signals be routed from point to point to achieve clean high-speed interconnects. This results in a signal route that is essentially a single trace that goes from the driver to each of the loads with no branches in the route.

Stub Length. Even with point-to-point routing it is rarely possible to avoid a short "stub" or connection from the load pin to the signal route. The maximum length of the stub is defined to assure that it will have no significant impact on signal quality.

Differential Pairs. Differential signal leads much have signal lengths that match to within the rule setting.

Length. It is possible to define minimum and maximum signal lengths for electrical rule classes.

Parallel Lengths. The maximum distance any two signals may be routed adjacent to each other may be defined. This is important to minimize signal crosstalk due to capacitive and inductive coupling of signals beside each other.

Via Limits. The maximum number of vias allowed in a signal route may be defined.

Layer Restrictions. It is possible to constrain which layers may be used for routing classes of signals.

Various signal routers are available. These include an interactive router, grid-based autorouters, and gridless shape-based autorouters. Autorouter algorithms have characteristics that are optimized for particular circumstances. For instance, a grid-based channel router is best for routing signals in a channel region that is not constrained by obstacles (such as a component pin padstack). A gridless maze autorouter is best for routing signals in a region that is of an irregular shape or that contains constraints. The high-speed router tools allow for multiple autorouting passes with each pass using a different type of autorouter.

Another feature of the layout application is the ability to add area fills. An area fill is typically used for power connections but may also be used for signals. An area fill will fill a used defined area with a selected metal layer. Connectivity constraints are used to allow the metal to connect to the intended net while avoiding pins or interconnect that is associated with other signals. Minimum width and spacing rules are adhered to for the metal layer created by the area fill. This feature allows for easy creation of power planes. Figure 11.24 shows the layout application shell.

11.7.5 Fablink

The final stages of a design process involve creating manufacturing files and documentation. Fablink also includes graphical drawing capabilities for adding text or polygon shapes and panelizing. Manufacturing data include:

Photoplotter Data. Photoplotter data files are used to create optical masks, screens, or stencils for patterning metal, dielectric, and silkscreen layers. Gerber format data of various types may be created with automatic or interactive creation of aperture tables. GDSII files may also be created and are primarily used for patterning layers used for MCM processing.

482   Chapter 11   Design Considerations

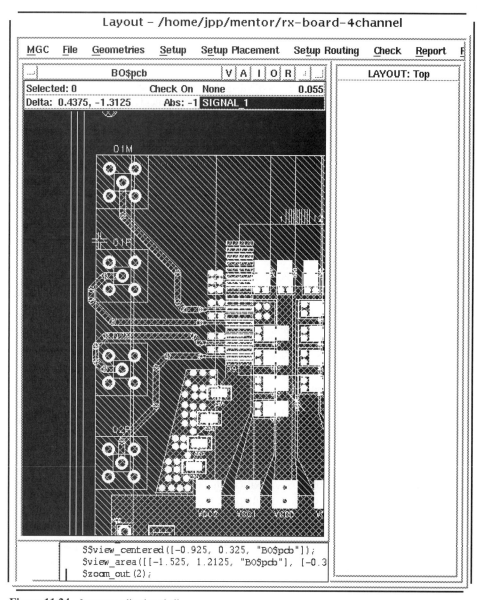

Figure 11.24   Layout application shell.

Drill Data. Files used to control drilling may be created that contain American Standard Code for Information Exchange (ASCII), Electronic Industry Association (EIA), or Extended Binary Coded Decimal Interchange Code (EBCDIC) descriptions of coordinates in Excellon or Trudrill formats. Drill tables may be created either automatically or interactively. An algorithm is used to minimize drill head movement to reduce manufacturing time. A simulation of the drilling process allows for the evaluation of the efficiency of created drill data.

Milling Data. Most PCBs require the milling of various shapes to allow for the attachment of hardware or compliance to mechanical structures. Milling files may be

11.7 PCB/MCM Board Design Process

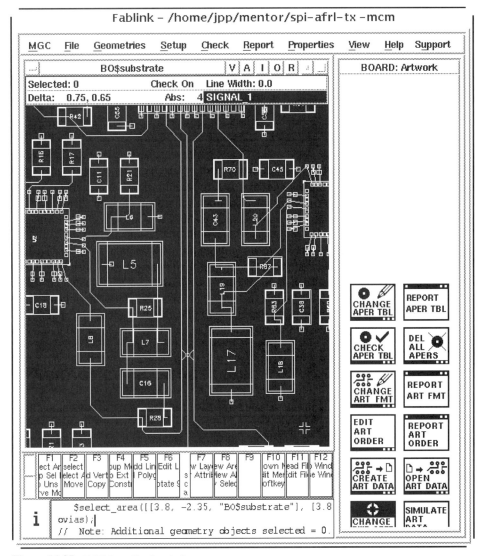

Figure 11.25 Fablink application shell.

created containing ASCII, EIA, or EBCDIC descriptions of coordinates in Excellon or Trudrill formats. Milling tables may be created either automatically or interactively. Simulation of the milling process allows for the evaluation of the milling procedure defined in the milling data.

Neutral File. The neutral file contains information about properties, attributes, and component locations. Neutral files are used as input for manufacturing machine interface programs. A neutral file is often used for the automatic creation of data for driving automatic test equipment, automatic component insertion equipment, surface-mount component placement equipment, wirewrap files, and others.

Figure 11.25 shows the fablink application shell.

11.7.6 Summary of Design Concepts

The application of EDA tools to the product development cycle has been described with a focus on applications available from Mentor Graphics Corporation. Many other EDA vendors supply products that are applicable to these various design tasks. The development of an electronic system involves many design and analysis steps. The application of EDA tools depends on the product development cycle selected, design expertise, development time, and specific requirements of the electronic system. It should be noted that this section contains an overview of the EDA tools and their application; a detailed description of the operation and interactions of these tools is beyond the scope of this text (and would surely fill many textbooks).

11.8 SUMMARY

This chapter has described the design process on two levels: the conceptual, requirements-driven process involving system partitioning and technology trade-offs and the more detailed process of reducing a design to practice.

Ian Ross, former President of Bell Labs, once told one of us (Schaper), "A group of competent engineers will do the thing right, but what's really important is doing the right thing." In other words, no amount of effort on detailed design, product introduction, or manufacturing cost reduction will make up for defining the wrong product. The up-front decisions on the trade-offs among competing product features and packaging requirements are key to product success in the marketplace.

REFERENCES

1. W. H. KNAUSENBERGER AND L. W. SCHAPER, Interconnection Costs of Various Substrates—The Myth of Cheap Wire, *IEEE Trans. Components, Hybrids, Manufact. Technol.*, pp. 261–263, Sept. 1984.
2. G. MESSNER, Cost-Density Analysis of Interconnections, *IEEE Trans. Components, Hybrids, Manuf. Technol.*, Vol. 10, pp. 143–151, June 1987.
3. R. TUMMALA, Multichip Packaging—A Tutorial, *Proc. IEEE*, Vol. 80, pp. 1924–1941, Dec. 1992.
4. W. J. JACOBI, N. J. TENEKETGES, AND L. A. WADSWORTH, Miniaturized, Low-Power Parallel Processor for Space Applications, presented at the AIAA/USU Conference on Small Satellites, Aug. 26, 1991.
5. E. E. DAVIDSON, in *Microelectronics Packaging Handbook*, Tummala and Rymaszewski, eds., New York: Van Nostrand Reinhold, 1989, p. 160.
6. B. S. LANDMAN AND R. L. RUSSO, On Pin versus Block Relationship for Partitions of Logic Circuits, *IEEE Trans. Computers* C-20, pp. 1469–1479, 1971.
7. H. B. BAKOGLU, *Circuits, Interconnections, and Packaging for VLSI*, Reading, MA: Addison-Wesley, 1990, p. 419.
8. TUMMALA AND RYMASZEWSKI, eds., *Microelectronics Packaging Handbook*, New York: Van Nostrand Reinhold, 1989, p. 873.
9. L. MORESCO, Electronic System Packaging: The Search for Manufacturing the Optimum in a Sea of Constraints, *IEEE Trans. Components, Hybrids, Manufact. Technol.*, Vol. 13, No. 3, pp. 494–508, Sept. 1990.
10. L. W. SCHAPER, Design of Multichip Modules, *Proc. IEEE*, Vol. 80, No. 12, p. 1958.
11. L. SCHAPER, S. ANG, Y. LOW, AND D. OLDHAM, The Electrical Performance of the Interconnected Mesh Power System (IMPS) MCM Topology, *IEEE Trans. CPMT*, Vol. 18, No. 1, pp. 99–105, Feb. 1995.
12. L. T. HWANG, et al., Measurement of High-Speed Signal Propagation Characteristics in Thin-Film Interconnections, *Proc. IEPS Conf.*, pp. 272–287, 1990.
13. S. SASAKI, et al., 3-D Electromagnetic Field Analysis of Interconnections in Copper–Polyimide Multichip Modules, *IEEE Trans. Components, Hybrids, Manufact. Technol.*, pp. 755–760, Dec. 1991.
14. R. SULHAN, et al., Thermal Modeling and Analysis of Pin Grid Arrays and Multichip Modules, in *Proc. SEMI-THERM Symp.*, pp. 110–116, 1991.
15. D. W. SNYDER, Thermal Analysis and Modeling of a Copper–Polyimide Thin-Film-on-Silicon Multichip Module Packaging Technology, in *Proc. Eighth IEEE SEMI-THERM Symp.* pp. 101–109.

BIBLIOGRAPHY

Books

COOMBS, C., *Printed Circuits Handbook*, New York: McGraw-Hill, 1996.

GARROU, P. AND IWONA, I., *Multichip Module Technology Handbook*, New York: McGraw-Hill, 1998.

HARPER, C., *Electronic Packaging and Interconnect Handbook*, New York: McGraw-Hill, 2000a.

HARPER, C., *High Performance Printed Circuit Boards*, New York: McGraw-Hill, 2000b.

HENNESSY, J. AND PATTERSON, D., *Computer Organization and Design*, San Francisco: Morgan Kaufmann, 1997.

SMITH, J., *Application-Specific Integrated Circuits*, Reading, MA: Addison-Wesley, 1999.

Software Application Manuals

Mentor Graphics Corporation, *PCB Products Overview Manual—Software Version 8.4_3*, 1995a.

Mentor Graphics Corporation, *A Guide to Design Process and Database Concepts—Software Version 8.4_1*, 1995b.

EXERCISES

11.1. The most effective exercise for the trade-off-related material in this chapter is for you to take a recent paper from an IMAPS, IEEE, or ASME electronic packaging conference and analyze it to see what trade-offs were made in solving the problem discussed. Usually, papers involving some form of module or subsystem development are the most effective for illuminating the trade-offs among materials and thermal, mechanical, and electrical performance issues. Study a paper and then present a review of the paper to the class as well as an analysis of the trade-offs involved.

11.2. Using a profit model as described in Section 11.5.2.1, develop an equation for the total sales of a product. Use the following terms:

T_i—time when the product is introduced into the market

T_p—time when peak sales are reached

T_e—end of the product lifetime

S_r—initial sales rate

T_s—total sales in dollars

11.3. Using an Internet search engine (www.google.com, www.lycos.com, etc.), locate a supplier of PCB CAD/EDA tools. Summarize the CAD/EDA tool capabilities and contrast them to those of Mentor Graphics (as described in this chapter).

Chapter 12

Radio Frequency and Microwave Packaging

FRED BARLOW AND AICHA ELSHABINI

12.1 INTRODUCTION AND BACKGROUND

An important question is "what is different about RF and microwave packaging"? To answer this question, we must consider the nature of the types of signals that are used in this type of circuit or system. The primary difference is the frequency of operation, which is often broken into a variety of ranges: radio frequency (RF) waves, microwaves, or millimeter waves.

12.1.1 Nature of High-Frequency Circuits

Radio frequency waves are described as frequencies from 300 kHz to ~0.3 GHz, the microwave range is from ~0.3 to 30 GHz, and the millimeter-wave range is from 30 to 300 GHz. Each of these frequency ranges is related to an equivalent range of wavelengths, as described in Chapter 6, which decreases as the frequency increases. The microwave and millimeter regions are broken into a series of designated bands, which are often used for different applications. These bands have both civilian designations and military designations, as illustrated in Figure 12.1. The civilian designations are the most commonly used to denote the bandwidth of a given application or circuit.

To understand the importance of wavelength and its impact on circuit design, consider the voltage on a short-circuit trace, assumed to be 5 cm long, in a low-frequency circuit. The signal of a sine wave generator connected to one end of the trace can be measured on the other end of the trace. If the trace is low loss, the difference between the generator voltage at one end and the measured voltage at the other end will be negligible.

Now consider that same line with a much higher frequency of operation such that the wavelength of the signal the generator feeds into the line is 10 cm in length. In air, this corresponds to a frequency on the order of ~6 GHz, which is close to the range used for cordless phones and many of the emerging wireless computer local area network (LAN) standards. Since the signals are sinusoidal in nature, this case produces a measured voltage that is exactly out of phase with the generator. So, if the generator is set to produce a wave with an amplitude of 1 V, the instant the voltage at the generator reaches its peak, the voltage

---

Advanced Electronic Packaging, Second Edition, Edited by Richard K. Ulrich and William D. Brown
Copyright © 2006 the Institute of Electrical and Electronics Engineers, Inc.

488 Chapter 12 Radio Frequency and Microwave Packaging

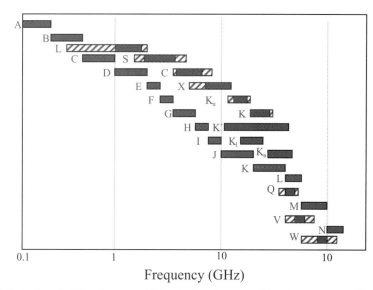

Figure 12.1 Military (solid) and commercial (hatched) designations of the microwave and millimeter spectrum. The solid areas of the commercial designations are the official designations, while the hatched areas designate commonly used terminology.

at the load is zero! This simple example illustrates just one impact of wave effects that must be considered for high-frequency circuits.

The primary difference between low-frequency circuits and high-frequency circuits is that the wavelength of the signal(s) of interest in a microwave or millimeter-wave system is on the order of the size of the circuit or the circuit elements. In this case, we must consider the electromagnetic effects in the system and not simply rely on circuit theory. Circuit theory assumes that the circuits are much smaller than the wavelength of the signals of interest.

A second difference between high-frequency circuits and low-frequency analog applications is that many high-frequency applications are often relatively narrow bandwidth systems. For example, consider the transmission of a TV signal via satellite. An individual TV channel has a bandwidth of approximately 6 MHz. Propagation of this signal is often performed at a much higher frequency, for example, on the order of 10 GHz, or more than a 1000 times higher in frequency. As a result, filters, amplifiers, and frequency conversion circuits used in this application normally are designed to operate over a narrow bandwidth (∼6 MHz) centered about a much higher propagation frequency (∼10 GHz).

12.1.2 Applications of High-Frequency Circuits

In many cases, these additional wave effects complicate the design of high-frequency circuits in comparison to similar low-frequency designs. However, several compelling reasons exist for the use of high-frequency signals, including:

- Transmission of the signals through the atmosphere
- Antenna efficiency
- Signal bandwidth
- Scattering of energy from objects for detection and tracking purposes; for example radar

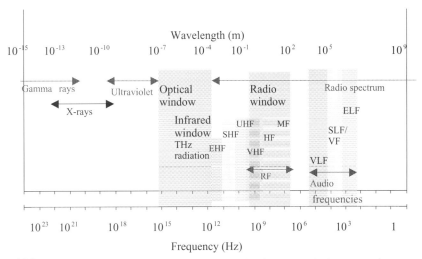

Figure 12.2 Electromagnetic spectrum illustrating the optical regions, RF and microwave regions, as well as ionization radiation regions.

First consider the effect of the atmosphere on signal propagation, as illustrated in Figure 12.2. Several windows exist that allow certain types of radiation to pass through the atmosphere with relatively little attenuation. Other frequencies are sharply absorbed by moisture in the atmosphere while still other frequencies propagate through the lower atmosphere but are reflected by the upper atmosphere and form the basis of short-wave radio, which is used for long-range terrestrial communication. Clearly, these effects can be exploited for a given application and must be carefully considered. For example, communication to a satellite via a 60-GHz carrier would not be ideal for most applications due to the loss in the atmosphere.

The second, and perhaps more important issue, is that of antenna efficiency. While the details of antenna design are beyond the scope of this text, in general, these devices are most efficient when they are on the order of the wavelength that they are receiving and/or transmitting [1–3]. This is why the antennas on most types of cell phones are all approximately the same length, since the frequency range of interest is similar even though the phones themselves may be of radically different design. Transmitting a signal at 100 Hz requires an unrealistically large antenna, which needs to be kilometers in length in order to efficiently radiate and receive that frequency range.

The third reason for the widespread use of high-frequency signals is the impact of signal bandwidth. It will be demonstrated later in this chapter that the design of circuits with large bandwidths is complicated by a number of factors. Engineers often express the bandwidth utilized in a system as a percentage of the center carrier frequency. So, for our previous example of a TV signal transmitted via satellite, the percentage utilization would be on the order of 0.06%, 6 MHz/10 GHz.

Now consider a microwave system intended to transmit TV signals, which occupy approximately 6 MHz. If the center frequency of the system is 600 MHz, each TV channel consumes 1% of the bandwidth. That same 1% bandwidth at 12 GHz allows transmission of 20 equivalent TV channels.

The final reasons for the use of high-frequency signals are range-finding, tracking, and remote-sensing applications, such as radar [4]. While these systems utilize a wide range of frequencies from the RF into the millimeter band, the sensitivity of the system to selected targets can be enhanced through the use of high-frequency signals.

12.1.3 Basic Concepts

Since the wavelength of the signals of interest is on the order of the size of the circuit elements and interconnects, wave phenomena play a key role and must be considered.

12.1.3.1 Impedance

The primary difference between low-frequency design and high-frequency design is the importance of impedance and controlling it throughout a circuit. For interconnects, the characteristic impedance is the quantity of interest, which, as described in Chapter 6, is the ratio of voltage and current amplitudes for either the positive-going wave or the negative-going wave. However, it is not the ratio of the total voltage and current amplitude, which will, in general, be the summation of positive and negative propagating waves interfering with each other.

The goal in most cases is to create a conjugate impedance match where the source and load impedances are complex conjugates of one another, since this will ensure maximum power transfer from the source to the load. The power source may be an antenna, and the load a filter, low-noise amplifier (LNA), or some other portion of the circuitry. The designer must take care to ensure that each element in the system is impedance matched to an adequate degree to ensure adequate signal strength throughout the system. Understanding the situations that require controlled impedance is the first step, since not all circuitry requires this more complex analysis and design procedure. In general, this subject becomes relevant when wave phenomena exist in a circuit or interconnects. This condition is met when the physical dimensions of the line or circuit are on the order of the wavelength of the signals in the circuit or system. So, for example, wave phenomena may be relevant even at very low frequencies for long transmission lines that are used to provide data connections from city to city. However, much higher frequencies are required before wavelike phenomena are relevant to the design process of an integrated circuit (IC).

The criteria that are used to evaluate the impedance match of a component, circuit, or system are the voltage reflection coefficient or the voltage standing-wave ratio (VSWR). As described in Chapter 6, these quantities are measures of how close a particular impedance is to the ideal match. Ideally, the voltage reflection coefficient should be zero, while an ideal value for the VSWR is one. VSWR values are often used as a performance specification for connectors and cables as well as circuit elements and system-level components. This is usually expressed as a ratio with respect to an ideal 50-Ω load and is a strong function of frequency.

12.1.3.2 Loss

The second primary difference between low- and high-frequency circuits is the fact that losses within the components at a given power level tend to increase with frequency. At low frequencies, the losses are primarily due to the conductor used with the circuit (or component), with a very small contribution from dielectric losses. As the frequency increases, the losses in the dielectric grow and, at some point, exceed the conductor losses. In addition, effects such as surface roughness and radiation loss can contribute to the total power dissipated within a high-frequency system. Since this power is either radiated away from the system (or component) or dissipated as heat, an important design goal is to minimize the losses within a given design.

The quality factor (Q) is often used to represent the losses in a given system (or component) as a function of frequency, where Q is the ratio of the time-averaged stored power times the radian frequency divided by the energy lost per cycle. Three principal parts make up the quality factor: conductor loss (Q_c), dielectric loss (Q_d), and radiation loss (Q_r). The total quality factor can be expressed as

$$\frac{1}{Q} = \frac{1}{Q_c} + \frac{1}{Q_r} + \frac{1}{Q_d} \tag{12.1}$$

Each component can be expressed in terms of the attenuation coefficient (α), which is also divided into its conduction (α_c), dielectric (α_d), and radiation (α_r) contributions [5, 6]. Each component can be expressed as

$$Q_r = \frac{\beta}{2\alpha_r} \quad Q_d = \frac{\beta}{2\alpha_d} \quad Q_c = \frac{\beta}{2\alpha_c} \tag{12.2}$$

where α is the attenuation coefficient and β is the propagation constant. The radiation losses are normally much smaller than the conductor and dielectric losses and, in some cases, may be neglected.

Conductor losses are usually expressed in terms of the surface resistance of a conductor. At frequency, the currents in a conductor are restricted to some portion of the outer skin of a thick conductor, and the electric field strength drops off as a function of depth from the conductor's surface. The depth at which the current density drops off to $1/e$ (~37%) of its surface value is the skin depth. The skin depth is of the form

$$\delta = \sqrt{\frac{2}{\omega\mu\sigma}} \tag{12.3}$$

where ω is the radian frequency ($2\pi^*$ freq.), μ is the permeability of the conductor (often $\mu = \mu_0 = 4\pi \times 10^{-7}$ H/m), and σ is the direct current (DC) conductivity of the conductor in ohms per meter. The important ramification of this effect is that conductors, which are much thicker than the skin depth, are normally a waste of metal, since only a very small amount of the current flows within the extra conductor material. Note that, as the frequency increases, the skin depth decreases, indicating that, at low frequencies, the skin depth may be so large that it is not relevant to most designers. However, in the microwave and millimeter-wave range, this effect is crucial since the current densities are normally far higher than one would expect if it is assumed that the current is uniform throughout the conductors cross-sectional area. A good rule of thumb for designers is to utilize conductors that are a few skin depths in thickness whenever practical in order to avoid excessive losses and or wasted metal thickness.

The surface resistance of a nonmagnetic conductor can then be expressed as

$$R_s = \frac{1}{\sigma\delta} = \sqrt{\frac{\pi f \mu_0}{\sigma}} \tag{12.4}$$

where δ is the skin depth in the conductor in meters, σ is the DC conductivity of the conductor material (in ohms per meter), $\mu_0 = 4\pi \times 10^{-7}$ H/m is the permeability of free space, and f is the frequency in question. Note that this expression contains an $f^{1/2}$ dependence of the surface resistance on the conductor losses, whereas the use of a simple DC conductivity or resistivity incorrectly implies no frequency dependence of the conductor losses. The conductor losses in many transmission line structures can be expressed as the product of this surface resistance and a geometric factor, which is described later in this chapter for each transmission line structure.

As mentioned above, surface roughness may also play a key role in the losses of a transmission line [7, 8]. This effect is due to localized current crowding in the conductor created by the rough bottom surface of a conductor, which will normally mirror the roughness in a substrate. This increased loss can be approximated as [9, 10]

$$\frac{\alpha_c}{\alpha_{c0}} = \left[1 + \frac{2}{\pi}\arctan\left[1.4\left(\frac{\Delta}{\delta}\right)^2\right]\right] \quad (12.5)$$

where α_c is the attenuation coefficient of the rough surface, α_{c0} is the attenuation coefficient of a perfectly smooth surface at DC, Δ is the surface roughness in micrometers and δ is the skin depth in micrometers. Note that the skin depth is frequency dependent and therefore the ratio of the attenuation constants will also exhibit strong frequency dependence, as one would expect.

12.1.3.3 Scattering Parameters

As described in Chapter 6, the voltages and currents in most high-frequency circuits are functions of time, but they are also a function of position at a given point in time. In other words, the voltage on a given trace within a circuit at any given point in time varies from point to point. As a result, most engineers find it more useful to consider scattering parameters, or "s parameters," rather than voltages and currents as a function of time [11, 12].

Scattering parameters are somewhat analogous to impedance in that a circuit or electrical element can be represented by its impedance or by its s parameters. The difference is that the impedance is the ratio of the voltage and current, while the s parameters are the ratio of the incident and reflected voltage waves. For a network with N ports, or electrical connections, there exists an $N \times N$ matrix of s parameters or N^2 s parameters than can be used to completely represent the electrical function of the network. Consider the electrical network in Figure 12.3, which has two electrical connections or ports and can, in general, have both incident and reflected voltage waves on each of the ports. The amplitude of the incident wave is denoted as V_i^+ and the amplitude of the reflected voltage wave is denoted as V_i^-, where i is the port number. This device may be a amplifier, filter, entire electrical system, or component. The s parameter matrix for this 2-port is 2×2 with components

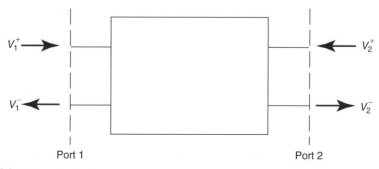

Figure 12.3 Two-port network.

S_{11}, S_{12}, S_{21}, and S_{22} of the form

$$\begin{bmatrix} V_1^- \\ V_2^- \end{bmatrix} = \begin{bmatrix} S_{11} & S_{12} \\ S_{21} & S_{22} \end{bmatrix} \begin{bmatrix} V_1^+ \\ V_2^+ \end{bmatrix} \tag{12.6}$$

Each of these parameters is determined by the expression

$$S_{ij} = \frac{V_i^-}{V_j^+} \bigg|_{V_k^+ = 0} \quad \text{for all } k \neq j \tag{12.7}$$

where i is the first subscript of the s parameter, j is the second subscript of the s parameter, and the voltage amplitudes are as defined in Figure 12.3 [13, 14]. This expression indicates that the s parameters are simply the ratio of incident and reflected voltage wave amplitudes, where the second subscript denotes the port being driven and the first subscript denotes the receiving port. For example, S_{11} is the ratio of the incident and reflected voltage waves for port 1. The requirement that $V_k^+ = 0$ for all $k \neq j$ simply means that all ports, other than the driving port, must be matched such that no incident voltage amplitude exists.

A key issue is the fact that s parameters are not meaningful unless the reference planes are carefully defined. By reference planes we mean the locations in the circuit or circuit element where the reflected and incident voltage waves are measured. In practice, these reference planes are defined through measurement calibration, which is described in more detail later in this chapter. Also keep in mind that all networks do not need to be 2-ports, and, in fact, the number of ports, N, can range from 1 to infinity. However, values of N from 1 to 4 are the most common in microwave and RF systems.

It is important to develop an intuitive feel for the meaning of these s parameters much like all electrical engineers have for impedance. Consider possible values for S_{11}, which range from zero to $-\infty$ decibels for a passive network (note values of $S_{11} > 0$ dB are possible with active devices). A S_{11} value of 0 dB means that the magnitude of the reflection is equal to the magnitude of the incident wave, since $S_{11} = 1 = 0$ dB (since $V_1^+ = V_1^-$). This situation corresponds to a perfect mismatch where none of the incident signal passes into or through the network. In contrast, a S_{11} value of $-\infty$ decibels corresponds to a perfect match, in which case $V_1^- = 0$. In reality, no impedance match can be perfect and typical measurement systems have noise floors around -60 to -70 dB. However, the decibel scale is logarithmic and -60 dB is not simply 6 times smaller than -10 dB. Using this scale, a designer would normally desire to have S_{11} as small as possible over the frequency range of the desired signals. In practice, values below -15 to -20 dB are considered a good match, since these values correspond to voltage reflection coefficients of ~ 0.17 and ~ 0.09, respectively.

Values of S_{21} are also normally expressed in decibels within the range from zero to $-\infty$ decibels for a passive network. The second subscript is the number of the driving port and the first subscript is the number of the receiving port. Therefore, S_{21} is a measure of the signal that is inserted into port 1 and travels out of port 2. An S_{21} value of 0 dB means that all of the power injected into port 1 comes out of port 2. The S_{21} values of -3 and $-\infty$ decibels correspond to half-power delivery to port 2 and no power delivered to port 2, respectively. As in the case of S_{11}, realistic values range down to the measurement noise floor of approximately -60 dB. Positive values of S_{21} are also possible for active circuits such as amplifiers. A positive value corresponds to gain created by the amplifier and values of $+5$ to $+30$ dB are common.

Consider the s-parameter data plotted in Figure 12.4 for a bandpass microwave filter. Note that this figure includes both the S_{11} and S_{21} data in decibels plotted versus frequency. It is unnecessary, in this case, to plot the S_{12} data since the device is reciprocal and this

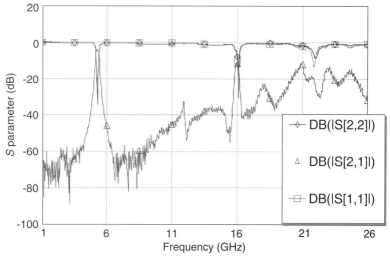

Figure 12.4 Example of measured s-parameter data for a prototype bandpass filter.

information is redundant. In fact, S_{11} and S_{22} are identical since this device is reciprocal. The S_{11} data approach 0 dB for signals out of band and are down 20 dB or more over the passband, which is centered at 5 GHz and repeats at higher frequencies for this particular filter type. This −30 dB S_{11} is a measure of the impedance match and corresponds to reflection coefficients of <0.09 and ~1 for the in-band and out-of-band regions, respectively. From these data, we can see clearly that this filter is reflective in that signals in band are transmitted and the out-of-band signals are reflected. This is also clear in the S_{21} data that show a small amount of loss in band, ~4 dB, but almost complete transmission of the signal through the filter. In the out-of-band regions, virtually no signal, actually less than −40 dB, is transmitted through the filter.

12.2 TRANSMISSION LINES

As was described in general in Chapter 11, a variety of transmission lines are used to convey signals from one point in a circuit or system to another. Although these transmission lines may appear as simple traces of metal, such as in the case for low-frequency circuits, they must be designed to guide the electromagnetic radiation from one point to the next with minimal loss or reflection of the signal. For purposes of discussion, we can divide transmission lines into two broad categories: system-level interconnects and planar transmission lines.

System-level interconnects include coaxial cables and rectangular waveguides that are used in applications where signals need to be conveyed from a card or module to another portion of the system. These transmission lines are generally capable of handling larger power levels than planar transmission lines because of their larger, more robust, construction and, in many cases, lower loss per unit length.

Planar transmission lines are much smaller in size and are used to convey signals from point to point on a chip or at the board or module level of a high-frequency system. In general, this type of interconnect exhibits a higher loss and lower power-handling capability than system-level transmission lines. However, a very large number of these structures can easily be fabricated in a relatively small volume. In addition, these planar transmission lines are normally used in situations where they are short in length so that the higher losses per unit length are not relevant.

The next few sections describe the variety of transmission lines in more detail and provide design and analysis equations that allow the reader to evaluate an existing transmission line performance as well as design transmission lines for a given application and material set. This analysis assumes that the materials involved are not magnetic in nature and that they are isotropic and homogenous. These are usually good assumptions, but they may not be true for all cases. Anisotropic, inhomogeneous, and/or magnetic materials require more involved analysis.

12.2.1 Transmission Line Modes

Each type of transmission line can support more than one form of electromagnetic propagation because of the different ways in which coupled E and B fields interact with particular geometries and materials that make up the transmission line. These different forms of electromagnetic propagation are called modes. Different modes propagate with different electrical characteristics, such as speed. However, it is usually undesirable to create multiple modes at one time in a given transmission line. Three basic types of modes that can be created are the transverse electromagnetic (TEM), transverse electric (TE), and transverse magnetic (TM). Each of these modes has one or more components transverse (perpendicular) to the direction of propagation, which we will define as z. For example, the TE mode's E component is transverse to the z direction and, therefore, does not vary in the direction of propagation. Likewise, for the TEM mode, the magnetic and electric fields do not vary in the direction of propagation. A more detailed presentation of this concept and its derivation from Maxwell's equations is given elsewhere [15]. Understanding these modes and when they can be excited in a given transmission line is important since each transmission line structure has a useful bandwidth, which is determined by the possible modes of propagation.

The TEM mode is the most important mode since it can, in general, exist at any frequency on a transmission line. However, not all transmission lines can support the TEM mode. Rules of thumb that can be used to determine if a guide can support the TEM mode are:

1. **Does it have at least two independent conductors?** Transmission lines composed of one conductor cannot support the TEM mode.
2. **Is it made up of a uniform dielectric?** More than one dielectric or nonuniform dielectric excludes the pure TEM case. Some transmission lines with nonuniform dielectrics may approach the TEM case. They are referred to as quasi-TEM.

If these conditions are met, then the transmission line supports the TEM mode and can be used from DC up to some maximum frequency.

The key issue to understand regarding the TE and TM modes is that, for a given transmission line configuration, different modes exist at different frequencies. There are an infinite number of TE and TM modes for any given transmission line structure, which are labeled as TE_{mn} and TM_{mn}, where m and n are integer counters that represent different modes. So, for example, in a given line, a TE_{10} mode may begin to propagate at a frequency we designate as f_{10}, while the TE_{11}, a different form of propagation, may not begin to exist until a much higher frequency f_{11}. This frequency dependence sets the useful range for a given guide based on the minimum frequency at which it will propagate any mode and the frequency at which the next highest mode of energy can exist.

The first mode with the lowest order, lowest cutoff frequency, is called the dominant mode and is the form of energy for which the transmission line is normally used. If the structure supports the TEM mode, then it will always be the dominant mode, since its cutoff frequency is by definition zero. One can see this later in the chapter by examining the structure of a stripline, which supports the TEM mode, versus a rectangular waveguide, which does not support the TEM mode. Clearly, the stripline can be used for a DC voltage, zero frequency, since the strip acts as one conductor and the two ground planes can act as the second current return conductor to complete the circuit. In contrast, the rectangular waveguide cannot possibly be used for a DC voltage since only one conductor exists, and, from basic circuit theory, we know that two conductors are required to complete a circuit.

12.2.2 System-Level Transmission Lines

Transmission lines are used to connect components at the chip level, package level, as well as at the system level. As you would expect, the types of structures used for these cases can be quite different. System-level transmission lines tend to be physically larger in size and are able to guide signals over large distances within a microwave system. The most common of these structures are rectangular waveguides and coaxial transmission lines.

12.2.2.1 Rectangular Waveguides

Waveguides are among the earliest types of transmission lines used to control the flow of electromagnetic energy within high-frequency systems. In fact, early radar systems were based on these guides, and, in many cases, the electrical components, such as filters and couplers, were fabricated from this type of guide. As illustrated in Figure 12.5, this type of guide in its simplest form is merely a hollow metal tube. Versions with rectangular as well as circular cross sections have their applications, with the rectangular design being the most widely utilized [16–18].

Since the waveguide structure is composed of one and only one conductor, it does not support the TEM mode and only TE and TM modes can exist. This type of guide can be filled with a dielectric, but it is most commonly utilized with a hollow air-filled center for minimum dielectric loss.

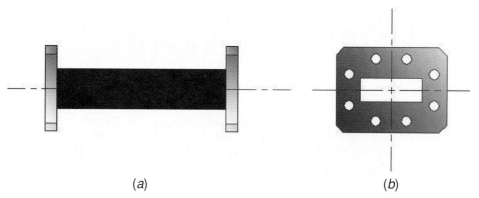

Figure 12.5 Rectangular waveguide: (*a*) side view; (*b*) end view. The center hollow section in (*b*) is the actual guide region with height a and width b, while the outer, smaller holes illustrate the typical mounting flange used to bolt these guides together.

The general solutions for fields that make up a TE mode of propagation in a rectangular waveguide can be found from the solution of the reduced Helmholtz equation and are of the form

$$H_z = A_{mn} \cos \frac{m\pi x}{a} \cos \frac{n\pi y}{b} e^{-j\beta z} \qquad H_y = \frac{j\beta n\pi}{k_c^2 b} A_{mn} \cos \frac{m\pi x}{a} \sin \frac{n\pi y}{b} e^{-j\beta z}$$

$$H_x = \frac{j\beta m\pi}{k_c^2 a} A_{mn} \sin \frac{m\pi x}{a} \cos \frac{n\pi y}{b} e^{-j\beta z} \qquad E_x = \frac{j\omega \mu n\pi}{k_c^2 b} A_{mn} \cos \frac{m\pi x}{a} \cos \frac{n\pi y}{b} e^{-j\beta z} \quad (12.8)$$

$$E_y = \frac{-j\omega \mu m\pi}{k_c^2 a} A_{mn} \sin \frac{m\pi x}{a} \cos \frac{n\pi y}{b} e^{-j\beta z} \qquad E_z = 0$$

where the direction of propagation is on the z axis, m and n are integer constants, and the subscripts denote the direction of the components. The β is the propagation constant and is of the form

$$\beta = \sqrt{k^2 - k_c^2} = \sqrt{k^2 - \left(\frac{m\pi}{a}\right)^2 - \left(\frac{n\pi}{b}\right)^2} \quad (12.9)$$

It is important to consider these expressions in order to understand the behavior of this type of transmission line. Note that if we do not constrain the value of k_c with respect to k, β could, in theory, be complex. Since β is itself the imaginary part of the complex propagation constant, a physically realizable wave cannot exhibit a complex value of β. Therefore, real propagating waves may only exist for values of k_c that are less than k. This important result means that only a certain range of values of k_c will produce propagating waves, and since the wavenumber is related to the frequency, we can define a cutoff frequency by

$$f_{c_{mn}} = \frac{k_c}{2\pi\sqrt{\mu\varepsilon}} = \frac{1}{2\pi\sqrt{\mu\varepsilon}} \sqrt{\left(\frac{m\pi}{a}\right)^2 + \left(\frac{n\pi}{b}\right)^2} \quad (12.10)$$

This mathematics means that an infinite number of TE modes of propagation exist since m and n can be any integer and that each mode will have a different cutoff frequency. For a particular waveguide with dimensions a and b, each mode of propagation will start propagating at its unique cutoff frequency and will exist for all frequencies higher than its cutoff frequency. Basic electrical engineering fundamentals indicate that this type of structure is not useful at DC since only one conductor exists, and therefore no possible return path for a DC exists. This effect can also be seen from the cutoff frequency equation above since no realistic combination of m, n, a, and b will create a waveguide with a cutoff frequency that approaches zero.

A similar solution exists for each TM mode of propagation, which also exhibits a unique cutoff frequency for a given set of physical dimensions and material used to construct a waveguide.

Since most applications desire only one mode of propagation, the net effect is that this type of guide is only useful over the frequency range in which the lowest order mode, and only the lowest order mode, propagates. This lowest order mode is normally called the dominant mode and rectangular waveguides are sometimes specified by their frequency range, X band, Ku band, and so on. For frequencies below the lowest order mode cutoff frequency, a given guide will not support energy transmission. The upper frequency is set by the fact that as the frequency is increased beyond the first cutoff frequency and the guide becomes useful, at some point the next mode will also begin to propagate. These different modes, in general, propagate with different velocities, and the use of more than one mode simultaneously results in significant signal distortion.

The applications for rectangular waveguides are for very high frequency and/or high-power signals that are difficult to handle with other planar structures or coaxial lines. Since these guides are air filled and made from a significant amount of thermally conductive material, not only are the losses generally very low, but the structure is very robust and can tolerate significant heating due to conduction losses without damage.

12.2.2.2 Coaxial Transmission Lines

A coaxial transmission line is composed of an inner conductor, or wire, and an outer cylindrical conductor, as shown in Figure 12.6. These two conductors are separated by a dielectric, and a protective coating is normally applied to the outside of the cable. This type of cable is commonly used for RF, as well as microwave applications, and can be found in a wide range of applications from cable TV connections to CB radios.

Coaxial cables can be obtained that will operate over a range from DC to as high as 100 GHz in some cases. Since the structure is TEM, no minimum cutoff frequency exists. This fact is often exploited by applications such as satellite TV receivers that use a coaxial cable to send power up to the rooftop dish and the same cable to send L-band signals containing the TV content down from the dish.

The characteristic impedance of the coaxial line can be expressed as

$$Z_0 = \sqrt{\frac{R + j\omega L}{G + j\omega C}} \qquad (12.11)$$

where the parameters R, L, G, and C are the resistance, inductance, conductance, and capacitance per unit length of the transmission line. Values for R, C, L, and G, can be found

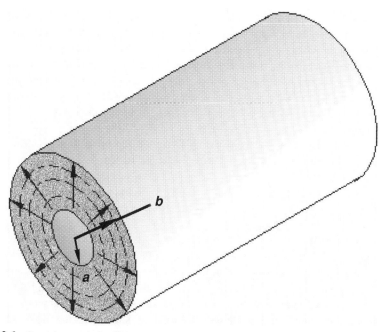

Figure 12.6 Coaxial transmission line.

from the expressions

$$R' = \frac{R_s}{2\pi}\left(\frac{1}{a} + \frac{1}{b}\right) \qquad C' = \frac{2\pi\varepsilon}{\ln(b/a)}$$

$$L' = \frac{\mu}{2\pi}\ln\left(\frac{b}{a}\right) \qquad G' = \frac{2\pi\sigma}{\ln(b/a)} \tag{12.12}$$

The cutoff frequency of the first higher order mode can be estimated as

$$f_c \approx \frac{ck_c}{2\pi\sqrt{\varepsilon_r}} \tag{12.13}$$

where k_c is the cutoff wavenumber of this mode and is of the form

$$k_c \approx \frac{2}{a+b} \tag{12.14}$$

12.2.3 Planar Transmission Lines

Large system-level interconnects that are useful at the board and system level are generally not suitable for integration into a module and certainly not into a semiconductor device. However, several kinds of transmission lines, which will be called planar transmission lines in this text, have been specifically developed for integration into compact chip-, board-, and module-level designs. The most important of these are the stripline, microstrip, and coplanar waveguide.

12.2.3.1 Stripline

Stripline transmission lines are composed of a metal strip buried in a dielectric material that is coated with two ground planes on its top and bottom surfaces, as shown in Figure 12.7 [19–21]. Many low-loss dielectric materials are used to fabricate this type of structure, including printed circuit boards, multilayer ceramics, and thin-film polymer coatings. The two ground planes must be maintained close to the same potential through the use of a large number of vias, which electrically tie these two planes together with minimum inductance.

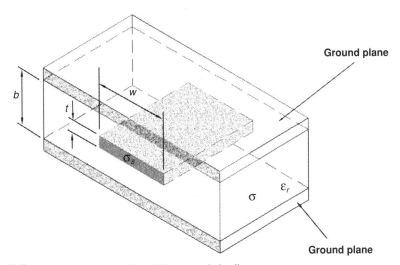

Figure 12.7 Cross-sectional view of a stripline transmission line.

The key attributes of a stripline that determine its electrical properties are the width of the strip (W), the dielectric thickness (b), the dielectric constant of the material, the loss tangent of the material, and the surface resistance of the conductors [22, 23].

For a given stripline geometry, the primary electrical characteristic of interest, the impedance, can be determined from the expression [24–30]

$$Z_0 = \frac{30\pi}{\sqrt{\varepsilon_r}} \frac{b}{W_e + 0.441b} \qquad (12.15)$$

where

$$\frac{W_e}{b} = \frac{W}{b} - \begin{cases} 0 & \text{for } \frac{W}{b} > 0.35 \\ \left(0.35 - \frac{W}{b}\right)^2 & \text{for } \frac{W}{b} < 0.35 \end{cases}$$

Similarly, a designer can determine the strip width needed for creating a given characteristic impedance once the dielectric constant and dielectric thickness (b) have been selected using the expression

$$\frac{W}{b} = \begin{cases} x & \text{for } \sqrt{\varepsilon_r} Z_0 < 120 \\ 0.85 - \sqrt{0.6 - x} & \text{for } \sqrt{\varepsilon_r} Z_0 > 120 \end{cases} \qquad (12.16)$$

where

$$x = \frac{30\pi}{\sqrt{\varepsilon_r} Z_0} - 0.441$$

The loss for a given stripline design can be calculated in two parts: the conductor loss α_c, and the dielectric loss α_d [31]. The total loss, neglecting radiation, is simply the sum of these two components:

$$\alpha_d = \frac{w\sqrt{\mu_0 \varepsilon_0 \varepsilon_r} \tan \delta}{2} \quad Np/m \qquad (12.17)$$

$$\alpha_c = \begin{cases} \dfrac{2.7 \times 10^{-3} R_s \varepsilon_r Z_0}{30\pi (b - t)} A & \text{for } \sqrt{\varepsilon_r} Z_0 < 120 \\[2ex] \dfrac{0.16 R_s}{Z_0 b} B & \text{for } \sqrt{\varepsilon_r} Z_0 > 120 \end{cases} \quad Np/m \qquad (12.18)$$

where

$$A = 1 + \frac{2W}{b-t} + \frac{1}{\pi} \frac{b+t}{b-t} \ln\left(\frac{2b-t}{t}\right)$$

and

$$B = 1 + \frac{b}{0.5W + 0.7t}\left(0.5 + \frac{0.414t}{W} + \frac{1}{2\pi} \ln \frac{4\pi W}{t}\right)$$

12.2.3.2 Microstrip

Microstrip transmission lines are very popular due to their simple construction and ease of fabrication in a variety of technologies. As illustrated in Figure 12.8, these transmission lines are composed of a thin conductor strip on top of a dielectric substrate and a ground

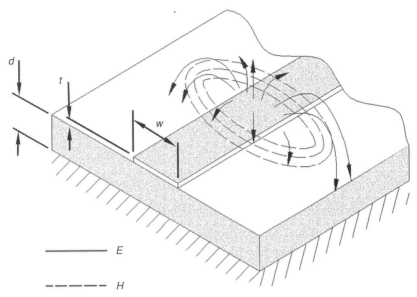

Figure 12.8 Microstrip transmission line and the electrical and magnetic fields associated with it.

plane on the bottom side of the dielectric substrate [32, 33]. The key physical parameters that impact the electrical performance of this type of transmission line are the metal thickness and composition as well as the dielectric properties and thickness.

Microstrip transmission lines do not support a pure TEM mode of propagation due to the inhomogeneous dielectric formed by the substrate and air above it. However, in many cases, the behavior approaches that of a TEM and is often referred to as a quasi-TEM transmission line. The analysis of this type of structure is more involved since the fields exist in the dielectric as well as in the air above the substrate [34–42]. The existence of fields in both the dielectric and the air above the dielectric is addressed through the effective permittivity. This quantity is used to transform the problem mathematically from the complex case illustrated in Figure 12.9a, where the fields must be analyzed in both the dielectric and the air. The ratio of the field density in air and the dielectric is a function of the dielectric constant. The effective permittivity allows a simple analysis of the structure, as shown in Figure 12.9b, where the air and substrate are replaced with a uniform dielectric that has a permittivity equal to the effective permittivity. This quantity can be calculated from the expression

$$\varepsilon_e = \frac{\varepsilon_r + 1}{2} + \frac{\varepsilon_r - 1}{2} \frac{1}{\sqrt{1 + 12d/W}} \quad (12.19)$$

Note that this is a function of the dielectric constant as well as the geometry of the microstrip line. Using this expression, the analysis equations for the microstrip case are of the form

$$Z_0 = \begin{cases} \dfrac{60}{\sqrt{\varepsilon_e}} \ln\left(\dfrac{8d}{W} + \dfrac{W}{4d}\right) & \text{for } \dfrac{W}{d} \leq 1 \\ \dfrac{120\pi}{\sqrt{\varepsilon_e}\,[W/d + 1.393 + 0.667\ln(W/d + 1.444)]} & \text{for } \dfrac{W}{d} \geq 1 \end{cases} \quad (12.20)$$

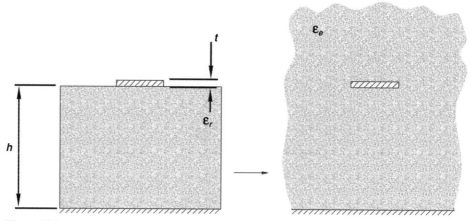

Figure 12.9 Concept of effective impedance, which transforms the idealistic microstrip from a combined dielectric air structure to a uniform effective dielectric constant.

Likewise a useful expression for the design of microstrip transmission lines is of the form

$$\frac{W}{d} = \begin{cases} \dfrac{8e^A}{e^{2A} - 2} & \text{for } \dfrac{W}{d} < 2 \\ \dfrac{2}{\pi}\left[B - 1 - \ln(2B - 1) + \dfrac{\varepsilon_r - 1}{2\varepsilon_r}\left\{\ln(B - 1) + 0.39 - \dfrac{0.61}{\varepsilon_r}\right\}\right] & \text{for } \dfrac{W}{d} > 2 \end{cases}$$

where

$$A = \frac{Z_0}{60}\sqrt{\frac{\varepsilon_r + 1}{2}} + \frac{\varepsilon_r - 1}{\varepsilon_r + 1}\left(0.23 + \frac{0.11}{\varepsilon_r}\right) \qquad B = \frac{377\pi}{2Z_0\sqrt{\varepsilon_r}} \qquad (12.21)$$

The losses for the microstrip case [43,44], as expressed in terms of the conductor and dielectric losses, respectively, are

$$\alpha_c = \frac{R_s}{Z_0 W} \quad \text{Np/m} \qquad \alpha_d = \frac{k_0 \varepsilon_r (\varepsilon_e - 1)\tan\delta}{2\sqrt{\varepsilon_e}(\varepsilon_r - 1)} \quad \text{Np/m} \qquad (12.22)$$

where

$$R_s = \sqrt{\frac{\omega\mu_0}{2\sigma_s}}$$

The first higher order mode cutoff frequency can be estimated as

$$f_c \cong \frac{c_0}{\sqrt{\varepsilon_r}(2w + 0.8h)} \qquad (12.23)$$

12.2.3.3 Coplanar Waveguides

Another common transmission line used as chip-, module-, and board-level interconnects is the coplanar waveguide. These guides are similar in construction to the microstrip with a thin metal strip on the top surface of the dielectric. However, in this case, the back-side ground plane present in the microstrip case is moved to the top surface and placed on either side of the conductive strip, as shown in Figure 12.10 [45]. The key parameters that affect the performance of this type of guide include the dielectric thickness, material properties, strip width (W), and gap between the strip and the ground planes (S). The gap on either side

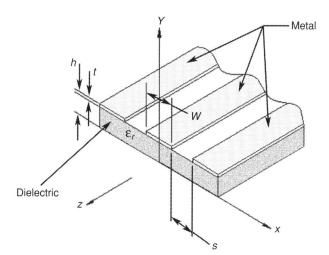

Figure 12.10 Coplanar transmission line.

of the strip is assumed to be symmetric, and ground planes extended out a distance that is usually much larger than the dielectric thickness.

Rather than closed-form design equations, as in the case of stripline, an analysis equation is often used to evaluate the impedance of a particular coplanar guide. Designs are often created by iterative solution of this analysis equation to find a suitable strip width and gap for a given substrate dielectric constant and thickness as well as a given metal thickness.

The analysis equation for coplanar waveguides can be expressed as

$$Z_0 = \frac{30\pi}{\sqrt{\varepsilon_r}} \frac{K(k')}{K(k)} \tag{12.24}$$

where the effective dielectric constant is of the form

$$\varepsilon_e = 1 + \frac{\varepsilon_r - 1}{2} \frac{K(k')}{K(k)} \frac{K(k_1)}{K(k_1')} \tag{12.25}$$

In these expressions,

$$k = \frac{a}{b} \qquad a = \frac{S}{2} \qquad b = \frac{S}{2} + W \qquad k_1 = \frac{\sinh(\pi a/2h)}{\sinh(\pi b/2h)} \tag{12.26}$$

Assume K is the complete elliptic function of the first kind and K' is its complement. The ratio K/K' can be approximated as

$$\frac{K(k)}{K(k')} = \begin{cases} \left[\frac{1}{\pi} \ln\left(2\frac{1+\sqrt{k'}}{1-\sqrt{k'}}\right)\right]^{-1} & \text{for } 0 \leq k \leq 0.7 \\ \frac{1}{\pi} \ln\left(2\frac{1+\sqrt{k}}{1-\sqrt{k}}\right) & \text{for } 0.7 \leq k \leq 1 \end{cases} \tag{12.27}$$

where

$$K'(k) = K(k') \qquad \text{and} \qquad k' = \sqrt{1-k^2}$$

The radiation losses are generally smaller than either the conduction or dielectric losses and in some cases may be neglected. The dielectric losses in a coplanar waveguide are identical to those given in Eq. (12.22) for the microstrip case. The conductor losses can be approximated as

$$\alpha_c = \frac{8.68 R_s \sqrt{\varepsilon_e}}{480\pi K(k_1) K'(k_1)(1 - k_1^2)} \left\{ \frac{1}{a} \left[\pi + \ln\left(\frac{8\pi a(1 - k_1)}{t(1 + k_1)}\right) \right] \right.$$
$$\left. + \frac{1}{b} \left[\pi + \ln\left(\frac{8\pi b(1 - k_1)}{t(1 + k_1)}\right) \right] \right\} \quad \text{dB/unit length} \quad (12.28)$$

where a, b, and k_1 are defined above. The metal is assumed to be of thickness t and possess a surface resistance R_s. This approximation assumes that the strip thickness t is greater than 3 times the skin depth and that the metal thickness is much less than the strip width or the gap. For this calculation, the individual values of K and K' are required. They may be approximated as

$$K(k) = \begin{cases} \frac{\pi}{2} \left\{ 1 + k^2 \left(\frac{1}{2}\right)^2 + k^4 \left(\frac{3}{2 \cdot 4}\right)^2 + k^6 \left(\frac{3 \cdot 5}{2 \cdot 4 \cdot 6}\right)^3 + \cdots \right\} & \text{for } 0 \leq k \leq 0.707 \\ \left\{ p + (p-1)\frac{k'^2}{4} + 9\left(p - \frac{7}{6}\right)\frac{k'^4}{64} + 25\left(p - \frac{37}{30}\right)\frac{k'^6}{256} + \cdots \right\} & \text{for } 0.707 \leq k \leq 1 \end{cases}$$

$$\text{where } p = \ln\left(\frac{4}{k'}\right) \quad (12.29)$$

12.2.3.4 Variations

Figure 12.11 illustrates some of the common variations of transmission lines from the basic designs described above. Many of these guides have niche applications that are appropriate due to fabrication constraints and/or specific application requirements. Variations on the traditional microstrip line include the buried microstrip, overlay microstrip, and suspended microstrip. Whole texts are available on this subject and the reader is referred to the references for further reading on this topic [47, 48].

The overlay microstrip (Fig. 12.11b) is a transmission line composed of a conventional microstrip with an additional dielectric layer placed above the metallic strip. This type of transmission line has found applications in tuning existing microstrip transmission line electrical properties [49] as well as in the fabrication of devices such as couplers and patch antennas.

The buried microstrip (Fig. 12.11a) is composed of a trench fabricated in a substrate and then filled with a dielectric material such as a polyimide [50, 51]. The microstripline is then placed on top of this dielectric fill. Recently, this approach has been used to create transmission lines on microwave monolithic integrated circuit (MMIC) devices [51].

The suspended microstrip is composed of a conventional microstripline without a ground plane on the back of the substrate that is then suspended over a ground plane either with the strip facing down toward the ground or facing away from the ground, as shown in Figure 12.11d. The case where the strip is facing the ground plane is often referred to as an inverted microstrip [37].

Another common variation on the principal types of transmission lines is the grounded coplanar transmission line [46]. This structure, shown in Figure 12.11c, is identical to the traditional coplanar line but with a ground plane added at the base of the substrate. This ground plane, in many cases, may be due to a housing or other metal structure used to support the substrate. In general, this ground plane tends to decrease the impedance of the transmission line in comparison to the coplanar line without a ground plane.

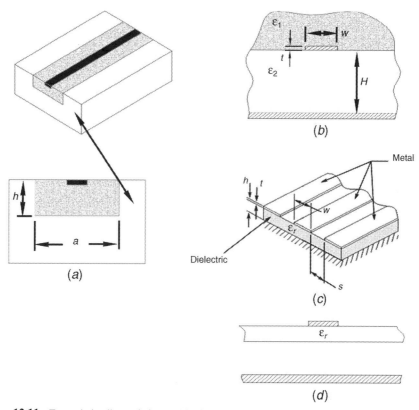

Figure 12.11 Transmission line variations: (*a*) buried microstrip; (*b*) overlay microstrip; (*c*) grounded coplanar; (*d*) suspended microstrip.

12.2.4 Discontinuities

An issue that often arises in microwave or millimeter-wave designs is the need to transition from one type of transmission line to another or from one impedance to a second impedance. Since the geometry of different types of transmission lines are normally radically different, optimization of a structure designed to interconnect the two requires considerable attention in order to minimize the reflected signal.

For example, consider a transition from a coaxial line to a coplanar waveguide. The center conductor of the coaxial line must be connected to the center conductor of the coplanar guide, and the outer ground ring of the coaxial line must be connected to the two outer coplanar ground planes.

Likewise, for transitions between impedances, a metal and dielectric structure must be designed to minimize the reflection caused by the abrupt change in the physical geometry needed to realize the two different impedances. A microstripline with an impedance of 50 Ω will have a wider conductor width than a 100-Ω microstrip on the same dielectric substrate. A gradual transition is normally used to connect these structures together in order to minimize the mismatch caused by the transition.

For purposes of discussion, we will divide the discontinuities into three categories: steps and bends within a transmission line family, transitions between families of transmission lines, and device interconnects.

12.2.4.1 Steps, Bends, and T Junctions

Within a transmission line configuration, such as a microstrip, the need often arises to create bends, steps in impedance, or T junctions. If these transitions are done in an arbitrary way, the discontinuity created by the abrupt change will often result in undesirable electrical effects, such as a capacitive loading of the line at that point in the design.

The most common of these discontinuities is the step change in impedance where one line impedance changes to a second line impedance. For example, a low-pass filter implemented in a distributed manner may utilize sections of low-impedance transmission lines (tens of ohms) connected to higher impedance lines (more than 100 Ω in some cases). The low-impedance lines are physically wider than the high-impedance lines implemented on the same substrate. Figure 12.12a illustrates a common transition that is used in these cases to minimize the impact of the physical transition through the use of a gradual change in the conductor width rather than an abrupt change [52–57]. It has been shown that an angle of 60° is optimum for most situations in order to minimize the impact of this transition.

Another common discontinuity is the bend or right angle in a planar transmission line. This structure is often required to route electrical signals from one place to another on either an IC or a substrate. In general, the ideal way to avoid these bends is to use gradual curves that do not contain a sharp radius of curvature. This type of gradual bend will produce very little, if any, discontinuity. However, in many cases, space prohibits the use of gentle bends and a right-angle bend is required in a transmission line. In this case, the optimum transition is composed of a mitered corner in a 90° bend, as illustrated in Figure 12.12b. It has been shown that in structures such as microstrip and stripline, the optimum miter is a section cut from the corner of the 90° bend with a width of $0.828W$, where W is the transmission line width [58–60].

Some circuits or circuit elements may also require a T junction. Such is the case for some power dividers, impedance-matching elements, and resonant structures. The T junction creates an inductive loading of the transmission lines that connect to form the T.

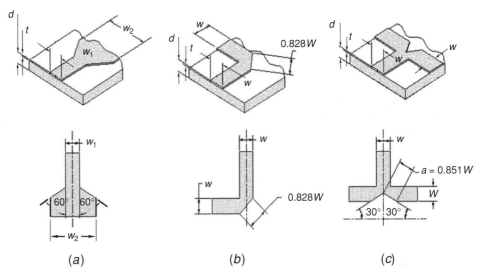

Figure 12.12 Common planar transmission line discontinuities: (a) optimized transition in strip width; (b) optimized 90° bend; (c) optimized T junction.

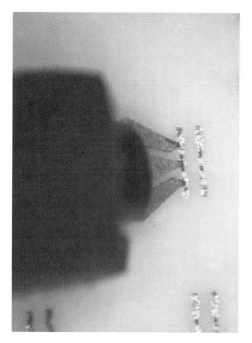

Figure 12.13 Ground–signal–ground (GSG) coaxial-to-coplanar probe.

The optimum way to minimize this discontinuity has been shown to be a right triangle cut from the T-junction metallization, as shown in Figure 12.12c. This cutout section is composed of an isosceles triangle with an angle of 30° and two equal side lengths of a, where the optimum value of a is 0.851 times the strip width [58–60].

12.2.4.2 Transitions

Transitions are often required between different types of transmission lines since the structures used to guide electromagnetic waves vary radically from one type to another. For example, consider the probe structure illustrated in Figure 12.13. This structure is a coaxial probe that is designed to allow connection between the coaxial line and a coplanar waveguide. The transmission line in this case is composed of gold on an alumina substrate that is terminated in a 50-Ω load, and the probe is composed of three metal contacts that are flexible in nature. The two outer probe contacts are connected internally in the probe to the outer ground electrode of the coaxial connector on the probe, while the inner probe contact is connected to the inner conductor of the coaxial connector. These transitions within the probe are carefully optimized to minimize the impedance mismatch at the interfaces and losses within the probes. The probe contacts are brought into contact with the coplanar waveguide and slightly deflected in order to create a physical contact between the probe tips and the coplanar waveguide metallization. This type of probe is available with bandwidths up to 110 GHz and can be used to create high-quality electrical contacts over this band.

In planar transmission line applications, such as those used in a substrate, it is often necessary to transition from one configuration of transmission line to another, for example, transition from a coplanar transmission line on the surface of the substrate to a stripline transmission line within the substrate. Figure 12.14 illustrates this type of transition in a low-temperature cofired ceramic (LTCC) substrate. In this case, the buried structure is just a

508 Chapter 12 Radio Frequency and Microwave Packaging

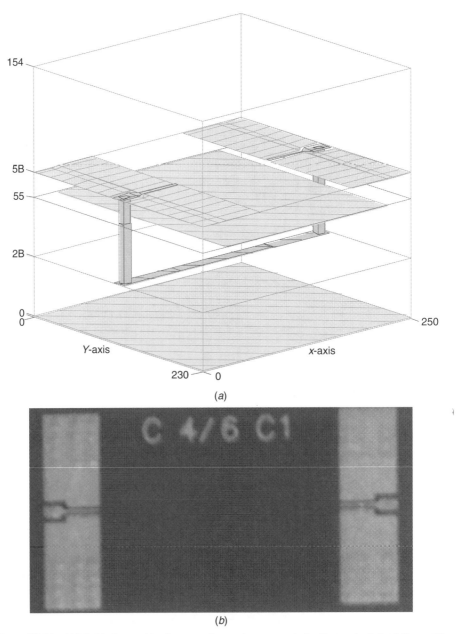

Figure 12.14 (*a*) Model of a transition from a surface coplanar transmission line to a buried stripline and then back to a surface coplanar transmission line. (*b*) Actual structure as fabricated in LTCC.

short section of stripline, but it may be a filter or other electrical component or circuit. Vias have been used to tie the buried stripline conductor to the coplanar center conductor as well as provide a low-inductance connection between the surface ground planes of the coplanar and buried ground planes of the stripline structure. In general, the design of these transitions are application specific and must be carefully optimized through the use of electromagnetic simulation.

12.2.4.3 Device Interconnects

An additional common discontinuity is an interconnect that is used to connect devices to a common substrate. For example, an RF IC or MMIC device will normally be mounted to a package lead frame or substrate and then electrically connected with wirebonds or solder bumps. In the case of solder bumps, the electrical connection is also the mechanical die-attach connection. The configuration of these interconnects is important in order to minimize the losses and impedance mismatch created by a poorly implemented transition from a controlled impedance interconnect on the substrate to a controlled impedance on the IC.

Wirebonds have historically been the method of choice for the interconnection of devices due to their high reliability and ease of automation. Trillions of wirebonds are made each year and the process is well developed with hundreds of references on the subject. However, wirebonds are intrinsically inductive in nature and care must be taken to account for the electrical impact of these structures in the signal path. The electrical effects of a wirebond, or set of wirebonds, have been modeled as a high-impedance transmission line with a T network at either end to account for the bond areas [61–63]. The characteristic impedance and effective dielectric constant of the transmission line created by this wire can be approximated as

$$Z_0 = \frac{\eta_0 u_0}{2\pi \sqrt{\varepsilon_w}} \qquad \varepsilon_w \approx \left[1 - \frac{k_0}{u_0}\left(\frac{\varepsilon_r - 1}{\varepsilon_r}\right)\right]^{-1/2} \qquad (12.30)$$

where

$$u_0 = \ln\left(\frac{h}{r_w} + \sqrt{\left(\frac{h}{r_w}\right)^2 - 1}\right) \qquad k_0 = \ln\left(\frac{\sqrt{(h/r_w)^2 - 1} + h_s/r_w}{\sqrt{(h/r_w)^2 - 1} - h_s/r_w}\right)$$

In this model, the wire is assumed to be suspended over a dielectric substrate of thickness h_s that is equipped with a ground plane on its bottom surface. This ground plane is a distance h from the wire, which has a radius of r_w and is located in free space with an impedance η_0.

Several methods are used to minimize the effects of this high-impedance connection, including the use of short wire lengths, multiple wirebonds, and ribbon bonds and/or recessing the die in a cavity. The first approach relies on the fact that the inductance of the wirebond is proportional to its length. Therefore, minimizing the length of the bond wire may be adequate in some cases to create acceptable electrical interconnects, particularly for RF ICs and the lower end of the microwave range.

The second method, the use of multiple bonds, has become popular in order to provide a closely matched interconnect over a wider bandwidth for high-speed digital applications [64–66]. Multiple bonds in parallel can be used to reduce the effective inductance or create a quasi-coplanar waveguide structure with lower impedance. The quasi-coplanar approach relies on a set of ground and signal pads on both the substrate and IC. The pads are arranged such that a ground pad is created on either side of the signal pad. Three wirebonds are then used to connect the signal pads and ground pads together, creating a structure similar in nature to a coplanar waveguide.

Ribbon bonds have been used for a variety of high-frequency interconnect applications in place of wirebonds. The wider conductor width results in significantly lower parasitic electrical inductance, which creates a lower impedance interconnect [67]. In addition, it has been demonstrated that ribbon bond first-level interconnects can be very reliable [68].

Cavities have been popular for decades in ceramic substrates since they can easily be formed in low- and high-temperature multilayer ceramics. This approach recesses the die such that its top surface is level with the surface of the substrate. This arrangement allows the use of very short wedge bonds with minimal loop heights, which make superior electrical connections when compared to the much larger loop heights required for a device mounted directly on the substrate surface.

The primary driver behind flip-chip interconnects has been the trend toward higher input/output (I/O) densities in modern digital ICs. However, this technology offers a number of advantages for the RF and microwave designer primarily due to the significant reduction in parasitic electrical effects associated with this type of interconnect. This type of interconnect can be created with epoxy or solder bumps and is appropriate for under-bump metallurgy (UBM) on an RF IC or MMIC device. The primary benefits are created by the very short interconnect path as well as the potential to create waveguide structures from multiple signal and ground bump interconnects.

A simple lumped-element electrical model for a flip-chip interconnect has been developed that expresses the interconnect as a shunt capacitance at either end of the bump and a series inductance for the bump structure itself [66]. The value of these lumped elements is a strong function of the materials used to create the bumps, the substrate, as well as the geometry of the structure. However, typical inductance values are on the order of 10 pH and typical capacitance values are on the order of 10 fF [69].

12.3 HIGH-FREQUENCY CIRCUIT IMPLEMENTATION

In any system, designers must consider the best way to implement the components or building blocks and then how to assemble them to form the desired electrical function. The building block approach is normally desirable for complex systems since it raises the overall system yield by allowing manageable pieces to be designed and built prior to incorporation into the final product. This implementation plan must consider the materials used as well as the logical use of ICs.

12.3.1 Material Considerations

12.3.1.1 Dielectric Losses

As described in Section 12.1.3.2, the total-loss picture for high-frequency circuits is more complex. In general, the dielectric losses become increasingly more important as the frequency increases and, at some point, begin to dominate conductor losses. Loss data for a given material are normally specified as the loss tangent ($\tan \delta$) and are a strong function of frequency for many materials. Loss information provided at 1 MHz may be of little value to a microwave designer since it is difficult to predict the losses at higher frequencies without direct measurement.

Table 12.1 provides some loss data for common dielectric materials at frequency. Note that, in general, the inorganic materials offer higher dielectric constants than the organic materials. Composites, such as ceramic-filled polymer substrates or boards, may allow for adjustment of the dielectric constant value over a wide range based on the amount of high-dielectric-constant fill material included in the polymer matrix. The desired dielectric constant is a strong function of the application. Higher values often allow for more compact circuit implementations since the size of many circuit elements is proportional to the wavelength. Since wavelength is inversely proportional to the square root of the substrate

Table 12.1 Common Dielectric Material Properties at High Frequencies and Room Temperature

Material	Dielectric constant (ε_r)	tan δ (1×10^{-4})	Typical surface roughness(μm)
Sapphire [-77]	9–12[a]	<0.002 (at 10 GHz)	<1
AlN [78]	8.29 (at 92.9 GHz)	4.6 (at 92.9 GHz)	1–10[b]
MgO [78]	9.8 (at 92.9 GHz)	<0.5 (at 92.9 GHz)	1–10[b]
99% Alumina [71]	9.9	0.43–7.1 (at 9 GHz)	1–10[b]
GaAs [73]	12.35 (at 9 GHz) 12.7 (at 70 GHz)	5–20 (at 10 GHz) ~60 (at 30 GHz)	<1
LTCC[c] [75, 76]	5.9–7.8	5–50 (at 10 GHz) <20 (up to 100 GHz)	1–10[b]
Quartz [72]	4.4	0.13 (at 9 GHz)	<1
FR-4	4–5	270 (at 10 GHz) [6]	<50[d]
BT	4.3	140 (at 10 GHz) [6]	<50[d]
Glass-reinforced PTFE [70]	2.2–2.6	9–22 (at 10 GHz) 4–5 (at 1 MHz)	(With rolled Cu) 2–4
Ceramic-filled PTFE [70]	3–10	13–27 (at 10 GHz)	(With rolled Cu) 2–4
Ceramic-filled thermoset [70]	6–10	22 (at 10 GHz)	(With rolled Cu) 2–4
Teflon [72]	2	2	—
BCB [74]	2.65	20 (at 10 GHz)	<1

Note: BT = bismaleinide triazine, PTFE = polytetrafluoroethylene, BCB = benzocyclobutene.
[a] Depending on orientation.
[b] As-fired roughness is typically 1–10 μm. Polished substrates can be produced with average roughness much less than 1 μm.
[c] Higher K dielectric tapes are available from several vendors.
[d] Dielectric roughness normally tracks applied metal roughness.

permittivity, selection of a high-permittivity material will reduce the circuit size. However, a designer should also carefully consider permittivity and its impact on the w/h ratio for a given transmission line. This is particularly important for higher power applications at the low end of the microwave frequency range. In this case, a lower dielectric constant produces wider conductors that reduce the conductor loss, which often dominates at this relatively lower frequency range.

Some polymer materials are not well suited to high-frequency designs due to a strong dependence of tan δ on frequency. For example, FR-4 and BT offer low losses at low frequencies due to thick, high-conductivity conductors and the fact that conductor losses are dominant in this range. However, the tan δ of these materials is a strong function of frequency, and, in the microwave and millimeter ranges, the loss is unacceptable for most applications. In addition, some polymers are prone to absorb water, which, even in small amounts, may dramatically increase the losses in a circuit. There are, however, a number of polymers that make attractive substrates over a wide frequency range, such as PTFE and PTFE composites. The surface roughness of these materials, as shown in Table 12.1, is often a strong function of the metallization that is used. Rolled copper foils generally offer smoother surfaces than electrodeposited copper films.

The losses in inorganic materials are generally low for low- to medium-dielectric-constant compositions. However, high-K ($K > 1000$ in some cases) ferroelectric materials can exhibit much higher losses. Dielectrics such as magnesium oxide (MgO), sapphire, and quartz offer extremely low losses into the millimeter wavelength range. The intrinsic roughness of these materials, as prepared, is often greater than a polymeric material. However, they can easily be polished to virtually any desired surface finish with average roughnesses of less than 1 μm in common use.

In the past few decades, the use of inorganic multilayer substrates, such as LTCC and high-temperature cofired ceramic (HTCC), has grown in popularity for high-frequency applications due to the need for higher circuit densities and the capability to integrate structures within and on top of these substrates. LTCC materials generally have an advantage in terms of losses due to the availability of formulations with very low loss tangents well into the millimeter range. The HTCC compositions are also relatively low in loss since they are largely composed of alumina or AlN. However, HTSC requires the use of refractory metals, which are high in resistivity compared to the noble metals that are compatible with LTCC.

12.3.1.2 Conductor Losses

As noted previously in Section 12.1.3.2, the losses are affected not only by the dielectric material selected but also by the conductors used within a substrate, module, or board. While these losses are more important at the lower end of the high-frequency spectrum, they must be carefully considered for virtually all applications. Table 12.2 gives some of the metals commonly used in high-frequency circuits and their properties. Three primary groups of metals are used: finish metals, good conductors, and adhesion metals. The finish metals, such as gold and nickel, are often used to provide corrosion-resistant surface finishes that can be wirebonded or soldered to. The good conductors, such as silver, gold, and copper, are used as the primary conducting medium due to their high conductivity. Adhesion layers, such as titanium and chromium, are used to provide a strong bond between the good conductors and a substrate, which, in thin-film form, rarely exhibit strong adhesion to most substrate materials. Adhesion metals are generally far less conductive than the group of good conductors, and, in some cases, these materials may be avoided, such as in the case of copper on laminate substrates and thick-film conductors.

Not only must the principal conductor that is used in a metallization be carefully considered, but also any adhesion layers. Since the skin effect will, in many cases, concentrate the current in the outer section of the conductor, a designer may inadvertently confine the

Table 12.2 Properties of Common Microwave Conductors

Material/configuration	Fabrication method	DC resistivity ($\mu\Omega$-cm)	Microstrip conductor attenuation α_c (dB/m)
Thin-film copper	Electrodeposited on quartz	1.9 [79]	1.7 at 4 GHZ [79]
			3.0 at 12 GHz [79]
	Electrodeposited on Al_2O_3		4 at 4 GHz [80]
Thin-film aluminum	Vacuum evaporation on quartz	2.73 [81]	2.1 at 4 GHZ [79]
			3.6 at 12 GHz [79]
Cr-Au	Vacuum evaporation on alumina	2.249 [81]	6 at 2 GHz [80]
			8 at 10 GHz [80]
Thick-film Au	Screen printed	2.6–6	
Thick-film Ag	Screen printed	<3	
Thick-film Cu	Screen printed	<3	

highest current density to a relatively low conductivity adhesion metal, such as titanium or chromium. This effect is most pronounced in planar transmission line structures, such as microstrip, where the adhesion metal layers, by necessity, must be on the bottom side of the conductor where the current densities are maximized.

12.3.2 Microwave Monolithic Integrated Circuits

Integrated circuits are the fundamental building blocks of most modern electronics due to the high degree of integration and low cost per electrical function in high-volume production. At high frequencies, these devices are often referred to as RF ICs MMICs to denote the unique design challenges associated with them [82]. The concept is, however, the same as that used in digital or low-frequency analog circuits—low-cost building blocks fabricated from semiconductor chips.

The key differences between low-frequency ICs and MMIC devices are the types of semiconductors used as well as the types of devices and the on-chip interconnects. MMICs are generally fabricated from groups III–V materials, such as GaAs, which offer semi-insulating substrate characteristics with superior high-frequency electrical performance. Metal–semiconductor field-effect transistors MESFETs, heterojunction bipolar transistors (HBTs), and high-electron-mobility transistors (HEMTs) are normally used in place of metal–oxide–semiconductor field-effect transistors (MOSFETs), or bipolar junction transistors (BJTs) due to their superior performance in the microwave and millimeter frequency ranges. The third primary difference is the use of controlled impedance interconnects and on-chip passive structures, such as matching networks, to fabricate the desired electrical function. The individual components on the surface of a MMIC device are often interconnected with microstrip or coplanar transmission lines rather than simple wire interconnects. These transmission lines use thin-film metal as the conductor and the semiconductor crystal as their dielectric.

Figure 12.15 illustrates a typical MMIC device (top left) as well as a number of single-chip packages that are used to protect these types of devices. As seen on the bottom right of the photo, this particular package is a land grid array package that is deigned to be soldered to a substrate or printed wiring board (PWB). MMIC devices are available in both packaged and bare die configurations.

12.3.3 MIC Technologies

In many cases, not all of the components required for a given circuit or system can be integrated easily onto MMIC devices. This may be due to the fact that they are difficult to integrate or that they consume so much die surface area that they are not cost effective to integrate on chip. Therefore, MMIC devices and additional elements are often integrated into a microwave integrated circuit (MIC). These MIC devices are modules or packages that contain one or more MMICs and discrete passive devices. The passive devices may be integrated into a substrate, which forms the base of the package or may be attached to the package/module substrate in a surface-mount configuration. A wide variety of MIC devices have been developed over many years, based on thick-film, thin-film, and laminate technologies [83]. An example of a microwave module developed by ANADIGICS for GSM cell phones is shown in Figure 12.16.

514   Chapter 12   Radio Frequency and Microwave Packaging

Figure 12.15   Top left is a GaAs-based MMIC chip. Bottom left and top right are typical single-chip packages used to protect these devices. As seen in the bottom right-hand side, these packages are designed as land grid arrays. (Courtesy of Triquint Semiconductors.)

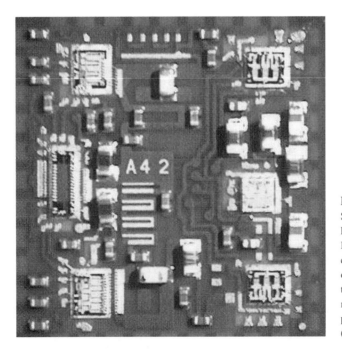

Figure 12.16
Second-generation GSM PowerPlexer (ANADIGICS, Inc.) module designed for use in cell phones. Normally, this type of module would be overmolded to protect the components. The module is shown without the protective covering for clarity. (Courtesy of ANADIGICS, Inc.)

12.4 LUMPED-ELEMENT COMPONENTS

While the design of passive devices has been covered in other sections of this text, the reader should be aware that significant care must be taken when designing lumped-element devices for high-frequency circuits. The key concerns are the undesired electrical parasitics that are built into the device as well as the overall size of the component [84–93]. To truly be a lumped-element device, the size of the component must be at the very least 10 times smaller than the wavelength of the signal(s) of interest. Parasitic reactance, in the form of undesired capacitance in an inductor and undesired inductance in a capacitor, causes these devices to self-resonate. In fact, these two issues are generally coupled in practice since the smaller the devices are in physical size, the lower the electrical parasitics will generally be.

With this in mind, the three key parameters a designer must consider are the self-resonant frequency (SRF) value and tolerance and the quality factor (Q). In many cases, these three metrics are mutually exclusive in that the desire to maximize SRF limits the achievable value.

12.4.1 Capacitors

As described in Chapter 9, there are two main designs of capacitor that are commonly used: the parallel plate and the interdigitated, as shown in Figure 12.17. The parallel-plate design (Fig. 12.17d) is attractive for higher value capacitors and is often used in conjunction with a high-dielectric-constant material. In this case, care should be taken to ensure that the

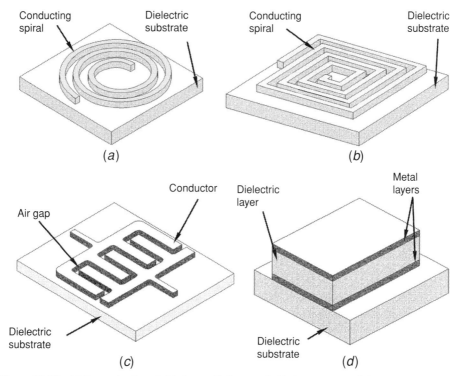

Figure 12.17 (a) Planar circular spiral inductor. (b) Square spiral inductor. (c) Interdigitated capacitor. (d) Thin film capacitor, (drawing is not to scale since the dielectric is normally very thin).

dielectric material is low in loss up to the desired frequency range and that the plate size and separation distance are much smaller than the signal wavelength. The interdigitated design (Fig. 12.17c) is useful for smaller values of capacitance and can be implemented in conjunction with planar transmission lines by simply etching the metallization to form the desired finger patterns.

While these two designs are common in high-frequency applications, many designs require very small values of capacitance. These values are often achievable by simply creating a small gap in a transmission line or by creating a small metal pad with a specified distance to a ground plane or embedded transmission line.

12.4.2 Inductors

Several types of inductors are commonly used in high-frequency circuits. These devices include spiral inductors as well as strip or ribbon inductors. Each configuration has its advantages and limitations, with the strip inductors generally finding use at higher frequencies than the spiral inductors.

12.4.2.1 Spiral Inductors

Flat spiral inductors, described in Chapter 9, are popular since they offer a wide range of achievable inductance with high Q factors. Two types of spiral inductors exist, square or circular, as shown in Figures 12.17 a and b, respectively. The circular inductor offers superior performance over the square spiral due, in part, to the increased resistance associated with the longer traces of the square spiral. Reports suggest that the circular spiral inductors offer 10 to 20% higher Q and SRF [91]. Limiting factors are size, acceptable Q factor, and maximum number of turns.

12.4.2.2 Strip or Ribbon Inductors

In many high-frequency circuits, small values of inductance are required. A range of small values can be fabricated by means of a small strip of metal. Since a metal surface or wire has some inductance, this effect can be exploited to create small inductance values in the middle of a transmission line. As illustrated in Figure 12.18, this type of inductor is normally fabricated using a high-impedance strip of metal between two controlled impedance transmission lines. The value of the inductance is given by

$$L = (2 \times 10^{-4} \ell) \left[\ln\left(\frac{\ell}{w+t}\right) + 1.193 + 0.2235 \frac{w+t}{\ell} \right] \quad (12.31)$$

where L is the inductance in nanohenrys per millimeter, w is the width of the strip, t is the metal thickness, and l is the length of the metal strip [84, 93]. The quality factor of a strip inductor is of the from

$$Q = \frac{2\pi f L}{K R_s l / [2(w+t)]} \quad (12.32)$$

where L is the inductance in nanohenrys per millimeter, w is the width of the strip, t is the metal thickness, l is the length of the metal strip, R_s is the surface resistance of the conductor, and f is the frequency of interest [84]; K is a correction factor that accounts for current crowding at the corners of the strip and can be calculated as

$$K = 1.4 + 0.217 \ln\left(\frac{w}{5t}\right) \quad \text{for } 5 < \frac{w}{t} < 100 \quad (12.33)$$

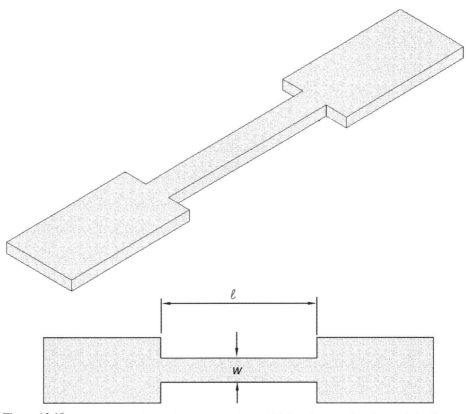

Figure 12.18 Configuration of a strip inductor made from a high impedance section of transmission line.

This analysis assumes that the inductor is not affected by a ground plane placed underneath or above the conductor strip. A ground plane, either above or below the inductor strip, reduces the effective inductance of the strip [89, 90]. This reduced inductance can be calculated from

$$L_{eq} = L \left[0.570 - 0.145 \ln \left(\frac{w}{h} \right) \right] \quad (12.34)$$

where L_{eq} is the inductance in nanohenrys per millimeter with a ground spaced a distance h away, w is the width of the strip, and L is the inductance without the ground plane [58, 93].

Most applications of strip inductors are between two controlled-impedance interconnects, such as stripline or microstrip transmission lines [94]. In these cases, it is convenient to express the inductance in terms of the strip's impedance,

$$L = \frac{Z_0}{2\pi f} \sin \left(\frac{2\pi l}{\lambda_g} \right) \quad (12.35)$$

where L is the inductance in henrys, Z_0 is the characteristic impedance of the strip, f is the frequency, l is the strip length, and λ_g is the wavelength in the dielectric. From examination of Eq. (12.35), it is clear that a high-impedance strip produces a greater inductance than a wider, low-impedance strip. Reducing the strip width is also important for the case of inductors located over or between ground planes, which is a common configuration. In these cases, the thinner strip exhibits a lower parasitic capacitance to ground than a wider strip.

The parasitic capacitance to ground for a strip inductor located above a ground plane is of the form

$$C = \frac{1}{2\pi f Z_0} \tan\left(\frac{\pi l}{\lambda_g}\right) \qquad (12.36)$$

where C is the termination capacitance to ground of the strip inductor in farads, Z_0 is the characteristic impedance of the strip, f is the frequency, l is the strip length, and λ_g is the wavelength in the dielectric.

12.4.3 Resistors and Terminations

Resistors are often required as two-port devices, a series resistor, or resistors to ground that are used for terminations. Generally, small values, <1 kΩ, are required for these applications. As in the case of the inductors and capacitors, the key concerns are not only the value and tolerance but also minimization of parasitic electrical effects. While the electrical response depends on the resistor design and materials used, low-value resistors tend to be inductive in nature [95]. This effect is best limited by creating resistors that are small in physical size.

12.5 DISTRIBUTED COMPONENTS

In many cases, lumped-element components such as the spiral inductors and parallel-plate capacitors described above are often inadequate at the higher end of the microwave range and in the millimeter-wave region. These devices often resonate due to parasitic effects in these frequency ranges and cease to provide the desired electrical function. In these cases, distributed components, based on electromagnetic principles, can be used in their place. These devices include filters [96–103], power dividers [104–113], and phase-shifting elements, but they can also implement the same basic capacitive and inductive effects. A good example of a distributed element is the strip inductor described above. Rather than creating all the inductance in one lumped spiral loop of metal, the natural inductance of a segment of transmission line is exploited instead.

These distributed components can be quite complicated to design, and whole texts have been written on this subject. The following discussion is intended to give the reader a basic understanding of the components and their key attributes.

12.5.1 Impedance-Matching Devices

A variety of impedance-matching devices are required for most microwave and millimeter-wave circuits. These devices, as described above, are used to minimize the reflection coefficient between the components within a system. Generally, these devices are created from transmission lines in the form of stubs of specific lengths or the impedance of a transmission line in series with a device [15]. As a packaging engineer, the key issue to understand is that these devices may often appear as simple metal traces on a MMIC or on a substrate. However, their lengths and widths have been precisely calculated to achieve a desired effect

12.5.2 Filters

Figure 12.19 illustrates a number of common distributed filter designs that are used in microwave and millimeter-wave circuits. The basic idea is that, at very high frequencies, the lumped-element capacitors and inductors do not behave as they would at lower frequencies since the signal wavelength is now on the order of the device [96]. However, transmission line structures can be used to produce similar behavior as that found in a lumped-element device and, therefore, create a filter response.

Figure 12.19a illustrates a band pass filter fabricated from a stripline or microstrip transmission line. Each section of the filter is designed as a coupled set of transmission lines that are capacitively coupled to the adjacent line. For a filter response of Nth order, there are N resonators and $N + 1$ coupled transmission line segments. While a detailed design of this type of filter is beyond the scope of this text, packaging engineers working at high frequencies should be aware that the length of the transmission lines controls the center frequency of the filter. Also, the width of each section of transmission line and the gaps between them are carefully designed to produce a particular filter response. Small

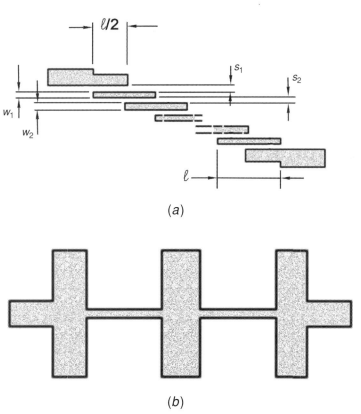

Figure 12.19 Planar microwave filters. (*a*) Top filter is designed to be a bandpass filter with the input as the far left and the output as the lower right. (*b*) Low-pass filter composed of microstrip transmission line segments.

deviations in any of these dimensions will alter the electrical response of the filter. Great care should be given to obtaining tolerance specifications from the designer in regard to the physical geometry of this type of filter to ensure that they are met in the final product.

The filter in Figure 12.19b is a low-pass filter formed by introducing several alternating segments of wide and narrow microstrip transmission lines [102, 103]. As in the band pass filter, the widths and lengths of the segments control the filter response, and while they are not as sensitive as in the bandpass case, large tolerances in these physical dimensions will radically alter the electrical response of the device.

These types of filters, and many variations on the theme, are commonly used in high-frequency circuits. It is not uncommon to implement these devices in or on a module-level substrate or printed circuit board rather than on chip since they consume a large amount of expensive chip area.

12.5.3 Power Dividers

Power dividers are used in microwave systems to split a signal into two parts or to combine two signals into one [104–107]. A divider can be fabricated from lumped elements; however, at higher frequencies, it is more common to fabricate these devices as distributed components from transmission lines. The two most common are the T junction [107] and the Wilkinson divider [104].

The T junction, as shown in Figure 12.20a, is simply a T-shaped structure fabricated from a planar transmission line, such as stripline or microstrip. The input line is normally designed to be the system impedance, such as 50 Ω, and the output arms are 100 Ω for an equal split divider. Care must be exercised to design this structure to prevent capacitive coupling between the two output arms.

The T junction offers simplicity and ease of fabrication since only a set of transmission lines are needed and this device can easily be fabricated in stripline or microstrip. However, this configuration has some limitations in that the output ports are not matched and it provides no isolation between the outputs. The lack of isolation refers to the fact that a signal induced at any output port will be transmitted to the input of the device as well as to the other outputs. While these may not be important for some applications, they may be unacceptable for other applications.

The Wilkinson divider uses a very similar T-like structure. However, the output arms for an equal split divider have impedances of $\sqrt{2}Z_0$, where Z_0 is the impedance of the input line, as shown in Figure 12.20b. This design also adds a resistor across the output transmission lines of the device with a value of $2Z_0$ ohms.

The benefit of the more complex Wilkinson design is that all the ports are matched to the system characteristic impedance and the outputs are isolated from each other. This device is somewhat more difficult to build, particularly at higher frequencies, since it requires a resistor, which can be difficult to implement without significant parasitics [105].

An example of the application of a power divider used to both split signals and combine them is often found in power amplifiers. This circuit splits the input into two parts that are then fed to identical power amplifiers. The output of the amplifiers is then fed back into a second power divider, which is operated as a combiner. That is, the two amplifier inputs are connected to the "output" terminals of the divider. The result is an amplifier that can handle twice the power rating of each of the individual amplifiers that are used to construct it. The circuit and variations on this scheme are often used for high-power applications, which are difficult to implement with a single amplifier device.

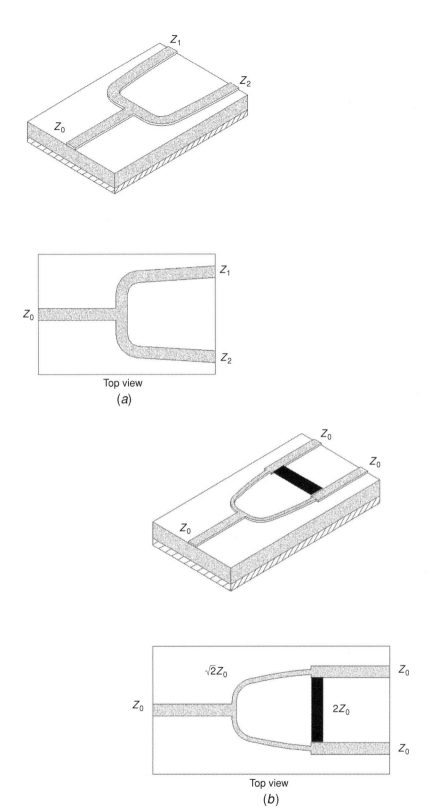

Figure 12.20 (*a*) T-junction power divider fabricated in microstrip. (*b*) Wilkinson power divider fabricated in microstrip. The dark black segment is a resistor between the two output transmission line segments.

12.5.4 Couplers

Couplers are four-port devices that can be used to combine or split a signal(s) in a number of ways. Figure 12.21 illustrates some of the more common designs that are used in microwave systems, including the branch line coupler, the rat race, coupled-line directional coupler, and Lange coupler. These devices can split signals equally or they can be designed to split the signals in an uneven manner, directing most of the signal to one output and a small portion of the signal to the second output port. An equal split device is often referred to as a "hybrid," not to be confused with the packaging terminology associated with this word. Each of these devices has four ports, which are labeled as shown in Figure 12.21c, and are often described as the input, coupled, through, and isolated ports. A signal connected to the input port is split between the coupled and through ports and, ideally, no signal exits the isolated port.

The performance of a coupler is given by the coupling coefficient C, the isolation I, and the directivity D, all normally expressed in decibels. The coupling, as described above, determines the percentage of power split between the coupled and the through ports. Ideally, the isolation is $-\infty$ decibels, representing no power leaving that port, while realistic values of I range from -50 to -60 dB. Recall that the decibel scale is logarithmic, such that -60 dB is much smaller than -50 dB, and the former approaches the noise floor of many measurement systems, which are often around -70 dB. Mathematically, these quantities can be expressed as

$$C = 10 \log \frac{P_1}{P_3} \qquad D = 10 \log \frac{P_3}{P_4} \qquad I = 10 \log \frac{P_1}{P_4} \qquad (12.37)$$

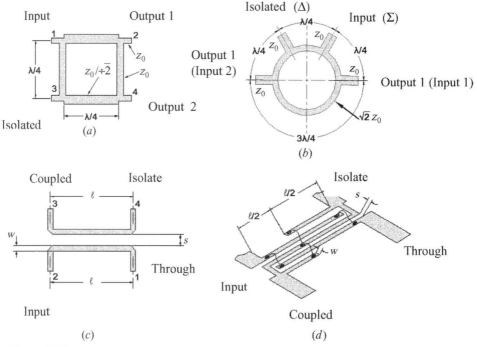

Figure 12.21 Typical couplers used in microwave and millimeter-wave circuits and packages: (*a*) branch line coupler; (*b*) rat race; (*c*) coupled-line device; (*d*) Lange coupler.

where P_1 is the power incident to port 1, P_2 is the power leaving port 2, P_3 is the power leaving port 3, and P_4 is the power exiting port 4 [15]. Note that the sum of the coupling coefficient and the directivity equals the isolation. Applications exist for a wide range of coupling values much like different inductor values are needed in electrical circuits. However, the isolation should, in general, be minimized.

The branch line coupler, shown in Figure 12.21a, is a hybrid that provides two outputs with equal power levels but phase shifted from each other by 90°. The devices are fabricated from a set of transmission lines in a boxlike configuration, with the vertical arms and input feed lines of the system of characteristic impedance (Z_0). The top and bottom transmission lines in the center box section are of impedance $Z_0/(2^{1/2})$. The entire structure is $\lambda/4$ in size, as measured from the center of the box arms, for some center frequency corresponding to λ in the transmission line used to fabricate the device. This requirement denotes a fundamental frequency limitation, since the device is only useful for a range of frequencies about this center frequency.

The rat race is also a hybrid that can be used to combine or divide signals, as shown in Figure 12.21b. A signal entering the input is split equally between the two output terminals with no phase difference, while a very small amount of leakage occurs at the isolated port. This configuration can be very useful for dividing signals in a number of applications. The device can also be used to provide the sum and/or difference of two signals by connecting each of the signals to the two "output" terminals. In this configuration, the sum of the signals appears at the input terminal and the difference appears at the isolated terminal. Rat race devices are most commonly implemented in microstrip; however, the basic concept can be applied to any planar transmission line configuration.

An even simpler coupler design is the directional coupler created by coupled transmission lines, as shown in Figure 12.21c. This device is fabricated from two transmission lines brought into close proximity to each other. The amount of coupling is a function of the spacing between the two lines as well as the type of transmission line configuration used (dielectric constant and geometry). The length of the coupled section of lines determines the center frequency of the coupler and corresponds to the length of the line being equal to $\lambda/4$. These devices are normally useful for some range of frequency about this center frequency.

In some cases, tight coupling, generally considered as more than -10 dB, is desired between the input and coupled ports. The couplers described so far are not effective in these cases since fabrication of them becomes very difficult. For example, consider the case of very tight coupling for the device in Figure 12.21c. The gap between the conductors must become very small since the coupling is inversely proportional to the gap size. As the gap goes to zero, these devices become impossible to fabricate with high yield and acceptable tolerance. The solution to this problem is the Lange coupler [113] shown in Figure 12.21d. This device uses the same basic concept as a set of coupled lines but uses several sets of coupled lines to increase the total coupling. Initially, these designs used wirebonds to connect the fingers as illustrated, but a growing trend is the use of a second metal layer to make these required connections.

12.6 SIMULATION AND CIRCUIT LAYOUT

A key aspect of high-frequency design is simulation and modeling. In many cases, every aspect of a circuit design, particularly the interactions between elements, cannot be easily calculated in a closed form. As a result, in recent years, the use of simulation tools has allowed designers to explore the behavior of design and/or circuit elements by reducing the need to fabricate and measure multiple iterations of the element or circuit in question. These tools can significantly speed up and improve product design cycles. However, the reader should be cautioned that these tools do not predict the behavior exactly and must be validated by prototypes that are fabricated and measured. Subtle differences between the simulator-predicted electrical behavior and reality are not uncommon due to simplifications made by the simulator and or manufacturing tolerances.

The design and simulation tools fall broadly into two categories of products: circuit simulators and field solvers. The first category of products are similar in function to low-frequency circuit simulators, such as SPICE, and they generally use closed-form models to predict the electrical behavior of elements, such as wirebonds, transmission lines, active devices, and other circuit elements. Products such as Ansoft Designer [114] and Microwave Office [115] include these types of tools. Designers can utilize these tools to simulate a high-frequency circuit and predict the s parameters and/or a host of other interesting electrical quantities. These simulations are most often frequency-domain oriented; however, time-domain products do exist and allow the user to predict these quantities over a chosen frequency range for a given design. These tools also generally include an intimate link to the physical structure of the circuit. This aspect is often not required at low frequencies but, as described above, is often critical to successful high-frequency designs. This integration of the physical and electrical structure allows designers to seamlessly integrate distributed physical transmission line structures with lumped circuit elements.

A key limitation of circuit-based simulators is that the user is restricted to the use of elements for which the s parameters or a model is known. In many cases, high-frequency designers may wish to use a structure or set of structures of their own design for which a model does not exist. Also, many complex structures and/or the electrical interactions between them are not easily condensed into a closed-form model. In these cases, a field solver that can calculate the electric and magnetic fields as a function of position and frequency within a dielectric and metal structure is invaluable. From these predicted fields, s parameters as well as the electrical currents and potential can be calculated and displayed for the designer. In most cases, one or more field solvers are integrated into a microwave circuit simulator that allows the designer to use the field solver for elements of the design and then import the results and use them in the circuit simulator. Another common approach is to use iterations of the fast circuit simulator to provide an optimized first-cut design prediction followed by a more accurate but considerably slower field solution of the geometry.

Field solvers are often grouped into planar three-dimensional (3D) and 3D varieties and use a variety of methods, including the method of moments or finite-element (FE) methods, to solve the electric and magnetic fields. The planar 3D solvers are designed for multilayer planar structures, such as are found in a multilayer ceramic substrate or on the top and bottom surfaces of a MMIC. These simulation tools are not well suited to true 3D structures, such as a wirebond; however, 3D solvers are designed to calculate the fields within a specific volume defined by the user. These solvers can also be used in conjunction with circuit simulators, and data can often be shared back and forth between the circuit model and the physical field model.

Field solvers use numerical methods to calculate the fields by breaking the geometry into small units and solving for the fields in these small areas. For most problems, this is a computationally intensive task that requires a high-speed computer and significant memory. Simulation times that last for hours and, in some cases, days are not unusual. Also, the designer must use great care in carefully defining the problem within the simulator so that the predicted electrical response correlates with reality.

12.7 MEASUREMENT AND TESTING

The measurement and testing of microwave and millimeter-wave circuits are quite different than the testing that is commonly done at low frequencies due to the additional complications associated with the electromagnetic nature of the signals of interest. Simple processes, such as probing of an IC, require more care at higher frequencies since the probe is launching or collecting an electromagnetic wave rather than a simple time-dependent voltage or current. Also, in many cases, the signals of interest may be very small, such as in the case of an input signal to a receiver.

12.8 FREQUENCY-DOMAIN MEASUREMENTS

Many of the applications at high frequency are fundamentally frequency-domain types of systems and consequently the determination of their electrical response as a function of frequency is very important. A variety of measurement systems are available that measure a given electrical response over a range of frequencies by repeatedly measuring the response at each frequency of interest. These systems are, or at least their sources are, often referred to as swept frequency sources since they are designed to rapidly make hundreds of measurements covering a broad frequency range in a manner of minutes.

12.8.1 Measurement Systems

The bulk of testing in the frequency range is done by measuring s parameters using either a vector network analyzer (VNA) or a scalar network analyzer (SNA). The primary difference between the instruments, as the name implies, is that the VNA measures both the magnitude and the phase information. Figure 12.22 illustrates a typical VNA used in the RF and microwave frequency ranges. This particular machine is limited to 26.5 GHz since it utilizes

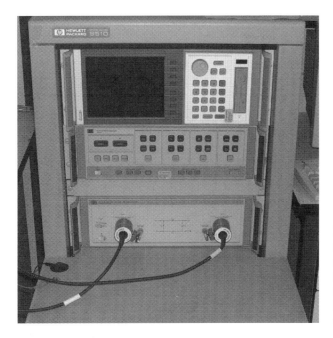

Figure 12.22 Typical VNA used for microwave and millimeter-wave measurements of s parameters, VSWR, and Γ. The two cables are coaxial transmission lines, one for each of the two measurement ports.

a 3.5-mm test set. However, similar units are available with bandwidths up to 100 GHz [116]. These instruments are generally two-port systems but can be used with supplemental switching systems to measure the properties of multiport devices. The response of a VNA is normally a log magnitude of the s parameter of interest in decibels, a phase plot of the s parameter of interest, or a Smith chart plot. A variety of other measurements can also be performed or calculated from the s parameter data, such as the voltage standing-wave ratio (VSWR).

The reader should recall that a key issue with s parameters is the reference plane location. That is, we are measuring the incident and reflected voltage waves, which are a function of position and time. Therefore, in order to use a VNA to measure a device, a careful calibration must be performed to set the reference planes at the desired location and account for any losses in the test system, cables, and probes. The reference planes are normally set at the input to the device under test (DUT), but, in some cases, it may be useful to set the reference plane inside the device. One example of setting the reference planes inside a device is the measurement of embedded components in a substrate where the electrical response of the device is of interest without the electrical response of the interconnect from the surface of the substrate to the internal device. A variety of calibration standards are used for this purpose, including an error model with some number of error parameters. In order to perform the calibration, an operator makes a series of measurements on known standards and then the VNA computes the error terms in the model and uses the error model to correct for losses and phase shifts within the test system. Popular calibration methods include the transmission reflect line (TRL), the short open-load thru (SOLT), and the line reflect reflect match (LRRM) [117–121].

12.8.2 Probing Hardware and Connectors

Stand-alone devices that provide coaxial connectors can be directly connected to a VNA. However, in many cases, the device or circuit to be tested has input and output connections that are made from planar transmission lines. Good examples of these types of components are MMIC devices to be tested in wafer form or a MIC substrate prior to application of external connectors. In these cases, a probe station and a set of probes are used to "launch" the signals into the planar transmission lines from the coaxial connectors provided by the VNA. One can think of these probes as coaxial-to-planar adapters that provide a low-loss impedance-matched transition between these different types of transmission lines. A typical probe station and probe are illustrated in Figure 12.23. This particular set of probes is designed for use up to 40 GHz. However, similar configurations can be used up to 110 GHz [122].

For device measurements on circuits and components that include connectors, a great variety of possible connector configurations exist. Since these devices are generally coaxial in nature, care must be taken to consider the bandwidth, losses, and power-handling capabilities of each connector. Since excitation of higher order modes is undesirable, each connector has a bandwidth from DC to some maximum. As noted in Table 12.3, 7-mm connectors are useful up to 18 GHz, 3.5-mm connectors work well to 26.5 GHz, 1.85-mm connectors are useful up to 67 GHz, and 1-mm connectors and cables operate effectively to 110 GHz. Also, some of these connectors are compatible and some are not. For example, 3.5-mm and SMA (Subminiature type A) connectors mate with each other, while a 7-mm connector will not mate with any connector other than itself and the sex of a 7-mm connector can be changed by turning the outer housing of the connector. The losses are a function

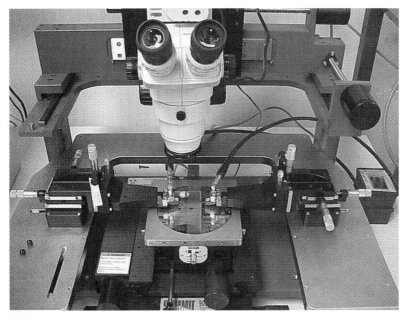

Figure 12.23 Typical probe station arrangement for on-wafer evaluation of components at high frequencies.

of the quality of the connector and include the possibility of an impedance mismatch, as measured by the connector VSWR, as well as losses in the connector materials. Higher quality connectors are made from airlines, which do not use a dielectric material between the conductors, but are air-filled instead. These airlines offer essentially no dielectric loss and are usually more carefully machined than lower quality connectors, resulting in very high levels of electrical performance. Great care must be taken in cleaning and measuring the protrusion of these connectors before each mating in order to ensure long service with minimal deviation in the electrical specifications.

12.9 TIME-DOMAIN MEASUREMENTS

Applications also exist for measurement data from the time domain since this type of measurement can often be gated in time to reveal the electrical response as a function of position inside a device or circuit. These measurements generally involve time-domain

Table 12.3 Properties of a Selection of Common Coaxial Connectors

Connector	Maximum bandwidth (GHz)	Compatibility
7 mm	18	Unisex; male and female interchangeable
SMA	18–26	3.5 mm and 2.92 mm
3.5 mm	26	SMA and 2.92 mm
2.92 mm/K	40	SMA and 3.5 mm
2.4 mm	50	1.85 mm
1.85 mm	67	2.4 mm
1 mm	110	None

reflectometry (TDR) and time-domain transmission (TDT). TDR utilizes a high-speed step response to excite the DUT, and then the reflection back from the device is measured and recorded for analysis purposes. Similarly, TDT utilizes that same high-speed pulse to excite one port of the DUT and then the response is measured at a second port on the device, thereby measuring the transmission of the step response through the device. In practice, most test systems employ TDR and TDT in one compact measurement apparatus, providing both the transmitted and reflected signal at once.

The advantage of these techniques is that, since the voltage step takes a finite amount of time to travel through the circuit and then be reflected or transmitted, one can analyze the signal as a function of time and correlate that with the position inside the circuit or component. This technique allows for not only the analysis of potential problems, but, in many cases, also the isolation of the undesired electrical response as a function of position [123, 124].

12.10 DESIGN EXAMPLE

In order to solidify the issues with high-frequency packaging described previously, consider the schematic in Figure 12.24. This is a simple wide-band amplifier with a bandwidth from 0.1 to 3.5 GHz [125]. While this circuit is very simple and not a complete system, it could be used as a building block for a variety of high-frequency applications and is appropriate to evaluate as a design example for educational purposes. The primary active device that will be implemented in this example is a single GaAs MMIC, (part no. TGA8061-SCC, available from TriQuint Semiconductor). This small 1.524-mm square chip of GaAs contains all of the active devices needed for the circuit's operation. However, a number of passive devices are needed to complete the operation of the circuit. These passive devices are difficult or not cost effective to implement on the GaAs MMIC, so external devices are needed in order

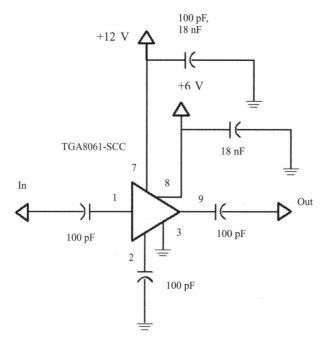

Figure 12.24 Schematic of a 0.1- to 3.5-GHz amplifier based on a Triquint MMIC.

12.10 Design Example   **529**

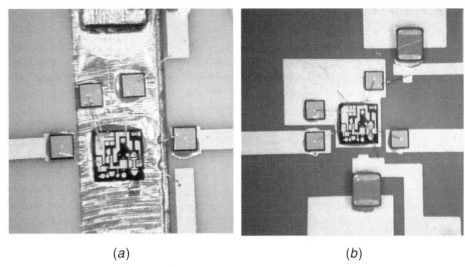

(a)   (b)

Figure 12.25 Simple packages fabricated for 0.1- to 3.5-GHz amplifiers: (*a*) fabricated using alumina and a kovar heat spreader; (*b*) design based on LTCC.

to implement these electrical functions. Two designs will be considered, one based on a traditional microwave package that uses a kovar base and thin film on alumina and a second based on LTCC.

Thermal management of this circuit is very important since the power density of the IC is nontrivial, with 1.2 W of power dissipated in a small footprint for a power density of 51.67 W/cm^2. In addition, since GaAs exhibits a lower thermal conductivity than Si, we must carefully consider the design of thermal paths to ground for this design. Both the kovar design and the LTCC design are mounted in a Cu housing that is designed to provide protection for the GaAs device and stress relief for the coaxial cable input and output. SMA connectors were chosen for this application due to their more than adequate bandwidth, compact size, and low cost.

The kovar design directly mounts the GaAs IC to a small kovar block that is mounted on the copper housing with a set of small machine screws, as shown in Figure 12.25*a*. Kovar was selected since it closely matches the coefficient of thermal expansion (CTE) of GaAs and the thermal stress is minimal between the die and the metal. Silver-filled conductive epoxy (Epoxy Technologies H20E) is used to mount the device to the kovar block and ensure a low electrical and thermal resistance to ground. This device, as is often the case for MMIC die, is an active back, which means the entire back surface of the die is a ground connection and must be connected to a low-inductance ground in order for the device to operate properly. Two alumina substrates with thin-film Ti/Cu/Ti/Au conductors are used to provide I/O connections as well as the power connections. The required capacitors are mounted directly to the substrate or kovar base and wirebonds are used to interconnect the components. The power connections are not controlled impedances and can easily be implemented with any conductor. However, the input signal and output signal must be impedance matched to 50 Ω up to at least 4 GHz, since the desired input and output impedances are 50 Ω, as is the impedance of the MMIC device. In order to design the transmission lines, a microstrip is selected due to its ease of fabrication in this type of thin-film structure. Since the substrate is 0.0635 cm thick and has a dielectric constant of 9.9, the *w/h* ratio can be calculated for a 50-Ω line from Eq. (12.21) as 0.96. This results in the requirement for a strip width of

0.0610 cm. In order to implement these microstriplines, the Ti/Cu/Ti/Au metal is deposited on both sides of the alumina substrates and etched on one side to form the desired conductive strips. These two substrates are then bonded to the kovar block with the same electrically conductive epoxy describe above. Once the substrates and die are bonded to the kovar, the capacitors are bonded in place and wirebonds are formed. The SMA connectors are then added to make physical connections to the ends of each microstrip transmission line. While this circuit is a simple prototype for illustration purposes, a similar hermetic housing would normally be used for high-reliability applications in order to protect the internal devices.

In contrast, the second design is based on an LTCC substrate that provides physical support for all the components as well as electrical interconnects. As shown in Figure 12.25b, the MMIC die and capacitors are bonded directly to the LTCC surface with conductive epoxy and then wirebonded to make electrical connections. The input and output feedlines are realized in microstrip. The design of this microstripline is based again on Eq. (12.21) using a substrate thickness of 60 mils (0.1524 cm) and a dielectric constant of 7.4 for this particular LTCC material. The result is a w/h ratio of 1.253 and a line width of 75 mils (0.191 cm). Note that, normally, the substrate thickness can be selected by the designer. In this example, the thickness was selected for ease of fabrication of the conductor trace as well as the mechanical properties of the substrate. Thermal management in the LTCC design is accomplished through thermal vias that connect the backside of the die to the metal housing in which the LTCC substrate is mounted. These thermal vias are solid filled Ag/glass cermets with a thermal conductivity of ~200 W/mK, compared to ~3 W/mK for the LTCC material itself. These vias greatly improve the thermal performance of the design compared to an LTCC design with no thermal vias.

Figure 12.26 illustrates the resulting performance of these amplifiers over a selected range. Note that virtually no difference exists in the electrical performance of the two designs. However, the kovar design has a thermal advantage with the die mounted directly to the metal block. The operating temperature of the kovar design was measured to be 63°C, while the LTCC design with thirty 10 mil-vias on 20 mils pitch was measured at 77°C. A larger number of smaller vias on a tighter pitch would reduce this difference. The main advantage of the LTCC design is that additional circuitry can easily be accommodated

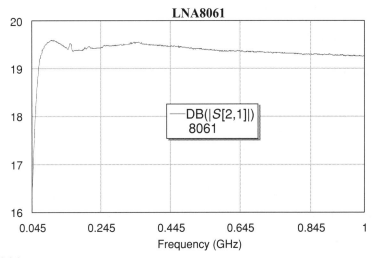

Figure 12.26 Typical s-parameter results for the 0.1- to 3.5-GHz amplifier, as measured up to 1 GHz.

in the substrate using stripline transmission lines embedded in the layers. The designer normally has at least 4–6 layers and, in some cases, 20 or more layers of circuitry that can be integrated into one substrate. In comparison, the thin film on alumina used in the kovar case is limited to only one circuit layer.

12.11 SUMMARY

This chapter describes the issues related to packaging of high-frequency circuits and components. These types of applications offer unique challenges that require consideration of the normal electrical effects associated with low-frequency circuits as well as electromagnetic phenomena. As a result, the electrical interconnects between components are not simply wires but must be evaluated in terms of their ability to guide electromagnetic energy from point to point within a circuit. In addition, electromagnetic phenomena are often exploited to build couplers, filters, and matching structures inside and on top of substrates and packages.

The key issues for packaging engineers working in this area are to develop a basic comprehension of the measurement techniques and terminology used to evaluate these types of circuits as well as a detailed understanding of the design and analysis required for the interconnects utilized in these systems.

REFERENCES

1. R. S. ELLIOTT, *Antenna Theory & Design*, New York: IEEE Press and Wiley, 2002.
2. W. L. STUTZMAN AND G. A. THIELE, *Antenna Theory and Design*, 2nd ed., New York: Wiley, 1997.
3. J. KRAUS AND R. MARHEFKA, *Antennas*, New York: McGraw-Hill, 2002.
4. M. I. SKOLNIK, *Introduction to Radar Systems*, 3rd ed., New York: McGraw-Hill, 2001.
5. H. P. HSU, On the General Relation Between α and Q (Correspondence), *IEEE Trans. Microwave Theory Tech.*, Vol. 11, No. 4, pp. 258–258, July 1963.
6. J. D. WELCH AND H. J. PRATT, Losses in Microstrip Transmission Systems for Integrated Microwave Circuits, *Northeast Electronics Research and Engineering Meeting (NEREM) Rec.*, Vol. 8, pp. 100–101, 1966.
7. M. El-Shenawee, Propagation Characteristics of Microstrip Transmission Line on Rough Dielectric Substrate Surface, *Proc. IEEE Antennas Propag. Soc.*, Vol. 1, pp. 190–193, 1997.
8. S. P. MORGAN, *J. Appl. Phys*, Vol. 20, p. 352, 1949.
9. E. O. HAMMERSTAD, ELAB Report STF44 A74169, University of Trondheim, Norwegian Institute of Techology, February, 1975.
10. E. O. HAMMERSTAD, AND O. JENSEN, Accurate Models for Microstrip Computer-Aided Design, *Microwave Symp. Dig.*, MTT-S, Vol. 80, No. 1, pp. 407–409, 1980.
11. Y. SATODA AND G. E. BODWAY, Three-Port Scattering Parameters for Microwave Transistor Measurement, *IEEE J. Solid-State Circuits*, Vol. 3, No. 3, pp. 250–255, Sept. 1968.
12. V. BELEVITCH, Topics in the Design of Insertion Loss Filters, *IRE Trans. Circuit Theory*, Vol., 2 , No. 4, pp. 337–346, Dec. 1955.
13. E. W. MATTHEWS, The Use of Scattering Matrices in Microwave Circuits, *IEEE Trans. Microwave Theory Tech.*, Vol. 3, No. 3, pp. 21–26, April 1955.
14. H. CARLIN, The Scattering Matrix in Network Theory, *IRE Trans. Circuit Theory*, Vol. 3, No. 2, pp. 88–97, June 1956.
15. D. M. POZAR, *Microwave Engineering*, 2nd ed., New York: Wiley, 1998.
16. L. PAGE AND N. I. ADAMS, Jr., Electromagnetic Waves in Conducting Tubes, *Phys. Rev.* Vol. 52, pp. 647–651, 1937.
17. K. S. PACKARD, The Origin of Waveguides: A Case of Multiple Rediscovery, *IEEE Trans. Microwave Theory Tech.*, Vol. MTT-32, No. 9, pp. 961–969, Sept. 1984.
18. G. C. SOUTHWORTH, Some Fundamental Experiments with Wave Guides, *Proc. IEEE*, Vol. 87, No. 3, pp. 515–521, March 1999.
19. R. M. BARRETT, Microwave Printed Circuits—A Historical Survey, *IEEE Trans. Microwave Theory Tech.*, Vol. 3, No. 2, pp. 1–9, March 1955.
20. R. M. BARRETT, Microwave Printed Circuits—The Early Years, *IEEE Trans. Microwave Theory Tech.*, Vol. 32, No. 9, pp. 983–990, Sept. 1984.
21. N. R. WILD, Photoetched Microwave Transmission Lines, *IEEE Trans. Microwave Theory Tech.*, Vol. 3, No. 2, pp. 21–30, March 1995.
22. S. B. COHN, Characteristic Impedance of Shielded Strip Transmission Line, *IRE Trans. Microwave Theory Tech.*, Vol. MTT-2, pp. 52–55, July 1954.
23. S. B. COHN, Problems in Strip Transmission Lines, *IRE Trans. Microwave Theory Tech.*, Vol. MTT-3, pp. 119–126, Mar. 1955.

24. H. Howe, Jr., *Stripline Circuit Design*, Dedham, MA: Artech House, 1974.
25. I. J. Bahl, and R. Garg, Designer's Guide to Stripline Circuits, *Microwaves*, Vol. 17, pp. 90–96, Jan. 1978.
26. H. A. Wheeler, Transmission Line Properties of a Stripline between Parallel Planes, *IEEE Trans. Microwave Theory Tech.*, Vol. MTT-26, pp. 866–876, Nov. 1978.
27. C. G. Montgomery, R. H. Dicke, and E. M. Purcell, eds., *Principles of Microwave Circuits*, New York: McGraw-Hill, 1948.
28. P. Bhartia, and I. J. Bahl, *Millimeter Wave Engineering and Applications*, New York: Wiley, 1984.
29. K. C. Gupta, R. Garg, and R. Chadha, *Computer-Aided Design of Microwave Circuits*, Dedham, MA: Artech House, 1981.
30. H. Howe, Jr., *Stripline Circuit Design*, Dedham, MA: Artech House, 1974, p. 125.
31. G. D. Vendelin, Limitations on Stripline Q, *Microwave J.*, Vol. 13, pp. 63–69, May 1970.
32. M. Caulton, J. J. Hughes, and H. Sobol, Measurements on the Properties of Microstrip Transmission Lines for Microwave Integrated Circuits, *RCA Rev.*, Vol. 27, pp. 377–391, Sept. 1966.
33. M. V. Schneider, Microstrip Lines for Microwave Integrated Circuits, *Bell Syst. Tech. J.*, Vol. 48, pp. 1422–1444, 1969.
34. G. I. Zysman and D. Varon, Wave Propagation in Microstrip Transmission Lines, in Dig. 1969 IEEE Int. Symp. on Microwave Theory and Techniques. Vol. 69, Issue 1, May 1969, pp. 3–9.
35. E. Denlinger, A Frequency Dependent Solution for Microstrip Transmission Lines, Vol. MTT-19, Jan. 1971.
36. R. Mittra and T. Itoh, Analysis of Microstrip Transmission Lines, in L. Young and H. Sobol, eds., *Advances in Microwaves*, Vol. 8, New York: Academic, 1974.
37. K. C. Gupta, R. Garg, and I. J. Bahl, *Microstrip Lines and Slotlines*, Dedharn, MA: Artech House, 1979.
38. K. C. Gupta, R. Garg, I. J. Bahl, and P. Bhartia, *Microstrip Lines and Slotlines*, 2nd ed., Norwood MA: Artech House, 1996.
39. E. O. Hammerstad, Equations for Microstrip Circuit Design, in *Proceedings of the European Microwave Conference*, Kent, UK: Microwave Exhibitors, 1975, pp. 268–272.
40. I. J. Bahl, and D. K. Trivedi, A Designer's Guide to Microstrip Line, *Microwaves*, Vol. 16, pp. 174–182, May 1977.
41. P. Pramanick, and P. Bhartia, CAD Models for Millimeter-Wave Finlines and Suspended-Substrate Microstrip Lines, *IEEE Trans. Microwave Theory Tech.*, Vol. MMT-33, No. 12, pp. 1429–1435, Dec. 1985.
42. H. A. Wheeler, Transmission Line Properties of a Strip on a Dielectric Sheet on a Plane, *IEEE Trans. Microwave Theory Tech.*, Vol. MIT-25, pp. 631–647, Aug. 1977.
43. M. Schneider, Microstrip Dispersion, *Proc. IEEE*, Vol. 60, Jan. 1972.
44. R. A. Pucel, D. J. Masse, and C. P. Hartwig, Losses in Microstrip, *Microwave Theory Tech. IEEE Trans.*, Vol. 16, Issue 6, pp. 342–350, June 1968.
45. C. P. Wen, Coplanar Waveguide: A Surface Strip Transmission Line Suitable for Nonreciprocal Gyromagnetic Device Applications, *IEEE Trans. Microwave Theory Tech.*, Vol. MTT-17, Issue 12, pp. 1087–1090, Dec. 1969.
46. G, Ghione and C. Naldi, Parameters of Coplanar Waveguides with Lower Ground Plane, *Electron. Lett.*, Vol. 19, pp. 734–735, 1983.
47. R. Mongia, I. Bahl, and P. Bhartia, *RF and Microwave Coupled-Line Circuits*, Boston: Artech House, 1999.
48. K. Gupta, R. Garg, I. Bahl, and P. Bhartia, *Microstrip Lines and Slotlines*, Boston: Artech House, 1996.
49. J. L. Prince, R. Santhinathan, O. A. Palusinski, and M. R. Scheinfein, Electrical Characteristics of Single Buried Microstrip Lines in the TEM Approximation, *Components, Hybrids, Manuf. Tech., IEEE Trans.* Vol. 11, Issue 3, pp. 279–283, Sept. 1988.
50. Seong-Ho Shin, In-Ho Jeong, Ju-Hyun Ko, Myung-Gyu Kang, Su-Jin Lee, and Young-Se Kwon, Monolithic Implementation of Air-Buried Microstrip Lines for High-Density Microwave and Millimeter Wave ICs, Electronic Components and Technology Conference, 2002, Proceedings, 52nd, 28–31 May 2002, pp. 1018–1020.
51. T. Ishikawa and E. Yamashita, Experimental Results on Buried Microstrip Lines for Constructing High-Density Microwave Integrated Circuits, *Microwave Guided Wave Lett. IEEE*, Vol. 5, Issue 12, pp. 437–438, Dec. 1995.
52. J. A. G. Malherbe, and A. F. Steyn, The Compensation of Step Discontinuities in TEM-Mode Transmission Lines, *IEEE Trans. Microwave Theory Tech.*, Vol. MTT-26, pp. 883–885, Nov. 1978.
53. M. A. Larson, Compensation of Step Discontinuities in Stripline, in *Proceedings of the Tenth European Microwave Conference*, Kent, UK: Microwave Exhibitors, 1980, pp. 367–371.
54. R. Chadha and K. C. Gupta, Compensation of Discontinuities in Planar Transmission Lines, *IEEE Trans. Microwave Theory Tech.*, Vol. MTT-30, pp. 2151–2155, Dec. 1982.
55. W. J. R. Hoefer, A Contour Formula for Compensated Microstrip Steps and Open Ends, *IEEE Intl. Microwave Symp. Dig.*, pp. 524–526, 1983.
56. I. Wolff, and W. Menzel, A Universal Method to Calculate the Dynamical Properties of Microstrip Discontinuities in *Proceedings of the Fifth European Microwave Conference*, Kent, UK: Microwave Exhibitors, 1975, pp. 263–L267.

57. T. C. Edwards, *Foundations for Microstrip Circuit Design* New York: Wiley, 1981, Chapter 5.
58. I. Bahl and P. Bhartia, *Microwave Solid State Circuit Design*, 2nd edition, Hoboken, NJ: Wiley, 2003.
59. R. Mehran, The Frequency-Dependent Scattering Matrix of Microstrip Right-Angle Bends, T-Junctions and Crossings, *AEU: Archiv. Elektronik Uber-tragungstechnik*, Vol. 29, pp. 454–460, 1975.
60. R. Mehran, Frequency-Dependent Equivalent Circuits for Microstrip Right-Angle Bends, T-Junctions and Crossings, *AEU*, Vol. 30, pp. 80–82, 1975.
61. F. Alimenti, U. Goebel, and R. Sorrentino, Quasi Static Analysis of Microstrip Bondwire Interconnects, *Microwave Symp. Digest*, 1995 IEEE MTT-S International, Vol. 2, May 16–20 1995, pp. 679–682.
62. H. Lee, Wideband Characterization of a Typical Bonding Wire for Microwave and Millimeter-wave Integrated Circuits, *IEEE Trans. Microwave Theory Tech.*, Vol. 43, No. 1, pp. 63–68, Jan. 1995.
63. A. Sutono, N. Cafaro, J. Laskar, and M. Tentzeris, Experimental Modeling, Repeatability Investigation and Optimzation of Microwave Bond Wire Interconnects, *IEEE Trans. Adv. Packaging*, Vol. 24, No. 4, pp. 595–603, Nov. 2001.
64. T. Krems, W. Haydl, L. Verweyen, M. Schlechtweg, H. Mabler, and J. Rudiger, Coplanar Bond Wire Interconnections for Millimeter-wave Applications, *Electrical Performance of Electronic Packaging*, Oct. 2–4 1995, pp. 178–180.
65. S. Yun and H. Lee, Parasitic Impedance Analysis of Double Bonding Wires for High-Frequency Integrated Circuit Packaging, *IEEE Microwave Guided Wave Lett.*, Vol. 5, No. 9, pp. 296–298, Sept. 1995.
66. T. Krems, W. Haydl, H. Mabler, and J. Rudiger, Millimeter-Wave Performance of Chip Interconnections Using Wire Bonding and Flip Chip, *Microwave Symp. Digest*, 1996 IEEE MTT-S International, 1995, Vol. 1, June 17–21 1996, pp. 247–250.
67. M. Hotta, Y. Qian, and T. Itoh, Resonant Coupling Type Microstrip Line Interconnect Using a Bonding Ribbon and Dielectric Pad, *Microwave Symp. Digest*, 1998 IEEE MTT-S International, Vol. 2, June 7–12 1998, pp. 797–800.
68. D. C. Guidici, Ribbon Wires Versus Round Wire Reliability for Hybrid Microcircuits, *IEEE Trans. Parts, Hybrids, Packaging*, pp. 159–162, June 1975.
69. D. Staiculescu, A. Sutono, and J. Laskar, Wideband Scaleable Electrical Model for Microwave/Millimeter Wave Flip Chip Interconnects, *IEEE Trans. Adv. Manufac. Part B: Adv. Packaging*, Vol. 24, No. 3, pp. 255–259, Aug. 2001.
70. Private communication with Rogers Corp.
71. R. A. Woode, E. N. Ivanov, M. E. Tobar, and D. G. Blair, Measurement of Dielectric Loss Tangent of Alumina at Microwave Frequencies and Room Temperature. *Electr. Lett.*, Vol. 30, No. 25, pp. 2120–2122, Dec. 1994.
72. R. G. Geyer, and J. Krupka, Microwave Dielectric Properties of Anisotropic Materials at Cryogenic Temperatures, *IEEE Trans. Instrumen. Measure.*, Vol. 44, No. 2, pp. 329–331, April 1995.
73. W. E. Courtney, Complex Permittivity of GaAs and CdTe at Microwave Frequencies, *IEEE Trans. Microwave Theory Tech.*, Vol. MTT-25, No. 8, pp. 697–701, Aug. 1977.
74. The Dow Chemical Company
75. D. Amey, S. Horowitz, and R. Keusseyan, High Frequency Electrical Characterization of Electronic Packaging Materials: Environmental and Process Considerations, 1998 International Symposium on Advanced Packaging Materials, pp. 123–128, Braselton, GA, March 1998.
76. The Ferro Corporation
77. M. Tobar, et al., Measurements of Low-Loss Crystalline Materials for High-Q Temperature Stable Resonator Applications, 1999 Joint Meeting EFTF–IEEE IFCS, Besancon, FR, pp. 573–576.
78. B. Komiyama, et al, Open Resonator for Precision Dielectric Measurements in the 100 GHz Band, *IEEE Trans. Microwave Theory Tech.*, Vol. 39, No. 10, pp. 1792–1796, Oct. 1991.
79. J. H. C. Van Heuven, A. G. Van Nie., and N. V. Phillips, Properties of Microstrip Lines on Fused Quartz, *IEEE Trans. Microwave Theory Tech.*, pp. 113–114, Feb. 1970.
80. S. Mahapatra and S. N. Prasad, A New Electroless Method for Low-Loss Microwave Integrated Circuits, *IEEE Trans. Components, Hybrids, Manufacturing Tech.*, Vol. CHMT-1, No. 4, pp. 428–431, Dec. 1978.
81. A. Elshabini and F. Barlow, Thin Film Technology Handbook, New York: McGraw Hill, 1997.
82. J. M. Mikkelson and L. R. Tomasetta, High-Density GaAs Integrated Circuit Manufacturing, *IEEE Trans. Semiconductor Manufact.*, Vol. 16, No. 3, pp. 384–389, Aug. 2003.
83. J. Hirshon, B. Kaplan, D. McElroy, and A. Pollino, Thick Film Hybrids: A Diode-Coupled Microelectronic Amplifier, *Components Parts, IEEE Trans.*, Vol. 11, No. 2, pp. 54–69, June 1964.
84. M. Caulton, S. P. Knight, and D. A. Daly, Hybrid Integrated Lumped-element Microwave Amplifiers, *IEEE J. Solid State Circuits*, Vol. SC-3, No. 2, pp. 59–66, June 1968.
85. D. A. Daly, et al., Lumped Elements in Microwave Integrated Circuits, *IEEE Trans. Microwave Theory Tech.*, Vol. MTT-15, pp. 713–721, Dec. 1967.
86. C. S. Aitchison, et al., Lumped-Circuit Elements at Microwave Frequencies, *IEEE Trans. Microwave Theory Tech.*, Vol. MTT-19, pp. 928–937, Dec. 1971.
87. R. S. Pengelly and D. C. Rickard, Design, Measurement and Application of Lumped Elements up to J-Band, in *Proceedings of the Seventh European Microwave Conference*, Kent, UK: Microwave Exhibitors, 1977, pp. 460–464.

88. G. D. ALLEY, Interdigital Capacitors and Their Application to Lumped-Element Microwave Integrated Circuits, *IEEE Trans. Microwave Theory Tech.*, Vol. MTT-18, No. 12, pp. 1028–1033, Dec. 1970.
89. A. GOPINATH AND P. SILVESTER, Calculation of Inductance of Finite-Length Strips and Its Variations with Frequency, *IEEE Trans. Microwave Theory Tech.*, Vol. MTT-21, pp. 380–386, June 1973.
90. R. CHADDOCK, The Application of Lumped Element Techniques to High Frequency Hybrid Integrated Circuits, *Radio Electron Eng.*, Vol. 44, pp. 414–420, 1974.
91. S. CHAKI, S. AONO, N. ANDOH, Y. SASAKI, N. TANINO, AND O. ISHIHARA, Experimental Study on Spiral Inductors, *IEEE MTT-S International Microwave Symposium Digest*, Vol. 2, 1995, Proceedings of the 1995 IEEE MTT-S International Microwave Symposium. Part 2 (of 3), May 16–20 1995, Orlando, FL, pp. 753–756.
92. F. W. GROVER, *Inductance Calculations*, Princeton, NJ: Van Nostrand, 1946; reprinted by Dover Publications, New York, 1962 and 2004.
93. F. E. TERMAN, *Radio Engineer's Handbook*, New York: McGraw, Hill, 1943, p. 51.
94. T. C. EDWARS, *Foundations for Microstrip Circuit Design*, 2nd ed., New York: Wiley, 1992.
95. W. GANGQIANG, V. RAJAGOPALAN, F. BARLOW, A. ELSHABINI, AND S. ANG, Effect of Design and Processing Parameters on Buried Resistors in LTCC Systems, Proceedings of the International Symposium on Microelectronics, Baltimore, Maryland, October, 2001.
96. G. MATTHAEI, L. YOUNG, AND E. JONES, *Microwave Filters, Impedance-Matching Networks, and Coupling Structures*, Norwood, MA: Artech House, 1980.
97. E. G. CRISTAL AND S. FRANKEL, Design of Hairpin-Line and Hybrid Hairpin-Parallel-Coupled-Line Filters, *Microwave Symposium Digest, GMTT Int.*, Vol. 71, No. 1, pp. 12–13, May 1971.
98. G. SAULICH, A Simple Method for Spacing the Adjacent Passbands of a Coupled-Line Filter, *Microwave Theory and Techniques, IEEE Trans.*, Vol. 28, No. 4, pp. 359–362, April 1980.
99. B. J. MINNIS, Printed Circuit Coupled-Line Filters for Bandwidths up to and Greater Than an Octave, *Microwave Theory Tech. IEEE Trans.*, Vol. 29, No. 3, pp. 215–222, March 1981.
100. C. CHO AND K. C. GUPTA, Design Methodology for Multilayer Coupled Line Filters, *Microwave Symposium Digest, IEEE MTT-S Intl.*, Vol. 2, 8–13 June 1997, pp. 785–788, 1997.
101. E. RIUS, G. PRIGENT, H. HAPPY, G. DAMBRINE, S. BORET, AND A. CAPPY, Wide- and Narrow-band Bandpass Coplanar Filters in the W-Frequency Band, *Microwave Theory and Tech., IEEE Trans.*, Vol. 51, No. 3, pp. 784–791, March 2003.
102. R. VAN PATTEN, Design of Improved Microwave Low-Pass Filters Using Strip-line Techniques, *IRE Int. Convention Record*, Vol. 5, pp. 197–207, March 1957.
103. R. LEVY, Tables of Element Values for the Distributed Low-Pass Prototype Filter, *Microwave Theory Tech., IEEE Trans.*, Vol. 13, No. 5, pp. 514–536, Sept. 1965.
104. E. J. WILKINSON, An N-Way Hybrid Power Divider, *Microwave Theory Tech., IEEE Trans.*, Vol. 8, No. 1, pp. 116–118, Jan. 1960.
105. Z. GALANI AND S. J. TEMPLE, A Broadband Planar N-Way Combiner/Divider, *Microwave Symposium Digest, MTT-S Intl.*, Vol. 77, No. 1, pp. 499–502, June 1977.
106. D. J. SOMMERS, Slot Array Employing Photoetched Tri-Plate Transmission Lines, *Microwave Theory Tech., IEEE Trans.*, Vol. 3, No. 2, pp. 157–162, March 1955.
107. L. I. PARAD AND R. L. MOYNIHAN, Split-Tee Power Divider, *Microwave Theory Tech., IEEE Trans.*, Vol. 13, No. 1, pp. 91–95, Jan. 1965.
108. J. A. GARCIA, A Wide-Band Quadrature Hybrid Coupler (Correspondence), *Microwave Theory Tech., IEEE Trans.*, Vol. 19, No. 7, pp. 660–661, July 1971.
109. J. LANGE, Interdigitated Stripline Quadrature Hybrid (Correspondence), *Microwave Theory Tech., IEEE Trans.*, Vol. 17, No. 12, pp. 1150–1151, Dec. 1969.
110. J. REED AND G. J. WHEELER, A Method of Analysis of Symmetrical Four-Port Networks, *Microwave Theory Tech., IEEE Trans.*, Vol. 4, No. 4, pp. 246–252, Oct. 1956.
111. W. A. TYRRELL, Hybrid Circuits for Microwaves, *Proc. IRE*, Vol. 35, pp. 1294–1306, Nov. 1947.
112. M. HOMO AND F. MEDINA, Multilayer Planar Structures for High-Directivity Directional Coupler Design, *IEEE Trans. Microwave Theory Tech.*, Vol. MlT-34, pp. 1442–1449, Dec. 1986.
113. J. LANGE, Interdigitated Stripline Quadrature Hybrid, *IEEE Trans. Microwave Theory Tech.*, pp. 1150–1151, Dec. 1969.
114. Ansoft Corporation
115. Applied Wave Research, Inc.
116. Agilent Technologies, Inc.
117. I. ROLFES, AND B. SCHIEK, LRR—A Self-Calibration Technique for the Calibration of Vector Network Analyzers, *Instrumen. Measure., IEEE Trans.*, Vol. 52, No. 2, pp. 316–319, April 2003.
118. F. PURROY, AND L. PRADELL, New Theoretical Analysis of the LRRM Calibration Technique for Vector Network Analyzers, *Instrument Measure., IEEE Trans.*, Vol. 50, No. 5, pp. 1307–1314, Oct. 2001.
119. W. WIATR, A Broadband Technique for One-Port Calibration of VNA, 12th International Conference on Microwaves and Radar, 1998, MIKON '98, Vol. 2, 20–22 May 1998, pp. 363–367.
120. E. W. STRID, AND K. R. GLEASON, Calibration Methods for Microwave Wafer Probing, *Microwave Symposium Digest, MTT-S International*, Vol. 84, No. 1, pp. 93–97, May 1984.
121. F. PURROY, AND L. PRADELL, New Theoretical Analysis of the LRRM Calibration Technique for Vector Network Analyzers, *Instrum. Measure., IEEE Trans.*, Vol. 50, No. 5, pp. 1307–1314, Oct. 2001.
122. Cascade Microtech, Inc.

123. A. M. NICOLSON, Broad-band Microwave Transmission Characteristics from a Single Measurement of the Transient Response, *IEEE Trans. Instrum. Meas.*, Vol. IM-17, pp. 395–402, Dec. 1968.
124. M. SWAMINATHAN, S. PANNALA, AND T. ROY, Extraction of Frequency Dependent Transmission Line Parameters Using TDR/TDT Measurements, Instrumentation and Measurement Technology Conference, 2001. IMTC 2001. *Proc. 18th IEEE*, Vol. 3, Nos. 21–23, pp. 1726–1730, May 2001.
125. H. KABIR, High Q Oscillator Based on HTS Sprial Resonator for RF and Microwave Applications, Masters Thesis, University of Arkansas, 2003.

EXERCISES

12.1. Design a 50-Ω microstrip transmission line for an LTCC substrate with a dielectric constant of 7.4 and a laminate board with a dielectric constant of 5. Assume both boards have identical thickness of 0.5 mm. Discuss the implications of using one board versus the other.

12.2. Design two 50-Ω stripline transmission lines for an LTCC substrate with a dielectric constant of 5.9. Assume that the dielectric for the first transmission line is 0.5 mm thick and for the second line is 0.25 mm thick.

12.3. Evaluate the conductor losses for the two transmission lines in problem 12.2. Compare and contrast your results for the two cases.

12.4. Calculate the dielectric losses in the two transmission lines considered in problem 12.1 in the X band. Assume that the polymer substrate is an FR-4 board with a $\tan \delta$ of 270×10^{-4} at 10 GHz and that the LTCC substrate has a $\tan \delta$ of 5×10^{-4} at 10 GHz.

12.5. Calculate the skin depth for a 50-Ω stripline transmission line operating in the Ku band based on copper foil and PTFE. Based on your calculations, what would you suggest as a metal thickness for this transmission line?

12.6. Consider an MMIC that has been attached to a substrate and requires wirebonding. Assume that the MMIC is fabricated from 100-μm-thick GaAs and that the substrate is 600-μm-thick alumina. What type of bond would you recommend and what would its impedance be?

12.7. Design a 1-nH inductor for operation in the X band. Assume that this inductor will be integrated into a stripline circuit fabricated in ceramic-filled PTFE. What type of inductor would you select and what would be the physical dimensions of this inductor?

Chapter 13

Power Electronics Packaging

ALEXANDER B. LOSTETTER AND KRAIG OLEJNICZAK

13.1 INTRODUCTION

The use of solid-state electronics to replace electromechanical mechanisms in consumer power electronics (e.g., cellular telephones, pagers, etc.) has resulted in tremendous improvements in product performance, functional density, reliability, and ease of manufacture. This can be attributed to novel and rapid advances in the miniaturization of devices, circuits, and systems by the semiconductor industry through advanced electronic packaging techniques. On the other hand, electronic packaging of high-power circuits and systems used in the space, avionics, military, industrial, consumer, factory automation, and automotive market sectors still rely largely on power hybrid circuit design and manufacturing techniques.

The power electronics industry is undergoing a "new" electronic revolution, a quest for converting packaging technologies across boundaries, from defense to commercial and from low-voltage, low-current to high-voltage, high-current applications. Power electronics miniaturization and packaging, the new electronic revolution, was truly set in motion by the advent of the power hybrid [1, 2]. Although impressive technological advancements continue in the realms of power semiconductor device fabrication and power circuit topology development, these could be considered incremental by some when compared to the advances required in power electronics integration. The increasing limitations on footprint, mass, and form factor require orders of magnitude of increase in integration while improving overall system reliability, performance, and cost. This change is a logical progression from that experienced by the microelectronics industry.

This chapter covers the fundamental knowledge required for an engineer to begin to work in the field of power electronics packaging, and key reference materials are presented for continued use. Where possible, practical examples are given that mimic real-world problems, decisions, assumptions, and simplifications. Throughout the chapter, current state-of-the-art technology as well as future trends are discussed.

13.2 SEMICONDUCTOR POWER DEVICE TECHNOLOGY

The core and most vital components of any modern-day power electronics circuit are the power semiconductor switches [3–12]. The fundamental purpose of the power switch is to

---

Advanced Electronic Packaging, Second Edition, Edited by Richard K. Ulrich and William D. Brown
Copyright © 2006 the Institute of Electrical and Electronics Engineers, Inc.

perform a wave-shaping function on the voltage and/or current signals in order to convert energy from one form to another [e.g., such as direct current (DC) to alternating current (AC) or AC to DC].

13.2.1 Ideal and Nonideal Power Switching

Power switches have three possible modes of operation:

1. *On State*. In this mode, an ideal power switch acts as a short circuit. In other words, it conducts current without a voltage drop across the switch. Realistically, however, all power switches have an on-resistance associated with their conduction mode, which results in a small voltage drop across the device and a small power loss within the device.
2. *Off State*. In this mode, an ideal power switch acts as an open circuit. In other words, it conducts no current and can block an infinitely large voltage. Realistically, power switches have a leakage current associated with their blocking mode that results in some amount of conduction through the device and a small power loss within the device.
3. *Commutation State*. This is the mode of operation where the power device is switching between the on and off states. This function is important for the determination of power-switching speeds, control strategies, and transient switching losses.

All three of these modes of operation contribute to the overall electrical losses of a power device. These electrical losses are observed as thermal energy that must be transferred from the power package. The thermal management strategy to accomplish the energy transference is one of the highest priorities when developing a power electronics package for a specific application. Figure 13.1 illustrates the typical switching waveforms of a power device.

The total average power loss of a power device can be determined by breaking the waveforms up into their three modes of operation and determining the energy loss of each

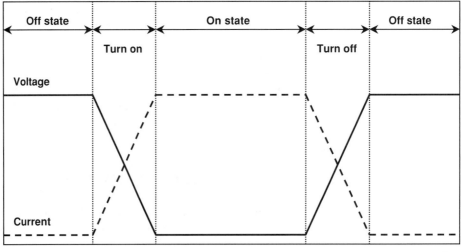

Figure 13.1 Typical Power Device Switching Waveforms.

13.2 Semiconductor Power Device Technology

mode [3–8], which are given by

$$W_{\text{on}} = \int_{\text{on-time}} i_{\text{on}} v_{\text{on}} \, dt \tag{13.1}$$

$$W_{\text{off}} = \int_{\text{off-time}} i_{\text{off}} v_{\text{off}} \, dt \tag{13.2}$$

$$W_{\text{transient}} = \int_{\text{turn-on}} i(t)v(t) \, dt + \int_{\text{turn-off}} i(t)v(t) \, dt \tag{13.3}$$

$$P_{\text{loss}} = \frac{W_{\text{on}} + W_{\text{off}} + W_{\text{transient}}}{T} \tag{13.4}$$

Example 13.1

Assume a nonideal power switch is rated by the manufacturer with an on resistance $R_{\text{on}} = 0.55\,\Omega$ and a leakage current $I_{\text{leakage}} = 25\,\mu\text{A}$ (both at 25°C). This power switch is used in a DC–DC converter to chop a 300-V DC input signal. Assume the switch has a 2.2-V drop while conducting and has a switch response as illustrated in Figure 13.1, with the following attributes:

> Duty cycle = 50%   Switching frequency = 20 kHz
> Turn-on time = 1 μs   Turn-off time = 1 μs

Assume the turn-on and turn-off voltage and current waveforms are linear ramp functions. Find the total average power loss, P_{loss}, of the switch.

Since the switching frequency is given as 20 kHz, the total time period T of the chopping circuit is found by

$$T = \frac{1}{f} = \frac{1}{20 \times 10^3 \text{ cycles/s}} = 50\,\mu\text{s}$$

Subtracting the turn-on and turn-off times from the time period gives the amount of time that the switch is on or off (but not transitioning) to be 48 μs. Duty cycle refers to the amount of time the switch is on compared to the amount of time the switch is off. Thus the on time and off time can be found as:

$$t_{\text{on}} = (T - t_{\text{turn-on}} - t_{\text{turn-off}})\,D = 24\,\mu\text{s}$$
$$t_{\text{off}} = (T - t_{\text{turn-on}} - t_{\text{turn-off}})\,(1 - D) = 24\,\mu\text{s}$$

The voltage drop across the switch during conduction is the on voltage, $v_{\text{on}} = 2.2$ V. The on current can be found by using the switch's on resistance and Ohm's law:

$$i_{\text{on}} = \frac{v_{\text{on}}}{R_{\text{on}}} = \frac{2.2\,\text{V}}{0.55\,\Omega} = 4\,\text{A}$$

So, from Eq. (13.1), the energy lost while the switch is on can be calculated:

$$W_{\text{on}} = \int_{0\,\text{s}}^{24 \times 10^{-6}\,\text{s}} (4\,\text{A})(2.2\,\text{V}) \, dt = (8.8\,\text{V} - \text{A})(t)\,|_{0\text{s}}^{24 \times 10^{-6}\,\text{s}} = 2.112 \times 10^{-4}\,\text{J}$$

The leakage current of the power switch is the off current, $i_{\text{off}} = 25\,\mu\text{A}$, and the total blocking voltage is the off voltage, $v_{\text{off}} = 300$ V. From Eq. (13.2), the energy loss while the switch is off can be calculated:

$$W_{\text{off}} = \int_{0\,\text{s}}^{24 \times 10^{-6}\,\text{s}} (25 \times 10^{-6}\,\text{A})(300\,\text{V}) \, dt = (0.0075\,\text{V} - \text{A})(t)\,|_{0\text{s}}^{24 \times 10^{-6}\,\text{s}} = 1.8 \times 10^{-7}\,\text{J}$$

Since the voltage and current waveforms are linear ramp functions, the turn-on and turn-off functions are found to be approximated by the following:

$$v_{\text{turn-on}} = \lfloor 300 - (3 \times 10^8 \, t) \rfloor \quad \text{V}$$
$$v_{\text{turn-off}} = (3 \times 10^8 \, t) \quad \text{V}$$
$$i_{\text{turn-on}} = (4 \times 10^6 \, t) \quad \text{A}$$
$$i_{\text{turn-off}} = \lfloor 4 - (4 \times 10^6 \, t) \rfloor \quad \text{A}$$

From Eq. (13.3), the energy loss due to switch transition can be calculated:

$$W_{\text{transient}} = \int_{\text{turn-on}} i(t)v(t)\,dt + \int_{\text{turn-off}} i(t)v(t)\,dt$$

$$W_{\text{transient}} = \int_{0\,\text{s}}^{1 \times 10^{-6}\,\text{s}} \left[(1.2 \times 10^9 \, t) - (1.2 \times 10^{15} \, t^2)\right] dt + \int_{0\,\text{s}}^{1 \times 10^{-6}\,\text{s}} \left[(1.2 \times 10^9 t) - (1.2 \times 10^{15} \, t^2)\right] dt$$

$$W_{\text{transient}} = 2 \times \left(\left[(6 \times 10^8) t^2\right] - \left[(4 \times 10^{14}) t^3\right]\right) \Big|_{0\,\text{s}}^{1 \times 10^{-6}\,\text{s}} = 4 \times 10^{-4} \quad \text{J}$$

From Eq. (13.4), the average power loss of the device can thus be calculated:

$$P_{\text{loss}} = \frac{W_{\text{on}} + W_{\text{off}} + W_{\text{transient}}}{T} = \frac{6.1138 \times 10^{-4}\,\text{J}}{50\,\mu\text{s}} = 12.23 \, \text{W}$$

Notice in the results of this example that the energy lost during the short turn-on and turn-off time is more than the energy loss of the device for the entire time it is actually on. Device switching speed is, therefore, an important characteristic to take into account when attempting to minimize overall power loss.

13.2.2 Power Diodes

Almost all semiconductor power devices are fabricated as vertical channel devices, where large numbers of current-carrying channels are formed and tied together in parallel in order to distribute the load through the device.

There are several types of common power diodes, including pn diodes, p-i-n diodes, and Schottky diodes. The two most important characteristics when using power diodes as switches in power converters are their blocking voltage capabilities and their reverse recovery time (which dictates switching speed). A pn diode is a p-type and n-type material fabricated to form a rectifying junction, while a p-i-n structure has a lightly doped intrinsic layer between the p and n regions. The Schottky diode forms a rectifying junction by depositing a specific work function metal on an n-type semiconductor material. The Schottky diode is a majority-carrier device (n metal with n-type semiconductor), so it has a very small reverse recovery time associated with it, making it ideal for fast switching. The disadvantage of the Schottky diode is that it also has a reduced voltage-blocking capability. Most Schottky diodes have implanted, heavily doped p regions to act as guard bands. These guard bands help to reduce current crowding and thus help to increase the voltage breakdown capability of the device. The vertical-channel pn and Schottky power diode structures are illustrated in Figure 13.2.

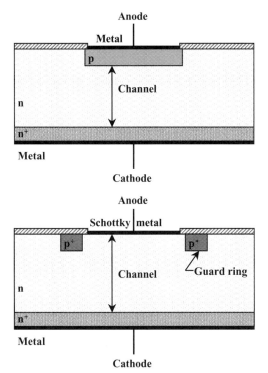

Figure 13.2 Cross section of (a) PN power diode and (b) Schottky power diode.

13.2.3 Thyristors

A thyristor is a multiple-junction device (p–n–p–n) that has a latching switch behavior. The thyristor is triggered into the on state by the application of a short-duration positive pulse current on the gate. Once the thyristor begins to conduct, it is latched into the on state and cannot be switched off by any gate signals. At this stage, it is conducting as a diode and can only be turned off as a diode turns off—when under the influence of the circuit, it stops conducting. The advantage of the thyristor is that it has very large voltage-blocking and current-carrying capabilities, but since it has three pn junctions, the device is slow, normally taking several microseconds to switch on and off. Thus it is only useful in low-frequency power circuits. A cross section of this device is illustrated in Figure 13.3.

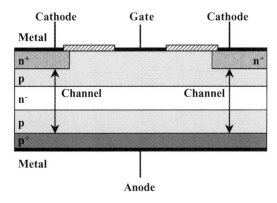

Figure 13.3 Cross section of a thyristor.

There is a category of thyristors that have become popular, referred to as the gate turn-off (GTO) thyristor. This device is similar to the thyristor, except that the device can be turned off by a negative gate current control. The disadvantage of the GTO is that the negative gate current required to turn the device off needs to be very large in magnitude (typically about one-third of the anode current being conducted through the device). This makes the required control circuitry complicated and costly when compared to other gate-controlled devices.

13.2.4 Power Bipolar Junction Transistors

Bipolar junction transistors (BJTs) are current-controlled devices that require a continuous gate current supply to keep them in the on state. The current gain in high-power BJTs is typically only about a factor of 10, and so, again, significant current is required from the control circuitry. BJTs can be connected in Darlington or triple Darlington configurations in order to increase the overall current gain. Most BJTs have been displaced by metal–oxide–semiconductor field-effect transistors (MOSFETs) and insulated gate bipolar transistors (IGBTs) for lower voltage applications (up to 1.2 kV) or GTOs for higher voltage applications (above 1.6 kV). A cross section of the power BJT is illustrated in Figure 13.4.

13.2.5 Power MOSFETs

Currently, the most common and popular power transistor is the MOSFET, a voltage-controlled device. It can switch substantial voltages and currents and is a fast device. By applying a bias to the transistor gate, charge builds up at the gate interface, opening a channel through the p body. The time required to charge and discharge this gate capacitance is the dominating factor in determining the switching speed of the device. Typically, the turn-on and turn-off time is on the order of 10 to 100 ns. Integral to the structure of the power MOSFET is a reverse body diode. Since the MOSFET is a voltage-controlled device, control circuits are relatively less complicated to implement (in comparison with GTOs or BJTs), making the power MOSFET a popular transistor choice. A cross section of the MOSFET is illustrated in Figure 13.5.

13.2.6 Insulated Gate Bipolar Transistors

The IGBT is essentially a MOS-controlled BJT. Thus, it has the voltage-blocking and current-carrying capacity of the BJT (which are higher than the MOSFET) with the ease

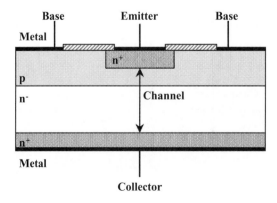

Figure 13.4 Cross section of a power BJT.

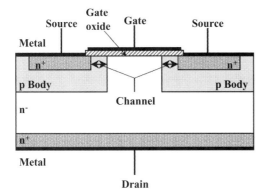

Figure 13.5 Cross section of a power MOSFET.

of low-voltage gate control. The major disadvantage of the IGBT is its slower switching speeds in comparison to the power MOSFET. A cross section of the IGBT transistor is illustrated in Figure 13.6.

13.2.7 Static Induction Transistors (SITs)

The SIT is essentially a vertical-channel junction field-effect transistor (JFET). These devices are popular in current silicon carbide (SiC) semiconductor research and development programs. In SiC technology, structures requiring gate oxides are currently far behind the development of other devices, thus making the SIT or JFET the most popular and advanced power device. The SIT is a normally-on, gate-controlled device. With no voltage applied to the gate, the drain and source are shorted through the body, and thus, the transistor is on. In order to turn off the transistor, a *negative* bias must be applied to the gates with respect to the source. This reverse voltage biases the p^+/n^- junctions, and the depletion regions at these junctions grow. As the negative gate voltage is increased, the depletion regions under the two gates grow toward each other, narrowing the channel, until the channel pinches off completely. A cross section of the SIT is illustrated in Figure 13.7.

13.2.8 Silicon Carbide Semiconductor Devices

As one observes the characteristics of current silicon-based device technology, the limiting factors of this technology are quickly becoming more prevalent and apparent. Clock speeds

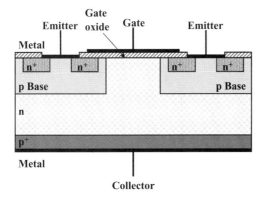

Figure 13.6 Cross section of an IGBT.

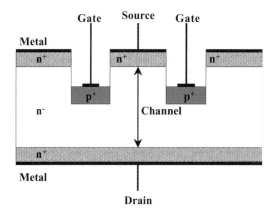

Figure 13.7 Cross section of a SIT.

of microprocessors are increasing, approaching the switching speed limitations of silicon-based devices; transistors are shrinking in size, resulting in higher power densities and increased packaging and heat-sinking complexities; and high-performance consumers (e.g., the U.S. military, NASA, etc.) are demanding electronics with higher efficiency coupled with extreme-environment operation [13, 14].

The past decade has seen an intense and steady increase in research on the viability of SiC-based device technology. This technology has the potential to solve many of the current limitations associated with silicon electronics, in particular the limitations of silicon device switching speeds, junction temperatures, and power density [15–17]. Silicon carbide power devices have recently begun to emerge on the commercial market, including the SiC Schottky, SiC SIT, SiC metal–semiconductor FET (MESFET), and SiC JFET [18–22].

Silicon carbide is a wide-bandgap semiconductor material capable of high temperature operation (theoretically up to ∼600°C). When compared to silicon-based devices, SiC devices possess a higher breakdown voltage (10 times that of Si), possess lower switching losses, are capable of higher current densities (approximately 3 to 4 times higher than Si or GaAs), and can operate at higher temperatures (approximately 5 times higher than Si). In short, SiC power devices have the potential to revolutionize the power electronics industry.

While extensive resources have been utilized to develop SiC semiconductor devices, relatively little research effort has been placed by industry or academic research institutions into developing adequate electronic packaging in order to provide for functional modules that will take full advantage of the capabilities of these devices. In particular, the market is just beginning to delve into the development of very high temperature packaging solutions (300°C+) that could allow the full utilization of SiC.

The interests in SiC technology can be traced to a comparison of the fundamental physics between it and silicon. Table 13.1 illustrates property comparisons between various semiconductor materials of interest to the power device community. Silicon carbide technology offers a higher electric field breakdown, which results in transistors with faster switching speeds (up to tens of gigahertz), smaller drive currents, and smaller on resistances, all contributing factors to reduced power losses and increased electrical efficiency over silicon devices.

The second major advantage of SiC electronics is their potential to operate in extremely harsh environments and, in particular, at high temperatures. This ability coupled with excellent electrical efficiencies gives SiC the potential to operate at very high power densities. Silicon carbide opens up a world of possibilities in electronics design and applications that, until recently, have been unthinkable [23–28].

Table 13.1 Comparison of Some Property Values of SiC with Other Semiconductor Device materials

Property	6H SiC	4H SiC	3C SiC	GaAs	Si
Energy bandgap (eV)	2.9	3.26	2.2	1.43	1.12
Electric field breakdown ($\times 10^6$ V/cm at 1 kV operation)	2.5	2.2	2.0	0.30	0.25
Thermal conductivity (W/m-K at room temperature)	490	370	360	55	150
Intrinsic carrier concentration, n_i (cm^{-3} at room temperature)	10^{-6}	8.2×10^{-9}	1.2	2.1×10^6	10^{10}
Electron mobility, μ_e (cm^2/V-s at room temperature)	330–400	700–980	800	8500	1400
Hole mobility, μ_h (cm^2/V-s at room temperature)	75	120	320	400	450
Saturated electron drift ($\times 10^7$ cm/s at $E > 2 \times 10^5$ V/cm)	2.0	2.0	2.5	1.0	1.0

Silicon carbide Schottky diodes were the first SiC power devices to reach the commercial market. In a Schottky diode, a significant voltage drop (and thus power loss) occurs due to the resistance in the bulk of the diode as current flows through the diode. The significant advantage of SiC Schottky diodes is the reduction of the on resistance in this area at high power, in comparison to Si or GaAs Schottky diodes. Analytical results of this comparison can be seen in Figure 13.8. Notice that SiC has a high specific resistance at low breakdown voltages. This is because the specific resistance in this region is dominated by the highly doped substrate resistance. The drift layer must be thicker and have a lower doping density

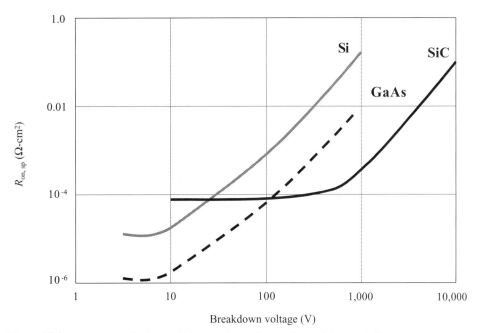

Figure 13.8 Comparison of Schottky diode operation utilizing various substrate materials.

in order to increase the breakdown voltage, thereby increasing the specific resistance of the diode. At high breakdown voltages, the breakdown field of the material becomes the dominant parameter—the breakdown field of SiC is an order of magnitude greater than Si or GaAs, and so SiC devices can operate at much higher voltage levels (approximately 10X) with equivalent resistance values. This improved power operation is possible even without the added benefit of high-temperature packaging.

While the physical properties and characteristics of SiC are in themselves impressive, it is the "end game" that drives the real interests and market directions. The specific potential consumers of SiC devices are the forces that guide the general direction of research and development. There are many potential markets for SiC devices, both long and short term, as well as a host of immediate markets.

Taking a look at the power density limitations of silicon leads one to immediately recognize that power electronics would benefit greatly from the implementation of SiC devices. Silicon carbide p-i-n diodes, Schottky diodes, power MOSFETs, MESFETs, SITs, JFETs, BJTs, and GTOs are all currently under development by research and manufacturing organizations (such as Northrop Grumman, SiCED, Cree, Rockwell Scientific, APT, and Semisouth), although few devices have yet been released into the commercial marketplace.

The U.S. military is interested in SiC power electronics for its hybrid-electric combat vehicles or naval warfare ships, which must utilize high-power converters and motor drives that are capable of operating within harsh environmental parameters or at high power densities. A second major area of interest to the military is in pulsed power applications (such as electromagnetic rail guns), which require very high power (especially high voltage) coupled with fast device responses.

Most solid-state microwave communications and radar electronics systems currently utilize GaAs devices, which offer superior performance over silicon. However, the U.S. military is always interested in increased performance capabilities. Again, high power densities and operating temperatures come into play. Coupling these characteristics with fast-switching capabilities makes SiC high-frequency devices ideal for military aircraft requiring lightweight, high-performance electronic radar systems, making this the third area of interest to the military. When SiC becomes more mainstream, these same performance advantages will make radio frequency (RF) SiC devices prime candidates for commercial communications industries such as air traffic control and weather radar stations, cell phone base stations, and television transmitting stations.

NASA is interested in SiC power converters for spacecraft and satellite applications to drive low-voltage digital and analog electronics from high-voltage power sources (such as the DC solar arrays). The advantage to NASA of an increased operational temperature range is a reduction in required heat sinks and heat exchangers, which are bulky, heavy components. This reduction would result directly in the saving of precious weight (along with its accompanying exorbitant launch costs). Silicon carbide electronics also offers NASA the opportunity for exploration of unique environments for which current technologies are not suited. This would include missions to high-temperature hostile environments, such as the surface of Venus. A Venus lander (and on-board electronics) would have to survive an atmosphere of sulfuric acid and a surface environment that exceeds 450°C at a pressure of almost 100 Earth atmospheres. This is perhaps one of the most difficult electronic packaging environments that one could imagine.

Energy companies are interested in SiC power electronics. Deep-earth exploration encounters hostile environments and extreme temperatures in which it is difficult to place silicon electronics. With SiC, it would be possible to send electronics down-hole, thereby improving motor drive control and efficiency, increasing exploration and sensors capabilities, and ultimately aiding in the discovery of previously hidden energy sources (such as large and wide but shallow oil reservoirs).

Power utility companies are also interested in SiC, specifically for very high voltage applications. Much of the power grid still operates on old electromechanical device technology due to the upper voltage limit of solid-state silicon technology. The current interests in overhauling the national power grid coincides with emerging SiC technology, which offers a significant improvement in semiconductor voltage limits. In 2004, Congress set up the National Center for Reliable Electric Power Transmission (NCREPT) for the sole purpose of developing solid-state electronic solutions that would improve the national power grid. NCREPT is located at the University of Arkansas, is managed by the Department of Electrical Engineering, and has a number of corporate and national partners [29].

These are all high-dollar niche markets requiring large capital investments and cutting-edge technology. When SiC becomes less expensive and more readily available, the entire commercial industry of motors and motor drives will open up. Just about any domestic or industrial electric motor in the world requires a power electronic drive—a drive that could be made smaller and more efficient with the use of SiC power devices. In order to achieve this, however, strides need to be made in packaging technologies.

13.3 COMMERCIALLY AVAILABLE POWER PACKAGES

Commercially available power packages can house either discrete power switches, several power switches (such as in "brick" configurations), or entire power circuit topologies (such as multichip power modules).

13.3.1 Discrete Power Device Packages

There is a wide range of commercially available packages used by industry to house discrete power devices [30–34]. The package of choice depends on a number of factors, including the size of the bare die device, the maximum power dissipation, and the circuit application. Of particular importance is the choice between a through-hole and surface-mount package. Surface-mount devices are smaller, but the maximum thermal dissipation from the package is highly dependent on the circuit substrate material. Through-hole components, although they take up more volume in an electronic system, can be mounted directly to heat sinks and thus can typically operate at an order-of-magnitude greater power densities.

A typical cross section of a surface-mount component is illustrated in Figure 13.9. The power die is mounted onto a chip carrier and wirebonded to internal posts. In later

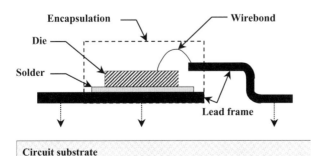

Figure 13.9 Cross section of a typical surface-mount package.

generation surface-mount components, the chip carrier is actually a metal bond plate to which the power device is soldered. This improves thermal dissipation by allowing heat to transfer from the device through the bottom of the package. The die is protected by encapsulation.

Improved thermal performance (i.e., higher power levels of operation) from discrete devices requires direct attachment to heat sinks. The first discrete power packages were the transistor outline (TO) packages for BJTs, which are referred to as "cans." A popular modern can package is the TO-3, illustrated in Figure 13.10. One of the disadvantages of the can package is that it is relatively large; however, it is more important is the heat sink attachment. The electrical connector leads of this package must mount *through* the heat sink, and this limits the size and types of heat sinks that can be mounted to the TO-3 package.

The more modern through-hole packages, such as the TO-220 and TO-247, are specifically designed to maximize thermal dissipation while minimizing volume. This package can be mounted directly to a large-surface-area heat sink with minimum impact on a circuit's footprint size. The assembly and mounting of such a device are illustrated in Figure 13.11.

The current advanced discrete-component packaging trends are toward developing small surface-mount power packages that are capable of dissipating significant amounts of thermal energy through an attached heat sink (without increasing the circuit size). An example of this is the DirectFET power package developed by International Rectifier, which is essentially a power flip-chip packaging method designed for easy surface mounting by applications houses. This technology is illustrated in the cross section of Figure 13.12. Since there are no wirebonds within the package, the complete surface area of all the die contacts are used for thermal dissipation. In addition, even though it is a surface-mount device, a heat sink can also be mounted directly to the top of the package.

The semiconductor power device companies are now moving toward even more advanced packaging techniques. They are developing bare die that can be solder bumped and then handled by standard surface-mount device (SMD) equipment. International Rectifier's FlipFET wafer-level packaging technology, illustrated in Figure 13.13, is an example. The wafer is passivated with a layer of SiN except for the areas where solder bumps are to be attached, which is coated with a thin under bump nickel–gold layer to protect the die from the solder and to seal the passivation.

Table 13.2 illustrates many of the common discrete-component packages. Smaller power diodes are usually placed in packages such as the small outline transistors SOT-23 and SOT-223, SMA/B/C, or TO-92. Power BJTs are usually found in cans, such as the TO-3 or TO-39. Power MOSFETs are found in a wide variety of packages, including dual in-line packages (DIPs), SO packages, Quadpaks (for special applications), DPaks, and the TO-220. IGBTs are manufactured in packages such as the TO-220 or TO-247. Finally, thyristors or very high power diodes are found in packages such as the TO-208 or TO-200.

13.3.2 Multichip Power Modules (MCPMs) and Completely Integrated Solutions

Even the largest of the power packages for discrete components can dissipate only a few watts of thermal energy (operation up to ∼200 W, electrically). When it is necessary to transfer very large amounts of thermal energy from the power die, a completely different strategy is required. This is where the "brick" strategy comes in. Multiple power die are packaged in common brick configurations, such as configurations used in a three-phase full bridge or a single-phase half bridge. In this manner, entire power stages of common power

13.3 Commercially Available Power Packages **549**

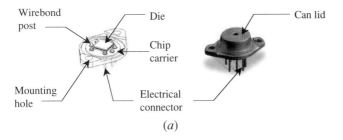

(a)

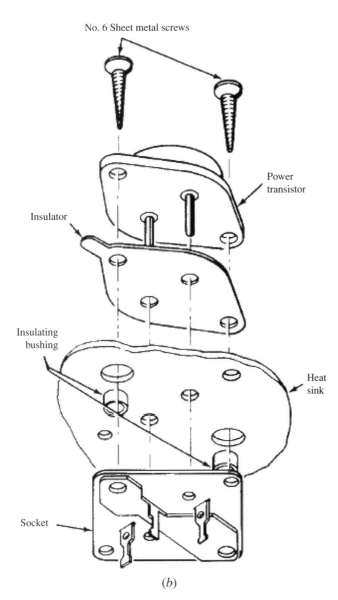

(b)

Figure 13.10 (a) Assembly of typical can power packages. (b) Motorola can power package.

550 Chapter 13 Power Electronics Packaging

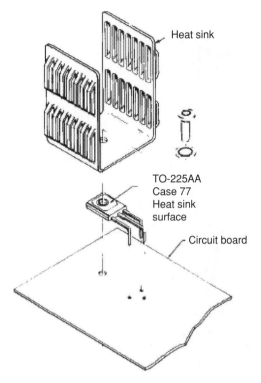

Figure 13.11 Assembly of a typical through-hole power package (Motorola).

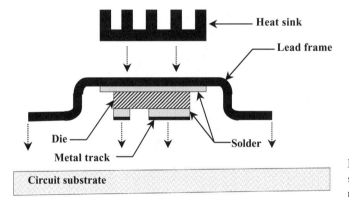

Figure 13.12 Cross section of a DirectFET power package.

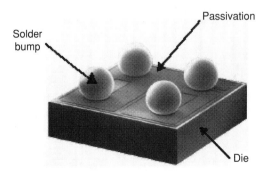

Figure 13.13 FlipFET wafer-level packaging technology.

Table 13.2 Common Discrete-Component Packages

Package	Package
DPAK	8-PIN DIP
D2PAK	TO-208
QuadPAK	TO-3
DirectFET	TO-39
SO-8	TO-92
SOT-23	TO-200
SOT-223	TO-220
SMA, SMB, SMC	TO-247

electronic circuits are packaged together in single modules that can then be mounted onto large heat sinks. The rest of the electronic circuit is then wired to the brick [35, 36].

Figure 13.14 illustrates the cross section of a typical power electronic brick. The entire theory of the package is built around spreading and dissipating large amounts of heat and in carrying very large amounts of current at high voltages. To this end, the package is typically built on an aluminum or copper heat spreader which serves as the base for the component. A high-thermal-conductivity direct-bonded copper (DBC) substrate is mounted onto the heat spreader. The DBC is a ceramic substrate manufactured from an advanced AlN or BeO ceramic (or sometimes the more common Al_2O_3) that has been chemically bonded through a high-temperature process with thick laminates of copper metallization. The DBC is a double-sided substrate in order to equalize thermal stresses within the ceramic. The top

552 Chapter 13 Power Electronics Packaging

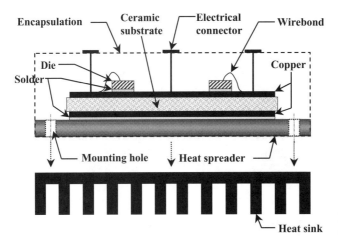

Figure 13.14 Cross section of a typical power brick.

layer of copper metallization is usually nickel and/or gold plated and then chemically etched [using standard printed circuit board (PCB) processing]. Multiple bare die power devices are mounted to it and wirebonded [37–41]. The brick is then usually filled with a gel for protection and the lid is mounted. Screw holes are tapped so the brick can be mechanically screwed down to a heat sink. Screws and pins on the lid provide for electrical connections (screws for high-power connections and pins for control connections) to the rest of the circuitry. More recently, metal matrix composites (MMCs) have been used as the heat-spreading baseplate material since they offer good thermal conductivity with the additional advantage of a closer coefficient of thermal expansion (CTE) match to the substrate, thus improving reliability [42–45].

It is only reasonable that if power bricks are available for common power electronics circuit configurations, then entirely integrated and packaged solutions have been developed for certain power applications. One of the most common of these consists of three-phase inverters and motor drives. The current state-of-the-art research thrust is in the development of fully integrated MCPMs—in other words, power modules that perform a complete function and are fabricated from bare die control *and* power devices[46–50]. This is a difficult undertaking due to the self-excluding requirements of the power devices and the control devices. The power die usually require large amounts of substrate surface area to spread thermal energy and thick metallization traces to carry large amounts of current. The control devices, on the other hand, require high-density packaging in order to minimize circuit size, which in turn requires thin metallization traces and tight pitch.

Table 13.3 illustrates examples of various bricks, MCPMs, and completely integrated solutions [51, 52].

13.3.3 Thermal Performance of Commercial Packages [53–58]

As discussed previously, a one-dimensional thermal network is similar to an electrical network, where heat transfer is given by the equation

$$q = hA\Delta T \tag{13.5}$$

13.3 Commercially Available Power Packages

Table 13.3 Examples of Various Bricks, MCPMs, and Completely Integrated Solutions

Package		Package	
MCM		IR PIIPM three-phase inverter	
MTP		Curamik power module	
SOT-227		UA three-phase inverter	
Half brick		PowerEx half brick	
Full brick		APEI, Inc. motor drive	

where

q = thermal power dissipated (W)
h = heat transfer coefficient
A = area involved in heat transfer
ΔT = temperature differential

Thermal resistance R_θ (in degrees Celsius per watt) can be derived from Eq. (13.5):

$$R_\theta = \frac{\Delta T}{q} = \frac{1}{hA} \qquad (13.6)$$

The basic industry standard thermal resistance model for power semiconductors is shown in Figure 13.15, where

$$q = \frac{T_j - T_c}{R_{\theta jc}} + \frac{T_c - T_s}{R_{\theta cs}} + \frac{T_s - T_a}{R_{\theta sa}} = \frac{\Delta T_{\text{Total}}}{R_{\theta \text{Total}}}$$

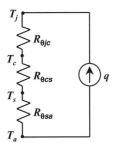

Figure 13.15 Power semiconductor thermal resistance model.

In this model equation, the key components of physical importance to the prevention of device failure are q and T_j, where T_j is the semiconductor device junction temperature and cannot exceed the maximum rated temperature for the device. Most power devices are rated to a maximum temperature of 150°C, but it is important to check a device's data sheet to verify this value. The other critical component in the equation is q, the amount of thermal energy (in watts) being released by the device. This power dissipation is the driving component.

The term $R_{\theta jc}$ is the thermal resistance between the device junction and the device case. Every component manufacturer lists $R_{\theta jc}$ in its data sheet. This is a value completely dependent on the die and the power package housing it.

The term $R_{\theta cs}$ is the thermal resistance of the interface between the device case and the heat sink. This is a variable that is widely dependent on heat sink mounting procedures and interface conditions. Figure 13.16 illustrates a magnified view of a typical interface between two materials. Since neither material is perfectly flat, an air gap forms at the interface, creating a thermal resistance. If a power package is simply mounted to a heat sink in this fashion, it is referred to as a dry mount. In order to reduce the thermal resistance at the contact, a thin coating of thermally conductive grease is typically applied to the surface of the heat sink. The grease fills in the air gaps between the two materials. It is important that only a *very thin* coating of thermal grease be applied so that a layer of grease is not formed between the two materials. This would actually increase the thermal resistance. Many times, it is undesirable to have an "electrically hot" heat sink, and so an electrical isolation pad is mounted between the power device and the heat sink. This isolation pad is usually a poor thermal conductor, which greatly increases the case-to-sink thermal resistance.

A final important factor in determining the case-to-sink thermal resistance is the force with which the heat sink is mounted to the power package. The heat sink is mounted with screws or clips, and a tight fit is required in order to obtain a good thermal contact. Typical contact thermal resistances are given in Figure 13.17.

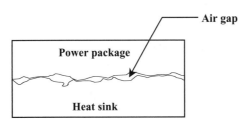

Figure 13.16 Power package–heat sink interface.

13.3 Commercially Available Power Packages

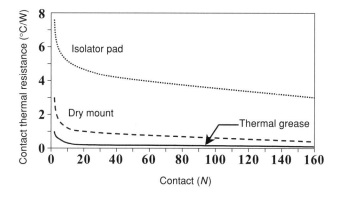

Figure 13.17 Typical contact thermal resistance with various interfaces.

Table 13.4 lists the thermal conductivity for a number of isolation pads or other materials that can affect the contact resistance at an interface. Given the thermal conductivity of a specific material, the thermal resistance can be found by applying the equation.

$$R_\theta = \frac{t}{kA} \tag{13.7}$$

Table 13.4 Thermal Conductivity for a Number of Isolation Pads and Other Materials

Material	k (W/m-°C)
Still air	0.033
Silicone grease	0.195
Mylar film	0.167
Mica	0.600
DeltaPad 175-6	0.433
DeltaPad 173-9	0.703
Type 120 joint compound	0.703
Type 121 joint compound	0.703
Type 126 joint compound	0.691
Types 152&D and 153&D	0.729
Type 151 DeltaCoate	0.820
DeltaPad 173-7	0.838
Types 152&A and 153&A	0.838
Types 152&B and 153&B	0.937
Types 152&C and 153&C	0.937
Type 156K	0.498
DeltaPad 174–9	1.4
PSA laminates	0.437–1.12
Filled silicone rubber	0.486
Anodized coating	7.0
Alumina (Al_2O_3)	34
Steel, carbon	47
Boron nitride	65
Beryllium oxide (BeO)	123
Aluminum type 1100	207
Copper, CDA110	394

556 Chapter 13 Power Electronics Packaging

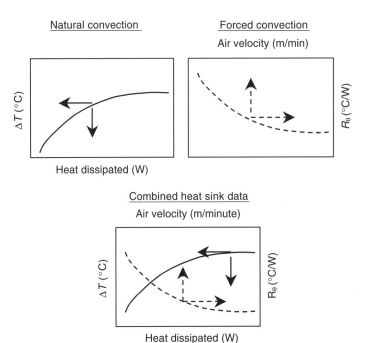

Figure 13.18 Graph form for manufacturers' heat sink data.

where

k = thermal conductivity (W/m °C)

t = material thickness (m)

A = cross-sectional area (m²)

The final resistance of the thermal model of Figure 13.15 is the heat-sink-to-ambient-environment thermal resistance, $R_{\theta sa}$. This property depends on the characteristics of the chosen heat sink and the cooling environment (forced air, forced liquid, etc.). Figure 13.18 shows the form of heat sink characteristic data provided by manufacturers. Natural convection data are given as a temperature delta between the sink and ambient environment versus dissipated thermal energy. Forced-convection data are given as sink-to-ambient-environment thermal resistance versus air velocity. Most manufacturers have this information combined into a single graph.

Example 13.2

Let us continue the problem of Example 13.1 except we will assume we have a power transistor that has an electrical loss of 8.25 W. This power is released as thermal energy and must be dissipated from the TO-247 package. The transistor manufacturer has given the thermal resistance between the junction and case to be 2.5°C/W and the device's maximum operating temperature to be 150°C. The transistor package is mounted on to a Wakefield 637 series forced-convection air-cooled heat sink with a thermal resistance property of 2°C/W [59]. The package must be electrically isolated, and it is decided that a beryllium oxide wafer (0.15 mm thick) with type 120 joint compound (25 μm) will be used at the case–heat sink interface (assume an otherwise perfect interface). Determine if this device and package will operate within their operating temperature parameters (assuming an ambient temperature of 25°C).

In the problem $R_{\theta jc}$ is given to be 2.5°C/W and $R_{\theta sa}$ is given to be 2°C/W. The ambient temperature T_a is given to be 25°C. Thus the only missing piece of information required to calculate the junction temperature is the thermal resistance at the case–heat sink interface.

A thin layer of thermal compound is applied between the package and the isolating wafer, and a second layer of thermal compound is applied between the isolating wafer and the heat sink. Since it is assumed to be an otherwise perfect interface, no other factors contribute to the thermal resistance at the interface. The thermal conductivities of the materials can be found in Table 13.3, where k of the type 120 joint compound is 0.703 W/m°C and k of the BeO wafer is 123 W/m°C. The thicknesses of both materials are given in the problem statement, 25 μm for each grease layer and 0.15 mm for the wafer. The missing component for determination of the case-to-sink thermal resistance is the cross-sectional area through which the thermal energy is being dissipated. This value can be determined by looking up the package size of a TO-247 package in any manufacturer's handbook (or website). The area of the TO-247 package mounted to the heat sink is 15.50 mm × 20 mm, and so A is calculated to be 3.1×10^{-4} m^2. The case-to-sink thermal resistance and the total thermal resistance can then be calculated as

$$R_{\theta cs} = \frac{t_{\text{grease}}}{k_{\text{grease}} A} + \frac{t_{\text{BeO}}}{k_{\text{BeO}} A} + \frac{t_{\text{grease}}}{k_{\text{grease}} A} = 0.233°\text{C/W}$$

$$R_{\theta \text{Total}} = R_{\theta jc} + R_{\theta cs} + R_{\theta sa} = 4.733°\text{C/W}$$

The junction temperature can then be calculated:

$$T_j = q R_{\theta \text{Total}} + T_a = 64.0°\text{C}$$

Thus, the power component will operate well beneath its maximum operating temperature of 150°C.

Example 13.3

Continue Example 13.2, but now assume that the heat sink is mounted with a DeltaPad 173-7, which is a greaseless interface. Looking up the properties of the DeltaPad 173-7 in Wakefield Engineering's data manuals shows the pad to have a thickness of 0.178 mm. From Table 13.4, it can be seen that this material has a thermal conductivity of 0.838 W/°C-m. Assume that each interface is non ideal and has a thermal resistance of 1°C/W. The resistances can be found and the junction temperature calculated:

$$R_{\theta cs} = \frac{t_{\text{pad}}}{k_{\text{pad}} A} + 2°\text{C/W} = 2.685°\text{C/W}$$

$$R_{\theta \text{Total}} = R_{\theta jc} + R_{\theta cs} + R_{\theta sa} = 7.185°\text{C/W}$$

$$T_j = q \cdot R_{\theta \text{Total}} + T_a = 84.3°\text{C}$$

The power component is still within its rated operating temperature range.

Example 13.4

Continue Example 13.2, but this time assume the package is not mounted to a heat sink at all. In this example, there are only two thermal resistances—junction-to-case and case-to-ambient-due to convection. The case-to-ambient can be calculated using the thermal convection equation and assuming free convection of air (Table 7.2). It can be estimated that the convection heat transfer coefficient h is approximately 5 W/m^2 °C. It can be assumed that the package encapsulation is not thermally conductive, the leads can be neglected, and so the area of thermal dissipation is the package backside where the heat sink would normally be attached. Thus, the area A remains unchanged from previous examples.

The heat transfer equations for convection are as follows:

$$q = hA\Delta T \qquad (13.8)$$

$$R_{\theta \text{convection}} = \frac{1}{hA} \qquad (13.9)$$

For this case, the thermal resistance due to convection and total resistance can be calculated:

$$R_{\theta \text{convection}} = \frac{1}{hA} = 645°C/W$$

$$R_{\theta \text{Total}} = R_{\theta jc} + R_{\theta \text{convection}} = 647.5°\text{C/W}$$

$$T_j = q \cdot R_{\theta \text{Total}} + T_a = 5367°\text{C}$$

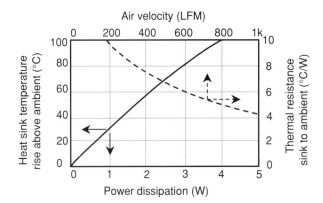

Figure 13.19 Wakefield engineering series 230 heat sink data.

Obviously, the component would fail in this example with these assumptions. It is important to note the ineffectiveness of thermal convection from such a small area. In reality, heat conduction through the package wirebonds, through the leads, and finally to the board actually has a much greater effect on the heat transfer rate than the convection factor. Thus, in a case such as this, realistically, the package leads cannot be neglected.

Example 13.5

Take the problem of Example 13.2, except, in this example, mount the power device to a Wakefield Engineering series 230 heat sink. Assume that the device is operating at a power level where half of the thermal power (of that calculated in Example 13.2) is dissipated. Also assume that the heat sink is mounted using the same BeO wafer of Example 13.2. The heat sink data are given in Figure 13.19. Find the junction temperature if the heat sink and package operate in a free-convection environment.

Reading the graph at 4.125 W of dissipation, the ΔT between the ambient environment and the heat sink is approximately 105°C. The resistances and junction temperature can be calculated as follows:

$$R_{\theta sa} = \frac{\Delta T_{sa}}{q} = 25.5°C/W$$
$$R_{\theta \text{Total}} = R_{\theta jc} + R_{\theta cs} + R_{\theta sa} = 28.2°C/W$$
$$T_j = q \cdot R_{\theta \text{Total}} + T_a = 141.5°C$$

The power device and package using the series 230 heat sink in a free-convection ambient environment is just within the operating parameters for maximum temperature.

Example 13.6

Continuing the problem of Example 13.5, if it is desired to use the same series 230 heat sink, determine if it is possible to operate the power device and package under the electrical conditions solved for in Example 13.2. If so, what are the conditions for heat sink cooling?

The maximum junction temperature is 150°C, the power dissipated is 8.23 W, and so the maximum allowable total thermal resistance and maximum allowable sink-to-ambient thermal resistance can be calculated:

$$R_{\theta \text{Total}} = \frac{\Delta T_{ja}}{q} = 15.2°C/W$$
$$R_{\theta sa} = R_{\theta \text{Total}} - R_{\theta jc} - R_{\theta cs} = 12.5°C/W$$

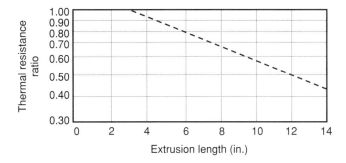

Figure 13.20
Natural-convection size selector (from 3-in. data).

Reading from the graph of Figure 13.19, forced-air-convection cooling of the heat sink with a minimum airflow of approximately 150 linear feet per minute (or 45 m/min) would be required to keep the sink-to-ambient thermal resistance below 12.5°C/W. So, yes, it is possible to operate the device at 8.25 thermal watts dissipation with a series 230 heat sink and still remain within the operational guidelines.

Power module bricks are usually mounted onto large extruded fin heat sinks. These heat sinks can be obtained from the manufacturers in varying lengths, and from the manufacturers' given data, the engineer must determine the proper length of the heat sink for the specific application. The heat sink manufacturer will normally give the sink-to-ambient thermal resistance for a particular length of the extruded heat sink.

Figure 13.20 (with data given for a 3-in. length) is used to determine the length of the extruded heat sink required under free-convection conditions. A simple formula is used in conjunction with the graph. The published heat sink data for a particular length are given as $R_{\theta \text{Data}}$ and the required thermal resistance is $R_{\theta \text{Desired}}$. Dividing the two, the thermal resistance ratio is calculated. The calculated ratio is found on the graph and the corresponding heat sink length can then be determined from

$$R_{\theta \text{Ratio}} = \frac{R_{\theta \text{Desired}}}{R_{\theta \text{Data}}} \tag{13.10}$$

Example 13.7

Your engineering firm is developing a three-phase motor drive. It has been decided that each phase of the power stage will be built with a separate IXYS MII 75-12 A3 IGBT 1.2-kV half-bridge rectifier (and each phase will have its own heat sink). The engineers on your team have determined that each half-bridge module will need to dissipate 100 W of thermal energy with a free-convection cooled Wakefield Engineering series 5772 heat sink. Your team manager asks you to determine the length required for each heat sink.

Data for the IXYS module can be found in an IXYS handbook at its website [60]. The junction-to-sink thermal resistance is given in the data sheet to be 0.66°C/W assuming a nonisolated interface and the use of a thermal grease compound. The maximum junction operating temperature is listed as 150°C. The ambient temperature for this example is assumed to be 25°C and the power dissipated is given as 100 W. The maximum required sink-to-ambient thermal resistance can be calculated as

$$R_{\theta \text{Total}} = \frac{\Delta T_{ja}}{q} = 1.25°\text{C/W}$$
$$R_{\theta sa} = R_{\theta \text{Total}} - R_{\theta js} = 0.59°\text{C/W}$$

Data for the series 5772 heat sink can be found in a Wakefield Engineering handbook or its website. The sink-to-ambient thermal resistance is given to be 0.71°C/W for a 3-in length. The thermal resistance

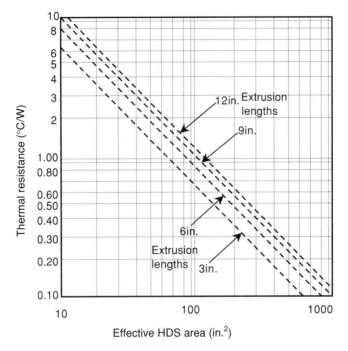

Figure 13.21 Forced-convection size selector.

ratio can thus be calculated:

$$R_{\theta \text{Ratio}} = \frac{R_{\theta sa}}{R_{\theta \text{Data}}} = 0.83$$

Locate the thermal resistance ratio of 0.83 on the graph of Figure 13.20, then read horizontally across to the line, and then read vertically down to the x axis to determine the length required. For this problem, the required minimum heat sink length is determined to be ~ 5 in.

The power module bricks must typically transfer very large amounts of heat, and the most efficient low-cost method of achieving this goal with the smallest available heat sinks is to attach a fan to the heat sink and cool with forced air convection. The curves of Figure 13.21 are utilized in this cooling strategy to determine the required length of extruded heat sink. In order to determine the required length, first, the desired thermal resistance is calculated. Next, the thermal resistance is located on the graph and the required heat dissipation surface (HDS) area is read off for *each* of the four extrusion lengths. The four resulting HDS areas are averaged together and divided by the chosen heat sink perimeter data in order to determine the extrusion length as given by

$$L_{\text{Extrusion}} = \frac{\text{HDS}_{\text{Average}}}{P} \quad (13.11)$$

Example 13.8

The three-phase motor drive project of Example 13.7 has taken a slight turn. The engineering team has decided that the heat sink lengths you calculated are too large for the project. It is required that the size and weight of the overall circuit be reduced. The team has been authorized by the engineering firm to investigate the use of fans and forced air convection for module cooling. The team manager has asked you to recalculate the required heat sink lengths, this time with forced-convection conditions.

The required sink-to-ambient thermal resistance remains the same at 0.59°C/W. The heat sink perimeter data are listed in the data sheet as $P = 99.79$ in. Reading the graph at 0.59°C/W yields the following information:

$$\text{HDS}_{\text{Average}} = 170 \, \text{in.}^2$$

HDS at 3 in. = 120 in.²
HDS at 6 in. = 150 in.²
HDS at 9 in. = 190 in.²

HDS at 12 in. = 210 in.²

From Eq. (13.11), the extrusion length required is

$$L_{\text{Extrusion}} = \frac{\text{HDS}_{\text{Average}}}{P} = 1.7 \, \text{in.}$$

The calculated heat sink extrusion length using a fan and forced air convection is approximately 1.7 in., which is only about one-third the length of a free air-cooling strategy. The slight additional cost of including a fan in the design will produce significant savings in the package volume and weight and greatly reduce the amount of required material. In the end, the savings from heat sink material purchases will greatly reduce the cost of the overall product.

13.4 POWER PACKAGING DESIGN METHODOLOGY

In designing power packages, the engineer must develop an overall system philosophy, select appropriate components (power switches, substrates, heat spreaders, heat sinks, etc.) to perform the desired functions, calculate power and thermal performance, and design topology layouts and module configurations.

13.4.1 Overall System Design Philosophies

There are a number of variables that need to be considered when designing a power system since they impact the power packaging strategy. The key from which all other decisions flow is the choice of semiconductor device. The criteria, which need to be considered when determining the device of choice, are as follows:

- Desired voltage and current operation levels [average, peak, root-mean-square (rms), and continuous operational levels]
- Switching frequencies (short rise and fall times, short delay times, high frequency)
- Losses and efficiency (conduction losses, switching losses, and leakage current)
- Thermal load due to device choice (directly related to losses)
- Cost of the semiconductor device(s) and impact on overall power circuit cost
- Reliability (capable of large numbers of temperature cycles, few thermal stress failures, and the ability to absorb voltage and current spikes or surges)
- Drive circuit complexity (easily controlled power devices are preferred)
- Modularity (power devices usable in a wide variety of applications with few circuit design changes are preferred)

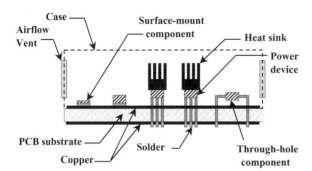

Figure 13.22 Cross section of a discrete-component power module on PCB.

- Size (weight and volume are important in many applications, in particular the size of the overall power circuit and packaging due to the power device chosen)
- Protection (the ability to design circuitry to protect or shut down the power device in case of a potential or inevitable catastrophic event is desirable)

The power circuit application and the choice of semiconductor device (or device technology) will directly impact the packaging technology. The power electronics design engineer essentially has five choices:

1. Select a completely integrated design solution. This is a "plug and play" option where the entire power circuit is simply a black box with a set of inputs and outputs that give the desired results.

2. Select discrete power devices for use on FR-4 or some other similar PCB material. This option can utilize either through-hole or surface-mount devices. Surface-mount devices optimize miniaturization, but due to the low thermal conductivity of PCBs circuit operation is limited to low thermal levels. Through-hole components greatly enhance the ability to dissipate thermal energy, but at the cost of increased package volume. This design philosophy is illustrated in Figure 13.22.

3. Select discrete power devices for use on ceramic hybrids. This option is restricted to surface-mount devices, but due to the high thermal conductivity of ceramic substrates, high levels of miniaturization can be obtained with the ability to dissipate significant amounts of thermal energy. Ceramic power hybrids are typically thick-film technology, thus restricting overall electrical power levels. Thick-film conductor traces are metal-filled pastes that have been fired in a furnace for solvent removal. The disadvantage of thick-film conductors is that they have approximately one-tenth the conductivity of pure metal, thus making them very lossy for power applications. As power levels become high, the electrical losses and current-carrying capabilities of thick-film conductors become impractical. The other advantage of ceramics is that they can survive and operate in more hostile environments than PCBs (such as high-temperature environments). This design philosophy is illustrated in Figure 13.23.

4. Separate the power and control stages and select a power module brick for the power stage. Typically, the control circuits of such power electronics systems are built on FR-4 or some other PCB material. The major advantage of this approach is that very high levels of electrical and thermal operation can be achieved. The obvious cost is that overall volume of the power package increases dramatically. The second disadvantage is that, due to the separation of circuit stages, significant parasitics are introduced, which increase electrical losses and reduce maximum obtainable switching frequencies. This design philosophy is illustrated in Figure 13.24.

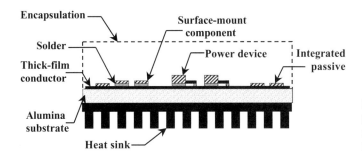

Figure 13.23 Cross Section of a Discrete-Component Power Module on Ceramic.

5. Select and design a complete custom MCPM. The advantage of this approach is that very high power density and excellent performance can be achieved with very high integration and miniaturization. This option is a necessity when high-performance integrated cooling strategies are required. The disadvantages are high cost, high complexity, long design and manufacture times, and the requirement for a wide range of high-dollar, in-house electronic packaging and processing equipment. One possible design philosophy for this approach is illustrated in Figure 13.25.

In this design, a DBC or IMS (insulated metal substrate) is used as the power substrate, similar to the brick design. Laminated to the power substrate is a multi layered control substrate fabricated from metallized polyimides or ceramics. The bare die and surface-mount control and passive components are then mounted onto the top of the control substrate and the components are wirebonded. The entire module is then encapsulated. Figure 13.26 presents photographs of such an MCPM during substrate processing.

13.4.2 Substrate Selection

After choosing the semiconductor device and the overall design philosophy, the next step is to choose the proper substrate. Substrate selection is highly dependent on the application. These materials should possess a number of properties, as follows:

- High thermal conductivity to help remove heat from the module
- CTE characteristics near those of the materials to be attached to the substrate

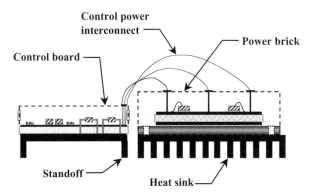

Figure 13.24 Cross section of a brick-based power module.

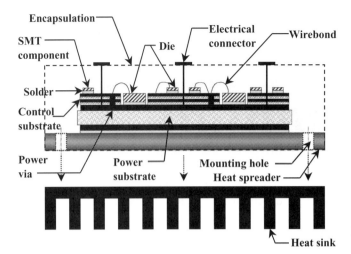

Figure 13.25 Cross section of a custom MCPM.

- Acceptable cost (or low cost for high-volume production applications)
- Good insulating characteristics for electrical isolation of circuit lines and components
- Low dielectric constant for low capacitive loading of circuit lines and low capacitive coupling
- High strength and toughness for ruggedness and "processability"
- High-dimensional stability
- Chemical and physical stability at the processing temperatures used for module fabrication and/or operation

PCB surface-mount or through-hole technology usually uses FR-4 where the main criterion is low cost. Ceramic technology usually uses Al_2O_3 where the main criteria are low cost and improved performance and miniaturization. Power module bricks usually use DBC with an AlN or BeO substrate, or an IMS, where the main criteria are high thermal conductivity, high current-carrying capacity, and high reliability with temperature cycling or electrical stressing. MCPMs can utilize all of these substrate technologies, as well as polyimide substrates for multilayer laminations or diamond and diamondlike carbon (DLC)

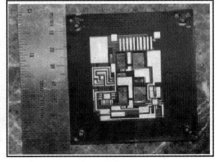

Figure 13.26 Power substrate and control substrate of an MCPM.

substrates for very high thermal conductivity. The copper metallization traces of high-performance substrates, such as in power brick or MCPM applications, are usually nickel or gold plated. This practice significantly reduces surface oxidation and contaminants, facilitating improved wirebonds and die attachments. This, in turn, reduces long-term corrosion and improves overall reliability. The specifics of all of these substrate technologies are discussed elsewhere in this book.

As shown in Figure 13.26, it is important to maximize the copper surface area in design and layout. This provides for the most efficient current flow through the module and also increases heat-spreading capability. Due to the copper metallization thickness, the minimum distance between traces is typically greater than 50 mils.

13.4.3 Baseplate and Heat Spreader Selection

Baseplate selection is a key factor when designing power packages using power module bricks or MCPMs, because this is the primary interface through which the heat will be removed. Thus, it is essential that the baseplate material have a high thermal conductivity, which allows for adequate thermal spreading. The second major characteristic of a baseplate material is high mechanical strength, since it provides the required structural support for the electronic circuits and modules [61].

A dilemma arises due to the fact that most of the materials that exhibit high strength and high thermal conductivity, key characteristics needed for baseplates, are metals (usually copper, or aluminum for lightweight applications). Metals are ideal except for their CTEs since the CTE of most metals is several times that of silicon. This mismatch can cause significant thermal stress reliability problems.

One potential solution to this problem is a class of materials referred to as MMCs. These materials are composed of a ceramic or graphite matrix that is injected with a metal filler. The resulting material combines the properties of the matrix with those of metals: high thermal conductivity, high strength, low thermal expansion, and low density. What is particularly attractive about MMCs is that their properties and characteristics can be tailored to specific applications by adjusting the material mixture recipes. In addition, some of these materials can also be formed into complex shapes, which allows for an extra degree of freedom in power module design. Several different materials are commercially available and their properties are given in Table 13.5 along with common device and substrate materials.

13.4.4 Die-Attach Methods [62–64]

Since most power semiconductors are vertical-channel devices, both the top and bottom surface connections must be electrically active. Therefore, the power die must be attached to the substrate with a metallurgical bonding process. There are three basic processes that can accomplish a metallurgical bonding:

1. **Welding**. This involves the melting together of two high-temperature metals, with or without filler metals. This process requires the application of localized high energy to create the bond. Power die are not attached to substrates by this method; however, it is often used for attaching lid packages.
2. **Brazing**. This process is a lower temperature process than welding but requires a higher temperature than soldering. The procedure utilizes a braze, or filler metal,

Table 13.5 Comparison of Material Thermal Properties

Material	CTE (ppm/K)	Thermal conductivity (W/m-K)
AlN	3.0–4.1	100–170
Al_2O^3 (96%)	6.4	35
AlSiC (60% Al)	12.6	240
AlSiC (63% SiC)	7.9	175 (minimum)
Aluminum	24	240
Be–BeO MMC	6.8	240
Copper	17	393
Cu–Mo	7.2	197
Cu–W (20% Cu)	7.0	248
Diamond	0.8–2.0	1000–2000
Graphite–Cu MMC	0–2.0 (directional)	356 (minimum)
Gold		317
Invar	3.1	11
Kovar	5.3	17
Molybdenum	4.9	143
Silicon	4.1	136
Silver		429
Solder		50

that typically has a melting temperature higher than 350°C but a lower melting temperature than the two metals being bonded. Fluxes are required to remove oxides and contaminants from the filler and the metals to be joined. These fluxes are typically corrosive. Brazing is under investigation for use in high-temperature electronics, and although it is not a standard die-attach method, it is often used in other packaging processes.

3. *Soldering*. In this process, two metals are joined by a filler metal with a melting temperature below 350°C. Typically, there is a temperature range between the metal's solid state and liquid state, referred to as the plastic region. If the joining metals are physically disturbed while cooling through this phase, the solder crystal can be disrupted, and a cold-solder joint will result (a solder joint with a high resistivity). This is the standard process for mounting power die.

The most desirable solder is a eutectic solder. This alloy transfers directly from the liquid state to the solid state without passing through the plastic region. The most common solders are lead–tin based, and the standard lead–tin eutectic solder is 63Sn–36Pb (63% tin and 36% lead), with a melting temperature of 183°C. Adding 2% silver to the alloy is a common practice, which strengthens the solder. This alloy is 62Sn–36Pb–2Ag. In the soldering process, it is desirable to heat the assembly to approximately the melting temperature plus an additional 30 to 50°C.

There are a large number of solders available on the commercial market, with a wide variety of melting temperatures, mechanical properties, and chemical compositions. Some of the more common solders, along with their physical properties, are listed in Table 13.6 (provided by Indium Corporation of America).

Table 13.6 Common Solders and Properties

Indialoy number	Composition	Liquidus °C	Solodus °C	Electrical conductivity % of IACS	Thermal conductivity W/cm°C @ 85°C	Thermal coefficient of expansion PPM/°C@20°C	Tensile strength PSI	Application notes
136	49Bi 21In 18Pb 12Sn	58	58	2.43	0.1	23	6300	Poor wettability but adequate for mechanical joining of metallic substrates *** corrosive type flux is used.
42	46Bi 34Sn 20Pb	96	96	—	—	—	—	Can be used on the same metallizations as SnPb based solder.
1E	52In 48 Sn	118	118	11.7	0.34	20	1720	Fair wettability on glass, quartz, and many ceramics. Good low-temperature malleability. Compensates for some difference in CTE.
281	58Bi 42Sn	138	138	4.5	0.19	15	8000	Good law melting point solder for electronics assembly or for applications where Cd and Pb are to be avoided. Also good for thermo-electric applications
290	97In 3Ag	143	143	23	0.73	22	800	Silver added to improve strength. Has nearly the wettability and low-temperature in malleability of Indium.
2	80In 15Pb 5Ag	154	149	13	0.43	28	2550	Especially useful for soldering against gold because it minimizes leaching. Good thermal fatigue
4	100In	157	MP	24	0.56	29	273	Pure indium. Soft, ductile metal. Good wettability on many surfaces including glazed ceramics, certain metallic oxides, glass, and quartz. Deforms indefinitely under load. Has no tendency to become brittle malding*** it valuable for cryogenic application

(*Continued*)

Table 13.6 (Continued)

Indialoy number	Composition	Liquidus °C	Solodus °C	Electrical conductivity % of IACS	Thermal conductivity W/cm°C @ 85°C	Thermal coefficient of expansion PPM/°C@20°C	Tensile strength PSI	Application notes
97	43 Pb 43Sn 14Bi	163	144	—	—	24	6400	Good general purpose step soldering alloy.
9	70Sn 18Pb 12In	167	154	12.2	0.45	24	5320	General purpose solder with good physical properties.
204	70In 30Pb	175	165	8.8	0.38	28	3450	Minimizes gold leaching characteristics. Good thermal fatigue properties.
Sn62	62Sn 36Pb 2Ag	179	179	11.9	0.5	27	7000	Good general purpose solder can be used on silver metallized*** surfaces to reduce scavenging.
205	60In 40Pb	181	173	7	0.29	27	4150	Minimizes gold leaching characteristics. Good thermal fatigue properties.
SN63	63Sn 37Pb	183	183	11.5	0.5	25	7500	Standard eutectic in-lead solder with wide application. Not recommended for use against silver or gold.
227	77.2Sn 20In 2.8Ag	187	175	9.8	0.54	28	6800	Can be used as a replacement for Sb63, and Sn60. Has similar melting point and equal or superior physical and mechanical properties Not for use over 100°C due to 118°C Sn/In eutectic.
201	91Sn 9Zn	199	199	15	0.61	—	7940	Recommended for soldering to aluminum using flux #3
7	50In 50Pb	210	184	6	0.22	27	4670	Minimizes gold leaching characteristics. Good thermal fatigue properties.
121	96.5Sn 3.5Ag	221	221	16	0.33	30	5620	Use when lead-based solders do not meet temperature, strength, or safety requirements. Not recommended against gold-plated surfaces.
206	60Pb 40In	231	197	5.2	0.19	26	5000	Minimizes gold leaching characteristics. Good thermal fatigue properties.

	Alloy						Notes
3	90In 10Ag	237	143	22.1	0.67	15	1650	Silver added to improve strength. Has nearly the wettability and low-temperature in malleability of indium.
133	96Sn 5Sb	240	235	11.9	0.28	31	5900	Used to join copper tubing for refrigeration and potable water systems. Good wettability with good creep resistance at elevated temperatures.
10	75Pb 25In	260	240	4.6	0.18	26	5450	Minimizes gold leaching characteristics. Good thermal fatigue properties.
150	81Pb 19In	275	260	4.5	0.17	27	5550	*Minimizes gold leaching characteristics. Good thermal fatigue properties.
182	80Au 20Sn	280	280	—	0.57	16	4000	Strong solder with excellent thermal fatigue resistance. Can be used against gold surfaces without flux in inert atmosphere.
165	97.5Pb 1.5Ag 1Sn	309	309	6	0.23	30	4420	Wide application in semiconductor assembly. Often used in reducing atmosphere such as hydrogen.
164	92.5Pb 5In 2.5Ag	310	300	5.5	0.25	25	4560	Particularly good thermal fatigue. Minimal gold leaching properties of indium-lead alloys. Often used in reducing atmosphere such as hydrogen.
171	96Pb 5Sn	312	308	8.8	0.23	30	4000	Highest melting temperature of the tin–lead system.

570 Chapter 13 Power Electronics Packaging

Quite often, the careful selection of solder can have a significant impact on the reliability of a product. There are a number of pitfalls that the power packaging engineer must be aware of, with some of the more profound being:

1. The copper metallization of many high-performance substrates are gold plated in order to reduce surface oxidation. Gold, however, is rapidly dissolved by the tin in standard solders, creating a brittle intermetallic compound at the solder joint. If the joint absorbs more than 3% gold, the risk of long-term failure becomes a significant reliability concern. The amount of gold absorbed into the solder joint is dependent on the amount of tin in the solder, the thickness of the gold on the pad or metallization trace, and the time and temperature of reflow. When the gold layer is thicker than 0.5 μm a solder substitute should be made. An indium- or gold-based solder is recommended in these cases.
2. Certain metallizations, such as nickel, are difficult to solder due to a thick layer of protective oxidation that forms on the surface. Solder pastes, which are often the solder mediums of choice for applying power die, usually contain less active fluxes, which makes it difficult to wet well to nickel. If nickel is the substrate metallization of choice, a solid solder ribbon, or preform, with a directly applied higher active flux should be used.
3. Copper metallization diffuses into indium, forming a brittle copper–indium intermatallic, resulting in weakened and failing joints. The diffusion occurs even at room temperature, and the process is referred to as solid-state diffusion. This is important because indium is a common solder alloy.
4. Aluminum is another metallization surface that is difficult to solder to, again because of the tenacious oxide that forms. It is not desirable to use tin–lead solders when metallic Al is involved because the difference in electropotential between the Al and solder can cause a galvanic coupling, accelerating corrosion and failure. Tin–zinc eutectic solder or a zinc–aluminum braze is recommended in these cases.

Table 13.7 outlines the compatibility issues and recommendations for a number of common metallizations, solders, and fluxes used in power packaging.

13.4.5 Wirebonding [65]

The majority of power die, once attached, must be electrically connected on the top side. In most cases, this connection is performed by wirebonding. Typical power connections are made from a set of parallel-bonded 5-, 10-, or 15-mil-diameter aluminum wire. The wirebonding concerns for power packages are:

- Current-carrying capacity
- Compatibility of the wire and pad
- Integrity of the wire with respect to fatigue
- Interdiffusion, which decreases bond strength, forms voids, and increases contact resistances
- Brittle intermetallic formation, which decreases bond strength

13.4 Power Packaging Design Methodology

Table 13.7 Flux and Solder Compatibilities

Base metal	Recommended indalloy flux	Recommended solder Indalloy number	Recommended solder Alloy	Incompatible solders
Gold (Au)[a]	#5R, #5RMA	4	100%In	Sn
#5R, #5RMA,		2	80In 15Pb 5Ag	Sn/Pb
silver (Ag)[b],		7	50In 50Pb	In/Sn
palladium (Pd),		10	75Pb 25In	Sn/Pb/In
platinum (Pt)		11	95 Pb 5In	Sn/Pb/Bi
		150	81Pb 19In	
		204	70In 30Pb	
		205	60In 40Pb	
		206	60Pb 40In	
		164	92.5Pb 5In 2.5Ag	
		182	80Au 20Sn	
		183	88Au 12Ge	
Copper (Cu),	#5R, #5RMA	42	46Bi 34Sn 20Pb	In
copper alloys		281	58Bi 42Sn	In/Pb
(brass, bronze)		104	62.5Sn 36.1Pb 1.4Ag	In/Sn
		105	63Sn 37Pb	In/Pb/Ag
		121	96.5 3.5Ag	
		133	95Sn 5Sb	
Tin (Sn),	#5R, #5RMA	1E	52 In 48Sn	c
solder plate (Sn/Pb)		106	63Sn 37Pb	
		121	96.5Sn 3.5Ag	
		281	58Bi 42Sn	
Nickel (Ni)	#4, #5RA	106	63Sn 37Pb	
		1E	52In 48Sn	
		204	70In 30Pb	Compatible with most solders
		205	60In 40Pb	
		206	60Pb 40In	
Aluminum (Al)	#3	201	91Sn 9Zn	Sn/Pb has poor corrosion resistance
		176	95Zn 5Al	
Stainless steel	#2	1E	52In 48Sn	Avoid Pb and Cd for food applications
		106	63Sn 37Pb	
		121	96.5Sn 3.5Ag	
Steel	#1	106	63Sn 37Pb	Compatible with most solders
		121	96.5Sn 3.5Ag	

[a] Indium-containing solder is good for operating temperatures below 125°C. For applications above 125°C, choose #182 (80Au 20Sn) or #183 (88Au 12Ge).

[b] When soldering to silver (Ag), it is recommended that the solder also contain some Ag. Such as #121 (96.5Sn 3.5Ag), #104 (62.5Sn 36.1Pb 1.4Ag), #151 (92.5Pb 5Sn 2.5 Ag).

[c] Avoid solders that contain indium when soldering to Sn or Sn/Pb. It is possible for localized pockets of the eutectic to form which melt at 118°C.

Table 13.8 Bonding Wire Constants for Several Metals

Metal	Wire length < 0.04 in. (0.10 cm)	Wire length > 0.04" (0.10 cm)
Aluminum	22,000	15,200
Gold	30,000	20,500
Copper	30,000	20,500
Silver	15,000	10,500
Other	9,000	6,300

Current-carrying capacity is perhaps the most important parameter in wirebond selection. The U.S. military specification standard MIL-M-38510 (Section 3.5.5.3) identifies the maximum allowable continuous current through a wirebond according to the following formula

$$I = \kappa d^{3/2} \tag{13.12}$$

where

I = maximum current (A)

d = wire diameter (in.)

κ = constant based on wire material and length, taken from Table 13.8

Example 13.9

Your power electronics team is building a half-bridge MCPM utilizing International Rectifier power MOSFETs. The engineering designs call for each power die source pad to be wirebonded with three 5-mil-diameter Al wires. You know that the packaged equivalent component of what you are using is the IRF740. Your manager has asked you to determine if the wirebonds called for in the engineering designs are proper specs.

You look up the IRF740 power MOSFET at International Rectifier's website and determine that the maximum current an IRF740 can conduct is 10 A. You cross reference the part and determine that the bare die equivalent component is the IRFC340. You then request a bare die data sheet for the IRFC340 to get the die layout, which is given in Figure 13.27. The die outline is International Rectifier's HEX-4 D22. Assuming that each bond connection is approximately twice the diameter of the wire, you graphically lay out the three bonds. With each wire at 5 mils diameter and each bond at 10 mils diameter, the total width of the bond pad has to be at least 30 mils in order to accommodate all three 5-mil wires. The source pad is actually 49 mils in width, leaving 19 mils of extra space, so this meets specification.

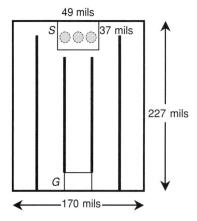

Figure 13.27 IRFC Bare Die Layout.

Equation (13.12) can be used to determine the maximum current for each wirebond. Noticing that the length of the source pad is 37 mils, it is safe to assume that the wirebond length will be greater than 40 mils. The value of the constant from Table 13.8 for an Al wire greater than 40 mils is 15,200. Thus

$$I = \kappa d^{3/2} = 5.37 \, \text{A}$$

Each wire can carry 5.37 A, so the three wires in parallel can carry a total of 16.11 A, which exceeds the maximum operation of 10 A specified by the device. Thus, the current-carrying capability meets the specifications. The wirebond design is sound and so you recommend to your manager to proceed as planned.

13.4.6 Thermal Design

Thermal analysis and heat sink selection for commercial packages were discussed in detail previously in this chapter. Custom-engineered MCPMs require a slightly different approach to thermal design, however. Thermal conductivity has been reviewed and so the discussion will pick up with thermal spreading [66–68].

A standard simple rule of thumb is that heat spreads through materials at a 45° angle, but with MCPM design, there is a requirement to perform a more detailed analysis. When materials are stacked together, such as the case with power modules, the degree of spreading changes as the thermal energy transfers from one material to the next. When two or more materials are stacked, the angle of thermal spreading and the surface length of the thermal effect (one-dimensional models) are calculated as follows [69]:

$$\alpha_a = \tan^{-1}\left(\frac{k_a}{k_b}\right) \tag{13.13}$$

$$L_2 = 2t_a \tan(\alpha_a) + L_1 \tag{13.14}$$

where

α_a = angle of thermal spreading through material a (°)
k_a = thermal conductivity of material a (W/m-°C)
k_b = thermal conductivity of material b (W/m-°C)
L_1 = length of thermal effect at interface 1 (m)
L_2 = length of thermal effect at interface 2 (m)

These parameters are illustrated in Figure 13.28.

Example 13.10

Your power electronics team has decided to build the half-bridge MCPM described in Example 13.9 with a modification to the structure. The original plan called for the power MOSFETs to be mounted

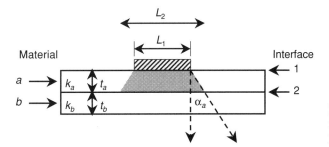

Figure 13.28 Thermal Spreading Through Stacked Materials.

574 Chapter 13 Power Electronics Packaging

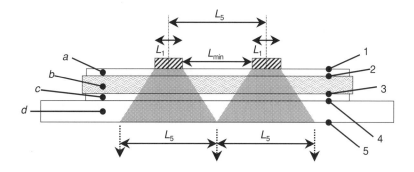

Heat sink attachment

Figure 13.29 Cross section of layout for example 13.10.

onto an Al_2O_3 substrate-based DBC mounted to a copper heat spreader. However, thermal performance was found to be lacking. So, a decision was made to go with an AlN substrate-based DBC. In an effort to improve reliability, the team has decided to replace the copper heat spreader with an AlSiC (63% SiC) heat spreader. The power MOSFETs are to be mounted side by side, as close together as possible. Your manager has asked you to estimate the minimum distance required between the two devices in order to obtain maximum thermal performance and miniaturization. The thicknesses of the copper metallization layers on the DBC are 0.30 mm on each side, the AlN substrate is 0.65 mm thick, and the heat spreader is 6.2 mm thick. Assume the effects of solder to be negligible.

You sketch the cross section of your layout, as illustrated in Figure 13.29, labeling the layer materials and interfaces. The minimum distance required between the two devices is found by determining the length of thermal spreading at the base of the heat spreader. Thermal spreading from the devices should not overlap if maximum thermal performance is desired. The parameter L_{min} is simply the final length of spreading minus the original length (which is the length of expansion throughout the package). In order to find the final length L_5, the angle of spreading through each material and the length at each interface must be determined.

The original length, L_1, is the width of the device (or length of the device). The width of the IRFC340 is 170 mils, or 4.318×10^{-3} m. The thermal conductivity of copper is $k_a = k_c = 393$ W/m-K. The thermal conductivity of AlN is $k_b = 170$ W/m-K, and for AlSiC (63% SiC) $k_d = 170$ W/m-K. The material thicknesses are $t_a = t_c = 0.30 \times 10^{-3}$ m, $t_b = 0.65 \times 10^{-3}$ m, and $t_d = 6.2 \times 10^{-3}$ m.

The angle of spreading through each material can be determined using Eq. (13.13):

$$\alpha_a = \tan^{-1}\left(\frac{k_a}{k_b}\right) = 66.6°$$

$$\alpha_b = \tan^{-1}\left(\frac{k_b}{k_c}\right) = 23.4°$$

$$\alpha_c = \tan^{-1}\left(\frac{k_c}{k_d}\right) = 66.6°$$

In order to determine the effect of thermal spreading through the heat spreader, the property of the material it is to be attached to must be known. In this case, we will assume an aluminum heat sink, which has a thermal conductivity $k_e = 240$ W/m-K, which gives

$$\alpha_d = \tan^{-1}\left(\frac{k_d}{k_e}\right) = 35.3°$$

The length of spreading at each interface can be determined by applying Eq. (13.14):

$$L_2 = 2t_a \tan(\alpha_a) + L_1 = 5.705 \times 10^{-3} \text{ m}$$
$$L_3 = 2t_b \tan(\alpha_b) + L_2 = 6.268 \times 10^{-3} \text{ m}$$
$$L_4 = 2t_c \tan(\alpha_c) + L_3 = 7.655 \times 10^{-3} \text{ m}$$
$$L_5 = 2t_d \tan(\alpha_d) + L_4 = 16.435 \times 10^{-3} \text{ m}$$

Thus, the minimum distance (one-dimensional analysis) between devices in order to have full thermal spreading is $L_5 - L_1$, which is approximately 12 mm.

In a more detailed analysis, solder interfaces would have a significant impact. Thermal energy would transfer laterally through the module in a two-dimensional analysis. In a true case, the heat would be transferred across the surface of the DBC as well, since it is convectively removed from the package by air. This effect would be reduced or altogether removed if the power module were to be encapsulated. Energy would also transfer laterally through the package and through the edges of the module. The procedure of this example makes the assumption that all heat transfer occurs through the heat sink beneath the devices *in a steady-state condition*. In other words, during startup, some heat would flow throughout the entire module (thus spreading outside the area indicated in Fig. 13.29) until some steady-state condition is reached; after that, all energy would flow directly to the heat sink within the prescribed spreading area.

Example 13.11

The MCPM developed by your team in Example 13.10 is almost ready for marketing. The product data sheets are being written, and your manager learns that the junction-to-case thermal resistance for the MCPM is unknown. You are assigned to help solve this problem by finding the resistance associated with the packaging *only*. You are asked to find the thermal resistance for a *single path* from a *single device* (not including the device) to the base of the heat spreader. Assume the solder is negligible. In order to simplify the problem, assume the power devices are square (i.e., assume length = width = 4.3 mm).

To solve this problem, the thermal resistance for each layer must be determined using Eq. (13.7). The difficulty here, however, is that the cross-sectional area under effect is not constant for each layer because of the thermal spreading effect. Thus, the average cross-sectional area should be used. The average area can be found by finding the average length of effect in a layer:

$$L_{a\text{AVG}} = \frac{L_2 - L_1}{2} + L_1 \tag{13.15}$$

$$A_{a\text{AVG}} = L_{a\text{AVG}}^2 \tag{13.16}$$

For this example, then, the following values are found:

$L_{a\text{AVG}} = 5 \times 10^{-3}$ m $\quad L_{b\text{AVG}} = 6 \times 10^{-3}$ m $\quad L_{c\text{AVG}} = 7.0 \times 10^{-3}$ m $\quad L_{d\text{AVG}} = 12.1 \times 10^{-3}$ m
$A_{a\text{AVG}} = 2.5 \times 10^{-5}$ m² $\quad A_{b\text{AVG}} = 3.6 \times 10^{-5}$ m² $\quad A_{c\text{AVG}} = 4.9 \times 10^{-5}$ m² $\quad A_{d\text{AVG}} = 1.46 \times 10^{-4}$ m²

The thermal resistances can be found for each layer as follows:

$$R_{\theta a} = \frac{t_a}{k_a A_{a\text{AVG}}} = 0.03°\text{C/W}$$

$$R_{\theta b} = \frac{t_b}{k_b A_{b\text{AVG}}} = 0.11°\text{C/W}$$

$$R_{\theta c} = \frac{t_c}{k_c A_{c\text{AVG}}} = 0.18°\text{C/W}$$

$$R_{\theta d} = \frac{t_d}{k_d A_{d\text{AVG}}} = 0.25°\text{C/W}$$

$$R_{\theta\text{Path}} = R_{\theta a} + R_{\theta b} + R_{\theta c} + R_{\theta d} = 0.39°\text{C/W}$$

The total thermal resistance of a single path from one device through the package is found to be 0.39°C/W.

There is an assortment of advanced cooling strategies for high-performance power electronic modules. These include liquid-cooled heat sinks, liquid cooling of the module baseplate, liquid spray cooling (essentially a phase change is achieved by "misting" power devices with a liquid coolant that evaporates upon contact), and the use of heat pipes and

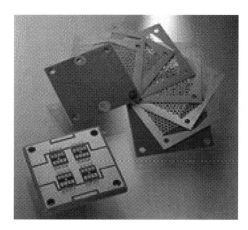

Figure 13.30 Curamik power substrate with integrated channels for liquid cooling.

radiators. Advanced materials and structures can be used, such as diamond and DLC substrates, or liquid-carrying microchannels that snake through a power substrate or baseplate. Figure 13.30 is a photograph of an integrated microchannel DBC substrate by Curamik.

13.4.7 Electromagnetic Interference (EMI) and Electromagnetic Compliance (EMC)

Power electronics engineers need to be aware of EMI issues for two reasons. First, EMI can cause interference with the pulse width modulation (PWM) or microcontroller chip that controls the operation of the power circuit, thus disrupting proper operation. Second, the majority of power electronics circuits contain significant electromagnetic transformers and inductors that can *put out* unwanted EMI and thus disrupt other electronics. For this second reason, commercial power electronics must comply with federal regulation standards that limit the risk of interference. The standard way of achieving this is by providing for some sort of shielding at one level or another within the power electronics package.

13.4.8 High-Temperature Power Electronics

Silicon-based power electronics are nearing their power density limits. The main reasons for this are the physical limitations associated with removing the heat from the power devices. If a limit is reached with regard to the heat flux from the power device, then the only way to increase power density is to increase the temperature limit of operation. This can be achieved through the use of SiC power devices.

There is a current heavy trend in power electronics research and development to pursue SiC devices in an attempt to improve power density. The theoretical limit of SiC operation exceeds 600°C, and so, once these devices are commercially available, it will be essential to develop high-temperature packages that can take advantage of these possibilities. There are numerous key technological issues and hurdles that must be faced:

1. Reliable high-temperature die-attach methods must be developed. Standard high-temperature solders rarely exceed 300°C.

2. As operational temperatures exceed 300°C, thermal stress issues will become a prime factor in determining long-term reliability. Mechanical and electrical integrity will have to be improved.

3. Oxidation and corrosion accelerates at high temperatures. This will have a significant impact on wirebond, solder joint, and interconnect reliability.

4. There are few high-temperature substrates. For temperatures in excess of 300°C, ceramics are the primary substrate choices. Curamik's DBC is rated to operate reliably up to 300°C.

5. Supporting control electronics will have to remain as silicon, thus severely limiting the usefulness of high-temperature operation, specifically with control integrated MCPM designs. Passives (following) are also a challenge.

6. There are few capacitors that can operate at high temperatures, and those that can (such as thick-film ceramics) are greatly limited in energy density. The majority of power electronics applications require high-energy electrolytic capacitors, which are typically rated to only 105°C.

7. High-temperature magnetic core choices for electromagnetic transformers and inductors are greatly limited at higher temperatures. The majority of standard core materials see a significant flux density drop at temperatures in excess of 200°C.

There are some organizations currently investigating high-temperature electronics and packaging. Honeywell has significant resources invested in its high-temperature MOS (HTMOS) silicon-on-insulator (SOI) technology. It offers a small assortment of high-temperature devices rated for operation up to 300°C, including op-amps, power MOSFETs, microcontrollers, and memory. Sandia National Labs has performed significant research on the development of high-temperature electronics and often hosts high-temperature electronics conferences.

13.5 SUMMARY

This chapter has provided a generalized overview of the packaging concepts an engineer needs to be aware of when designing for power electronics systems. Current and future state-of-the-art power-switching technologies were discussed. Methodologies for determining power switch losses were presented as well as thermal management strategies to dissipate these losses. The student was asked to design for power package selection (if using commercially available packages) and heat sink selection. Engineering methods for designing custom power electronics packages were also discussed, including designing for layout, thermal dissipation, current conduction, power interconnects, materials choices, and reliability.

REFERENCES

1. M. HARRIS, An Overview of the DoD Needs for Power Electronics and Power Electronics Packaging, *Adv. Microelectr*, Vol. 24, No. 1, pp.17–19, Jan./Feb. 1997.

2. JOHNSON, WRIGHT, et al., Recent Progress and Current Issues in SiC Semiconductor Devices for power Applications, *IEEE Proc. Circuits Devices Syst.*, Vol. 148, No. 2, pp. 101–108, April 2001.

3. MOHAN, UNDELAND, AND ROBBINS, *Power Electronics*, Hoboken, NJ: Wiley, 2003.

4. M. BROWN, *Power Supply Cookbook*, Newton, MA: Butterworth Heinemann, 1994, Chapter 3, pp. 25–94.

5. J. KASSAKIAN, M. SCHLECHT, AND G. VERGHESE, *Principles of Power Electronics*, Reading, MA: Addison Wesley, 1992, Part I pp. 9–139.

6. BOSE, *Modern Power Electronics and AC Drives*, Upper Saddle River, NJ: Prentice Hall, 2002.

7. BIRMAN, *Power Supply Handbook*, Flushing, NY: Kepco, 1965.

8. PRESSMAN, *Switching Power Supply Design* 2nd ed, New York: McGraw-Hill, 1998.
9. S. SZE, *Physics of Semiconductor Devices*, New York: Wiley, 1981.
10. EDWARDS-SHEA, *The Essence of Solid-State Electronics*, New York: Prentice Hall, 1996.
11. J. BALIGA, *Modern Power Devices*, Malabar, FL: Krieger, 1992.
12. J. BALIGA, *Power Semiconductor Devices*, Boston, MA: PWS Publishing Company, 1996.
13. M. HARRIS, An Overview of the DoD Needs for Power Electronics and Power Electronics Packaging, *Adv. Microelectro.*, Vol. 24, No. 1, pp.17–19, Jan./Feb. 1997.
14. KIRSCHMAN, *High-Temperature Electronics*, New York: IEEE Press, 1999.
15. JOHNSON, WRIGHT, et al., Recent Progress and Current Issues in SiC Semiconductor Devices for Power Applications, *IEEE Proc. Circuits Devices Syst.*, Vol. 148, No. 2, pp. 101–108, April 2001.
16. CHOYKE, MATSUNAMI, PENSL, *Silicon Carbide: A Review of Fundamental Questions and Applications to Current Device Technology, Volumes I and II*, Pittsburgh, PA: Akadamie, 1997.
17. G. HARRIS, *Properties of Silicon Carbide*, Washington DC: Howard University & INSPEC, 1995.
18. MITLEHNER, BARTSCH, BRUCKMANN, DOHNKE, WEINERT, The Potential of Fast High Voltage SiC Diodes, *IEEE Inspec.*, pp. 165–168, 1997.
19. SINGH, COOPER, et al., SiC Power Schottky and PiN Diodes, *IEEE Trans. Electron Devices*, Vol. 49, No. 4, pp. 665–672, April 2002.
20. SINGH AND KUMAR, A New 4H-SiC Lateral Merged Double Schottky (LMDS) Rectifier with Excellent Forward and Reverse Characteristics, *IEEE Trans. Electron Devices*, Vol. 48, No. 12, pp. 2695–2700, Dec. 2001.
21. WANG AND WILLIAMS, Evaluation of High-Voltage 4H-SiC Switching Devices, *IEEE Trans. on Electron Devices*, Vol. 46, No. 3, pp. 589–597, March 1999.
22. SINGH, RYU, et al., 1500 V, 4 Amp 4H-SiC JBS Diodes, ISPSD 2000, Toulouse, France, pp. 101–105, May 22–25, 2000.
23. J. HORNBERGER, A. B. LOSTETTER, K. J. OLEJNICZAK, S. MAGAN LAL, AND A. MANTOOTH, A Novel Three Phase Motor Drive Utilizing Silicon on Insulator (SOI) and Silicon-Carbide (SiC) Electronics for Extreme Environment Operation in the Army Future Combat Systems (FCS), 37th International Symposium on Microelectronics (IMAPS 2004), Long Beach, CA, November 2004.
24. A. LOSTETTER, J. HORNBERGER, S. MAGAN LAL, K. OLEJNICZAK, A. MANTOOTH, AND A. ELSHABINI, Development of Silicon-Carbide (SiC) Static-Induction-Transistor (SIT) Based Half-Bridge Power Converters, 36th International Symposium on Microelectronics (IMAPS 2003), Boston, MA, October 2003.
25. A. LOSTETTER, T. MCNUTT, J. HORNBERGER, S. MAGAN LAL, AND A. MANTOOTH, The Application of Silicon-Carbide (SiC) Semiconductor Power Electronics to Extreme High-Temperature Extraterrestrial Environments, Proceedings of the 2004 IEEE Aerospace Conference, MT, March 2004.
26. A. LOSTETTER, K. OLEJNICZAK, AND A. ELSHABINI, Silicon Carbide Power Die Packaging in Diamond Substrate Multichip Power Module Applications, 34th International Symposium on Microelectronics (IMAPS 2001), Baltimore, MD.
27. TRIVEDI, SHENAI, AND NUEDECK, High-Speed Switching Performance and Buck Converter Operation of 4H-SiC Diodes, *IEEE Inspec.*, pp. 69–78, 2000.
28. T. FUNAKI, J. C. BALDA, J. JUNGHANS, A. S. KASHYAP, F. D. BARLOW, H. A. MANTOOTH, T. KIMOTO, AND T. HIKIHARA, SiC JFET DC Characteristics Under Extremely High Ambient Temperatures, *IEICE Electron. Express.*, Vol. 1, No. 17, pp. 523–527, Dec. 10, 2004.
29. University of Arkansas, Department of Electrical Engineering website, http://www.eleg.uark.edu/.
30. SERGENT AND HARPER, *Hybrid Microelectronics Handbook*, 2nd ed., New York: McGraw-Hill, 1995.
31. ELSHABINI AND BARLOW, *Thin Film Technology Handbook*, New York: McGraw-Hill, 1998.
32. Kyocera's Electronic Packaging website, http://global.kyocera.com/prdct/semicon/.
33. Motorola, *Power MOSFET Transistor Device Data*, Motorola Inc., 1996.
34. International Rectifier website, http://www.irf.com.
35. International Rectifier, *Hexfet Designer's Manual*, El Seguedo, CA: International Rectifier, 2000.
36. International Rectifier, *IGBT Designer's Manual*, El Seguedo, CA: International Rectifier, 2000.
37. Toshiba, *Direct Bond Copper (DBCu) AlN Substrates*, Toshiba Product Data.
38. Curamik website, http://www.curamik.com/.
39. J. SNOOK AND R. VISSER, Direct Bond Copper (DBCu) Technologies, *IEEE Industries Applications Conference*, 1995.
40. Transene Company Inc., *Watts Nickel Plating Solution*, Transene Company Inc. Product Data.
41. Transene Company Inc., *Bright Electroless Gold*, Transene Company Inc. Product Data.
42. Brushwellman, *Beryllium Metal-Matrix Composite*, Brushwellman Data Sheet M08.
43. D. JECH AND J. SEPULVEDA, Advanced Copper-Refractory Metal Matrix Composites for Packaging Heat Sink Applications, International Symposium on Microelectronics, pp. 90–96, 1997.
44. U. KASHALIKAR, P. KARANDIKAR, L. FELTON, AND K. JAYARAJ, *Metal Matrix Composites for Thermal Management*, Foster-Miller Data Sheets.
45. M. NAGAI, Y. YAMAMOTO, K. TANABE, Y. KUMAZAWA, AND N. FUJIMORI, Composite Diamond Heat Spreader, International Symposium on Microelectronics, pp. 302–306, 1997.
46. E. PORTER, S. ANG, K. BURGERS, M. GLOVER, AND K. OLEJNICZAK, Miniaturizing Power Electronics Using Multichip Module Technology, *Intl. J. Microcircuits Electro. Packaging*, Vol. 20, No.3, 3rd Qt., pp. 397-401, 1997.

47. C. Cotton and D. Kling, Design and Development Challenges for Complex Laminate Multichip Modules, *Intl. J. Microcircuits Electr. Packaging*, Vol. 20, No. 3, 3rd Qt., pp. 339-344.
48. E. Porter, S. Ang, K. Burgers, M. Glover, and K. Olejniczak, Power Electronics Combines with MCM Technology to Create Multichip Power Modules, International Symposium on Microelectronics, pp. 78–84, 1997.
49. A. Lostetter, K. Olejniczak, W. Brown, and A. Elshabini, The Utilization of Diamond and Diamond-Like Carbon Substrates for High-Performance Power Electronic Packaging Applications, 2001 European Power Electronics Conference, Graz, Austria.
50. A. Lostetter, R. Hoagland, J. Webster, F. Barlow, and A. Elshabini, Integrated Power Modules (IPMs), A Novel MCM Approach to High Power Electronics Design and Packaging, *Intl. J. Microcircuits Electr. Packaging*, Vol. 21, No. 3, pp. 274–278, 3rd Q., 1998.
51. APEI, Inc. website, http://www.apei.net.
52. Powerex Semiconductors website, http://www.pwrx.com/.
53. Boley and Weiner, *Theory of Thermal Stresses*, Mineola, NY: Dover, 1960.
54. Gere, *Mechanics of Materials*, 5th ed., Pacific Grove, CA: Brooks/Cole, 2001.
55. Burgreen, *Elements of Thermal Stress Analysis*, Jamaica, NY: C. P. Press, 1971.
56. F. Incropera and D. Witt, *Introduction to Heat Transfer*, 2nd ed., Wiley, 1990, Appendix A, pp. A1–A30.
57. T. Kitahara and K. Nakai, Thermal Analysis for High-Performance Small Devices, *Intl. J. Microcircuits Electr. Packaging*, Vol. 20, No. 3, 3rd Qt., pp. 274–281.
58. D. Hopkins, J. Pitarresi, and J. Karker, Thermal Impedance and Stress in a Power Package Due to Variations in Layer Thickness, International Symposium on Microelectronics, pp. 72–77, 1997.
59. Wakefield Thermal Solutions website, http://www.wakefield.com/.
60. IXYS Corporation website, http://www.ixys.com/.
61. M. Gordon, B. Chandron, and R. Ulrich, Thermal-Mechanical Issues for Diamond-Based MCMs, *Adv. Microelectr.*, Vol. 25, No. 1, Jan./Feb. 1998.
62. Indium Corporation of America, *Research Solder Kits*, Indium Corporation of America (Indalloy) Solder Data Sheets, Form No. 97504 R2.
63. Indium Corporation website, http://www.indium.com/.
64. Williams Advanced Materials website, http://www.williams-adv.com/.
65. Harman, *Wire Bonding in Microelectronics: Materials, Processes, Reliability, and Yield*, New York: McGraw-Hill, 1998.
66. N. Nguyen, Properly Implementing Thermal Spreading Will Cut Cost While Improving Device Reliability, International Symposium on Microelectronics, pp. 383–388, 1996.
67. Godbold, Sankaran, and Hudgins, Novel Designs in Power Modules, IEEE IAS Meeting, p. 911–915, 1995.
68. A. Lostetter, F. Barlow, and A. Elshabini, Thermal Evaluation and Comparison Study of Power Baseplate Materials, *Adv. Microelectr.*, Vol. 25, No. 1, pp. 25–27, Jan./Feb. 1998.
69. B. Sonuparlak and M. Lehigh, Design and Fabrication of Thermal Management Materials for High Performance Electronic Packages, International Symposium on Microelectronics, pp. 377–382, 1996.

EXERCISES

13.1. In Example 13.1, it was assumed a resistive load was used. Now make the assumption that you are switching an inductive load. The waveforms may look like Figure 13.31. Calculate the estimated transient turn-on power loss. Assume a period of 100 μs.

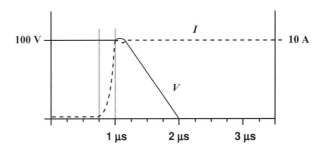

Figure 13.31

13.2. Several MMCs are listed in Table 13.5 as possible alternatives to copper and aluminum heat spreaders. Do some research into today's R&D market and try to find an even better alternative. Explain.

13.3. You are looking at designing a brick and have the choice of using an Al_2O_3 DBC, AlN DBC, or Be-BeO DBC. Discuss the advantages and disadvantages of each.

13.4. You are designing a power package for long-term reliability. Discuss the advantages and disadvantages between using Au or Al wirebonding.

13.5. International Rectifier specifies the junction-to-case thermal resistance as $R_{\theta jc} = 1.0°$ C/W for the IRF740. Verify this is correct through calculation.

13.6. Find the die outline of an International Rectifier HEX-5 D26. What are the length and width?

13.7. Wakefield has three heat sink designs specifically for "full brick" modules. These are the series 557, 558, and 559 heat sinks. If you wanted to maximize thermal performance, disregarding other design factors, which series would you select and why? Find the thermal resistance with an airflow of 300 ft/min. Find the power dissipation when a temperature differential of 40°C is across the heat sink.

13.8. The Wakefield Engineering series 510, 511, and 512 high-fin-density heat sinks are designed specifically for power module applications. Find the $R_{\theta sa}$ of a series 510-9M heat sink under forced-convection conditions of 100 ft^3/min.

13.9. You are building a DC–DC buck converter using a single IRF540. The transistor will be attached to a heat sink with a $R_{\theta sa} = 2.0°$C/W. You have calculated that the transistor will be dissipating 35 W. Should you use an isolation pad? Why or why not?

13.10. A DBC substrate has copper metallization on both the top and bottom sides of the ceramic. Why is this? Why isn't the metallization on one side only?

13.11. Calculate the maximum current an Al wirebond with 15 mils diameter and 200 mil length can carry.

13.12. You have been given the two-module designs shown in Figure 13.32 to analyze:

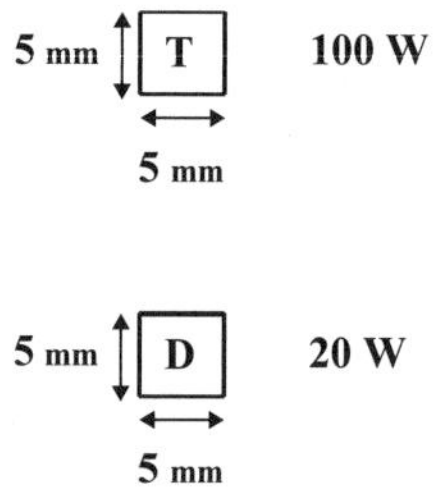

$$5\text{ mm} \updownarrow \boxed{\text{T}} \quad 100\text{ W}$$
$$\overset{\leftrightarrow}{5\text{ mm}}$$

$$5\text{ mm} \updownarrow \boxed{\text{D}} \quad 20\text{ W}$$
$$\overset{\leftrightarrow}{5\text{ mm}}$$

Figure 13.32

1. AlN DBC
   Cu = 10 mils
   AlN = 12 mils
   Cu = 10 mils

2. AlN DBC
   Cu = 12 mils
   AlN = 10 mils
   Cu = 10 mils

Which module design would you recommend? Explain through calculation. Assume a copper heat spreader.

13.13. Design a half brick that will house a half-bridge power stage capable of 600 V/10 A power conversion (Fig. 13.33). You may use any die you wish from any manufacturer. Include a layout design, materials and cross-sectional design, wirebond requirements, $R_{\theta jc}$, and heat spreader design.

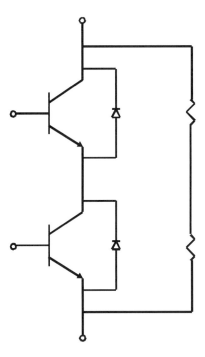

Figure 13.33

13.14. For your design of exercise 13.13. design a heat sink attachment that will maximize the module's operability. Assume 10 kHz switching frequency and a purely resistive load of $R_{L1} = R_{L2} = 90\,\Omega$. Find maximum T_j.

13.15. For your design of exercise 13.14, find the maximum switching frequency that keeps T_j below 150°C, assuming $T_{ambient}$ = room temperature. What thermal management modifications could you make to improve thermal performance? Investigate forced air convection or liquid cooling. With these methods, what is the maximum power you could dissipate from your bare die?

Chapter 14

Multichip and Three-Dimensional Packaging

JAMES LYKE

14.1 INTRODUCTION

Multichip packaging approaches seem like natural extensions of single-chip packaging, but they generally involve new concepts and technologies. This chapter provides an overview of the evolution of multichip packaging and identifies a number of basic two- and three-dimensional (2D, 3D) schemes. Both "classic" and modern multichip packaging approaches are reviewed.

The decision processes used to identify the need and type of multichip packaging approaches is then undertaken, and the various factors leading to the definition of a particular set of options are discussed. Multichip packaging is usually driven by some system-level constraints, and these constraints are discussed.

14.1.1 Brief History of Multichip Packaging

Establishing an effective means to combine a number of discrete electronic components within a single unifying package structure has been a perennial pursuit. Indeed, this pursuit predated and motivated the development of the first integrated circuits (ICs) by Kilby in 1958 [1]. Even before the invention of the transistor, cordwood assemblies and other innovative techniques were examined in development programs dating back, at least, to World War II. With the advent of the monolithic IC in 1961 (Kilby's earlier work required the components formed on a common substrate to be connected together through the addition of manually placed discrete wires), it might have briefly seemed that the need to worry about multicomponent integration diminished, but limitations in process yield and the diversity of components of the early IC processes motivated development of the first "chip-and-wire" hybrid microcircuits in the 1960s [2]. The need for these assemblies, perhaps the first "recognizable" multichip modules (MCMs), was largely driven by the needs of the aerospace industry. Military/aerospace demand for these hybrids arose from the performance requirements of dimensionally constrained platforms, such as missiles. The emergence of military

---

Advanced Electronic Packaging, Second Edition, Edited by Richard K. Ulrich and William D. Brown
Copyright © 2006 the Institute of Electrical and Electronics Engineers, Inc.

standards to specify their methods of construction and qualification reflected a reduction to practice.

Over the next two decades, hybrid microcircuits continued to evolve and mature to become the dominant class of MCM technology. These hybrids were based on ceramic thick- or thin-film substrates. Originally driven by aerospace applications, the more modern hybrids became commonly used across a wider range of commercial applications (including consumer products), especially high-power amplifiers and converters. In parallel, in the 1980s, computer mainframe developers mobilized the development of hybrids of a much more elaborate form. In this case, performance (not miniaturization) became a driving factor in development. Despite the Moore's law trend of doubling transistor counts every 18 to 24 months, some researchers sought to magnify performance gains by combining many ICs within a single package. One of the more well-known examples is IBM's thermal conduction module (TCM), which combined up to 100 ICs per module and employed about nine such modules per board [3].

In the mid/late 1980s, several abortive, high-profile attempts were undertaken to revive the concept of monolithic wafer-scale integration (WSI) [4]. At the same time, new advancements in materials, processes, and computer-automated design began to converge to form a technology base for improved MCM approaches. In addition to more aggressive forms of the earlier ceramic hybrid microcircuits, which were eventually given the industry nomenclature MCM-C, thin-film substrate approaches gave rise to a fundamentally new type of MCM. Denser and theoretically greater in performance (as measured in signal time-of-flight and parasitic latency reduction), these MCM-D (D standing for "deposited") technologies promised to revolutionize the packaging of more complex systems. Even the laminate printed wiring board (PWB) technology, staple of the throwaway digital wristwatch and four-function calculator, evolved into a powerful chip-on-board (COB) MCM technology, also referred to as MCM-L. This simple classification is summarized in Table 14.1.

The 1990s seemed to mark the "coming of age" of the MCM, as evidenced by the introduction of dozens of ceramic, thin-film, and laminate MCM substrate technologies. The pace of development was further fueled by large-scale initiatives in the government from the Defense Advanced Research Projects Agency (DARPA) and the Department of Commerce (e.g., the NIST "dual-use" Advanced Technology Program [5]). A number of significant but less visible development programs were undertaken by other government/military organizations, ranging from logistics centers to aircraft avionics upgrade programs. The developmental boom led to proclamations of imminent demand and commercial success. Unfortunately, some of the predictions were based on a set of technically flawed premises. One of these was the claim that MCM methods were necessary in systems operating at elevated clock frequencies. Since clock frequencies increased in a predictable way due to Moore's law, some notional targets were proposed as the point in the future when MCMs might be necessary in computers and other complex systems. By the mid-1990s (when

Table 14.1 Industry Categorization of MCMs by Substrate Type

Substrate technology	Dielectric
MCM-C	Alumina (Al_2O_3), AlN
MCM-D	Thin-film polymer or SiO_2
MCM-L	PWB materials (e.g., FR-4, BT resin)

these targets were converged upon in typical ICs), the overselling of MCM approaches led to complacency and criticism. Not only were MCMs not required, but they added expense rather than value in commodity high-performance applications. Nevertheless, the frenetic development of MCM approaches and prototype demonstrations continued. During the latter half of the 1990s, a slowing occurred as the diversity and number of MCM suppliers dwindled. As the out-year funding of previously justified government programs tapered off, a more-or-less downward spiral resulted in a shakeout in the industry. Failing to consider the market realities and increased competition from emerging ball grid array (BGA) and chip-scale package (CSP) approaches, the *real* MCM market was apparently significantly less than originally expected. Only the most competitive schemes in companies survived a downturn in demand for the new MCMs that continued through the late 1990s.

Far from disappearing, however, the modern MCM has undergone an interesting transformation. In consumer appliances, fine-line PWBs, CSPs, 3D chip stacks coexist in low-cost cellular telephones. Hybrid microcircuits continue to flourish. Even the once celebrated thin-film MCM-D approaches that suffered a sharp decline continue to exist and even thrive in niche applications. What has not occurred is the widespread demand for vast quantities of large-format MCM packages studded with dozens of complex ICs. Today's MCMs are present but understated, often similar in size and shape to other single-chip packages used to populate a PWB.

14.1.2 Motivations for Multichip Packaging

There are basically two reasons for multichip packaging approaches: dimensional constraints and performance. Multichip packaging often allows ICs and other discrete components to be placed in much closer proximity, reducing volume and mass. The closer proximity can lead to performance benefits, since time of flight between components is reduced, and the parasitic capacitance and inductance associated with the interconnection manifold are decreased.

These benefits are depicted generically in the simplistic diagram shown in Figure 14.1. A decade ago, when most conventional single-chip packaging was based on dual-in-line, quadruple flat packages, or pin grid array (PGA) packages, the analogy represented a sharp delineation of the benefits of single- and multichip packaging. Figure 14.2 illustrates a typical example from the mid-1990s. The left photograph in Figure 14.2 represents the physical package body of a Texas Instruments TMS320C30 digital signal processor (DSP) packaged

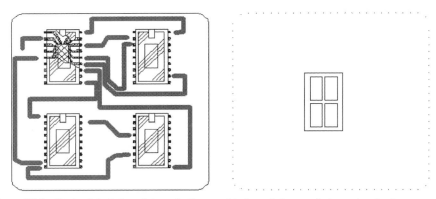

Figure 14.1 Notional depiction of size reduction possible through the use of advanced packaging.

586 Chapter 14 Multichip and Three-Dimensional Packaging

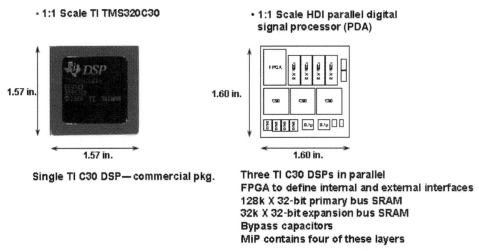

Figure 14.2 Real-world example of size reduction through the use of a MCM approach.

as a single die within a PGA package. Using an aggressive form of MCM technology, three TMS320C30 die were copackaged into an MCM, along with a field-programmable gate array (FPGA) and two banks of memory and other discrete components. The floorplan of the resulting MCM is approximately the same size as the package of a single IC die in a PGA package [6]. The resulting compact MCM floorplan is shown in the right illustration of Figure 14.2. Such spatial compression ratios were representative in the thin-film MCM approaches developed in the mid-1990s. With the advent of more recent technologies, such as fine-line PWBs and chip-scale packaging, the delineation is not always as crisp as Figure 14.2 might lead one to believe.

14.1.2.1 Application Sectors

Aerospace systems have been the traditional drivers for high-performance, embedded systems with demanding requirements for miniature electronics. Space systems, for which each fielded pound of mass represents a significant cost investment, typically seek to minimize the volume displaced by electronics. Advanced packaging approaches permit a greater concentration of functionality without a corresponding mass penalty. Space systems represent a problematic application sector, since the total demand for electronics by volume is insignificant. Submunitions and other ordnance applications represent a higher demand, not only for miniaturized electronics, but also for ruggedized electronics, capable, in some cases, of withstanding more than $10,000\,g$ acceleration. Soldier systems represent another application class in which mass represents a compromise in the quantity of weapons and supplies that can be effectively carried in the battlefield. Other classes of military platforms, such as aircraft, ground vehicles, and naval vessels, represent a potential application base where reduced size and mass are desirable. In aircraft, the performance of complex subsystems (such as fire control radar) often represent a more significant driver for efficient packaging than the desire for mass reduction.

Whereas aerospace systems often benefit from both performance improvement and mass reduction, commercial systems that drive multichip packaging typically concentrate on only one of these two benefits. High-end test equipment, medical equipment, and high-performance computers resort to multichip packaging often to gain a performance improvement. In many cases, size and mass reductions, while desirable, are a secondary concern. On the other hand, consumer applications, which are intensively cost driven, place a premium on miniature devices (e.g., cellphones, camcorders, PDAs) that are functional and compact.

While applications tend to focus on reduced mass or increased performance, other benefits accrue from multichip packaging. At the same time, there are distinct disadvantages to multichip packaging approaches, which is why they are not ubiquitous approaches to packaging electronics systems.

14.1.2.2 Advantages

Besides the benefits of reduced size and mass, along with performance improvements due to the compression of distance between components, multichip packaging offers other potential advantages. In some cases, multichip packaging results in improved reliability due to the systematic reduction of interconnect outside the package. As shown in Figure 14.3, the number of intervening structures necessary to achieve system-level contacts are reduced when otherwise separate components are combined into a common package. Figure 14.3a demonstrates that nine structures are involved in connecting signals between two separate single-chip packages mounted onto a PWB. Employing one MCM approach (Figure 14.3b)

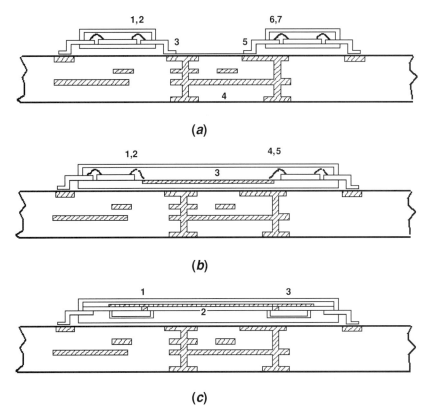

Figure 14.3 Chip-to-chip interconnections; (a) conventional single-chip packages; (b) typical "chip-and-wire" MCM-C; (c) patterned overlay MCM.

results in the elimination of two structures, whereas another MCM approach (Figure 14.3c) yields still further reduction (to three total interconnection structures).

Cost benefits are also possible in some cases. For military/aerospace systems, in which parts are often subjected to expensive qualification processes, multichip packaging can actually improve costs, since the ensemble of components is treated as a single assembly. In this manner, the cumulative cost of individual qualification tests is replaced by that of a single qualification test.

14.1.2.3 Disadvantages

Multichip packages have disadvantages that are, in some cases, significant enough to disqualify them from many applications. Although the amount of material is decreased in these assemblies compared to conventionally packaged versions, the cost is not necessarily reduced. As such, an MCM is often more expensive to create than the corresponding collection of singly packaged ICs mounted onto a PWB, displacing a larger board area. This cost increase is due to the use of denser but more exotic interconnection technologies, especially in the lower volume assemblies typically built in some of the more elaborate MCM processes.

Since components are not always individually tested before being committed to multichip assembly, the lower yield resulting from the "known good die" problem (discussed later) creates additional expense due to the need to repair or discard defective assemblies. Finally, the existence of another level in the packaging hierarchy (discussed next) can be an unwelcome complication and risk in a program development.

14.2 PACKAGING HIERARCHY AND TAXONOMY

In this section, the context and definition of particular multichip packaging approaches are more precisely defined. Multichip packaging represents a sort of modulation to the hierarchy of structures that combine the electronic components of a system together. The terminology of multichip packaging is outlined. Then, the different approaches are related in terms of the taxonomy, parts of which are of only academic interest. Included in the taxonomy are 3D approaches, which have represented some of the most creative but least successful advanced packaging developments of the last decade.

14.2.1 Hierarchy

Like a complex organism, which has a circulatory system, respiratory system, and so on, a complex system or platform, such as a spacecraft, airplane, or automobile, contains an electronic system. This electronic system combines together all of the discrete components through packaging and interconnection structures. This set of structures is hierarchical, with levels to this hierarchy as suggested in Figure 14.4.

14.2.2 Anatomy of an MCM

An MCM is usually defined as two or more nontrivial electronic components (usually ICs) that share a common interconnecting substrate and package. Figure 14.5 identifies a number of features that exist in a typical MCM. As shown later, not all of these features exist in

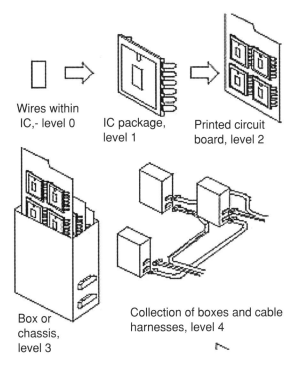

Figure 14.4 Packaging hierarchy.

all MCM configurations. The single most important defining characteristic of the MCM is the substrate. In Figure 14.5, the substrate is divided into two sections: the mechanical substrate and the interconnecting substrate. The mechanical substrate provides a physical support structure onto which the various components of an MCM are mounted. It is typically desirable that the mechanical substrate have a physical robustness consistent with the expected operating environment of a finished MCM assembly. It is usually desirable that the mechanical substrate have adequate thermal conductivity to promote efficiency heat transport from components to the package exterior. It is possible to have an interconnecting substrate that is distinct from the mechanical substrate, as is typically the case in thin-film MCMs. The role of the interconnecting substrate is to provide a supply of signal and power conductors sufficiently dense and available in the required locations to accomplish all routing connections between the terminals of mounted components.

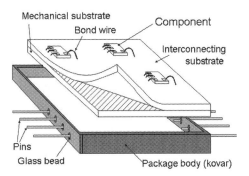

Figure 14.5 Features of a typical MCM.

590 Chapter 14 Multichip and Three-Dimensional Packaging

Components are mounted onto the interconnecting substrate using one of a number of element attachment approaches that are described later (wirebonds are shown as a representative example). Substrates can be mounted into a separate package structure, as shown in Figure 14.5. In many cases, the construction of the package is integral to the substrate, and a separate structure is not necessary. In this example, the package serves as a container for the MCM and supports the necessary "services": physical protection in transport of power, signal, and thermal energy. In this case, a number of pins are shown penetrating a Kovar package body. Since Kovar is electrically conductive, it is necessary to isolate power and signal pins using, for example, glass beads. In more modern constructions employing Kovar, ceramic inserts with patterned conductors for pin terminals are used, since glass beads do not typically support high contact densities (i.e., <0.050 in.). Not shown in Figure 14.5 is the package lid that is used to seal the assembly. Hermetic sealing is typically preferred, particularly in military/aerospace applications, and can be accomplished through the use of solder, seam welding, or laser welding.

Not all multichip packages follow the template shown in Figure 14.5. In the case where no substrate exists, the resulting assembly is sometimes referred to as a multicomponent package (MCP). Examples of the latter include physically separated ICs that are mounted onto electrically isolated lead frame paddles (in the case of plastic-encapsulated microcircuits) and simple chip stacks. An example of a four-chip MCP is shown in Figure 14.6. In

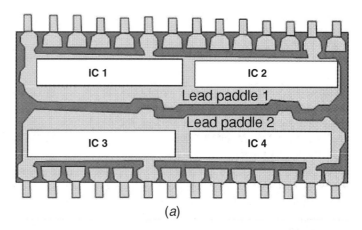

(a)

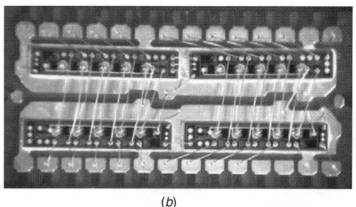

(b)

Figure 14.6 Example a MCP: (*a*) floorplan containing four identical IC die on two electrically isolated lead frame paddles; (*b*) photograph revealing intrapackage bonding scheme.

fact, since most 3D assemblies are die stacks that follow this method, most die stacks are classified as MCPs.

14.2.3 Planar MCM Approaches

Like PWBs and ICs, the interconnections of MCMs are generated using planar fabrication methods. The industry established a classification system based on the type of substrate materials used, and these designations were mentioned previously. To be more complete in defining the variety of MCM approaches, it is necessary to supplement the industry nomenclature with information about the arrangement of the substrate and the ways that components can be attached to them.

14.2.3.1 Substrates

MCMs that employ a ceramic substrate are referred to as MCM-Cs. Two broad classes of MCM-C technologies exist. The most mature of these is the high-temperature, cofired ceramic (HTCC) technologies, and a typical process flow is provided in Figure 14.7 [2]. Cofired substrates typically are based on a number (1 to 100) of single plies stacked together while still in a "green" state (unfired). Conductive patterns are screened onto each layer before firing using inks containing refractory materials capable of withstanding >1000°C temperatures. Via connections between the layers are formed by punching holes in the plies with a mechanical tool and filling the vias with conductive paste. The multilayer stackup is cofired, resulting in a complex structure that serves both as a mechanical and an electrical substrate. Components can be mounted directly onto the substrate. Often, the same structure can serve as a final package. In this case, components within the interior of the MCM are framed with a seal ring. Terminal pins and the seal ring are attached using a brazing process.

The most common material used in HTCC assemblies is alumina (Al_2O_3). Other materials include aluminum nitride (AlN) and berillium oxide (BeO). Aluminum nitride is particularly attractive due to its superior thermal conductivity and reasonably close match in thermal expansion coefficient to silicon. Figure 14.8 illustrates a large-scale, double-sided (components could be mounted on both sides) AlN substrate developed for a U.S. Air Force

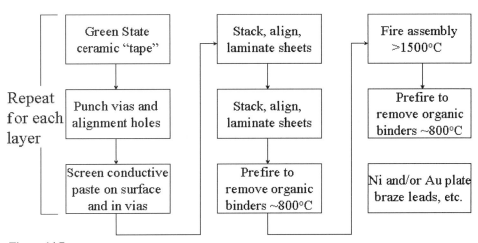

Figure 14.7 Typical HTCC process.

Figure 14.8 MCM-C module based on an aluminum nitride substrate (shown before component mounting).

application. This illustration demonstrates a number of typical features of MCM-C assemblies. Clearly visible around a number of component attachment sites (the photograph was taken before component assembly) is the seal ring and a large number of pins brazed to the substrate. The pins are temporarily secured using nonconductive tiebars around a perimeter in this integral package–substrate configuration.

A second, popular MCM-C technology is low-temperature, cofired ceramic (LTCC). LTCC approaches are attractive due to lower firing temperatures, which permit the use of nonrefractory materials, such as copper, which have a higher electrical conductivity. All ceramic approaches, which are considered midrange in cost (between laminate and thinfilm), have a common problem in that they shrink during the firing process. Due to this characteristic, it is necessary not only to compensate for shrinkage in design but also to carefully control many variables in the firing process, such as temperature, humidity, and even the loading factors of other furnaces used in firing substrates. In the past, practical problems were experienced in using cofired ceramic substrates for stacked applications, where a number of similar substrates with high pin counts were stacked. In these applications, shrinkage variability between individual substrates posed challenges in forming connections due to run-out errors accumulated across a row of contacts when comparing the edges of two different substrates. Fortunately, shrinkage control appears to be less of a problem in most applications today due to more careful attention to a wider range of process control variables and starting material quality.

The second class of planar MCMs, based on industry standard nomenclature, is referred to as the deposited, thin-film MCM or MCM-D. MCM-D represented an eclectic class of approaches, usually based on polymer dielectrics with copper metallization. However, a number of other approaches employed aluminum metal, and some used a very large scale integration (VLSI) arrangement of aluminum and SiO_2. Technically, these approaches, for the most part, are considered as having an interconnect substrate that is clad to a mechanical substrate, as suggested in Figure 14.5. The interconnecting substrate contains several layers of copper/polyimide (Cu/PI) interconnect structure connected together by interlayer vias.

The PI layers are typically spun on in multiple cycle buildups (to prevent pinhole defects) or they may be laminated one by one as each metal layer is added. The typical thickness of spun layers is usually 1 to 5 µm, whereas laminates are typically 20 40 µm. The laminates have the advantage of improved planarity and increased thickness. Thickness is considered an advantage when characteristic impedances above ~30 Ω are desired, along with lower resistance cross sections. The dielectric materials are commonly a PI (such as kapton). A number of researchers have explored alternative formulations, such as benzocyclobutenes (BCBs), which have lower permittivity and reduced loss tangent.

To improve adhesion and reduce potential reaction with dangling hydroxide bonds that might be present in the dielectric, barrier metals are almost always used with copper. Examples of barrier metals include nickel, titanium, and chromium. The copper layers are almost always electroplated to several micrometers thickness for improved ductility and reduced resistance. Most constructions employ staircase vias, although filled vias (which can be directly stacked) have also been explored.

Because MCM-D substrates employ dielectrics with reduced permittivity (dielectric constant) and reduced resistivity interconnections, they are considered to have higher performance. The first-order time constant as a measure of performance can be summarized as the product of resistance (R) and capacitance (C) per unit area:

$$\tau \cong RC = \frac{\rho}{A} \varepsilon C_0 \tag{14.1}$$

where ρ is resistivity (lower for copper than tungsten), ε is permittivity (lower in PI than ceramic), and C_0 represents capacitance in vacuum. The thicker, plated conductors, the superior conductivity of copper, and the lower permittivity of polymer dielectrics (e.g. ~3 vs. ~9) give MCM-Ds a strong performance advantage over most other MCM classes.

MCM-D approaches are based on photolithographic methods used in VLSI, so linewidths <10 µm are realizable, whereas, with other approaches, linewidths are typically >50 µm. Also, via densities are far greater in MCM-D technologies, which increases wireability and gives MCM-D a greater capacity to route signals. Consequently, two layers of a typical MCM-D technology can support the effective wiring capacity of many layers of MCM-C technology.

The General Electric (GE) high-density interconnect (HDI) process is an unusual representative of the MCM-D class. The typical process flow is illustrated in Figure 14.9. The process integrates ICs and other components face up into the recesses of a planarizing mechanical substrate (Figure 14.9a). The substrate composition can be ceramic, glass, metal, or plastic. The recesses are, in this case, preformed and must be within ~10 µm so that all components, after placement (die attach), create a nominally planar assembly (Figure 14.9b). The MCM-D interconnect system is created after this step. It is initiated by laminating a glue-clad Kapton® sheet (Figure 14.9c), that is cured, laser drilled to form vias, and then metallized. The metal (Ti–Cu–Ti) is then patterned and etched to form a

594 Chapter 14 Multichip and Three-Dimensional Packaging

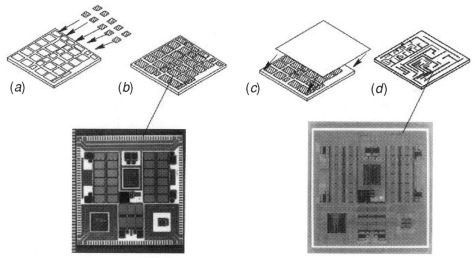

Figure 14.9 Simplified HDI process flow: (*a*) components introduced into recesses of mechanical substrate; (*b*) planar substrate with integrated components; (*c*) formation of patterned overlay; (*d*) completed HDI substrate.

patterned single layer. The lamination, drilling, metallization, patterning, and etch steps are repeated as required to create a finished, interconnected substrate (Figure 14.9*d*).

Formation of a patterned overlay system is complicated by many factors. One important factor is component alignment, which can change even after placement due to drift in the curing steps associated with die attach. To combat this problem, GE's HDI process employs a novel *adaptive lithography* system, which is shown in Figure 14.10. The GE HDI system, as

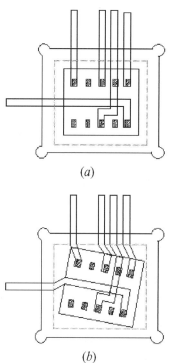

Figure 14.10 Adaptive lithography: (*a*) desired alignment and interconnect pattern; (*b*) misaligned die and corrected lithography generated at run time.

originally invented, forms all conductor patterns using direct-write lithography, as opposed to mask-based patterning. In adaptive lithography, some portions of the lithography near components are generated dynamically. In particular, "moat" regions are identified around each component in the floorplan of an MCM that exploits adaptive lithography. During module construction, the locations of each die within particular substrates are registered upon placement and curing of the die-attach adhesive. Using these fiducial cues (from the registration), translation/rotation misalignments can be compensated through algorithms that generate jogs in conductors within the adaptive moat regions. Direct-write lithography is necessary, in this case, to compensate for the misalignments, since corrected lithography is not deterministic.

A third class of MCM packaging approaches, based on substrate type, is defined by the industry as laminate or MCM-L. MCM-L approaches are based on PWB technologies in which components are directly attached to a substrate. This approach is also referred to as COB. MCM-L approaches were originally considered to be the simplest and lowest cost MCM category. Before the advent of microvia PWB technologies, it was impractical to consider a combination of very complex ICs and PWBs due to limited wiring capacity. Significant recent progress in microvia PWBs, however, has dramatically improved the interconnect density of board technologies, making MCM-L approaches more competitive than originally imagined, since they can implement more complex designs. The typical feature sizes of MCM-L technologies are approaching those of MCM-C technologies. Still, the most common use of microvia PWBs are as substrates for single-chip packages, especially BGA configurations.

In addition to the so-called MCM-C/D/L substrate classes, a number of hybridizations have been commonly practiced. For example, high-performance MCMs in supercomputers have often employed a combination of MCM-C and MCM-D. The tandem combination is powerful, since the higher capacitance MCM-C layers are useful for power rail holdup and provide structural support, while the MCM-D layers supply a more intensive level of interconnections for signal routing. At the present time, IBM holds the record for substrate complexity based on this construction approach. One module, designed for G4/G5 server applications, is reported to contain as many as 100 individual wiring layers [7]. Furthermore, MCM-D/L approaches, in which rigid mechanical substrates do not exist, define a class of foldable/flexible MCMs, sometimes referred to as *chip on flex* (COF). Even within a particular classification, different types of substrate approaches can be mixed, such as the combination of through-hole and microvia.

14.2.3.2 Substrate Configurations

The relationship of components to substrates in an MCM defines a substrate arrangement or substrate configuration. A number of potential substrate configurations are shown in Figure 14.11. Substrates can be used in a manner similar to PWBs, in which components are mounted onto the assembly. Such configurations are referred to as patterned substrate approaches. Figure 14.11a represents the most commonly used substrate configuration. This particular example illustrates distinct interconnecting and mechanical substrates. A variation of this configuration sometimes used for power devices is referred to as the *recessed patterned substrate* and is shown in Figure 14.11b. In this case, access holes are formed through the interconnecting substrate, resulting in a more intimate substrate contact and, therefore, lower thermal resistance. A fundamentally different substrate configuration results when the interconnection manifold is created over the substrate that contains the components, as shown in Figure 14.11c. This approach is referred to as patterned overlay.

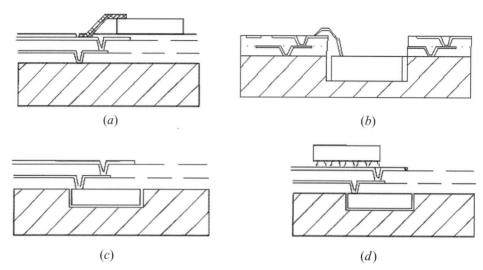

Figure 14.11 MCM substrate configurations: (*a*) patterned substrate; (*b*) recessed patterned substrate; (*c*) patterned overlay; (*d*) combination.

In this case, components are recessed into a planarized substrate, which serves as a starting surface onto which multiple levels of interconnects are created. Examples of patterned overlay (*chips-first*) approaches include the GE HDI process [8] and, most recently, the Intel *bumpless build-up layer* (BBUL) technology [9], which has been proposed as a next-generation packaging system for Intel's future gigascale ($>10^9$-transistor) microprocessors. Normally, the interconnect system is patterned as it is built onto the substrate. Prebuilt overlays can also be laminated onto the substrate and then joined to the components using a variation of the same process [10]. It is conceivable that both approaches can be combined, which can be done as suggested in Figure 14.11*d*.

14.2.3.3 Element Attach

Component-attach or element-attach approaches are defined as the explicit schemes for which components are mechanically and electrically attached to the substrate. A number of approaches are shown in Figure 14.12. Even today, the vast majority of components are attached to packages and substrates using wirebonding technologies (Figure 14.12*a*). Wirebonding was the first large-scale chip attachment approach. It has withstood the test of time despite premature speculations regarding its demise. It is important to put this statement in perspective, since the vast majority of components are relatively simple (<200 terminal connections). Such components are typically not *input/output (I/O) bound* and do not require extremely aggressive pitches for their bond pads. The evolution of wirebonding to accommodate more aggressive components, such as FPGAs, is very impressive. In 2003, the practical state-of-the-art in wirebonding was ~50 μm center-to-center pitch, and research to improve this limit continues.

Another approach, still practiced but far less common, is the tape-automated bonding (TAB) approach (Figure 14.12*b*). In TAB, a flexible circuit tape intended for reel applications, contains an inner lead bond pattern for an IC die framed by an outer lead bond pattern for a package or substrate. There are those who believe that TAB is superior to wirebonding

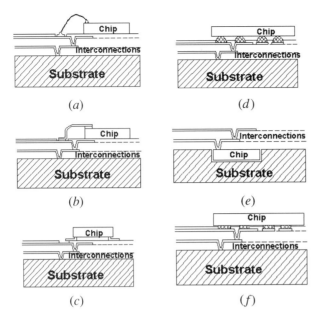

Figure 14.12 Element-attach methods: (a) wirebond; (b) TAB; (c) flip-TAB; (d) flip-chip; (e) patterned overlay; (f) conductive adhesive.

because components are loaded onto reels and attached rapidly using a gang bonding approach. For MCMs, TAB is attractive since premounting ICs permits the pretesting of components before they are mounted onto substrates. One disadvantage of TAB is the need to provide a relatively large footprint for the space displaced by the outer lead bond-out. A flip-TAB method (Figure 14.12c) is used to reduce the footprint created by the outer lead bond pattern. For reasons not totally clear, TAB has never presented a serious threat to wirebonding.

Flip-chip attachment (Figure 14.12d), once called controlled collapse chip connection (C4), is based on the use of a fine-pitch solder ball array formed onto the bond pads of ICs, which are inverted and mounted to the substrate by solder reflow. Flip-chip has become increasingly popular in FPGAs, whose terminal counts now exceed 1000 in some cases. Unlike wirebonds, which are typically restricted to perimeter connections, flip-chip approaches have an advantage of being able to exploit the entire die surface for contacts. Of course, higher contact density impacts substrates through increased wiring demand.

Patterned overlay (Figure 14.12e) defines not only the substrate approach but also the chip attachment approach. It is defined as an approach in which interconnects are metallurgically joined to component bond pads as part of the intrinsic process used to form interconnects. An increasing number of chip scale-packaging approaches employ a similar method, under the label *wafer-level packaging*, which essentially operates the same way. One difference between wafer-level approaches and more generalized patterned overlay approaches is that the former is applied to entire wafers, as opposed to heterogeneous arrangements of individual components.

Another method that is gaining popularity is the use of conductive adhesives (Figure 14.12f), which permits rapid attachment of components to substrates at low temperatures. In conductive adhesives, conductive particles are loaded into a polymer matrix, which, upon curing, forms a conductive bridge. The technique is commonly applied in low-cost, low-performance applications. Recently, the technique has been used in much more complex assemblies, such as the attachment of grid array assemblies to substrates. At present, the approach appears to be improving the capability to interconnect tight-pitch assemblies with high yield, but it is presently still limited to applications at or above 0.5 mm center-to-center pitch.

14.2.3.4 Package Body Styles

A final package, however simple or complex, represents the transitional structures between the MCM and the board (or other higher level assembly) into which it integrates. The basic package body styles (e.g., dual-in-line package, PGA, quad-flat package, BGA, etc.) have already been reviewed in previous chapters. Multichip packaging does not alter these basic concepts, since MCMs can be packaged the same way. The degree of spatial efficiency in package body styles can have an impact on the effectiveness of MCM approaches if improved performance is desired. For example, as shown in Figure 14.13a, substrates can be strategically engineered to provide component mounting surfaces on both sides for at least the MCM-C and MCM-L approaches. In general, integral package constructions (where the substrate and the package structures are combined) are preferred. Figure 14.13a achieves a higher component density by using both sides of a substrate for components and by employing an integral package construction. Both of these objectives are difficult to achieve in MCM-D approaches. Another useful integral package construction for patterned substrate approaches is shown in Figure 14.13b. This is recognized as simply an extension of approaches commonly used today in BGA packages. It is possible to achieve integral package structures with patterned overlay. Figure 14.13c demonstrates the extension of the patterned overlay structure to form the attachment service for a BGA. Figure 14.13d demonstrates the conversion of a patterned overlay MCM to an integral, surface-mount package.

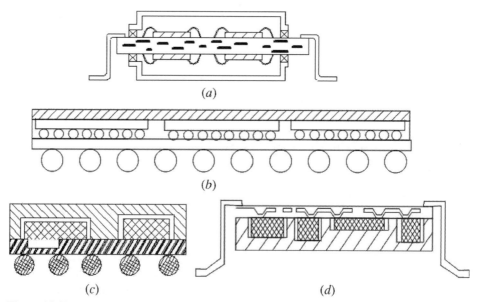

Figure 14.13 Examples of package body styles: (a) double-sided MCM-C quad-flat package; (b) flip-chip patterned substrate BGA; (c) patterned overlay BGA; (d) patterned overlay quad-flat package.

14.3 THREE-DIMENSIONAL SYSTEMS

Planar MCM approaches are limited in that, even when chips are placed in a perfect tile (no gaps between any chip), the highest density achievable is a single circuit layer. While this is an impressive density, especially in modern IC processes, one can do far better by exploiting the third spatial dimension, analogous to high-rise buildings in areas with limited surface real estate.

There are a great many schemes for doing 3D packaging, but almost all approaches can be divided into stacking and folding approaches. In *stacking* approaches, a number of planar elements are arranged by placing one on top of another, as in a stack of playing cards. In this case, the word "element" has a different meaning than in the previous discussion and can, in fact, refer to (1) individual ICs, (2) MCM substrates, or (3) packages containing ICs or MCMs. In *folding* approaches, one can fold a planar assembly with a flexible substrate into a more compact shape, as in the case of some wallet designs with accordion-folded credit card holders.

14.3.1 Defining Characteristics of 3D Systems

The potential diversity of 3D packaging approaches can be vast and confusing. A number of distinguishing characteristics are useful in sorting out this diversity, and a few of these key features are discussed here.

14.3.1.1 Large-Scale Join Configurations

Electronic assemblies that are densely packaged can be joined in one of three basic orientation styles, as shown in Figure 14.14. In the first method, shown in Figure 14.14a, the edges of a stacked electronic assembly engage a planar surface orthogonally at regular pitch spacings. This style is referred to as an *edge–plane* connection, since contacts formed on the edge of the spaced assemblies are formed on two contacts that exist in a planar surface. Edge–plane connections are commonly found in motherboard-to-daughtercard assemblies. The second orientation style (Figure 14.14b) takes advantage of the planar surface area of two different assemblies that are mounted so that at least part of the overlapping surfaces can be dedicated to contact distribution. This style is referred to as a *plane–plane* connection. It has the advantage of greater surface area for contacts. In mezzanine assemblies, which can be viewed as a stack sequence of plane–plane connections, 3D arrangements are also possible. Finally, two sets of regular assemblies consisting of a sequence of stacked circuit layers can be abutted so that their edge surfaces come into contact. This style is referred to

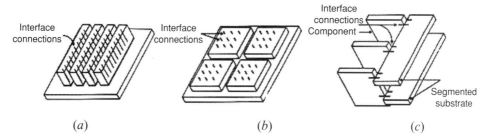

Figure 14.14 Join configurations for 3D Assemblies: (*a*) Edge-plane; (*b*) Plane–plane; (*c*) Edge-edge.

as *edge–edge* connection. As shown in Figure 14.14c, the engagement angles can be varied, which produces an arrangement such that signals from a layer of one set can be broadcast to any or all layers of the second set. Such arrangements can be useful in communication routing applications.

14.3.1.2 Input/Output Location

Another distinguishing characteristic of stacked 3D assemblies is the manner in which conducting input/output (I/O) terminals are arranged on each of the constituent stacked elements (boards, MCMs, packages, die). There are two basic possibilities, as shown in Figure 14.15. The first of these two methods applies to edge–plane connections (Figure 14.14a) and involves the arrangement of conducting terminals on the edge surfaces of stacked elements. This approach is very compact, since no additional vertical headroom is taken up by interconnect structures. However, since this approach is limited to only perimeter contact schemes, it is difficult to build structures in this fashion with very many I/O. This disadvantage can be mitigated by adding multiple rows of conductors on each edge surface, although this technique results in increased complexity of the elemental assemblies. The second method, which applies to plane–plane connections, involves the distribution of I/O contacts across the top and bottom surface area of an assembly, as shown in Figure 14.15b. This scheme then permits the formation of 3D stacks with very high theoretical I/O densities when a mezzanine-style construction is involved. A slight variation of this scheme is shown in Figure 14.15c. In this scheme, I/O pads are restricted to the border region of a planar surface. Usually, the placing of components is confined to the resulting space within the annular frame. This scheme is simpler to implement since it avoids some of the obvious practical problems in comingling the placement of I/O pins and components. The plane–plane (face-to-face) contact geometry provides the greatest amount of overlap geometry between two elements of a 3D assembly, and this overlapping region can be densely populated with contacts between the two elements. A disadvantage occurs in cases where a number of layers are stacked together, one atop another, using this "plane–plane" scheme. Such a *many-layered* 3D assembly must support not only the routing requirements between adjacent layers but also the signals that pass through to nonadjacent layers. In a five-layer stack of plane–plane connected elements, it must be the case that the contacts unique between element 2 and element 4 must traverse through element 3, for example. Such routes are referred to as *pass-through* connections, since they are not strictly useful to the element containing the contacts for these connections. Under some circumstances, many pass-through contacts may be required, and the probability and quantity are likely higher in stacks having more elements. Conceivably, pathological cases could be defined where the number of "pass–through" connections dominate the total signal count on most

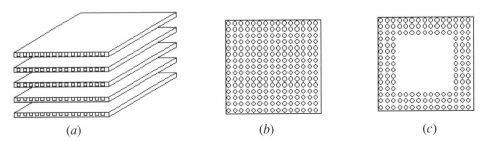

Figure 14.15 I/O configurations: (*a*) perimeter; (*b*) areal; (*c*) border.

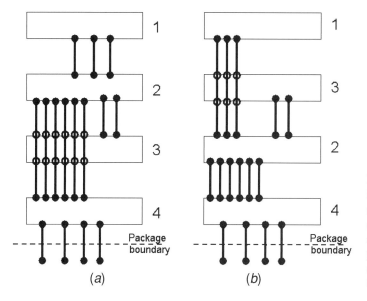

Figure 14.16
Reducing pass-through connections in a vertical stack of elements: (*a*) original configuration (six pass throughs in layer 3); (*b*) reordered configuration (three pass throughs).

elements. Under these conditions, it may be more advantageous to repartition the design or reassign the elements within a stack to produce an ordering that may reduce the number of pass-through connections. An example where reordering is used to reduce pass throughs is shown in Figure 14.16. In some cases, it may be advantageous to exploit the edge–plane connection scheme as an alternative approach or even to apply some combination of both basic methods.

14.3.1.3 Number of Layers in a Stacked Assembly

The way that components are stacked results in a limited number of active layers that can be combined in a single 3D assembly. For example, the most common chip-on-chip stacks, based on wirebonded die, are limited to two layers, with a smaller number extending to a limit of three layers. Extending these approaches to a higher number of layers is possible by using double-sided substrates and/or combining back-to-back flip-chip schemes with wirebonding. At least one result based on these techniques has reported up to eight layers.

A number of techniques are iterable, that is, stackable to an indefinite limit. These include so-called sugar-cube approaches, neopackaging [11], 3D high-density interconnect, and some mezzanine-based MCM stacking systems. Early stacks based on sugar-cube approaches have reported as many as 128 layers. The advantage of such approaches is that very high functional densities are possible. The main disadvantages are lower yield and difficult thermal management.

14.3.1.4 Demountability

Three-dimensional systems can also be characterized by their ability to be readily disassembled for easy access of individual layers within a system. As the complexity of 3D MCM approaches increases, so does expense and probability of failure. *Demountable* 3D assemblies not only offer the advantage of layer repair but also provide additional modularity and the possibility of interchange between 3D technologies. The most common examples

602 Chapter 14 Multichip and Three-Dimensional Packaging

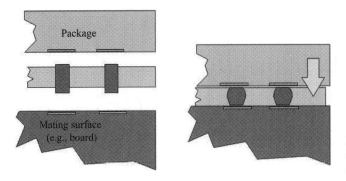

Figure 14.17 Interposer in noncompressed and compressed conditions.

of demountable assemblies use fuzz button or elastomeric interposers using the border I/O scheme (Figure 14.14c) to provide compliant contact mechanisms between layers, as typified in (Figure 14.17). Interposers (based on mechanical spring arrays, fuzz buttons, or elastomeric) operate by bringing matching grids on two flat surfaces into conduction by compression. The compliance of the interposer compensates for nonplanarities. *Nondemountable* systems simply refer to 3D assemblies in which layers are glued or fused together. The assemblies are sometimes repairable, particularly when thermoplastics are used.

14.3.2 Die and Package Stacks

Die/package stacks seek to exploit the floorplan of a substrate in the most effective way, not unlike the way that parking garages or skyscrapers attempt to maximize utilization of relatively small amounts of real estate. These approaches typically do not feature a special substrate but rely on relatively simple connection schemes, such as multiple tiers of wirebonds. Once again, it is possible to identify a simple taxonomy for approaches. There are four basic elements or dimensions to the die-stacking taxonomy: (a) layer configuration, (b) connective relationship, (c) role of substrate, and (d) type of element attach used.

The layer configuration identifies the relationship of the shapes of individual elements making up a stacked IC or package. Figure 14.18 outlines the three basic possibilities. In all of these approaches, the goal remains the same: Expose all necessary bond pad surfaces of each stacked element and do so in the minimal amount of volume. One approach in stacking die is to arrange smaller die above larger die. This approach (Fig. 14.18a) is reminiscent of the familiar "towers of Hanoi" game and can be referred to as a *telescopic* arrangement. The second approach (Fig. 14.18b), useful in assemblies based on regular components, is referred to as a homogenous arrangement. Shown in the illustration is the introduction of spacers, which provide headroom for wirebonding. When wirebonding is not employed, the spacers may not be necessary. The remaining possibility arranges die to expose different parts of die at different levels for easy wirebond access. This scheme is referred to as a staggered arrangement.

The connective relationship between the elements of a stacked die or package are, in many cases, a very sophisticated system of point-to-point connections. For this case, without a more elaborate interconnecting substrate, there are two basic possibilities, as shown in Figure 14.19, using wirebonding maps for illustrative purposes. In the first case, (Fig. 14.19a), bond pads at one vertical level of a stack are completed by connection to a

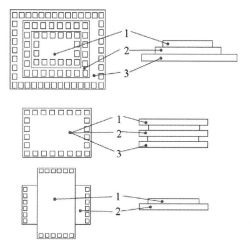

Figure 14.18 Three-dimensional die layer configurations: (*a*) telescopic; (*b*) homogeneous; (*c*) staggered.

vertical nearest neighbor. This scheme amounts to an *intrastack* or *relative* bonding. In the illustration, all of the connections are shared, meaning that each bond pad from a different level connects to a bond pad that connects to yet another level. This is not a strict limitation, as individual bond pads can be dedicated to separate connections between each layer. Dedicated pads are minimized due to the obvious impact on reducing the amount of die surface area available for overlap and for placement of active circuitry. Another application of relative connections occurs when a vertical stack is placed very close to another die or vertical stack sharing a common mechanical substrate. In this case, wirebonds can bridge across and connect, similar to the approaches used in multicomponent packages [12]. The second case (Fig. 14.19*b*) is referred to as *absolute* or direct substrate connection. This case does not employ intermediate connections but directs that all outer bond connections be made to a common substrate. This approach is ultimately limited in the number of layers that can be accommodated. Each additional layer increases the average wirebond length and substrate footprint while reducing yield.

A number of practical examples of wirebonded die stacks based on combinations of layer arrangements and bond-out schemes are shown in Figure 14.20. The most dominant

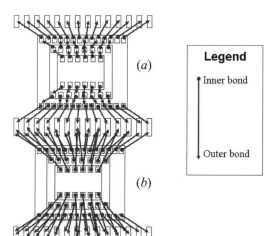

Figure 14.19 Three-dimensional die bond-out schemes: (*a*) relative or intrastack; (*b*) absolute or direct substrate.

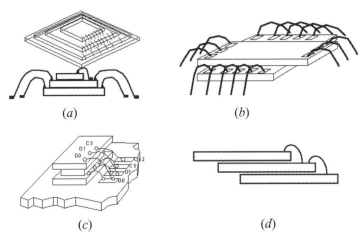

Figure 14.20 Three-dimensional IC stacking methods based on wirebonding: (*a*) telescoping; (*b*) staggered bond-out; (*c*) spacer; (*d*) staggered staircase.

form of 3D packaging is a two-die chip stack using a telescoping or staggered arrangement and direct substrate bonding. By 2003, over 100 million assemblies of this construction had been produced, with a growth in use of 50% per year [13]. Figure 14.20 demonstrates the aforementioned absolute and relative bonding approaches in different applications. Figure 14.20*a* shows a longitudinal cross section of a stack that employs both relative and absolute bonding approaches. In Figure 14.20*b*, two identical die are stacked by rotating them to produce a staggered stack. Clearly, the long edges in this example contain no bond pads, simplifying the bonding by eliminating the possibility of entangling wirebonds of components on two different levels. On the other hand, Figure 14.20*c* requires *bond-over-bond* connections. Managing *bond-over-bond* connections can be considerably complex, requiring different types/thicknesses of bond wires based on the length and/or orientation of individual bond wires. Figure 14.20*c* also demonstrates the role of spacers in a homogeneous stack. In this case, identical memory die are stacked and all connections of data pins (e.g., D0 and D1) are direct bonded to common substrate termini, while the chip-select pins (e.g., CS1 and CS2) are connected to distinct termini. These ideas lead to simple schemes for stacking existing die effectively to produce dense memory modules. Finally, Figure 14.20*d* demonstrates an unusual stair-step arrangement of die using a combination of die staggering and relative die bonding, which was first demonstrated by LETI in a dense MCM application [14].

The third element of the die/package stack taxonomy considers the element-attach approach. In the examples described so far, only wirebonding has been considered. TAB approaches are also used, most famously perhaps in the 3D Plus method. In this method, TAB frames are stacked, potted, and then cut and polished to form flat-edge surfaces, which are then patterned. Tessera has described a Micro-Z ball stack method [15] (Fig. 14.21) in which individual die are joined bond-side down onto individual accessory substrates. Interior ports are cut into the substrate and individual die bonds are formed through this portal. The finished layers are then stacked using BGA connections patterned on the border of each substrate. The BGA of the final level becomes the contacts of the primary package.

Flip-chip technology has also been used and proposed to make stacked IC arrangements. The most natural method involves a simple chip-on-chip stack in which one or more small die are face-down bonded onto a larger die, creating a simple two-level stack. Such arrangements were studied extensively in the early 1990s by AT&T. More recently, the use of flip-chip and wirebond in the same stack has been proposed. In one scheme, a flip-chip die is attached

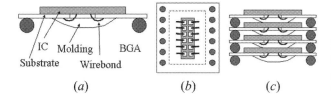

Figure 14.21 Tessera's Micro-Z ball stack technology: (*a*) cross section of single layer; (*b*) plan view; (*c*) four-die stack.

to a substrate. Then a simple die stack, such as a two-die or three-die chip stack, can be simply glued onto the mounting surface created by the backside of the mounted flip-chip component. It is also conceivable that face-to-face flip-chip mounting arrangements can be substituted for the spacers shown in the Figure 14.20*c* concept. Such a scheme could increase the effective density by replacing the otherwise inert spacers with active silicon.

Die based on overlay patterning approaches (at die or wafer level) can be directly stacked without spacers. In this case the patterned overlay provides two functions: bond pad repatterning (for die of dissimilar size and shape) and insulation between stacked levels. The bond pad repatterning step provides a means to route surface pads to one or more edges. The stacked assembly must be completed by forming an edge overlay to connect layers together. A prime example of this scheme is an approach introduced by Irvine Sensors Corporation referred to as Neo-Stacking [11, 16]. GE's 3D extensions of HDI, described later, have also been used in the same way to produce dense stacks of heterogenous components.

The final element of a stacking taxonomy for die and packages considers the role of the substrate. Most of the schemes discussed so far (except the Micro-Z ball approach) use a single substrate. One way to increase density is to use the single mounting substrate as a mirror plane and replicate the bottom of the substrate with another stack. This approach is straightforward in cofired ceramic packages where two cavities are formed [17]. Using this scheme, die stacks containing as many as eight layers have been formed.

14.3.3 MCM Stacks

MCM stacks have a distinct advantage over die stacks in that collections of components of dissimilar size and process can be readily cointegrated. One example of an MCM stacking method is illustrated in Figure 14.22. This method is based on a set of identically sized

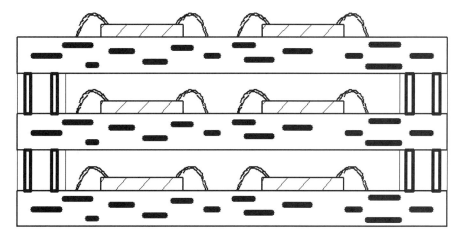

Figure 14.22 MCM-C stack using interposers.

606 Chapter 14 Multichip and Three-Dimensional Packaging

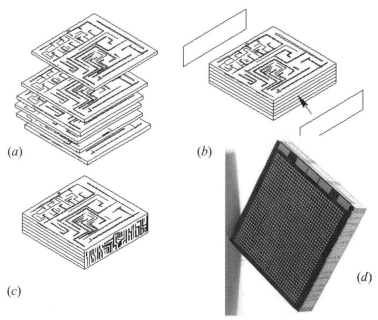

Figure 14.23 Three-dimensional HDI process: (*a*) specially prepared 2D HDI layers stacked; (*b*) edge overlays formed; (*c*) finished stack; (*d*) example of a four-layer 3D HDI module containing 32 floating-point processors, memory, and four Myrinet (gigabit/sec duplex) ports.

MCM-C substrates in which conductors are patterned on the top and bottom surfaces with land grid arrays. Compliant interposers are introduced in the borders of each MCM to connect signals vertically.

Patterned overlay approaches are particularly suited for 3D assemblies, since the component layers of the stack are planar [18]. A process flow and typical example module are shown in Figure 14.23. In this flow, a number of planar modules are combined to form a stack. The process used to form these layers is similar to that shown in Figure 14.9 except that the substrates must contain edge conductors, either built into the substrate or added through a special metallization process. This modification then permits the set of stacked modules to be interconnected using an edge overlay system. The process follows the Figure 14.9 planar process, except that the individual substrate edges are now the components (adaptive lithography again can play an important role in maintaining alignment during fabrication). Once again, Kapton® is laminated to the assembly (Fig. 14.23*b*) and patterned to create the interconnections between layers (Fig. 14.23*c*). A typical module built this way is shown in Figure 14.23*d*.

A still more advanced 3D example is shown in Figure 14.24. This approach is a modular (demountable) 3D system referred to as *highly integrated packaging and processing* (HIPP). It combines a number of special MCM assemblies that are called *segments* in pairs (Fig. 14.24*a*). These segments contain MCMs, are patterned with land grid arrays (LGAs), and are jacketed by efficient thermal management structures. The segment pairs are sandwiched onto thin multilayer printed wiring structures to form a clamshell (Fig. 14.24*b*). The sandwich employs interposers (not shown) to ensure a uniformly compliant (positive) contact between each of the 1225 conductors on the mated LGAs. A number of such clamshells can be efficiently stacked (Fig. 14.24*c*) and then combined with separate dedicated wiring board structures for power and signal distribution.

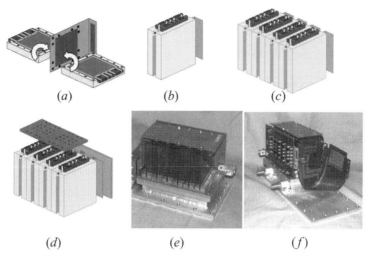

Figure 14.24 HIPP 3D approach: (*a*) MCM pair and fanout translation circuit; (*b*) formation of clamshell; (*c*) combination of clamshells; (*d*) integration of power and signal distribution planes; (*e,f*) photographs of experimental system.

14.3.4 Folding Approaches

An alternative method for creating 3D assemblies is by converting planar assemblies to 3D through folding. This is done with individual die using an oversized patterned overlay. By extending the overlay beyond the floorplan of an individual component, a "flap" is formed by the excess, which can be folded around the body of the die. If the flap is patterned, it can be used to produce a convenient connection surface for another similarly patterned component [19]. This arrangement then creates another method for vertical signal transport and any of the Figure 14.18 stacking approaches can be employed. Another method involving folding produces a planar floorplan of components that is simply folded into a 3D assembly. An early example, based on HDI, is shown in Figure 14.25. Folded assemblies have the

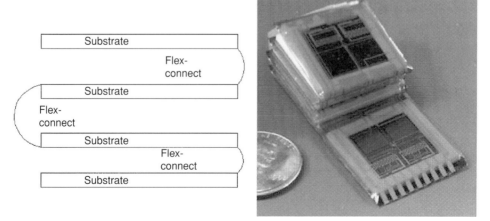

Figure 14.25 Folded flex-based MCM.

clear advantage of eliminating some of the process steps and complexities associated with stacking and connecting the various levels of a 3D system.

14.4 OPTIONS IN MULTICHIP PACKAGING

Circuit design for multichip packaging has several options as well as constraints. This section examines both as well as the trades required for specific applications.

14.4.1 Yield/Known Good Die

The yield of a multichip assembly is directly related to components simply as

$$Y = \prod_i y_i \qquad (14.2)$$

where y_i is the yield of each component. In assemblies with N identical components, module yield is at best y^N. So, the number of components and the degree to which their yield can be controlled drive the yield of MCM assemblies.

This problem is referred to as the *known good die* (KGD) problem, and it has often been regarded as a fundamental barrier to the success of MCM approaches. Substrate yield is another factor. It is a function of surface area and the number of layers. A simple model for the combination of these factors can be postulated:

$$Y = \left(\prod_i y_{\text{chip},i}\right) \left(\prod_j y_{\text{layer},j}\right)^a \qquad (14.3)$$

where the $y_{\text{chip},i}$ is the yield of a particular component, $y_{\text{layer},j}$ is the yield of a particular interconnection layer per unit area, and a is substrate area. The yield of modules and substrates is not the only factor in assembly yield; even perfect substrates can suffer yield loss at later stages of assembly. This makes it necessary, even with strong provisioning for testability, to consider schemes for redundancy/defect tolerance/repair. Repair can introduce a tremendous penalty in terms of physical overhead. Repairability in dense MCMs usually requires *separability* (the ability to release parts of a tightly fused assembly), which drives contradictory goals. It may be necessary to add margins far in excess of 100% to achieve the level of serviceability required for effective repair. The alternatives involve designing assemblies for defect tolerance/full redundancy. In certain, highly regular structures, redundancy at the substrate or subsubstrate level is relatively simple to enact. More complex designs, such as random logic or processors, require more careful consideration. In some cases, redundancy implies not only full replication of large portions of circuitry but also an effective means of selecting and repressing the good and bad copies, respectively, of the replicated functions. The use of reconfigurable logic (e.g., FPGAs, analog programmable arrays) can be advantageous in these circumstances, since this creates essentially a large pool of computational resources that can be shaped and reshaped as circumstances dictate.

14.4.2 Process Compatibility

One of the clearest advantages of MCMs over monolithic systems is in the ability to combine components from divergent processes within a single assembly. The objective of mixing technologies can be complicated by special requirements, such as air bridge structures in microwave ICs, which are subject to contamination. Another important category of devices is comprised of microelectromechanical system (MEMS) devices, which not only are subject to similar issues but also have requirements of access portals to sample the environment.

14.4.3 Density Metrics in 2D and 3D Packaging

The most relevant density metric in 2D (MCM) packaging is substrate efficiency, which is a measure of the fraction of a substrate actually covered by active components and can be written $a_{si}/a_{substrate}$. As a surface metric, substrate efficiency has little meaning for 3D approaches. For 3D approaches, it may be better to consider the amount of active silicon area in a unit volume (cm^2/cm^3). In this metric, one would find that a stack of raw (noninterconnected, unthinned) wafers would demonstrate a value of approximately 20 cm^2/cm^3. While it will soon be shown that this "reference standard" can be exceeded, it is true that even the densest typical electronic systems, including those that employ MCMs, very rarely exceed 1 to 2 cm^2/cm^3 "in practice." This was one of the central problems in the application of MCM technology in the 1990s. MCMs are capable of realizing significant improvements in localized density, but sometimes only for a few components in a complex electronic systems design. Under these conditions, the density advantage of MCMs in "spot applications" is mitigated. Using several MCMs in a single board is not always an improvement due to floorplan overhead involving MCM lead frames. In these real-world cases, MCMs are not a worthwhile proposition in dimensionally constrained systems. In 3D systems, floorplanning overhead can be absorbed within the assembly, and even the simplest stacking schemes offer a powerful advantage over complex MCMs involving only planar layouts. Estimates of the scalability of selected 3D approaches in terms of density are summarized in Table 14.2.

14.4.4 Wiring Density

The wiring density or wiring capacity of MCM substrates is a measure of the amount of physical wiring that is maximally available to interconnect components together. It is generally specified in terms of unit length per unit area [21]:

$$W_c = \frac{e_{wire}m}{p} \qquad cm/cm^2 \qquad (14.4)$$

where m is the number of wiring layers, p is center-to-center pitch (in terms of unit length), and e_{wire} is wiring efficiency. The latter term is a function of a number of factors, such as routing algorithm efficiency and area removed by via occlusion. Generally, substrates with higher wiring capacities are capable of accommodating more complex assemblies. MCPs and stacked packages have no wiring capacity. They must rely on arrangements where the proximity of pads on dissimilar die is sufficient to permit completion of intrapackage connections.

Table 14.2 Comparison of Volumetric Density of Different Packaging Approaches

Approach	Estimated scaling limit (cm^3/cm^3) and I/O density	Comments
Stacked packages	10 cm^2/cm^3, indefinite, very low	Tall stacks, limit of 0.5 mm package body thinness, I/O scaling bad
Telescopic	8 cm^2/cm^3, ~4, low	Requires a choreography of telescoping substrates
Spacers	8 cm^2/cm^3, very low	Stacks much more complex above four layers
Staggered die	30 cm^2/cm^3, very low	I/O scaling bad unless a complex angled backplane can be devised
µZ-Ball	17 cm^2/cm^3 [20], moderate	Significant lateral headroom
Folded	200 cm^2/cm^3, moderate	Extreme density assumes thinning low-interconnect density, extra lateral overhead for folding axes
Stacked interposers	5 cm^2/cm^3, moderate	Separable (good) but compliant interposer limits density
Patterned overlay die stacks (3D HDI, Irvine sensors)	>200 cm^2/cm^3, low to high	Possible problems with handling thinned die; highest I/O requires ultrathin substrates
Stacked TAB frames 3D plus)	20 cm^2/cm^3, low	Low-cost approach
Stacked wafer schemes (HRL)	10 cm^2/cm^3, high	Limited by spring structure and rigidity requirement of silicon
3D IC with integral feedthroughs (Tezzaron)	>1600 cm^2/cm^3, extremely high	Extremely problematic processing of feedthroughs and yield above four layers

14.4.5 Input/Output

The relation of terminal count to gate count in systems is often described through a simple relationship known as Rent's rule [22]:

$$T = AG^p \qquad (14.5)$$

where T is the number of terminals or I/O, A is a constant referring to the number of terminals per "block," G is the number of gates or blocks, and $0 < p < 1$ is Rent's exponent. Although, originally, this simple relationship was informally documented as an empirical observation to certain types of circuit modules in computer systems in the early 1960s, Rent's rule has been found to be germane to most of the recursive elements in a hierarchy such as that shown in Figure 14.4. For example, using Rent's rule, Donath [23] developed an analytic distribution of interconnection lengths in integrated circuits, which is compared in Figure 14.26 against a typical distribution in an actual IC [24].

Rent's rule demonstrates the importance of power law relationships in understanding certain aspects of packaging complexity, but it requires careful interpretation. For example, it explains the steady increase in I/O for complex ICs, such as FPGAs, as gate counts have increased progressively according to Moore's law. This growth rate in I/O is dependent

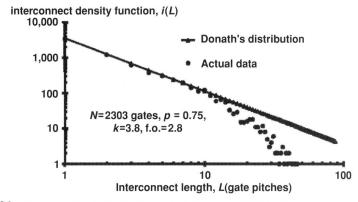

Figure 14.26 Interconnect length distribution in a typical modern VLSI circuit.

on component type, as suggested in Figure 14.27. Different types of components give rise to different Rent's exponents. Regular structures, such as memories, possess a low Rent's exponent and, therefore, have a very slow growth in I/O. On the other hand, processors and complex application-specific integrated circuits (ASICs) lack such regularity in structure and, therefore, drive higher I/O for a comparable number of gates.

It has been reported that the Rent's exponent of many electronic systems falls in the range $0.65 < p < 0.75$ [23]. This range for Rent's exponent becomes problematic for 2D systems, however, as the complexity of systems increases. It is straightforward to show this by comparison to an arbitrarily large 2D system. Assume that a large 2D system is divided into four equal square partitions, with pins spaced equidistant in each partition. Assume each partition has t pins. For each $k \times k$ arrangement of t-pin blocks, Rent's rule takes on the form

$$T = AG^p \qquad 4\left(\frac{1}{4}k\right)t = t(k^2)^p \qquad (14.6)$$

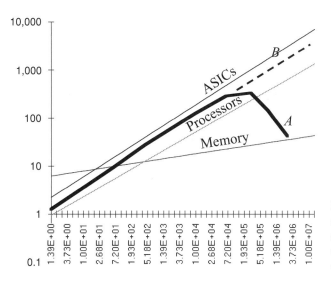

Figure 14.27 Rent's rule for different component types and system classes: A, dimensionally constrained; B, performance constrained.

Since Rent's rule applies recursively, we observe that each partition has t pins and there are k^2 partitions. By definition, there are t pins per block. Since only the blocks on the perimeter contribute to the global I/O count, the exposed partitions contribute one of four edges, and the global system has four edges. Obviously, the only solution for arbitrarily large k is given in the limit $p = 0.5$, corresponding to the maximal p in infinite 2D systems. For 3D systems (which is assigned as an exercise), it can be shown that the corresponding $p = \frac{2}{3}$, which closely corresponds to the previously observed range for most real-world systems.

The analysis of real systems, which may contain mixtures of circuit types each having a different local p, is not so easily captured by a single power law expression. At least one group of researchers has developed heterogeneous Rent's rule expressions to deal with such mixtures [25]. Cases are easily contrived, however, where monotonic increases in gate count do not result in monotonic increases in I/O. Two such cases are demonstrated graphically in Figure 14.28. In these diagrams, rectangular and circular blocks represent a number of gates, while lines represent one or more terminals per line. In Figure 14.28a, the encircled partitions (dotted lines) cut across a smaller number of terminals than the sum of terminals emerging from each block. Obviously, this reduction is due to the combination of terminals within a block. In Figure 14.28b, a collapse of the single partition onto the star produces an initial nonmonotonic change in the number of lines cut due to the transition between two different regular grids. This type of delineation, a departure from Rent's rule behavior, may correlate with the boundary of a system and is referred to as a *containment*. Without further elaboration or proof here, it is claimed that dimensionally constrained systems represent well-defined containments and may be recognized by a marked nonmonotonicity in the terminal-to-gate relationship, as suggested in case A of Figure 14.27. Almost any electronic "appliance," such as a personal computer, which contains billions of electronic devices, represents not only a system but also a containment. On the other hand, a partition that appears more consistent with an extension of Rent's rule behavior is indicative of a portion of a performance-driven system. In the following sections, these classes of systems are further described by defining distinct packaging approaches.

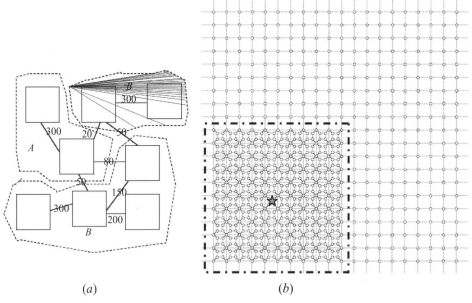

Figure 14.28 Examples of nonmonotonic pin-to-gate relationship.

14.4.6 Electrical Performance and Substrate Selection

In high-performance systems, signal-bearing conductors require low latency and distortion to support high clock frequencies. The propagation delay for digital signals can be roughly specified as

$$t_{prop} = t_{flight} + t_{rise} \tag{14.7}$$

In this expression, t_{flight} is the delay component due to time of flight,

$$t_{flight} = \frac{l\sqrt{\varepsilon_0 \varepsilon_r}}{c} \tag{14.8}$$

where ε_0 is the permittivity of free space, ε_r is the dielectric constant, l is conductor length, and c is the velocity of light. The second component of propagation delay is the rise time, which is roughly given as $t_r \propto rc$. The conductors for power and ground, however, benefit from high capacitance, which helps to keep supply rails insensitive to sudden surges related to simultaneous switching noise while reducing the alternating current (AC) impedance of the grounding system.

One way to address these conflicting requirements is to place discrete capacitance near each component. Often, two types of capacitors are used, one to address impedance reduction and a second to provide instantaneous storage. This latter capacitor could be replaced by a microbattery, although research in microscale batteries is still relatively immature. Another method involves the use of integrated capacitance in the substrate itself. Integral passives are described in detail elsewhere in this book, and this is one of their most important applications. A simpler scheme that has often been used in high-performance MCMs involves the use of a two-tier substrate design, specifically a combination MCM-C/D substrate. The MCM-C layers of the substrate provide a medium with 2 to 3 times the capacitance, ideally suited for power and ground conductors, while the MCM-D layers support high speed and high density for signal distribution.

14.4.7 Thermal Management

The most obvious potential problem in a large-scale, high-performance system is the ability to remove heat from an assembly. The first key principle is to avoid circuitry with excessive power, and a great deal of DARPA and commercial research is currently on-going in power-aware and reduced-power circuitry [26]. Even relatively low power circuits, however, can result in unacceptably high power densities when dozens of circuit layers are compressed in an ultradense packaging system.

Thermal management involves the control of temperatures of materials within a package. Usually, thermal management targets the buildup in temperature distributions within a package that, without special attention, could result in hot spots. These hot spots can have a severe impact on reliability due to both the activation of failure modes in components and disruptions in physical structures inside the package due to differential expansion of dissimilar materials at their interfaces. Thermal management is an important consideration in MCM packaging approaches, since physical size decreases faster than the power density. In fact, power density may increase in MCMs due to reduced interconnection latencies and, therefore, faster cycle times. While thermal management problems are expectedly worse for 3D electronics, even monolithic ICs can sustain unacceptably high power density, and the trends continue to become worse, based on the International Technology Roadmap for

Semiconductors (ITRS) [27]. The situation is compounded even more in 3D systems, which, unlike most planar approaches, have buried components.

In any thermal management approach, the central objective is to efficiently shunt heat from each power generation source within a package to an external surface, from which heat can be rejected into a host system. The requirements for a viable thermal management approach in advanced packaging are stringent. Solutions must (1) be compatible with the mechanical and electrical requirements, (2) not significantly impact the density advantage, and (3) be affordable.

Thermal management usually involves techniques based on material and geometry considerations. Material approaches focus on harmonizing thermal expansion coefficients to the maximum degree possible and maximizing thermal conductivity. Geometry approaches focus on engineering low-resistance pathways from heat-generating surfaces to appropriate heat-rejecting surfaces. Examples of techniques include (1) thinning components, (2) introducing heat spreaders, (3) increasing cross sections, and (4) shortening paths to reduce thermal resistance.

There are two basic cases for thermal management in 3D systems: few layer and many layer. In few-layer systems, heat transport is vertically directed (as in single ICs), since the component stack is mounted flat. In this case, it is necessary to conduct heat through the stack. Since, in most cases, a polymer will be the dominant material, this could create a problem. The total traversal distance is short, thereby mitigating the high thermal resistance of the polymer. In this circumstance, two basic approaches can be considered. First, the stack can be arranged in a way that places the most dissipative substrates closest to the heat removal surface. A heat removal surface must always be assumed to exist, since heat generated *must* be directed somewhere, unless a phase change or thermal storage block is included in the construction of the system. Second, it is possible to consider loading the polymeric materials with insulators that conduct heat better than Kapton®.

In many-layer systems, heat transport is more isotropic. In this case, the two previously discussed approaches apply, but the heat removal "surface of choice" may be different. It is also possible to consider the inclusion of other heat removal structures, such as periodic heat spreader layers, which act as lateral thermal shunts, or vertical columns or slugs of highly conductive materials that can be inserted throughout the lateral extent of the entire ensemble of an ultradense MCM assembly.

14.4.7.1 Special Considerations

When the operating duty factors within a package are sporadic, it is possible to consider thermal storage as an alternative to rejecting heat. Such approaches might be referred to as adiabatic, since they need not rely on thermal management interfaces at the package boundary.

Other special cases exist in which it is important to cool and/or heat components within a package to control the temperatures of selected components within a target range, independent of the ambient temperature. Examples include cooled optical sensors, precision oscillators, and circuits where thermal drift impacts performance. Again, materials and geometries can play an important role, but not necessarily in the same way as in conventional thermal management. It may, for example, be necessary to impede thermal transport for cooled devices to minimize the work required to maintain a desired temperature.

In addition to passive measures, active cooling components can be introduced into packages. For example, thermoelectric coolers are commonly used in laser diode assemblies, and flow-through systems can be engineered to deliver coolant directly to the backsides of hot components.

14.4.8 Testability

Design for test (DFT) is an important consideration in the development of most complex ICs. Since a number of complex ICs are often combined, it is important to test them thoroughly. This is not easily achieved since the components are often from a mixture of suppliers and the approaches employed to test these devices are not universally interchangeable. Intellectual property and special test access facilities are often protected as highly proprietary. Hence, users have only a limited window. To complicate matters further, die are not always available for a desired part that is otherwise freely available as a prepackaged component. Attempts in the 1990s to establish an infrastructure for KGD were only partly successful. As such, projects that try to perform an "MCM conversion" on an arbitrary brassboard (prebuilt prototype design) are rarely as straightforward as a piecewise miniaturization exercise. Instead, it is usually necessary to convert the brassboard to another format. The conversion process is itself error prone. An effective testability scheme is required to ensure that the fidelity of the converted design is consistent with the original prototype.

Despite these challenges, testability is usually achievable by exploiting any existing "hooks" in predesigned components and adding other features that permit an enhanced level of observability and controllability. In all-digital designs, a thorough incorporation of boundary scan is ideal, and a joint testability action group (JTAG) is commonly used to configure complex components, such as FPGAs. Ideally, a set of components employing boundary scan can be isolated, tested individually, and then tested in controllable subsets. Developing test code even for throroughly documented digital designs is complex, and techniques such as automatic test pattern generation (ATPG) are less applicable in complex systems, and ad hoc methods are usually required. Packaging can provide some additional options, such as the incorporation of special test points or additional diagnostic circuitry.

14.4.9 System in a Package Versus System on a Chip

Dimensionally constrained systems often have very strict limitations in size, mass, and power and require the highest functional densities. Examples include pacemakers, cellular telephones, and submunitions. For these systems, it is often possible to collapse the packaging hierarchy into one or two levels (Fig. 14.4). The case of a collapse to a single-level hierarchy can be referred to as a *system on a chip* (SOC). Two-level packaging hierarchies, generally referred to as MCMs, can, in some cases, be referred to as a *system in a package* (SIP). There is no clear delineation between the two, except that SIP designs tend to be smaller and simpler than some of the traditional MCMs and are more easily integrated within a conventional PWB design without special attention to mounting considerations. In other words, SIPs can be treated like normal singly packaged ICs, whereas MCMs often require special consideration in assembly.

14.5 EMERGING TRENDS IN DENSITY SCALING

From Table 14.1, it appears that only certain 3D processes can scale to the point beyond about 50 layers/cm thickness due to overhead material and/or geometry limitations. Patterned overlay technology, as mentioned previously, traditionally involves a mechanical substrate. The first mechanical substrates were ceramic, but later, other materials were used. It is not necessary to have a mechanical substrate at all, as previously described in the COF

Figure 14.29 COF examples based on patterned overlay high-density interconnect (HDI); (*a*) memory module: (*b*) TRAM.

processes. Figure 14.29 illustrates two modules using a COF construction. Figure 14.29*a* demonstrates a memory module using COF, which eliminates >90% of the mass of a normally packaged approach. Figure 14.29*b* illustrates the transmit–receive antenna module (TRAM) design, which forms a very lightweight antenna structure [28]. These modules were produced without a mechanical substrate; such COF structures actually exploit the flexible interconnections as a mechanical substrate. Initially, the die are placed and glued face down onto a stretched sheet of Kapton® (PI) dielectric. The sheet is then processed as a patterned overlay, like a standard HDI module, using a series of laminations, whose number is dictated by wiring density demands. The Kapton® sheet is patterned and metallized to connect the component bond pads and more levels of interconnect can be added as required, as summarized in the simplified process sequence shown in Figure 14.30.

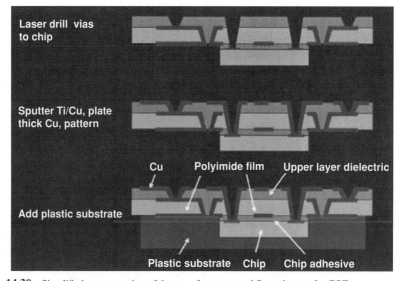

Figure 14.30 Simplified representation of the use of prepatterned flex substrate for COF.

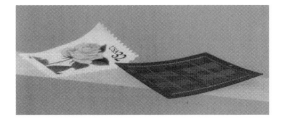

Figure 14.31 Paper-thin HDI MCM based on postthinned substrate and die.

In this scheme, module thickness is substantially reduced, as it is defined simply as the thickness of the die, glue, and interconnection manifold. For an unthinned die (about 500 μm) and three levels of electrical interconnect, a module built this way would be about 625 μm thick. If such modules could be stacked with a 15-μm glue line between each module, the resulting density would be 16 cm^2/cm^3. It is possible to do better by:

1. Using thinned IC die
2. Using fewer levels of interconnections
3. Using thinner interconnect layers
4. Using thinner glue lines

Handling thin die is problematic, however. It turns out that a better method may be to build HDI modules and then thin the entire module, die included, to realize a thinner module assembly. In doing this, it may be better to use a sacrificial substrate material, which is ablated away as the module is ground flat. The process is illustrated in Figure 14.31. This construction method was used in the fully functional memory module built in 1998 using backgrinding approaches to realize an ~75-μm-thick substrate (which if stacked with a 5-μm glue line yields a density of 125 cm^2/cm^3).

There are several basic concepts for creating a dense 3D system based on stacks of these ultrathin substrates. Three methods are discussed in more detail in the following sections. Each approach, in principle, supports arbitrarily thin layers and is extensible to many layers of substrates and each is, in principle, pretestable prior to stacking.

14.5.1 Method 1: For Regular and/or Low-Pincount Assemblies

The simplest and lowest risk approach for ultrathin stacks involves creating a number of thin platelets that can be separately tested and stacked, like playing cards. All signals and power conductors are mapped to an identical border grid on each substrate. Although this border grid is identical in each substrate, the circuitry and specific signal mappings to the grid can be different on each substrate. For example, one could map an entire VersaModule Eurocard (VME) bus to the border grid, and each platelet could represent a bus card. The platelets, when stacked, are initially isolated electrically. The entire laminate can then be processed using PCB techniques to connect together each pad of the border array using barrel vias in which tiny holes are mechanically drilled through every substrate at each pad location. The holes are then metallized through an electroplating process that connects each substrate's ith conductor (for all i in the border array). The finished stack can then be mounted onto a PCB using headers, as shown in Figure 14.32, or using a BGA, which requires bump metallurgy only on the topmost substrate.

618 Chapter 14 Multichip and Three-Dimensional Packaging

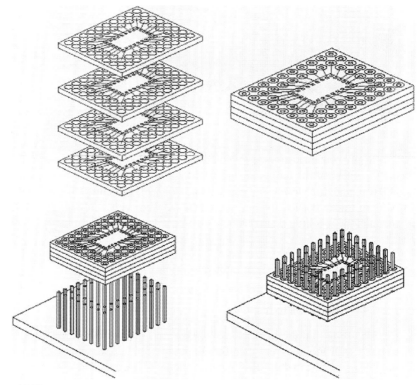

Figure 14.32 Low-cost, "thickness agnostic" stacking scheme for simple modules.

This form of stacking is probably the easiest to implement, since it can be tested on thicker substrates and gradually "shrunk" by thinning each substrate uniformly in subsequent prototyping runs. Furthermore, all connections are simple, are exposed for testability, and do not require sophisticated buildups or edge metallization as do the other approaches described. However, this simple stacking scheme does suffer from two disadvantages. First, the ultimate connection density is quite limited. Mechanical PCB vias cannot be very densely placed (e.g., >10 mils pitch), and the number of nets in the entire stack cannot exceed the number of pads in the border array. Second, the border array adds a higher lateral size overhead than other approaches discussed. Nevertheless, as an incremental strategy, it is hard to beat this scheme for allowing initial work in substrate thinning to be validated in a representative prototype. Simple test assemblies and, indeed, very dense digital storage and/or integrated processing systems alike can be readily integrated using such an approach. The very highest performance multiprocessors, however, tend to have an excessively high pincount, which results in larger substrates due to pincount overhead. Such high-performance designs would benefit from one of the subsequently described schemes.

14.5.2 Method 2: For Moderately Complex Pincount Assemblies

In this approach, the "thinness" of substrates permits a very powerful technique to be applied, namely the idea of *spanning vias*. With spanning vias, it is possible to create feedthroughs lithographically that penetrate entire substrates. In this ultrathin regimen, the

14.5 Emerging Trends in Density Scaling

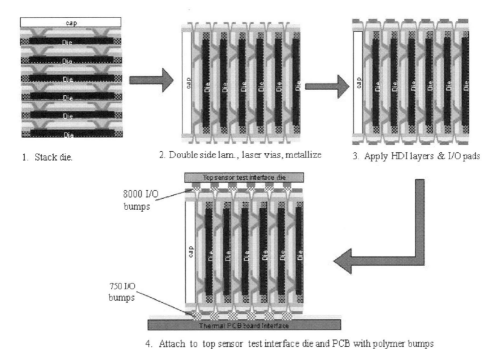

Figure 14.33 Summary of 3D process used in high-speed (megaframe-per-second) imaging application: (1) die are stacked; (2) edge overlays applied on one or more edge surfaces; (3) final layer patterns formed to accommodate imaging array; (4) 3D assembly mounting to imaging array and PWB/flex accessory plane.

thickness of an entire substrate can be comparable to the thickness of a single dielectric layer. Pretested substrates are added sequentially and "riveted" to the previous buildup. With this approach, a great density of interconnections can be formed from layer to layer. The advantage of this scheme is that it allows many thousands of connections between substrates. The disadvantages of the scheme are that (1) it is not repairable in any obvious way and (2) nonadjacent connections are subject to the same limitations illustrated in Figure 14.16.

14.5.3 Method 3: For Moderately Complex Pincount Assemblies

This method is similar to method 2 and, in fact, can be applied *after* method 2 is completed. This method involves forming an edge-metal HDI system on a stacked assembly, which provides a better solution for cases where the I/O communications are more complex. In the application shown in Figure 14.33, a megaframe imager is formed by stacking multiplexers and interconnecting the top and bottom edge surfaces via an applied layer (or layers of HDI interconnects).

14.5.4 Issues in Ultradense Packaging

Unthinned silicon die are typically about 500 μm thick and, in the bare condition, are fragile and easily breakable until transferred into a package. Thinning die down to about 125 μm is sometimes done en masse (millions per month) at the wafer level for making low-profile packages, such as thin small outline packages (TSOPs). Thinned die are far more fragile, and special handling is required to prevent breakage. Using such die in an MCM assembly is problematic, especially when a number of different die types, thinned and unthinned, are used in a single substrate. It is for this reason that embedding unthinned die into a patterned overlay MCM and then thinning the entire MCM may result in improved manufacturability. This embodiment eliminates the need to further handle bare die in this most fragile state. Finally, it is clear that upon even further thinning of silicon (<50 μm), it is often seen that the silicon takes on a different behavior. The behavior is more like a foil that appears to be able to bend without breaking, as suggested in Figure 14.31, which demonstrates the flexibility of a 1-μm-thick silicon wafer.

There are three basic approaches for die thinning: (1) mechanical/chemical polishing, (2), epitaxial lift-off, and (3) backside removal. The first method involves using only mechanical ablation or some combination of mechanical ablation and chemical polishing. Manual grinding involves mounting die face down onto a working disc with a releasable substance (such as wax). The disc is lowered onto a platen containing a slurry that spins underneath the disc. Grinding proceeds by removing material from the back of the silicon die and can be done using a sequence of grinding polishes of successive finer grit sizes (analogous to proceeding from coarse to fine grits in using sandpaper). Slurries can also be chemically active to further accelerate material removal. Manual grinding is done routinely, and simple table-top machines can be calibrated to remove specific quantities of silicon. The problem with this technique in doing extreme thinning is that the starting wafer thicknesses fluctuate and, experimentally, thinning below 100 μm is problematic since calibration on one wafer may not work on subsequent wafers, resulting in die that are completely ground away or are, in other cases, too thick. In this case, it is more desirable to have in situ means of monitoring die thickness.

A second method is referred as epitaxial liftoff (ELO). Originally, it involved the use of groups III to V semiconductor materials containing a heterostructure with an engineered release layer. This release layer is usually about 1 μm or less from the top surface of the material, which can contain active circuitry. ELO is performed by immersing a sample in a wet etch that is generally inert to every part of the heterostructure but the release layer. The etching rate is initially quite slow but eventually accelerates due to the natural tendency of the sample to curl, which increases the etching cross section. Even though the curling occurs and the etching process appears to empirically exceed the theoretically projected etch rate, the process is still quite slow. The sample, when released, is foillike and tends to bond due to van der Waals forces onto substrates. ELO is an interesting but seemingly limited approach. It would probably never be practical to carry out this process on samples much larger than a small IC die, due to the very small etch cross section presented by the lateral, exposed edge of the release layer. Furthermore, the handling of successfully lifted die is difficult and the process is not easily scaled.

The third method, backside removal, is an alternative to ELO. Like ELO, it requires the existence of a release layer, although this release layer more accurately serves the role of an etch stop. The approach can be used with silicon on insulator (SOI), which is being used by several leading foundries. The approach involves using the first method to remove most of the bulk material from the backside of a die or wafer. Following this step, the samples are placed in a XeF_2 chamber. The sublimation of this material produces an etching effect that prefers silicon about 100,000 times more than SiO_2, and it is believed that this effectively allows the removal of the backside material quite effectively, leaving only the top silicon and the residual backside oxide, which can probably be left in place or easily removed (if care is taken to protect the topside oxide, which is an essential part of the native interconnection system in silicon components). This technique may represent a preferred method of preparing ultrathin silicon, but it is still an unproven method. Furthermore, it may be necessary to reengineer circuit design rules for components built this way, due to the potential effect of the work function relationships of the removed material. This effect, if it exists, is not well understood. If it exists, it could be compensated for by either (1) design-only changes or (2) backside deposition of another material to introduce a compensation effect.

Substrate alignment is another problem in stacked assemblies. The need for substrate alignment is driven by interconnection density. With very low interconnection counts between substrates, larger contact regions are permissible. Such contact regions are forgiving of misalignment, and a variety of methods can be used to align layers. The first stacking method described previously, for example, can only support this low-precision requirement. As the interconnect densities between substrates increase, however, the physical contact dimension decreases, so the need for increased alignment precision between substrates is more critical. With the patterned overlay HDI process, the use of adaptive lithography can compensate for misalignments in contact geometries on the order of 25 µm. The second stacking method is an example of a process where such flexibility might be applied. Of the three stacking approaches discussed, the third stacking method has the most severe demands on alignment. It is necessary to control the surface topography of the interconnection manifolds very tightly. For this reason, it is desirable to use substrates as opposed to bare die, since the edges of substrates can be polished after stacking to create a flat working surface for further processing. Further increases in contact density are not likely possible without precision alignment, as even adaptive lithography is limited to small displacements that can be compensated for by simple adjustments confined to a single plane of a complex interconnection manifold. It is conceivable that more elaborate adaptation systems can be devised, but such elaborations will likely have impacts on throughput and scalability due to the complexity of the run time algorithms required to support the generation of dynamic jogs in lithography. Obviously, any of these adaptive approaches are limited to systems that employ direct-write (vice-mask-based) lithography.

14.6 SUMMARY

This chapter addresses packaging configurations in which multiple components are combined. The combination is typically done for reasons of performance or miniaturization (or both). The principal approaches include MCMs and 3D assemblies, especially stacked ICs.

Besides introducing a number of definitions and classification schemes for these concepts, drivers for MCMs were discussed, including I/O and thermal management. Some interesting future possibilities, such as the use of paper-thin packages and COF were discussed. Other subjects, such as reliability, were not addressed in depth, as Chapter 16 has been dedicated to the subject of reliability in packaging.

One important lesson to learn in the use of multichip approaches is an objective assessment of the problems in packaging that are being addressed in a system design. In the 1990s, MCM technologies were considered an inevitable solution to many problems. Few of the developers of substrates in the early 1990s considered the advent of BGA and CSP approaches and the subsequent evolution of the high-density PWB industry and surface-mount technology, which mitigated many of these problems adequately without the considerable expense of MCMs. Even though the expense of MCMs was a problem to be solved through the economies of scale, the scale never emerged, in most cases, to supply the evidence necessary for this proof of principle. Instead, less attractive, cruder methods, such as simply stacking two die inside of a package, became the dominant counters to complex MCMs in many assemblies. Still, the increasing sophistication of surface mount rivals that of the most complex MCMs of the previous decade. The landscape of the MCM has receded somewhat, to be replaced with that of the CSP, stacked CSP, and SIP. The challenges still exist, in some cases, and are as complex as before. Most of the concepts of MCMs from the 1990s did not really die but have reemerged in ways not originally expected. For this reason, it is very important to understand not only the most popular schemes but also the entire range of MCM/3D technologies.

REFERENCES

1. S. AUGARTEN, *State of the Art*, New Haven: Ticknor & Fields, 1983.
2. J. J. LICARI AND L. R. ENLOW, *Hybrid Microcircuit Technology Handbook*, Park Ridge, NJ: Nyoes Publications, 1988.
3. A. J. BLODGETT AND D. R. BARBOUR, Thermal Conduction Module: A High-Performance Multilayer Ceramic Package, *IBM J. Res. Develop.*, Vol. 26, pp. 30–36, 1982.
4. D. E. MEYER, Wafer Scale Integration? Time to Get Serious, *Semicond. Int.*, Vol. 9, p. 32, 1986.
5. Information on the NIST Advanced Technology Program.
6. S. CANNON, J. LYKE, J. STAGGS, D. WATSON, AND D. FULLER, A Parallel C30 Architecture for Miniaturized 3D Monolithic Packaging, Transputer Research and Applications Conference (NATUG-7), 1994.
7. G. A. KATOPIS, W. D. BECKER, T. R. MAZZAWY, H. H. SMITH, C. K. VAKIRTZIS, S. A. KUPPINGER, B. SINGH, P. C. LIN, J. J. BARTELLS, G. V. KIHLMIRE, P. N. VENKATACHALAM, H. I. STOLLER, AND J. L. FRANKEL, MCM Technology and Design for the S/390 G5 System, *IBM J. of Res. Develop.*, Vol. 43, pp. 621–649, 1999.
8. C. W. EICHELBERGER, R. J. WOJNAROWSKI, R. O. CARLSON, AND L. M. LEVINSON, HDI Interconnects for Electronic Packaging, SPIE Symposium on Innovative Science and Technology, Symp. Proceedings, 1988.
9. E. BOGATIN, Chips-First Technology: Has Its Time Finally Come? *Semicond. Int.*, p. 46, 2002.
10. R. A. FILLION, R. J. WOJNAROWSKI, B. GOROWITZ, W. DAUM, AND H. S. COLE, Conformal Multichip-on-Flex (MCM-F) Technology, International Microelectronics and Packaging Society MCM-95, Denver, CO, 1995.
11. K. D. GANN, Neo-Wafer 3D Packaging, International Microelectronics and Packaging Society, 3D Packaging Advanced Technology Workshop, San Diego, CA, 1998.
12. J. H. DAY, Packaging Options Stack up for Stacking Active Devices, in *Electronic Engineering Times*, 2001, pp. 85, 94.
13. Stacked Die CSPs Piling up Gains, in *Printed Circuit Design and Manufacture*, 2003, p. 4.
14. J. C. LYKE AND G. A. FORMAN, Space Electronics Packaging Research and Engineering, in *Microengineering Aerospace Systems*, H. Helvajian, ed., El Segundo, CA: Aerospace Press, 1999.
15. V. SOLBERG AND I. OSORIO, 3D Packaging Solution for High-Performance Memory, *Adv. Packaging*, pp. 23–25, 2003.
16. K. GANN, Neo-Stacking Technology, *HDI Mag.*, 1999.
17. K. STURCKEN, S. KONECKE, AND K. MASON, Bare Chip Stacking, *Adv. Packaging*, pp. 17–19, 2003.
18. J. C. LYKE, R. J. WOJNAROWSKI, R. SAIA, G. A. FORMAN, B. GOROWITZ, AND M. CICCARELLI, Development and Application of Practical Three-Dimensional Hybrid WSI Technology, XVIII Government Microcircuit Applications Conference (GOMAC), Las Vegas, NV, 1992.
19. E. BOGATIN, Origami-Style Structure Simplifies Packaging Efficiency, *Semicond. Intl.*, pp. 50, 2003.
20. J. C. DEMMIN, Stacked CSPs: Issues and Results, *Electr. Packaging Produc.*, pp. 26–28, 2003.
21. D. SERAPHIM, R. LASKY, AND C.-Y. LI, *Principles of Electronic Packaging*, New York: McGraw-Hill, 1989.
22. H. B. BAKOGLU, *Circuits, Interconnections, and Packaging for VLSI*. Reading, MA: Addison-Wesley, 1990.
23. W. E. DONATH, Placement and Average Interconnection Lengths of Computer Logic, *IEEE Trans. Circuits Syst.*, Vol. CAS-26, pp. 272–277, 1979.
24. J. A. DAVIS, *A Hierarchy of Interconnect Limits and Opportunities for Gigascale Integration*. Atlanta, GA: Georgia Institute of Technology, 1999.
25. P. ZARKESH-HA, J. DAVIS, W. LOH, AND J. MEINDL, On a Pin versus Gate Relationship for Heterogeneous Systems: Heterogeneous Rent's Rule, IEEE Custom Integrated Circuit Conference, 1998.
26. *Information on the DARPA Power-Aware Computation/Communications Program*.
27. *International Technology Roadmap for Semiconductors*, Semiconductor Industry Association, 1999.
28. J. LYKE, K. AVERY, AND P. BREZNA, BMDO/AFRL Partnership: Advances in Data Handling Systems for Space Experiment Control, *AIAA J. Spacecraft Rockets*, Vol. 39, pp. 481–488, 2002.

EXERCISES

14.1. Prepare a table headed with columns labeled MCM-C, MCM-D, and MCM-L. Prepare the following rows: density, cost, speed, and thermal performance. For each row, identify the relative descriptor that applies to each technology.

14.2. Prove the corresponding Rent's exponent for two- and three-dimensional systems: two dimensional, $p = \frac{1}{2}$; three dimensional, $p = \frac{2}{3}$.

14.3. Identify at least two advantages of the Figure 14.11d construction and identify at least two different methods for attaching such a module to a PWB.

14.4. Calculate the wiring capacity (in./in.^{2}) of a PWB having 4-mil lines and spaces and 12 layers. For this board, it is estimated that vias will occupy about 10% of the available area and routing efficiency is 80%.

Chapter 15

Packaging of MEMS and MOEMS: Challenges and a Case Study

AJAY P. MALSHE VOLKAN OZGUZ, AND JOHN PATRICK O'CONNOR

15.1 INTRODUCTION

Microelectromechanical systems (MEMS) and microoptoelectromechanical systems (MOEMS) packaging are much different from conventional integrated circuit (IC) packaging. Many MEMS/MOEMS devices must interface to the environment in order to perform their intended function, and the package must be able to facilitate access with the environment while protecting the device. The package must also not interfere with or impede the operation of the MEMS/MOEMS device. For example, the die attachment material should be low stress and low outgassing, while also minimizing stress relaxation over time that can lead to scale factor shifts in sensor devices. Many devices are application specific, requiring custom packages that are not commercially available. Devices may also need media-compatible packages that can protect the devices from harsh environments in which the MEMS/MOEMS device may operate. Techniques are being developed to handle, process, and package the devices such that high yields of functional packaged parts will result. This chapter reviews and discusses packaging challenges that exist for MEMS/MOEMS and exposes these issues to new audiences from the IC packaging community. Additionally, in the light of the challenges, a case study for packaging design considerations and guidelines for a specific and successful MOEMS, the Digital Micromirror Device is discussed.

15.2 BACKGROUND

Today's artificially intelligent systems are fabricated using submicro-scale devices in micro- and meso-scale packages, primarily using electronic signals and the logic of programming to interface to the macroscopic world. The progress in this arena began in the 1940s with the invention and commercialization of the silicon transistor. Over the years, the packaging of these devices, which started out as an "afterthought" processing step to protect the device and connect it to the outside world, ultimately created its own standards and an industry unto itself.

Advanced Electronic Packaging, Second Edition, Edited by Richard K. Ulrich and William D. Brown
Copyright © 2006 the Institute of Electrical and Electronics Engineers, Inc.

15.2.1 Mixed Signals, Mixed Domains, and Mixed Scales Packaging: Toward the Next-Generation Application-Specific Integrated Systems

With the advent of a new type of knowledge-based economy in which the next product must steal the market of the earlier version to maintain the leadership, application-specific soft features of products–along with fast, dense, and inexpensive traits—became the guideline for designing next-generation electronic products. These guidelines implicitly result in three key features for the next generation *application-specific* systems-on-a-chip (SOAC) or systems-on-a-package (SOAP): mixed signals, mixed environmental domains, and varying scales of devices and packaging components (see Fig. 15.1). The collision of these variables makes the packaging a critical part of the device. During the past few decades, various research laboratories and corporations around the world have been busy researching and developing novel MEMS, with an ultimate goal of combining microelectronics, nanotechnology, biotechnology, optics—you name it—to enable diverse functionality in a single chip or a package. For the packaging community, this progress poses new development and manufacturing challenges for integrating mixed signals (electrical, optical, chemical, neural, magnetic, acoustic, etc.), mixed ambient domains (vacuum, hermetic, aqueous, optical, etc.), and last, but not the least, various sizes and shapes of devices.

These new developments only add to what needs to be juggled by the packaging engineer. Such professionals must be more multifaceted than ever. The challenges faced by the packaging industry will continue to dominate in a range of technical fields, such as dicing, releasing, handling, interconnection, by-product management, materials, manufacturing process, methods for reliability testing, modeling and simulation techniques, standards, professional development of today's packaging community, and, of course, interdisciplinary and integrated education at universities to foster a new set of leaders in packaging technology. On the background of these challenges and ongoing developments, this chapter details packaging challenges for MEMS and discusses technical approaches for packaging MOEMS through a case study.

15.2.2 Microelectromechanical Systems

Microelectromechanical systems consist of mechanical devices and machined components ranging in size from a few microns to a few hundred microns. They can be the mechanical

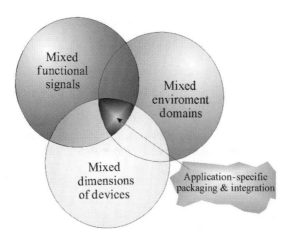

Figure 15.1 Integration of signals, domains, and scales for application-specific systems.

interconnects of microsystems and can also transduce signals from one physical domain to another, such as mechanical-to-electrical, electrical-to-mechanical, electrical-to-chemical, and so forth. These devices are broadly categorized as either sensors or actuators. MEMS sensors are devices such as pressure sensors, accelerometers, and gyrometers that perceive an aspect of their environment and produce a corresponding output signal. MEMS actuators are devices that are given a specific input signal on which to act and a specific motion or action is produced. For example, a microengine is connected to a gear train, which moves in response to an electrical input signal. Other examples of MEMS actuators are microengines, microlocks, and optical discriminators [1, 2].

Sensors can be thought of as being passive, waiting on a signal from the environment to elicit a response, while the user activates actuators. The sources of motion for MEMS elements, such as gears or microlocks, are usually electrostatics, thermoactuation, wobble motors, inertia, and with limited success, even microsteam-engines [1]. These "motors" provide the mechanical source input required to activate the actuators and can be used with a gear train to raise or lower a micromirror for a digital light processing application [2]. These microengines produce the motion of a linear actuator or activate a weapon safety system that requires motion of the mechanical locking elements [3–5]. Just these few examples show the diversity of interfaces that MEMS packaging engineers face between signal domains.

In state-of-the-art manufacturing, MEMS are primarily fabricated using manufacturing processes and tools borrowed from the microelectronics industry. Many of these processes and tools are used directly, while some have been modified to meet the specific needs of MEMS [6]. The development cycle for MEMS is long and often several design cycles are required to comply with design specifications [7]. This is because, in many cases, the products, as well as the technologies, are being developed simultaneously. MEMS design and fabrication are more like that of application-specific integrated circuits (ASICs). There are currently relatively few generic parts that can be rearranged to create any one of thousands of different devices in the manner that resistors, transistors, and capacitors can be arranged in IC design [7]. This can be problematic for developing MEMS packaging since every device is different in fabrication and application. Research laboratories are developing libraries of different parts that can be reassembled in various configurations [8, 9]. Researchers are also working on CAD (computer-aided design) tools for MEMS that will facilitate solid modeling of these MEMS devices, leading to reduced design cycle times. This allows faster design times and increased modularity of the devices. Further, it is important to note that MEMS devices and the packaging community face challenges in understanding complex material properties in static as well as dynamic conditions. One example is tribology, particularly when interfacing dissimilar materials across signal, domain, and scale boundaries. Any modeling at these scales is subject to input parameter accuracy. Materials do not necessarily obey the physical behavior observed at common length scales, and their properties in the system depend on microstructure, residual stress, and the like, which can vary due to conditions ranging from set process parameters to the skill of the operator to control the process parameter window.

As in MEMS fabrication, MEMS packaging borrows many processes and toolsets. However, extra care must be taken since MEMS dice need to interface with the environment for sensing, interconnection, and/or actuation [10]. MEMS packaging is application specific, and the package allows the physical interface of the MEMS device to the environment [10]. In the case of a fluid mass flow control sensor, the medium flows into and out of the package. Harsh environments may create different challenges for the packaging of MEMS [10]. These types of packaging are referred to as media-compatible packaging. In addition to challenges related to the environment of the MEMS chip and interfacing it with the environment, challenges also exist inside the MEMS package with die handling, die attach, interfacial stress, and outgassing [11]. These new challenges in the field of MEMS packaging need immediate research and development efforts. The reason for this attention is that devices can easily be destroyed by assembly and packaging processes. If a device no longer performs its intended function, whether it be rotary actuation or pressure sensing, it is simply defined as failure.

To date, most of what is known about MEMS packaging remains proprietary and published literature is scarce. A disproportionality exists between the resources spent on the packaging of MEMS and the time spent researching MEMS packaging. Currently, the cost of MEMS packaging typically accounts for 60% or more of the sale price of the device [12]. In contrast with IC packages, which are considered commodity items, many MEMS packages will be specialty items that will generate higher profit margins. This paradigm shift is creating new opportunities for small, low-volume packaging companies.

MEMS packaging is extremely important for the viability of MEMS success. New devices are continually being developed that have great promise for miniaturizing existing sensors and actuators, as well as novel devices not previously possible. Currently, it is the packaging of these devices that is limiting their market applications. MEMS devices that are currently available, such as pressure sensors and accelerometers, are devices that have been relatively simple to package and, therefore, have commercial viability. MEMS is a few billion dollar market today and is looming on the edge of revolutionizing the entire world with products that are both smaller and less expensive than current products in the sensor and actuator regime. MEMS has been called the second silicon revolution, with benefits far beyond what has been seen with the first silicon revolution in integrated circuit technologies. Fabrication techniques, equipment, and new technologies are evolving every year that advance the field of MEMS even further. It is of the highest importance that MEMS packaging stay abreast of the fabrication so that new systems can be both created and integrated into widespread commercial and strategic applications.

15.3 CHALLENGES IN MEMS INTEGRATION

There is a philosophical difference between the motivation for packaging ICs and packaging MEMS. The goal of IC packaging is to provide physical support and protection for the chip, provide an electrical interface to active chip(s) in the system, supply signal, power, and ground interconnections, and allow heat dissipation [13]. Also, a package must effectively isolate the chip physically from its environment. MEMS devices, on the other hand, often are intimately interfaced with their environment [10]. Another issue is the media compatibility of the MEMS package. MEMS devices may need to operate in diverse environments or media, such as under automobile hoods, with intense vibrations, and in saltwater, strong acids, or alkaline or organic solutions. For some pressure sensors, media compatibility has been achieved by protective parylene coatings and silicone gel fills [10]. Overall, the package, while performing detection or actuation, must be able to withstand the environment(s).

Another challenge in MEMS packaging is the effect of packaging parameters on reliability. The package is part of the complete system, and all aspects of the system must function together, and must be compatible with each other. This determines which materials and what design considerations and limitations become important. One of the main scientific challenges of MEMS is the issue of material properties. The properties of the materials depend on how they are used, processed, the heat treatments to which the materials are subjected, and even the specific pieces of equipment used during fabrication. For instance, using a die-attach material with a large coefficient of thermal expansion (CTE) mismatch with silicon can induce undesirable stress levels on the device. Not all the materials used react in the same way to these parameters, so compromises must be made. Some materials may be hard to obtain with research and development production run numbers. Low quantities of materials are used, and suppliers are reluctant to sell small quantities or develop new products for limited markets, such as protective media compatibility coatings [11]. One favorable aspect about the materials used in microsystems is that the material properties generally get better at the microscale. This is due to a decrease in the number of defects encountered in the materials. The defect density remains about the same as in the macroscale, but since the MEMS devices are so small, the chance of a killer defect occurring in a device is reduced.

Packaging of MEMS dice is application specific, and, hence, desired process steps can vary significantly. For example, a pressure sensor package will be dramatically different than a package for a micromirror array. Thus, it is important to classify MEMS dice by their packaging requirements and then develop the packaging standards and related knowledge base. Different standards would apply for different types of MEMS applications since they operate in such a variety of different physical domains. For example, microfluidic devices have requirements that are vastly different than electrostatic actuators. The device and the package should be designed concurrently with the application and the environment in mind from the project conception.

Table 15.1 summarizes the different techniques that can be used to meet the various packaging parameter challenges. These solutions are not complete or foolproof but are techniques that have been exercised and provide direction for further developments. The first column of Table 15.1 lists are the packaging domains where challenges exist. The

Table 15.1 Current Packaging Parameters, Challenges, and Suggested Possible Solutions for MEMS

Packaging parameters	Challenges	Possible solutions
Release etch and dry	Prevention of device stiction, washing away of parts during release, must release parts individually after dicing.	Freeze-drying, supercritical CO_2 drying, roughening of contact surfaces, coatings, or processes that reduce surface tension, use of dimples, develop a dicing process in which parts can be released before dicing, possibly wafer cleaving or laser sawing.
Dicing/cleaving	Eliminating contamination caused by cooling fluid and particulates during wafer sawing.	Release dice after dicing, cleaving of wafers, laser sawing, wafer-level encapulation.
Die handling	Damages top die face contact region.	Fixtures that hold MEMS dice by sides rather than top face, such as collets that fit existing pick-and-place equipment.
Stress	Abating performance degradation and resonant frequency shifts, curling of thin-film layers, misalignment of device features.	Low-modulus, low-creep die-attach material, annealing, die-attach materials with CTE similar to that of silicon.
Outgassing	Stiction, corrosion, outgassing of organic solvents from polymeric die-attach materials.	Low outgassing epoxies, cyanate estares, low-modulus solders, new die-attach materials, removal o f outgassing vapor.
Testing	Applying nonelectric stimuli to devices, testing moving device features before release, unable to release parts before dicing.	Electrical test structures to mimic nonelectrical functions, modify (where possible) wafer-scale probers to do nonelectrical tests, cost-effective, high-throughput, and parallel packaged devices test systems.

second column is the list of challenges that must be overcome for efficient batch processing of MEMS with high-yield margins. The final column is a list of possible solutions that are being used or could be used to meet the challenges in the center column. The following section goes into much more detail in discussing these challenges and possible solutions [14].

15.3.1 Release and Stiction

The process of freeing the die by etching oxide so it becomes operational is referred to as release of the die. Releasing of the MEMS dice is an important step in MEMS packaging. Typically, the polysilicon features are surrounded by silicon dioxide during fabrication, which protects the features and prevents them from becoming damaged or contaminated. This oxide is etched away, freeing the devices, so they become operational. This is typically done using an HF etch, which is selective between SiO_2 and Si [3]. The challenge exists in economical and efficient releasing of the MEMS parts. It is most economically done in wafer form as a batch process, but this leads to near certainty that contamination of the devices and destruction of the parts during the dicing of the released wafer. The cooling fluid used in dicing will obliterate the tiny mechanical devices. The most inefficient time is after dicing since each chip is released individually rather than the entire wafer at once. The MEMS features, however, do remain protected throughout the potentially lethal dicing stage, and, currently, this is the only way to obtain functional devices at the die level.

During and after the release, there is the possibility of a beam-type element being pulled down onto the substrate due to capillary forces. This is referred to as stiction. Stiction occurs from the capillary action of the evaporating rinse solution in the crevices between the elements, for example, between cantilevers and the substrate [15]. As the device dries, surface tension pulls its features into contact with the substrate. Unfortunately, stiction can render the MEMS devices useless after all other resources have been invested in them. Preventing stiction from occurring after release is a major challenge. Some methods that are effective are freeze-drying and supercritical CO_2 drying [16]. These methods remove the liquid surface tension from the drying process, preventing stiction from occurring. Although not yet foolproof, these methods have been used with some success.

MEMS parts can also be placed in an additional bath that applies a monolayer film, which reduces stiction on the surface of the devices. There are coatings of different compositions that can be used and have been proven successful in preventing stiction [17]. There are issues, however, on the stability and integrity of these coatings. Little is understood about how these films age, and if they will still be present on the devices after a year or several years. Also, these films can be scrubbed off by the friction between moving parts, where stiction is a major issue. Additional research is needed on the subject of these coatings that includes stability studies over time.

Another method that can be used as a processing step is to roughen the surface of the devices to minimize contact area between features [18]. A major problem here is that the very areas susceptible to stiction are very difficult to process in this manner. The close tolerances and gaps between the features and the substrate make it impossible for dry etching methods to be effective. Stiction can also be reduced in the design process by using dimples in regions of the device where stiction may be a problem [19,20]. These small protrusions on the bottom of an element can greatly reduce the contact area between the MEMS device element and the substrate. These are simple cuts in the sacrificial oxide layers that allow polysilicon to form a bubble on the bottom side of the device feature.

15.3.2 Dicing

Another challenge in MEMS packaging is dicing the wafer into individual dice. Dicing is typically done with a diamond saw a few mils thick. This requires that coolant flow over the surface of the very sensitive dice, along with silicon and diamond particles that are generated during sawing, and which are all deadly to the devices. These particles, combined with the coolant, can contaminate and even destroy the devices and, thereby, negatively impacting the yield. The fluid simply washes the features off the surface of the wafer. Contaminants can also get into the crevices of the remaining features, causing those devices to fail. An alternative to dicing is wafer cleaving. Wafer cleaving, also known as the scribe-and-break technique, is common in III-V semiconductor lasers and also has applications in MEMS [21]. Cleaving does not require coolant and does not generate nearly as many particles as sawing. However, cleaving is difficult to master and is highly dependent on the crystal structure and direction of the wafer. Laser dicing and wafer-level encapsulation are two other methods that may decrease the hazards of dicing. Laser dicing is very accurate and generates few particles, but may be expensive and slower than diamond sawing.

15.3.3 Die Handling

Die handling is another area of MEMS system fabrication that requires additional development. Because of the delicate surface features of MEMS, these dice cannot be handled from the top surface using vacuum pick-up heads as is done in traditional IC die assembly. The vacuum heads may damage the surface features of MEMS devices. The MEMS dice must be handled by their edges, which requires new infrastructure for automated handling. Handling chips by the edges is more difficult than by the top surface because of a greatly reduced surface area in the pick-up region and increased dexterity requirements of the pick-and-place equipment. Effective edge handling equipment could simply be collets that fit into existing pick-and-place equipment and allow the dice to be picked up by the edges using vacuum or mechanical clamping. These MEMS die-handling fixtures could be fingers or clamps that delicately handle the MEMS dice by their edges. In order to handle the high volumes of MEMS chips, die handling fixtures and methods that handle the chips by the edges must become commonplace in intermediate- to high-volume MEMS packaging houses.

15.3.4 Wafer-Level Encapsulation

Wafer-level encapsulation mitigates many of the challenges associated with MEMS packaging, such as dicing, handling, and encapsulation. In this technique, a capping wafer is bonded to the top of a device wafer and, when diced, each MEMS chip effectively has a protective lid attached to it. There are several methods that can be used to bond the capping wafer to the device wafer. The wafer bonding can be done using direct bonding, however, the required temperature is about 1000°C [22] and is incompatible with integrated CMOS (complementary metal–oxide–semiconductor) control circuitry. Glass-frit or anodic bonding is more commonly used because the processing temperature is between 450 and 500°C [10]. However, the glass frit may cause stress in the die if the glass used has a CTE sufficiently different from that of silicon. Anodic bonding requires high voltage, which can also be a disadvantage for integrated systems combining MEMS and IC devices on a single chip [3]. These wafers can also be bonded in a vacuum to produce a vacuum environment inside each device chip, which may be desirable for some applications. In an important development, recent papers address the possibility and effects of pulsed, as well as continuous-wave, laser-assisted *selective* wafer bonding [23–25]. For example, a continuous-wave (CW) carbon dioxide (CO_2) laser at a wavelength of 10.6 μm and a focal spot size of 0.5 mm in diameter was employed to bond silicon-to-silicon [26] and gallium arsenide-to-silicon in a selective vacuum environment using eutectic lead–tin solder as the intermediate bonding material. The bonding temperature was less than 200°C and the bonding strength was more than 3 MPa, which is comparable with other bonding methods. Although wafer-level encapsulation has many advantages, it may also add some cost to device fabrication. However, a cost analysis of fabrication, packaging, and testing may result in a reduction of the overall manufacturing costs due to greatly reduced packaging and testing costs.

15.3.5 Stress

When polysilicon is deposited, a great deal of stress is created in the film. Most of this stress can be annealed out at a temperature of around 1000°C and is most effective if the polysilicon is deposited amorphously and then annealed to form a polycrystalline structure [27]. This creates the lowest stress arrangement with the fewest defects in the polysilicon. The second source of stress is the die-attach material at the interface between the MEMS die and the package substrate. Depending on the die-attach material and CTE mismatch between the package and the chip, interfacial stress can develop within a MEMS package [28]. Die stress can be reduced by simply using a low (modulus) stress die-attach material. However, many low (modulus materials) stress attaches are also high in outgassing, which can cause other detrimental effects on the MEMS. Also, low-stress die-attach materials may creep, leading to stress relaxation.

A major drop in reliability may be caused by excess stress in the package. This stress can be caused by stress-inducing fabrication processes, CTE mismatches between the die attach/substrate, and/or die attach/die, lid sealing, or shrinkage during the attach curing. The results of stress are that the devices may deform, gear teeth may become misaligned, tensile stress may cause the resonant frequency to increase and result in device breakage, and excessive compressive stress causes long beam elements to buckle. Packaging stresses can induce both offset and scale factor shifts in sensors. In addition, the use of hard solders, such as AuSn or AuSi, can put excessive stress on the delicate components and cause the features, as well as the die itself, to warp or fail [29, 30].

Stress effects also worsen as chip size increases. MEMS chip sizes may be larger than many IC chips since the feature sizes are larger and the devices typically require more die area. This stress can be reduced by using lower modulus die-attach materials that deform as the chip and package expand and contract [29]. These low-modulus die-attach materials may also allow stress relaxation, which may be a poor choice because a change in the stress state can lead to changes in device performance. These variations in stress lead to changes in device characteristics, such as resonant frequency, offset voltage, and scale factor. Although high stress may be undesirable, it is also undesirable for the stress-state to change over time [31].

15.3.6 Outgassing

For die attachment, either polymer or metal solders can be used. When a polymer, such as an epoxy or cyanate ester, is used for die attach, water and organic vapors are generated as they cure [32]. The emission of these water and/or organic vapors is called outgassing. Outgassing also occurs after die attachment when the die is in a package. The water and organic vapors generated in outgassing may then redeposit on the features, in crevices, and on bond pads. This leads to device stiction and corrosion. Stiction prevents the operation of the moving parts, while corrosion can affect the electrical conduction paths by increasing impedance and promoting electrical failures. Die-attach materials with a low Young's modulus, like epoxies, also allow the chip to move during ultrasonic wirebonding, resulting in low bond strength, which has been documented in certain pressure sensors [11]. These poor wirebonds create reliability issues, as well as increased impedance, due to the poor attachment. Possible solutions to outgassing challenges include very low outgassing die-attach materials with sufficiently high Young's modulus and/or the removal of outgassing vapors during die-attach curing. For solving the problem of outgassing after die attach, when the die is in the package, special mechanisms/techniques can be adapted so that there is chemisorption or physisorption of outgassed materials. One such technique that can be used to remove water and/or organic vapor is the use of gettering materials.

15.3.7 Testing

The testing of MEMS devices is also an issue. No one wants to package a bad chip; it is too expensive and time consuming. MEMS devices are difficult to test for several reasons. One is that, in the unreleased state, the devices are not free to move, so operational testing of moving features is impossible. In this state, only electrical measurements, such as sheet resistance and conductivity, can be made. After releasing, devices can be tested for proper mechanical operation; however, this is after significant time and processing investments in the part. Also, individual dice are time consuming to handle and are not well suited for automatic test probers, as are wafer-scale parts. At the current time, the most efficient testing is done after packaging when the full operation of the parts can be tested, and the parts are easy to handle in packaged form. Any electrical testing that can be done prior to releasing the devices should also be used to eliminate die that have electrical faults from the process stream.

15.3.8 State-of-the-Art in MEMS Packaging

The present state-of-the-art in MEMS packaging combines MEMS with ICs and utilizes advanced packaging techniques to create complex MEMS. Two different approaches being used are multichip modules (MCMs) and monolithic integration. Monolithic integration is the placing of CMOS and MEMS on the same chip. The challenge is that the processing steps for CMOS and MEMS are not compatible. For instance, high-temperature annealing used in MEMS processing destroys the diffusion profiles and aluminum interconnects used in the CMOS devices. There are three main methods that have been used to create the monolithic integration of CMOS and MEMS: (1) Electronics First (University of California, Berkeley), (2) MEMS in the Middle (Analog Systems), and (3) MEMS First (Sandia National Laboratories) [3].

One of the most recent efforts in monolithic integration has been Sandia's MEMS First effort in which the MEMS are first fabricated in an etched trench, then covered with a sacrificial oxide. After the trench is filled completely with SiO_2, the surface is planarized. This flat surface serves as the starting material for the CMOS foundry. The sacrificial oxide covering the MEMS is removed after the CMOS devices are fabricated. This protects the MEMS devices from the CMOS processing steps [3, 33]. An alternative approach to monolithic integration is the use of MCMs [34]. An IC and a MEMS dice can be placed in the same package, either in two- (2D) or three-dimensional (3D) fashion, as opposed to being on the same die, thereby eliminating these processing incompatibilities. However, parasitic effects are ultimately reduced, placing the control electronics as close as possible to the MEMS devices.

Advanced packaging techniques, like MCM and flip-chip technologies, are being actively pursued for use with MEMS [35]. The logic behind the use of MCM technology is that several different MEMS sensors and actuators, or a combination, can be combined into a single package, forming complex systems that can perform several functions in a single package. These MCM systems should be modular and any number of them can be constructed from the available MEMS sensors and actuators [34]. This opens the already nearly limitless range of applications for systems of much greater complexity.

A downside to MCMs use is the apparent added packaging expense. Further, signal loss is especially noticeable with some capacitive devices. For some of these devices, the capacitance changes being sensed are less than a femtofarad. A single 100 μm by 100 μm bond pad can add a picofarad of capacitance, thereby swamping the desired signal change [36]. In these applications, one may put detection circuitry right next to the MEMS device in order to decrease the effects of parasitic capacitance. The apparently greater cost of packaging is due to the issue of known good die (KGD), or pretested dice, that are known to work before being placed into the MCM. MCMs also have larger package sizes that do cost more than a single-chip package, but fewer packages are needed. Also, depending on the MCM requirements, the substrate can be a major part of the packaging costs if it requires multiple layers and high-density signal lines. Stress can also be increased from the larger package dimensions [13]. The disposability of MCM-packaged systems becomes much less palatable and the cost of rework more acceptable.

The next state-of-the-art process is the idea of lab-on-a-chip. This is the concept of several sensors and actuators on a single chip, or in the same package using MCM technology, forming a mixed signal system that fulfills a function or group of functions. The objective is to realize the size, weight, and cost reductions of microsystems that create significant benefits over many larger systems. Mixed signal in this sense refers to input and/or output signals that can be mechanical, electrical, magnetic, optical, biological, acoustical, chemical, and so forth. There are several directions researchers are going with the concept of lab-on-a-chip. There are applications involving chem-lab-on-a-chip, where the object is to create several sophisticated chemical sensors on a single chip [36]. Another application is an optical bench-on-a-chip, which utilizes semiconductor lasers, beam splitters, movable mirrors, lenses, and the like to make a miniaturized version of an optical bench [37]. Also, efforts have been directed toward a DNA (deoxyrionucleic acid) lab-on-a-chip, where the polymerase chain reaction (PCR) is carried out to amplify DNA and then various separation techniques are used to analyze the DNA [38].

Due to the size of MEMS devices, it is best to have noncontact signal interfaces with the environment. Even very small contact forces can destroy microdevices; however, they can generally survive amazingly high noncontact force levels, such as inertial forces [39]. Mechanical signal interfaces, in the micronewton range, between MEMS devices, such as gear trains, however, are commonplace. The lab-on-a-chip concept capitalizes on the many possibilities and advantages of MEMS. It is conceivable that an entire control panel of sensors could be reduced to hand held size. This would have tremendous value in spacecraft, fighter jets, and other strategic applications where space and weight considerations are critical [35].

15.3.9 Future Directions

The role of packaging is to provide a compact housing as well as an interface between a device and the outside world. It should protect the chip, while letting it perform its intended function cleanly with very little attenuation or distortion of the *signals*, such as electrical and/or mechanical and/or optical in the given environment, and do so at a low cost. The packaging and assembly, as well as the materials used, are integral parts of a microsystem. The total cost of processing and assembly must be taken into account when designing and fabricating a device. Currently, packaging is a significant portion of the total cost of a MEMS device. MEMS packages will be specialty items that generate higher profit margins, creating new opportunities for small, low-volume packaging companies. Traditional front-end and back-end packaging become blurred and united in the development of the system. MEMS packaging breaks the paradigms of traditional packaging and is an exciting field ready for additional widespread research and development of new applications.

Currently, there is scattered research being performed related to MEMS packaging, for example, in the areas of die attach and outgassing. However, focused efforts on packaging issues, such as stiction, stress, and outgassing, are essential for reliable implementation of the technology.

In the future, the field of packaging and integration will be required to consider not only the packaging of individual devices but also the seamless integration of electrical (ICs) and/or mechanical (MEMS) and/or optical (integrated optics) and/or chemical devices. Generic packaging schemes, "breadboards/templates" should be developed so that multiple users can select a common package design. MEMS device designers should not be required to also develop custom packages where more generic packages would serve their purposes well. The advanced packaging and integration concepts need to address the packaging of individual devices, as well as multisignal devices, for synergistic response in the desired environment. In particular, for MEMS devices, based on the current knowledge base, we believe that the packaging and integration could be realized in three ways: hybrid thick film, multichip module (MCM), and monolithic integration. Hybrid thick film is the integration of passive components, such as resistors, that can be fabricated using thick-film technologies. MCM is another integration regime that takes existing chips from different signal domains and integrates them into a single package. This increases silicon efficiency, decreases the effects of electrical parasitics, and creates systems that are more compact. Monolithic integration is the most compact form of system integration but creates the most processing challenges. With CMOS and MEMS processing incompatibilities already mentioned, it takes creative processing solutions to create functional systems.

The future of MEMS is most assuredly the marriage of mechanical devices to control electronics and optics. Also, there will be interaction between micromechanical, optical, chemical, biological, and electrical systems to perform sensing and actuation tests and analysis in many domains never before possible. Success of this development in these domains will result in allowing application-specific systems and product designers to deliver next-generation engineered systems with novel features such as self-organization for new interfaces, self-repair, self-powdering, multifunctionality and reconfigurability, on-demand information-to-knowledge conversion, and devices that are surely heading in the direction of "biomimetic packaging."

15.4 PACKAGING CONSIDERATIONS AND GUIDELINES RELATED TO THE DIGITAL MICROMIRROR DEVICE

The integration of micromechanics and semiconductor electronics has led to microsystems, that are finding their way into a broad range of applications involving optics, electronic, mechanical, thermal, fluidic, magnetic, and electromagnetic signals and related phenomena. Due to the photonics and communication revolution, of particular interest is their implementation and functionality in optical applications.

15.4.1 Introduction and Background to MOEMS and Particularly DMD Devices

Optical MEMS or MOEMS combine micromechanics with electooptics (mass, force, displacement with photons and electrons). The many applications of optical MEMS can be divided into two general categories:

1. Sensing optical signals to acquire, process, and communicate information
2. Controlling optical signals for image processing and generation

They are used to sense displacement, acceleration, flow, temperature, and pressure. They are also widely used in the fiber-optics networking industry for information communication. MOEMS mirrors, shutters, actuators, and fiber aligners are employed in optical switches, tunable filters and lasers, attenuators, add/drop multiplexers, and performance monitors.

MOEMS have found significant use in display systems to generate images onto a screen for real-time viewing or to present images on recording media. They can also be used to scan for information, such as bar code readers, and to serve as read–write devices for optical storage applications. Two such MEMS devices for displays are the grating light valve, which is fabricated of thin deformable silicon ribbons, and the digital micromirror device, which consists of an aluminum array of movable micromirrors integrated on top of a silicon chip.

MOEMS are in their infancy and are receiving considerable attention due to the huge growth opportunities the technology provides. Increased research and development efforts will increase and expedite their delivery into the above and into new industries. New materials and processes will be developed and will improve existing applications and enable new applications. Some new applications include miniaturized optical spectrometers and spectrally sensitive arrays for infrared imaging. Industry is also exploiting the miniaturization of these electromechanical microsystems in the fields of biology and medicine. MOEMS are assured to observe considerable growth in both their application and technology as we move forward. Designing, fabrication, packaging, testing, and cost remain important areas for further progress.

MOEMS packaging encompasses the design, fabrication, assembly, and test of a reliable and functional "home" for a MEMS device. In light of the theme, this section of the chapter focuses on packaging considerations and guidelines related to the Digital Micromirror Device (DMD), an optical MEMS developed by Texas Instruments, currently and primarily used in projection display markets such as conference room projectors, large venue/digital cinema, and home entertainment systems.

The DMD device is a reflective light modulator and its primary use is in projection display markets. Its function is to reflect light from an illumination source (lamp), through an optical configuration (color wheel, prism, projection lens), and onto a display screen. The device is composed of an array of highly reflective aluminum mirrors that project a digital

15.4 Packaging Considerations and Guidelines Related to the Digital Micromirror Device

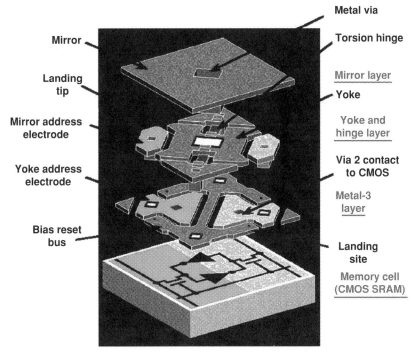

Figure 15.2 DMD pixel stackup.

image. The number of mirrors in the array is determined by its resolution (SVGA; 800 × 600; XGA; 1024 × 768; or SXGA; 1280 × 1024), and the active array size is determined by the resolution and the mirror size and pitch. Each mirror represents an active pixel on the screen image. Using the DMD device to digitize light in this fashion leads to a Digital Light Processing (DLP) system.

The micromirror array is fabricated using surface micromachining layering technologies (repetitive deposition, patterning, and removal processes [36,37]. The array rests on top of and is electrically connected to a silicon substrate utilizing static random-access memory (SRAM) CMOS circuitry. A stack-up of a single DMD pixel is illustrated in Figure 15.2. Each mirror is supported by a metal via that rests on a torsion hinge and is addressed to a specific SRAM cell for independent control. The mirrors can be rotated $\pm 10°$, representing either an on or off pixel. Mignardi [38] provides a more complete overview of the DMD pixel and its fabrication.

Figure 15.3 shows four scanning electron microscope (SEM) images of the mirror array. Figures 15.3*a* and *b* illustrate the mirror length scale by comparison to an ant foot and salt crystal, respectively. Figure 15.3*c* illustrates a stuck mirror, and Figure 15.3*d* shows the yoke structure under a removed mirror. The mirrors are 17 μm square with a 0.8-μm gap separating them. The metal via at the center of the mirrors is evident in each image. These illustrations provide an understanding of micromirror sensitivity to foreign particles and contamination.

15.4.2 Parameters Influencing DMD Packaging

Figure 15.4 identifies parameters that influence the DMD package. The initial driver is *market*, what products can employ the technology, then *schedule*, when are the products

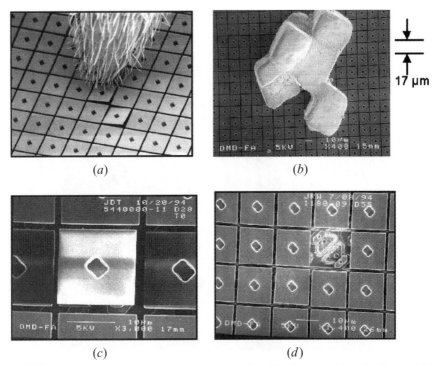

Figure 15.3 SEM images of the DMD mirror array: (*a*) ant foot, (*b*) salt crystal, (*c*) stuck mirror, and (*d*) missing mirror.

needed, and finally what is the *cost* structure for the product. Projection display markets currently supported are:

Conference room projectors

Large venue/digital cinema

Home entertainment systems

These markets have different schedule, cost, product life, and optical requirements; therefore, packaging of the DMD device is highly influenced.

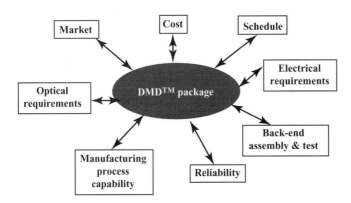

Figure 15.4 Parameters influencing the DMD.

15.4 Packaging Considerations and Guidelines Related to the Digital Micromirror Device

From a performance standpoint, the key metric is *optical image quality,* and every aspect of the package, its geometry, size, material, piece part quality, structural/thermal design, fabrication, assembly and testability, are influenced. The key image quality characteristics that influence the DMD package are:

Brightness (ANSI lumens)

Roll-off/uniformity

Contrast ratio

On screen blemishes

Image focus

Characteristics 1 and 2 influence the temperature distribution throughout the package. Higher lumen levels and high uniformity lead to increased thermal energy dissipation within the package.

A dark border with minimal light artifacts around the perimeter of the projected image is desired (3). This can be addressed at the system or package level. When volume (and arguably cost) drive the design, it is addressed at the package level. Blemishes, particles, and scratches on the package optical interface (window) must be controlled to prevent these defects from being projected onto the screen image (4). Finally, image focus (5) must be tightly controlled; therefore, DMD array planarity is critical. Array planarity is sensitive to package flatness and parallelism, as well as the die-attach process.

The package must provide an *electrical* interface between the DMD device and the external environment. One of three data interface configurations can be used, each of which employs different speed and signal-level control parameters:

Single data rate low-voltage CMOS (SDR–LVCMOS); double data rate low-voltage CMOS (DDR–LVCMOS); double data rate low-voltage differential signal (DDR–LVDS)

To achieve higher speed and lower power operation, the DDR–LVDS electrical interface is desired. The signal speeds are up to 50, 100, and 400 MHz, for three listed configurations, respectively. Increasing signal speed leads to tighter substrate impedance matching tolerances. Using the LVDS configuration places tighter tolerances on substrate transmission line impedance. Substrate material, internal routing, and plating are affected by the data interface configuration and must also be accounted for.

Reliability considerations strongly influence package design. Douglass [39] provides a detailed overview of how elevated temperatures accelerate specific DMD package failure modes, and Grimmett [40] identifies thermal requirements. Of significant importance is controlling the array temperature and package temperature gradient, therefore, thermal dissipation is a high design priority.

Structural integrity must be addressed through material selection to minimize CTE-induced stress. In addition, the package/projector system mechanical interface must be designed to prevent excessive package stress concentrations [41]. Mirror array functionality is sensitive to contamination, moisture, and stiction, and the package must provide protection from these elements, as well as a stable operational environment.

Manufacturing process limitations must be accounted for in the design of packaging piece parts. Piece part parameters that are sensitive to manufacturing process variation must be identified. Prudent judgement in specifying tolerances for these parameters is important. Controlling tolerances too tightly will drive up manufacturing costs; failure to control parameters tightly enough can lead to poor performance and reliability. Finally, it

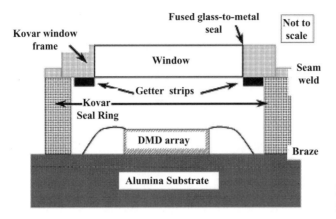

Figure 15.5 Hermetic DMD package cross section.

is required that the package be *assembled and tested* cost effectively. Every effort must be taken to make the package compatible with existing assembly and test equipment, fixtures, and tooling. Failure to do so will drive up costs and cause schedule delays.

15.4.3 DMD Package Design

With this introduction, we can summarize what functions the DMD package must provide:

- Electrical interconnect
- Heat dissipation
- Support and protection
- Flat/parallel DMD interface to system optics
- Transparent optical interface
- Reliable micromirror operation
- Manufacturing process compliance
- Cost-effective assembly and test

Figure 15.5 illustrates a cross-sectional view of the DMD package (not to scale). The package contains two primary piece parts, a *substrate* to which the DMD array is mechanically bonded and electrically attached, and a transparent *window* "lid," which provides an optical interface for the micromirror array. To design these two-piece parts, consideration must be given to the parameters outlined in Figure 15.4 and to the list provided above.

15.4.3.1 DMD Die Size

The initial step in designing the package is to determine the DMD die size, which is composed of three elements:

- Active DMD array
- Bond pad region
- Die "light shield"

15.4 Packaging Considerations and Guidelines Related to the Digital Micromirror Device

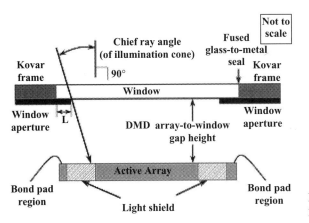

Figure 15.6 DMD aperature and light shield design.

Recall that the active array size is a function of array resolution and mirror size/pitch. The number, size, and pitch of the bond pads, and their required circuitry, determine the bond pad region size. To determine the die light shield size, optical image quality is considered, and, in particular, border contrast. There are two design goals: Only light reflected from the active array is to be projected on the screen image (no light artifacts outside array image) and the projected array image perimeter must have a sharp, well-defined border.

The first goal is addressed at the package design level by placing an absorptive metal aperture on the bottom (inside) surface of the window. The second is addressed by designing a light shield around the active array on the DMD die. These two parameters are coupled and must be sized simultaneously. Figure 15.6 provides a cross-sectional view of the DMD window and die (not to scale). It illustrates two of the three primary parameters that influence the aperture and light shield size:

Illumination profile (chief ray angle)

DMD array-to-window gap height

Window aperture-to-DMD die alignment

The chief ray angle is the angle at which incident light enters the package, measured from the package normal. The array-to-window gap height is dependent upon many parameters, optical image quality being of primary concern. The closer the window is to the active array, the more sensitive the on-screen image is to window defects. Reducing this gap height requires controlling window defects to tighter tolerances and leads to increased window manufacturing costs. Increasing the gap height makes the projected image less sensitive to window defects but the package size increases (affecting cost and volume), and the package headspace volume increases. The third parameter that influences aperture and light shield size is aperture-to-die alignment capability. There are aperture size and positioning process tolerances, die placement tolerances, and window-to-substrate placement tolerances that influence their relative position.

Knowing the active array size, the illumination optical profile (chief ray angle), the desired array-to-window gap height, and the aperture-to-die alignment tolerance, the window aperture and die light shield size (and, therefore, total die size) can be quantified. They must be sized so that the window aperture does not shadow the active array (the entire active

642 Chapter 15 Packaging of MEMS and MOEMS: Challenges and a Case Study

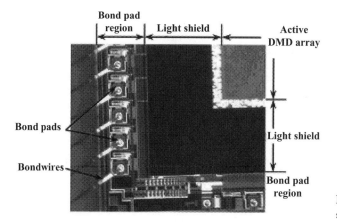

Figure 15.7 DMD die light shield.

array must be illuminated), and so that from the perspective of the projection lens pupil, the bond pads and bondwires are not visible. Figure 15.6 provides a conceptual illustration of these two requirements.

Figure 15.7 illustrates a photograph of a DMD die (lower left-hand corner). It shows the active array, the die light shield, and bond pad region. The light shield is fabricated from a highly absorptive metal and, therefore, provides significant contrast between the active array and border area. Having the light shield located on the die allows it to be in focus with and perfectly aligned to the active array. Therefore, it provides a sharp, high-contrast perimeter to the active array projected image.

Figure 15.8 illustrates how the window aperture, die light shield, and active array are positioned from a vertical perspective. The edge of the window aperture falls within the die light shield region. It is evident that the window aperture prevents light from directly

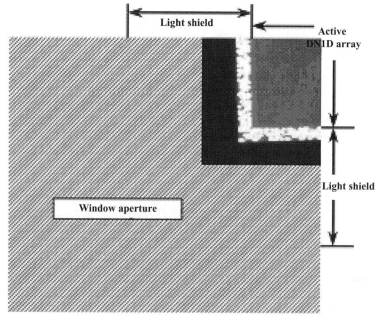

Figure 15.8 DMD window aperature.

15.4 Packaging Considerations and Guidelines Related to the Digital Micromirror Device

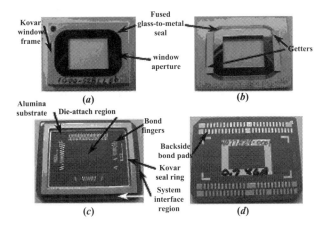

Figure 15.9 DMD package piece parts: (*a*) window (top view), (*b*) window (bottom view), (*c*) substrate (top view), and (*d*) substrate (bottom view).

striking the bond pads and wires. With the die size and bond pad definition complete, the package can be sized and designed.

15.4.3.2 Package Piece Parts

Figure 15.9 illustrates photographs of the DMD package piece parts. The window (top and bottom views, 15.9*a* and 15.9*b*, respectively) and substrate (top and bottom views, 15.9*c* and 15.9*d*, respectively) are shown. Providing a controlled and stable operating environment within the package headspace is a requirement for reliable mirror operation. Package material selection plays an important role in accomplishing this. The substrate and window materials must address the following:

- Low permeability to fluids
- Compatibility with package materials
- Compatibility with headspace materials
- Sufficient thermal conductivity
- Sufficient dimensional stability
- Sufficient structural strength

15.4.3.3 Substrate Design

The substrate is fabricated from multiple layers of laminated alumina (Al_2O_3) ceramic sheets with internal tungsten circuitry. Tungsten (melting point ~3400°C) is required to withstand the ceramic manufacturer cofire process temperatures (~1500°C). Ceramics are widely employed in hermetic packaging [36, 42–44]. They do not retain or absorb moisture and are impermeable to fluids. They exhibit excellent chemical stability when exposed to water, acids, and solvents, and they have excellent dimensional stability during and after manufacture processing.

The thermal conductivity of alumina does not approach that of metals but is significantly higher than that of organic materials. The coefficient of thermal expansion is similar to that of ASTM-F15 (kovar), glass, and silicon, three additional materials used in the DMD package (Table 15.2). Ceramics are brittle and sensitive to excessive stress concentrations and this must be considered during design.

Table 15.2 DMD Material Properties at 25°C

Material	Modulus (GPa)	CTE $\times 10^{-6}$ (ppm/°C)	Thermal cond. (W/m-°C)
Alumina	310	5.4	17
Silicon	130	2.6	148
ASTM-F15 - Kovar	138	6.5	16.3
Glass	62	3.3	1.118

The substrate internal bond fingers and external backside contacts are electrolytically plated with nickel/gold (Fig. 15.9c and 15.9d, respectively). All exposed substrate surfaces are either alumina or gold plated. Gold plating is required to prevent oxidation on any metal surfaces for compatibility with constituents in the package headspace, and for wirebond and window/substrate seam weld integrity. The location of the internal bond fingers depends on the DMD die size and die bond pad layout. Their size, pitch, positioning, and tolerance must be in compliance with substrate and wirebond process capability. The external backside contacts must be compatible with the electrical interconnect technology used between the DMD package and system interface board and in compliance with substrate process capability.

A nickel-plated kovar frame ("seal ring") is brazed onto the top surface of the substrate and serves as the package "sidewalls" (Fig. 15.5 and Fig. 15.9c). The ceramic substrate surface area to which the seal ring is brazed has a layer of nickel-plated tungsten. During the braze process, a copper/silver shim fuses the nickel-plated seal ring to the nickel-plated area of the substrate. After braze, all metal surfaces on the substrate are nickel/gold plated. This seal ring provides a surface for attaching the window to the substrate via a seam weld process (discussed elsewhere in this chapter). The similarity in thermal expansion coefficients between alumina and kovar (Table 15.2) minimizes sealing stress concentrations.

Image quality is sensitive to the relative position between the DMD active array and system optics. Therefore, additional parameters of importance for the substrate are flatness of the region to which the DMD die is bonded (die-attach region—Fig. 15.9c), and the parallelism of this region with respect to the substrate system interface region (Fig. 15.9c). The die-attach region exists over the center of the substrate, and the system interface region lies over specific areas just along the outside perimeter of the seal ring and aligns the package to the system optics. Controlling the flatness and parallelism of these two regions improves the DMD active array-to-system optics alignment.

Finally, as illustrated via Figure 15.3, any source of contamination caused by the substrate must be controlled. The substrate manufacturing cycle involves exposure to high-temperature furnaces, plating and grinding processes, and many different equipment and inspection operations, after which the substrates must be shipped in containers from the manufacturer to the DMD assembly house. From a design perspective, all substrate edges are chamfered to minimize particle generation during these processing steps. In addition, there are numerous cleaning stages to which the substrates are exposed, so particles and contamination sources are minimized.

15.4.3.4 Window Design

Recall the material requirements outlined for the window in the previous section. The window assembly contains a nickel/gold-plated kovar frame that is fused to glass (Figs. 15.5, 15.9a, 15.9b). Again, gold plating is required to minimize oxidation, for headspace compatibility, and for the window-to-substrate seam weld process (discussed

15.4 Packaging Considerations and Guidelines Related to the Digital Micromirror Device

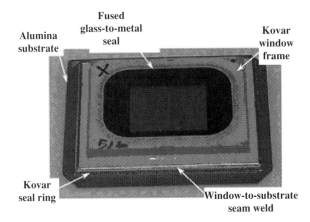

Figure 15.10 Hermetic DMD package assembly.

elsewhere in this chapter). After cooling from the elevated fuse process temperature, a compression seal is provided. This is evident by comparison of the thermal expansion coefficients of glass and kovar (Table 15.2). Glass is a brittle material and is sensitive to excessive stress concentrations, and this must be accounted for in the design. Like ceramics, glasses do not retain or absorb moisture and are impermeable to fluids. They are also chemically stable when exposed to a broad range of environments.

One of the primary objectives of a projection display system is to get the maximum amount of available light at the image plane (on screen). The DMD package is a single component in this system, and it must be designed as efficiently as possible with regard to achieving this objective. From an optical design standpoint, this and the image quality parameters must be considered when designing the window. The design is focused on controlling parameters in the visible spectrum (photopic—400 to 700 nm). The window glass thickness and index of refraction must be selected appropriately. Glass surfaces transmit, reflect, and absorb light. Any loss of illuminant energy passing through the window results in a loss in screen luminance (brightness); therefore, antireflective (AR) coatings are tailored to optimize performance in the photopic spectrum. Knowing the DMD array-to-window gap height (Fig. 15.10), analytical techniques are used to quantify how tightly window defects (blemishes, scratches, digs, stains) and particles must be controlled.

The photopic properties of the window aperture must also be tailored. Recall the objective of the window aperture is to provide a dark border around the active array projected image. It is designed to prevent light from striking areas inside the package other than the active array and to prevent any stray reflected light inside the package from escaping beyond the aperture opening. Therefore, the window aperture is designed to be opaque (controlled photopic transmittance).

Considering light in the photopic spectrum incident on the package and striking the top surface of the aperture, the aperture must be absorptive enough to prevent light from being reflected and projected onto the screen. In direct conflict with this is that absorbing energy on the window aperture increases package operational temperatures due to the high thermal impedance of the glass window. Therefore, there is a trade-off in specifying the photopic reflectance of the window aperture. It must absorb as much photopic energy as possible without subjecting the package to excessive thermal loads.

Depending upon the projection system, the amount of energy outside the photopic spectrum incident on the package varies, but, in general, because it is not visible, it does not affect the aperture design. Manufacturing process limitations require that the photopic and radiometric [ultraviolet (UV): 100 to 400 nm, infrared (IR): 700 to 1000 nm] reflectance values of the aperture be similar. As with the substrate, any source of contamination on the window piece part must be controlled, and there are many cleaning and inspection process steps to minimize contamination.

15.4.3.5 Package Size

Package size can be driven by either the window or substrate size. Inspection of Figures 15.9c and 15.9d provides insight on the required substrate size. From a topside perspective (Fig. 15.9c), the substrate must accommodate the entire DMD die, provide sufficient real estate for bond fingers, wirebonds, the seal ring, and a large enough system interface region. From a bottom side perspective (Fig. 15.9d), there must be sufficient area for a heat sink interface and the backside contacts. The larger of the top-side and bottom-side requirements will typically drive the substrate and package size. An exception to this is if the window footprint drives the substrate size.

Considering window size, a starting point is the required aperture size previously discussed. Depending on the window height and chief ray angle, there is a specific length requirement (L) between the aperture edge and window glass-to-metal (kovar) fuse line to prevent blocking the incoming light cone (Fig. 15.6). In addition, making this distance too small could cause window frame reflections on the image plane. Finally, there is a length requirement between the glass-to-metal seal and the window-to-seal ring seam weld edge. During seam weld, the package is exposed to elevated temperatures and controlling this distance will minimize thermal stress at the glass-to-metal seal. Based on the above discussion, the larger of the substrate or window size will determine the overall package size. The final design must be verified via structural and thermal finite-element modeling.

15.4.3.6 Headspace Getters

Recall that mirror array functionality is sensitive to contamination, moisture, and stiction. Therefore, the package must be designed to provide reliable mirror operation over the intended life of the product. Inspection of Figures 15.5 and 15.9b indicates getters are attached to the window's bottom surface and are located inside the package headspace. The getters, typically using a proprietary process, provide headspace environmental control by absorbing moisture and controlling headspace chemistry. Permeation and chemical analysis and experimentation are used to design/size the getters.

15.4.4 DMD Hermetic Package Assembly

Inspection of Figures 15.5 and 15.10 helps highlight the process steps for the hermetic package assembly. After the DMD die singulation process is complete, the die is bonded down to the substrate using a thermally conductive and structurally compliant adhesive. Recall that image quality is of paramount importance. Therefore, die height and parallelism with respect to the system interface region (Fig. 15.9c) must be tightly controlled (die attach is critical). Electrical connection between the DMD die and substrate is then made via wirebond.

After additional process steps, the window is attached to the substrate. The window is placed on top of the seal ring, and the window aperture is aligned to the DMD die. The window frame and substrate seal ring are then seam-welded together, providing a hermetic

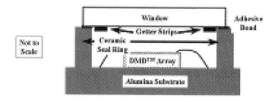

Figure 15.11 Nonhermetic DMD package cross section.

seal. The piece part manufacturing and assembly process steps, identified in Figures 15.5 and 15.10:

Window glass-to-metal fuse

Seal ring braze-to-substrate

Window frame-to-seal ring seam weld

are designed to provide a hermetic enclosure that is capable of meeting helium leak rates of less than 5×10^{-8} cm^3 when exposed to repeated thermal cycles between -55 and $125°$C. There are many more process steps involved in the package assembly, each of which is important. Certainly, moisture, contamination, and particle control are essential throughout the assembly process.

15.5 FUTURE PACKAGING CHALLENGES

Section 15.4 of this chapter describes, in some detail, the design considerations and features of the hermetic DMD package. As is typical of other MEMS, packaging accounts for a significant percentage of the total product cost. Hermetic sealing materials and processes do not lend themselves to low-cost and high-throughput conditions. A significant challenge for our packaging team is to lower the cost associated with packaging, including raw material and assembly costs.

Figure 15.11 illustrates a non-hermetic package (not to scale) that focuses on reducing package costs. It has piece parts similar to the hermetic design; however, a solid glass window is bonded directly down to a ceramic seal ring. Therefore, it does not require a fused window, nor does it require the seal ring braze or window-to-seal ring seam weld processes. Therefore, the raw material and assembly costs are reduced. This design is subject to moisture permeation that affects mirror reliability, and a more thorough understanding of package headspace chemistry as a function of operating environment is required. Additionally, the window-to-substrate adhesive interface is much more sensitive to temperature excursions and mechanical loadings, and the work is in progress to bring this design to the marketplace.

Figure 15.12 illustrates a DMD packaged at the wafer level (not to scale). This concept is under consideration and presents many formidable technical challenges. However, it does

Figure 15.12 Wafer-level DMD package cross section.

provide significant cost advantages. Packaging of the entire wafer would be conducted in one process step, where the window would be attached directly to the DMD die (silicon) in wafer form. The glass-to-silicon interface may be hermetic (fuse process via localized heating) or nonhermetic (adhesive). The ideal process would require that the mirror array be released at the wafer level and tested for functionality. Glass would then be attached to the die and the package would be singulated.

The micromirror array would be exposed to an unprotected environment only in a wafer-level-class cleanroom. During all downstream packaging operations, the array is protected, and fallout due to assembly contamination would be minimal. A reduction in yielded material costs and improved throughput would be realized. The unyielded window costs would be reduced compared to either the hermetic or nonhermetic packaging approaches previously discussed, and the substrate to which the die is attached could be low cost. Inspection of this design indicates that the die-to-window gap height is very small and could pose serious image quality concerns. These will have to be addressed via glass quality. In addition, window aperture and getter processing concerns must also be addressed.

Acknowledgments

The author (APM) would like to acknowledge the National Science Foundation (NSF) and the Defense Advanced Research Projects Agency (DARPA) for their financial support of MEMS packaging research. Also, the author (APM) acknowledges technical contributions from colleagues, students, and collaborators from Sandia National Laboratories. Also, recognition is given to the multidiverse engineering team, the many technicians, support and business personnel, and the manufacturing areas of the DMD device, who have worked on this product for years. It is a significant accomplishment for the team to have brought this to market. There are many individuals, too numerous to mention by name, who, through meetings or individual discussions, have provided insight for much of the information provided, and the author (JPO) would like to acknowledge their contributions. The author (JPO) would specifically like to highlight the efforts of J.W. Liu and Brad Haskett of the DMD package design team. Without their continuous efforts to understand and improve the DMD package design, and their many discussions with the author, this contribution would not be possible.

REFERENCES

1. J. J. SNIEGOWSKI, Multi-level Polysilicon Surface-MicromachiningTechnology: Applications and Issues (Invited Paper), ASME 1996 International Mechanical Engineering Congress and Exposition, Proc. of the ASME Aerospace Division, November 1996, Atlanta, GA, AD-Vol. 52, pp. 751–759.
2. J. J. SNIEGOWSKI, Moving the World with Surface Micromachining, *Solid State Tech.*, pp. 83–90. Feb. 1996.
3. Sandia National Laboratories Introductory MEMS Short Course, June 29–July 1, 1998.
4. J. J. SNIEGOWSKI AND E. J. GARCIA, Microfabricated Actuators and Their Application to Optics, Proceedings of Photonics West '95.
5. D. L. HETHERINGTON AND J. J. SNIEGOWSKI, Improved Polysilicon Surface-Micromachined Micromirror Devices Using Chemical-Mechanical Polishing, International Symposium on Optical Science, Engineering, and Instrumentation, SPIE's 43rd Annual Meeting, San Diego, CA, July 1998.
6. B. KLOECK AND N. F. DE ROOIJ, Mechanical Sensors, in *Semiconductor Sensors*, S. M. Sze, ed., New York: Wiley, 1994, pp. 153–199.
7. PEETERS, Challenges in Commercializing MEMS, *IEEE Computat. Sci. Eng.*, pp. 44–48, 1997.
8. Sandia National Laboratories, Albuquerque, NM, Micromachine Initiative web page, http://www.mdl.sandia.gov/Micromachine/summit5.html.
9. MCNC, Research Triangle Park, NC, Cell Library web page, http://mems.mcnc.org/camel.html.
10. D. J. MONK, T. MAUDIE, D. STANERSON, J. WERTZ, G. BITKO, J. MATKIN, AND S. PETROVIC, Media Compatible Packaging and Environmental Testing of Barrier Coating

Encapsulated Silicon Pressure Sensors, Solid State Sensor and Actuator Workshop, Hilton Head, SC, June 2–6, 1996, pp. 36–41.

11. D. J. MONK AND M. K. SHAH, Packaging and Testing Considerations of Bulk Micromachined, Piezoresistive Pressure Sensors, Motorola Sensor Products Division.

12. H. DE LAMBILLY, R. GRACE, AND K. SIDHU, Package or Perish, Proceedings of the 1996 Sensors Expocon, 1996, p. 275.

13. W. D. BROWN, ed., *Advanced Electronic Packaging: With Emphasis on Multichip Modules*, New York: IEEE Press, 1999, pp. 3–8, 16–22, 35–41.

14. A. P. MALSHE, C. O'NEAL, S. B. SINGH, W. P. EATON, W. D. BROWN, AND W. M. MILLER, Challenges in the Packaging of MEMS, *Intl. J. Microcircuits Electr. Packaging*, Vol. 22, No. 3, p. 233, 1999.

15. R. LEGTENBERG, A. C. TILMANS, J. ELDERS, AND M. ELWENSPOEK, Stiction of Surface Micromachined Structures after Rinsing and Drying: Model and Investigation of Adhesion Mechanisms, *Sensors and Actuators, A: Phys.* Vol. 43, Nos. 1-3, pp. 230–238, 1994.

16. G. T. MULHERN, D. S. SOANE, AND R. T. HOWE, Supercritical Carbon Dioxide Drying of Microstructures, the 7th International Conference of Solid-State Sensors and Actuators, Transducers '93, pp. 296–299, 1993.

17. A. P. MALSHE, S. SINGH, K. HANKINS, J. J. YOUNG, B. S. PARK, S. N. YEDAVE, AND W. D. BROWN, Effects of Packaging Process Steps on the Functionality of MEMS Devices: Investigation of Electrical Interconnection on Lubricated MEMS, Proceedings of IMAPS 2000, September 2000.

18. M. R. HOUSTON, R. MABOUDIAN, AND R. T. HOWE, Ammonium Flouride Anti-stiction Treatments for Polysilicon Microstructures, Digest of Technical Papers, 1995 International Conference on Solid-State Sensors and Actuators, Transducers '95, paper 45, 1995.

19. W. C. TANG, T-C. H. NGUYEN, AND R. T. HOWE, Laterally Driven Polysilicon Resonant Microstructures, 1989 International Workshop on Micro Electromechanical Systems (MEMS '89), p. 53, 1989.

20. M. H. KIANG, O. SOLGAARD, K. Y. LAU, AND R. S. MULLER, Electrostatic Combdrive-Actuated Micromirrors for Laser-Beam Scanning and Positioning, *J. Microelectromech. Syst.*, Vol. 7, No. 1, pp. 27–37, March 1998.

21. Dynatex Inc. Santa Rosa, CA, web page, http://www.dynatex.com.

22. J. B. LASKY, S. R. STIFFLER, F. R. WHITE, AND J. R. ABERNATHY, Silicon-on-Insulator (SOI) by Bonding and Etch-Back, Technical Digest, 1985 IEEE International Electron Device Meeting, pp. 684–687, 1985.

23. C. LUO, L. LIN, AND M. CHIAO, Nanosecond-Pulsed Laser Bonding with a Built-in Mask for MEMS Packaging Applications, Proceedings of the 11th International Conference on Solid-State Sensors and Actuators, Munich, Germany, June 10–14, 2001.

24. U. M. MESCHEDER, M. ALAVI, K. HILTMANN, CH. LIZEAU, CH. NACHTIGALL, AND H. SANDMAIER, Local Laser Bonding for Low Temperature Budget, Proceedings of the 11th International Conference on Solid-State Sensors and Actuators, Munich, Germany, June 10–14, 2001.

25. M. J. WILD, A. GILLNER, AND R. POPRAWE, Advances in Silicon to Glass Bonding with Laser, *Proc. SPIE*, Vol. 4407, pp. 135–141, 2001

26. Y. TAO, A. P. MALSHE, AND W. D. BROWN, Laser-assisted Selective Bonding for Wafer-level & Chip-scale Vacuum Packaging of MEMS and Related Micro Systems, Proceedings of IMAPS 2002, Denver, CO, September 2002.

27. H. GUCKEL, D. W. BURNS, C. R. RUTIGLIANO, D. K. SHOWERS, AND J. UGLOW, Fine-sgrained Polysilicon and Its Application to Planar Pressure Transducers, Technical Digest, The 4th International Conference on Solid-State Sensors and Actuators, 1987 pp. 277–282.

28. M. L. KNIFFIN AND M. SHAH, Packaging for Silicon Micromachined Accelerometers, *Intl. J. Microcir. Electr. Packaging*, Vol. 19 No. 1, pp. 75–86, 1996.

29. B. I. CHANDRAN, Determination and Utilization of AuSn Creep Properties for Bonding Devices with Large CTE Mismatches, University of Arkansas, 1996.

30. M. D. BROWN, Investigation of Flip-Chip Die Attachment to Diamond Heat Spreaders", University of Arkansas, 1998.

31. A. C. MCNEIL, A Parametric Method for Linking MEMS Package and Device Models, 1998 Solid State Sensors and Actuators Workshop, Hilton Head, SC, '98, June 1998, pp. 166–169.

32. I. Y. CHIEN AND M. N. NGUYEN, Low Stress Polymer Die Attach Adhesive for Plastic Packages, *Electr. Eng.*, pp. 41–46, Feb. 1995.

33. J. H. SMITH, S. MONTAGUE, J. J. SNIEGOWSKI, J. R. MURRAY, and P. J. MCWHORTER, Embedded Micromechanical Devices for the Monolithic Integration of MEMS with CMOS, Technical Digest, The 1995 International Electron Device Meeting, pp. 609–612.

34. J. T. BUTLER, V. M. BRIGHT, AND J. T. COMTOIS, Advanced Multichip Module Packaging of Microelectromechanical Systems, Transducers '97, 1997 International Conference on Solid-State Sensors and Actuators, Chicago, June 16–19, 1997, pp. 261–264.

35. J. LYKE, Packaging Technologies for Space-Based Microsystems and Their Elements, *Packaging Tech.*, pp.133–180.

36. S. T. PICRAUX AND P. J. MCWHORTER, The Broad Sweep of Integrated Microsystems, *IEEE Spectrum*, Vol. 35, No. 12, pp. 24–33, Dec. 1998.

37. J. H. SMITH, M. S. RODGERS, J. J. SNIEGOWSKI, S. L. MILLER, D. L. HETHERINGTON, P. J. MCWHORTER, AND M. E. WARREN, Micro-electro-optical Devices in a Five-level Polysilicon Surface-micromachining Technology, Proceedings of Micromachined Devices and Components IV, *Proc. SPIE*, Vol. 3514, pp. 42–49, Sept. 1998.

38. A. T. WOOLEY, D. HADLEY, P. LANDRE, A. J. DE MOLLO, R. A. MATHIES, AND M. A. NORTHRUP, Functional Integration of PCR Amplification and Capillary

Electrophoresis in a Microfabricated DNA Device, *Anal. Chem*, Vol. 70, p. 158, 1998.
39. T. G. BROWN AND B. S. DAVIS, Dynamic High-G Loading of MEMS Sensors: Ground and Flight Testing, Materials and Device Characterization in Micromachining, *Proc. SPIE*, Vol. 3512, pp. 228–235, Sept. 1998.
40. N. MALUF, *An Introduction to Microelectromechanical Systems Engineering*, Artech House, 2000.
41. P. R. CHOUDHURY, *Handbook of Microlithography, Micromachinging, and Microfabrication, Vol. 2: Micromachining and Microfabrication*, SPIE Press, 1997.
42. M. A. MIGNARDI, From ICs to DMDTMs, *Texas Instrum. Techn. J.*, Vol. 15, No. 3, pp. 56–63, 1998.
43. M. R. DOUGLASS, Lifetime Estimates and Unique Failure Mechanisms of the Digital Micromirror Device (DMDTM), 1998 IEEE International Reliability Physics Proceedings, Cat. No. 98CH36173.
44. J. D. GRIMMETT, Thermal Analysis of a Light Reflecting Digital Micromirror Device, 1997 ISPS Proceedings, pp. 242–247.
45. M. N. VARIYAM, Behavior of a MOEMS Package (TI DMDTM) under Mechanical and Thermal Loads, 2000 ASME International Mechanical Engineering Congress and Exposition Proceedings, MEMS Vol. 2, pp. 601–608.
46. J. LAU, C. P. WONG, J. L. PRINCE, AND A. L. LONDON, *Electronic Packaging—Design, Materials, Process, and Reliability*, 3rd ed., New York: McGraw Hill, 1998.
47. R. R. TUMMALA, E. J. RYMASZEWSKI, AND A. G. KLOPFENSTEIN, *Microelectronics Packaging Handbook—Semiconductor Packaging Part II*, 2nd ed., Chap Hall: Kluwer Academic, 1999.
48. G. R. BLACKWELL, *The Electronic Packaging Handbook*, Boca Raton, FL: CRC Press, 2000.

EXERCISES

15.1. List and describe MEMS packaging challenges and possible solutions to address these challenges. Use of an example is advised for elaboration.

15.2. Identify the major differences in IC and MEMS packaging, which make MEMS packaging application specific.

15.3. What is wafer-level packaging? Write a note on the role of material interfaces in wafer-level packaging of MEMS devices, particularly for MOEMS packaging. Use of an example is advised for elaboration.

15.4. Describe fundamental challenges in the packaging of DMD devices.

15.5. Identify differences in packaging requirements for micromirror verses microfluidic devices. Highlight differences taking into consideration signals, environments, and scales from device-to-system-level integration.

15.6. Design a package for wafer-level encapsulation of DMD devices by further improving the package described in the text. It is important to note that the fundamental signal, temperature, and interfacing requirements remain essentially the same.

Chapter 16

Reliability Considerations

RICHARD ULRICH

16.1 INTRODUCTION

This chapter will address the science and modeling of failure mechanisms for packaged microelectronics by considering the following issues:

- What mechanisms might cause components in a system to fail?
- How can the probability of failure be estimated without having to wait out the expected lifetime of the system under field conditions?
- If we know the probability of failure from every failure mechanism for every component in the system, what is the probability that the system as a whole will fail?
- How can we decrease the probability of the various failure mechanisms, thereby increasing the reliability of the system as a whole?

What we broadly call *microelectronics* can be broken down into two distinct areas: integrated circuits (the "front end") and packaged systems (the "back end"). This book addresses the latter, but the reliability issues for both are driven by the same trend: miniaturization. Over the past 40 years, feature sizes on integrated circuits (ICs) have been shrinking at a geometric rate, as described empirically by Moore's law. At the same time, the size of the average IC has increased slightly. The combined result is that each successive generation of ICs has a larger number of more delicate structures. Both of these factors exacerbate reliability issues. A similar trend is underway for packaged systems, but the situation is not as severe because the dimensions are larger and the component densities are smaller. The finest features on any circuit board are always at least one or two orders of magnitude larger than those on a typical IC and are about equal to IC feature dimensions from about 20 years ago. But, as is the case for ICs, board-level linewidths, layer thicknesses, and component sizes have all been dropping steadily over the decades. The result is, of course, the same: an increase in the types and frequency of failures. In fact, many of the same failure mechanisms that have long been present at the chip level are starting to show up at the packaging level.

16.1.1 Definitions

This chapter addresses a number of concepts for which most people have an intuitive feel but may not have ever thought about quantitatively. First of all, what is *failure* itself? If you

---

Advanced Electronic Packaging, Second Edition, Edited by Richard K. Ulrich and William D. Brown
Copyright © 2006 the Institute of Electrical and Electronics Engineers, Inc.

turn on your computer and it does not boot, we would all agree that this is a failure. But if your CD burner suddenly requires that you tap the door with your finger to make it work, but it operates perfectly every time you do that, has it failed? Is it worth taking your computer out of service for a week to have that fixed? Failure is a matter of definition, but it must be firmly established in order to be used in the quantitative failure analyses that follows in this chapter. It does not much matter what level of poor performance is deemed a failure, but for quantitative analysis it has to be consistently defined. If it has not failed, it is considered to be *operational*. *Failure mechanisms* are the stresses that cause a component to move from operational to failed. These stresses include electrical, chemical, and mechanical processes.

If the concept of failure is sufficiently defined, the next step is to consider what is failing. A *system*, such as a computer or cell phone, is made of *components*, such as packaged ICs, surface-mount capacitors, connectors, clock crystals, and entire circuit boards. Systems do not fail, although that is where the problem is noticed, the problem usually arises in a tiny part of a single component, such as a ball bond or a solder joint or a single interconnect. One useful definition of a component is it is the smallest unit that can be replaced to provide a repair. You could replace a packaged chip but not a ball bond. Either systems or components can be tested, evaluated, and tracked to establish their reliability behavior, and we will use the term *unit* to mean either one.

Although they are linked through similar types of root causes, *yield* is a different concept from *reliability*. The two meet on the loading dock of the manufacturing plant. The fraction of units that start out in the manufacturing flow and end up tested and fully functional on the loading dock is the yield. Reliability analysis starts at this point and assumes that, to start with, 100% of the systems are 100% operational. Certainly the same failure mechanisms can lead to either or both a loss of yield or reliability. For instance, a manufacturing fault that results in that broken solder ball would cause a loss of yield if the ball broke just before final test at the factory but would cause a loss of reliability if it happened a few days later after being purchased.

More formally, reliability as applied to microelectronics, or anything for that matter, has two equivalent definitions:

The probability that a specific unit will be operational for a given period of time

The fraction of a group of units manufactured together that are operational for a given period of time

Since reliability is a probability or a fraction, it is a number between zero and one. The "given period of time" part means that reliability is a time-dependent quantity; a reliability that is not associated with a time is meaningless. If a unit made by your company has a reliability of 90% at 8 years, then any one of these units has a 90% chance of still working at the end of 8 years or, expressed the other way, 90% of the units manufactured will be operational after 8 years by your definition of "operational." The proper symbology is

$$R(8 \text{ years}) = 0.90$$

For reliability measurements, time-equals-zero is after a final test at the factory and the reliability of anything at that time is 100% and can only remain the same or else decrease as time progresses. In mathematical terms,

$$R(0) = 1$$
$$\frac{dR(t)}{dt} \leq 0 \quad \text{at all values of } t$$

From an economic point of view, reliability considerations are driven by two extremes of requirements:

Cost-Driven Reliability The unit's reliability should be sufficient to maximize the manufacturer's profit. Examples include desktop computers, home entertainment systems, automotive electronics, and almost any other consumer system. This does not mean that reliability is ignored; unhappy customers are bad for profit.

Performance-Driven Reliability No reasonable amount of money or effort will be spared to maximize the device's reliability. This includes safety-related applications such as automobile airbag triggers, pacemakers, aircraft control systems, and security systems as well as units that cannot be replaced or repaired during operation such as those in satellites, missiles in flight, and undersea sensors.

Accelerated Testing The process of inflicting severe stress on a system in order to accelerate the time to failure. For instance, 1 out of every 500 computers might be taken off the end of the assembly line and put in an 85/85 chamber (85°C and 85% relative humidity) for 1000 h to check for weak points in the system that would fail at longer times in the more benign environment in an office. This unit would not be then put back to be sold. There are many types of accelerated testing designed to stimulate various failure mechanisms.

Reliability Metrology The mathematics of failure prediction. Failure rate information from short times in the field or from accelerated testing might be used to predict the failure patterns on units in service in the more distant future.

16.1.2 Patterns of Failure

It is very important to be able to predict the reliability versus time for a given product since this provides information regarding future repair costs under warranty, the competitive position of the product compared to others, and the cost benefits of additional spending on reliability enhancement. The classic failure pattern for a population of almost any manufactured article, including clocks, cars, pens, and hot water heaters as well as packaged microelectronic devices, follows the well-known shape of the *bathtub curve* as shown in Figure 16.1.

The bathtub curve is actually the superposition of three independent failure rates as shown in Figure 16.2. *Early failures*, also called *patent failures* or *infant mortality*, are out-of-spec manufacturing faults that are able to pass the final test at the factory but fail quickly due to normal usage stresses. The units are operational within their performance

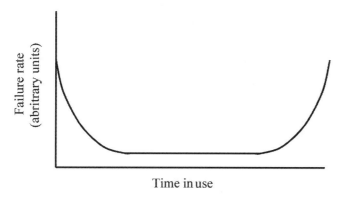

Figure 16.1 Failure rate for a population of devices: the bathtub curve.

654 Chapter 16   Reliability Considerations

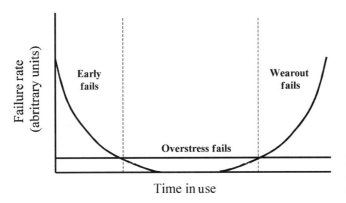

Figure 16.2 Bathtub curve decomposed into its three contributions.

specification when shipped but are close to failure and may even be harming the manufacturing yield. For example, a solder ball may be cracked as a result of a manufacturing error but not all the way through, so it passes final test at the plant. But a dozen temperature cycles due to being turned on and off in the user's office might complete the crack in a short time. These faults represent the costliest type of failure since they occur early in the unit's expected lifetime but may be detectable through careful yield analysis. Increased severity and length of burn-in testing can uncover more of these parts before shipment, but at the risk of causing early overstress failures in components that would have otherwise operated reliably over their expected lifetime.

After the early fails are culled out, the components will settle into a long period of *overstress failures* or *intrinsic failures* which are caused by high-level stresses beyond those expected in normal usage, resulting in instant failure. These overstresses are not necessarily the fault of the user, they may be caused by the failure of some other associated component in the system. For instance, a voltage regulator may drift out of spec on the high side, causing failure of another board, or a cooling fan may fail, resulting in extremely high device temperatures, or the part may simply be dropped on the floor, causing a mechanical overstress. Electrostatic discharge is a common mode of overstress failures for ICs and passives. In each case, the devices were within specifications when shipped but were never intended to tolerate these levels of stress. Since the failures are random, their frequency is flat with time.

Finally, *wearout failures* will begin to be significant as normal failure mechanisms begin to take their toll on units that were manufactured as well as they could be. This is caused by the cumulative effect of low-level stresses from normal everyday usage. Examples include corrosion, mechanical fatigue, or dendrite growth. As in the case of all three general failure modes shown in Figure 16.2, the devices were operational when shipped.

Can you guess when the manufacturer sets the warranty to expire in Figure 16.1? This is usually set to be just before wearout failures come into play to maximize the perceived length of warranty but minimize the number of devices that have to be repaired or replaced. A warranty that is too long would result in a sizable fraction of the products having to be replaced. This underscores the economic value of determining the reliability of a component versus time.

16.1.3 Coverage in This Chapter

In this chapter we will mainly address the technical aspects of reliability and spend little time on the economic aspects. This is not to say that the economic aspects are unimportant;

they are the main driving force for reliability studies and the implementation of efforts to enhance reliability. The reason for its cursory treatment here is that the economics of reliability are a function of the product's intended usage, the manufacture's economic goals, and the intended level of product reliability, and these are very application specific. On the other hand, the technical aspects are common to all treatments of reliability, cost driven or performance driven, and are necessary to be completely understood before any economic analysis can be performed. We will not address the issue of software reliability, only hardware, although software errors and conflicts are by far the most common cause of your computer crashing. Unlike hardware failures, the system can generally fully recover from a software problem.

16.2 FAILURE MECHANISMS

Failure mechanisms typically occur on some small portion of the lowest component level, but the results are observed at the highest system level. This means there are many more causes of failure than there are symptoms of failure and, as a result, extensive failure analysis may be necessary before you can know for sure what went wrong.

What sorts of things can fail in microelectronic packaging (Figure 16.3) and how can these failures be prevented? The mechanisms that account for nearly all of the failures in microelectronic equipment were identified long ago. The major effort these days is not in discovering new failure mechanisms but in figuring out when the usual ones will occur, estimating how long they will take to cause failure under field conditions, and how to prevent them in units with continuously shrinking feature sizes. In this section we will discuss the failure mechanisms. Estimating the time to failure will be covered in the next section.

Failure is caused by some sort of stress—electrical, chemical, or mechanical—that results in irreversible hardware damage, although that damage may be microscopic. These stresses can be divided into two broad categories: *overstress mechanisms* in which the stress in a single event is sufficient to cause failure and *wearout mechanisms* in which a lower level of stress is delivered over a long period of time until cumulative damage results in unit failure. A leading cause of cell phone failure is being dropped in water, which is an example of chemical overstress.

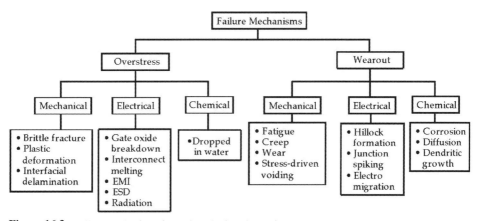

Figure 16.3 Failure mechanisms for packaged microelectronics.

16.2.1 Corrosion

The progressive decrease in metal feature sizes has rendered bondwires, bondpads, and board-level interconnects increasingly susceptible to corrosion failure. For instance a single 8×8 mil bondpad consists of only about 10^{-7} g of Al of which about 5% actually participates in the bond itself. A comparable amount of chloride, phosphoric acid, or organic acids along with a few monolayers of water can result in corrosion that can destroy that bond pad, interrupting the electrical continuity of these structures. Thus, microscopic amounts of water and ions can cause corrosion failure of metal components in microelectronic packaging.

Corrosion is an electrochemical reaction that occurs when metal comes into contact with water and certain dissolved ions. The result of the reaction is that electrically conductive metal atoms, with valance of zero, are oxidized to nonconductive positive valances. These metal ions either form a nonconductive oxide crust on the remaining metal (such as rust on iron) or become soluble in the water phase and leave the conductor. Either way, conductive metal is lost, leaving less behind to carry signals or power until the connection is totally cut. The exclusion of either ions or water will remove the possibility of corrosion occurring. If these are both present, corrosion is made possible, but the actual rate of corrosion depends on several further factors.

Thermodynamic Stability of Metal Being Corroded The more tightly a metal holds on to its electrons the slower it will corrode. That is why gold does not corrode at all, it is more thermodynamically stable as a zero valance metal than as an oxidized gold ion. On the other hand, most metals such as Al, Cu and Fe are more stable as ions in an aqueous solution than they are as solid metals. For this reason, gold, platinum, silver, and, occasionally, copper is found in the earth as a metal while aluminum, iron, and nickel are found only as oxide ores that must be reduced to metals by smelting. Gold and platinum are examples of *noble metals* because they resist corrosion in almost any ambient environment due to their inherent thermodynamic stability. Figure 16.4 is a list of the thermodynamic oxidation potential of some common metals, with the ones used in microelectronics shown in bold print. The voltage scale is relative to the oxidation of hydrogen gas at 0 V, and the higher on the scale, the more stable the metal to corrosion. This is only a general listing of relative resistance to corrosion because there are other factors involved, as explained below.

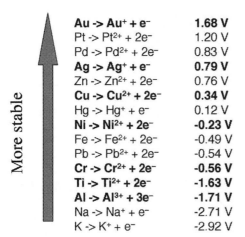

Figure 16.4 Relative thermodynamic stability of metals against corrosion.

Notice that the most common conductor used to form tiny chip-level interconnects, Al, is one of the least thermodynamically stable.

An unfortunate aspect of this relative thermodynamic stability of these metals is that if you put two of them into physical contact, electrons will want to flow from the less stable to the more stable metal, resulting in corrosion of the less stable one. This is how a battery works; it is essentially a corrosion cell, which is why it contains acids. This problem can arise when you have a 1-μm-thick Al bondpad with a 25-μm-thick Au bondwire ball-bonded on top. Corrosion of the Al bondpad will be accelerated due to contact with the gold relative to those with no bondwires attached. This can be a good reason to use Al wire on Al bondpads.

Availability of Moisture There has to be liquid water in order for significant corrosion to occur. For microscopic systems, such as metal objects on the order of microns in size, a few molecular layers on the surface may suffice. Hermetic packages made of metal and ceramic will prevent water from reaching the devices inside (assuming that no water was packaged in the enclosure in the first place), but molded plastic will allow enough moisture to diffuse into the package to provide plenty for corrosion. Molded plastic is orders of magnitude more porous to moisture than ceramic and metal enclosures, but it may be good enough and will cost much less. Which to use depends on the intended reliability and the target price for the system.

Presence of Ions The hydrogen ion is one of the most reactive ions for corrosion and is even present at neutral pHs. That is why acids tend to cause corrosion. Almost any other type of dissolved ion will promote corrosion through a number of less effective, but significant, mechanisms. Cl^-, Na^+, CO_3^{2-}, and other common species are ubiquitous in the environment, even in a cleanroom, and are very difficult to completely exclude. Stringent measures to decrease ionic contamination in molding compounds and cleanroom chemicals by manufactures since about the 1970s led to greatly increased reliability in all kind of electronics. If you take a bare die or multichip module (MCM) substrate with exposed thin-film Al bondpads or interconnect metallization and simply touch it with your finger, the metal will dissolve before your eyes—there are that many salty ions on your skin.

Condition of Oxide Layer Upon exposure to air, all metals except Au spontaneously form a layer of oxide on their surface. For most metals, this oxide is a few monolayers thick, less than about 10Å. However, in the case of Al the oxide can grow in air to be several hundred angstroms of a tighly adhearant surface layer. Despite its low thermodynamic stability, aluminum is considered to be relatively corrosion resistant because this thick oxide blocks the corrosion reactions. You have never in your life touched a piece of Al, only Al_2O_3.

Halide ions, especially chloride and fluoride, catalytically remove the oxide layer exposing the bare metal to corrosion. The halides therefore do not attack the metal directly; they just remove the protective layer. To return to the battery analogy, common Zn-based flashlight batteries contain added chloride in the paste electrolyte that prevents the formation of an oxide layer over the internal zinc electrodes, which would decrease their efficiency in a battery. So now you can see why hydrochloric acid is so corrosive to many metals such as Fe, Cu, and Al. Hydrochloric acid is an aqueous solution containing a very reactive ion (H^+) and a halide ion (Cl^-) that attacks the protective oxide. It does not attack gold at an appreciable rate, however, because gold does not depend on an oxide layer to block the corrosion reaction, and it is thermodynamically stable against corrosion.

In practice, the fabrication metals are ranked something like this:

Au, Pt will not corrode even in severe usage conditions

Ag

Cu susceptible to corrosion only if halides and acid present

Ni, Ti, Al very sensitive to moisture, halides, and acids

Corrosion of metallization, including bondpads, interconnects, and printed wiring board (PWB) traces, is always spotlike and highly localized and is never seen as a general thinning even over areas as small as a square centimeter. The corrosion site will appear as a spot or small region of missing metal, possibly surrounded by corrosion products. This is because corrosion begins at a flaw in a coating, either a passivation flaw (crack or a pinhole) or a thin area of metal oxide. Bondpad corrosion is also easy to spot and can be so severe that the bond may separate leaving the bondwire hovering over the remains of the bondpad. In some cases bondpad corrosion in molded plastic may show up as an intermittent failure as the bond loses and regains contact with uncorroded bondpad metal when the package undergoes cyclic thermal strains. If an opening is found and microscopy cannot distinguish it from other possible causes (electromigration, stress-induced cracking), then surface analysis may indicate the presence of the contaminant that caused the corrosion to occur such as chlorides or phosphates.

The probability of corrosion failure in microelectronics can be decreased by several strategies. The overall goal is to exclude moisture and/or ions.

Encapsulation in Hermetic Packages Most polymers provide a very small amount of protection against the intrusion of moisture. Epoxies are the most common in the electronic system since it is the main component in most IC package molding compounds, FR-4 circuit boards, and many underfills. But epoxies will absorb as much as a couple percent of their own mass of moisture, and this water can penetrate millimeter in less than a day. More exotic polymers such as Teflon and liquid-crystal materials can do much better but are not yet common in the fab lines, mainly due to processing complications. Hermetic packages, where the chip sits in a cavity completely surrounded by only metals and ceramics, keep moisture out very effectively and remove almost all concerns about corrosion, but at a high cost, sometimes more than the chip itself. Because of their expense and large size relative to molded plastic parts, hermetic packages are uncommon in consumer electronics and are even disappearing from military equipment.

Cleanliness in and Around Assembly Areas Prior to the mid-1980s most die-attach, wirebonding, package seal, and molding operations were done in workshop-type areas by persons wearing no masks or gowns. Now most companies have moved assembly operations into class 1000 or even class 100 cleanrooms.

Lowering the Ion Levels in Materials Contamination levels in molding compounds have dropped by at least a couple of orders of magnitude since the early 1970s and this made it possible to achieve decent corrosion reliability in plastic packaging.

Cleaning the Parts Prior to Encapsulation Rinses with deionized water are known to be effective in removing ions from surfaces and, by analyzing the rinse water, the level of contamination of the parts can be estimated and this information used for quality control in the processes upstream. If the ionic contaminants are held in place by organic contaminants, such as oil from a worker's finger or a backstreaming pump, the water rinse may not be able to wet the surface and remove the ions. It may be necessary to remove the greases first with an organic solvent. There has also been some research into using plasmas or ozone to oxidize the organics into volatile gases. The advantage of plasma or ozone is that they can penetrate occluded cavities better than liquids and there are no organic solvents leftover that create a disposal problem. The drawbacks are that plasma and ozone can damage some materials. For instance, they will both form a thick, flaky black oxide on Ag.

16.2.2 Mechanical Stress

This section covers a large range of failure phenomena, which are covered in more detail in Chapter 8, Mechanical Considerations. The cause of mechanical stresses are usually traceable to coefficient of thermal expansion (CTE) mismatch between dissimilar materials that are tightly bonded together. Table 16.1 shows the CTE range for several materials of interest in packaging. In general, CTEs are a function of the material's composition and atomic structure. As a result, the measured values can span a large range for polymers, which can have a highly variable structure, and a tighter range for metals and ceramics, which have a smaller number of crystal arrangements or else are altogether amorphous. Polymers make up a large fraction of the mass of packaged microelectronics and typically have much higher CTEs than metals or Si chips. Ceramic fillers, such as alumina, are frequently added to polymers to lower their CTE to better match either Si or Cu. In fact, the main purpose of having woven glass fibers in the epoxy that makes up a PWB is to lower the overall CTE of the board to match that of the Cu interconnects at about 17.

Even with CTE-matched materials, assembly processes can build in significant stresses from the start. Catastrophic mechanical failures, such as broken bond balls, broken bondwires, cracked die, and broken connectors, that occur early in the product's life are usually

Table 16.1 CTE Ranges for Common Materials in Electronic Systems

Material	CTE range (ppm/°C)
Ceramics	1–10
SiN	2.8
Si	2.6
Borosilicate glass	4
Al_2O_3	6–8
Quartz	8.1
Metals	10–25
steel	11–13
Cu	17
Al	23–25
Solder	25–30
Polymers	20–100
FR4	16–18
Filled epoxies	15–30
Polyimide	7–30
Nylon	22–110

indicative of manufacturing faults that left behind a combination of high stress and mechanically weak points in the unit. Fatigue-based mechanical failures, such as creep and wear, are caused by long-term or repetitive forces that are well below the tensile failure point of the materials involved. These failures are usually seen much later in the product's life and are less likely to be due to manufacturing faults.

16.2.3 Electrical Stress

Electrostatic discharge (ESD) is the most common electrical failure mechanism in microelectronic packaging by far. Static electricity is caused when two different insulating materials come into contact, they exchange electrons resulting in an equilibrium charge difference between them and then are suddenly separated, taking the charges with them. If the materials are insulators, such as human skin or plastics or cloth, the charge does not dissipate immediately and remains on the surface of the object. When this surface comes near a metallic conductor, the charge can jump over as a spark, resulting in a very high voltage surge but with very low current. Thousands of volts are common, more than enough to cause irreversible breakdown of thin insulating structures such as capacitor dielectrics, gate oxides, and even insulation between layers in the circuit board. It is even possible to melt metal conductors in small areas. This is overstress failure for sure and usually results in instantaneous failure. The prevention strategy is to avoid ESD reaching the components in the first place by employing careful grounding practices for personnel and work-in-progress in the manufacturing areas, and perhaps building in robust structures in EDS-prone areas and, of course, stringent destructive ESD testing of sample products.

16.2.4 Techniques for Failure Analysis

Although you can do a considerable amount of testing yourself, all microelectronics manufacturers have a captive failure analysis lab where you can send a part in order to have the failure mechanism diagnosed. They generally work on parts that were returned from the field, but they may also do in-house work for the reliability R&D effort. Some of their tools are listed below.

Optical Microscopes This will always be the most important failure analysis tool. Light microscopes will resolve down to about the wavelength of light, which is around 1 μm, which amounts to a magnification of about 1000×. Attempts to magnify above this will not bring out more detail; you will just be magnifying blur. Chip-level linewidths are now well below this, and it takes an scanning electron microscope (SEM) to see them at all, but almost all packaging-level structures are easily visible. In some cases it may be necessary to section a part in order to see it in cross section. This is done by encasing the part in a hard plastic, cutting it in two with a fine-toothed saw, and polishing the exposed face to reveal a cross section that can be inspected under the microscope or with surface analysis equipment. A skilled technician can cross section through a single 3-mil bond ball.

Electron Microscopes SEMs can achieve much higher magnifications than light microscopes, resolving down to perhaps 100 Å or so but require more sample preparation. The sample cannot be larger than about a square inch, so no large packages can be inserted into the microscope intact, and polymers pose a problem since they may outgas and either decrease the vacuum level or else contaminate the inside of the vacuum chamber with organic residues. Also, the sample must be electrically conductive, so it may be necessary to give polymers and ceramics a thin coating of evaporated gold before they can be imaged.

Surface Analysis   Auger, X-ray photoelectron spectroscopy (XPS), secondary ion mass spectroscopy (SIMS), energy-disperisive X-ray spectroscopy (EDX), and other microanalytical methods are especially valuable for failure analysis wherever corrosion is suspected. It is often difficult to tell the difference between an interconnect corrosion failure and electromigration or stress cracking (described below) from visual or SEM images because they are so similar in appearance. Surface analysis techniques can indicate the presence of corrosion products such as aluminum oxides and, possibly, can find the contaminants that caused the corrosion. The most common request that a reliability worker will make to the surface analysis lab is "what's that spot?"

Package Opening   Hermetic packages can generally be opened without damaging the contents either by mechanical means (cutting it open) or by melting the hermetic seal and pulling the package apart. However, corrosion usually occurs in molded plastic packages, not in hermetic packages, and it is first necessary to get the molding compound out of the way before you can see the failure site. This procedure is called *decapsulation* and can be accomplished by putting heated droplets of anhydrous sulfuric acid on the top of the package and letting them eat their way down to the chip. As long as the acid is truly water free and as long as you renew the droplet every couple of minutes, you will get a smooth well all the way down to the chip. Once it reaches the chip's surface, the acid will not attack the ceramic portions but may slowly etch the metal. So as soon as the chip becomes visible, the etching action should be stopped with a water flood. This is somewhat of an art, but every fab site will have at least one technician that is skilled at decapsulation.

X-Ray and Ultrasonic Imaging   Plastic packages can be X-rayed with enough resolution that individual 1-mil bondwires and conductors are clearly visible so that mechanical and corrosion-induced breakage can be seen. Generally, you will want two orthogonal angles on the part so you can really see what is where. Sound-based "microscopes" are very good at detecting delaminations of all sorts.

16.3   ACCELERATED TESTING

The purpose of accelerated testing is to make failure mechanisms happen in a much shorter time than they would in field usage. This is accomplished by subjecting the components to exaggerated mechanical, environmental, and/or electrical conditions. It is critical that the exaggerated stresses do not cause failure mechanisms that would not happen in normal usage conditions. There are several purposes to accelerated testing, including to determine what failure mechanism will happen for the unit, to estimate the time in normal usage until those failures occur, and to compare the performance of two or more materials or processes. The second purpose requires an acceleration transform, which is the ratio of time to failure for a given mechanism in field usage to that in accelerated testing. Acceleration transforms are rarely known with much precision.

For instance, let us say your company has for many years used the same epoxy molding compound for single-chip packages, but along comes another company that says it has a better formulation. You are given the task of determining whether or not your company's packaged ICs will perform just as well and just as long with the new compound as it did with the old compound. You might start out by going to your company's failure analysis lab and learning what the normal failure mechanisms are for the devices in the old compound. Let us say there are two:

- Cracking of the die due to mechanical stresses from the molding compound. The average time until this happens in normal field usage has been estimated from failure analysis of returned parts to be 6.5 years.
- Input/output (I/O) pad corrosion caused by intrusive moisture reacting with contaminants from either the molding compound or leftover from the chip's manufacturing process. The average time to failure in normal field usage by this mode is 4.0 years.

Obviously, your company cannot wait 6.5 or 4.0 years for you to give your report, so you need to speed things up a little. For the mechanical testing you might choose to utilize a temperature shock test where a representative number of parts, say 100, are dunked in a liquid at 150°C for 10 min and then dunked in another liquid at −65°C for 10 min for perhaps 1000 cycles of this. After every 100 cycles or so you might test the parts for die cracking by checking the electrical performance of the part or by X-raying them. You might find that the average number of cycles before die cracking with the new compound is 700. Since the same parts molded in the old compound lasted only 550, you would judge that the time to failure due to chip cracking would be longer for the new compound. In fact, you could estimate the time in the field as:

$$\left(\frac{700 \text{ cycles with the old compound}}{550 \text{ cycles with the old compound}}\right)(6.5 \text{ years with the old}) = 8.3 \text{ years with the new}$$

For the corrosion failure mechanism, you might take 100 parts packaged in the new compound and subject them to an atmosphere of 130°C and 85% relative humidity in a highly accelerated stress test (HAST) chamber. You could test them every 3 h or so for bondpad corrosion until enough of them have failed to determine an average lifetime at test conditions. Let us say that the average time to failure by bondpad corrosion for the new compound in the HAST chamber was 18 h. You did not have time to perform an identical HAST test on parts packaged in the old compound so you have to estimate the average lifetime another way. A quick search of the literature indicates that 130/85 HAST conditions has an acceleration transform of 5.63 h of HAST equals 1 year of field stress for bondpad corrosion. Applying this acceleration transform gives

$$(18 \text{ h at HAST conditions})\left(\frac{1 \text{ year in field until failure}}{5.63 \text{ years HAST until failure}}\right)$$
$$= 3.2 \text{ years expected time to failure}$$

So the expected corrosion lifetime with the new molding compound is 3.2 years as opposed to 4.0 years with the old. Certainly these calculated times to failure are estimates at best, but the test results indicate that the new molding compound should perform better than the old with regard to the die cracking problem but will not perform as well with regard to pad corrosion. Relative results are always more accurate than absolute predictions of failure time.

There are dozens of accelerated test procedures, and they can be grouped into three categories that parallel the types of failure mechanisms they induce:

Environmental Tests Simultaneous high temperature and high humidity to induce corrosion failures. High temperature can also be used alone.

Mechanical Testing Cyclic temperature changes to exacerbate CTE mismatch problems or else direct mechanical bending of assemblies.

Electrical Testing Constant high voltage or pulsed electrostatic discharge testing.

Units subjected to accelerated testing are not usually tested in more than one manner and are never returned to the product flow for later sales. They may, however, go to the failure analysis lab if unusual failure patterns are seen. Accelerated testing can be used for qualification of new products or for routine evaluation of the manufacturing process for old products.

16.3.1 Accelerated Environmental Testing

Accelerated environmental testing involves subjecting the units to high-temperature and/or high-humidity conditions, possibly combined with electrical bias. Environmental testing began in the 1960s when reliability researchers put components into boiling water or in a steam autoclave. Later in that decade, Western Electric developed the 85/85 test where units were placed in an 85°C and 85% relative humidity environment, and a simultaneous electrical bias could also be applied to the units during testing to simulate operation. During the 1970s a common criteria for passing reliability testing was 1000 h of 85/85 (1000 h = 6 weeks). At that time no one could be sure how 1000 h of 85/85 related to lifetime in the field, but 85/85 lifetime provided a benchmark for decreasing contamination in the fab and in the packaging materials. The 85/85 test provided a valuable comparison for the development of different molding compounds, manufacturing alternatives, and IC materials.

Having available a quantitative environmental testing method such as 85/85 enabled companies to identify the causes of moisture-related failures and improve this type of reliability. Since the late 1960s contamination levels have been aggressively reduced in molding compounds and in the fab, resulting in products that became more reliable with each new generation. As 85/85 lifetimes began to far exceed 1000 h in the early 1980s, it became apparent that the time to failure by this test was becoming too long to wait out. This led to a set of more aggressive test conditions that fall under the acronym of HAST, highly accelerated stress test. These involve higher temperatures, up to perhaps 140°C, and unsaturated humidities, generally 85%. In order to achieve these conditions, the HAST chamber has to be pressurized, causing it to be much more expensive to purchase and to operate than 85/85. The 1980s also saw the development of semiempirical correlations for acceleration factors—the ratio of time to failure in usage conditions to the time to failure in the HAST chamber. Since so much older data was in the form of 85/85 results and since 85/85 is much more consistent than actual usage conditions, many companies choose to relate HAST data to expected 85/85 lifetimes instead of expected field lifetimes.

There are two major families of standards for accelerated testing. MIL-STD is a collection of test and evaluation procedures that are often specified for military parts and was written by the Rome Air Development Center of the U.S. Department of Defense. The most common test procedures for plastic parts are given in the JESD-22 and JESD-26 listings, which were set by the Joint Electronic Devices Engineering Council (JEDEC) of the Electronic Industry Association. However, most companies use their own proprietary test procedures, which are usually derived from combinations MIL-STD, JEDEC, and their own experience. While there is no definitive set of accelerated reliability tests, most facilities gravitate toward similar specifications for HAST, 85/85, temperature shock, temperature cycle, ESD, and so forth with regard to conditions and testing time.

16.3.1.1 Saturated versus Unsaturated Accelerated Environmental Testing

If accelerated environmental testing is conducted using boiling water at 1 atm, then the test is at 100°C since that is the boiling point of water at 1 atm of pressure. This is a test under saturated conditions since there is liquid water present. It does not matter if the part is actually immersed in the boiling water or is in the 100% relative humidity (RH) vapor stream directly above it. Obviously, if the part is directly immersed, it will become saturated with water, but it will also become saturated if it is in the vapor stream just above the boiling liquid since this vapor stream is in thermodynamic equilibrium with the liquid phase. A part in this vapor stream will "sweat" as moisture condenses on it and runs off so it is completely wet. Based on this, the 1-atm boiling water conditions would be labeled a 100/100 test: 100°C and 100% relative humidity. Saturated testing can be run at higher temperatures than this by using a "pressure cooker." For instance, 120/100 can be achieved by boiling water at a total pressure of about 2 atm.

Environmental testing is no longer done at saturated conditions because it is not representative of actual usage, and the failures induced are, therefore, not representative of what would be expected in humid, but unsaturated, conditions. It is acceptable to exaggerate humidity for testing, but saturation is fundamentally different and unrealistic. In addition, saturated testing results in poor reproducibility because those portions of the unit that are wetted by condensation or dripping water will fail at a much faster rate. Finally, it is not possible to put bias on parts in saturated testing. These problems convinced reliability workers that unsaturated testing is preferred, and this lead to the development of 85/85 testing that was the de facto standard between about 1970 and 1990.

To reach unsaturated humidity at temperatures above 100°C, it is necessary to pressurize the chamber, up to around 3 atm in practice. This means that HAST chambers must have strong walls and bolted-down doors to resist the large mechanical forces brought about by pressurization. Of course, there must also be safety features to prevent the door from being opened while the chamber is under pressure; at 3.0 atm a 12-in.-diameter door would have almost 5000 lb of force pushing against the latches. Test results have shown that reliability testing in HAST chambers is not a function of the total pressure up to at least several atmospheres. Humidity in accelerated environmental testing is usually not set above about 85% in order to provide some safety margin against moisture condensation. Although higher pressures would enable the chamber to reach higher temperatures at 85% RH, it usually is not set above about 150°C because this would exceed the glass transition temperature of many polymer materials used in electronic packaging. Above T_g the CTE of polymers typically is much higher than at normal operational temperatures, and this could lead to unrepresentative mechanical failures.

16.3.1.2 Acceleration Transforms for Environmental Testing

If all you need to do is compare the moisture performance of two different molding compounds over the same chip, then you just mold them up in the two compounds, put them in the HAST chamber, and see which one lasts longer. A more generalized approach would be to determine an acceleration transform that tells you how a certain number of hours in a certain temperature/RH/bias condition translates into expected lifetime in actual usage conditions or to archived 85/85 conditions. This kind of acceleration transform was developed by Pecht in his classic 1986 study Comprehensive Model for Humidity Testing Correlation (24th Annual Proceedings of the Reliability Physics Symposium 1986, p. 44, 1989). He proposed that the acceleration transform between various values of temperate and relative humidity used during testing should take the form:

$$\text{Time to failure} \propto \frac{1}{\text{RH}^n} \exp\left[\frac{\Delta E_a}{kT}\right] \qquad (16.1)$$

where

> RH = relative humidity
> n = empirical constant
> ΔE_a = empirically found energy of activation for failure mechanism, eV
> k = Boltzmann's constant = 8.62×10^{-5} eV/K
> T = absolute temperature, K

This equation utilizes the standard Arrhenius temperature dependency familiar to chemists, along with an empirical relative humidity term. Pecht found values of n and ΔE_a by fitting this equation to 61 reported cases of time to failure as a function of T and RH in accelerated testing and recommended $n = 2.66$ and $\Delta E_a = 0.79$ eV. Later work added another 26 cases to the database, and the constants were updated to $n = 3.00$ and $\Delta E_a = 0.90$ eV. Actually, these slight changes make little difference, and it should be kept in mind that this is far from an exact science.

Let us say your company has been using a 120/75 test when certifing the parts as passed if they survive 200 h. But your quality control is getting so good that all the parts are now passing 200 h of 120/85, and you want to start using a more stringent test. What is the equivalent time at 140/85? To calculate this, simply take the ratio of the above equation for the two different relative humidities and temperatures:

$$\frac{\text{Time to failure at } T_2 \text{ and RH}_2}{\text{Time to failure at } T_1 \text{ and RH}_1} = \left(\frac{\text{RH}_1}{\text{RH}_2}\right)^n \exp\left[\frac{\Delta E_a}{k}\left(\frac{1}{T_2} - \frac{1}{T_1}\right)\right] \quad (16.2)$$

If we use $n = 3.00$ and $\Delta E_a = 0.90$ eV, then this ratio is 0.198, so 200 h at 120/75 is equal to $(0.198)(200) = 40$ h at 140/85. Based on $n = 3.00$ and $\Delta E_a = 0.90$ eV, Table 16.2 lists time to failures equivalent to 1000 h of 85/85.

The idea of relating any test conditions to ambient conditions—office, jungle, or otherwise—is a bit tenuous and has not been thoroughly tested due to the extreme lengths of time that would be involved. Also, units are usually electrically biased during testing to simulate actual use, and this may cause heating that produces local temperatures in excess of the chamber's setpoint, which may influence the results. Environmental testing is an inexact science, and all results should be treated as such. Direct comparisons of the same units

Table 16.2 Equivalent Times to 1000 h 85/85

Temp. (°C)	RH (%)	Time to failure (h)	
85	85	1,000	6 weeks
110	85	149	6.2 days
120	85	74	3.1 days
130	85	39	1.6 days
140	85	21	0.88 days
25	40	3,410,000	389 years (office conditions)
35	95	81,500	9.3 years (jungle conditions)

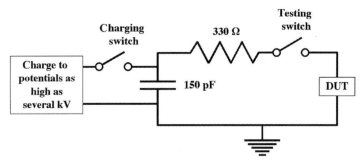

Figure 16.5 Schematic for electrostatic discharge testing, protocol IEC-61000-4-2, Human Body Model.

with only one difference in the same chamber are the most accurate, such as the same chip with the two molding compounds mentioned earlier. The next most accurate acceleration transforms would be between various testing conditions, while transforms from any testing condition to field conditions are the least accurate.

16.3.2 Electrostatic Discharge Accelerated Testing

There are several standards for ESD testing, but the most commonly used is the IEC-61000-4-2 Human Body Model, which simulates the sudden release of charge from a person to an electronic device. Testing is performed by attaching the device under test (DUT) to a charged capacitor and resistor in series as shown in Figure 16.5. The protocols for the Human Body Model stipulate a 150-pF discharge capacitor, a 330-Ω resistor, and a switch that can close in less than 1 ns. The capacitor is charged to anywhere from 1 kV up to, perhaps, 10 to 20 kV. After charging, the power supply is disconnected and the capacitor is discharged through the switch to the DUT.

Other methods are available, such as MIL-STD-883 method 3015.7 and the Electronic Industry Association of Japan (EIAJ) IC-121. Both were developed as an aid in understanding the precautions necessary for packaging and handling ICs. This method tests each package pin, against other groups of pins and classifies the device according to the lowest voltage for which failure occurs.

16.3.3 Other Accelerated Tests

In addition to the widely used temperature/humidity tests described in the previous section, there are others such as:

Temperature Cycling MIL-STD-883 method 1010 and JEST-22 method A104 describe tests designed to create mechanical stresses in units by subjecting them to slowly changing temperatures. The cycling is done in the gas phase with a transition time of at least 5 min and a dwell time of at least 10 min at each temperature to ensure that the entire part reaches isothermal conditions at each temperature extreme. Temperatures of −65 to +150°C are typical. Since this cycling is done relatively slowly (compared to thermal shock as described below), the temperature at every point of the unit is almost the same, just changing slowly all together from one extreme to the other; 1000 cycles is common.

Temperature Shock MIL-STD-883 method 1011 and JEST-22 method A106 describe tests designed to create higher mechanical stresses in units than temperature cycling by subjecting them to rapidly changing temperatures by moving the parts between two liquids. The cycling is done between temperatures similar to those used in temperature cycling but with a more rapid transition time, perhaps as little as a few seconds since they move from one liquid to the other in seconds. This cycling is done quickly enough that the part is not isothermal as its temperature changes, inducing added stresses due to thermal gradients within the units. The liquids used in temperature shock testing are inert, chemically stable, and nontoxic such as Fluorinert from 3M; 10 to 2500 cycles are common.

Temperature Soak These tests involve simply storing the part at high temperature in dry conditions, up to 150°C for something like 1000 h (6 weeks), far in excess of normal operating temperatures for electronics that are rarely higher than about 60°C. Since this is a constant-temperature test, there is no CTE mismatch issue only the acceleration of thermally activated failure mechanisms, although it is possible that the initial heat-up causes a CTE-induced failure.

Pressure Cooker Saturated steam at 120°C for, typically, 50 to 200 h. This requires a total pressure of about 2 atm.

Salt Spray MIL-STD-883 method 1009 simulates a seacoast environment. The test is conducted in an environmental chamber using a flowing salt fog at 35°C for 24, 48, 96, or 240 h.

In addition there are various mechanical tests for vibration resistance, g force, drop shock, and many others.

16.3.4 Test Structures

There are basically two types of unit you can subject to accelerated testing: the actual product (single-chip package, MCM, PWB, rack, computer, etc.) or a test structure that is designed to fail by specific mechanisms and would be expected to show a very narrow spread of time to failure. You test the actual product when you want to find the weak spots in its design, out-of-spec materials, or manufacturing faults, but you use test structures when you want to investigate the specific mechanism. For instance, you might have the task of testing the moisture reliability of a new molding compound relative to an old compound for use with a microprocessor chip. You could mold up 100 microprocessors in the old compound and 100 in the new and put them into HAST testing, but that might cost a small fortune just for the chips. Besides, the microprocessor chip is so complex that there may be many corrosion failure modes going on at once. Since you are testing the molding compound and not the IC, it might be better to use a chip that has very specialized testing structures as shown in Figure 16.6.

This chip has only one level of metal, no diffusions, no active devices, no passivation layer, and requires only one masking level. The corrosion snake has only one failure mode and is easily tested by simply looking for electrical continuity. The moisture sensor should normally be an open circuit but will show some conductivity when moisture reaches it through a package, especially if there are ions dissolved in that water. Other types of test structures are available for measuring mechanical strain, bonding errors, die sawing faults, and many others.

668 Chapter 16 Reliability Considerations

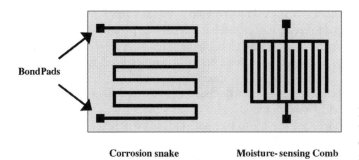

Figure 16.6 Corrosion test chip containing a corrosion snake and a moisture sensor.

16.4 RELIABILITY METROLOGY

Metrology means "methods of measurement and quantitization." Reliability metrology then refers to the measurement and mathematical modeling of reliability and the resulting patterns of failure. In this section we will look at the time dependency of failure for components, and the systems they make up, by any arbitrary failure mechanism. We will make no distinction between what the failure mechanisms actually are—that is one of the convenient aspects of this sort of analysis. However, these methods may be somewhat easier to correlate with test data if one failure mode dominates the observed failure patterns. The methods we will look at are not dependent on the complexity of the devices being tested or the failure modes, making these modeling approaches very general. The basic principles that you will see in this section were established decades ago in other industries but were readily applicable to microelectronics so there is much literature available. The techniques are applicable to both field failure data and failure during accelerated lifetime testing.

16.4.1 Failure Rate, MTBF, and FITs

The most common general expressions for quantitizing electronic failures are closely related: the failure rate, the mean time between failures (MTBF), and failures in time (FITs). They are defined as follows:

$$\text{Failure rate} = \frac{\text{total number of failures}}{\text{total number of device hours}}$$

So if you have 500 devices on test for 1000 h, this is 500,000 device-hours of testing. If you experience four failures during that test run, the failure rate is then 8×10^{-6} failures per device-hour or, which is commonly expressed as eight failures per million device-hours. The MTBF is the reciprocal of this, or 125,000 h between failures. Failures in time is the number of failures per billion device-hours, so, for this example, it is 8000 FITs. These are obviously very crude ways of expressing failure patterns, mainly because it utilizes two assumptions that may not be very good: Failure rate is constant with time, and a small number of failures represents the rate for a potentially very large population. Various statistical treatments can be used to mitigate both of these, as will be shown in Section 16.4.3.

16.4.2 Reliability Functions

Reliability functions are shaped to predict failure versus time, enabling the reliability engineer to use a limited amount of failure data to predict behavior out into the future. As a

Table 16.3 Failure Rate Data for ROMs in Commercial Controller

Days of field usage	Number of ROMs failed in each 50-day period
50	45
100	60
150	78
200	85
250	82
300	78
350	81
400	85
450	88
500	78
550	73
600	68
650	75
700	65
750	60
800	60
850	50
900	52
950	45
1000	40

practical example, let us say you are going to monitor the failure rate of a read-only memory (ROM) packaged in a new type of molding compound. Eighteen hundred of these ROMs go out into field usage in a commercial controller, and you will follow them by looking for failures in equipment returned to your company for repairs. The repair shop keeps track of the units that failed due to the ROM, regardless of what went wrong with the individual ROMs, and supplies you with the data in Table 16.3.

That is all you have to work with, and your task is to pull some information out of this data, such as average expected part lifetime and some information on the failure modes' behavior, whatever these modes may be. The first thing you might notice about the data is that the number of ROMs failing per 50 days of usage is decreasing just a little with time. That might make sense because there are fewer ROMs left in service as time goes by to fail. You will also notice that if you add up the number that have failed at 1000 days, the total is 1348, or 74% of the 1800 that went out. It would be very rare that you would have the luxury of waiting for all the units to fail before you begin data analysis, so it is normally the case that you will be working with less than 100% failure data. One more thing that should jump out at you from this data is that these ROMs have a serious problem if they are failing this fast this soon, but it is easier to demonstrate this methodology with a higher failure rate.

Reliability functions are used to ascertain some information about the failure mechanism that is operating. There are four main ones, but we will start with the two most basic:

$$R(t) = reliability\ function$$

This is the fraction of the original devices that is still operating at time t. This is the same as the $R(t)$ defined in the introduction to this chapter.

$$F(t) = \text{cumulative failure function}$$

This is the fraction of the original devices that has failed by time t.

Both of these are a number between zero and unity, and they always have a length of time associated with them. Since a component either operates properly or does not, by whatever definition of "operational" has been set, then the sum of these two functions is always unity:

$$R(t) + F(t) = 1$$

The value of these two functions can easily be calculated from the tabulated data shown in Table 16.3. For instance,

$$F(150 \text{ days}) = \frac{\text{number failed at 150 days}}{\text{total original number of devices}} = \frac{45 + 60 + 78}{1800} = 0.102$$

$$R(150 \text{ days}) = \frac{\text{number operating at 150 days}}{\text{total original number of devices}} = \frac{1800 - (45 + 60 + 78)}{1800}$$
$$= 1 - F(150) = 0.898$$

Therefore, for a large number of these packaged ROMs in field usage, you would expect that 10.2% of the original devices will fail by 150 days of usage and that a given ROM has an 89.8% chance of still being operational at this time (which, by the way, is not very good). Remember that you are dealing with statistics here, so that the bigger your test population, the more accurate the resulting predicted values of $F(t)$ and $R(t)$ will be for a larger population.

Figure 16.7 gives the plots of $F(t)$ and $R(t)$ for the example data. Notice that in this and all subsequent plots discrete data is shown as discrete data points and calculated values from continuous mathematical functions are shown as continuous lines. Also, the discrete points for 150 days are plotted at time equals 125 days, the midpoint between 100 and

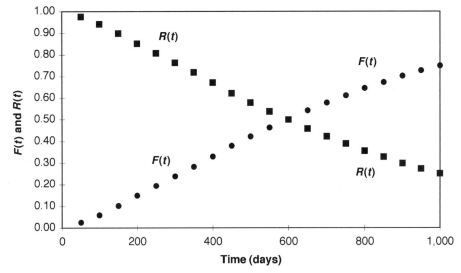

Figure 16.7 Cumulative failure function and reliability function for the example data.

150 days, since this actually represents phenomena that occurred somewhere between these two measurement times.

For reliability calculations, a population of devices, components, or systems is fully functional to start with and eventually they will all fail. Thus,

$$F(0) = 0 \quad \text{and} \quad F(\infty) = 1$$
$$R(0) = 1 \quad \text{and} \quad R(\infty) = 0$$

What is the probability that a new ROM will fail by 700 days? It is about 58%. What is the probability that a new ROM will fail by 800 days? It is about 64%. What is the probability that a new ROM will fail between 700 and 800 days? It is $64 - 58 = 6\%$.

The third of four reliability functions we deal with in this chapter is the *failure density function*, $f(t)$. This is defined as the time derivative of the cumulative failure function and has the units of reciprocal time:

$$f(t) = \frac{dF(t)}{dt} = -\frac{dR(t)}{dt} \tag{16.3}$$

$$F(t) = \int_0^t f(t)\,dt \tag{16.4}$$

Physically, the failure density function is the fractional rate that the *original* devices are failing at a given time. The higher the failure density function, the more rapidly the original population is failing. Since all devices will eventually fail, the time integral of the failure density function from zero to infinity is unity:

$$F(\infty) = \int_0^\infty f(t)\,dt = 1$$

The failure density function is related to the rate that the original population is failing at a given time, so this function does not have to be zero at $t = 0$, unless there are no failures occurring at this time, but it must go to zero eventually when there are very few or, possibly, none left to fail.

$$f(\infty) = 0$$

The failure density function is the time derivative of the cumulative failure distribution and, for discrete data such as in Figure 16.7, this derivative can be expressed in finite-difference form, in this case the backwards difference form:

$$f(t) = \frac{dF(t)}{dt} = \frac{F(t) - F(t - \Delta t)}{\Delta t} \tag{16.5}$$

Taking numerical derivatives can result in considerable scatter in the results since small dispersions in the y direction can lead to large departures in the resulting slopes.

At 800 days the value of the failure density function is

$$f(800 \text{ days}) = \frac{dF(t)}{dt} = \frac{F(800) - F(750)}{50} = \frac{0.645 - 0.612}{50} = 6.67 \times 10^{-4} \text{ day}^{-1}$$

The failure density function for the example data is shown in Figure 16.8. If you look at the data in Figure 16.8, it seems that the failure rate increases to a rough maximum around 400 days and then decreases beyond that. But keep in mind that, as time goes by and devices fail and drop out of the test, the number of devices that are *on test* decrease. Therefore, there are fewer devices to fail, so the failure density function must decrease, eventually reaching zero.

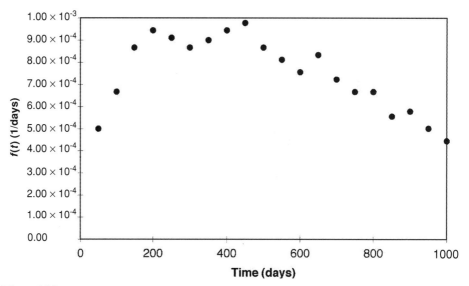

Figure 16.8 Failure density function for the example data.

The failure density function, $f(t)$, represents a rate of failure at a given time t based on the *initial* population of devices. Thus, at large enough times, a failure density function shows a negative slope as the number of devices on test drops to a small fraction of the original population even though the rate of failure per device *still on test* may be as high as it was at the start of the test. This means that a curve of $f(t)$ does not give much intuitive information on the rate of device failure for those still on test. A better measure of the failure rate of those devices still working is the *hazard rate*, $h(t)$, which is defined as the rate of failure based on the steadily decreasing number of devices that are still on test (not the constant original number) and have, therefore, not yet failed:

$$h(t) = \frac{\text{rate of failure}}{\text{number of devices still operating}} = \frac{f(t)}{R(t)} = \frac{f(t)}{1 - F(t)} \quad (16.6)$$

This also has units of reciprocal time. The hazard rate at 800 days is

$$h(800 \text{ days}) = \frac{f(800)}{R(800)} = \frac{6.67 \times 10^{-4}}{0.355} = 1.88 \times 10^{-3} \text{ days}^{-1}$$

The form of the hazard rate for the example data (Fig. 16.9) indicates that, whatever the failure mechanism is, it gets worse as the devices get older. That implies that it probably is not a random overstress failure but probably some sort of wearout mechanism. If it were random overstress, the hazard rate would be a flat line. Remember that the final data point for the hazard rate is based on the failure rate of only 492 devices that made it past 950 days, not on the 1800 that started. Once the surviving population fraction, $R(t)$, becomes small, random fluctuations in the failure rate can have a large effect on the measured hazard rate so the points can scatter significantly near the right end. In statistical calculations, the more data you have the more certain your results are. When nearly all of the devices have failed, there is little data coming in from which to calculate current failure rates and, as a result, the hazard rate at long times tends to become very scattered. Table 16.4 shows the four functions calculated for all available times.

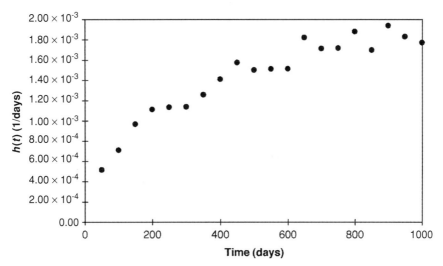

Figure 16.9 Hazard function for example data.

Table 16.4 Failure Rate Data for ROMs in a Commercial Controller

Time (days)	Number of fails per time unit	Results of numerical analysis			
		$F(t)$	$R(t)$	$f(t)$	$h(t)$
50	45	0.025	0.975	5.00×10^{-4}	5.13×10^{-4}
100	60	0.058	0.942	6.67×10^{-4}	7.08×10^{-4}
150	78	0.102	0.898	8.67×10^{-4}	9.65×10^{-4}
200	85	0.149	0.851	9.44×10^{-4}	1.11×10^{-3}
250	82	0.194	0.806	9.11×10^{-4}	1.13×10^{-3}
300	78	0.238	0.762	8.67×10^{-4}	1.14×10^{-3}
350	81	0.283	0.717	9.00×10^{-4}	1.25×10^{-3}
400	85	0.330	0.670	9.44×10^{-4}	1.41×10^{-3}
450	88	0.379	0.621	9.78×10^{-4}	1.57×10^{-3}
500	78	0.422	0.578	8.67×10^{-4}	1.50×10^{-3}
550	73	0.463	0.537	8.11×10^{-4}	1.51×10^{-3}
600	68	0.501	0.499	7.56×10^{-4}	1.51×10^{-3}
650	75	0.542	0.458	8.33×10^{-4}	1.82×10^{-3}
700	65	0.578	0.422	7.22×10^{-4}	1.71×10^{-3}
750	60	0.612	0.388	6.67×10^{-4}	1.72×10^{-3}
800	60	0.645	0.355	6.67×10^{-4}	1.88×10^{-3}
850	50	0.673	0.327	5.56×10^{-4}	1.70×10^{-3}
900	52	0.702	0.298	5.78×10^{-4}	1.94×10^{-3}
950	45	0.727	0.273	5.00×10^{-4}	1.83×10^{-3}
1,000	40	0.749	0.251	4.44×10^{-4}	1.77×10^{-3}

16.4.3 Weibull Distribution

To be able to compare failure patterns of devices in field usage or in accelerated testing, semi empirical reliability distribution functions are used to reduce the observed, and often incomplete, failure data to just one or two parameters. These are presumed shapes of the four reliability distribution functions, $F(t)$, $R(t)$, $f(t)$, and $h(t)$, based on various assumptions. These can be used to compare the failure patterns of different units by comparing just these couple of parameters instead of all the individual failure data points. As an example, let us look at one of the most widely used distributions, the Weibull (pronounced "Y-bull") distribution. For the Weibull distribution the failure density function is

$$f(t) = \frac{\beta}{\lambda}\left(\frac{t}{\lambda}\right)^{(\beta-1)} e^{-(t/\lambda)^\beta} \tag{16.7}$$

In principle, if you are given any one of these four functions, you can derive the other three:

$$F(t) \int_0^t f(t)\,dt = \int_0^t \left[\frac{\beta}{\lambda}\left(\frac{t}{\lambda}\right)^{(\beta-1)} e^{-(t/\lambda)^\beta}\right] dt = 1 - e^{-(t/\lambda)^\beta} \tag{16.8}$$

$$R(t) = 1 - F(t) = e^{-(t/\lambda)^\beta}$$

$$h(t) = \frac{f(t)}{R(t)} = \frac{f(t)}{1 - F(t)} = \frac{\frac{\beta}{\lambda}\left(\frac{t}{\lambda}\right)^{(\beta-1)} e^{-(t/\beta)^\beta}}{1 - \left[1 - e^{(t/\lambda)^\beta}\right]} = \frac{\beta}{\lambda}\left(\frac{t}{\lambda}\right)^{(\beta-1)} \tag{16.9}$$

The Weibull distribution is a two-parameter model:

λ = The *lifetime parameter*, a measure of the average time to failure. This has the units of time and, as we will see, is equal to the time at which 0.632 of the population has failed.

β = The *shape parameter*, a measure of how the failure frequency is distributed around the average lifetime or, in other words, a measure of the shape of the hazard function. This is dimensionless and is usually between about 0.5 and 2.0.

The idea is to use measured lifetime data, such as that in our example case, and calculate values of λ and β for use in comparison to other product's lifetime data or to correlate λ and β for various products and processes, and you do not have to compare large amounts of raw failure data. The following three figures show these three functions. The failure density function itself is shown in Figure 16.10 plotted with the time axis expressed as time divided by the lifetime parameter, or t/λ and for a λ of unity. As you can see, as the shape parameter increases the time at which the majority of failures appear.

The cumulative distribution function, $F(t)$, is shown below in Figure 16.11. It is clear from this figure that λ, the lifetime parameter, is the time at which 0.6321 of the units have failed (this is equal to $1 - e^{-1}$), and β, the shape, influences how the failure rate is distributed around λ.

Finally, we can get a physical sense for the shape parameter if we plot out the hazard rate for the Weibull distribution as shown in Figure 16.12. The shape parameter β is related to the distribution of the hazard rate. A shape parameter of unity means that the chances of device failure is constant in time as you might expect from random overstress failures. A smaller β indicates that there is a higher chance of early failure, perhaps due to manufacturing defects,

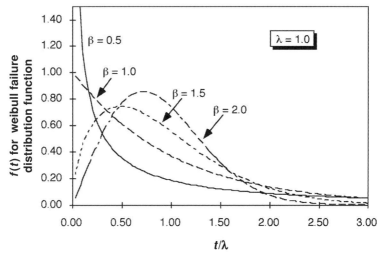

Figure 16.10 Weibull failure density function.

and a $\beta > 1$ means that wearout failures are causing an increase in the failure rate at later times.

So how do you use this correlation to determine λ and β? The best way is to curve fit the rawest, least processed data you have, which is nearly always the cumulative failure distribution, $F(t)$. In the case of the Weibull distribution this is

$$F(t) = 1 - e^{-(t/\lambda)^\beta} \qquad (16.10)$$

So if you plot your observed cumulative failure data $F(t)$, such as from Table 16.4, against this function with some sort of least-squares fit procedure you would get something like Figure 16.13.

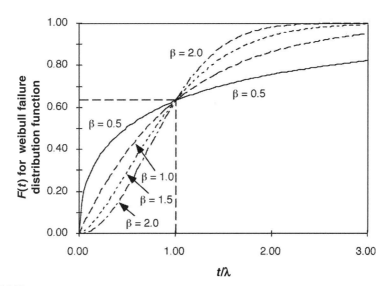

Figure 16.11 Cumulative failure distribution from Weibull distribution function.

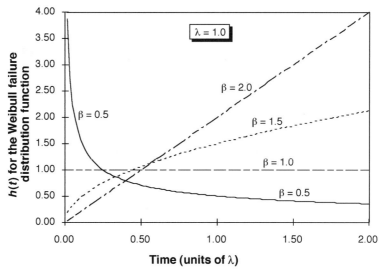

Figure 16.12 Hazard distribution from Weibull distribution function.

Three different curve fits are shown in Figure 16.13, each with λ 782 days but with different values of β. Clearly the best fit is at a shape parameter of 1.35. This fit is perhaps better than might be expected from real data, but remember that $F(t)$ data is always monotonic and will always show less scatter than $f(t)$ and $h(t)$, which are calculated from numerical derivatives. Figure 16.14 shows the functional curve fit for the hazard rate.

The hazard rate rises with time indicating that, whatever the failure mechanism is, the longer the device lives the more likely it is to fail. This means that the failure mechanism is a wearout failure; if it were random overstress the hazard rate would be flat, and if it were an infant mortality failure the hazard rate would be decreasing with time. Therefore, while

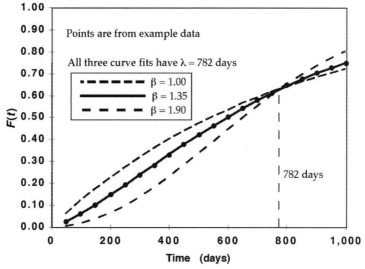

Figure 16.13 Example's failure data used to determine lifetime parameter and shape parameter of Weibull distribution.

16.4 Reliability Metrology

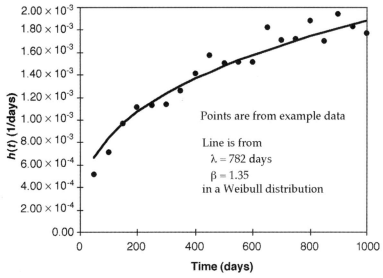

Figure 16.14 Comparison of experimental and curve-fit hazard rate for example data.

this sort of analysis does not identify the failure mode, it does give some clues about its behavior. (See Fig. 16.15).

16.4.4 Normal Distribution

The normal distribution is the one you are used to using for averages and standard deviations. You may never have thought about it as a mathematical model for failure patterns, but it is very effective for that, when used correctly. The independent variable is time, and we are

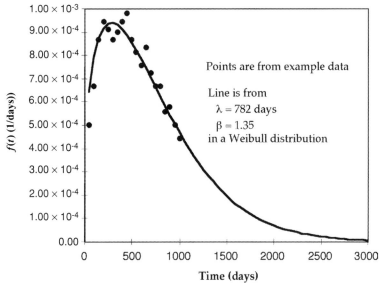

Figure 16.15 Weibull curve fit for failure density function.

interested in the time to failure for microelectronic units. If you start with some number of units, which are all functional at $t = 0$, and you measure the time to failure for each and every one, then you can calculate an average lifetime and a standard deviation for those units in the familiar way:

$$t_{\text{avg}} = \frac{\text{sum of times to failure for each unit}}{\text{initial number of units}} = \frac{\sum t_i}{N} \tag{16.11}$$

$$\sigma = \sqrt{\frac{\sum (t_i - t_{\text{avg}})^2}{N}} \tag{16.12}$$

The average lifetime is the lifetime parameter, and the standard deviation is the shape parameter, filling the same role that λ and β did for the Weibull distribution. Equations (16.11) and (16.12) only work if every single unit has failed, not just most of them, otherwise you will get an unrealistically low calculated average. This is a surprisingly common error in practice. However, it is also possible to calculate average lifetimes and standard deviations before all the units have failed, by curve fitting as we did for the Weibull distribution. First, let us understand how the normal distribution can be cast as a set of reliability functions; then we will look at how to use it.

The normal distribution works best for wearout failures, and is not usable for early failures or for overstress failures, a fact that will become more clear when we look at the curves. It is most suitable when the time to failure is "normally distributed" around an average value, and this distribution defines the standard deviation. The failure density function for the normal distribution is

$$f(t) = \frac{1}{\sigma\sqrt{2\pi}} \exp\left[-\frac{1}{2}\left(\frac{t - t_{\text{avg}}}{\sigma}\right)^2\right] \tag{16.13}$$

This is the familiar *bell-shaped curve*, and is shown in Figure 16.16 for an average time to failure of 100 months and a standard deviation of 16 months. These numbers correspond to another variable that is normally distributed around an average of 100 and has standard deviation of 16, that is, intelligence quotient (IQ).

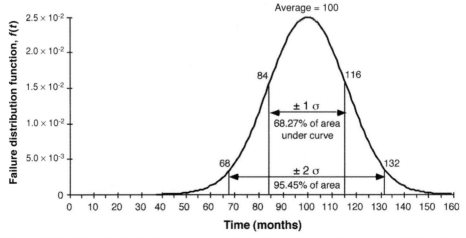

Figure 16.16 Failure distribution function for normally distributed failures with an average of 100 months and a standard deviation of 16 months.

16.4 Reliability Metrology

It is no accident that IQ values are normally distributed, along with lots of other things if they are truly random, such as the as-manufactured length of erasers, the weight of people, and the frequency with which you step on cracks on the sidewalk. According to the mathematical relationship between the reliability functions, the other three are:

$$F(t) = \int_0^t \frac{1}{\sigma\sqrt{2\pi}} \exp\left[-\frac{1}{2}\left(\frac{t - t_{\text{avg}}}{\sigma}\right)^2\right] dt \equiv \Phi\left(\frac{t - t_{\text{avg}}}{\sigma}\right) \quad (16.14)$$

$$R(t) = 1 - \Phi\left(\frac{t - t_{\text{avg}}}{\sigma}\right) \quad (16.15)$$

$$h(t) = \frac{\frac{1}{\sigma\sqrt{2\pi}} \exp\left[-\frac{1}{2}\left(\frac{t - t_{\text{avg}}}{\sigma}\right)^2\right]}{1 - \Phi\left(\frac{t - t_{\text{avg}}}{\sigma}\right)} \quad (16.16)$$

The integral in Eq. (16.14) cannot be solved in closed form and is commonly given the symbol Φ for brevity. Values are available in tables and as preprogrammed functions in many software packages. The cumulative failure density function and the hazard rate for the normal distribution are shown in Figure 16.17, and it is easy to understand why the shape of this distribution is suitable only for wearout failures. In order to use this distribution,

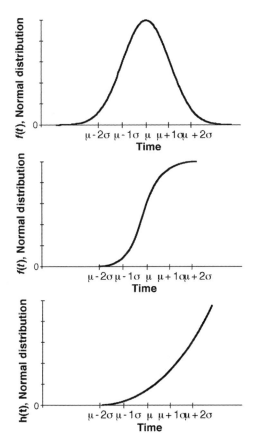

Figure 16.17 Distribution functions for the normal distribution.

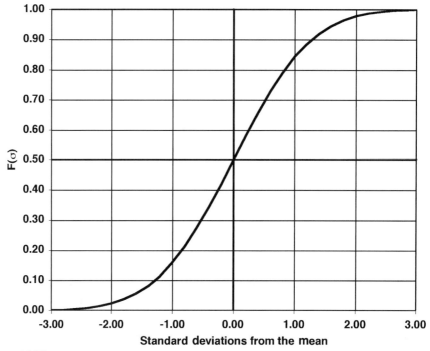

Figure 16.18 Cumulative failure function for the normal distribution.

the average needs to be at least three standard deviations away from zero or the correct bell shape cannot be used. The normal distribution is best used for wearout failures since the hazard rate rises rapidly within two standard deviations of the mean. Notice also that the normal distribution is defined relative to the mean failure time, not time $= 0$. This distribution works best for wearout failures where the first fails some period of time and then rises rapidly after that.

Figure 16.18 shows the cumulative failure function, $F(t)$, for the normal distribution in more detail. The way to calculate the average and standard deviation for normally distributed failure data before everything has failed is similar to what we did for the Weibull distribution. In both cases you plot the known $F(t)$ versus t data with various guesses of the two parameters, average and standard deviation in this case, until it fits the shape of the curve from the distribution as well as it can be made to. If no values of average and standard deviation can make the data fit, then the failures are not normally distributed.

16.4.5 Failure Distributions and the Bathtub Curve

The well-known bathtub curve was described in the introduction of this chapter as being the superposition of three different failure curves, as shown in Figure 16.19. No single reliability distribution can model the entire shape of the classical bathtub curve with just two parameters. Using the Weibull distribution, you would have to use three different curves, one with $\beta < 1$, one with $\beta = 1$, and one with $\beta > 1$ (see Fig. 16.10), but you will rarely need to model all three; they are usually studied separately for a given product. The Weibull distribution can handle all three sections, and the normal distribution is fine for wearout

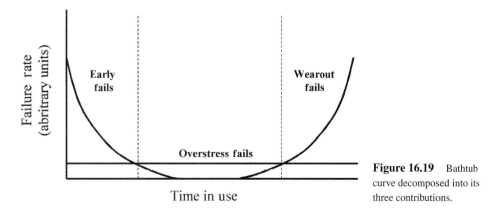

Figure 16.19 Bathtub curve decomposed into its three contributions.

failures that start after a significant portion of the expected lifetime is over. Other types of distributions are also available such as the lognormal and exponential.

16.5 FAILURE STATISTICS FOR MICROELECTRONIC SYSTEMS

The reliability of the overall system can be calculated from the known or estimated reliabilities of the individual *criticality one* components that make up that system. A criticality one component is one that, if it fails, the entire system fails. For instance, let us say you are asked to estimate the reliability of a certain circuit board for your company. The time period of interest is 6 years. This particular circuit board consists of three chips, and all three chips must function properly for the circuit board as a whole to function within acceptable limits. The circuit board is the system, the chips are the components. Nothing else on the board will fail at a significant rate. The probability that each of the three ICs will fail within 6 years was measured using accelerating life testing in your company's reliability lab. They found that the only significant failure modes were underfill cracking and that the estimated probabilities of failure in 6 years for each of the three components (chips) are:

   Component A: 0.30 chance of failure
   Component B: 0.40 chance of failure
   Component C: 0.50 chance of failure
   Component A: $1 - 0.30 = 0.70$ chance of working successfully
   Component B: $1 - 0.40 = 0.60$ chance of working successfully
   Component C: $1 - 0.50 = 0.50$ chance of working successfully

So the reliability of the system over the 6-year time interval is

   (probability of A operating)(probability of B operating)(probability of C operating)
   $= (1 - 0.30)(1 - 0.40)(1 - 0.50) = 0.210$

The circuit board will have a reliability of 21% over 6 years or, conversely, a 79% chance of failure over 6 years. In other terms, $R(6 \text{ years}) = 0.21$ and $F(6 \text{ years}) = 0.79$.

Now, what if further testing indicated that one of the chips also had a bondpad corrosion failure mode due to contamination during manufacture, but that the probability of failure by that mode in 6 years was only 1%. Then the reliability of the system is a function of the four possible failure modes:

$$(0.70)(0.60)(0.50)(0.99) = 0.208$$

The additional low probability failure mode make little difference in the reliability of the overall system. This indicates that the failure rate of the system is dominated by the most probable failure modes.

In this example the 6-year time interval was stated at every opportunity in order to emphasize that the reliability of a component or of a system is defined over a given time interval. A reliability, estimated or measured, that is not associated with a definite time span is meaningless. We can generalize this example to get the reliability for a nonredundant system where every component must work for the system as a whole to work:

$$R_{\text{sys}} = \prod_{i=1}^{N} R_i = \prod_{i=1}^{N} (1 - F_i) \qquad (16.17)$$

where

R_{sys} = probability that the system will work over a certain time interval

N = number of components in system

R_i = probability that component i will work over certain time interval

F_i = probability that component i will fail over certain time interval

If each component has the same probability of operating properly over a certain time interval, then

$$R_{\text{sys}} = \left(R_{\text{comp}}\right)^N = \left(1 - F_{\text{comp}}\right)^N \qquad (16.18)$$

where

R_{comp} = probability that component will work over certain time interval

F_{comp} = probability that component will fail over certain time interval

Reliability is usually written as $R(t)$ to indicate that the value of R is a function of time. I have left off the (t) for brevity, but keep in mind that a value of reliability is not meaningful without an associated time.

This equation is plotted in Figure 16.20 for various values of N and R_{comp} along with typical ranges of the number of components that would be expected for various systems and components.

How does this relate to the reliability of a real microelectronic system? A modern cell phone has about 500 passive components on the main circuit board. If the set of passives alone has a reliability over 3 years of 97%, then the average reliability of an individual passive can be calculated as follows:

$$R(t) = (R_i)^{500} = 0.97$$
$$R_i = 0.99994 = 99.994\%$$

Because modern electronic systems are so reliable as systems and contain a large number of components, the reliability of individual components is very high.

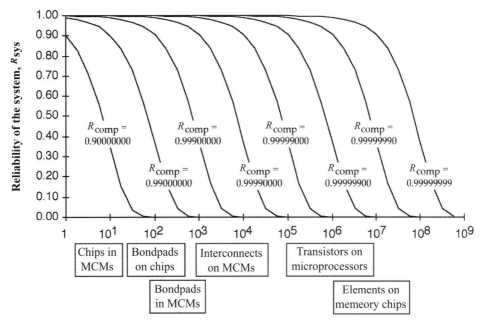

Figure 16.20 Reliability of system consisting of N nonredundant components, each having the same individual component reliability R_{comp}.

16.5.1 Predicting Failure in Components That Have Multiple Failure Modes

The same sort of logic applies to the case of multiple failure modes for a single component. If a component—or a system for that matter—has several different failure modes, then each mode is like a nonredundant subcomponent, and the situation is treated like a system of nonredundant components. For instance, if a single chip has a 7.0% chance of failing in 8 years due to bondpad corrosion and a 10.0% chance of failing in 8 years due to thermal overload and a 12% chance of failing in 8 years due to solder joint breakage, then the chip will fail if any of the three failure mechanisms occur:

Bondpad corrosion occurs, 7.0% chance *or*

Thermal overload occurs, 10.0% chance *or*

Joint breakage occurs, 12.0% chance

Again, this is the same as saying that the probability of no chip failure is the probability of:

Bondpad corrosion does not occur, $100\% - 7.0\% = 93\%$ chance *and*

Thermal overload does not occur, $100\% - 10.0\% = 90\%$ chance *and*

Joint breakage does not occur, $100\% - 12.0\% = 88\%$ chance

The probability of no chip failure at 8 years, its reliability, is then

$$(0.93)(0.90)(0.88) = 0.737 = 73.7\%$$

16.6 INDUSTRIAL PRACTICE OF RELIABILITY SCIENCE FOR MICROELECTRONICS

As you can see from the contents of this chapter there are a number of different testing and data analysis methods available for application in reliability science for microelectronics. A specific company tends to use the conditions and data analysis techniques that it used the year before, perhaps with minor modifications to cope with changing circumstances. For instance, most companies used some form of 85/85 testing for years before moving to the more expensive HAST chambers only when test times became too extreme to wait out. The exact standards used at a given company are, then, an evolved situation based on many factors.

The type of task you will perform most often as a reliability engineer is to qualify a new product, procedure, or material with respect to one that has been in production or use for some time. This is usually quite straightforward in that you simply treat two populations of parts with the different procedure or material and put them on test in the way that your company has used before. When there is no test that was used before, be conservative in coming up with new tests. The more extreme a test is relative to usage conditions, the less likely it is to stimulate the same failure mechanisms. As a reliability engineer, especially a new one, you should familiarize yourself with the various testing and analysis procedures as best as you can, both inside and outside your company. Some sources for this are given in the Bibliography.

BIBLIOGRAPHY

Books

Ayyub, M., and McCuen, H., *Probability, Statistics, & Reliability for Engineers*, Boca Raton, FL: CRC Press, 1997.

Chan H., and Englert P., eds., *Accelerated Stress Testing Handbook: A Guide for Achieving Quality Products*, IEEE Press, 2001.

Crow D., and Feinberg A., *Design for Reliability*, Boca Raton, FL: CRC Press, 2000.

Dia S., and Wang M., *Reliability Analysis in Engineering Applications*, New York: Van Nostrand Reinhold, 1992.

Grosh D., *A Primer of Reliability Theory*, New York: Wiley, 1989.

Limnios N., and Nikulin M., eds., *Recent Advances in Reliability Theory: Methodology, Practice, and Inference*, Birkhauser, 2000.

Pecht M., *Integrated Circuit, Hybrid, and Multichip Module Package Design Guidelines—A Focus on Reliability*, New York: Wiley, 1994.

JEDEC Standard, Test Method A110, Highly-Accelerated Temperature and Humidity Stress Test (HAST), Electronic Industries Association, July 1988.

Conference Proceedings

The *Annual Proceedings of the Reliability Physics Symposium*, is published by IEEE after each conference and contains the results of some of the best basic research in this area. To order these through your interlibrary loan office, describe the volume as ***th Annual Proceedings of the Reliability Physics Symposium where "***" plus 1962 = the year of the conference. For instance, the 18th Annual Proceedings were from the $18 + 1962 = 1980$ conference.

Articles and Periodicals

Pecht M., Comprehensive Model for Humidity Testing Correlation, 24th Annual Proceedings of the Reliability Physics Symposium 1986, p. 44, 1989.

Electronic Device Failure Analysis, ASM International,

EXERCISES

16.1. Estimate the number of the following in a personal computer:

- Boards
- Chips
- Chip-level I/O

- Transistors (do not forget the memory chips)
- Purely mechanical devices

Be sure and justify your counts. Many of your estimates may be low.

16.2. Your company makes 1000 circuit boards in a production run. Of these 800 pass final testing prior to shipment and, 1 month after shipping, 200 of those that shipped have failed. What is the numerical value of $R(0)$?

16.3. You make 1800 of a certain product in one lot. The yield was 85% and the reliability at 6 months after shipment was 92%. How many of these units were still working 6 months after shipment?

16.4. Your company manufactures circuit boards that have a 5-year warranty. If one fails during that time, you will replace it for free with a new circuit board and a new 5-year warranty. Which of the following two cases will cost your company more money: a customer's MCM fails after 1 year or a customer's MCM fails after 4 years? Why?

16.5. Your company has for years been performing HAST testing at 120/80 and deeming a part "passed" if it survived 300 h. What is the equivalent test time in hours at 140/85? Use $n = 2.60$ and $E_a = 0.82$ eV.

16.6. What HAST testing time at 130/85 is equivalent to 500 h of 120/80 if you use an activation energy of 0.83 eV and a humidity exponent of 2.50?

16.7. Parts are failing due to corrosion and are showing an average lifetime of 6.5 years at an operating temperature of 40°C. What would be their average lifetime at 60°C if the activation energy for this failure mechanism is 0.72 eV?

16.8. 800 units go onto temperature storage test at 120°C in a dry nitrogen chamber for 600 h. During this time, there are seven failures.

  a. What is the failure rate, MTBF, and FITs?

  b. What would be the estimated FITs at the actual operating temperature of 35°C if you use an activation energy of 0.9 eV?

16.9. Your company is trying to develop a correlation for its temperature/RH environmental test data. The failure lab tested a large number of corrosion test structures at various conditions to determine the time to failure. Here are the results:

Temp. (°C)	RH (%)	Average time to failure (h)
85	85	6200
85	95	3300
100	85	2800
100	95	1500
120	85	1200
120	95	520
130	85	610
130	95	420
140	85	310
140	95	210

What are the values of n and ΔE_a (in electron volts) that best fit this data using the Pecht equation? Use some sort of best-fit routine to get the best values and make a plot showing that your values give a good match.

16.10. If a type of unit fails in an average time of 180 h during a temperature soak at 120°C, what would its average lifetime have been at its operating temperature of 60°C?

16.11. For the following temperature shock data, what is the value of the four reliability functions at 80 cycles? There were initially 900 devices on test.

Number of temperature shock cycles	Number of fails per time unit
20	62
40	48
60	56
80	50
100	46
120	79
140	60
160	58
180	62
200	34
220	28
240	43
260	69

16.12. If you draw the hazard function versus time for units that fail exclusively by overstress failure (failure rate is constant and independent of number on test), does the line have a positive, zero, or negative slope? Why?

16.13. You are a reliability engineer for a company that makes chips packaged in a ball grid array (BGA). The units you make are found to be undergoing early failures where the failure density function is approximately linear. At $t = 0$, the failure density function is 10^{-3}/week and at $t = 40$ weeks half of the units have failed. What is the equation for the four reliability functions with t in weeks?

16.14. If the Weibull distribution is used to model the failure rate of a population of units and it is found that the lifetime parameter is λ days and the shape parameter is β, what is the time at which 50% of the units will fail?

16.15. You are tracking the failure rate of 1200 MCMs in temperature shock accelerated testing.

Number of temp. shock cycles	Number of fails
10	120
18	78
26	58
35	47
50	65
56	26
66	35
75	28
81	18
105	68
121	34
142	40
160	35
172	20
188	26
211	32
227	20
238	14
250	12
270	20

a. Plot the data to determine the Weibull parameters for this data.
b. Plot the hazard rate versus time for this data along with the hazard rate from these two Weibull distribution parameters.
c. What is the probability of a new part failing between 130 and 200 cycles? Calculate this from the Weibull parameters.

16.16. 1800 microcontrollers packaged in molded plastic are subjected to high-temperature storage testing. The failure results are give in the table below. Clearly, a normal distribution is best for fitting this since the fails begin after a "long" time.

a. Back out the best-fit average and standard deviation for the time to failure by plotting the failure data against the cumulative failure distribution for the normal distribution. If you use a spreadsheet, or MathCAD or computer code, hand that in. If you hand in a spreadsheet, include the formula sheet (tools:preferences:view:formulas) so that the equations you used are visible.
b. Hand in a plot showing the discrete data points and your model's curves for both $f(t)$ and $h(t)$.
c. At what time would you expect 95% of the devices to have failed?

Time (days)	Number of fails per time increment
0	0
40	0
80	3
120	1
160	0
200	4
240	0
280	2
320	1
360	0
400	3
440	0
480	2
520	1
560	14
600	25
640	9
680	14
720	24
760	40
800	35
840	86
880	64
920	120
960	144
1,000	119
1,040	144
1,080	153
1,120	164
1,160	148
1,200	125

16.17. Two thousand memory modules went into service. Over the next 4 years there was the following failure rate. How many weeks until 90% will have failed? Use the normal distribution to calculate this.

Time (weeks)	Total number of failures	Time (weeks)	Total number of failures
20	0	260	50
40	2	270	22
60	0	280	35
80	0	290	48
100	3	300	100
120	0	310	68
140	0	320	82
160	1	330	112
180	0	340	133
200	5	350	86
210	0	360	64
220	1	370	142
230	2	380	155
240	22	390	115
250	16	400	108

16.18. For an average lifetime of 8.5 years with a standard deviation of 1.3 years, how long until 95% fails?

16.19. For an average lifetime of 6.4 years with a standard deviation of 9.8 months, how long until 35% fails?

16.20. If the failure pattern of some parts are described by an average lifetime of 75 weeks and a standard deviation of 12 weeks, how long will it take for 10% of the parts to fail?

16.21. Someone gives you the Weibull parameters (λ and β) for a certain product, and β is a fairly large value, say bigger than 3 or so. However, your company does not use Weibull, you use the normal distribution. How would you convert λ and β into average and standard deviation?

16.22. The reliability function for the exponential analysis is $\exp(-t/\gamma)$. Starting with this, derive the other three functions for the exponential distribution: $F(t)$, $f(t)$, and $h(t)$.

16.23. You have a gate array with 15,000 gates and 80 bondpads. The possible failure modes over 10 years are:
- Bondpad corrosion, 0.045% chance of failure per bondpad
- Thermal failure, 5.0×10^{-4}% chance of failure per gate

What is the reliability of this gate array over 10 years? What is its chance of failure in 10 years?

16.24. The individual chips in a memory module have been tested for failure patterns. The Weibull distribution was used in the data analysis and each memory chip has these parameters: $\lambda = 600$ months, $\beta = 1.50$. If the memory module contains 15 of these chips, what is the reliability of this assembly at 36 months?

16.25. You have an MCM with 6 chips and 900 bondpads. The possible failure modes over 12 years are:
- Bondpad corrosion, 0.05% chance of failure per bondpad
- Chip cracking, 3.0 % chance of failure per chip

What is the reliability of this MCM at 12 years?

16.26. You have an MCM with 6 chips. Each chip has 100 bondpads. The possible failure modes over 12 years are:
- Bondpad corrosion, 0.05% chance of failure per bondpad
- Chip cracking, 3.0% chance of failure per chip

Everything has to work for the MCM to be considered operational. Out of an initial run of 400 of these MCMs, how many would you expect to still be working at 12 years?

16.27. The individual chips in an memory module have been individually tested for failure patterns. The Weibull distribution was used in the data analysis. This group of chips has these parameters: $\lambda = 600$ months, $\beta = 1.50$. If the memory module contains 15 of these chips, what is the reliability of this assembly at 36 months?

Chapter 17

Cost Evaluation and Analysis

TERRY R. COLLINS, SCOTT J. MASON, AND HEATHER NACHTMANN

17.1 INTRODUCTION

No discussion of high-density packaging can be considered complete without an assessment of manufacturing costs. As new technologies are introduced to manufacture high-density packages, managing the costs to produce these products are of great interest to the manufacturer. This chapter overviews the costs that must be considered by both the user and manufacturer in either the purchase or production of high-density electronic packages.

This chapter initially describes the product cost analysis process with particular emphasis on direct and indirect costs. Traditional and activity-based costing techniques are presented to provide the reader with the ability to analyze product costs. Next, fixed and variable costs are used to support a discussion of linear break-even analysis. The next section of the chapter discusses popular forecasting models such as the moving average, and the single (simple), double (Holt's method), and triple (Winter's method) exponential smoothing methods. The last forecasting method discussed is the least-squares regression.

A comparative analysis based on the concept of the time value of money is presented in the decision-making process to invest or purchase in a product or process. Capital project evaluation and replacement analysis are further discussed in detail in this chapter. Finally, different techniques are presented to perform a sensitivity analysis to aid in addressing uncertainty in the decision-making process.

17.2 PRODUCT COST

Conventional product cost analysis methods may be readily used to describe the costs associated with the production of high-density electronic packages. Most manufacturing costs can be classified as either a direct or indirect cost. In situations where the cost is easily tracked and assigned to a cost driver, this is considered a direct cost. A cost driver is any output measure that causes cost [1]. In contrast, when a cost is very difficult to economically identify the cost for a specific cost driver, this cost is classified as an indirect cost.

Advanced Electronic Packaging, Second Edition, Edited by Richard K. Ulrich and William D. Brown
Copyright © 2006 the Institute of Electrical and Electronics Engineers, Inc.

17.2.1 Direct Costs

Direct costs can be further stratified into categories of *direct labor* and *direct material costs*. Direct labor is the cost of all "hands-on" activities associated with the manufacture of a product. The completion of direct labor activities usually impart some added value to the product being manufactured. Typical direct labor activities include fabrication procedures, assembly, testing/troubleshooting, inspection, and rework/repair efforts that can be traced directly to the manufacture of a specific product.

Similarly, direct material consists of the costs of all materials that are included in the end product that is being produced. Examples include printed wiring boards (PWBs), chips, other system components, and materials used in interfacing the finished high-density electronic package.

17.2.2 Indirect Costs

Traditionally, *indirect manufacturing costs*, or *factory overhead*, are all costs that are not easily identifiable or economically feasible to track [1]. *Indirect costs* are normally classified as overhead by the electronics packaging manufacturer. Two types of manufacturing overhead are *indirect labor* and *indirect material*. In electronics packaging there are some types of labor and material used in the production facility that are spread out over many different products and processes, making it virtually impossible to track. Examples of indirect labor include costs for management/supervision, accounting, engineering, labor costs to support cleanroom facilities, and the like. An example of indirect material is when materials are partially consumed during the production process, such as solder. These manufacturing costs are usually considered to be indirect, even though a portion of these materials may be resident in the end product or its subassemblies. Other components of overhead costs include building and equipment depreciation, rent, insurance, utility costs, and the like.

17.2.3 Traditional Volume-Based Costing

Unlike direct costs, indirect or overhead costs are difficult to charge directly to products. Product costing requires an understanding of cost pools and cost drivers. A cost pool is a group of individual costs that is allocated to products using a single cost driver. A traditional costing system assumes that overhead cost is driven by a single volume-based cost driver. These traditional volume-based costing (VBC) systems may provide distorted product cost information when products are diverse in size, complexity, material requirements, and/or setup procedures [2]. VBC systems group all overhead costs into a single cost pool and divide the total overhead cost by a volume-based cost driver such as direct labor hours. This provides a volume-based overhead rate that allocates overhead cost to products based on their direct labor hour requirements. This overhead rate only takes direct labor requirements into account when allocating overhead costs to products and does not consider other product characteristics that may affect overhead consumption such as size or complexity. In general, VBC systems do not provide adequate information regarding the effects that modifying services, product lines, production volume, capacity, or outsourcing procedures have on profitability [3].

The VBC process is described below:

Step 1. Calculate the total cost pool costs, C_i, for cost pool i

$$Ci = \left(\sum_{j=1}^{n} X_{ij} \right) \quad (17.1)$$

where

X_{ij} = overhead cost element j for cost pool i

n = total number of overhead cost elements

$i = 1$

Step 2. Compute the total cost driver levels, D_i, for cost pool I.

Step 3. Calculate the VBC overhead rates, R_i, for cost pool I:

$$Ri = \left(\frac{C_i}{D_i}\right) \tag{17.2}$$

Step 4. Calculate VBC overhead cost for each product:

$$OH_k = \left(\sum_{i=1}^{m}(R_i \times CDC_{ik})\right) \tag{17.3}$$

where

CDC_{ik} = product k consumption of cost driver for cost pool i

m = total number of cost pools

Step 5. Calculate VBC total product cost for each product:

$$PC_k = (DL_k + DM_k + OH_k) \tag{17.4}$$

where

DL_k = direct labor cost for product k

DM_k = direct material cost for product k

Step 6. Compute the VBC profit (loss), PL_k, from the profit (loss) analysis for product k:

$$PL_k = (P_k - PC_k) \tag{17.5}$$

where

P_k = selling price for product k

As an example, consider a VBC system that allocates overhead costs based on direct labor hours developed for an electronics packaging facility. Production of ball grid array (BGA) packages requires expenditures of $39,000 in overhead cost and 300 h of direct labor. Additional product information is provided in Table 17.1. All prices/costs are per 1000 units.

Table 17.1 BGA Package Data

	BGA-S	BGA-S+
Number of units produced	50	10
Selling price	$1300	$3200
Direct labor cost	$120	$300
Direct material cost	$500	$1200
Direct labor hours (DLH) per unit	4	10

The VBC total product cost and profit (loss) for the BGA-S ($k = 1$) and BGA-S+ ($k = 2$) packages were calculated using Equations (17.1) through (17.5).

$$C_1 = \$39,000$$

$$D_1 = 300$$

$$R_1 = \left(\frac{\$39,000}{300}\right) = \$130 \text{ per direct labor hour}$$

$$OH_1 = (\$130 \times 4) = \$520$$
$$OH_2 = (\$130 \times 10) = \$1300$$

$$PC_1 = (\$120 + \$500 + \$520) = \$1140$$
$$PC_2 = (\$300 + \$1200 + \$1300) = \$2800$$

$$PL_1 = (\$1300 - \$1140) = \$160$$
$$PL_2 = (\$3200 - \$2800) = \$400$$

From the above calculations, a profit of $160 and $400, respectively, resulted for each BGA-S and BGA-S+ package sold.

17.2.4 Activity-Based Costing

Activity-based costing (ABC) was developed to provide more accurate overhead cost analyses based on the premise that overhead costs should not be allocated solely on a volume basis [4–7]. ABC analysis identifies business activities that are performed to develop, produce, and sell a product. Overhead costs associated with these activities are identified and grouped into multiple cost pools based on similar cost drivers. For example, all activities associated with shipping and handling may be driven by the number of shipments made. Therefore, all overhead costs associated with these activities may be grouped into a shipping and handling cost pool with the number of shipments as a cost driver.

ABC systems contain multiple cost pools and multiple cost drivers. The total overhead cost in each cost pool is divided by its corresponding cost driver, which results in an ABC overhead rate for each cost pool. Product cost driver consumption data is collected for each cost driver. For example, the number of shipments performed for each product is tallied. To calculate a product's overhead expenses: (1) the overhead rate for each cost pool is multiplied by the product's consumption of the corresponding cost driver; and (2) these values are totaled. These multiple cost pools and drivers allow for a more inclusive analysis of indirect product costs by accounting for product diversity. It is important to note that the amount of required data collection and analysis associated with development of an ABC system can be much greater than that spent to develop an analogous VBC system. Managers and cost analysts need to be aware of these increased resource requirements and be committed to the development and use of an ABC system [8].

The ABC process is described below:

Step 1. Calculate the total cost pool costs, C_i, for cost pool i:

$$C_i = \left(\sum_{j=1}^{n} X_{ij}\right) \tag{17.6}$$

Table 17.2 ABC System Information

Cost pool	i	Total cost pool cost	Cost driver	Total cost driver level	BGA-S cost driver level	BGA-S+ cost driver level
Purchasing	1	$15,000	Number of purchases	50	20	30
Utilities	2	$8,000	Square feet of facility	2000	800	1200
Shipping	3	$6,000	Number of shipments	25	10	15
Accounting	4	$10,000	Number of orders	30	10	20
Total		$39,000				

where

X_{ij} = overhead cost element j for cost pool i

n = total number of overhead cost elements

i = total number of cost pools

Step 2–5. See Steps 2 to 5 of VBC process.

Note, the total number of cost pools is the primary developmental difference between the VBC and ABC systems, where the VBC system is restricted to $i = 1$.

Detailed analysis divided the $39,000 total overhead cost into four cost pools and drivers and determined the cost driver levels consumed by each product, which are provided in Table 17.2. This information and Eqs. (17.6) and (17.2) through (17.5) are used to calculate ABC total product cost and profit (loss) for the BGA-S ($k = 1$) and BGA-S+($k = 2$) packages.

$$R_1 = \left(\frac{\$15,000}{50}\right) = \$300 \text{ per purchase}$$

$$R_2 = \left(\frac{\$8000}{2000}\right) = \$4 \text{ per squarefoot}$$

$$R_3 = \left(\frac{\$6000}{25}\right) = \$240 \text{ per shipment}$$

$$R_4 = \left(\frac{\$10,000}{30}\right) = \$333.33 \text{ per order}$$

$$OH_1 = \frac{(\$300 \times 20) + (\$4 \times 800) + (\$240 \times 10) + (\$333.33 \times 10)}{50} = \$298.67$$

$$OH_2 = \frac{(\$300 \times 30) + (\$4 \times 1200) + (\$240 \times 15) + (\$333.33 \times 20)}{10} = \$2406.67$$

$PC_1 = (\$120 + \$500 + \$298.67) = \$918.67$

$PC_2 = (\$300 + \$1200 + \$2406.67) = \$3906.67$

$PL_1 = (\$1300 - \$918.67) = \$381.33$

$PL_2 = (\$3200 - \$3906.67) = -\$706.67$

The ABC analysis shows that the company is making a higher profit on each BGA-S ($381.33) package than indicated by its previous VBC system and is actually losing $706.67 on each BGA-S+ package produced and sold. The company should look into the

696 Chapter 17 Cost Evaluation and Analysis

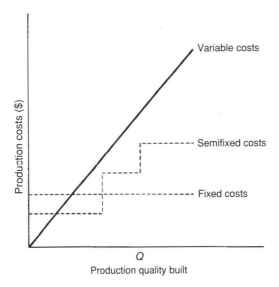

Figure 17.1 Fixed, variable, and semifixed costs [9].

feasibility of increasing the price of the BGA-S+ package or eliminating this product from its product mix.

17.3 BREAK-EVEN ANALYSIS

In high-density packaging production, it is appropriate both to consider and classify manufacturing costs on the basis of their relationship to production quantities. *Fixed costs* are those costs that are considered to be functionally independent of the quantity that is produced. Examples include setup costs, cost to program component insertion equipment, test equipment programming costs, and costs to either purchase or design and build production fixtures to facilitate electronic testing. *Variable costs* are those costs that increase with the quantity produced on a "per unit" basis. Some examples are per unit assembly costs, testing costs, costs for rework/repair, and as discussed earlier, direct labor and material costs.

Semifixed costs vary with consecutive groups of units that are produced. Periodic recalibration of production machines/processes and the performance of scheduled maintenance at regular intervals are examples of semifixed costs. Fixed, variable, and semifixed costs are illustrated in Figure. 17.1. Once these costs are determined, the manufacturer can identify the break-even point for the product, the point where revenues equal expenses.

17.3.1 Linear Break-even Analysis

Break-even analysis concepts have been overviewed in some detail [10,11]. The procedure addresses the question of what production quantity is required to recover the initial fixed and variable costs associated with the startup of production. This is an appropriate cost analysis tool to consider for high-density packaging processes because of the large fixed costs associated with this type of manufacturing. The following notation applies:

Let

Q = number of units built or sold

P = sales price per unit

I = income

F = fixed costs
V = variable cost per unit for a specific level of production
R = profit at a production level of Q units

Where

$$I = Q \times P \quad (17.7)$$
$$C = \text{total costs} = F + Q \times P \quad (17.8)$$
$$R = I - C \quad (17.9)$$

The relationship between these costs is illustrated by Figure 17.2. The break-even point is seen to occur where total income equals total costs. From this illustration, it is apparent that all costs and income are assumed to be linear functions of the quantity produced. Losses occur to the left of the break-even point, whereas the region to the right of this point depicts profits earned. The break-even quantity may be expressed as

$$\text{Total income} = \text{total costs}$$
$$QP = F + QV \quad (17.10)$$
$$Q = \frac{F}{P - V} \quad (17.11)$$

Consider the following example. Fixed costs associated with the fabrication of a specific high-density product are known to be $35,000. Variable costs associated with the manufacture of this product are also known to be $57.00 per unit. If each product can be sold for $75.00, how many units must be produced to recover the fixed costs associated with production?

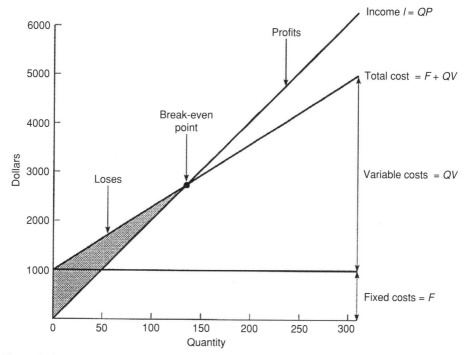

Figure 17.2 Linear break-even analysis [9].

Using Eq. (17.11), we find that

$$Q = \frac{F}{P - V}$$
$$= \frac{\$35,000}{\$75 - \$57}$$
$$= 1945 \text{ units}$$

What is the profit margin at a production level of 3000 units?

$$\text{Income} = I = QP = 3000 \text{ units at } \$75/\text{unit} = \$225,000$$
$$\text{Costs} = C = F + QV = \$35,000 + 3000 \text{ units at } \$57/\text{unit} = \$206,000$$
$$\text{Profit} = R = I - C = \$225,000 - \$206,000 = \$19,000$$

17.3.2 Piecewise Linear Break-Even Analysis

The assumption of both cost and income linearity with production quantity is rarely true in actual manufacturing applications. In practice, it is difficult to express these relationships in any form of continuous equation. Where price breaks occur for larger order quantities and semifixed costs are known to exist, piecewise linear break-even analysis provides a method for addressing these additional cost variables.

Consider the following example. Suppose a process exists that has an initial fixed cost of $1000. A semifixed cost of $1500 is incurred at the 300th unit produced. The sales price is $15 per unit for the first 500 units and drops to $2.50 per unit above a quantity of 500 units. The costs and income as a function of production quantity are illustrated in Figure 17.3. As seen from this illustration, four break-even points exist and occur at 140, 300, 425, and 575 units of production, respectively.

The following example illustrates piecewise linear cost/income calculations. Suppose 125 units are produced. The first 75 units are to be sold for $20 each. The next 25 are to be sold for $15 per unit. The last 25 units are to be sold for $10 each. Fixed costs are known to be $100. A semifixed cost of $75 is incurred at unit 100 and every 100 units thereafter. Variable costs are $8 per unit for the first 85 units and increase to a level of $15 per unit thereafter. What is the level of profit if all 125 units are produced?

$$\text{Total income} = I = Q_1 P_1 + Q_2 P_2 + Q_3 P_3 \quad (17.12)$$
$$= 75(\$20) + 25(\$15) + 25(\$10)$$
$$= \$1500 + \$375 + \$250 = \$2125$$
$$\text{Total costs} = F = F' + Q_4 V_1 + Q_5 V_2 \quad (17.13)$$
$$= \$100 + \$75 + 85(\$8) + 40(\$15)$$
$$= \$1455$$
$$P = I - C = \$2125 - \$1455 = \$670$$

17.4 LEARNING CURVE RELATIONSHIPS

Microelectonic packages can be configured many different ways. Each time a new microelectronic package configuration is run the cost of production normally decreases with increasing levels of production. These reductions result from the human aspect of learning

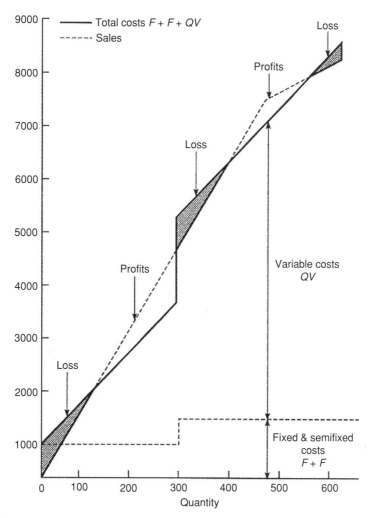

Figure 17.3 Piecewise linear break-even analysis [9].

as operators familiarize themselves with production processes and procedures. Therefore, an understanding of the learning curve concept is important when investigating cost considerations for microelectronic packages. Knowing the cumulative amount of time to produce 1000 units associated with a predetermined learning curve percent will have a dramatic effect on the cost to produce the 1000 units.

Errors are often found in the operation of equipment and production that are corrected in the early stages of production. This is also true of design and production process documentation and test procedures. Correction of errors in test procedures can result in dramatically reduced troubleshooting and rework times. The net effect is that costs usually decrease at a decreasing rate in the early stages of the production run. The effect can be described with the use of product improvement or learning curves. Product improvement curves have been described in detail in the literature [9–12]. Such curves may be mathematically described by the following relationship:

$$Y = KX^N \tag{17.14}$$

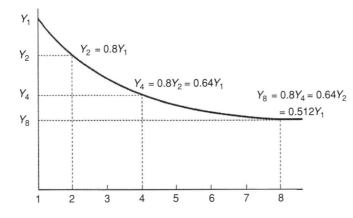

Figure 17.4 An 80% product improvement curve [9].

where

K = time to build the first unit

X = cumulative total units manufactured

N = negative exponent that determines percent decrease in Y each time X is doubled

Y = cumulative average time to build X units or time to build the Xth unit

Using log–log paper, the curve is represented by

$$\log Y = \log K + N \log X \tag{17.15}$$

The way in which the term, Y, is defined determines the type of product improvement curve. If Y is described as the cumulative average time per unit to build X units, the curve is designated as a *cumulative average curve*. If Y is described as the time to build the Xth unit, the curve is designated as a *unit curve*.

The relationship specified by Eq. (17.14) is illustrated in Figure 17.4. Each curve has an associated improvement percentage. The improvement percentage illustrated is 80%. This means that Y decreases by a factor of 80% (0.8) each time the production quantity X is doubled.

17.4.1 Determining Exponent Values for Improvement Rates

The improvement curve percentage is indirectly described by the exponent in Eq. (17.14). If the curve percentage is known, it is possible to determine the value of this exponent by solving a set of simultaneous equations.

Assume a learning curve with an improvement percentage of 80%. Consider two points (X_1, Y_1) and (X_2, Y_2) where $X_2 = 2X_1$. From Eq. (17.14),

$$Y_1 = K X_1^N \qquad Y_2 = K X_2^N$$

$$\frac{Y_1}{Y_2} = \frac{K X_1^N}{K X_2^N} = \frac{X_1^N}{X_2^N} = \left(\frac{X_1}{X_2}\right)^N$$

If $X_2 = 2X_1$, then $Y_2 = 0.8Y_1$ and

$$\frac{Y_1}{Y_2} = \frac{Y_1}{0.8Y_1} = \left(\frac{X_1}{X_2}\right)^N = \frac{1}{0.8} = 1.25 = \left(\frac{X_1}{2X_1}\right)^N$$

$$1.25 = \left(\frac{1}{2}\right)^N$$

$$N \log(1/2) = \log(1.25)$$

In some cases the learning curve is not given and has to be calculated. If the negative exponent is known, the learning curve again can be determined by solving a set of simultaneous equations:

$$\frac{Y_{\text{c.a.}} \ 2}{Y_{\text{c.a.}} \ 1} = \frac{K(X^2)^N}{K(X_1)^N} = 2^N \tag{17.16}$$

where $Y_{\text{c.a.}}$ is the cumulative average hours per unit for any amount of X units produced. Then, to calculate the percent learning curve from the example above

$$N = 0.80 \text{ or } 80\% \text{ learning curve}$$
$$2^N = 2^{-0.322}$$

These calculations can be repeated for different improvement percentages. Table 17.3 presents exponent values for common improvement percentages. If another percentage is used in learning curve calculations Eq. (17.14) can be used to determine the exponent value.

If Eq. (17.14) is slightly modified to represent the time element in the learning curve relationship, the following equation results. The cumulative average time (Y) to build X units is now represented by T_x. The initial unit time (K) is represented by T_i:

$$T_x = T_i X^N \tag{17.17}$$

where

T_x = total direct hours required to build a group of X consecutive manufactured units

T_i = total direct hours to build the initial first unit

Table 17.3 Exponent Values for Common Improvement Curve Percentages[a]

Curve percentage	N
65	−0.624
70	−0.515
75	−0.415
80	−0.322
85	−0.234
90	−0.152
95	−0.074
100	0.00

[a]Based on Eq. (17.14).

17.4.2 Learning Curve Examples

The following examples illustrate these relationships using Eq. (17.14) to (17.17). Assuming that it takes a cumulative average of 50 production hours per unit for the first 250 units, what would be the cumulative average product improvement curve if 500 units are produced at a cumulative average of 35 production hours per unit? Equation (17.15) can be used to set up the problem.

$$\log Y_{\text{c.a.}} = \log K + N \log X$$
$$\log 50 = \log K + N \log 250$$
$$\log 35 = \log K + N \log 500$$
$$\log 50 - \log 35 = N(\log 250 - \log 500)$$
$$N = \frac{\log 50 - \log 35}{\log 250 - \log 500}$$
$$= \frac{0.15491}{-0.30103}$$
$$= -5.14$$
$$2^N = 2^{-.514} = 70\% \text{ learning curve}$$

In the next example, assume that the first unit, T_i, requires 530 h to produce and the learning curve is known to be 75%. It follows that the hours required to produce the 200th unit would be

$$T_x = T_i X^N$$
$$T_{250} = (530)(200)^{-0.415}$$
$$= 58.8 \text{ h}$$

In some situations the amount of time required to produce the initial unit is not know. Therefore, in most cases, a standard level of performance is targeted for the Xth unit. For example, a company expects the 500th unit to take 75 production hours at a 95 % learning curve. Using this information, T_i can be approximated as:

$$T_i = \frac{T_x}{X^N}$$
$$T_i = \frac{75}{500^{-0.152}}$$
$$= 193.30 \text{ h}$$

Similar to the approximation calculations for the initial unit, an approximation can be determined for the cumulative average work hours:

$$T_{\text{c.a.}} = \frac{T_i X^N}{1 + N}$$

where $T_{\text{c.a.}}$ is the cumulative average hours for the Xth unit.

So, for the given information in the example above, the cumulative average hours for the 250th unit is

$$T_{\text{c.a.}} = \frac{T_i X^N}{1 + N}$$
$$= \frac{193.3(250)^{-.152}}{1 - 0.152}$$
$$= 101.49 \text{ h}$$

17.5 FORECASTING MODELS

Forecasting is the study of historical data to discover the underlying tendencies and patterns. This knowledge is used to project the data into future time periods as forecasts. Some typical questions that people seek to answer via forecasting include the following: "What demand can/should we expect to see from customer A for product B next month?" and "Has this demand been constant over time, or is there any sort of trend in the data?"

The combination of new forecasting techniques, software, and powerful desktop computing platforms has empowered managers to utilize very sophisticated data analysis techniques. Forecasting software packages are readily available to the corporate world with very low setup cost. An understanding of these techniques is now essential, as inaccurate forecasts can lead to bad (costly) decisions.

Forecasting is not an exact science. Dr. George E. P. Box is credited with the saying, "All models are wrong... some are useful." Predictions as to future outcomes rarely are precisely on the mark. The forecaster can only endeavor to make the inevitable errors as small as possible.

Given this fact, why should a company endeavor to forecast at all? All organizations operate in an atmosphere of uncertainty. Despite this fact, decisions must be made that affect the future of the organization. Educated guesses or "gut" feelings are more valuable than uneducated guesses. In fact, these are "simple" forecasts. Two basic types of forecasts exist: long term and short term. Long-term forecasts help set the general course of an organization and are typically the focus of top management. In contrast, short-term forecasts help design immediate strategies and are typically the focus of first-line and midmanagement.

In order to choose the most appropriate forecasting approach, various questions must be answered, including "What level of detail is required?" "Is the answer needed very soon or later in the future?" and "To what extent are qualitative and quantitative methods appropriate?" Regardless of the approach employed, the results of any chosen forecasting method must facilitate the decision-making process of the organization's managers. In sum, companies strive to produce a forecast that is accurate, timely, and understandable by management. The forecast should help to promote better decisions.

During the forecasting process, companies extend past experiences into the uncertain future. The implicit assumption in forecasting is that the conditions that generated past data are indistinguishable from the conditions of the future, except for those variables explicitly contained in the forecasting model. As with all business endeavors, forecasting has both costs and benefits associated with it. The costs of forecasting include personnel, training, hardware, software, and other factors (such as time). However, the benefits of forecasting may include reduction of inventory levels and more accurate production scheduling. A company should only continue with its forecasting effort provided the benefits of forecasting outweigh the costs of performing the forecast.

The following sections describe three common forecasting models for time-series data: moving averages, exponential smoothing, and least-squares regression. Each model is predicated on the assumption that time-series data is being analyzed and forecasted. A time-series is a set of points whose values are realized over time. For example, y may describe a series of values $y(1), y(2), \ldots, y(t)$ that have been observed from time 1 up to the present time t. The goal in forecasting is to estimate the values for y at various points in the future, such as time $t+1, t+2$, and so on. In the following sections, $f(t)$ denotes the forecasted value for y in time period t.

Forecasting time-series data is a three-step procedure. First, one must select a model that computes forecasts from historical (i.e., time series) data. Next, forecasts are produced for existing, known data to establish the accuracy of the chosen forecast method. This is accomplished through the use of various different error measures. Forecast accuracy is judged by comparing $y(t)$ and $f(t)$. The *residual* or *forecast error* $e(t)$ in period t is calculated as

$$e(t) = f(t) - y(t) \tag{17.18}$$

In other words, a residual is the difference between a forecast value and its actual (observed) value. Positive (negative) forecast errors are an indication that the current forecast is over- (under-) estimating the actual data values. Obviously, one endeavors to have as little forecast error as possible. The following list of forecast error measures is not meant to be exhaustive; rather, four common approaches to measuring forecast error are presented for the reader's reference. In each case, n denotes the total number of forecasted data values.

17.5.1 Mean-Squared Error (MSE)

MSE measures forecast accuracy by averaging the square of each residual:

$$\text{MSE} = \frac{\sum_{t=1}^{n} [y(t) \quad f(t)]^2}{n} \tag{17.19}$$

A penalty is associated with "large" forecast errors due to the squaring of the residual. One caution for the reader is that a forecasting technique that produces moderate errors may be preferable to one that usually has small errors but occasionally yields extremely large ones.

17.5.2 Mean Absolute Deviation (MAD)

MAD measures forecast accuracy by averaging the magnitudes of forecast errors, where magnitude is defined as the absolute value of each forecast error:

$$\text{MAD} = \frac{\sum_{t=1}^{n} |y(t) - f(t)|}{n} \tag{17.20}$$

MAD is a very useful error-measuring approach, as the computed MAD is measured in the same units as the original data series.

17.5.3 Mean Percentage Error (MPE)

MPE measures forecast accuracy in terms of percent error:

$$\text{MPE} = 100 \times \frac{\sum_{t=1}^{n} \frac{y(t) - f(t)}{y(t)}}{n} \tag{17.21}$$

The MPE approach is useful when one needs to see if a forecasting method is biased (i.e., consistently forecasting high or low).

17.5.4 Mean Absolute Percentage Error (MAPE)

MAPE measures forecast accuracy in terms of absolute percent error:

$$\text{MAPE} = 100 \times \frac{\sum_{t=1}^{n} \frac{|y(t) - f(t)|}{y(t)}}{n} \qquad (17.22)$$

Employing MAPE for the forecast error measure is useful when the size of the forecast variable is important in evaluating forecast accuracy.

Finally, if the accuracy of the forecast for existing data is acceptable, then the chosen forecasting approach is employed. Otherwise, various forecasting model parameters can be adjusted or refined for further accuracy improvements, or the chosen forecasting model can be discarded in favor of another forecasting model approach. At this time, the whole three-step process begins again with model selection.

17.5.5 Moving Average

Simple or naïve forecasting models assume that recent periods are the best predictors of the future. For example, if 300 units were sold in June [i.e., $y(\text{June}) = 300$], then a simple forecast for July would be $f(\text{July}) = 300$. A generic version of this simple model is

$$f(t+1) = y(t) \qquad (17.23)$$

where $y(t)$ is the most recent observation and $f(t+1)$ is the forecast for one period in the future.

In general, averaging methods use a form of a weighted average of past observations to smooth short-term fluctuations in the data. The underlying assumption of averaging methods is that the fluctuations in past values represent random departures from some smooth "curve." If there were no random deviations, then the future observations would follow the curve. Therefore, averaging methods try to find the curve by smoothing the fluctuations such that the forecast is based on it. One averaging approach is to simply compute the average of all past observations:

$$f(t+1) = \frac{1}{t} \sum_{i=1}^{t} y(i) \qquad (17.24)$$

This approach is known as *simple averaging*. For example, if historical sales data revealed that $y(\text{March}) = 400$, $y(\text{April}) = 450$, $y(\text{May}) = 250$, and $y(\text{June}) = 300$, then the sales forecast for July [i.e., $f(\text{July})$] is computed as

$$f(\text{July}) = \tfrac{1}{4}(400 + 450 + 250 + 300) = 350 \text{ units} \qquad (17.25)$$

Alternately, a *moving-average* forecast with a time window of n periods considers only the n most recent observations. A moving-average forecast requires less historical data than does a simple averaging approach, as the one must only retain the most recent n periods of data. However, now an additional decision must be made with regard to the appropriate value for the moving-average time window n. An n-period moving-average forecast is calculated as

$$f(t+1) = \frac{1}{n}[y(t) + y(t-1) + \cdots + y(t-n+1)] \qquad (17.26)$$

For example, the simple or naïve forecasting model given at the beginning of this section is in fact a moving-average forecast with $n = 1$.

Using the same sales data above from March to June, a two-period ($n = 2$) moving-average forecast is computed as

$$f(\text{July}) = \tfrac{1}{2}\left[y(\text{June}) + y(\text{May})\right] = \tfrac{1}{2}[300 + 250] = 275 \text{ units} \quad (17.27)$$

It should be noted that when forecasting p periods into the future,

$$f(t + p) = \hat{y}(t + 1) \quad (17.28)$$

as no additional observation data is available for future periods at time period t.

As the chosen value for n becomes larger, more observations are used in the forecast computation. This, in turn, increases the smoothing effect and little attention is given to fluctuations in the data. Moving-average forecasts typically work best with stationary or stable data. Even though moving-average approaches work better than simple averaging for data with trend or seasonality, moving-average forecasts typically do not handle trend or seasonality very well.

17.5.6 Forecasting Sales Based on Historical Data

A review of historical sales data uncovers the sales, in thousands of units, listed in Table 17.4. In this table, the results are given from four different forecasting approaches: naïve, simple averaging, moving average with $n = 2$ [MA(2)], and MA(4), as well as the forecast residuals and the calculation of four different measures of forecast error: MSE, MAD, MPE, and MAPE. These forecasts are also presented graphically in Figure 17.5, along with the time-series data being forecast.

The reader should note how the simple models mimic the shape of the time series being forecasted, but lag the actual time-series data by one time period. Simple averaging, which is equivalent to an MA(1), continues to stabilize its forecast as more and more historical data points are added to the overall average. Comparing MA(2) to MA(4) demonstrates the

Table 17.4 Comparison of Forecasting Historical Sales with Four Different Approaches

		Naïve		Simple avg.		MA(2)		MA(4)	
Period t	Sales $y(t)$	$f(t)$	$e(t)$	$f(t)$	$e(t)$	$f(t)$	$e(t)$	$f(t)$	$e(t)$
1	7,000								
2	12,000	7,000	−5,000	7,000	−5,000				
3	22,000	12,000	−10,000	9,500	−12,500	9,500	−12,500		
4	33,000	22,000	−11,000	13,667	−19,333	17,000	−16,000		
5	9,000	33,000	24,000	18,500	9,500	27,500	18,500	18,500	9,500
6	17,000	9,000	−8,000	16,600	−400	21,000	4,000	19,000	2,000
7	22,000	17,000	−5,000	16,667	−5,333	13,000	−9,000	20,250	−1,750
8	37,000	22,000	−15,000	17,429	−19,571	19,500	−17,500	20,250	−16,750
9	11,000	37,000	26,000	19,875	8,875	29,500	18,500	21,250	10,250
10	12,000	11,000	−1,000	18,889	6,889	24,000	12,000	21,750	9,750
11	31,000	12,000	−19,000	18,200	−12,800	11,500	−19,500	20,500	−10,500
12	40,000	31,000	−9,000	19,364	−20,636	21,500	−18,500	22,750	−17,250
13		40,000		21,083		35,500		23,500	
	MSE	2.05×10^8		1.61×10^8		2.37×10^8		1.23×10^8	
	MAD	12090.9		10985.3		14600.0		9718.8	
	MPE	16.4%		−7.8%		19.5%		20.2%	
	MAPE	75.1%		52.1%		80.0%		52.7%	

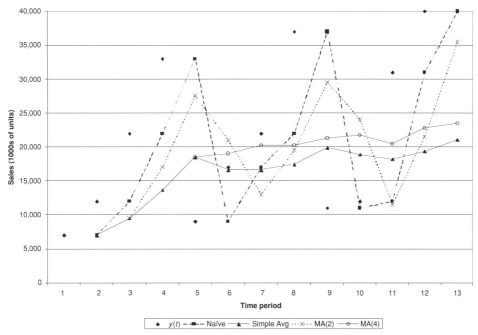

Figure 17.5 Forecasts resulting from four forecasting techniques in Table 17.4.

fact that as the size of the moving-average window n increases, the overall smoothing effect increases and little attention is given to fluctuations in the data. Finally, examining the four error measures presented suggests the most appropriate of the four simple and averaging techniques for forecasting historical sales is MA(4), as it produces the forecast with the best performance in terms of the error measures of interest.

17.5.7 Exponential Smoothing

In contrast to averaging methods that place equal weight or importance on each historical data point under consideration, *exponential smoothing* forecasts are based on averaging or smoothing past observations in a decreasing manner. While exponential smoothing forecasts are still a weighted average of all observations, more (less) weight is placed on the most (least) recent observations. As was the case with the size of the moving-average window n, an additional decision must be made with regards to the appropriate value(s) for the smoothing parameter(s). Three primary exponential smoothing models are applicable to a variety of different data types: simple (single) exponential smoothing, Holt's method (double exponential smoothing), and Winter's method (triple exponential smoothing).

17.5.7.1 Simple (Single) Exponential Smoothing

The simple exponential smoothing forecast for $f(t+1)$ is a weighted average of the newest observation and the last forecast:

$$f(t+1) = \alpha y(t) + (1-\alpha) f(t) \tag{17.29}$$

In this expression, α is a smoothing parameter with $0 \leq \alpha \leq 1$. In other words, an exponential smoothing forecast is a weighted average of all observations, with exponentially decreasing weights for past observations. This type of forecast can also be thought of as the old forecast plus a portion of the forecast error in the old forecast, as simple algebraic manipulation can show that

$$f(t+1) = f(t) + \alpha\,[y(t) - f(t)] \tag{17.30}$$

When using the exponential smoothing forecasting technique, the smoothing parameter α is key, as it determines the extent to which the most recent observation affects the forecasted value. For example, again assume that actual sales in June were 300 units [i.e., y(June) $=$ 300]. Further, let f(June) $=$ 350 units. The following calculations show the forecast results from three different specified values for α:

- If $\alpha = 0.1$, then f(July) $= 0.1y$(June) $+ 0.9f$(June) $= 0.1(300) + 0.9(350) = 345$.
- If $\alpha = 0.6$, then f(July) $= 0.6y$(June) $+ 0.4f$(June) $= 0.6(300) + 0.4(350) = 320$.
- If $\alpha = 0.9$, then f(July) $= 0.9y$(June) $+ 0.1f$(June) $= 0.9(300) + 0.1(350) = 305$.

Therefore, the speed at which past values lose their importance in an exponential smoothing forecast depends on α. Consider the previous examples where unit sales are being forecasted. Smaller values for α cause a new forecasted value to be closer in magnitude to the last forecasted value [e.g., f(July) $= 345$ when $\alpha = 0.1$], while larger α values produce forecast values that are closer to the last sales data [e.g., f(July) $= 305$ when $\alpha = 0.9$].

If the person responsible for producing the forecast desires to have stable predictions with random variations in the data smoothed out, a small value of α should be selected. However, if the forecaster wants to respond to real changes in the underlying data quickly, a larger value of α should be selected. An approach that is commonly employed by software packages is to calculate a number of forecasts for the test data using a number of different values for α. The "best" α is then selected as the one that produces the minimum forecast error measure. Finally, as was the case with moving-average forecasting, the simple exponential smoothing forecast for p periods into the future is given by

$$f(t+p) = f(t+1) \tag{17.31}$$

17.5.7.2 Holt's Method (Double Exponential Smoothing)

In some circumstances, data values consistently increase (or decrease) over time. When this is the case, the data is said to contain a trend. One can adjust any simple forecasting model by taking into account the change that occurs between periods. When a significant trend exists, simple exponential smoothing forecasts lag behind the actual observations. Therefore, to handle the linear trend in the data, one can employ Holt's method. Holt's method is often referred to as double exponential smoothing as it estimates not only the data's level (i.e., the function of simple exponential smoothing) but also the data's trend.

In Holt's method, two smoothing constants are required: α, for smoothing the base level (or intercept), and β, for smoothing the trend (or slope) estimate. As was the case with α, $0 \leq \beta \leq 1$. The forecast prescribed by Holt's method is the aggregate of two parts: the latest estimate of base-level demand at time t is defined as $L(t)$, while the latest estimate of trend (either growth or decline) is denoted as $T(t)$. Therefore, our most recent forecast is given by

$$f(t) = L(t) + T(t) \tag{17.32}$$

Trend can be simply defined as $L(t) - L(t-1)$. To estimate the trend or slope, it is clear that $T(t) = \beta$(latest trend in base-level forecast) $+(1-\beta)$ (the most recent estimate of the trend). Using this definition, it follows that

$$T(t) = \beta\,[L(t) - L(t-1)] + (1-\beta)T(t-1) \tag{17.33}$$

Similarly, the base-level estimate $L(t) = \alpha$ (the latest observation) $+ (1-\alpha)$(the most recent forecast). This equates to

$$L(t) = \alpha y(t) + (1-\alpha) f(t-1) \qquad (17.34)$$

Using our definition of $f(t)$, we now have that

$$L(t) = \alpha y(t) + (1-\alpha)[L(t-1) + T(t-1)] \qquad (17.35)$$

Further, the double exponential smoothing forecast for period $t+1$ is given by

$$f(t+1) = L(t) + T(t) \qquad (17.36)$$

However, Holt's method forecast for p periods into the future is given by

$$f(t+p) = L(t) + p^* T(t) \qquad (17.37)$$

(i.e., the current base-level estimate plus p periods of trend).

17.5.7.3 Forecasting Sales Data Revisited

Continuing our previous sales data forecasting example, we now use the same historical sales times-series data and determine both the single and double exponential smoothing forecast for the data. Table 17.5 displays the resulting forecasts, residuals, and error measure calculations for $\alpha = \beta = 0.2$, while Figure 17.6 shows the single and double exponential smoothing forecasts graphically.

The reader should note the superior performance of Holt's method (double exponential smoothing) for this historical data time series. This technique results in superior error measure performance, especially for MPE. This suggests a trend exists in the historical

Table 17.5 Comparison of Single and Double Exponential Smoothing Forecasting of Historical Sales

		Single exponential smoothing		Double exponential smoothing			
Period t	Sales $y(t)$	$f(t)$	$e(t)$	Level $L(t)$	Trend $T(t)$	$f(t)$	$e(t)$
1	7,000	22,000		12,000	0		
2	12,000	19,000	7,000	13,400	280	12,000	0
3	22,000	17,600	−4,400	14,464	436.8	13,680	−8,320
4	33,000	18,480	−14,520	15,617	580	14,901	−18,099
5	9,000	21,384	12,384	14,234	787.5	16,197	7,197
6	17,000	18,907	1,907	18,199	822.9	18,022	1,022
7	22,000	18,526	−3,474	18,922	803.1	19,022	−2,978
8	37,000	19,221	−17,779	19,624	782.9	19,725	−17,275
9	11,000	22,776	11,776	20,881	877.6	20,407	9,407
10	12,000	20,421	8,421	21,491	824.1	21,759	9,759
11	31,000	18,737	−12,263	21,600	681.0	22,315	−8,685
12	40,000	21,190	−18,810	22,062	637.3	22,281	−17,719
13		24,952				22,700	
		MSE	1.34×10^8				1.21×10^8
		MAD	10248.7				9132.8
		MPE	15.5%				2.5%
		MAPE	54.4%				43.5%

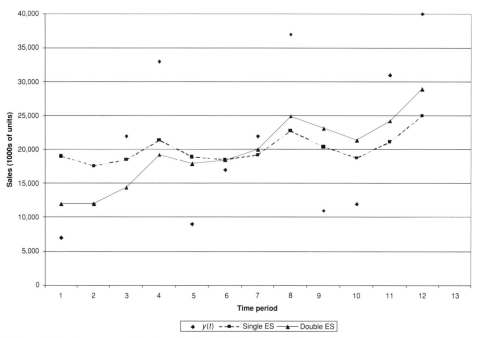

Figure 17.6 Forecasts resulting from single and double exponential smoothing techniques.

sales data. The final step in the forecasting analysis will be to determine if any other pattern(s) exist within this time series of interest.

17.5.7.4 Winter's Method (Triple Exponential Smoothing)

Certain types of data, in addition to containing a trend, are characterized by seasonal changes. Winter's method (triple exponential smoothing) adjusts forecasts for both linear trend and seasonality effects using a third smoothing parameter γ. Again, $0 \leq \gamma \leq 1$. Using this approach, one can now estimate the seasonality of the underlying data along with the base level and trend. First, the length of the season must be determined, such as a year. Let N represent the number of periods (e.g., $N = 4$ quarters) before the seasonal pattern begins to repeat. Further, define seasonality factor $S(t)$ as the average amount that the demand in period t of the season is above or below the overall average. For example, $S(3) = 1.25$ signifies that the demand in the third period of the season is 25% above the average demand level.

Clearly, actual (past) observations include both trend and seasonal effects. If we define the actual sales in period t as $y(t)$, then the deseasonalized (or level) sales for period t is given by $y(t)S(t)$. Therefore, the seasonality factor estimate

$$S(t) = y(t)/L(t) \qquad (17.38)$$

To estimate seasonality, we note that $S(t) = \gamma$ (the current estimate of the seasonality factor) $+ (1 - \gamma)$ (the last season's seasonality factor). It follows that

$$S(t) = \gamma[y(t)/L(t)] + (1 - \gamma)S(t - N) \qquad (17.39)$$

The trend estimate under Winter's method is the same as was used in Holt's method (double exponential smoothing):

$$T(t) = \beta[L(t) - L(t-1)] + (1-\beta)T(t-1) \tag{17.40}$$

However, the base-level estimate requires the use of deseasonalized demand. The base-level estimate $L(t) = \alpha$ (the latest deseasonalized demand) $+ (1-\alpha)$ (the most recent forecast). Therefore,

$$L(t) = \alpha[y(t)/S(t-N)] + (1-\alpha)[L(t-1) + T(t-1)] \tag{17.41}$$

Once all three components have been estimated (level, trend, and seasonality), the forecast for period $t+1$ using Winter's method equates to (the base-level estimate + the trend estimate) × (the seasonality factor estimate). Therefore,

$$f(t+1) = [L(t) + T(t)]S(t+1-N) \tag{17.42}$$

Finally, the Winter's method forecast for p periods into the future is given by

$$f(t+p) = [L(t) + pT(t)]S(t+p-N) \tag{17.43}$$

17.5.7.5 Forecasting Sales Data Using Winter's Method

Finally, we conclude our sales data forecasting example by using Winter's method (triple exponential smoothing) to forecast the data. Table 17.6 displays the resulting forecasts, residuals, and error measure calculations for $\alpha = \beta = \gamma = 0.2$, while Figure 17.7 shows the triple exponential smoothing forecasts graphically.

Table 17.6 Triple Exponential Smoothing Forecasting of Historical Sales

		Triple exponential smoothing				
Period t	Sales $y(t)$	Level $L(t)$	Trend $T(t)$	Seasonality $S(t)$	$f(t)$	$e(t)$
1	7,000			0.3784		
2	12,000			0.6486		
3	22,000			1.1892		
4	33,000	18500	0.0	1.7838		
5	9,000	19557	211.4	0.3947	7,000.0	−2,000
6	17,000	21057	469.0	0.6804	12,822.9	−4,177
7	22,000	20920	348.0	1.1617	25,597.9	3,598
8	37,000	21163	327.0	1.7767	37,938.3	938
9	11,000	22765	582.0	0.4124	8,483.1	−2,517
10	12,000	22205	353.6	0.6524	15,885.3	3,885
11	31,000	23384	518.7	1.1945	26,206.1	−4,794
12	40,000	23625	463.1	1.7600	42,468.1	2,468
13					9,934.7	
14					16.017.1	
15					29,879.0	
16					44,839.7	
				MSE	1.07×10^7	
				MAD	3047.2	
				MPE	−3.5%	
				MAPE	17.8%	

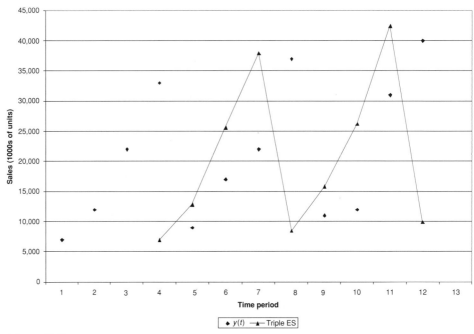

Figure 17.7 Forecasts resulting from triple exponential smoothing.

As expected, the seasonal pattern evident in the data is most appropriate forecast using Winter's method. Using the approach, MSE is reduced by an order of magnitude when compared to Holt's method (double exponential smoothing) forecasting that does not consider data seasonality. Additional experimentation varying the three smoothing parameters α, β, and γ can result in additionally superior forecast error measure performance.

17.5.8 Least-Squares Regression

When the data under consideration depends on or is at least correlated with certain factors, it may be a good idea to undercover the relationship between the data and those factors explicitly. In this case, the forecast is a dependent variable that is a function of one or more independent variables. For example, housing sales are a function of mortgage interest rates. Similarly, product sales may be a function of income level and stock market levels.

In least-squares regression, a dependent variable of interest $y(t)$ is forecast (e.g., sales) based on its relationship to the independent variable $x(t)$ (e.g., price). In other words, the relationship between x and y is used to forecast y assuming that the value of x is known. In this forecasting approach, the squared difference between the y value estimated for a given point x and its actual y value is determined for each point in the original data set. The sum of these squared differences is called the *residual sum of squares*. Next, the sum of the squared differences between the actual y values and the average of the y values is calculated (i.e., the *total sum of squares*). The smaller the residual sum of squares is compared with the total sum of squares, the larger the value of the coefficient of determination (R^2), which is an indicator of how well the equation resulting from the least-squares regression analysis explains the relationship among the variables in the forecasting model.

The primary assumption in least-squares regression is that the underlying relationship between x and y is linear (i.e., straight line). With this in mind, define y as the sum of some intercept or level term a and the product of a slope or trend term b and x:

$$y = a + bx \qquad (17.44)$$

Therefore, using this least-squares regression equation, an estimate or forecast can be computed for any value of x. Both a and b can be estimated using the following formulas [13]:

$$a = \bar{y} - b\bar{x} \qquad (17.45)$$

$$b = \frac{\sum_{i=1}^{n} x_i y_i - \bar{x} \sum_{i=1}^{n} y_i}{\sum_{i=1}^{n} x_i^2 - \bar{x} \sum_{i=1}^{n} x_i} \qquad (17.46)$$

Further, Microsoft Excel's LINEST function can also be used to estimate a linear relationship between x and y by estimating values for both a and b. For additional information on linear regression, the reader is referred to Neta and co-workers [14].

17.5.8.1 Forecasting Sales Data Using Least-Squares Regression

Figure 17.8 displays the least-squares regression forecast for the historical sales data presented above. In this figure, the regression equation (computed using Microsoft Excel) is given as

$$y = 1549x + 11015 \qquad (17.47)$$

Therefore, the sales in any time period x can be directly estimated using this equation. However, as $R^2 = 0.2317$ in the figure, it follows that least-squares regression forecasting of the

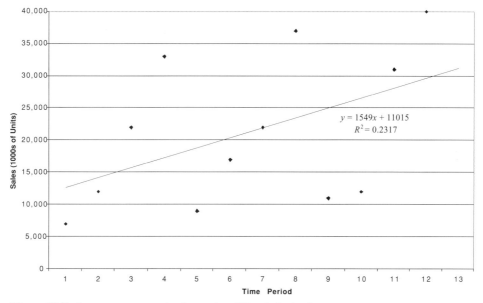

Figure 17.8 Least-squares regression forecasting of historical sales data.

historical sales data time series may not be the most appropriate course of action, especially given the evidence of data seasonality discussed in the previous section on Winter's method (triple exponential smoothing).

17.6 COMPARATIVE ANALYSIS

It is often the case that a decision must be made as to whether or not a proposed investment or purchase should occur. A fundamental understanding of the *time value of money* is paramount for engineers who strive to make fiscally responsible investment or purchase decisions. Money has time value as banks, businesses, and people are willing to pay interest for the use of money. A dollar today may or may not be worth a dollar one year from now. This section discusses some of the basic economic concepts typically used by engineers to justify or nullify capital investments. The reader is referred to any one of the various engineering economic analysis textbooks currently in print for a more detailed discussion on any one of the following topics.

Simple interest is interest that is computed based on the original amount of money borrowed. Assume a funding source offered to loan a person P dollars for n years at an annual interest rate of i percent. The amount of interest the person would owe the funding source after one year is calculated as

$$\text{Total interest} = iPn \qquad (17.48)$$

Therefore, the amount of money owed by the person at the end of the loan (based on simple interest) would be

$$P + iPn = P(1 + in) \qquad (17.49)$$

Under this scenario, one states the investment or purchase has a *present worth* or *present value* of P dollars. This signifies the investment's value in today dollars. However, in reality the most common form of interest accrual on a loan is *compound interest*. Under this method, any interest that is owed at the end of year t and not paid off in full in that year is added to the unpaid loan principal. Therefore, this unpaid interest in year t will be subject to additional interest charges in future time periods. Assuming the same lending source above made a similar loan arrangement for P dollars under a compound interest scenario, the amount of money owed by the person at the end of the loan can be expressed as

$$\text{Total amount owed under compound interest} = P(1+i)^n \qquad (17.50)$$

Note than in this equation, the implicit assumption is that no money is paid back to the lending sources until the end of the loan period (i.e., n years). Therefore, the equation is calculating the *future worth* of a present-day amount, assuming a specific interest rate i and time period n. This can also be expressed in a common economics functional notation as

$$F = P(F|P, i, n) = P(1+i)^n \qquad (17.51)$$

This expression is commonly referred to as "find the future (value), given the present (value)." In fact, many economics books provide lookup tables containing the value of the factor $(F|P, i, n)$ (as well as other factors discussed in this section) for various values of i and n.

Another calculation frequently of interest to the practicing engineer is to determine the *annual worth* (A) of a proposed investment or purchase. In this case, the equivalent annual return or payment amount can be calculated from an investment's present or future worth.

In the case of compound interest, a purchase's annual worth equates to the level annual payment required to pay back the borrowed money within the allotted time period of n years. Annual worth can be calculated from present worth P as

$$A = P(A|P, i, n) = P\left[\frac{i(1+i)^n}{(1+i)^n - 1}\right] \quad (17.52)$$

The reader should note that this equation can also be solved for P in terms of A. Further, annual worth can be calculated from future worth F as

$$A = F(A|F, in, n) = F\left[\frac{i}{(1+i)^n - 1}\right] \quad (17.53)$$

The time value of money, specifically with regards to calculating a proposed investment or project's present worth, future worth, and annual worth, is the basis for most capital project selection decisions made in businesses today. For additional information on economic analysis, the reader is referred to Newnan and co-workers [15].

17.6.1 Capital Project Selection and Evaluation

Often an available business or investment "opportunity" must be evaluated to determine whether or not it is financially justifiable. Companies are in business to make money. Therefore, any evaluation of a potential project or opportunity must consider whether or not the proposed engagement will be profitable. Any one of the three economic techniques described above can be directly applied in this type of project selection/evaluation problem.

Assume a business opportunity is available for selection by a company, called alternative 1. Alternative 1 has an initial cost to the company of P_1 dollars and a project lifetime of 5 years.

After 5 years, alternative 1 will have a book or salvage value of S_1 dollars. If the company selects alternative 1, in each of the next 5 years, the company will receive a guaranteed income cash flow of I_1 dollars. This may be generated by interest payments to the company or return of profits generated by the investment. Finally, the company's stated cost of capital (i.e., the minimum interest rate percentage the company will earn on its money if it invests it elsewhere, like a bank) is 8%.

Using these facts, the present worth of alternative 1 can be calculated. There are three different cash flow streams to consider: the initial outflow associated with alternative 1's purchase price of P_1 dollars, the annual cash inflow of I_1 dollars, and the future salvage value of S_1 dollars. The present worth of any proposed project or business opportunity is the sum of the present worth of cash inflows (i.e., income) minus the sum of the present worth of cash outflows (i.e., expenses). The present worth of the cash inflows can be determined from the annual worth equation (17.51) after a little algebraic manipulation:

$$P = A\left[\frac{(1+i)^n - 1}{i(1+i)^n}\right] \quad (17.54)$$

For alternative 1, $A = I_1$, $i = 8\%$ per year, and $n = 5$ years. Substituting these values into Eq. (17.53), we see that the present worth of the cash inflows for alternative 1 is

$$\text{PW(CashInflows)} = I_1\left[\frac{(1+0.08)^5 - 1}{0.08(1+0.08)^5}\right] = I_1\left(\frac{1.4693 - 1}{0.1175}\right) = 3.9927(I_1) \quad (17.55)$$

Now, the present worth of alternative 1's salvage value of S_1 dollars in year 5 must be determined. For this calculation, Eq. (17.50) is applicable after slight algebraic manipulation:

$$P = \frac{F}{(1+i)^n} \qquad (17.56)$$

Therefore, it follows that

$$\text{PW(Salvage Value)} = \frac{S_1}{(1+0.08)^5} = \frac{S_1}{1.4693} = 0.6806(S_1) \qquad (17.57)$$

The present worth of alternative 1's cash outflow is simply P_1 dollars, as this expense is incurred today (i.e., at the present time). Therefore, the present worth of alternative 1 equals

$$\begin{aligned}\text{PW(alternative 1)} &= \text{PW(Inflows)} - \text{PW(Outflows)} \\ &= [3.9927(I_1) + 0.6806(S_1)] - P_1\end{aligned} \qquad (17.58)$$

From this expression, it follows that this business opportunity should only be pursued if

$$I_1 \geq \frac{P_1 - 0.6806(S_1)}{3.9927}$$

as this would lead to a positive present worth for alternative 1 (i.e., a profitable business venture).

It is important to note the consistency in time units that is maintained throughout these three inputs, as the cash inflows occur annually, an annual percentage rate is given, and the useful life of alternative 1 is also expressed in years. One must always ensure that economic problem inputs maintain their dimensional consistency to ensure valid results are produced during the analysis.

17.6.2 Replacement Analysis

Another decision that must be made in the future is whether or not to replace an existing piece of equipment or other asset with a new one and if so, when should this replacement occur. In order to help answer this question, an asset's *minimum cost life* can be calculated to determine the number of years at which the annual ownership and operating costs are minimized. By keeping an asset for the number of years prescribed by the minimum cost life, then replacing it at that time, a company can minimize the cost of owning and operating the asset.

Consider a recently acquired asset for which the same company paid $3000. The asset has a useful life of 6 years and no salvage value. The asset's annual operating, maintenance, and repair costs are estimated to be $500 in the first year of ownership. After this first year, these costs are expected to rise by $500 per year. Finally, no cash inflows are currently associated with this asset. A typical replacement analysis question is how many years should this asset be kept by the company before it is replaced.

Two primary cash flows need to be considered in this minimum cost life analysis. Obviously, an annual cash outflow is associated with the operating, maintenance, and repair expenses. However, the cost of purchasing the replacement asset must also be considered on an annual basis, as these funds could have been invested otherwise. The total annual cost of this asset must be determined for each potential year that it could be replaced (i.e., years 1 through 6, its useful life). The year in which this total annual cost is minimized is the minimum cost life of the asset.

Assuming the company's cost of capital is again 8%, the total annual cost of the asset in year i equals the sum of the annualized purchase price and the annualized operating, maintenance, and repair costs. For example, in year 2, the annualized purchase price can be found using Eq. (17.51) as

$$\text{AnnualCost (Year 2)} = P\left[\frac{i(1+i)^n}{(1+i)^n - 1}\right] = \$3000\left[\frac{0.08(1+0.08)^2}{(1+0.08)^2 - 1}\right] = \$1682.31 \tag{17.59}$$

Further, the annual cost in year 2 of the asset's operating, maintenance, and repair expenses is found in two steps. First, the present worth of these expenses is calculated using Eq. (17.50) as

$$\begin{aligned}\text{PW(Expenses)} &= \text{PW(Year 1 Expenses)} + \text{PW(Year 2 Expenses)} \\ &= \frac{\$500}{(1+i)^1} + \frac{\$1000}{(1+i)^2} = \frac{\$500}{1.08} + \frac{\$1000}{1.1664} \\ &= \$462.96 + \$857.34 = \$1320.30\end{aligned} \tag{17.60}$$

Next, this present worth of $\$1320.30$ is annualized over 2 years using Eq. 17.51:

$$\begin{aligned}\text{AnnualExpenses(Year 2)} &= P\left[\frac{i(1+i)^n}{(1+i)^n - 1}\right] \\ &= \$1320.30\left[\frac{0.08(1+0.08)^2}{(1+0.08)^2 - 1}\right] = \$740.38\end{aligned} \tag{17.61}$$

Therefore, the year 2 total annual cost can be calculated as

$$\text{TotalAnnualCost(Year 2)} = \$1682.31 + \$740.38 = \$2422.69 \tag{17.62}$$

Performing similar calculations for each potential year of asset replacement gives the total annual costs shown in Table 17.7.

From the calculations, the asset's minimum cost life is 4 years, as $2108 in year 4 is the minimum total annual cost. Therefore, the asset should be replaced at the end of the fourth operating year in order to minimize the total cost of owning and operating the asset.

17.7 SENSITIVITY ANALYSIS

Uncertainty occurs when there is a possibility of error as a result of having less than total information about the existing environment. This possibility of error exists in all measurements and estimates to some degree. Representing uncertainty can be extremely

Table 17.7 Example Minimum Cost Life Calculations

Year	Annual purchase cost	Annual expenses	Total annual cost
1	$3240	$500	$3740
2	$1682	$740	$2423
3	$1164	$974	$2138
4	$906	$1202	$2108
5	$751	$1423	$2175
6	$649	$1638	$2287

important in economic analysis where accurate information is not readily available. The accuracy of a cash flow estimate is related to the amount of available information and expertise. Estimates that are developed with little or no historical information and based on guesswork have a high degree of uncertainty. Estimates are presumed to be more certain when they are based on high-quality data and reliable expertise.

It is important to quantitatively handle this uncertainty to minimize the risk of making poor economic decisions. Sensitivity analysis is a quantitative technique that measures the sensitivity of a decision to changes in the values of one or more input parameters. This allows the analyst to explore the degree to which each input parameter affects the measure of performance. In addition, the extent to which a parameter value must change before another alternative becomes preferable can be determined. It is recommended to perform sensitivity analysis when conditions of uncertainty exist for one or more parameters involved in the analysis. Two methods of sensitivity analysis that are employed frequently in practice are the single parameter procedure and the optimistic–pessimistic approach. There are also multiparameter procedures that consider the possibility of interaction of various parameters [16].

17.7.1 Single-Parameter Sensitivity Analysis

The single-parameter procedure considers the sensitivity of the performance measure caused by changes in a single parameter. The procedure is simply to vary uncertain parameter values one at a time and observe the effect on the performance measure. For example, the analyst may know that the useful life of new equipment is uncertain. In order to understand how this uncertain parameter affects the economic value of this equipment, the analyst will deviate the equipment life from the initial estimate and observe the effect on the measure of performance, such as the present worth of the equipment investment.

For example, purchase of a burn-in oven is being considered. The oven can be purchased for $100,000. It is assumed that the oven has a useful life of 5 years and a salvage value of $20,000. The annual cash flows (receipts – disbursements) are predicted to be $30,000. The company's cost of capital is 8%. Equations (17.54) and (17.56) are used to calculate the investment's present worth:

$$P = -\$100,000 + \$30,000 \left[\frac{(1+0.08)^5 - 1}{0.08(1+0.08)^5} \right] + \frac{\$20,000}{(1+0.08)^5} \quad (17.63)$$
$$= \$33,392.97$$

Assuming our parameter estimates are accurate, the equipment purchase is a good investment since the present worth P is greater than zero. Due to the uncertainty surrounding our parameter estimates, it is important to investigate how sensitive the results are to changes in the initial parameter estimates. To perform sensitivity analysis on the uncertain parameters (project life, annual cash inflows, and cost of capital), the initial values of each parameter is deviated one at a time, and the effect on our present worth performance measure is observed. The results of the single-parameter sensitivity analysis are presented in Figure 17.9. Figure 17.9 suggests that P is most sensitive to changes in annual cash inflows and least sensitive to changes in the cost of capital. The equipment investment will be profitable as long as either the project life is at least 67% of the initial estimate, the annual cash inflows

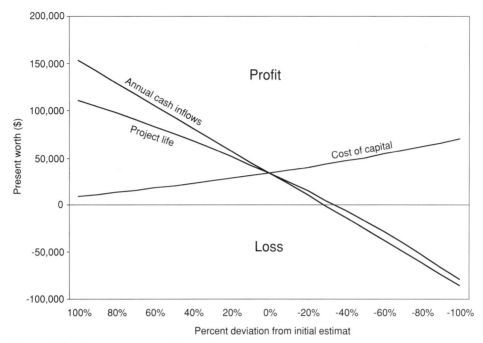

Figure 17.9 Single-parameter sensitivity analysis.

do not decrease by more than approximately 27%, or the cost of capital does not increase by more than 140%.

17.7.2 Optimistic–Pessimistic Sensitivity Analysis

The optimistic–pessimistic approach involves changing the estimates of one or more parameters in favorable (optimistic) and unfavorable (pessimistic) outcome directions to determine the effect on the results of the study. The initial values are kept as the expected or most likely outcome. Table 17.8 contains optimistic, most likely, and pessimistic estimates for the burn-in oven purchase decision.

Equations (17.52) and (17.53) are used to find the optimistic, most likely, and pessimistic annual worth values of the investment. The investment is profitable under the optimistic and most likely conditions, but the pessimistic condition leads to an unprofitable

Table 17.8 Optimistic–Pessimistic Estimates

	Optimistic estimates	Most likely estimates	Pessimistic estimates
Purchase price	$100,000	$100,000	$100,000
Project life	8	5	3
Annual cash inflows	$50,000	$30,000	$20,000
Salvage value	$20,000	$20,000	$20,000
Cost of capital	8%	8%	8%

Table 17.9 Annual Worth for All Combinations of Estimate Conditions

		Annual cash inflows		
Annual worth		Optimistic	Most likely	pessimistic
Project Life	Optimistic	$34,478.82	$14,478.82	$4,478.82
	Most likely	$28,363.48	$8,363.48	−$1,636.52
	Pessimistic	$17,357.32	−$2,642.28	−$12,642.70

result (i.e., a negative annual worth).

$$A(\text{optimistic}) = -\$100,000 \left[\frac{0.08(1+0.08)^8}{(1+0.08)^8 - 1}\right] + \$50,000 + \$20,000 \left[\frac{0.08}{(1+0.08)^8 - 1}\right]$$
$$= \$34,478.82$$

$$A(\text{most likely}) = -\$100,000 \left[\frac{0.08(1+0.08)^5}{(1+0.08)^5 - 1}\right] + \$30,000 + \$20,000 \left[\frac{0.08}{(1+0.08)^5 - 1}\right]$$
$$= \$8363.48$$

$$A(\text{pessimistic}) = -\$100,000 \left[\frac{0.08(1+0.08)^3}{(1+0.08)^3 - 1}\right] + \$20,000 + \$20,000 \left[\frac{0.08}{(1+0.08)^3 - 1}\right]$$
$$= -\$12,642.70$$

These results are useful if the estimate conditions for both uncertain parameters, project life and annual cash inflows, are perfectly correlated, for example, optimistic project life occurs with optimistic annual cash inflows. If the estimate conditions are not correlated, the analyst must consider all combinations of all estimate conditions. Table 17.9 contains the annual worth results for all of these combinations.

These results suggest that purchasing the burn-in oven is profitable under all combinations when either annual cash inflow or project life is in the optimistic condition and when both parameters are in the most likely condition. The analyst can feel fairly confident that the investment should be made based on these results.

17.8 SUMMARY

This chapter presents numerous methodologies and techniques to address the cost, forecasting, and economic evaluation issues for microelectronic packaging. Manufacturers of microelectronic packages are continually striving to lower production costs, increase profits, and identify opportunities to predict with some degree of accuracy future demand trends. Therefore, it will be advantageous to develop an understanding of the cost considerations for the manufacturer of high-density electronic packages.

The volume-based and activity-based costing techniques allow electronics manufacturers to determine costs based on the traditional volume-based approach, or by the business activities that dictate the cost to manufacture electronic packages. Several examples have been provided to illustrate the application of VBC and ABC in an electronic packaging situation. Linear and piecewise break-even analysis is a useful technique to develop a relationship between production quantities and manufacturing costs. Learning curves for new products and processes are of significant importance when estimating the costs to manufacturer mass quantities of high-density electronic packages. Predicting the demand for microelectronic packages is extremely difficult considering the rapid change in technology and competition among manufacturers. Various forecasting models are presented with associated illustrations to provide a more in-depth understanding of how to predict product demand.

No cost evaluation is complete without considering engineering economic analysis. This chapter has provided a basic understanding of the different economic analysis tools and how they are applied to microelectronics packaging. Several examples are provided to illustrate the fundamentals of engineering economy. Finally, in the electronic packaging environment there are varying levels of uncertainty when dealing with highly unstable manufacturing parameters. Sensitivity analysis has been presented in this chapter to address the variable levels of uncertainty and to provide an approach to determine the most desirable outcome.

REFERENCES

1. C. Horngren, G. Sundem, and W. Stratton, *Introduction to Management Accounting*, 12th ed., Upper Saddle River, NJ: Prentice Hall, 2002.
2. R. Copper and R. S. Kaplan, How Cost Accounting Distorts Product Costs, *Management Account.*, Vol. 69, No. 10, pp. 20–27, 1988.
3. M. Walker, Attribute Based Costing: For Decision Making, *Manage. Account.*, Vol. 77, No. 6, pp. 18–22, 1999.
4. R. Cooper, The Rise of Activity-based Costing—Part One: What Is an Activity-based Cost System? *J. Cost Manage.*, Vol. 2, No. 2, pp. 45–54 1988.
5. R. Cooper, The Rise of Activity-based Costing—Part Two: When Do I Need an Activity-based Cost System? *J. Cost Manage.*, Vol. 2, No. 3, pp. 41–48, 1988.
6. R. Cooper, The Rise of Activity-based Costing—Part Three: How Many Cost Drivers Do You Need and How Do You Select Them? *J. Cost Manage.*, Vol. 2, No. 4, pp. 34–46, 1989.
7. R. Copper, The Rise of Activity-based Costing—Part Four: What Do Activity-based Cost Systems Look Like? *J. Cost Manage.*, Vol. 3, No. 1, pp. 38–49, 1989.
8. M. Morrow and T. Connnolly, Practical Problems of Implementing ABC, *Accountancy*, Vol. 113, No. 1205, pp. 76–80, 1994.
9. E. Malstrom, *What Every Engineer Should Know about Manufacturing Cost Estimating*, New York: Marcel Dekker, 1981.
10. E. Malstrom and R. Shell, A Review of Product Improvement Curves, *Manuf. Eng.*, Vol. 82, No. 5, 1979.
11. E. Malstrom, ed., *Manufacturing Cost Engineering Handbook*, New York: Marcel Dekker, 1984.
12. T. Landers, W. Brown, E. Fant, E. Malstrom, and N. Schmitt, *Electronics Manufacturing Processes*, Englewood Cliffs, NJ: Prentice Hall, 1994.
13. J. R. Canada W. G. Sullivan, and J. A. White, *Capital Investment Analysis for Engineering and Management*, 2nd ed., Englewood Cliffs, NJ: Prentice Hall, 1996.
14. J. Neter, M. H. Kutner, C. J. Nachstheim, and W. Wasserman, *Applied Linear Statistical Models*, 4th ed., Chicago: Irwin, 1996.
15. D. G. Newnan, J. P. Lavelle, and T. G. Eschenbach, *Essentials of Engineering Economic Analysis*, 2nd ed., New York: Oxford University Press, 2002.
16. R. T. Clemen and T. Reilly, *Making Hard Decisions with Decision Tools*, 2nd ed., Pacific Grove, CA: Duxbury, 2001.

EXERCISES

17.1. A company produces ball grid array (BGA) and quad flat pack (QFP) packages. Presently, the company uses a single cost driver (number of direct labor hours) to allocate its overhead costs. Based on unexpected loss of profits during the last year, management is concerned that the current volume-based costing (VBC) system may be distorting product costs and has asked you to investigate. As a result, you are considering the possibility of implementing an activity-based costing (ABC) system. Your preliminary analysis divides the company's overhead costs into four major cost pools—purchasing, utilities, shipping, and accounting. Associated with each cost pool, you have identified an appropriate cost driver. The data you have collected is summarized below. All prices/costs are per container of 1000 units. Note that 360 h of direct labor were expended and the company pays an average hourly labor rate of $18.

Product	Direct labor hours (hours/container)	Material cost ($/container)	Sales volume (containers)	Selling price ($/container)
BGA	3	400	100	850
QFP	2	370	30	600

Cost pool	i	Overhead cost	Cost driver	Total cost driver level	BGA cost driver level	QFP cost driver level
Purchasing	1	$11,640	Number of purchases	70	40	30
Utilities	2	$5,600	Square feet of facility	900	600	300
Shipping	3	$8,000	Number of shipments	49	32	17
Accounting	4	$7,680	Number of orders	45	27	18
Total		$32,920				

a. Determine the product costs per container of the BGA and QFP packages using the current volume-based costing method.

b. Determine the product costs per container of the BGA and QFP packages using an ABC approach.

c. Observing the results of (a) and (b), what conclusions can you draw? What are your recommendations?

17.2. A cost analyst is studying a manufacturing operation for the production of microelectronic packages. Setup, equipment, and facilities modifications required to initiate production total is approximately $400,000. As production begins, recurring costs are known to be $75 per unit. If each unit can be sold for $90, how many units must be initially produced to recover the initial investment associated with production? What is the margin of profit at a level of production equal to 50,000 units?

17.3. Suppose the production scenario in the previous problem is to be reconsidered using additional cost information. It is realized that equipment maintenance cost will be an additional $25,000. This cost is to occur after the production of the first 25,000 units and every additional 25,000 units thereafter. The manufacturer also plans to reduce the sales price to $85 per unit after a production level of 100,000 units. What is the level of profit in this operation at a level of production equal to 200,000 units?

17.4. A particular assembly process is known to have a 70% cumulative average product improvement curve. It is also known that it took 60 h to make the first 100 units. The work standards indicate that the run time should be 0.60 h per unit; 300 units have been scheduled for production. What is the negative exponent for this problem? How many hours were spent in building the first unit? How many hours were expended to build the 50th unit? How many units were produced before the standard time was reached?

17.5. A microelectronic package manufacturer has just received a rather sizable order for a specialized microelectronic package, the AAA-1. The AAA-1 product is completely new and requires additional setup and modifications to the existing production process. It is determined that existing data cannot be used to determine the learning curve, so a test run is prepared and the following data is collected. The first unit required 100 h and the 20th unit took 64 h. Based on the available data determine the learning curve for the AAA-1 product.

17.6. What are the two primary types of forecasts? Under what conditions should each of these forecasts be made? For what is each type of forecast useful? What levels of company personnel should focus on each of these two types of forecasts?

17.7. Consider the following sales data for 2 years of quad flat pack package sales:

	Year 1				Year 2			
	Q_1	Q_2	Q_3	Q_4	Q_1	Q_2	Q_3	Q_4
Sales	7,054	12,109	22,870	32,231	8,922	16,226	21,590	38,815

Forecast quad flat pack sales in Q_1 of year 3 using simple averaging, moving average with a time window of 2 quarters, and moving average with a time window of 4 quarters. Calculate appropriate error measures, commenting on the utility of each of the three average-based forecasting approaches for forecasting quad flat pack sales data.

17.8. Using the quad flat pack sales data given in Exercise 17.7, forecast quad flat pack sales in Q_1 of year 3 using both single and double exponential smoothing techniques with $\alpha = \beta = 0.2$. Compare the quality of these two forecasts in terms of appropriate error measures.

17.9. A packaging firm needs to borrow $1,000,000 to fund various corporate initiatives. Assuming an annual interest rate of 8% and a 10-year loan term, calculate the firm's loan payment assuming (a) equal annual payments (i.e., 10 equal payments) and (b) equal monthly payments (i.e., 120 equal payments). What is the total amount paid by the firm to borrow the $1,000,000 under each type of loan payment plan?

17.10. A electronics packaging firm is trying to decide whether or not to purchase an automated line loading machine that is advertised to help load products more efficiently into the front of the manufacturing process. The line loading machine costs $200,000 and is expected to produce $45,000 in annual savings over its 5 years of useful life. Assuming the firm's cost of capital is 8%, what is the minimum acceptable salvage value S for the machine that will ensure the firm would be making a good (i.e., profitable) decision if it decided to buy the line loading machine?

17.11. A packaging firm just purchased a new burn-in oven for $80,000. The oven has a useful life of 10 years and no salvage value. The asset's annual operating, maintenance, and repair costs are estimated to be $15,000 in the first year of ownership. After this first year, these costs are expected to rise by $4000 per year. Finally, no cash inflows are currently associated with this asset. When should this oven be replaced by the firm, assuming a cost of capital of 10%?

17.12. An engineering department in a manufacturing company requested a replacement of an old (worth nothing) ball grid array packaging machine with a new one; currently the new machine costs $C = \$38,000$. According to the manufacturer, the new machine's useful life and salvage values are expected to be $n = 7$ years and $F = \$7000$, respectively. The net annual cash flows are predicted to be $12,500, and the company's minimum attractive rate of return is $i = 12\%$. The engineering department is uncertain about the purchase date, which affects the machine costs as it is changing with the time and market situation. Perform a single-parameter sensitivity analysis showing the effect of $\pm 100\%$ changes in the initial machine capital cost, life, salvage value, and annual net cash flow based on the PW value and discuss their impacts on the purchasing decision.

17.13. Assume that in Exercise 17.12 the company confirmed the purchase in which the machine costs $38,000 and its useful life is 7 years. However, the annual cash flow and salvage value are still uncertain. The following table contains the optimistic, most likely, and pessimistic estimates for the ball grid array packaging machine investment. Based on these information, perform an optimistic–pessimistic sensitivity analysis by calculating the annual worth values for all combinations assuming that the estimate conditions are not correlated.

	Optimistic estimates	Most likely estimates	Pessimistic estimates
Purchase price	$38,000	$38,000	$38,000
Project life	7	7	7
Annual cash inflows	$12,000	$9,000	$7,500
Salvage value	$5,800	$4,200	$3,500
Cost of capital	12%	12%	12%

Chapter 18

Analytical Techniques for Materials Characterization

EMILY A. CLARK, INGRID FRITSCH, SEIFOLLAH NASRAZADANI, AND CHARLES S. HENRY

18.1 OVERVIEW

This chapter covers in detail some highly useful analytical techniques used to characterize materials employed in electronics fabrication technology. Knowing the composition and structure of materials are helpful in the synthesis, design, and development of electronic and photonic modules, as well as in the analysis of their failure. Each section in this chapter focuses on a different technique and includes a summary to serve as a quick reference for the reader, a description of the basic principles of the physical and chemical phenomena involved, an overview of the instrumentation, and one or more real examples of how the technique has been used in electronics-related work. Some common problems that are encountered and their solutions are also considered. First, X-ray diffraction (XRD) and Raman spectroscopy provide examples of techniques capable of analysis of the bulk structure and composition of materials. Next, physical imaging techniques, such as scanning electron microscopy/energy dispersive X-ray spectroscopy (SEM/EDX), confocal microscopy, scanning force microscopy (SFM), and scanning tunneling microscopy (STM) methods are presented. Finally, surface analytical techniques, including Auger electron spectroscopy (AES), X-ray photoelectron spectroscopy (XPS), and secondary ion mass spectrometry (SIMS) are examined. Table 18.1 summarizes some of the important aspects of these techniques.

18.2 X-RAY DIFFRACTION

X-ray diffraction (XRD) is a powerful and nondestructive technique that was developed primarily to provide crystallographic characterization of solid materials. XRD can attain not only structural information of bulk materials and thin films but also the composition of crystallographic phases present in a sample, the extent of defects, the thickness of layers (in a multilayer thin-film structure), stresses developed during manufacturing of thin films, size and orientation of grains, and the lattice match between interlayers in a multilayer structure.

Advanced Electronic Packaging, Second Edition, Edited by Richard K. Ulrich and William D. Brown
Copyright © 2006 the Institute of Electrical and Electronics Engineers, Inc.

Table 18.1 Analytical Techniques Employed in Development and Analysis of Materials Used in Multichip Modules

Technique	Acronym	Incident probe	Analyzed phenomenon	Depth of analysis	Spatial resolution	Detection limit	Applications
X-ray diffraction	XRD	X-rays (1–3 kW, 18 kW for rotating anodes; 1.54056 Å for Cu anode)	Angles of X rays relative to incident beam and sample that are diffracted by sample	100 Å–1000's Å	>10 μm	1%	Crystal structure, texture, orientation, phase analysis, elemental composition
Raman spectroscopy	RS	Photons (UV or visible radiation)	Photons	<50 Å for surface-enhanced RS much thicker (cm) for transmission RS through liquids or solids that are mostly Raman transparent	2 μm with RS microscopy	0.50%	Phase identification, functional groups (bonded structures)
Scanning electron microscopy	SEM	Electron beam (0.3–300 keV)	Electrons	—	5 Å	—	Morphology, topography
Auger electron microscopy	AES	Electron beam (0.5–10 keV)	Secondary electrons from sample	1–30 Å	100 Å with scanning Auger microscopy (SAM)	0.05 atomic%	Surface layer defects, elemental analysis and distribution, bonding information

Technique	Abbreviation	Incident probe	Detected signal	Depth resolution	Lateral resolution	Detection limit	Information
X-ray photoelectron spectroscopy	XPS	X-rays (>1 keV; 1486 eV for Al K_α radiation)	Electrons emitted by sample	5–100 Å	3 μm with imaging XPS	0.01 atomic%	Surface elemental composition and oxidation state, bonding information
Energy dispersive X-ray analysis	EDX or EDS	Electron beam (1–30 keV)	X rays emitted by sample		1 nm–0.5 μm	0.1 mol%	Elemental composition near-surface
Secondary ion mass spectrometry	SIMS	Ion beam (0.5–30 keV)	Secondary ions ejected by primary ions from sample	10's Å (with static 1000's Å–μm SIMS) to microns (dynamic SIMS)	600 Å	<0.0001 mol%	Trace elemental, isotope and molecular composition, bonding information
Scanning probe microscopy (e.g., scanning tunneling microscopy, atomic force microscopy)	SPM (STM, AFM)	Cantilever or conducting tip	Current (STM) or force (AFM)	0.1 Å	0.1 Å	Each atom	Morphology, topography surface defects

18.2.1 Summary

- What XRD does
  - The angle of reflection of X rays from a sample is related to the crystal structure and composition of the material.
- Types of analyses
  - Lattice parameter measurements
  - Crystallite size and distribution
  - Texture analysis—preferred orientation of crystallites in a material
  - Internal and residual stress measurements
  - Coefficient of thermal expansion measurement
  - Phase and composition identification
  - Film thickness measurements
- Performance
  - Depth of analysis 100 to 1000's Å
  - Spatial resolution >10 μm
  - Detection limit 1%

18.2.2 Basic Principles

Polycrystalline materials contain many grains that are randomly oriented so they form a mosaic structure. Each grain contains atoms in arrangements having crystallographic planes. Figure 18.1 illustrates different planes in the face-centered cubic structure of MgO. Any lattice plane can be described by a sequence of numbers called Miller indices. Three are needed for cubic structures and four for hexagonal structures. Lattice planes provide a reference grid to which the atoms in the crystal structure may be referred. They sometimes coincide with layers of atoms but not always. The distance between two successive planes for a given set of Miller indices is called the interplanar spacing, d, and depends on the orientation of the plane through and the composition of the crystalline grain. The magnitude

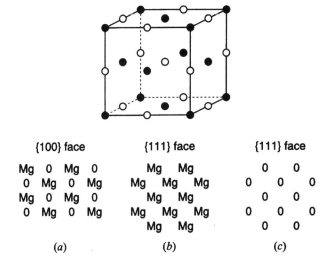

Figure 18.1 Arrangement of atoms in a face-centered cubic unit cell, as for MgO. The solid circles are Mg atoms and the open circles are O atoms. The d spacing is between lattice planes, which contain coplanar atoms. Different planes may be sliced from this unit cell. (*a*) A plane with a (100) face contains both Mg and O atoms. A plane with a (111) face contains only (*b*) Mg or (*c*) O atoms. (*Reprinted with permission from* [2]. *Copyright* 1989 *John Wiley & Sons, Ltd.*)

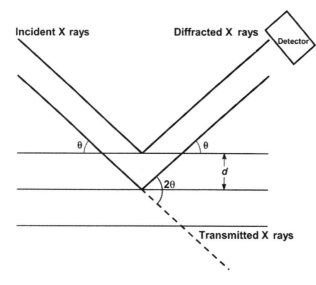

Figure 18.2 Geometric illustration of the X-ray diffraction experiment. Each horizontal line represents a cross section of a lattice plane.

of d is on the order of the wavelength of X rays. This characteristic of crystalline material makes X rays suitable for studying a material's crystallographic features.

When X-rays fall on a set of lattice planes in a crystalline material, "constructive" interference takes place, and the beam gets reinforced only when Bragg's condition of diffraction is satisfied. Mathematically, Bragg's condition is defined as

$$n\lambda = 2d_{hkl} \sin \theta \qquad (18.1)$$

where n is the order of diffraction, λ is the wavelength (in Å), d is interplanar spacing (in Å) and θ is the angle of incidence or diffraction of the X rays in degrees. Figure 18.2 shows a geometrical demonstration of these parameters. Note that for practical considerations, the angle, 2θ, is sometimes measured, instead of θ. To obtain the 2θ value, the transmitted X-ray incident beam serves as the reference point for the angle measurement of the diffracted beam.

In cubic crystals, the interplanar spacing is related to the lattice parameter, a, which is defined as the length of the cube side using Eq. (18.2).

$$d_{hkl} = a/(h^2 + k^2 + l^2)^{1/2} \qquad (18.2)$$

where *hkl* indicates the Miller indices of a given set of planes. A list of the relationships between a and d for different lattice planes is given in a variety of texts [1, 2].

In noncrystalline materials, destructive interference of X rays takes place because they are out of phase with respect to each other. This effect results in no beam reinforcement. Hence, little or no diffraction is observed.

18.2.3 Instrumentation

The components for an XRD instrument include an X-ray generator, a goniometer, sample stage, detector, optics, and a computer (for data collection and analysis). There are two main types of X-ray sources, sealed tubes (1 to 3 kW) and rotating anodes (18 kW). When electrons that are produced at a heated cathode (e.g., tungsten) are accelerated with a large voltage and collide with a metal anode target (e.g., copper, molybdenum, chromium, and cobalt),

X rays are produced. At appropriate accelerating voltages, line spectra are superimposed on a continuum of X rays emitted from the target. Copper, for example, exhibits lines at 1.54056 Å ($K_{\alpha 1}$), 1.54439 Å ($K_{\alpha 2}$), and 1.39222 Å (K_β). To obtain monochromatic radiation for diffraction applications, a filter is employed to remove most of the continuum and undesired lines in the spectrum. Two types of devices are used, called goniometers, that allow independent control over the positions of the sample stage and detector ($\theta/2\theta$ configuration) or of the X-ray tube and the detector (θ/θ configuration). Dual-goniometer systems are available to allow two different configurations to be set up at the same time, while sharing the same source, which facilitates analysis on a variety of samples without reconfiguring. Sample stages may offer pressure and temperature control, others allow tilting and rotation of the sample (Eulerian cradles). Point detectors, linear position-sensitive detectors, and charge-coupled devices are available for detecting the diffracted X rays. The latter two offer speed advantages over point detectors, which collect diffracted intensity only one angle at a time.

The optics are arranged to maximize the signal-to-noise ratio for characterizing different types of samples. Debye optics are used in single-crystal instruments and area detector systems for analysis of small single-crystal specimens and sample mapping or microdiffraction with high spatial resolution. This setup includes a spot source of X rays and pinhole collimation. Optics used with thin films are Bragg–Brentano and Seemann–Bohlin geometries. The Bragg–Brentano type (Figure 18.3a), which is primarily used for the analysis of both preferentially and randomly oriented polycrystalline films, incident X rays, which are collimated by a slit, impinge on the specimen at an angle θ. The diffracted beam is also collimated using a receiving slit before its detection. In this geometry, the detector moves to an angle twice that of the sample tilt, and crystallographic information from planes parallel to the surface are obtained. The sensitivity of this method is highest when the receiving slits, the specimen, and the focal point of the X-ray source lie on a circle. In this condition, X rays are approximately focused on the receiving slits. This is known as the *parafocusing* condition. In the Seemann–Bohlin geometry (Figure 18.3b), the specimen is fixed at a small angle of about 5 to 10° with respect to the incident beam of X rays. The detector is allowed to move along a circle and record the diffracted X rays. Since the incident angle in this

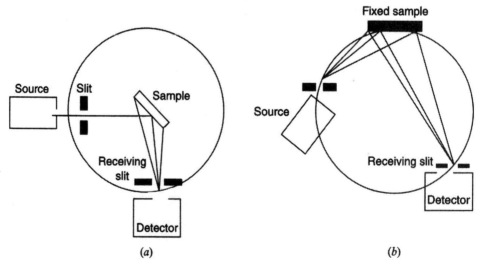

Figure 18.3 Configuration of optics for XRD of materials: (*a*) Bragg–Brentano and (*b*) Seemann–Bohlin.

method is small, a large volume of material becomes exposed to X rays, which leads to good sensitivity. Only planes with the correct orientation diffract X rays. The Seemann–Bohlin geometry is usually used for polycrystalline films containing randomly oriented crystallites.

18.2.4 Practical Considerations and Applications

A wide variety of analyses can be performed on materials with XRD. A few examples are described herein.

18.2.4.1 Crystal Structure and Phase Determination

A simple X-ray diffraction experiment involves the exposure of a sample to monochromatic X rays at a known angle θ, and the collection of the diffracted beam at 2θ. This analysis is referred to as a $\theta/2\theta$ measurement. In such an experiment, the detector is mechanically linked to the sample rotation such that the detector moves through 2θ while the sample moves through θ. A plot of the resulting intensity of the diffracted beam versus 2θ can then be used to locate the angular positions where the strong peaks occur. The peaks of intensity occur at angles where the Bragg condition is met. This information is extremely valuable because the angular positions (2θ) at which diffraction occurs provide information about the spacing between a particular set of planes (d), and therefore the geometry of the crystal lattice, whereas the intensity of the diffracted X rays provides information about the number and kind of atoms within the crystal. If a sample contains multiple phases (i.e., different types of crystals of the same or different composition), then the intensity is also proportional to the relative amount of a given phase.

Basically, crystal structure determination involves three major steps. First, based on some knowledge of the approximate chemical composition of the substance, an educated guess must be made to determine to which of the seven crystal systems (cubic, hexagonal, tetragonal, orthorhombic, rhombohedral, triclinic, monoclinic) the unknown structure belongs. With this assumption, Miller indices are assigned to each of the observed peaks. If a good correlation between the assumed crystal form and the angular peak positions is obtained, one would proceed to calculate lattice parameters. The second step involves the computation of the number of atoms per unit cell from the lattice parameters, chemical composition, and measured density. The third step is to obtain atomic positions in the unit cell based on the relative intensities of the diffracted peaks.

Determining the crystal structure of an unknown sample is really the task of a crystallographer. However, materials scientists are often involved in phase identification of semiknown substances. For example, varying the synthesis parameters and conditions for the development of new materials, such as ceramic superconductors, requires phase identification using X-ray diffraction. In this case, a $\theta/2\theta$ measurement is made on the sample under investigation. Using Bragg's law [Eq. (18.1)], corresponding d spacing of all crystallographic planes can be calculated. A knowledge of the d spacings and a comparison with those given for known phases in a library, provided by the Joint Committee on Powder Diffraction Standards (JCPDS), often leads to a positive identification of unknown substances. Figure 18.4 shows data from such a library for iron. Computers facilitate such a library search and comparison. Consequently, a complete identification of phases is possible within a matter of minutes.

Fe
Iron
Iron, syn
Hanawalt 2.03/X 1.17/3 1.43/2 0.91/1 1.01/1 0.83/1 0.00/1 0.00/1 0.00/1 0.00/1

Lambda 1.5405 Sys. Cubic			d	Int	h	k	l	d	Int	h	k	l
Sg Im3m	PS cl 2.00		2.026	100	1	1	0					
a 2.8664	b	c	1.433	20	2	0	0					
α	β	τ	1.170	30	2	1	1					
			1.013	10	2	2	0					
A 2.8664	C	z 2	0.9064	12	3	1	0					
D_x 7.875	Dm	V23.55	0.8275	6	2	2	2					
F (N) 225.2	M(20) 999.9	I/Ic										
d-sp Not given												
Int Diffractometer												
Total d's 6												
Color Gray, light gray metallic												
Temp X-ray pattern at 25 C.												

Figure 18.4 American Society of Testing Materials (ASTM) X-ray diffraction card for elemental iron, where the fourth line is the reference text for the information in the card; lambda is the wavelength; a, b, c are lattice parameters of the unit cell; α, β, τ are angles of the unit cell; C and A are lattice parameters in Å; D_x is the density calculated in g/cm^3; D_m is the measured density; V is the volume in Å^3. (*Reprinted with permission from Total Access Diffraction Database. Copyright* 1993 *Philips Analytical B.V.*)

18.2.4.2 Measurement of Strain in Epitaxial Films

For characterization of strain in epitaxial films, a geometry very similar to Bragg–Brentano is used with the exception that a perfect single crystal is placed near the focal point of the incident X-ray beam (source) to collimate and monochromatize the incident X rays. A diffractometer used in this manner is referred to as a double-crystal diffractometer, and it allows high angle resolution measurements. In this setup, the detector is fixed near a known angular position (2θ) for the (hkl) planes of interest, and the receiving slits are selected as large as possible to accept a large range of 2θ. When the specimen is rotated slightly (about 2θ), all asymmetric reflections of the planes that are not parallel to the surface can be measured. The resulting plot of intensity versus 2θ, referred to as a *rocking curve*, indicates very small deviations in d spacing due to strain. This is possible because narrow peaks are expected from strain-free specimens.

18.2.4.3 Texture: Preferred Orientation of Crystallites in Materials (Pole Figures)

Yttrium-stabilized zirconium oxide is of interest as a dielectric or substrate for superconducting materials. There is great interest in depositing the material on to substrates in a thin-film form that is suitable for patterning. However, thin films of a material do not always possess the same properties of bulk single crystals because the structure of the film is usually less ordered. To evaluate the texture, or preferred orientation of crystallites in a material, pole figures can be obtained. In these measurements, a diffraction angle is chosen that corresponds to reflection from planes possessing the Miller indices of interest. The optics are set at that angle. The sample is positioned first in a horizontal position, so that the tilt angle, Ψ, is zero. While diffracted X rays are monitored at the detector, the sample is tilted by a

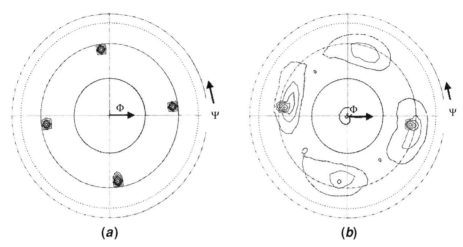

Figure 18.5 Comparison of texture (as evaluated from pole figures) for the (111) crystal orientation of a single crystal of yttrium-stabilized zirconium oxide and of a film of the same material formed by deposition onto an aluminum substrate from a laser ablation process.

small angle and rotated through 360° (the Φ angle). The sample is tilted more and rotated again. This process is continued through as much as $\Psi = 90°$. Pole figures (contour or three dimensional) that plot the diffraction intensity at the fixed angle of θ as a function of Ψ and Φ angles are produced from the data. Figure 18.5 compares pole figures that were obtained for a single crystal of cubic Y-stabilized ZrO_2 and for a film that was laser ablated onto a substrate for optics that were fixed at an angle that reflects off of the (111) plane. Intensity of diffracted radiation is very high and narrow at four different locations at 45° tilt for the single crystal. The narrow peaks are indicative of extremely high order of the reflection planes, as is expected from a single crystal. The 45° tilt angle for the (111) is consistent with a crystal that is oriented with its (100) face horizontally in the starting position. The pole figure for the film formed by laser ablation also shows four regions of peak intensity at 45° tilt, which indicates that there is some preferential ordering of the crystallites in the film similar to that of the single crystal. However, the intensities are much lower and broad, suggesting that there is a significant degree of disorder within the film.

18.2.4.4 Reflectometry: Determination of Thickness of Thin Films

An X-ray instrument is also capable of obtaining the thickness of thin films based on reflection of the radiation. The same laws of physics are obeyed by the X rays as those by visible light. When the incident X rays arrive at the film of a sample (at a grazing angle), interference fringes are produced from X rays reflected from the surface and those reflected from the interface of the film and the substrate. The thickness of the film can be calculated with the following equation:

$$\text{Thickness} = \frac{\lambda}{2 \left| \sin \frac{2\theta_{n+1}}{2} - \sin \frac{2\theta_n}{2} \right|} \quad (18.3)$$

where λ is the wavelength of the X rays and n is the nth interference fringe maximum. Figure 18.6 shows interference fringes from a reflectometry measurement of a thin film of gold with a chromium adhesion layer on a glass substrate, yielding a film thickness of 443 $\pm$ 3 Å. Other techniques available for thickness measurements of thin films are atomic force

734 Chapter 18 Analytical Techniques for Materials Characterization

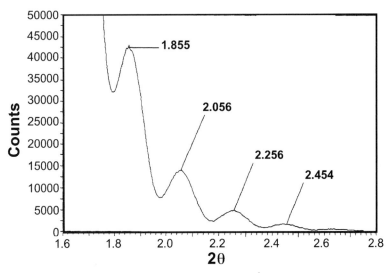

Figure 18.6 X-ray reflectometry of a thin film of gold (with a <10 Å chromium adhesion layer) on a glass substrate, where the wavelength of X rays is 1.542 Å (average of $K_{\alpha 1}$ and $K_{\alpha 2}$). The thickness of the film obtained from the data is 443 ± 3 Å.

microscopy (AFM) and profilometry [3,4]. Although they can measure local thicknesses of films on patterned surfaces, which X-ray reflectometry cannot do, they require a step edge across which to scan the probe tip. In addition, X-ray reflectometry is better suited for smaller thicknesses than profilometry. Not only can X-ray reflectometry measure film thicknesses of up to several hundred nanometers, it can also assess surface *and* interface roughness.

18.3 RAMAN SPECTROSCOPY

Raman spectroscopy involves a phenomenon whereby the frequency of light is changed when it is scattered by molecules. The frequency changes occur as a result of coupling between incident radiation and the vibrational energy levels of molecules. However, Raman scattering is well-established for electronic and rotational energy levels as well. This technique provides information on molecular/crystal structure, microcrystallinity (especially in semiconductors), strain effects, bond characterization (from bond polarizability), defects, and structural disorders, and composition of functional groups in organic materials. Although the focus of this section is on bulk and thin and thick films of solids, it should be noted that Raman spectroscopy is also capable of analyzing liquids and gases.

18.3.1 Summary

- What Raman spectroscopy does
  - Measures frequency change of light scattered by sample relative to incident light.
  - Provides information on vibrational frequencies of chemical bonds present in the sample (functional group analysis).
  - Analysis of phase transition.

- Types of analysis
  - Depth profiling for transparent materials
  - Spot analysis
  - Line scan or imaging when coupled with Raman microscopy
- Performance
  - Lateral resolution 1 μm with microfocus instrument
  - Spectral resolution 1 cm^{-1}
  - Depth probed varies from microns to millimeters and is material dependent

18.3.2 Basic Principles

In Raman spectroscopy, a monochromatic source of intense radiation in the visible to near-infrared frequencies provides photons that cause a momentary distortion (polarization) of electrons distributed around bonds in molecules of the sample. The molecules increase in energy equal to the energy of the photons, to a nonquantized, "virtual state" between the ground and first electronic excited states. The radiation that emits when the bonds return to their normal states is called scattered light. Three types of scattering can result and are shown in Figure 18.7. Most of the scattered photons have the same energy as the incident photons, a case known as Raleigh scattering. About 0.001% of the scattered light has energy that is different than the incident radiation, which is Raman scattering. In this case, the excitation frequency has been affected by the frequency of a bond vibration. Stokes emission occurs when the photon is emitted at a lower frequency than the incident radiation, indicating that the molecule relaxes to an energy level that is higher than its starting level. Anti-Stokes emission occurs when a photon is emitted with a higher frequency than the incident radiation, indicating that the molecule relaxes to an energy level that is a lower level than its starting level. Stokes emission is more favored than Anti-Stokes emission. A Raman spectrum displays the difference in frequency of the scattered radiation from the incident radiation, usually in units of reciprocal centimeters (cm^{-1}) [5], which corresponds to the vibrational mode transition.

18.3.3 Instrumentation

The basic components of a Raman spectrometer are a light source, focusing and collection optics (often at 90° from the source or filters are used to remove wavelengths of the laser source), and a detector. In order to achieve detectable Raman scattering, lasers are often

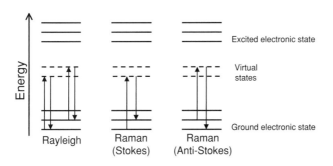

Figure 18.7 Energy transfer involved in Stoke's, anti-Stoke's, and Rayleigh scattering.

used to provide the necessary incident intensity. Lasers also offer a single wavelength from which the energy shift of scattered light can be measured. Common laser sources used in Raman spectroscopy are: argon ion (488.0 or 514.5 nm), krypton ion (530.9 or 647.1 nm), helium/neon (632.8 nm), diode laser (782 or 830 nm), and Nd/YAG (1064 nm). Radiation in the near-infrared wavelength region have the advantages of minimal photocomposition of the sample and interferences from fluorescence.

Different types of Raman spectrometers are available [6]. Dispersive instruments employ double grating systems to reduce the interfering radiation from reaching the photomultiplier tube (PMT) detector but suffer from transmission losses. Fourier transform and multichannel types take spectra in about 1/20th of the time of the dispersive instruments. Either cooled germanium transducers or charge-coupled device (CCD) detectors are used, which are more sensitive than PMTs to the 782-nm radiation of diode lasers. CCDs are less sensitive to 1064-nm radiation of the Nd/YAG lasers.

18.3.4 Practical Considerations and Applications

Raman spectroscopy may be used in a variety of applications. The following sections cover composition and structure of diamond, GaAs, and silicon carbide that are of interest to the materials, packaging, and semiconductor industries [7, 8].

18.3.4.1 Thick- and Thin-Film-Analysis

Many different types of thick and thin films of materials can be analyzed using Raman spectroscopy. Even a submonolayer of a chemical species on a surface may be sufficient signal for analysis under the right conditions. Raman analyses of free-standing diamond and diamond deposited on a variety of substrates are illustrated here. These materials possess interesting properties, one of which is excellent heat conduction for use in the electronics industry [9].

Carbon forms several different crystal structures, which influence Raman spectra significantly. In the diamond form, carbon atoms are sp^3 hybridized and located in tetrahedral positions of a cubic unit cell with a bond length of 1.54 Å, exhibited by a single Raman peak at a wavenumber of 1332 cm^{-1}, whereas in graphite the carbon atoms are sp^2 hybridized and form a crystal structure that is hexagonal with bond lengths of 1.42 Å, exhibited by a Raman peak at 1580 cm^{-1}. Highly oriented pyrolytic graphite (HOPG), however, is disordered and exhibits a peak shifted from 1580 cm^{-1} with an extra peak at 2719 cm^{-1}. Figure 18.8 compares Raman spectra of different quality and types of (a) diamond and (b) graphitic carbon. As is shown in the figure, mixtures of sp^2 and sp^3 hybridized atoms can be found in many carbon-based materials, such as diamond-like carbon.

18.3.4.2 Stress Measurement

Diamond films deposited via chemical vapor deposition (CVD) processes generally are highly stressed due to the presence of high thermal energy and high concentration of hydrogen during formation. Raman spectroscopy has been successfully used to qualitatively evaluate and compare residual stresses in diamond films grown under different conditions. Microcrystallinity and lattice strain cause the Raman line to broaden and shift. Depending on the direction of the Raman shift, one can even determine the compressive or tensile nature of the stresses. For example, Englert et al. [10] determined the built-in stress in silicon-on-sapphire (SOS) by evaluating Raman spectra like those shown in Figure 18.9.

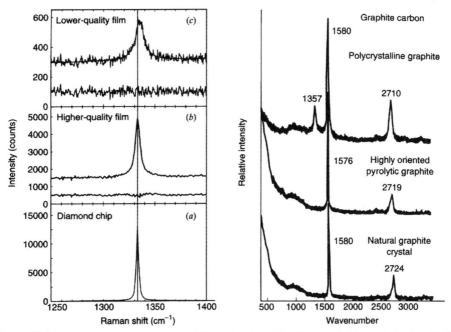

Figure 18.8 Comparison of Raman spectra of diamond and graphite. (*Reprinted with permission from* (1) [8]. *Copyright* 1991 *American Physical Society and* (2) [7]. *Copyright* 1989 *Materials Research Society.*)

18.3.4.3 Defects and Structural Disorder

Raman spectra of materials having lattice defects (i.e., structural disorders) exhibit pronounced line broadening compared to ordered structures. Hang et al. [11] studied the Raman spectra of GaAs that was damaged by polishing for 60 min with 0.3-μm grit. The results are shown in Figure 18.10. Consider the spectra shown in the figure for (100) GaAs.

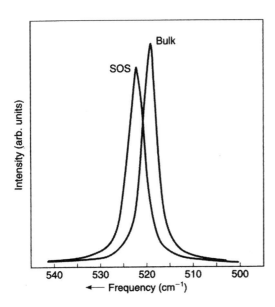

Figure 18.9 Raman spectra showing shift due to built-in stress in silicon film. (*Reprinted with permission from* [10]. *Copyright* 1980 *Elsevier Science.*)

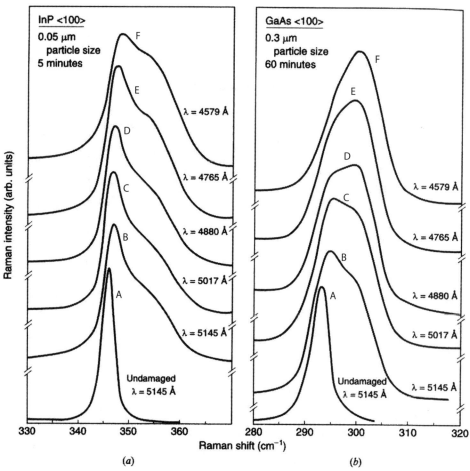

Figure 18.10 Raman scattering from undamaged and damaged materials. (*Reprinted with permission from* [11]. *Copyright* 1988 *American Institute of Physics.*)

Spectrum A was taken of the undamaged material using a probe beam wavelength of 514.5 nm, while spectra B through F were taken of damaged samples using various wavelengths. The broad shoulder in spectrum B reflects the damage caused by the polishing process. Because the shorter wavelength incident beams penetrate further into the sample, by using different wavelengths, depth-dependent information can be obtained. From the figure, it is clear that the damage caused by polishing decreases with distance from the surface.

18.3.4.4 High Spatial Resolution with Raman Microscopy

Raman spectra of 2-μm sized spots are possible by adapting an optical microscope with objective lenses through which the laser beam passes, and scattering is detected with a highly sensitive CCD detector. Figure 18.11 illustrates the use of Raman microscopy to characterize 140-μm diameter silicon carbide fibers that had been fabricated by CVD of silicon- and carbon-containing compounds on 30-μm diameter core carbon fibers. The microstructure of the deposited layers is highly dependent upon the CVD conditions (pressure, supply rate, and composition of the gases, as well as temperature). Figure 18.11*a* shows optical micrographs

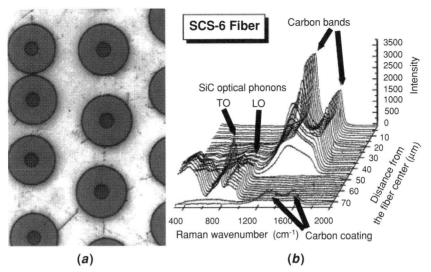

Figure 18.11 Characterization of CVD-deposited SCS-6 silicon carbide (SiC) fibers: (*a*) optical micrograph of a polished cross section of fibers in a metal matrix, and (*b*) Raman microscopy maps at 2-μm increments from the center of one carbon fiber toward its surface. Assignments of peaks are shown in figure; TO = transverse optic phonon; LO = longitudinal optic phonon of silicon carbide polytypes. (*Reprinted with permission from* [105]. *Copyright* 2001 *Kluwar Academic Publishers.*)

of a polished cross section of a number of fibers that had been embedded in a metal matrix. The core can be distinguished from three surrounding layers. The one adjacent to the core is 1 to 2 μm thick and the external one is ∼3 μm thick. Spectra taken by Raman microscopy at 2-μm intervals from the center of one of the fibers to its surface reveal the composition of each layer (Figure 18.11*b*). Three main regions show distinctively different Raman spectra. The first, 15 to 20 μm from fiber center, exhibits strong carbon bands with narrow widths, indicative of a well-ordered, graphitic carbon, crystallite core. The positions and widths of the peaks in the spectra of the next 25 μm suggest that both carbon (less ordered) and SiC are present in small crystallite structures. Spectra from the outer 25 μm exhibits only SiC peaks that are initially sharp and narrow, indicating stoichiometric quantitites of Si and C with larger crystallite structure, and peaks broaden toward the surface due to diminishing crystallite size. Broad carbon bands reappear at the surface of the fiber, indicating a thin disordered carbon film.

18.3.4.5 Enhancement of Raman Spectroscopy

Two forms of enhancing the signal from Raman scattering have been employed. One is resonance Raman spectroscopy, where the incident laser energy is near that of an absorption peak of an electronic transition state of the sample. Enhancements of 10^2 to 10^6 are possible. This approach requires a tunable laser. The major disadvantage of resonance Raman is interference with fluorescence.

Surface-enhanced Raman spectroscopy (SERS) dramatically improves the sensitivity of a surface analysis. SERS involves collecting spectra from molecules or films adsorbed to the surface of colloidal or roughened metals [6]. These spectra show enhancement factors as great as 10^{15} [12, 13]. The large enhancement originates from excitation at the metal surface and resonance effects [14].

18.4 SCANNING PROBE MICROSCOPY

Scanning probe microscopy (SPM) includes many different types of high-resolution instruments that are used to measure surface topography and other surface properties. All SPM instruments have a sharp probe that scans at a very close distance over the surface of the sample. A very important characteristic of this class of techniques is the ability to produce images from samples in many environments (ambient air, vacuum, or liquid) [15]. This section focuses on two kinds of SPM techniques, scanning tunneling microscopy (STM) and the scanning force microscopy (SFM) known as atomic force microscopy (AFM).

18.4.1 Summary

- What SPM does
  - STM—Monitors the tunneling current to determine distance between sample and tip primarily of conducting solids.
  - SFM—Uses attractive and repulsive forces between atoms of sample and atoms of tip. Samples can be either insulating or conducting solids.
- Types of analyses
  - Three-dimensional images of physical and chemical phenomena
  - Three-dimensional images of topography
- Performance
  - STM—Atomic-level resolution
  - SFM—Atomic-level resolution for depth and 1 nm lateral resolution

18.4.2 STM Principles and Instrumentation

In STM, a quantum mechanical tunneling current is produced by a bias voltage applied between the sample and a conducting tip (usually metallic) located 50 to 100 Å away from the sample. The sample is attached to a piezoelectric transducer to facilitate scanning of the tip across it (Figure 18.12). Feedback control maintains a constant distance between the probe tip and the sample. There are many possible materials used in the fabrication of probe tips. These include tungsten, sharpened electrochemically to atomic dimensions, and Pt–Ir, made by stretching a wire to an extremely small diameter. The Pt–Ir probe is more resistive to oxidation and more suitable for imaging large structures. Other materials include silicon [16], GaAs [17], and Pt–Cr [18].

In STM, images are produced from the dependence of the magnitude of the tunneling current on the mean distance between the tip and the surface under investigation. Mathematically, this relationship is of exponential form. The precision with which the image is reproduced depends on the sensitivity of the piezoelectric scanner to current variation across the surface. Tunneling current also depends on the local density of states, and this feature is helpful in performing spectroscopic analyses of the materials in terms of their electronic properties.

In terms of spatial resolution, STM has atomic-level resolution in the lateral direction, as well as in height. In fact, STM is by far the technique of choice for surface roughness measurements. The technique is truly nondestructive in nature since the probing tip is kept a few angstroms away from the sample surface Although STM images have recently been

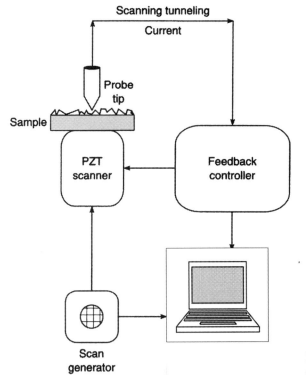

Figure 18.12 Major components of a scanning tunneling microscope.

collected for insulators [19,20], the analysis is much easier with electrically conductive samples.

18.4.3 SFM Principles and Instrumentation

SFM takes advantage of the attractive and repulsive van der Waals forces between atoms of the probe tip and the surface to produce an image, rather than utilizing a tunneling current. A common form of SFM is atomic force microscopy (AFM). In SFM, a sharp tip is constructed on a flexible cantilever beam using semiconductor mass production methods and etching of silicon, silicon oxide, or silicon nitride. [Focused-ion beam (FIB) milling can be used to fabricate very thin cantilevers (<1 µm) [20, 21].] A tip radius of about 400 Å is used to allow for high lateral resolution. The tip is scanned over a stationary sample, or the sample is scanned under a stationary tip. A piezoelectric scanner is responsible for correcting the distance between the tip and sample to maintain a constant force. The most common way to measure tip movement, is to shine a beam of light from a laser off a reflective coating on the backside of the cantilever to a position-sensitive photodiode (PSPD). The minute deflections of the tip result in large deflections in the position of the laser beam on the PSPD. The output of the PSPD is the input to a feedback loop that controls the piezoelectric scanner (Figure 18.13) [15]. A lateral resolution of about 5 nm is achieved with SFM and, if an extremely sharp tip scans an extremely flat sample, atomic-level lateral resolution can be achieved.

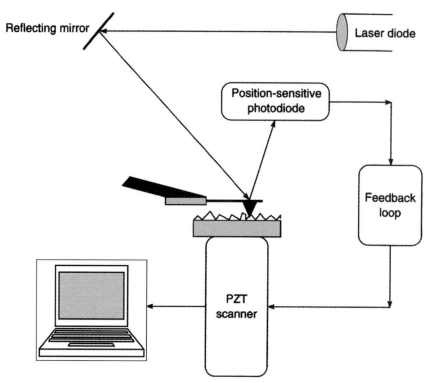

Figure 18.13 Major components of a scanning force microscope.

18.4.4 Practical Considerations and Applications

One of the important applications of SFM is in the characterization of surfaces of materials and thin films. Characterizations of interest include defect detection, measurement of surface roughness, and determination of crystallite size. These characteristics of a sample can be correlated with film deposition parameters. A typical SFM output showing surface roughness is given in Figure 18.14. In addition to an image of the scanned area, statistics are provided for the scanned area, as well as for a smaller specified area within the image area [22]. In this case, a diamond film was deposited onto a WC–Co substrate in a hot filament chemical vapor deposition (CVD) reactor using a gas mixture of 1% methane in hydrogen.

SFM can also be used to examine c-axis grains of YBCO superconducting films with thicknesses as small as 200 Å. Another application of SFM is in the analysis of thin-film magnetic disk read/write head poles that are typically recessed by approximately 300 Å below their surrounding surface. Surface scratches on such films due to polishing are easily detected. SFM can also be used on materials that are extremely soft, such as polyethylene and biological samples.

STM is very useful in a number of different applications ranging from spectroscopic evaluation of Si (111) 7 × 7 reconstruction to a roughness measurement of metallic surfaces. In the spectroscopic mode, one can increase the bias voltage between the tip and the sample and measure the corresponding change in tunneling current. From these data, a dI/dV versus V plot at a given site on the sample surface can be made. This experiment provides information on the local density of states of a sample.

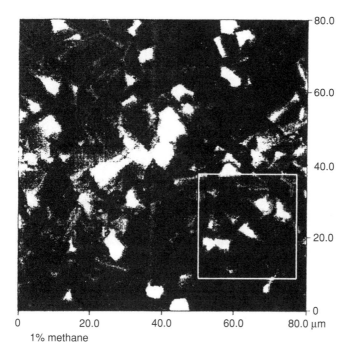

Figure 18.14 SFM roughness analysis for a diamond-coated WC-Co tool insert deposited with 1% methane (*Reprinted with permission from* [22].)

Figure 18.15 shows an example of an STM image with atomic resolution of an annealed material of Pd islands deposited onto a TiO_2 single crystal (Ti^{4+} form present) that is of interest in catalysis. In catalysis, understanding the structure and composition of the *surface* of the catalyst material is important. X-ray photoelectron spectroscopy was used to help determine the composition of the surface. Annealing produced an oxygen monolayer with a hexagonal arrangement between the Pd islands and reduced Ti^{2+} and Ti^{3+} species that terminate the surface. The bright spots in Figure 18.15a correspond to titanium atoms located in the higher, twofold bridge sites or nearly atop sites over oxygen atoms, and the darker spots correspond to titanium atoms located in the lower, threefold hollow sites of the oxygen atoms. There is no titanium atom in the middle of the pinwheel because that is an energetically unfavorable position exactly atop of an underlying oxygen atom.

744 Chapter 18 Analytical Techniques for Materials Characterization

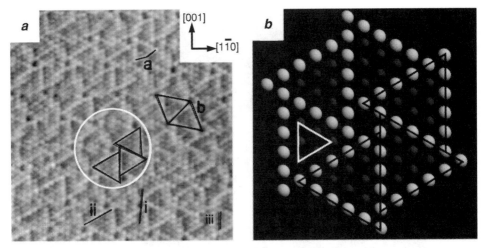

Figure 18.15 (*a*) An atomic resolution STM image (105 Å × 105 Å) of a pinwheel structure on a surface of a material consisting of ~14 monolayers of Pd deposited onto the (110) face of a single crystal of TiO_2, followed by annealing in ultrahigh vacuum at 573 K for 25 min, 773 K for 20 min, and 973 K for 20 min. The circle marks out a pinwheel, the triangles indicate parts of the pinwheel, and the lines *a* and *b* show a bend and the unit cell for the superstructure, respectively. (*b*) Schematic of idealized location of atoms in the pinwheel. (*Reprinted with permission from* [106]. *Copyright* 2002 *the Americal Chemical Society.*)

In addition to STM and SFM, there are many other SPM techniques that provide further characterization of samples. Scanning electrochemical microscopy (SECM) uses an ultramicroelectrode to perform and evaluate spatially resolved electrochemistry, which is of interest from corrosion to biological samples. The actions of electroactive species between the tip and the sample can also provide information about sample topography and conductivity [23]. Magnetic force microscopy (MFM) can image the magnetic properties of samples. Scanning thermal microscopy detects spatial variations in temperature, thermal conductivity, or thermal diffusivity of a surface [20]. In addition to characterizing surfaces, SPM has also been used to pattern materials on the nanometer and atomic scale by applying various voltages to a conductive tip [24–28].

18.5 SCANNING ELECTRON MICROSCOPY AND ENERGY DISPERSIVE X-RAY SPECTROSCOPY

The single most important feature of SEM, compared to optical microscopic methods, is the great depth of field that is possible. In addition to revealing three-dimensional morphological characteristics of a surface, an energy-dispersive spectrometer (EDX or EDS) attachment can provide information about the chemical composition of a specimen.

18.5.1 Summary

- What SEM and EDX do
    - SEM—Secondary and backscattered electrons are detected from a sample rastered with a focused primary electron beam, resulting in a surface image with a great depth of field.

- EDX—The energy of X-rays emitted from a sample that is irradiated with the primary electron beam lead to elemental analysis and mapping (for $Z > 4$, beryllium)
- Types of analysis
  - Particle size and distribution analysis
  - Thickness and feature size measurements
  - Morphology
  - Elemental analysis: spot, line, and two-dimensional images
- Performance
  - SEM
    - $500,000\times$ magnification
    - Lateral resolution of 1 to 50 nm
    - Depth of analysis: few nanometers to few micrometers
  - EDX
  - Lateral resolution of ~ 0.5 μm
  - Limit of detection $\sim 0.1\%$ weight
  - Sample depth 0.02 μm

18.5.2 Basic Principles

When an electron beam strikes the surface of a sample, a number of processes occur (Figure 18.16). The interaction of the electron beam with a specimen can produce secondary electrons, backscattered electrons, Auger electrons, X rays, and even visible light. Depending on the selection of the detector, one may obtain images containing different information about the sample. SEM analysis involves detection of secondary and backscattered electrons and EDX detects the X-ray emission.

Backscattered electrons (BSE) exit the specimen at random angles and with the same energy of the primary beam (typically 20 keV), after the incident electrons undergo elastic collisions with the nuclei of atoms in the sample. The area over which backscattered electrons emerge from the sample can be several microns and is a factor in limiting spatial resolution in SEM of BSE. The backscattered electron yield is higher for elements with higher atomic numbers. The dependence of the backscattering phenomenon on atomic number, Z, can produce an image contrast between low Z elements and high Z elements.

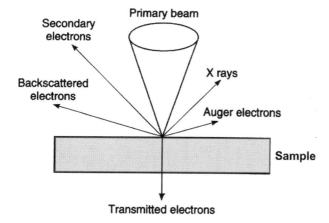

Figure 18.16 Interactions of a sample with a primary beam of electrons in a scanning electron microscope.

Secondary electrons (SE) are produced by inelastic collisions of electrons of several kiloelectron volts with those of the atoms in a sample under investigation. In this process, some of the energy of the electrons in the impinging beam is transferred to the electrons of the sample atoms. If the energy transferred is sufficiently high to overcome the work function of the materials, electrons of a few electron volts exit the sample from a region slightly larger in area than the incident beam and can be detected for high-resolution imaging purposes. Secondary electrons from atoms deeper than 50 to 500 Å suffer additional inelastic collisions and never reach the detector. The yield of secondary electrons decreases slowly with increasing beam energy after reaching a peak at around 1 keV and shows very little variation over the full range of elemental atomic numbers [29].

In EDX, X rays emitted from the sample are used for chemical analysis and elemental mapping. The production of X rays results from transfer of energy from primary beam electrons to sample atoms resulting in the ejection of inner-shell electrons. The excited atom returns to its ground state by emitting an X-ray photon or Auger electron of an energy that is characteristic of the element. Information about the spatial distribution of elements in a sample can be acquired with a spatial resolution of about 0.5 μm. X rays produced by atoms as deep as 1 to 2 μm beneath the surface can be detected.

18.5.3 Instrumentation

An SEM instrument has an electron beam source, optics that focus the beam, and a detector for electrons that emerge from the sample surface. The EDX instrumentation is an attachment to the SEM, which is simply a detector for X rays emitted from the sample. SEM and EDX analyses take place with the sample in a chamber that is under vacuum.

18.5.3.1 SEM Instrumentation

A general schematic of an SEM instrument is illustrated in Figure 18.17. Typical instruments maintain a vacuum of 10^{-7} torr, which is necessary to minimize interactions between gas-phase particles and the electron beam, emitted electrons, and radiation emitted from the sample. It is also sufficient to minimize contamination of the sample. More recently, environmental SEM instruments have been made available, which operate at higher pressures (a few torr), to accommodate wet samples and diminish charging of insulating samples, but with degraded spatial resolution. Typical pumps for SEM include a roughing pump (mechanical) to provide the initial vacuum to the sample chamber and a liquid-nitrogen-trapped diffusion pump or a turbomolecular pump to achieve higher vacuum. An electron gun, which has a V-shaped filament, usually made of tungsten (often coated with thoria or lanthanum hexaboride, LaB_6, to lower the work function and, thereby, decrease the operating temperature) emits a beam of electrons (primary beam) that is accelerated by a voltage in the range of 1 to 30 keV, and directed through a column containing a condenser lens, stigmator, objective lens, and scanning coils in sequential order. The lenses form a fine electron beam (5 to 200 nm) that is focused onto the sample in a specimen chamber at the opposite end of the column. The purpose of the scanning coils is to allow the primary electron beam to scan across a very small area of the specimen surface. Either secondary or backscattered electrons are collected using a proper detector and then used to form an image. Secondary electrons can be selectively removed from the total electrons that reach the detector by applying a slight negative bias to the transducer housing. The image is formed digitally and is displayed on a computer monitor.

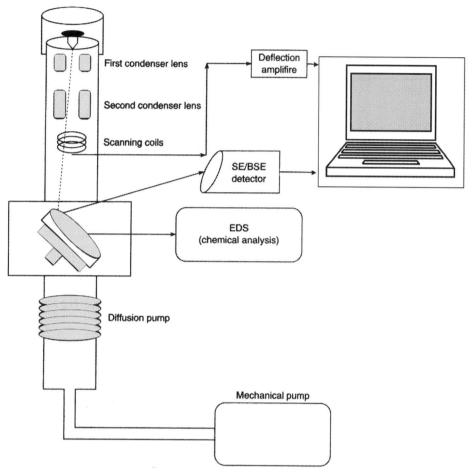

Figure 18.17 Schematic that illustrates components of a scanning electron microscope. (Types of pumps will vary.)

18.5.3.2 EDX Instrumentation

The main components of an EDX system, which is usually an integral part of a scanning electron microscope, consists of a semiconductor transducer, such as a Si(Li) (or lithium-drifted silicon) detector, a field-effect transistor, an amplifier, a multichannel analyzer, and aquisition and display electronics. Figure 18.18 shows a schematic of an EDX system. The heart of the instrumentation is the solid-state Si(Li) detector. Basically, an incoming X-ray photon is absorbed, a highly energetic photoelectron that is produced loses its kinetic energy through the elevation of several thousand electrons in the silicon to the conduction band, thereby increasing the conductivity. In short, the absorbed photon is converted into a current pulse that is proportional to the energy of the absorbed photon. The pulse is converted to a voltage, amplified by a field-effect transistor, separated according to its energy in a multichannel analyzer, and stored in the proper channel. This information is used to form a spectrum and is displayed.

Almost all elements with $Z > 4$ (beryllium), if they are present in sufficient quantity, can be detected using modern EDX instruments that are either windowless or possess ultrathin windows (diamond). In older instruments, a beryllium window before the detector

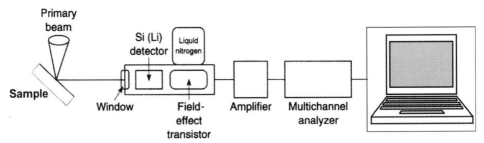

Figure 18.18 Schematic diagram of an energy dispersive X-ray spectrometer.

limits analysis to elements with $Z \geq 11$ (sodium), because the window prevents the low-energy X rays from the lighter elements to reach the detector. This limitation can be crucial because it prevents analysis for some of the most common elements found in materials: carbon, oxygen, and nitrogen.

18.5.4 Practical Considerations and Applications

Instrumentation and sample preparation for SEM and EDX are complicated by the need to perform the analysis with a charged particle beam of high energy and the presence of a vacuum. These issues and applications for SEM and EDX are considered below.

18.5.4.1 SEM Sample Handling and Preparation

When samples are analyzed in traditional SEM instruments with high vacuum systems, they must not contain, be contaminated with, or be attached to the sample holder with materials that can outgas extensively and degrade the vacuum. These include volatile oils from fingerprints and adhesives from tape on samples or sample holders. In addition, samples easily charge when irradiated by an electron beam and, therefore, must be grounded to draw away the charge. Insulating samples may be first coated with very thin carbon or metal films that do not obstruct the features of the sample. Environmental SEM that operates at higher pressures minimizes the need for such special sample handling and conductive coatings.

18.5.4.2 EDX Accelerating Voltage Selection

Voltage selection in EDX is critical and depends greatly on the type of sample under investigation. High voltages lead to less intense characteristic peaks for light elements due to absorption by the heavier elements of the X rays produced by the light elements. A typical example of this occurs in the analysis of carbon in steel. As the accelerating voltage is increased, the characteristic peak for carbon in the EDX spectrum decreases in magnitude. Measurements for light elements are best when the electron beam voltage is low and take-off angle for X-ray measurement is high (i.e., small path length through the sample to minimize absorption). In thin-film analysis, the accelerating voltage should be very close to the critical ionization potential of the line of interest. For bulk sample analysis, one should select as high a potential (25 to 30 keV) as possible in order to maximize the number of X-ray signals produced.

18.5.4.3 Applications

The advancement of electronics technology, to a great extent, depends on, not only the proper selection of bulk and thin-film materials but also the study of their failures in order to correct them, and thus, avoid them in the future. For example, the surface roughness of is an important consideration when selecting a substrate. SEM offers a qualitative description of rough surfaces. It is also useful in observing corrosion of metallization systems and in examining substrates during processing to determine if the various processing steps are yielding linewidths and spacings as specified by the design. The SEM-EDX system is successfully used in the analysis of Pb–Sn eutectic solder that is commonly used as the attachment solder in electronic assembly. Other alloys, such as Pb–In, are also used to provide improved thermal fatigue characteristics of solder bumps for flip-chip mounting [30]. The quantitative chemical analysis capability, along with the fine probe size of SEM-EDX, makes it the method of choice for analyzing such alloys. In most electronics technologies, aluminum wire bonding is used to provide electrical interconnection between integrated circuits (ICs) and the substrate. As the functionality of ICs increases, the input/output (I/O) requirement also increases. SEM, with its excellent depth of field, can be employed to analyze bond mesh, spacing, and placement.

Figure 18.19 shows an example of an SEM image with the corresponding EDX spectrum for Ag_2Se nanowires of 32 nm diameter [31]. Ag_2Se in its different phases possesses a number of properties (e.g., narrow bandgap semiconductor, superionic conductor, high Seeback coefficient, and low lattice thermal conductivity) that make it of interest for thermoelectric applications, photochargeable secondary batteries, and photosensitization of thermochromic materials. Template-directed synthesis produced the one-dimensional materials that promise enhancement of the properties and new applications. SEM easily allows these nanowires to be visualized. EDX analysis is limited to about a 0.5-μm spot, larger than an individual wire, and thus a strong copper signal from the Cu grid, which was used to support the wires, appears in the spectrum. However, quantitative analysis of the silver and selenium EDX peaks indicate the desired 2 to 1 stoichiometric ratio for Ag_2Se.

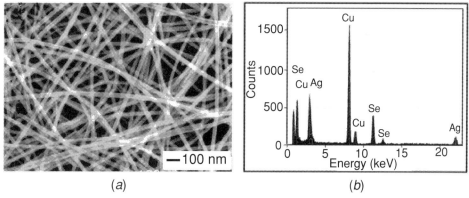

Figure 18.19 (a) SEM image and (b) corresponding EDX spectrum of Ag_2Se nanowires on a copper support grid. (*Reprinted with permission from* [31]. *Copyright* 2001 *American Chemical Society.*)

18.6 CONFOCAL MICROSCOPY

Confocal microscopy is a three-dimensional (3D) imaging technique in which light is focused on a very small section of the sample, and the reflected light is focused through a pinhole before reaching the detector [32]. Its ease of use and nondestructive nature make it an attractive technique for collecting high-quality 3D images. Although the spatial resolution is not as high as in SEM and scanning tunneling or force microscopies (STM or SFM), confocal microscopy offers advantages to those techniques. In contrast to SEM, it does not require a vacuum or low pressure. Unlike STM and SFM, it does not need samples that are fairly flat nor does it have the added complexity of physically scanning a minute probe tip, which is difficult to reproducibly construct, across a sample surface.

18.6.1 Summary

- What confocal microscopy does
  - Focused beam of light is scanned across sample. Reflected light is focused through a pinhole onto the detector.
- Types of analyses
  - Nondestructive, noncontact topography measurements of surfaces
  - Three-dimensional imaging
  - x-z imaging
- Performance
  - z resolution down to 5 nm
  - Lateral resolution down to 0.2 µm

18.6.2 Basic Principles

Confocal microscopy provides several advantages over traditional optical microscopy: depth discrimination, improved resolution, reduced blurring from scattered light, and improved signal-to-noise ratio [33, 34]. Analysis of surfaces by confocal microscopy is accomplished by spatial filtering of unfocused light. A schematic diagram of the instrument is shown in Figure 18.20 [32]. Light is focused onto the sample surface through a lens. The reflected light passes through the lens again, and a pinhole is placed in front of the detector to allow only focused light to reach the detector. In most cases, the optics are stationary, and the stage is moved (by computer control) to scan the focused beam of light over the sample.

18.6.3 Instrumentation

Confocal microscopy analysis can be performed at atmospheric pressure, greatly simplifying the instrumentation. The main parts of the microscope are the light source, the lens, and the detector. The light source can be a laser or white light (such as from xenon or halogen lamps) [35]. The objective lenses used are the same as those used in conventional optical microscopes [36]. Photomultiplier tubes (PMT) and charge-coupled device (CCD) cameras are the most commonly used detectors [32, 37]. A computer is used to facilitate building the 3D image from data collected by the detector.

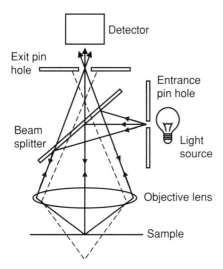

Figure 18.20 Schematic diagram of a confocal microscope. Solid lines represent focused light, and dashed lines represent out-of-focus light [32].

Some confocal instruments contain a Nipkow disk between the beam splitter and the objective lens. The opaque disk contains a series of rectangular holes, positioned in a spiral pattern around the center of the disk. When the disk rotates, the holes scan every part of the sample in a raster pattern. Thus, the image can be created by monitoring the brightness of each image element (the light through each individual hole) with the detector [33].

There are many different imaging modes possible with confocal microscopy. Z-series imaging involves collecting a series of images at different levels on the z axis. Changes in fine focus (performed by a computer-controlled stepping motor) are coordinated with image acquisition at each step. A computer is used to convert the z-series images into a 3D image. Another mode is x-z imaging, which produces an image of a vertical slice of the sample. This analysis is conducted by performing a line scan across the x axis at varying z-axis depths. Most confocal imaging is performed in reflection mode, but, for transparent samples, transmission imaging is possible [38].

18.6.4 Practical Considerations and Applications

Confocal microscopy has a very wide range of applications, imaging samples as different as semiconducting devices and biological cells. Its packaging applications include monitoring warpage, lead coplanarity, laser markings, film thicknesses, and surface and contact roughness [39].

An example of determining the thickness of a layer of material is shown in Figure 18.21. Poly(dimethylsiloxane), PDMS, was spin-coated onto a glass substrate and cut with a razor blade to reveal an edge of 22 µm. Such images were used to analyze the dependence of layer thickness on spin rate. The figure also reveals details of the roughness of the edge of the PDMS layer that resulted from the cutting. PDMS is of interest in interfacing microfabricated devices to fluids because it can be easily molded and forms good seals with oxidized silicon and with itself.

Figure 18.22 shows a confocal image of a 50-µm diameter microcavity, formed by reactive ion etching, through two sets of alternating layers of gold (~1000 Å with a 10-Å Cr adhesion layer) and polyimide (4 µm thick), with a total depth of 8 µm. Devices similar

752 Chapter 18 Analytical Techniques for Materials Characterization

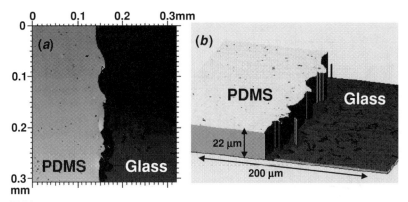

Figure 18.21 Confocal microscope image of PDMS layer that was spin coated on glass substrate at 4000 rpm, showing roughness of the PDMS edge and demonstrating topographical imaging capability of confocal microscopy.

to these have been used for chemical assays and sensors [40–43]. The metal layers are thinner than the spatial resolution of the technique and cannot be distinguished from the polyimide layers. The image provides dimensions of the microcavity and shows misalignment of the layers that occurred during a modified version of the fabrication procedure. It also provides information about the smoothness of the cavity edges.

18.7 AUGER ELECTRON SPECTROSCOPY

Auger electron spectroscopy (AES) provides elemental and chemical state analysis of the topmost 5 to 100 Å of the surface of a solid, with absolute detectability as low as 100 ppm for most elements [44, 45]. AES is one of three complementary ultrahigh vacuum (UHV) surface characterization techniques that are described in this chapter. X-ray photoelectron spectroscopy (XPS) and secondary ion mass spectroscopy (SIMS) are the other two methods and are presented in Sections 18.8 and 18.9, respectively.

18.7.1 Summary

- What AES does
  - Ultrahigh vacuum (UHV) surface analytical technique.
  - Primary electron beam ultimately results in ejection from the sample of Auger electrons whose energies provide elemental and chemical information.

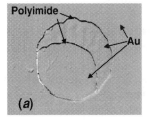
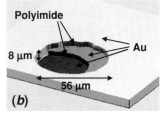

Figure 18.22 Confocal microscope images of microcavity fabricated into alternating layers of gold and polyimide: (*a*) top-down view showing misalignment of layers and (*b*) a three-dimensional image [107].

- Types of analyses
  - Point or averaged area
  - Line scan
  - Depth profiling; inert gas sputtering or bevel sectioning
  - Imaging; scanning Auger microscopy (SAM)
- Performance
  - Spatial resolution down to 4 nm but typically ~10 nm
  - Limit of detection down to 0.05 atomic percent but typically 0.1 atomic percent
  - Depth of analysis 5 to 75 Å

18.7.2 Basic Principles

Analysis of surfaces by AES is based on the phenomenon that occurs when an incident primary electron beam bombards a surface and produces a series of electron cascade events that ejects an Auger electron, which is detected. Analyses have been performed on a wide range of solid samples, including insulators, conductors, and superconductors. Advances in AES have led to scanning Auger microscopy (SAM), which can map the elemental composition of surfaces with a spatial resolution of less than 30 nm [44, 46–48]. Another newer adaptation, angle-resolved AES (ARAES), provides both distribution and structural information because it relates the direction in which Auger electrons are ejected from the surface to the layered or crystal structure of the sample [44, 45, 47, 48].

An Auger electron is one possible product of the fundamental processes that are involved when an incident electron beam (primary electrons) interacts with the surface of a sample, as illustrated in Figure 18.16. The basic concepts of atomic structure must be learned to understand the processes involved in AES [49]. Some of these concepts are also relevant to the XPS process (Section 18.8).

18.7.2.1 Atomic Theory

The electrons surrounding the nucleus of each atom are differentiated according to a set of quantum numbers $n, l,$ and s. The first is sometimes called a shell and is related to the average distance of electrons in that shell from the nucleus, where $n = 1, 2, 3, 4,$ and so forth. On average, it takes more energy to remove an electron from a shell close to the nucleus ($n = 1$) than far from the nucleus ($n = 4$). Likewise, the binding energy of an electron that is closer to the nucleus is greater than one farther away. The second quantum number is the orbital angular momentum, l, which has values from 0 to $n - 1$. It is also common to denote l values with the letters s, p, d, f, respectively. For a given n, lower l values correspond to higher binding energies. The third quantum number is the electron spin momentum, s, which can have values of $+\frac{1}{2}$ or $-\frac{1}{2}$. Sometimes it is more convenient to use total electronic angular momentum, j, which is the vectorial summation of l and s. Each element has an electronic configuration that can be described with these quantum numbers, as shown in Table 18.2. Electrons involved in transitions in AES and XPS can be designated with this notation [44].

In AES, it is common practice to designate principal electron energy levels with the letters $K, L, M,$ and N instead of principal quantum numbers 1, 2, 3, and 4, respectively. Subscripts of the letters, as the 1 in L_1 and the 3 in M_3, further designate the total

Table 18.2 AES and XPS Notation for Electron Energy Levels

Quantum numbers			Subscript for AES	Electron designation in AES	Photoelectron designation in XPS
n	l	j			
1	0	1/2	1	K	$1s_{1/2}$
2	0	1/2	1	L_1	$2s_{1/2}$
2	1	1/2	2	L_2	$2p_{1/2}$
2	1	3/2	3	L_3	$2p_{3/2}$
3	0	1/2	1	M_1	$3s_{1/2}$
3	1	1/2	2	M_2	$3p_{1/2}$
3	1	3/2	3	M_3	$3p_{3/2}$
3	2	3/2	4	M_4	$3d_{3/2}$
3	2	5/2	5	M_5	$3d_{5/2}$
...	...	...	...	...	...

electronic angular momentum, j, which is the vectorial summation of the orbital angular, $l = 0, 1, 2, 3 \ldots$ and spin momenta, $s = \pm\frac{1}{2}$. Table 18.2. summarizes the notation [44].

18.7.2.2 Auger Process

Figure 18.23 shows a schematic of the AES process for an atom in solid silicon; it involves three different electrons. Energy from an incident electron beam is transferred to the first electron, which lies in one of the core electron shells. If the transferred energy is high enough, it will cause that electron to be ejected, and the atom will become ionized. A second electron, which resides in an outer-lying shell (higher potential energy and lower binding energy), drops into the hole left behind. Excess energy that is produced from this electron transition, equal to the difference in binding energies for the two levels, is released in one of two competing processes: Either an X-ray photon is produced (this is the basis of EDX) or the energy is transferred to a third electron, the Auger electron that also resides in an outer-lying

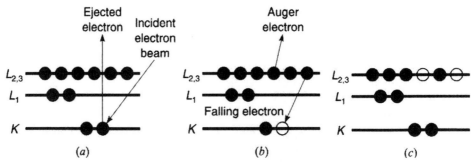

Figure 18.23 Schematic diagram of the emission process of an atom that produces the Auger electron $KL_{2,3}L_{2,3}$. (a) An incident electron has sufficient energy, E_p, to create a hole in the K level by ejecting a K electron, which has a binding energy E_{b1}. For this process to occur, $E_p \gg E_{b1}$. (b) The vacancy in the K level is filled by an electron that falls from the $L_{2,3}$ level, which has a lower binding energy, E_{b2}. The excess energy, $E_{b1} - E_{b2}$, that results from this process is released in the form of either a photon or ejection of an Auger electron with a kinetic energy of $E = E_{b1} - E_{b2} - E_{b3}^*$. The asterisk indicates the binding energy of the Auger electron for an atom in the presence of a hole in $L_{2,3}$. (c) The final state is doubly ionized.

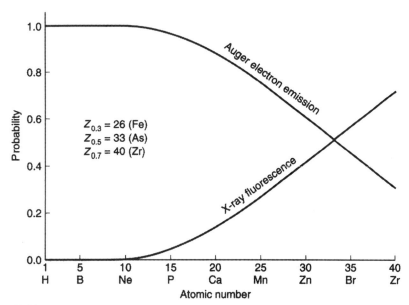

Figure 18.24 Relative probabilities for emission of Auger electrons and X-ray photons (the EDX process) for relaxation of an atom in which a hole is created in the K level. (*Reprinted with permission from* [44]. *Copyright 1990 John Wiley & Sons, Ltd.*)

orbital, and is ejected. Figure 18.24 shows the relative probabilities of relaxation by these two processes for K shell ionization, where the Auger process dominates for elements of low atomic number up to As ($Z = 33$) [44, 50]. For L and M initial vacancies, the Auger process dominates for all elements. Ejection of an Auger electron leaves the atom in a final state that is doubly charged (Figure 18.23c). Each kind of Auger electron is named after the transitions that occur, such as KLL for the example in Figure 18.23. The kinetic energy, E, of the Auger electron is detected by the AES instrument, and is defined by the binding energy of the three electrons, E_{b1}, E_{b2}, and E_{b3}, and the hole–hole repulsion energy, U [50, 51].

$$E = E_{b1} - E_{b2} - E_{b3} - U \qquad (18.4)$$

Because the binding energies involved in the electron cascade are specific for different elements, the kinetic energy of the Auger electron can be used to identify the species on the surface. Note that U is only 0 to 10 eV and usually will not complicate the identification process. Also, with higher atomic number, more electron transitions, and thus, types of Auger electrons are possible. Figure 18.25 is one of many kinds of charts that illustrate the number and relative probabilities of Auger electron energies for different elements [52].

The number of Auger electrons of the same kinetic energy, E, that are detected is proportional to the intensity, $N(E)$, of a peak at that kinetic energy in the Auger spectrum. Figure 18.26 illustrates derivative [$dN(E)/dE$ versus E] and nonderivative [$N(E)$, or $EN(E)$, versus E] representations of Auger data for a contaminated Mo sample [53]. Note that C, O, and N are common in airborne contaminants. The nonderivative form has several advantages, including proportionality to the number of Auger electrons produced, better signal-to-noise ratio, higher resolution, and ease of evaluating peak shape and overlapping peaks [54].

A peak in a nonderivatized spectrum (Figure 18.26a) exhibits tailing, called the "loss tail," which occurs on its low-energy side and extends to zero, causing a large background

756 Chapter 18 Analytical Techniques for Materials Characterization

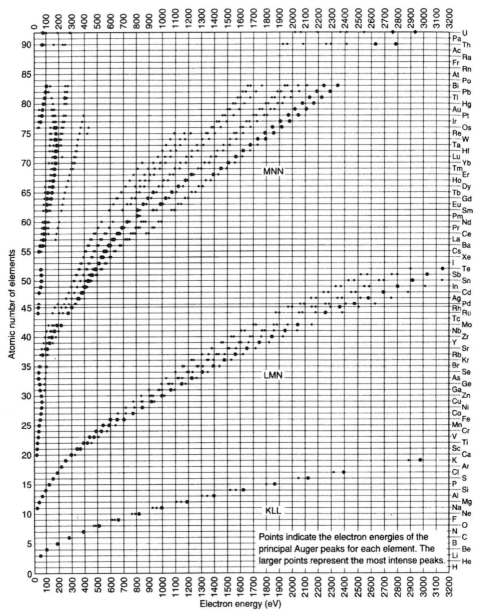

Figure 18.25 Auger electron energies for different elements. Each dot represents the electron energy for principal Auger peaks. Larger dots indicate more intense peaks. (*Reprinted with permission from* [49]. *Copyright* 1995 *Physical Electronics, Inc.*)

signal [50, 54]. This is due to Auger electrons that inelastically scatter within the sample before escaping, and thus, have lost some of their initial kinetic energy. The extent of inelastic scattering is a function of the material through which the electron travels and the electron energy. The average distance that the electrons travel before suffering an inelastic event is the mean free path, typically a few angstroms long. For example, 95% of the Si *LMM* Auger electrons from oxidized Si are produced in the top 15 Å of material [54]. This

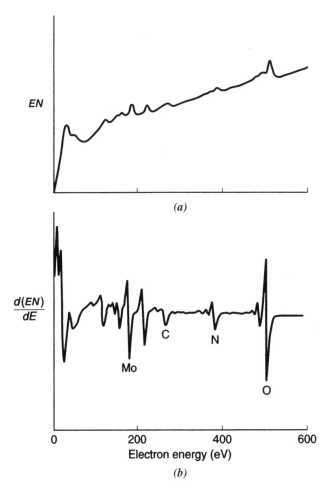

Figure 18.26 Standard representations of Auger spectra for a contaminated Mo surface: (a) $EN(E)$ and (b) $dEN(E)/dE$. (*Reprinted with permission from* [53]. *Copyright* 1989 *John Wiley & Sons, Ltd.*)

is why AES is a surface-specific analytical tool. Only those electrons that escape *without* inelastic scattering are useful for identifying elements in the sample.

18.7.3 Instrumentation

The basic AES setup includes an analysis chamber with UHV system, a higher pressure introductory chamber with vacuum system, a sample manipulator and transfer arm, an electron gun, an electron analyzer, an ion gun and differential pump, and computer control and data acquisition systems. There are a variety of other accessories for specialized applications, such as controlled-temperature sample stages and sample fracture or cleaving devices. The major instrument components are described further below.

18.7.3.1 Vacuum Systems

AES requires a UHV (10^{-9} to 10^{-10} Torr) system to maximize the mean free path so that electrons in the incident beam reach the sample and so that Auger electrons from the sample reach the detector. This low pressure can be accomplished with the following pumps, which are used in combination with roughing or turbo pumps, or both: diffusion pump, cryo pump (10^{-11} Torr), or ion pump (also used in conjunction with a titanium sublimation pump), preceded by a mechanical pump and turbomolecular pump.

18.7.3.2 Electron Beams

The incident electron beam is created using the same kinds of electron guns and focusing elements that are used in SEM analysis, described in Section 18.5.3. Thermionic sources are operated at 10 to 30 keV and field emission sources at 30 to 60 keV. Spot sizes can be as low as 10 nm and beam currents are normally from 0.1 to 10 nA [44, 51]. The electron beam may be rastered across the sample, which provides the capability to perform elemental line scans and two-dimensional elemental images of surface features.

18.7.3.3 Electron Energy Analyzers and Detectors

Auger electron detection usually involves an electrostatic electron energy analyzer and an electron detector. The principal detector for AES is the cylindrical mirror analyzer (CMA) [51], which is shown in Figure 18.27. It consists of two concentric cylinders with radii r_1 and r_2. Electrons enter from the source (the sample) and pass through a mesh-covered aperture in the inner cylinder, which is grounded. The angular spread of the electrons that pass through ultimately defines the energy resolution of the CMA. The electrons with an energy E_0, where $E_0 = (K \text{ eV})/\ln(r_2/r_1)$ and K is a characteristic constant, are deflected from the outer cylinder held at $-V$ and focused onto either an entrance slit to a second CMA analyzer on the same axis or to an electron detector [44]. When the potential on the outer cylinder of the CMA is scanned, the electrons with different energies are scanned onto the detector, directly giving the energy distribution of electrons. However, because the CMA transmission varies as E, the recorded distribution is not $N(E)$ but $EN(E)$, as in Figure 18.26. An advantage of the CMA is the very high solid angle of acceptance, which provides for greater transmission. Its disadvantage is the minimal working space around the sample.

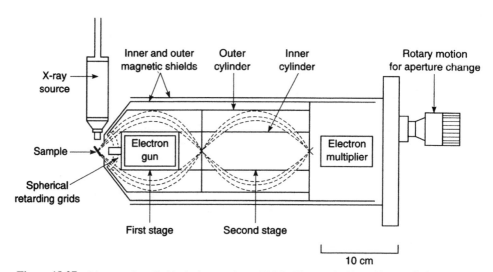

Figure 18.27 Diagram of a cylindrical mirror analyzer (CMA). The retard grids and inner cylinder are grounded for AES. (*Reprinted with permission from* (1)[55]; *copyright* 1974 *Elsevier Science; and* (2) [44]; *copyright* 1990 *John Wiley & Sons, Ltd.*)

Also, if the sample surface is moved slightly away from the focal point of the analyzer, it will cause a significant decrease in signal intensity and a shift in peak position. To eliminate this latter problem, a double-pass CMA is used as shown in Figure 18.27 [55, 56] in which the electron energy in the second stage is then defined by the entrance and exit slits and not by the sample position. The CMA can be operated in combination with channeltron, Cu–Be electron multiplier, and position-sensitive electron multiplier detectors.

18.7.3.4 Ion Guns

An AES instrument usually has an ion gun for sputter cleaning surfaces and for depth profiling. Samples that are contaminated with a variety of chemical species will produce surface analyses that are not representative of the sample's composition. Thus, samples are either cleaved in the UHV or cleaned with a short exposure to an ion beam. Extended exposure to an ion beam will cause extensive sputtering of the surface, and if alternated with AES analysis, can be used to obtain an elemental analysis as a function of sample depth. Ion beams of He^+, Ne^+, Ar^+, and Xe^+ are used, with Ar^+ and Xe^+ being the most common. Depending on how the ions are formed, the final beam energy lies between 0.5 and 5.0 keV, focuses to spot sizes of 0.5 mm to $\sim$10 μm with beam currents from 30 μA to 1 nA, and rasters over areas of 1 mm $\times$1 mm to 10 mm $\times$10 mm [44]. When an ion gun is in use, it is necessary to differentially pump the source volume, which is at $\sim 10^{-4}$ Torr, so that the chamber pressure remains near 2×10^{-7} Torr.

18.7.4 Practical Considerations and Applications

It is important to consider a number of different aspects regarding samples (handling, preparation, mounting, charging), data analysis and quantification, parameters and settings for data acquisition for a variety of samples, as well as features (two-dimensional imaging, depth analysis, and angle-resolved detection), and limitations of AES. Several AES applications that are relevant to the semiconductor and packaging industries are used as examples to illustrate some of these considerations.

18.7.4.1 Sample Handling

Generally, greater precautions must be taken for samples analyzed by AES than those for standard SEM (see SEM Section 18.5.4.1), primarily because of the UHV conditions and the high sensitivity of AES to surface contaminants. Descriptions of sample handling can be found in ASTM E 1078-85 [57]. AES samples are usually mounted at the edges with metal screws or under metal masks that are screwed into the holder (that minimize charging). The sample surface that is to be analyzed must be carefully protected so that contaminants or unwanted reactions with air are minimized. Air-sensitive samples can be transferred into the AES instrument via a glovebag, glovebox, or transfer flask (filled with a nonreactive gas such as argon or nitrogen) at the entrance port to the sample introductory chamber. Also, surfaces may be cleaned inside the UHV chamber by heating and/or bombardment with ions, or samples may be mechanically fractured (for bulk analysis) to expose a clean surface. Sputter cleaning is usually performed long enough for the Auger signal of the unwanted element to disappear, such as a carbon signal from inorganic samples. However, sputter cleaning suffers from the same complications as depth profiling, as described below.

18.7.4.2 Sample Charging

Both insulators and conductors may be analyzed by AES, but there are limitations due to charging of the surface. Because the Auger process causes emission of electrons from the sample, it can result in a positively charged surface, just as in XPS analysis. Because AES also involves an incident beam of electrons, the surface may also become negatively charged. Charging is a problem because it causes shifting of peaks in the spectrum (just as it does in XPS analysis; see Figure 18.34), and may include peak broadening or splitting, making identification of elements and chemical states difficult, if not impossible. Conducting samples or conducting parts of samples do not suffer charging problems when they are shorted to the sample holder, which is grounded. However, analysis of insulators and their charge compensation is much more complicated. The simpler approaches for minimizing sample charging in AES [44,48] include lowering the beam energy and beam current, tilting flat samples at grazing angles (30°) to the electron beam, surrounding the sample with a metal mask or foil that is grounded, and using a positive ion beam (e.g., He^+ or Ar^+) of very low energy.

18.7.4.3 Sample and Data Analysis

AES can be performed in four ways. One is the point (or area) mode in which the electron beam is rastered over a specified area of the sample while Auger electron energy is scanned by the analyzer. This results in a spectrum like that in Figure 18.26, which is an average over the sample area that is rastered. The second way is the depth profile, where an ion beam sputters away one atomic layer at a time with intermittent AES analyses in the point mode. The electron energy analyzer may either collect entire spectra for each time increment or measure only the signal at specified peak energies for different elements. The signal for an element is plotted as a function of time (and thus, depth, if the sputtering rate is known). The third mode is the line scan, which involves a series of point analyses along a line across the sample, while the electron energy analyzer monitors the signal at the peak energy of a specific element. The fourth type of analysis is the mapping mode, which is a series of line scans that provide an image of the spatial distribution of an element on the sample in two dimensions. Examples of these modes are shown throughout this section.

AES spectra can be evaluated qualitatively to determine the presence of different elements and chemical states and quantitatively to obtain percent composition. The number of peaks and their peak positions are determined by the identity of the atoms on the surface. The main ones are summarized in Figure 18.25. Multiple peaks can be useful for verifying the presence of an element, especially when one peak overlaps with that of a different element. The line shape and shift of the peaks in the Auger spectrum may be used to ascertain chemical information. Although chemical shifts in XPS can be smaller than those in AES, the AES lines are much broader, making it generally harder to distinguish chemical states. Also, AES line shape is sensitive to transitions where valence electrons are involved. For example, in sodium fluoride, the *KLL* transition for F is narrow. However, in polymers containing F, the valence electrons in $L_{2,3}$ are involved in bonding orbitals, which causes the *KLL* line to broaden [44].

Quantification in AES is largely based on comparisons to standards rather than basic, first-principles calculation of Auger peak intensities [58,59]. This is because numerous factors contribute to the intensity of an Auger peak and many are difficult to measure or compute. These include the density of the element, the probability of an Auger transition, and the ionization cross section, which are a function of beam voltage. Other factors are more instrumental in nature, such as the beam current, transmission efficiency of the analyzer, and electron detector efficiency. The latter two are also dependent on the kinetic energy of Auger electrons. Backscattering of secondary electrons within the sample is another important factor that creates more detected Auger electrons. This effect depends on the incident beam angle, voltage, and sample material.

Sensitivity factors for each element i, S_i, are used to account for the unknown parameters in the measurement. The Auger background-subtracted peak height (PH) for non-derivatized spectra, or peak-to-peak height (PPH), for derivatized spectra, of the element is divided by the sensitivity factor and then divided by the sum of the total signal/sensitivity ratio to calculate the relative atomic percent present, as shown in Eq. (18.5) [51]:

$$\%i = 100 \times \frac{(PH/S_i)}{\sum(PH/S_n)} \qquad (18.5)$$

Sensitivity factors are found in numerous texts [44, 48] and when combined with Eq. (18.5), can provide a fair estimate of sample composition. However, for best quantitative results, they are best determined from standards of similar composition as the sample, with the same energy analyzer, at the same energy resolution, at the same primary electron beam energy, and at the same orientation to the beam and energy analyzer. It is also important that the line shapes are similar for standards and samples. The standards are usually either pure elements or pure compounds. Thus, the more complicated the sample, the more difficult it is to obtain appropriate standards and perform accurate quantitative analyses. New methods are continuously being developed to enhance the quantitative analysis of AES of a large variety of samples [45].

The effects of charging, as described previously, must also be carefully considered in the quantification of insulating samples or samples that are patterned with both insulating and conducting materials. Charge compensation may help to eliminate this problem. However, the standards used to obtain the sensitivity factors must be treated in the same manner.

18.7.4.4 Sample Damage

The incident electron beam in AES can damage both organic and inorganic samples [44, 48]. Any chemical changes that occur in the sample during analysis result in inaccurate quantification of its composition. Damaging processes include electron-stimulated desorption of elements, such as chlorine and oxygen, and heating of the sample, which can result in temperature increases of several hundred degrees. Beam damage can be determined by following changes in the AES spectra either over an extended irradiation time or at different beam current densities. Damage can be minimized by operating the AES instrument at lower beam current densities (which may be offset by longer analysis times to achieve desired signal-to-noise ratios) and exposing the samples to the beam for a minimum amount of time. One useful technique is to irradiate the area of interest for analysis only after establishing the instrument parameters on a different part of the sample.

18.7.4.5 Scanning Auger Microscopy (SAM)

Elemental distributions on a surface are of interest in patterned materials, such as in the microfabrication of electronic circuits, the analysis of surface corrosion, and evaluation of heterogeneity in thin films. Scanning Auger microscopy instruments can provide this information with a routine spatial resolution of under 30 nm [44–46, 48, 60, 61], which is comparable to SIMS imaging resolution, but an order of magnitude better than XPS. Scanning Fourier transform infrared and Raman microscopies are limited by the wavelength of light.

Figure 18.28 is an example of the imaging capabilities of SAM [62]. A gold array of lines, patterned on a silicon-nitride-coated Si wafer, was modified with a single monolayer of

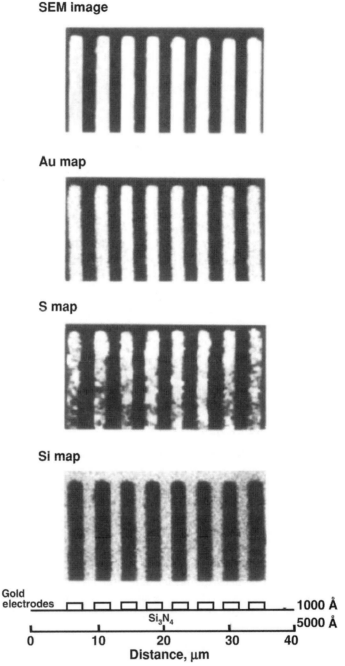

Figure 18.28 Scanning electron microscopy, AES element maps, and cross-sectional schematic of a gold band electrode array on silicon nitride, which has been selectively modified with a monolayer of organic molecules, (1-mercapto-3,6-dithiaheptanyl)octamethylferocene. (*Reprinted with permission from* [62]. Copyright 1991 *American Chemical Society.*)

an organic species, (1-mercapto-3,6-dithiaheptanyl)octamethylferrocence, which is known to attach to the gold surface through the sulfur atom (the mercapto group). The SAM analysis produces Au and S signals in registry, but not the Si. There is no S detected on the Si portions of the array, thus demonstrating that the molecules selectively modified the Au. These maps were difficult to obtain because of low signal, electron-beam-induced degradation, and drift of the instrument during the 2 to 4-h analysis time for the $\sim 30 \times 30$ µm region at 0.1-µm resolution. Surface charging was minimized by using low beam currents, 0.1 to 5 nA, and a sample tilted at 40 to 60° relative to the electron beam. Low beam voltages of 8 to 10 kV were used to minimize sample damage, although higher voltages would have reduced charging because of permeation through the silicon nitride layer into the silicon semiconductor. Other problems of an overlapping Auger Au peak with the strongest S peak had to be solved.

There are several issues that should be addressed in SAM analysis that can lead to artifacts. If the sample is not flat, then shadowing of parts of the sample may occur and provide false elemental maps. If the sample is patterned with insulator and conductor materials, then peak shifts in the insulating areas due to charging may move the Auger peak of interest out of the detection window, resulting in a false elemental map. Other issues are demonstrated by the example in Figure 18.28. The increased beam current density that results from a smaller spot size for high spatial resolution analyses may increase sample damage. Longer acquisition times are often necessary to achieve desired signal-to-noise ratio for the smaller electron beam spot size. Thus, line scans, and especially two-dimensional maps (which are a sequence of line scans that are displaced from each other), can take hours of acquisition time. Because of these lengthy analyses, any drift in sample position or vibrations affect the final result.

18.7.4.6 Depth Profiling and Ion Beam Bevel Sectioning

Thin-film technology has led to the construction of a number of electronic and optical devices that consist of multiple layers of thin films. There are two main methods to evaluate the composition, thickness, and interdiffusion of the thin films. One involves alternating analysis and sputtering of a sample to obtain a depth profile perpendicular to the surface of the sample. An example is given in Figure 18.29 [63], which illustrates changes in the distribution and depth of elements in Ag (200 nm)/Al (80 nm) bilayer structures on SiO_2 substrates during annealing. Ag is of interest in integrated circuit technology because of its bulk conductivity. In the example, AES aids in assessing the thermal instability of Ag films and the mechanisms responsible for it. The figure shows how Al, which starts at the Ag/SiO_2 interface, diffuses to the surface and combines with oxygen to form a thin aluminum–oxide film. Ag, in contrast, shifts away from the surface. These changes have been related to a decrease in resistance, an increase in surface passivation, and an improvement of adhesion of Ag to SiO_2.

Another method for obtaining composition as a function of depth also uses sputtering, but only initially, to create a bevelled edge through the multiple layers. SAM can then be used to scan this edge and obtain a spatial distribution of the composition. An example of ion beam bevel sectioning is shown in Figure 18.30 [64]. A sample with alternating layers of GaInAs and InP in 220-Å periods was beveled with Xe^+ ions at 2 keV. The SEM image in Figure 18.30 shows how the bevelling expands the layers to 250- to 300-µm periods for easy microscopy analysis. The periodicity of the P line scan across this edge is in registry with the SEM image and demonstrates the linearity of the bevel. However, there may be preferential sputtering of the P because the magnitude of the line scan never flattens out.

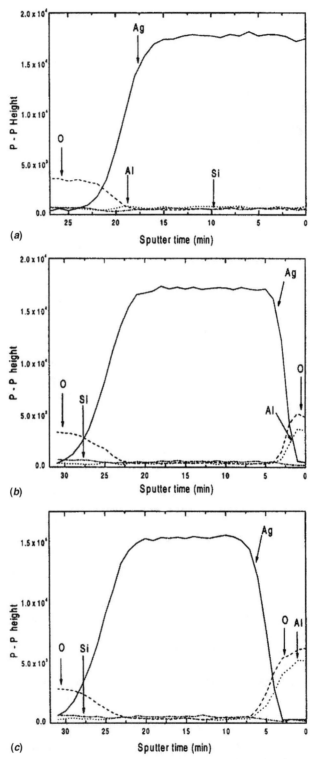

Figure 18.29 Auger depth profiles of Ag/Al bilayer on SiO_2, illustrating the diffusion of Al through the Ag layer during annealing. Profiles are shown for the following samples: (*a*) as deposited, (*b*) annealed at 400°C, and (*c*) annealed at 700°C in argon. (*Reprinted with permission from* [63]. *Copyright* 2001 *American Institute of Physics.*)

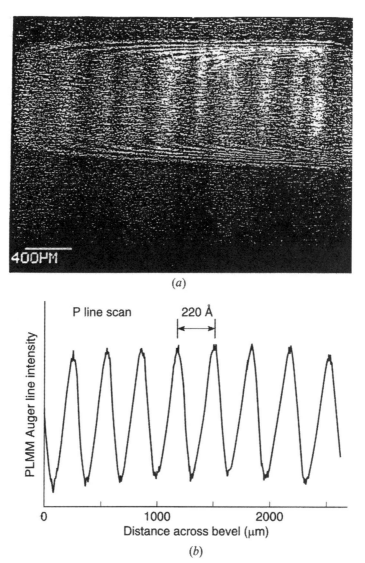

Figure 18.30 Depth analysis of P by AES, using the beveled method. (*a*) SEM image of ion beam bevel section sputtered into a 220 Å period superlattice of GaInAs/In. (*b*) AES line scan for P across the beveled section. (*Reprinted with permission from* [64]. *Copyright* 1995 *EDP Sciences.*)

This example also illustrates the utility of a small analysis area. XPS is less well equipped to handle such samples because of its lower spatial resolution.

Although sputtering is a valuable tool, there are a number of issues that can lead to artifacts and, thus, must be considered [44, 54, 64, 65]. These include atomic mixing, nonplanar removal of material, damage- or radiation-induced diffusion, preferential sputtering, surface segregation and migration, redeposition of sputtered material, and surface roughening. The sputtering behavior of the material should be studied before attempting analysis. A range of samples must be produced for calibration and analyzed by other analytical methods, as well. One way to study the sputtering behavior is to compare the AES analyses of known sputtered and vacuum-cleaved samples [64].

18.7.4.7 Angle-Resolved Auger Electron Spectroscopy

Auger electrons are emitted anisotropically and engage in scattering within the sample. These phenomena result in an anisotropic angular distribution from the surface that can be related to the geometric structure of the surface region. Angle-resolved Auger electron spectroscopy (ARAES) is the technique that monitors this angle distribution [44, 45, 47, 48]. The resulting data can be used to understand the atomic structure near a solid surface and the fundamental interactions of electrons with matter. Applications of this technique include depth profiling of polycrystalline solid surfaces and structure analysis of monatomic adlattices, multiple atomic layers, single crystals, and layered crystals.

18.8 X-RAY PHOTOELECTRON SPECTROSCOPY

X-ray photoelectron spectroscopy (XPS), also known as electron spectroscopy for chemical analysis (ESCA), is a sensitive technique that nondestructively analyzes the elemental composition and chemical state of the top 5 to 100 Å of a surface with a sensitivity of 0.1% of a monolayer [66–70]. XPS is often considered the "workhorse" ultrahigh vacuum (UHV) technique for surface analysis because of its ease of use, ability to provide information on surface chemistry, applicability to insulators (as well as conductors) without excessive difficulty in establishing charge compensation and a small amount (if any) damage to organic materials.

18.8.1 Summary

- What XPS does
  - UHV surface analytical technique
  - Irradiation of sample with X rays results in emission of photoelectrons whose energy is related to binding energy of the electrons, providing elemental and chemical information with minimal damage
- Types of analyses
  - Spot
  - Line scan
  - Depth profiling; inert gas sputtering or angle-resolved technique
  - Imaging; scanning X-ray beam or scanning electron detection aperture
- Performance
  - Spatial resolution down to 3 μm but typically ~10 μm
  - Limit of detection down to 0.01 atomic%
  - Depth of analysis 5 to 100 Å

18.8.2 Basic Principles

In XPS, electrons are emitted from a sample surface when the energy of the incident X rays ($h\nu$) exceeds the binding energy (E_b) of the electrons [66, 70]. These electrons are called photoelectrons. They originate from specific orbitals in an atom and can be designated by notation, such as $1s, 2s, 2p_{1/2}, 2p_{3/2}$, which is explained in more detail in Section 18.7.2.1. Each element in a given oxidation state may emit electrons from one or more orbitals that

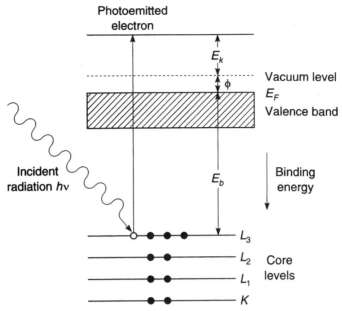

Figure 18.31 Schematic of the XPS process for an atom, resulting in emission of a $2p_{3/2}$ (or L_3) electron. An incident X-ray photon is of enough energy, $h\nu$, to cause the photoemission of the electron at the $2p_{3/2}$ level in this atom. The electron obtains a kinetic energy of $1/2\ mv_e^2$, which can be measured. The binding energy of that electron may be calculated from: $h\nu$ - the kinetic energy - work function of the spectrometer. The K, L_1, L_2, and L_3 electron energy levels in the atom are the same as $1s$, $2s$, $2p_{1/2}$ and $2p_{3/2}$, respectively. (*Reprinted with permission from* [72]. *Copyright* 1997 *Prentice-Hall.*)

have binding energies distinct from other elements and other oxidation states. Figure 18.31 schematically illustrates the XPS process.

XPS spectrometers measure the kinetic energy ($\frac{1}{2}mv_e^2$) of the ejected electrons [71, 72], which is related to the incident X-ray energy and the binding energy by Eq. (18.6),

$$h\nu = \tfrac{1}{2}mv_e^2 + E_b + q\Phi \qquad (18.6)$$

where ν is the frequency of the incident X rays, h is Planck's constant, m is the mass of the ejected electron, v_e is the velocity of the electron, E_b is the binding energy of the electron, and $q\Phi$ is the work function. The work function depends on both the surface composition (organic, inorganic, metallic, etc.) and the spectrometer and represents a small energy barrier that must be overcome before the electron can escape from the surface. The work function is much smaller than the binding energy and, for most cases, is ignored. The binding energy is calculated by subtracting the measured kinetic energy from the known incident X-ray energy. The XPS spectrum contains peaks at binding energies that are specific for the elements present in the sample. The peak intensity is proportional to the number of electrons detected at that binding energy. Figure 18.32 shows an XPS survey spectrum for silver. Multiple peaks in the spectrum arise from electrons that are elastically ejected from different electron energy levels (i.e., $3d$, $4f$, etc.) within the atom. The peak intensities are not the same because the probability for photoejection from each orbital (the photoionization cross section) is different.

An XPS spectrum, such as in Figure 18.32, exhibits a large background that extends from each peak toward higher binding energies. This results from detection of

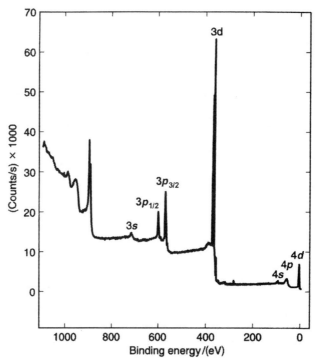

Figure 18.32 XPS survey spectrum of a sputtered Ag surface obtained with monochromated Al $K\alpha$ X rays at 1 eV steps. Each set of peaks is identified according to the electron energy level from which the photoelectrons that produced the peak originated. The doublet for the $3p$ photoelectrons can be clearly observed, the lower intensity peak corresponds to $j = \tfrac{1}{2}$ and the peak of larger intensity corresponds to $j = \tfrac{3}{2}$. This is a standard reference sample used for the calibration of XPS spectrometers.

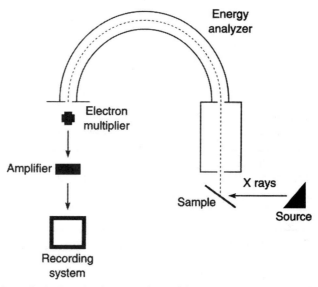

Figure 18.33 Generalized schematic of the core of an XPS instrument. The X-ray source is a cathode of solid Mg or Al. The escaping electrons first pass through a slit system before entering the energy analyzer. The electrons that make it through the analyzer are multiplied at the detector, whose signal is electronically transferred to a computer.

inelastically scattered photoelectrons. These electrons have lost energy by colliding with atoms or molecules in the sample before escaping from the surface. Their kinetic energy is lower than, and not discrete like, elastically ejected photoelectrons.

Another phenomenon that affects the spectrum is the spin–spin coupling that is described by the total angular momentum, j (see Section 18.7.2.1) [73]. In the p shell, values of both $\frac{1}{2}$ and $\frac{3}{2}$ are possible for j. The result of having multiple angular momenta is the appearance of a doublet (see Fig. 18.32). This occurs in $d(j = \frac{3}{2}$ and $\frac{5}{2})$ and $f(j = \frac{5}{2}$ and $\frac{7}{2})$ shells, too. The ratio of intensities of the two peaks is fixed, as is the ratio between all j numbers. The larger of the two peaks corresponds to the angular momentum possessed by the larger number of electrons.

XPS provides not only elemental information but also chemical state information during the same analysis. Different chemical states for a given element are exhibited by different shifts in binding energy of its XPS peaks [67–69, 74, 75]. The chemical state is a measure of electron density that surrounds the nucleus. The electron density depends on the oxidation state and the surrounding elements. The electron density increases with either increasingly negative charge (more reduced state) on an atom (e.g., S^{2-} has more electron density than S^0) or if the nucleus attracts electrons from neighboring atoms (e.g., Cl in NaClO has a higher electron density than in $NaClO_4$, where the larger number of O atoms pull electrons away from Cl with more strength). Thus, less energy is required to remove an electron and a lower binding energy is observed. The electron density decreases with either increasingly positive charge (more oxidized state) on an atom (e.g., Fe^{3+} is more oxidized that Fe^{2+} or Fe^0) or if neighboring atoms pull the electrons away [e.g., C attached to F in the compound $CF_3(CF_2)_5(CH_2)_2SH$ has less electron density than the C attached to H in that compound]. Thus, more energy is required to remove an electron, and a higher binding energy is observed. Table 18.3 illustrates this point. It shows binding energies for four elements in different compounds.

18.8.3 Instrumentation

Figure 18.33 shows the basic set of components in an XPS instrument. They are: at least one X-ray source, an energy analyzer, and an electron detector located in a UHV ($<10^{-8}$ torr) chamber. The UHV system prevents gaseous molecules from scattering the incident X rays and minimizes the rate of surface contamination during analysis [68, 74]. Because XPS is sensitive to only the uppermost layers of a surface, any contamination of the surface can alter the results. (Gaseous molecules can readily adsorb onto surfaces at pressures of 10^{-6} torr

Table 18.3 Binding Energies for Al, Si, Cu, and Sn in Different Compounds [44]

Material/energy level	Binding energy (eV)	Material/energy level	Binding energy (eV)
Al $2p$	72.85	Cu $2p$	932.67
AlAs $2p$	73.6	Cu_2O $2p$	932.4
AlN $2p$	74	CuO $2p$	933.8
Al_2O_3 (sapphire) $2p$	74.1	$CuCl_2$ $2p$	934.4
Si $2p$	99.7	Sn $3d$	484.87
Si_3N_4 $2p$	101.9	SnO $3d$	486.9
SiO_2 $2p$	103.4	SnO_2 $3d$	486.6

and form a monolayer in approximately 1 s under these conditions.) Additional discussion on UHV systems is given in the AES section (Section 18.7.3). Optional equipment can include an ion sputtering source for surface preparation and depth profiling, an aperture system for control of analysis area, and a movable stage that allows for small (5 μm) two-dimensional movements, rotation in the plane of the sample, and tilt to be used in acquiring angle-resolved spectra.

18.8.3.1 X-ray Sources

The X rays used in XPS are generated by applying a large voltage between an anode and a cathode. The anode is an electron source consisting of a filament (e.g., thoriated tungsten). The cathode is constructed from a metal that produces X rays when struck with high-energy (typically 15 kV) electrons generated at the anode. Only Mg and Al are both inexpensive and provide the requisite intense, narrow radiation band (<1 eV) to be of practical use. The energies of X rays from Mg and Al anodes are 1253.6 and 1486.6 eV, respectively.

The X rays are frequently passed through a monochromator [68, 70, 74] because this improves energy resolution and removes the Bremsstrahlung continuum background and satellite peaks. Signal-to-noise ratio is enhanced. Monochromatic X rays provide these advantages at a cost—lower incident X-ray intensity translates into fewer ejected electrons and a lower sensitivity.

18.8.3.2 Electron Energy Analyzers and Detectors

After electrons have been ejected from the surface, they must be separated according to their kinetic energy (velocity) before reaching the detector. This is usually accomplished with a concentric hemispherical energy analyzer (CHA). The analyzer consists of two, hemispherically shaped, charged plates. As electrons pass between the plates, they are attracted to the bottom plate (+V) and repelled from the top plate (−V). Only electrons with the proper velocity pass between the plates and reach the detector. The potential between the two plates is ramped, allowing electrons with different velocities to be separated. Channeltrons, microchannel plates, or resistive anode plates can be used for detection.

18.8.4 Practical Considerations and Applications

Calibration of the XPS instrument, considerations of the effects of XPS on the sample, and careful data examination and quantification must be addressed to ensure proper XPS analysis. These are described in the following sections along with several examples of applications of XPS in the electronics industries.

18.8.4.1 Charge Neutralization

The photoinduced ejection of electrons from a sample results in the eventual buildup of positive charge on the sample, unless replacement electrons are supplied. The accumulated charge causes the apparent binding energy of the electrons to change. Insulating and semi-conducting samples are subject to this effect. Without a method to neutralize the charge, XPS is limited to analyzing conducting samples that are grounded to the sample holder. The most common method for neutralizing charge is to flood the surface with an electron flood gun. In Figure 18.34, two spectra from a quartz sample excited with an Al X-ray source are shown. The first spectrum was collected while using an electron flood gun. The

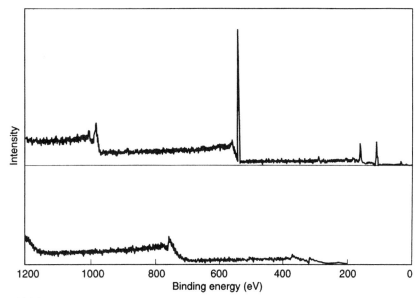

Figure 18.34 XPS spectra of a quartz sample, showing the effects of sample charging on peak intensity and location. The top spectrum was taken during charge-compensation with an electron flood gun. The bottom spectrum was taken with the flood gun turned off. (*Reprinted with permission from* [44]. *Copyright* 1990 *John Wiley & Sons, Ltd.*)

second spectrum was taken under the same conditions but without charge compensation. Two considerations should be noted when charge neutralization takes place with an electron gun. The flood of electrons (1) may damage fragile samples such as organic compounds and polymers (described further below) and (2) exposes the whole sample and neighboring samples on the same sample holder to the electrons, not just the specific area being analyzed.

18.8.4.2 Sample Damage

XPS is considered a nondestructive technique. It is less destructive than AES (Section 18.7) and secondary ion mass spectroscopy (SIMS, Section 18.9). However, sample damage can occur in XPS in some cases. The damage typically comes from three sources; the vacuum, the X rays, and the electrons from the flood gun, if one is used. The UHV requirements for XPS can volatilize materials from the surface and change the composition. For example, it has been shown that waters of hydration may be removed during the time of the analysis [76]. X rays may damage the sample by inducing chemical reactions between species on the surface. One of the first examples is the reduction of Pt(IV) compounds [77]. Other important materials that can be damaged are halogenated organic compounds, which are important as photoresists in microfabrication. Electron flood guns typically damage the surface by causing a variety of reactions. This again is especially true among organic and reactive inorganic materials.

18.8.4.3 Data Analysis

XPS spectra can be evaluated qualitatively to determine the presence of different elements and their chemical state and quantitatively to obtain the relative atomic concentration of those elements. As with other surface analysis techniques, such as AES (Section 18.7), elements on the sample surface are identified by examining the number of peaks and peak positions in the spectrum. Generally, XPS more easily distinguishes between chemical states than AES.

Correct identification of chemical states in a sample can be achieved by proper calibration of the XPS spectrometer. This is often performed with either silver or gold standards. Shifts caused by instrumental variation or charging can be corrected by comparing the binding energy of the strong peaks of the Ag $3d_{5/2}$ or Au $4f_{7/2}$ transitions to the predicted energies of 368.2 and 84.0 eV, respectively.

It is common practice in XPS to quantify the chemical composition of a surface [72]. The peak intensity or area in an XPS spectrum, which corresponds to the number of escaping photoelectrons for a given element, i, is directly proportional to the number of atoms of that element present on the surface. The proportionality constant is called the sensitivity factor, S_i, and one is determined for each photoelectric transition of a given electron orbital in a given element. This factor is a function of the photoelectric cross section, the angular efficiency factor for the instrumental configuration, the efficiency of photoelectron production, the area of the sample, efficiency of the detector, and the mean free path of the photoelectrons. Standard tables of sensitivity factors are available [78].

The relative atomic percent concentration, $\%i$, for an element i is calculated using a relationship that is similar to that in Eq. (18.5) for AES. In summary, the area under the most intense peak of a single photoelectric transition for element i in a given set of peaks is chosen (instead of its peak height) and is normalized by its corresponding sensitivity factor. The resulting ratio is divided by the sum of the ratios of one selected peak area to its sensitivity factor for each the elements, n, present on the sample surface.

An example of a qualitative analysis with XPS in an electronics application is given by Figure 18.35. This figure shows two survey spectra for polypropylene (PPP) packaging films. The PPP films are used to provide a barrier against the diffusion of oxygen and water. During processing, the films must be heat-sealed to the chip. The sample with spectrum (a) shows a loss of sealing after heating. A spectrum for a sample that shows the desired behavior is shown in (b) for comparison. The failed material contains Ti on the surface. (Note that peaks due to Auger electrons also appear in the spectra.) XPS is a good technique for this analysis because it has the ability to identify only the surface species and analyze a "soft" polymeric sample without significantly altering the surface.

18.8.4.4 XPS Imaging

Imaging (or mapping) in XPS allows a two-dimensional image to be obtained showing the location of elements on a surface. An XPS image is obtained using one of two instrumental methods. One method places an aperture between the sample and the energy analyzer. The aperture sizes typically range from 30 μm and up. The surface is flooded with X rays and the optics scanned so that the electrons that reach the aperture are from a specific location on the surface. The size of the aperture and the scanning speed determine the resolution and speed of imaging. It takes more time to collect a reasonable signal with a smaller aperture because fewer electrons reach the analyzer. The other method uses a focused X-ray beam. A map is obtained by moving the sample very slowly under the beam. The size of the beam can be changed to improve the speed or resolution of the measurement. High-resolution measurements require more time to allow a reasonable amount of signal to be measured. This second method generally exhibits lower resolution than the first. Resolution for XPS imaging is presently limited to ~10 μm, however, as the technology matures, this limit

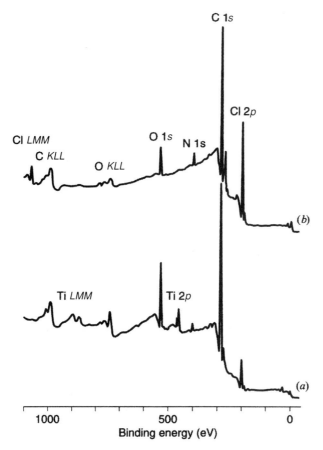

Figure 18.35 Survey spectra of a polypropylene (PPP) packaging film that had been sealed to a chip during processing. The sample producing the top spectrum had maintained adhesion to the sample, while the bottom sample lost adhesion. The adhesion failure is caused by the presence of Ti at the interface. (*Reprinted with permission from* [44]. *Copyright 1990 John Wiley & Sons, Ltd.*)

will be reduced significantly. For both instrumental setups, electron counts are taken at a discrete energy that is specific for the element of interest.

Figure 18.36 illustrates how XPS imaging can reveal chemical heterogeneity of a polymer surface [79]. A polymer blend of poly(vinyl chloride) (PVC, with molar mass of 77.3 kDa) and poly(methyl methacrylate) (PMMA, with molar mass of 75 kDa) was investigated. It was formed by pipetting a solution of a 50 : 50 mixture in tetrahydrofuran solvent onto a Teflon substrate, allowing it to dry, and pealing the resulting film from the substrate. There is a 2-eV shift in binding energy of the C $1s$ signal from PVC (287 eV for C–Cl) to PMMA (289 eV for C=O) due to bonding of carbon with elements of different electronegativities. These peaks can be used to distinguish between PVC and PMMA. XPS signals from elements more unique to the polymers were used for imaging, chloride for PVC and oxygen for PMMA. Because of contamination from substances in the air, the surface oxygen was higher than expected for PMMA. (Interferences from air contaminants can also be a problem in the C $1s$ region.) From Cl $2p$ and O $1s$ images like those in Figure 18.36, it was found that the air side of the polymer film is enriched in PMMA (60 to 70%) and heterogeneous. (Fourier transfer infrared spectroscopy was used to show that heterogeneity extends into the bulk material.) When the molar mass of PMMA increased, surface segregation of PMMA was found to increase as well. Imaging of the substrate side of the film (not shown) indicated 80 to 90% PVC enrichment and a homogeneous distribution.

Cl 2p image   **O 1s image**

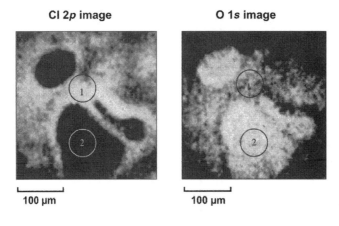

100 μm   100 μm

Figure 18.36 Cl 2p and O 1s XPS images of a PVC/PMMA polymer blend. The inverse correlation of bright areas in the Cl 2p (PVC enrichment) and the O 1s (PMMA enrichment) images indicates the phase separation and heterogeneity at the surface of the polymer blend. (*Reprinted with permission from* [79]. *Copyright* 2000 *Society for Applied Spectroscopy.*)

18.8.4.5 Depth Analysis

Depth analysis can be achieved in two ways. One is depth profiling, which uses an argon ion gun to sputter layers off the surface. Analysis is performed at regular intervals between sputtering steps. Elemental and chemical state variation with sputtering time is displayed in a way similar to that in Figure 18.29 for AES depth profiling. The depth resolution depends on the energy of the incoming ions, the material being sputtered, and the time spent sputtering the sample. Depth profiling is good for most inorganic materials but is not as reliable for polymers and other organic materials. The high energy of the Ar^+ ions can change bonding of carbons in polymers, preventing an accurate analysis of the surface. Depth profiling may be used successfully with polymers if the presence of one or more elements other than carbon is of interest. The AES section (18.7) and the SIMS section (18.9) provide further discussion of ion beams and effects on a sample.

Angle-resolved XPS (ARXPS) provides a nondestructive approach to depth analysis. It involves tilting the sample relative to the incident beam. As the tilt varies from a perpendicular orientation relative to the X rays (0° takeoff angle) to grazing angles (large takeoff angles), fewer photoelectrons originate from the bulk and more originate from closer to the surface. Information cannot be obtained from depths greater than the escape depth of the emitted photoelectrons, however. Figure 18.37 illustrates the use of ARXPS to investigate

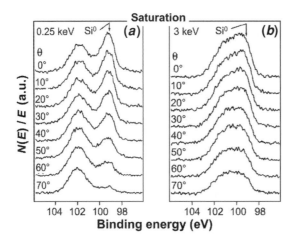

Figure 18.37 Nondestructive depth analysis with ARXPS in the Si 2p region of silicon nitride layers that were formed by N_2^+ ion implantation at (*a*) 0.25 keV and (*b*) 3 keV at the saturation dose into Si (100) single crystals. The takeoff angles (θ) for each spectrum are indicated in the figure. (*Reprinted with permission from* [108]. *Copyright* 2002 *American Chemical Society.*)

the chemical composition of silicon nitride films grown on silicon through low-energy nitrogen ion (N_2^+) implantation. Ion beam methods for growing films hold promise for modifying surfaces of materials without adhesion problems or changing bulk properties. Five states of silicon, Si^0, Si^1, Si^2, Si^3, and Si^4 (centered at 99.1, 99.9, 100.7, 101.5, and 102.3 eV, respectively) can be deconvoluted from narrow scans in the Si $2p$ region, representing species bonded to zero, one, two, three, and four nitrogens. The Si^4 form is the most desired in nitride films, Si_3N_4, formed by chemical vapor deposition and thermal methods. The narrow scans of the Si $2p$ region taken at different takeoff angles indicate higher oxidation states of silicon toward the surface for silicon implanted with N_2^+ at 0.25 keV. Nitridation of silicon at low N_2^+ energies is a main effect. The distinctive variation of silicon oxidation state with depth might not have been observed if depth profiling with Ar^+ ions were performed instead of ARXPS. The oxidation states are fairly evenly distributed with depth when silicon is treated with N_2^+ at 3 keV, suggesting a mixed layer due to "knock-on" mechanisms that are enhanced with increased ion beam energies. The N_2^+ ion beam at these high energies behaves similar to an Ar^+ depth profiling beam, and an increased Si^0 content is attributed to preferential sputtering of nitrogen.

18.9 SECONDARY ION MASS SPECTROMETRY

Secondary ion mass spectrometry (SIMS) provides elemental, isotopic, and molecular information about the surface and near surface of solid samples [80–88]. The molecular information that SIMS provides distinguishes it from AES and XPS. Detection limits are in the ppb–ppt range, which are two to four orders of magnitude better than those of XPS or AES [69, 83, 84, 88, 89], allowing for trace analysis.

18.9.1 Summary

- What SIMS does
  - UHV surface analytical technique.
  - Primary ion beam causes sputtering of the sample surface. Secondary ions from the sample consisting of molecules, fragments, and elements are detected.
  - Isotopes can be distinguished from each other.
  - Hydrogen can be detected.

- Types of analyses
  - Small area
  - Depth profiling
    - Dynamic mode—Use primary ion beam to erode sample
    - Static mode—Need second ion beam to etch sample
  - Imaging. Ion microscope or scanning microprobe

- Performance
  - Spatial resolution down to 50 nm
  - Limit of detection of ppb to ppt
  - Depth of analysis is 2 Å to several micrometers

18.9.2 Basic Principles

The SIMS process is shown schematically in Figure 18.38. A primary ion beam is focused onto a sample target, giving rise to elastic and inelastic collision events that cause ejection (or sputtering) of sample particles [83, 85, 87, 90, 91]. Only the topmost two or three monolayers of the sample are sputtered, making SIMS one of the most surface-sensitive analysis techniques [88]. A majority of the particles are neutral atoms or clusters and a much smaller fraction are ionized atoms, molecules, and molecular fragments (providing chemical bonding information) of negative and positive charge [91]. The fraction of these secondary ions can vary from 10^{-1} to 10^{-6} [85]. The extent of ionization depends on the surrounding material (the matrix), surface conditions, ionization probability, and incident beam properties (flux, type, and energy). Because of the influence of these many factors, the ionization process is far less understood than that involved in XPS and AES [85,86], rendering quantification difficult. Additional ions may be produced by postionization of the sputtered neutral species using a laser, ideally ionizing all of the sputtered species and producing a more quantifiable analysis [83–86, 92].

Either positive or negative ions are focused onto the entrance of an analyzer that separates them by their mass-to-charge (m/z) ratio. The separated ions reach the detector and a spectrum is formed by plotting the ion counts versus m/z. The m/z value is simply the sum of the masses of the individual atoms of the ion, divided by the charge on the ion. If the charge is -1 or $+1$, the m/z equals the ion's mass. Calculations of m/z values for ion fragments must involve the exact masses of the isotopes, not the average masses that are listed in the periodic table. For example, naturally occurring carbon consists of 98.89% ^{12}C (12.0000 amu) and 1.11% ^{13}C (13.003 amu), where the superscript corresponds to the

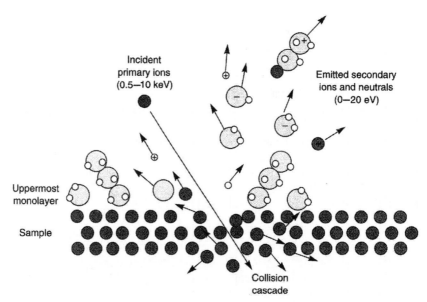

Figure 18.38 Schematic of processes that occur when an incident primary ion beam bombards a sample surface. This example shows an ordered sample that is either contaminated or modified with approximately a monolayer of other compounds. If the energy of the primary beam is great enough, recoil atoms in the sample generate a collision cascade. Ions and neutral species of whole molecules and molecular fragments, adducts of whole molecules and molecular fragments with other fragments and atoms, and individual atomic species are sputtered from the surface with lower energy than that of the primary ion beam.

18.9 Secondary Ion Mass Spectrometry 777

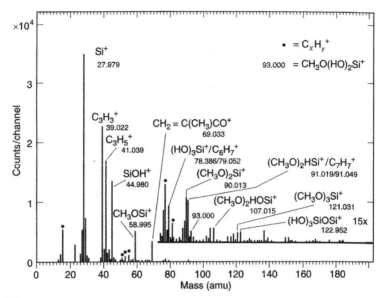

Figure 18.39 Example of a SIMS spectrum, which was obtained using a TOF mass analyzer, showing molecular positive ions. The sample is a silicon wafer that was modified with a 0.5 vol% solution of methacryloxysilane $(CH_3O)_3Si$—$(CH_2)_3$—O—CO—$C(CH_3)$=CH_2 in a 1 : 1 ratio of isopropanol to water. (*Reprinted with permission from* [93]. *Copyright* 1995 *John Wiley & Sons, Ltd.*)

mass number (i.e., the total number of protons and neutrons for that isotope). Compounds having large numbers of carbon atoms have a significant probability that a ^{13}C will occur, and thus, will yield mass spectra that show a series of peaks for each ion fragment, one for the ion containing only ^{12}C isotopes, and others of one mass-unit increments for ions that contain increasing substitutions of ^{13}C for ^{12}C. Some elements have more than one isotope with high probabilities, such as bromine, whose average mass is 79.90 amu, and the exact isotopic masses are 78.9183 for ^{79}Br at 50.87% abundance and 80.9163 for ^{81}Br at 49.13% abundance. Thus, SIMS analysis of compounds containing Br results in mass spectra showing peak doublets, separated by two mass units. The isotopic abundances, which are unique to each element, can be used to determine the exact species present in a mass spectrum.

An example of a SIMS spectrum is shown in Figure 18.39. The sample is a silicon wafer surface that was chemically modified with an organosilane reagent [methacryloxypropyltrimethoxysilane, $(CH_3O)_3Si$—$(CH_2)_3$—O—CO—$C(CH_3)$=CH_2] to improve adhesion to a subsequently deposited polymeric layer. The spectrum shows the strengths of SIMS to detect both atomic species (Si^+, m/z 27.979) and molecular fragments [CH_2=$C(CH_3)CO^+$, m/z 89.033] from the modified surface. Molecular fragments indicate that some complete hydrolysis (($HO)_3Si^+$, m/z 78.986) and incomplete hydrolysis [$(CH_3O)_2HOSi^+$, m/z 107.015] has occurred. Incompletely hydrolyzed groups are weakly bound and may result in poor adhesion between the silicon wafer and polymeric layers [93].

There are two different SIMS modes: dynamic and static. Dynamic SIMS is inherently destructive. A high current primary ion beam is used to rapidly erode the sample surface with rates of 10 to 50 monolayers per second [88]. This mode is typically used to obtain elemental analysis as a function of depth on the scale of microns or in the bulk of inorganic samples. Because of the violent sputtering process, dynamic SIMS provides little or no chemical or molecular information. In the static mode, an extremely low beam current is used so that, in a given analysis time, no primary ion strikes the same molecule or surface region twice. This allows for less fragmentation and sampling of only the uppermost layer of molecules, thus preserving the molecular structure and bonding information. As a result, static SIMS is commonly used to evaluate surface contaminants (submonolayer quantities), organic coatings, and polymers. Another valuable feature of static SIMS is that the signal for a surface-bound species is linear with coverage, up to one monolayer. Its disadvantages include a higher sensitivity to surface topography and an inability to provide reasonable depth profiling rates. Thus, when depth profiling is desired, static SIMS instruments are outfitted with a second ion gun capable of high beam currents, which allows for alternating depth erosion and analysis.

18.9.3 Instrumentation

The basic SIMS instrument consists of a UHV chamber with ion gun, focusing optics, sample holder, secondary ion extraction optics, mass analyzer, and ion detector. Figure 18.40 shows a schematic of these components in a time-of-flight (TOF) SIMS [85]. UHV pressures of 1×10^{-8} Torr or less are required to minimize contamination of the surface and to ensure efficient generation and detection of secondary ions. The requirements for dynamic SIMS are less stringent than static SIMS because the much higher erosion rates can exceed the contamination rates. Typical UHV vacuum systems are described in the AES section (18.7.3.1). Options illustrated in Figure 18.40 are an electron flood gun for charge compensation, a postionization laser to ionize neutrals and thereby enhance sensitivity and quantification, and a raster for moving the ion beam across the surface for spatially defined analysis and for evaluating areas larger than the beam diameter. The ion beam pulsing unit is specific to a TOF instrument.

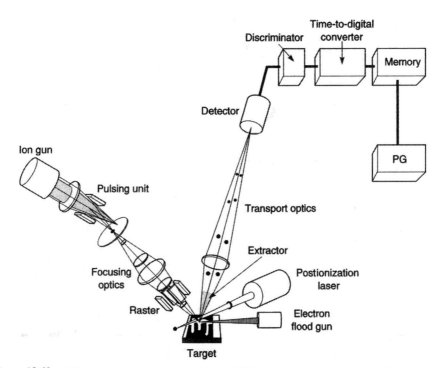

Figure 18.40 SIMS schematic for an instrument with a TOF mass analyzer, laser postionization, and charge compensation. (*Reprinted with permission from* [85]. *Copyright* 1993 *American Chemical Society.*)

18.9.3.1 Ion Guns

Primary ion beams in SIMS may consist of Ar^+, Xe^+, Ga^+, O_2^+, and Cs^+ (and less commonly, He^+, Ne^+, and Kr^+) in the energy range of 0.5 to 30 keV. A SIMS instrument is typically configured with two ion guns, one to enhance negative secondary ion emission and the other to enhance positive secondary ion emission. One can vary beam current, voltage, size, and electronegativity of primary ions to vary ionization processes and depth resolution. Ion current densities for dynamic and static SIMS are >1 μA/cm^2 and <1 nA/cm^2 (and 1×10^{12} ions/cm^2 dosage), respectively [87]. Spot sizes of O_2^+ and Cs^+ ion sources are typically from 30 to 75 μm, but smaller diameters of 0.5 and 0.15 μm have been achieved, respectively [88]. A Ga liquid metal ion gun (LMIG) is employed to obtain better than 100-nm lateral spatial resolution [85]. Mass filtering mechanisms are employed to remove neutrals, impurities, and multiply charged species from primary ion beams [88]. Ion guns are partly described elsewhere in the context of accessories for AES (Section 18.7) and XPS (Section 18.8) instruments for sample preparation and depth-profile analyses.

18.9.3.2 Mass Analyzers

Static and dynamic SIMS place different demands on the mass analyzer. There are three main kinds of mass analyzers used in SIMS and shown in Figure 18.41; quadrupole and double-focusing magnetic sector (typically used for dynamic SIMS, although also used for static SIMs), and time of flight (TOF, used for static SIMS). Secondary ion optics transport ions from the sample surface to the entrance of the mass analyzer. A series of electrostatic lenses and apertures perform this task. The mass analyzer largely determines the mass resolving power (MRP), mass range, detection limit, and acquisition time. The MRP is a measure of how well an instrument can separate two peaks of similar mass and is defined as the mass divided by the width of the peak ($m/\Delta m$).

Quadrupole Mass Analyzers By far the least expensive and most compact and widely used is the quadrupole mass analyzer (QMA) [84,85] shown in Figure 18.41a. The QMA consists of four conducting rods that are arranged as shown in the figure. The pairs of opposite rods have like charge. Direct current (DC) and antternating current (AC) (radio frequency, RF) voltages are applied to the rods so that the charge on the rods alternates. Ions are repelled from one set of rods and attracted to the other set, and then switch when the charges switch. This induces a frequency in the ions passing along the long axis between the rods. The frequency depends on the m/z ratio of the ions and on the ratio of the RF and DC voltages. Those ions whose m/z is just right so that they have a stable path pass through and are detected. The QMA detects one mass at a time and must scan a range of masses to obtain a spectrum. The mass extent is up to $m/z \sim 300$ to 2000 and the MRP is often unity [94], but no better than 250 [88]. The disadvantages of a QMA include relatively low ion collection efficiency (~ 0.1 to 10% of secondary ions) [94] and a spectrometer transmission that decreases with increasing mass. Figure 18.42 shows a SIMS spectrum that was obtained using a QMA. The purpose of the analysis was to evaluate, in the manufacturing of power devices, the effect of overetching a SiO_2 film that covers Al pads. The etching solution was aqueous $NH_4F + HF$. This procedure typically produces a problem, called "spotted" Al pads. The sample for the SIMS analysis in the figure was Al–1%Si that was deposited by sputtering onto a silicon wafer. The spectrum was taken after etching the sample in the aqueous $NH_4F + HF$ solution for 180 s [95]. The peaks at m/z that correspond to F^+, AlF^+, and AlO^+ are consistent with the formation of aluminum fluoride during overetching, which is followed by hydrolysis to Al_2O_x. Consequently, a dry etching process, which is generally

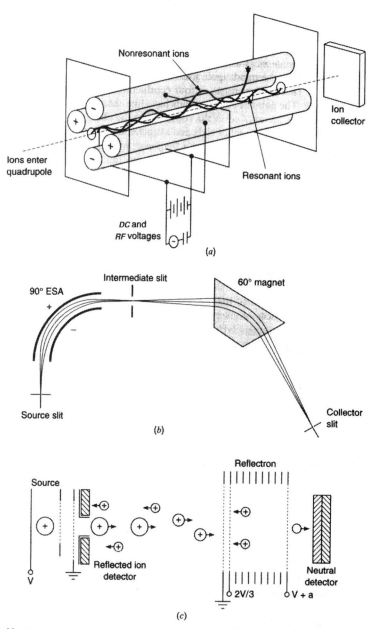

Figure 18.41 Diagrams of mass analyzers: (*a*) quadrupole and (*b*) magnetic sector. (*Reprinted with permission from* [96]. *Copyright* 1997 *Prentice-Hall.*) and (*c*) TOF. (*Reprinted with permission from* [97]. *Copyright* 1993 *American Chemical Society.*)

more easily controlled, would be preferred for such samples. For the illustrated application, the QMA is adequate. However, its low mass resolution results in ions of similar masses being unresolved. For example, $^{14}N^+$ (m/z 14.0025) would not be resolved from $^{28}Si^{2+}$ (m/z 13.9879). Likewise, $^{16}O^{2+}$ (m/z 31.9893) would not be resolved from $^{32}S^+$ (m/z 31.9715).

Double-Focusing Magnetic Sector Analyzers Another means of analyzing masses is with a double-focusing magnetic sector spectrometer [88,96], which consists of an

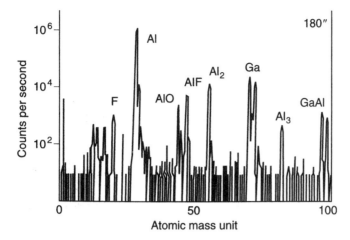

Figure 18.42 SIMS spectrum, using a quadrupole mass analyzer of an Al surface that was etched with $NH_4F + HF$ solution for 180 s. Note presence of AlO^+, AlF^+, and F^+ species. (*Reprinted with permission from* [95]. *Copyright* 1995 *Springer-Verlag.*)

electrostatic analyzer for energy filtering and a magnetic sector for mass separation. When ions pass through a magnetic sector, they travel in a trajectory with a radius defined by the charge, mass, accelerating voltage, and magnitude of the magnetic field. Typically, the magnetic field is scanned to allow ions of different m/z to pass through. However, there is a velocity spread in the ions when they enter the magnetic sector, resulting in a broadening of the peak in the spectrum. To provide velocity focusing, an electrostatic analyzer is placed in series with the magnetic sector. When ions of different velocities enter the radial electric field at right angles, they experience different trajectories, resulting in a highly focused ion beam.

A double-focusing mass spectrometer of the Nier–Johnson type is shown in Figure 18.41*b* [96]. Like the quadrupole mass spectrometer, only a single mass is analyzed at a time, and, thus, most of the secondary ions that are generated go undetected. Unlike QMAs, the resolving power of a double-focusing magnetic sector can exceed 100,000, but a more practical limit is 10,000. However, higher mass resolution is accompanied by a loss in sensitivity. The mass range goes up to 10,000 to 15,000 amu. Sector mass spectrometers are sensitive to changes in the environment and current in the power supply, and thus, must be calibrated periodically.

Time-of-Flight Mass Analyzers TOF analyzers have several advantages over double-focusing magnetic sectors and QMAs. Secondary ions are generated by a pulsed (1 to 10 ns) primary ion beam, not a continuous one, and thus, allow static SIMS to be performed easily. In addition, the entire mass spectrum can be collected in a few microseconds [86] and in a single pulse. Spectra from several pulses can be averaged to obtain better signal-to-noise ratios. A full spectrum obtained in a few primary ion pulses is powerful for determining surface composition and minimizing sample damage. A 10^4-fold improvement in detection over previous mass analyzers is possible [81]. Also, TOF has the ability to detect very high masses, so that polymers and proteins may be analyzed. The key component in a TOF system is a drift tube that is typically 1 to 2 m long. Ions enter the tube at the same energy. Lighter ions travel at a faster velocity, v, than heavier ions to maintain the same energy. Thus, the ions separate themselves according to mass and are detected at the end of the drift tube. Sophisticated electronics monitor the timing between the primary ion pulse and arrival time at the detector, so that a rise in detector signal and the mass of the ion can be correlated. Routine mass analysis of more than 10,000 Da is typical, although the mass range of TOF is theoretically infinite.

By doubling the length of the drift tube through the use of a reflector with a two-stage electrostatic field gradient, the MRP improves. The first gradient decelerates the ions and a retarding field gradient reverses their direction of travel and focuses them back onto a detector, near the opening of the drift tube. Figure 18.41c shows one type of reflectron TOF analyzer [97]. The MRP is about 10,000 and essentially constant over the mass range, decreasing only slightly for lower masses (≤ 25 amu) [88]. This configuration allows the possibility for detection of neutrals, as the figure indicates. The transmission does not vary with mass resolution (i.e., high mass resolution without loss of sensitivity).

Figure 18.43 shows TOF–SIMS spectra for a silicon wafer treated with a CHF_3 plasma, followed by UV/ozone treatment [85]. This analysis was motivated by the need to evaluate contamination that occurs during production of some semiconductor devices. Several features are demonstrated. First, because some species easily form negative ions and others form positive ions, it is important to collect mass spectra in both modes to get a full understanding of the composition of the surface. Second, TOF–SIMS easily detects the higher mass polymeric species $C_n F_m$ at $> m/z$ 300 that are formed on pure Si (Figure 18.43a). Third, the value of high MRP is demonstrated. Figure 18.40b reveals nine different peaks of approximate m/z 56. One is $^{56}Fe^+$, from 2×10^{10} atoms/cm^2 Fe contamination in this sample. The quantification of this species is accurate to about 20% on an area of 30×30 μm^2. The other peaks at m/z 56 show further contamination with organic species.

18.9.3.3 Detectors

Modern detectors in mass spectrometry systems use electron multipliers with high-speed electronics and computers. When ions strike the electron multiplier, across which a high potential is maintained, secondary electrons are produced. These collide with the walls of the multiplier and produce more electrons, which are counted. Dual-channel plates or resistive anodes are commonly used for image detection.

18.9.4 Practical Considerations and Applications

Practices in sample handling, ion beam effects on the sample, depth profiling, imaging, data analysis, and possible artifacts (and correction of artifacts) are described below. Some of these issues are similar to those in AES and XPS, Sections 18.7 and 18.8, respectively. For example, because SIMS analyses are carried out under UHV conditions, samples must be treated with the same care as those for AES and XPS. Other issues are quite different for SIMS, however. For example, damage to the sample must occur to produce ions for analysis. Static SIMS produces less damage than dynamic SIMS. In addition, only a mass spectrometry technique can distinguish between different isotopes on surfaces.

18.9.4.1 Sample Charging

The bombardment of a sample by an ion beam (usually of positive ions) and subsequent sputtering of secondary ions causes charging of the surface. Conducting samples can be grounded to compensate for the charge. However, SIMS analysis of insulating samples (with resistivities above 10^8 Ω) [83] requires charge neutralization, usually achieved with electrons that are extracted from a hot W filament or a LaB_6 filament. The latter is more expensive, but allows for higher spatial resolution and longer life [88].

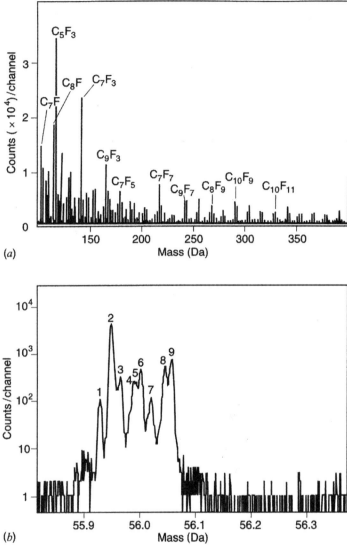

Figure 18.43 (a) TOF–SIMS negative ion spectrum over a large range of m/z of a silicon wafer treated with CHF_3 plasma. (b) TOF–SIMS positive ion spectrum over a narrow range, showing advantages of high mass resolution, for the same wafer after UV/ozone treatment. Identity of the peaks are: $1 = {}^{56}Fe^+$; $2 = {}^{28}Si^{2+}$; $3 = {}^{28}SiCO^+$; $4 = {}^{28}SiCNH_2^+$; $5 = {}^{29}SiC_2H_3^+$; $6 = {}_{28}SiC_2H_4^+$; $7 = C_3OH_4^+$; $8 = C_3NH_6^+$; $9 = C_4H_8^+$. (*Reprinted with permission from* [85]. *Copyright* 1993 *American Chemical Society.*)

18.9.4.2 Sample and Data Analysis

Spectral Interpretation (Qualitative Analysis) Spectra of known samples and consideration of simple fragmentation mechanisms help to elucidate the identity of the m/z peaks in an unknown sample. Mass resolving power and the natural isotopic distributions that produce characteristic multiplet peaks for molecular ions are important in interpreting spectra. In addition, because peak position, m/z, in a spectrum is inversely dependent upon the charge of the ion, the possibility of multiply charged species should also be considered.

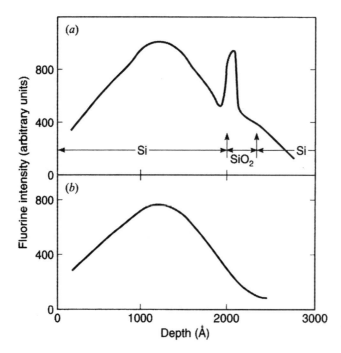

Figure 18.44 Depth profiles of F implanted into 2000 Å of Si on a SiO$_2$ substrate. (*a*) Ionization efficiency of F is enhanced at the Si/SiO$_2$ interface. (*b*) Laser postionization removes such matrix effects. (*Reprinted with permission from* [92]. *Copyright* 1992 *Butterworth-Heinemann.*)

Quantitative Analysis Although SIMS can detect species in a broad range of concentrations—majority, minority, and trace—caution must be exercised in quantification. Even more difficult is to obtain quantitative information as a function of depth or lateral position. Figure 18.44 shows an example of the matrix effect on ion yield during a depth analysis of F in a sample consisting of a silicon film on a silicon dioxide substrate. The SIMS data in Figure 18.44*a* show an enhanced F$^+$ ion yield from the oxygen at the SiO$_2$/Si interface. However, if the species sputtered from the sample by the primary ion beam are completely ionized with laser postionization, which is matrix independent, the profile of F$^+$ is smooth across the SiO$_2$/Si interface (Figure 18.44*b*).

For well-understood samples under strictly controlled conditions, some empirical techniques may be used that employ the relative sensitivity factor, RSF, which is calculated from Eq. (18.7) [98],

$$\text{RSF}_{x/\text{ref}} = (i_{x/\text{iref}})(C_{\text{ref}} f_{\text{ref}} / C_x f_x) \tag{18.7}$$

where x is the analyte, ref is the reference species (internal or external standard), i is the signal intensity, C is the elemental concentration, and f is the isotopic abundance. An unknown concentration C_x can be obtained by rearranging this equation after RSF is determined from a known concentration of the species x.

18.9.4.3 Depth Profiling

In dynamic SIMS, analysis as a function of depth is performed with the same primary ion beam that erodes the sample. This is possible because the primary ion beam produces a high sputtering rate. In contrast, static SIMS must use a depth-profiling beam in alternating sequence with the analysis beam.

Several factors must be addressed in depth profiling. The effect of changing matrix on ion yield is described in Section 18.9.4.2 and is demonstrated by Figure 18.44. The limitations of using depth profiling with polymers and the mixing of layers during depth profiling are described elsewhere in this chapter. Mixing reduces depth resolution and causes a sharp interface between two materials to appear broad. High sputter rates provide depth profiling of thick layers and improved detection limits because the rate of ion ejection is high, they are not appropriate for thin films, such as 1000-Å implantation layers in semiconductor materials. In SIMS, lower energy primary ions of 2 keV Xe^+ and Cs^+ provide better resolution than O_2^+ [88].

18.9.4.4 SIMS Imaging

Spatial resolution in SIMS is much better than XPS and only slightly worse than AES. There are two modes for imaging [83–88, 98, 99]: ion microscope and scanning microprobe. In an ion microscope, a large area is irradiated and ion optics must be used to preserve the spatial relationship of the desorbed ions from the sample to the detector. This system functions analogously to an optical microscope, except that particles are mass selected. The lateral resolution of microscope imaging is inferior to scanning microprobe analysis, and typically, no better than 500 nm. Scanning microprobe involves rastering a small spot ion beam across the sample while recording the secondary ion intensity. The diameter of the primary ion beam limits the spatial resolution, which can be as small as 50 nm [83, 100]. Partly for this reason, spatial resolution during dynamic SIMS analyses, which use larger ion beams with higher ion current, is much poorer than in static SIMS. Also, analysis of molecular ions typically has worse spatial resolution than atomic species.

Figure 18.45 shows SIMS images that were obtained to evaluate the high-temperature corrosion of β-NiAl after 2 h in $^{16}O_2$ and an additional 2 h in $^{18}O_2$ at 1200 °C [101]. This example illustrates the advantage of SIMS over other techniques in the independent detection of two different isotopes of oxygen. It also shows the use of SIMS' high spatial resolution for following nonuniform growth of oxides. The analyses indicate that a thin film of oxide (containing $^{16}O^-$) is initially formed over the sample, followed by continued growth at the ridges (containing $^{18}O^-$).

18.9.4.5 Other MS Techniques for Analysis of Solids

Other ionization methods have been implemented in conjunction with mass spectral analysis to study surfaces of materials, and which have imaging capabilities. These include laser desorption mass spectrometry (LD–MS), where the ion beam is replaced by a pulsed laser beam. The mass analyzer is either a TOF [97] or a Fourier transform mass spectrometer [102, 103]. The ionization process is different than SIMS and generally results in less fragmentation of molecules. However, the analysis is deeper and over a larger area because of the laser fluence and diameter. It can also be used to analyze nonconductors. Detection limits for LD–MS are in the 1 to 100 ppm range.

Sputtered neutral mass spectrometry (SNMS) [104] involves using an RF argon plasma to ionize sputtered neutral species. Because the ionization process is separate from the sputtering process, SNMS is semiquantitative. The area of analysis is large, at about 5 mm, which makes high spatial resolution impossible. Limits of detection for SNMS are in the 10 ppm–ppb range.

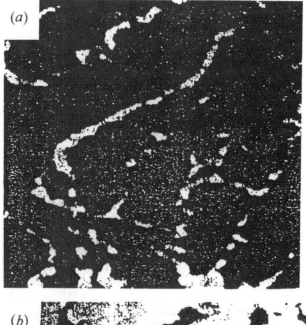

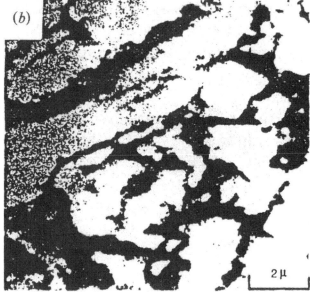

Figure 18.45 SIMS images of β-NiAl for (a) ^{16}O and (b) ^{18}O isotopes of oxygen, showing oxide formation. Sample was heated at 1200°C for 2 h in ^{16}O$_2$ followed by 2 h in ^{18}O$_2$. High concentrations are white. (*Reprinted with permission from* [101]. *Copyright* 1993 *Institute of Materials, IOM Communications, Ltd.*)

REFERENCES

1. B. D. CULLITY, *Elements of X-ray Diffraction*, Reading, MA: Addison-Wesley, 1978.
2. A. R. WEST, *Solid State Chemistry and Its Applications*, New York: Wiley, 1989.
3. M. P. NAGALE AND I. FRITSCH, Individually-Addressable, Submicron Band Electrode Arrays: 1. Fabrication from Multilayered Materials, *Anal. Chem.*, Vol. 70, pp. 2902–2907, 1998.
4. M. P. NAGALE AND I. FRITSCH, Individually-Addressable, Submicron Band Electrode Arrays: 2. Electrochemical Characterization, *Anal. Chem.*, Vol. 70, pp. 2908–2913, 1998.
5. J. D. INGLE AND S. R. CROUCH, *Spectrochemical Analysis*, Upper Saddle River, NJ: Prentice Hall, 1988.

6. D. A. SKOOG, F. J. HOLLER, AND T. A. NIEMAN, *Principles of Instrumental Analysis*, 5th ed., Austin: Harcourt Brace, 1998.
7. D. S. KNIGHT AND W. B. WHITE, Characterization of Diamond Films by Raman Spectroscopy, *J. Mater. Res.*, Vol. 4, pp. 385–393, 1989.
8. J. W. AGER, III, D. K. VEIRS, AND G. M. ROSENBLATT, Spatially Resolved Raman Studies of Diamond Films Grown by CVD, *Phys. Rev. B (Condensed Matter)*, Vol. 43, pp. 6491–6499, 1991.
9. M. MERMOUX, B. MARCUS, L. ABELLO, N. ROSMAN, AND G. LUCAZEAU, In situ Raman Monitoring of the Growth of CVD Diamond Films, *J. Raman Spectros.*, Vol. 34, pp. 505–514, 2003.
10. T. ENGLERT, G. ABSTREITER, AND J. PONTCHARRA, Determination of Existing Stress in Si Films on Sapphire Substrates Using Raman Spectroscopy, *Sol. St. Electr.*, Vol. 23, pp. 31–33, 1980.
11. Z. HANG, H. SHEN, AND F. H. POLLAK, Raman Study of Polish-Induced Surface Strain in <100> and <111> Gallium Arsenide and Indium Phoshide, *Proc. SPIE-Intl. Soc. Opt. Eng.*, Vol. 118, pp. 524, 1985.
12. J. T. KRUG, II, G. D. WANG, S. R. EMORY, AND S. NIE, Efficient Raman Enhancement and Intermittent Light Emission Observed in Single Gold Nanocrystals, *J. Am. Chem. Soc.*, Vol. 121, pp. 9208–9214, 1999.
13. S. P. MULVANEY AND C. D. KEATING, RAMAN SPECTROSCOPY, *Anal. Chem.*, Vol. 72, pp. 145R–157R, 2000.
14. B. PETTINGER, G. PICARDI, R. SCHUSTER, AND G. ERTL, Surface-Enhanced and STM-tip-enhanced Raman Spectroscopy at Metal Surfaces, *Single Molecules*, Vols. 5–6, pp. 258–294, 2002.
15. M. G. HEATON AND F. M. SERRY, Scanning Probe/Atomic Force Microscopy: Technology Overview and Update, *Digital Instruments*, Veeco Metrology Group, 2001.
16. A. CHAND, M. B. VIANI, T. E. SCHAFFER, AND P. K. HANSMA, Microfabricated Small Metal Cantilevers with Silicon Tip for Atomic Force Microscopy, *J. Microelectromech. Syst.*, Vol. 9, pp. 112–116, 2000.
17. S. HEISIG, O. RUDROW, AND E. OESTERSCHULZE, Optical Active Gallium Arsenide Cantilever Probes for Combined Scanning Near-Field Optical Microscopy and Scanning Force Microscopy, *J. Vac. Sci. Technol., B*, Vol. 18, pp. 1134–1137, 2000.
18. D. W. LEE, T. ONO, AND M. ESASHI, Cantilever with Integrated Resonator for Application of Scanning Probe Microscope, *Sens. Actuators, A*, Vol. A83, pp. 11–16, 2000.
19. K. BOBROV, A. J. MAYNE, AND G. DUJARDIN, Atomic-Scale Imaging of Insulating Diamond through Resonant Electron Injection, *Nature*, Vol. 413, pp. 616–619, 2001.
20. M. A. POGGI, L. A. BOTTOMLEY, AND P. T. LILLEHEI, Scanning Probe Microscopy, *Anal. Chem.*, Vol. 74, pp. 2851–2862, 2002.
21. J. MCCARTHY, Z. PEI, M. BECKER, AND D. ATTERIDGE, FIB Micromachined Submicron Thickness Cantilevers for the Study of Thin Film Properties, *Thin Solid Films*, Vol. 358, pp. 146–151, 2000.
22. M. TAHER, Design Construction and Optimization of a Hot Filament Chemical Vapor Deposition (HFCVD) Reactor for Diamond Coating of Cemented Carbide Tool Inserts, MS Thesis, Fayetteville: University of Arkansas, 1995.
23. M. V. MIRKIN AND B. R. HORROCKS, Electroanalytical Measurements Using the Scanning Electrochemical Microscope, *Anal. Chim. Acta*, Vol. 406, pp. 119–146, 2000.
24. S. E. MCBRIDE AND G. C. WETSEL JR., Nanometer-scale Features Produced by Electric-field Emission, *Appl. Phys. Lett.*, Vol. 59, pp. 3056–3058, 1991.
25. C. T. SALLING, I. I. KRAVCHENKO, AND M. G. LAGALLY, Atomic Manipulation for Patterning Ultrathin Films, *J. Vac. Sci. Technol.*, Vol. 13, pp. 2828–2831, 1995.
26. R. YANG, D. F. EVANS, AND W. A. HENDRICKSONG, Writing and Reading at Nanoscale with a Scanning Tunneling Microscope, *Langmuir*, Vol. 11, pp. 211–213, 1996.
27. F. P. ZAMBORINI AND R. M. CROOKS, Nanometer-Scale Patterning of Metals by Electrodeposition from an STM Tip in Air, *J. Am. Chem. Soc.*, Vol. 120, pp. 9700–9701, 1998.
28. P. KRUSE AND R. A. WOLKOWA, Gentle Lithography with Benzene on Si(100), *Appl. Phys. Lett.*, Vol. 81, pp. 4422–4424, 2002.
29. J. B. BINDELL, Scanning Electron Microscopy, in *Encyclopedia of Materials Characterization: Surfaces, Interfaces, Thin Films*, C. R. Brundle, C. A. Evans, Jr., and S. Wilson, eds., Boston: Butterworth-Heinemann, 1992, pp. 70–84.
30. T. M. MOORE AND R. G. MCKENNA, *Characterization of Integrated Circuit Packaging Materials*, Boston: Butterworth-Heinemann, 1993.
31. B. GATES, Y. WU, Y. YIN, P. YANG, AND Y. XIA, Single-Crystalline Nanowires of Ag_2Se Can Be Synthesized by Templating against Nanowires of Trigonal Se, *J. Am. Chem. Soc.*, Vol. 123, pp. 11500–11501, 2001.
32. S. W. PADDOCK, An Introduction to Confocal Imaging, in *Confocal Microscopy, Methods and Protocols*, Vol. 122, S. W. Paddock, ed., Totowa, NJ: Humana Press, 1999, pp. 1–34.
33. S. INOUE, Foundations of Confocal Scanned Imaging in Light Microscopy, in *Handbook of Biological Confocal Microscopy*, J. B. Pawley, ed., 2 ed., New York: Plenum, 1995, pp. 1–18.
34. M. GU, *Principles of Three-Dimensional Imaging in Confocal Microscopes*, NJ World Scientific, 1996.
35. V. CHEN, Non-Laser Light Sources, in *Handbook of Biological Confocal Microscopy*, J. B. Pawley, ed., 2nd ed., New York: Plenum, 1995, pp. 99–109.
36. H. E. KELLER, Objective Lenses for Confocal Microscopy, in *Handbook of Biological Confocal Microscopy*, J. B. Pawley, ed., 2nd ed., New York: Plenum 1995, pp. 111–126.

37. J. Art, Photon Detectors for Confocal Microscopy, in *Handbook of Biological Confocal Microscopy*, J. B. Pawley, ed., 2nd ed., New York: Plenum, 1995, pp. 183–196.
38. S. W. Paddock, T. J. Fellers, and M. W. Davidson, Introduction to Confocal Microscopy, Imaging Modes, *Nikon Microscopy*, vol. 2003, 2003.
39. NanoFocus μSurf® C Product Data Sheet, Germany: NanoFocus, 2003.
40. Z. P. Aguilar, W. R. Vandaveer IV, and I. Fritsch, Self-Contained Microelectrochemical Immunoassay for Small Volumes using Mouse IgG as a Model System, *Anal. Chem.*, Vol. 74, pp. 3321–3329, 2002.
41. W. R. Vandaveer IV and I. Fritsch, Measurement of Ultrasmall Volumes Using Anodic Stripping Voltammetry, *Anal. Chem.*, Vol. 74, pp. 3575–3578, 2002.
42. Z. P. Aguilar and I. Fritsch, Immobilized Enzyme Linked DNA-Hybridization Assay with Electrochemical Detection for *Cryptosporidium parvum* hsp70 mRNA, *Anal. Chem.*, Vol. 75, pp. 3890–3897, 2003.
43. W. R. Vandaveer IV, D. J. Woodward, and I. Fritsch, Redox Cycling Measurements of a Model Compound and Dopamine in Ultrasmall Volumes with a Self-Contained Microcavity Device, *Electrochim. Acta*, Vol. 48, pp. 3341–3348, 2003.
44. D. Briggs and M. P. Seah, *Practical Surface Analysis by Auger and X-ray Photoelectron Spectroscopy*, Vol. 1, 2nd ed. Chichester: Wiley, 1990.
45. N. H. Turner and J. A. Schreifels, Surface Analysis: X-ray Photoelectron Spectroscopy and Auger Electron Spectroscopy, *Anal. Chem.*, Vol. 68, pp. 309R–332R, 1996.
46. J. M. Cowley, Electron Microscopy, in *Surface Imaging and Visualization*, A. T. Hubbard, ed., Boca Raton, FL: CRC Press, 1995, pp. 131–156.
47. D. Frank, Angle-Resolved Auger Electron Spectroscopy, in *Surface Imaging and Visualization*, A. T. Hubbard, ed., Boca Raton, FL: CRC Press, 1995, pp. 1–22.
48. C. Klauber, X-ray Photoelectron and Auger Electron Spectroscopy, *Microbeam Anal.*, Vol. 4, pp. 341–368, 1995.
49. K. D. Childs, et al., *Handbook of Auger Electron Spectroscopy*, Eden Prairie: Perkin-Elmer Corp., 1995.
50. M. Sancrotti, Qualitative and Quantitative Surface Analysis via X-ray Photoemission and Auger Electron Spectroscopies, *Surface Rev. Lett.*, Vol. 2, pp. 859–883, 1995.
51. N. H. Turner, Auger Electron Spectroscopy, in *Surface Imaging and Visualization*, A. T. Hubbard, ed., Boca Raton, FL: CRC Press, 1995, pp. 33–44.
52. L. A. Harris, Analysis of Materials by Electron-Excited Auger Electrons, *J. Appl. Phys.*, Vol. 39, p. 1419, 1968.
53. J. T. Grant, Methods for Quantitative Analysis in XPS and AES, *Surf. Interface Anal.*, Vol. 14, pp. 271–283, 1989.
54. Y. E. Strausser, Auger Electron Spectroscopy, AES, in *Encyclopedia of Materials Characterization: Surfaces, Interfaces, Thin Films*, C. R. Brundle, C. A. Evans, Jr., and S. Wilson, eds., Boston: Butterworth-Heinemann, 1992, pp. 310–323.
55. P. W. Palmberg, Combined ESCA/ Auger System Based on the Double Pass Cylindrical Mirror Analyzer, *J. Electron Spectrosc.*, Vol. 5, p. 691, 1974.
56. P. W. Palmberg, Combined ESCA and Auger Spectrometer, *J. Vac. Sci. Technol.*, Vol. 12, pp. 379–384, 1975.
57. Active Standard: E1078-02 Standard Guide for Specimen Preparation and Mounting in Surface Analysis, in *Book of Standards*, Vol. 03.06, ASTM International, 2003.
58. C. J. Powell and M. P. Seah, Precision, Accuracy, and Uncertainty in Quantitative Surface Analysis by Auger Elelctron Spectroscopy and X-ray Photoelectron Spectroscopy, *J. Vac. Sci. Technol. A*, Vol. 8, pp. 735–763, 1990.
59. S. Mroz, Physical Foundation of Quantitative Auger Analysis, *Prog. Surf. Sci.*, Vol. 46, pp. 377–437, 1994.
60. M. A. Baker and J. E. Castle, Scanning Auger Spectroscopy, *Mater. Sci. Technol.*, Vol. 2B, pp. 219–239, 1994.
61. H. Oechsner, Scanning Auger Microscopy, *Vide Couches Minces*, Vol. 274, pp. 141–150, 1994.
62. J. J. Hickman, D. Ofer, C. Zou, M. S. Wrighton, P. E. Laibinis, and G. W. Whitesides, Selective Functionalization of Gold Microstructures with Ferrocenyl Derivatives via Reaction with Thiols or Disulfides: Characterization by Electrochemistry and Auger Electron Spectroscopy, *J. Am. Chem. Soc.*, Vol. 113, pp. 1128–1132, 1991.
63. G. F. Malgas, D. Adams, P. Nguyen, Y. Wang, T. L. Alford, and J. W. Mayer, Investigation of the Effects of Different Annealing Ambients on Ag/Al Bilayers: Electrical Properties and Morphology, *J. Appl. Phys.*, Vol. 90, pp. 5591–5598, 2001.
64. D. K. Skinner, The Role of Auger Electron Spectroscopy in the Semiconductor Industry, *Microsc. Microanal. Microstruct.*, Vol. 6, pp. 321–343, 1995.
65. E. Darque-Ceretti and M. Aucouturier, Surface and Interface Analysis in Materials Science, Method Section, Difficulties and Applications, *Analysis*, Vol. 23, pp. 49–58, 1995.
66. M. Seah and D. Briggs, A Perspective on the Analysis of Surfaces and Interfaces, in *Practical Surface Analysis by Auger and X-ray Photoelectron Spectroscopy*, M. Seah and D. Briggs, eds., New York: Wiley, 1983, pp. 1–16.
67. S. Thomas and A. Joshi, Failure Analysis of Microelectronic Materials Using Surface Analysis Techniques, presented at 28th National SAMPE Symposium, Anaheim, CA, 1983.
68. L. Muhloff, Photoelectron Spectroscopy in Semiconductor Surface Characterization, *Proc. Electrochem. Soc.*, Vol. 90, pp. 173–189, 1990.

69. T. M. Duc, Surface Analysis Applied to Polymers and Biological Samples, *Surf. Rev. Lett.*, Vol. 2, pp. 833–858, 1995.
70. P. M. A. Sherwood, X-Ray Photoelectron Spectroscopy, in *The Handbook of Surface Imaging and Visualization*, A. Hubbard, ed., New York: CRC Press, 1995, pp. 875–888.
71. P. M. A. Sherwood, Photoelectron Spectroscopy, in *Spectroscopy*, Vol. 3, B. J. Straughan and S. Walker, eds., London: Chapman and Hall, 1976.
72. D. R. Chopra and A. R. Chourasia, X-ray Photoelectron Spectroscopy, in *Handbook of Instrumental Techniques for Analytical Chemistry*, F. Settle, ed., Englewood Cliffs, NJ: Prentice Hall, 1997, pp. 809–827.
73. D. Briggs and J. Riviere, Spectal Interpretation, in *Practical Surface Analysis by Auger and X-ray Photoelectron Spectroscopy*, D. Briggs and M. P. Seah, eds., New York: Wiley, 1983, pp. 87–140.
74. J. C. Riviere, Instrumentation, in *Practical Surface Analysis by Auger and X-ray Photoelectron Spectroscopies*, D. Briggs and M. P. Seah, eds., New York: Wiley, 1983, pp. 17–84.
75. T. L. Barr, Electron Spectroscopic Analysis of Multicomponent Thin Films with Particular Emphasis on Oxides, in *Multicomponent and Multilayered Thin Films for Advanced Microtechnologies: Techniques, Fundamentals and Devices*, O. Auciello and J. Engemann, eds., Netherlands: Kluwer Academic, 1993, pp. 283–309.
76. S. Evans and M. D. Scott, Chemical and Structural Characterization of Epitaxial Compound Semiconductor Layers Using X-ray Photoelectron Diffraction, *Surf. Interf. Anal.*, Vol. 3, pp. 269, 1981.
77. R. G. Copperthwaite, The Study of Radiation-Induced Chemical Damage at Solid Surfaces Using Photoelectron Spectroscopy: A Review, *Surf. Interf. Anal.*, Vol. 2, p. 17, 1980.
78. C. D. Wagner, L. E. Davis, M. V. Zeller, J. A. Taylor, R. M. Raymond, and L. H. Gale, Empirical Atomic Sensitivity Factors for Quantitative Analysis by Electron Spectroscopy for Chemical Analysis, *SIA, Surf. Interf. Anal.*, Vol. 3, pp. 211–225, 1981.
79. K. Artyushkova, B. Wall, J. Koenig, and J. E. Fulgham, Correlative Spectroscopic Imaging: XPS and FT-IR Studies and PVC/PMMA Polymer Blends, *Appl. Spectrosc.*, Vol. 54, pp. 1549–1558, 2000.
80. A. Benninghoven, F. G. Rudenauer, and H. W. Werner, *Secondary Ion Mass Spectrometry, Basic Concepts, Instrumental Aspects, Applications and Trends*, New York: Wiley, 1987.
81. J. C. Vickerman, A. Brown, and N. W. Reed, *Secondary Ion Mass Spectrometry Principles and Applications*. Oxford: Clarendon, 1989.
82. R. G. Wilson, F. A. Stevie, and C. W. Magee, *Secondary Ion Mass Spectrometry*, New York: Wiley, 1989.
83. P. K. Chu, Dynamic Secondary Ion Mass Spectrometry, in *Encyclopedia of Materials Characterization: Surfaces, Interfaces, Thin Films*, C. R. Brundle, C. A. Evans, Jr., and S. Wilson, eds., Boston: Butterworth-Heinemann, 1992, pp. 532–548.
84. B. Katz, Static Secondary Ion Mass Spectrometry, in *Encyclopedia of Materials Characterization: Surfaces, Interfaces, Thin Films*, C. R. Brundle, C. A. Evans, Jr., and S. Wilson, eds., Boston: Butterworth-Heinemann, 1992, pp. 549–558.
85. A. Benninghoven, B. Hagenhoff, and E. Niehuis, Surface MS: Probing Real-World Samples, *Anal. Chem.*, Vol. 65, pp. 630A–640A, 1993.
86. N. Winograd, Ion Beams and Laser Postionization for Molecule-Specific Imaging, *Anal. Chem.*, Vol. 65, pp. 622A–629A, 1993.
87. J. A. Gardella, Jr., Secondary Ion Mass Spectrometry, in *Surface Imaging and Visualization*, A. T. Hubbard, ed., Boca Raton; FL: CRC Press, 1995, pp. 705–712.
88. M. A. Ray, E. A. Hirsch, and G. E. McGuire, Secondary Ion Mass Spectrometry, in *Handbook of Instrumental Techniques for Analyical Chemistry*, F. Settle, ed., Englewood Cliffs, NJ: Prentice Hall, 1997, pp. 829–847.
89. R. A. Hockett, Ultratrace Impurity Analysis of Silicon Surfaces by SIMS and TXRF Methods, in *Handbook of Semiconductor Wafer Cleaning Technology*, W. Kern, ed., Park Ridge, NJ: Noyes, 1993, pp. 537–592.
90. H. J. Borg and J. W. Niemantsverdriet, Applications of Secondary Ion Mass Spectrometry in Catalysis and Surface Chemistry, *Catalysis*, Vol. 11, pp. 1–50, 1994.
91. W. Husinsky and G. Betz, Fundamental Aspects of SNMS for Thin Film Characterization: Experimental Studies and Computer Simulations, *Thin Solid Films*, Vol. 272, pp. 289–309, 1996.
92. S. G. Mackay and C. H. Becker, Surface Analysis by Laser Ionization, SALI, in *Encyclopedia of Materials Characterization: Surfaces, Interfaces, Thin Films*, C. R. Brundle, C. A. Evans, Jr., and S. Wilson, eds., Boston: Butterworth-Heinemann, 1992, pp. 559–570.
93. H. W. Werner and H. van der Wel, Some Applications of Secondary Ion Mass Spectrometry in the Micro-Electronics Industry, *Anal. Methods Instrum.*, Vol. 2, pp. 111–121, 1995.
94. D. Briggs, Polymer Surface Characterization by XPS and SIMS, in *Characterization of Solid Polymers, New Techniques and Developments*, New York: Chapman & Hall, 1994, pp. 312–360.
95. S. Pignataro, Surface Analysis in Microelectronics, *Fresenius J. Anal. Chem.*, Vol. 353, pp. 227–233, 1995.
96. I. Vidavsky and M. L. Gross, High-Resolution Mass Spectrometry of Volatiles and NonVolatives, in *Handbook of Instrumental Techniques for Analytical Chemistry*, F. Settle, ed., Eaglewood Cliffs, NJ: Prentice Hall, 1997, pp. 589–608.
97. R. J. Cotter, Time-of-Flight Mass Spectromety for the Structural Analysis of Biological Molecules, *Anal. Chem.*, Vol. 64, pp. 1027A–1039A, 1992.
98. I. Gay and G. H. Morrison, Chemical Imaging Using Ion Microscopy, in *Surface Imaging and Visualization*,

99. J. M. CHABALA, K. K. SONI, K. L. LI, K. L. GAVRILOV, AND R. LEVI-SETTI, High-Resolution Chemical Imaging with Scanning Ion Probe SIMS, *Intl. J. Mass Spectr. Ion Process.*, Vol. 143, pp. 191–212, 1995.
100. M. J. GRAHAM AND R. J. HUSSEY, Analytical Techniques in High Temperature Corrosion, *Oxidation Metals*, Vol. 44, pp. 339–374, 1995.
101. R. PRESCOTT, D. F. MITCHELL, AND M. J. GRAHAM, in *Microsopy of Oxidation, Proc. Int. Conf.*, S. B. Newcomb and M. J. Bennett, eds., London: Institute of Materials, 1993, pp. 455–462.
102. R. B. CODY, A. BJARNASON, AND D. A. WEIL, Applications of Laser-Desorption-Fourier Transform Mass Spectrometry to Polymer and Surface Analysis, in *Lasers and Mass Spectrometry*, D. M. Lubman, ed., New York: Oxford University Press, 1990, pp. 316–339.
103. J. R. SCOTT, L. S. BAKER, W. R. EVERETT, C. L. WILKINS, AND I. FRITSCH, Laser Desorption Fourier Transform Mass Spectrometry Exchange Studies of Air-Oxidized Self-Assembled Monolayers on Gold, *Anal. Chem.*, Vol. 69, pp. 2636–2639, 1997.

A. T. Hubbard, ed., Boca Raton, FL: CRC Press, 1995, pp. 45–53.

104. Mass and Optical Spectroscopies, in *Encyclopedia of Materials Characterization: Surfaces, Interfaces, Thin Films*, C. R. Brundle, C. A. Evans, Jr., and S. Wilson, eds., Boston: Butterworth-Heinemann, 1992, pp. 527–531.
105. Y. WARD, R. J. YOUNG, AND R. A. SHATWELL, A Microstructural Study of Silicon Carbide Fibres through the Use of Raman Microscopy, *J. Mater. Sci.*, Vol. 36, pp. 55–66, 2001.
106. R. A. BENNETT, C. L. PANG, N. PERKINS, R. D. SMITH, P. MORRALL, R. I. KVON, AND M. BOWKER, Surface Structures in the SMSI State; Pd on (1x2) Reconstructed $TiO_2(110)$, *J. Phys. Chem. B*, Vol. 106, pp. 4688–4696, 2002.
107. M. ETIENNE, E. A. CLARK, S. R. EVANS, W. SCHUHMANN, AND I. FRITSCH, Feedback-Independent Pt Nanoelectrodes for Shearforce-Based Constant-Distance Mode Scanning Electrochemical Microscopy, submitted to *Anal. Chem.*, 2005.
108. C. PALACIO AND A. ARRANZ, Chemical Structure of Ultathin Silicon Nitride Films Grown by Low-Energy (0.25–5 keV) Nitrogen Implantation: An Angle-Resolved X-ray Photoelectron Spectroscopy Si 2p Study, *J. Phys. Chem. B*, Vol. 106, pp. 4261–4265, 2002.

EXERCISES

18.1. A diffracted X-ray beam is observed from the (220) planes of a thin film of iron at a 2θ angle of 99.1°. When a copper tube, with $\lambda = 1.5418$ Å, is used to produce X rays, calculate the lattice parameter of the iron.

18.2. Given the data in Figure 18.6, calculate the thickness of the gold film if the wavelength used had been $\lambda = 1.392$ Å.

18.3. A sample shows a Raman shift of 700 cm^{-1} when excited by a 632.8-nm He–Ne laser. What are the wavelengths of the Stokes and anti-Stokes Raman lines?

18.4. Describe why depth of field in a scanning electron microscope is the best among all available types of microscopy.

18.5. An accelerating Voltage of 8 keV was used for the EDX analysis of a thin film. Approximately what accelerating voltage should be used for the EDX analysis of a bulk sample of the same composition as the thin film?

18.6. In Figure 18.19, the peak at ~9 keV is the Cu K_β peak. (These X-ray photons are generated as electrons relax from the M shell to the K shell of Cu). The peak at ~8 keV is the Cu K_α peak. Based on your knowledge of atomic energy levels, from which electron relaxations do you think these X-ray photons arise? Draw a diagram to explain why the K_α peak occurs at a lower energy than the K_β peak.

18.7. Describe how the energy resolution in the EDX technique is measured and compared between two different instruments.

18.8. In order to evaluate the surface roughness of an oxidized silicon wafer, one would select SFM rather than STM. Why? Compare/contrast these two techniques.

18.9. Select a proper technique to evaluate the hydrogen content of a diamond film and describe advantages/disadvantages of the selected method with at least one other technique.

18.10. Why is it not possible for AES to detect H or He?

18.11. Calculate the kinetic energy of a peak in an AES spectrum for a *KLL* electron for C. The binding energy of a $1s$ electron is 290 eV, that of a $2s$ electron is 20 eV, and that of a $2p$ electron is 10 eV.

18.12. Given the information about carbon in Exercise 18.11, what binding energies would you expect peaks in an XPS spectrum to have? Assume that the X-ray source is a 1486.6-eV monochromated Al K_α.

18.13. In an XPS spectrum of C it may be possible to observe the *KLL* Auger peak in addition to the XPS peaks. Why would changing the source from Al K_α to Mg K_α allow you to identify which are due to the Auger process and which are due to the XPS process?

18.14. Which will show an XPS peak at a higher binding energy? (a) the Ti $2p$ peak of Ti or TiO_2? and (b) the $1s$ peak of C or Si? Why?

18.15. In a typical XPS spectrum, there is a background that extends from each peak toward higher binding energies. It is especially noticeable for large peaks. The background is due to inelastically scattered photoelectrons. Explain why the background appears at higher binding energies and not lower binding energies.

18.16. A thin-film microfabricated feature of 4 μm × 4 μm is not performing as it should. An analysis of its surface composition is needed to find out if it is contaminated from the processing or mixed with underlying materials due to diffusion at high processing temperatures. What analysis technique(s) and accessories (if any) would you choose and why?

18.17. Features of 1 mm × 1 mm feature on a microfabricated circuit were electrodeposited with a binary metal alloy. The desired concentrations are 20% Ni and 80% Sn. What technique(s) would you use to determine its actual composition?

18.18. A semiconductor is not performing well. You suspect that *trace* inorganic contaminants may be the cause. What technique(s) would you use to identify the trace elements and why? Identify some of the instrumental components that would make this possible.

18.19. A high-purity silicon is analyzed by SIMS using an oxygen ion beam. What ions would you assign to the following m/z values? (Note that the ion beam species can combine with the sample. Also, you will need a list of isotopic masses or at least a periodic table to answer this question.) $m/z = 15.995, 27.9769, 43.972, 55.9538, 71.9488, 87.9438$.

18.20. An engineer wants to know the composition of a YBCO superconducting thin film so that she can determine if the composition can explain the film's conducting properties. She has access to an AES instrument to perform the analysis. Why would it be important for her to sputter through the uppermost atomic layer(s) of the sample first, before AES analysis?

18.21. Si_3N_4 is a material that can be deposited as a thin film using plasma-enhanced chemical vapor deposition (PECVD) from SiH_4 and NH_3 gases. The parameters of the deposition must be carefully controlled and the gasses must be pure to achieve the proper 3-to-4 ratio of Si-to-N in the material. Not only is the ratio important but also the oxidation state of the Si is important. XPS analysis is often used to evaluate the quality of Si_3N_4 films. A spectrum of a particular sample shows a very large peak at 101.9 eV. Smaller peaks appear at 99.6 and 103.4 eV. What can you conclude about this sample? (Information in Table 18.3 is helpful in solving this problem.)

Index

A

Absolute substrate connection/bonding, 603
Accelerated testing, 661–667
 acceleration transforms, 664
 electrostatic discharge testing, 666
 HAST and 85/85 testing, 663
 reliability metrology, 668
 saturated vs. unsaturated, 663
 temperature cycling and shock, 666
 test structures for, 667
Accelerators, 91
 for epoxy cure, 50–52
Acetone, 96
Activation energy, 88
Active cooling of components, 614
Active devices in systems, 351–352
Active metal brazing, 182
Adaptive lithography, 594, 595
Additive processes, 90, 94
 embedded passives 357, 362, 363, 371
 metallization, 93
Adhesion, 10, 30, 49, 95
 interfacial, 31
 PCB layers, 131
 promoter, 49
Adhesion promoters, 95, 101
Adhesives, 5
Advanced Technology Program (ATP), 584
Aerospace systems, 586
Agglomeration, 91
Ag2Se nanowires, 749
Air
 bridge structures, 609
 bubbling, 96
 cooling, 282
Aligner, 95
Alignment patterns, 95
Alkaline
 aqueous-based liquid, 95
 developers, 93
 electroless copper bath, 91
Alloys, 82, 101
 sputtering targets, 84
Alternating current impedance, 613
Alumina, *see* Aluminum oxide
Aluminum, 71, 81
Aluminum nitride, 35, 40, 71, 591

Aluminum oxide, 37, 71, 150, 153, 591
 tape-cast, 40
Ambient gas, 81
Amorphous, 86
Analytical techniques, 726
Angle of incidence, 85
Angle-resolved Auger electron spectroscopy, 753, 774
Anisotropic, 98
Anode, 84, 86, 90
Anodization, 155, 356, 370
Anti-Stokes emission, 735
Application-specific integrated circuits (ASICs), 449, 611
 design, 465
Applied electric field, 90
Artwork for PCB's, 112
Aspect ratio, 87
Assembly, 389, 390, 395
Atmospheric pressure CVD (APCVD) 88, 89
Atomic beam, 81
Atomic force microscopy, 727, 734
Atomic weight, 92
Attenuation coefficient, 491
Auger electron microscopy, 726, 752–766
 angle-resolved, 753, 766
 sensitivity factors, 761
Autocatalyzed reduction of metal, 90
Automatic test pattern generation (ATPG), 615
Average kinetic energy, 85
Avogadro's number, 92
Axial force, 313

B

Back-to-back flip-chip schemes, 601
Backgrinding, 617
Backplane, 443
Backscattered electrons (BSE), 745
Backside
 bonding, 101
 deposition, 621
 oxide, 621
 removal, 620
Backstreaming, 82
Backward traveling wave, 228
Ball grid array (BGA), 585, 595, 604, 617, 621
Bandwidth, 203, 489
Bare die testing, 17
Barium titanate, 364–367
 in polymer thick film dielectrics, 370

Barrel vias, 617
Barrier layer metal, 60, 593
Base pressure, 85
Basket filament, 83
Baths for plating, 91
Bathtub curve, 653, 680
Beam, 81
    atomic and molecular, 81
    ion, 98
    lead, 101
Bed of nails tester, 457
Belt furnace, 101
Bending strength, 40
Benzene, 50
Benzocyclobutene (BCB), 46, 71, 72, 364-370, 593
Beryllia, *see* Beryllium oxide
Beryllium oxide, 35–39, 43, 71, 151, 591
Binders, 36, 40
    organic, 36, 39
Binding energy, 767, 769
Biot-Savart law, 195
Bis-aryl diazide, 93
Bisbenzo-cyclobutenes, 71
Bismaleimide triazine (BT) resin, 67
Boat filament, 83
Boiling, 283
Boltzmann constant, 78
Bond-over-bond connections, 604
Bond
    pads, 596, 597
    repatterning, 605
    stress, 326
Bonding, 5, 603
    covalent, 35
    dipole interaction, 45
    eutectic, 5, 58
    flip-chip, 5, 6, 58–62
    forces, 35, 45
    gang, 7
    hydrogen, 45
    inner and outer lead, 7
    intrastack, 603
    ionic, 35, 45
    primary covalent, 45
    relative, 603
    secondary, 45
    silicon-oxygen, 52
    solid-state, 98
    tape-automated, 4–6, 50, 58, 60, 90, 98
    thermocompression, 7, 59
    thermosonic, 59
    ultrasonic, 59
    wire, 5, 6, 58, 98
    Van der Waals, 45
Bondpads, 98, 99
Border
    array, 617, 618
    grid, 617

Boron nitride, 35, 41
    diamond and hexagonal structure, 41
Borosilicate glass, 42
Boundary scan, 615
Bragg's condition, 729
Brassboard, 615
Brazed metal ring, 98
Brazing, 7, 98, 101, 591
Break even analysis, 696
Brighteners, 93
Brittleness, 98
Brittle rupture, 318, 320
Buffers, 91, 96
Bulk precipitation, 91
Bump pad, 102
Bumping, 61
    adhesive, 421
    copper pillar, 421
    gold stud, 421
    metallurgy, 617
    polymer bump, 421
    solder plate, 420
    solder print, 420
Bumpless build-up layer (BBUL) technology, 596
Buried components, 614

C

CAD/CAM, 107
Calcium oxide, 37
Calibration methods, 526
Camber, 154
Capacitance, 200
    interconnects, 188
Capacitors, 31, 363. *See also* Passives and Embedded passives
    decoupling, 381–384
    dielectric materials, 364–369
    electrical performance, 374–376
    lumped, 515
    parasitic properties, 376
    paraelectrics and ferroelectrics, 365
    self-resonance frequency, 375
    sizing, 367
    temperature coefficient of capacitance (TCC), 365
    value range in systems, 354
Capillary head, 99
Carbides, 35
Carboxylic acids, 91–95
Carrier mobility, 32
Casting, 37
    clip, 37
    tape, 37, 40
Catch pads, 448
Cathode, 84, 86, 90
CCl4 plasma, 96
Cell phones, 17
    number of components in, 352–354
Cell potential, 92

Index **795**

Centerline average, 154
Ceramic capillary, 99
Ceramic chip carriers, 93, 101
Ceramics, 29, 149
   amorphous, 35
   cleanroom environments, 166
   crystalline, 35
   design considerations, 178
   firing, 175
   glass, 35
   High temperature cofired ceramics (HTCC), 70, 163, 180
   inserts, 590
   lamination, 173
   Low temperature cofired ceramics (LTCC), 70, 163
   material choices, 150
   polycrystalline, 35
   reasons for substrates, 150
   screen printing, 156–159
   substrates, 70, 149, 584
   thin films on, 155
Cermets, 357–362
Chamber, 82
   glass bell jar, 82
   stainless steel, 82
Chamber volume, 80
Characteristic impedance, 71, 214, 220, 239, 593
   coaxial waveguides, 498
Charge, 32
   electronic, 32
   ionic, 32
   neutralization, 770, 782
Charge accumulation, 86
Chemical etch, 97
Chemically adsorbed (chemisorption), 81
Chemical polishing, 620
Chemical reactions, 88
Chemical vapor deposition (CVD), 42, 77, 88
   atmospheric pressure, 88, 89
   low pressure, 89
   plasma-enhanced, 89
   processes, 84
Chemisorption, 81
Chip carrier, 4, 10, 33, 99
   ceramic, 14
   polymeric film-based, 57
   TAB, 57
Chip-on-board (COB), 4, 15, 43, 67, 584, 595
Chip-on-chip stack, 601, 604
Chip-on-flex (COF), 595, 615, 616, 621
Chip-scale package (CSP), 14, 585, 621
   ceramic carrier, 14
   compliant members, 14
   lead frame, 14
   lead-on-chip, 14
   mechanical protection, 14
   resin mold, 14
   silicon-based rigid laminate, 14
   space transformers, 14
   stacked, 622
   tape carrier-flexible laminate, 14
Chip stacks, 590
   homogeneous, 604
Chip-to-bond pads, 98
Chlorine-based plasmas, 98
Chloroethylene, 43
Circuit, 37
   flex, 68
   hybrid, 37
Circuit board, 33, 77
Circuit board fabrication, 95, 96
Circuit support and protection, 3
Clamshell, 606
Cleanrooms, 106
   ceramic processing, 166
Clock distribution, 452
Coalesce, 81
Coaxial waveguides, 498
Co-deposition, 82
Coefficient of thermal expansion (CTE), 30–33, 47, 54, 66, 339, 659
   ceramics, 165
   printed wiring boards, 110, 129
Co-fired metallization, 40, 41
Cohesive strength, 49
Cold-wall reactor, 88, 89
Collision-free path, 78
Colorants in polymers, 50, 52
Common substrate, 583
Comparative cost analysis, 714
Complexant, 91
Complexed radical, 96
Complimentary metal oxide semiconductor (CMOS), 21
Component
   attach, 596
   density, 598
   mounting surfaces, 598
Compression head, 99
Concentration gradient, 92
Conducting materials, 86
Conduction
   heat, 251, 257
Conductive
   adhesive, 597
   bridge, 597
   particles, 597
   paste, 591
Conductivity, 21, 92, 593
   anisotropic, 63
   electrical, 67, 491
   thermal, 21, 34–42, 50, 54, 66
Conductor
   films, 89
   foils, 89
   losses, 512
   materials, 32, 69
   metals, 77

Confocal microscopy, 750
Conformal coating, 77
Conformal films, 84
Connection density, 618
Connectors, 90
  copper-based, 90
  electronic, 90
  pad-to-pad, 90
  spring-to-pin, 90
Constant current plating, 92
Consumer applications, 587
Contact density, 597
Contact printing, 95
Contact resistance
  electrical, 192
  thermal, 261
Contamination, 85
  surface, 85
Continuous-feed reactor, 88
Contract review, 125
Controlled collapse chip connect (C4), 12, 61, 101, 597
Convection, 88, 252, 556
Conventional circuit boards, 89
Conveyor belt, 88
Coolant, 79
Cooling, 282
Coplanar waveguides, 501
Copolymerization, 44
Copolymers, 44
Copper, 71
  etching, 114–116
Copper-based connectors, 90
Copper-plated substrate, 94
Copper plating bath, 91
Copper/polyimide (Cu/PI) interconnect structure, 593
Copper sulfate, 96
Copper sulfate/sulfuric acid bath, 92
Cordierite, 42
Cordwood assemblies, 583
Corningware, 42
Corrosion, 67, 656
Corrosion-resistant metal pads, 77, 98
Cosine law, 82
Cost
  estimation, 20, 691
  of interconnectivity, 439
  profit model, 463
Coulombs, 92
Couplers, 522
Crazing, 48
Creep, 311
Cross-linked negative photoresist, 95
Cross-linking in polymers, 45, 93
  covalent, 45, 46
  density, 48, 52
Crosstalk, 6, 21, 226
CrSi, 357, 360
Cryogenic pump, 80, 81

Crystal structure, 731
Crystalline materials, 728
Crystallinity, 86
Curing, 45, 597
  carboxylate, 53
  curing agent, 50, 52
Curling, 620
Current, 8
  alternating, 33
  direct, 33
  junction leakage, 8
Curvature, 314
Cutoff frequency, 499
CVD-deposited film, 88
CVD reactors, 88
Cyclized polyisoprene (rubber), 93
Cylindrical
  magnetron sources, 86
  mirror analyzer, 758

D

Dangling hydroxide bonds, 593
Daughterboards, 443
DC
  diode discharges, 86
  diode sputtering, 85
  sputtering, 87
Debye optics, 730
Decapsulation, 661
Decoupling capacitors, 71, 202, 381–384, 453
Defects and structural disorder, 737
Defect tolerance/full redundancy, 608
Defense Advanced Research Projects Agency (DARPA), 584
Deflocculants, 39
Deformation zone formation, 48
Degree
  of cure, 48
  of planarization, 49, 66
  of polymerization, 43
Delamination, 48, 49
Delta I noise, 196, 233
Demountable 3D assemblies, 601
Department of Commerce, 584
Deposited thin-film substrates, 71
Deposition, 85
  high rate, 84
  ion-beam-assisted deposition (IBAD), 87
  line-of-sight, 81
  plasma-enhanced, 85
  sequential, 82
  sputter, 85
  thin film, 84, 85
  vapor-phase, 77
Deposition chamber, 82
Deposition techniques
  anodization, 155, 356, 370
  on ceramics, 155–161

Design, 21, 437, 468
    ASIC, 465
    for test (DFT), 615
    rules in ceramics, 154, 178
Desmear, 119
Desorbed, 81
Desorption, 78
Developer, 95
Development cycles, 460
Diamond, 43
Die attach, 5, 48, 50, 101, 593
    adhesive, 595
    flip-chip, 101
Die attachment, 405, 423
Dielectric, 32, 50, 89
    losses, 200, 223, 500, 510
    strength, 24, 32, 40, 66
Dielectrics, 71
    benzocyclobutene (BCB), 364-370
    breakdown, 68
    capacitors, 369
    ferroelectrics, 365–367
    in PCBs, 127
    interlayer, 72
    intermetallic, 50
    materials, 593
    metallization, 72
    paraelectrics, 365–367
    polymer, 593
    substrate materials, 71
    thick film, 162
Dielectric constant 21, 200, 363–371, 593, 613
    capacitors, 363, 369
    high frequency, 511
    in-phase, 33
    loss, 19, 37
    out-of-phase, 33
    relative, 32
    thin films, 77, 89
Die/package stacks, 501, 602–604
Die thinning, 620
Differential expansion, 613
Diffusion
    barrier, 10
    processing, 88
    pump, 80
    region, 92
    surface, 87
Digital light processing (DLP), 636
Diglycidyl ether of bisphenol-A (DGEBA), 51
Diluent gases, 89
Dimensional stability, 110
    constrained platforms/systems, 583, 609, 612, 615
Dimple filament, 83
Diode sputtering systems, 86
Dipolar complexes, 92
Direct
    bonded copper on ceramic, 181

chip attach (DCA), 13, 15
    metallization, 121
    substrate connection/bonding, 603, 604
    write lithography, 595
Direct current sputtering, 85
Direct-drive pumps, 79
Direct-drive, dual-stage, oil-sealed pumps, 79
Discrete electronic components, 583
Dispersants, 39
Dispersive instruments, 736
Displacements, 300
Disposable emulsion mask, 95
Dissipation factor, 31, 152, 364
Distortion, 613
    energy, 318-320
Distributed circuit analysis, 209, 518
Doctor blading, 36, 39
Double-crystal diffractometer, 732
Double-sided substrates, 601
Drilling, 594
Drivers, 205, 235
Dry etching, 96, 97
Dry-film photoresist, 94, 95
Dry pressing, 40
Dual-in-line package (DIP), 4, 36, 585
Dual-stage pumps, 79
Ductile, 98
Ductile rupture, 318
Ductility, 24, 593
Ducts, flow in, 275
Dyes, 95

E
E-glass, 67, 109
Economics of scale (EOS), 622
Edge
    bead, 94
    conductors, 606
    metal HDI system, 619
    overlay, 605, 606
    plane connections, 600, 601
Efficiency, 609
    wiring, 609
    routing algorithm, 609
85/85 testing, 663
Elastic modulus, 48, 308
Electrical
    conductivity, 592
    modeling, 187
Electric arc welding, 98
Electric field, 86,
    applied, 90
Electrical properties of materials
    ceramics, 152
    dielectrics, 363
    metals, 190
    resistors, 162, 361
Electrochemical, 90

Electroless
  Cu, 120–122
  nickel plating, 60
  plating, 77, 89–94
Electrolyte, 90, 92
Electromagnetic
  field analysis, 451
  interference, 576
Electromigration, 191
Electron
  beam evaporation, 83
  cyclotron resonance, 89
  cyclotron resonance (ECR) microwave plasma source, 87
  dispersive X-ray spectroscopy (EDX), 754
  emitting filament, 84
  energy analyzer, 758
  flood gun, 779
  tube, 37
Electronic
  connectors, 90
  packaging, 102
Electroplating
  copper, 60
  techniques, 77, 89, 90, 94
Electrostatic
  discharge, 660
  fields, 84
Element attach, 596
Embedded passives, 349. See also Passives, Resistors, Capacitors
  capacitors, 363
  decoupling, 381
  definitions, 354
  dielectrics, 369
  electrical characterization, 373
  inductors, 371
  motivation and problems, 378
  paraelectrics and ferroelectrics, 365
  parasitic properties, 361
  resistors, 358
  sizing, 360, 367
  systems, 586
  thick films, 363, 370
  value range in systems, 354
Emulsion, 95
Encapsulants, 7
Encapsulation, 425. See also Sealing
  non-hermetic, 427
Energy, 33, 81
  activation, 88
  average kinetic, 85
  kinetic, 81, 84
  power, 590
  signal, 590
  surface, 81
  thermal, 81, 590
  threshold, 85
  translational, 33

  vibrational, 33
  volume, 81
Energy dispersive X-ray analysis, 727, 744
Energy of bombarding ions, 85
Engineered release layer, 620
Environmental
  protection, 1
  SEM, 746
Epitaxial liftoff (ELO), 620
Epoxy, 7, 30, 46–50
  bisphenol, 51
  polyfunctional, 67
Equipment, 404
ESCA, see X-ray photoelectron spectroscopy
Etch
  chemistries, 95
  processes, 88
  rate, 96
  recipes, 95
  uniformity, 96
Etchant solution, 96, 97
Etchback, 119
Etching, 77, 85, 93, 96
  cross section, 620
  dry, 96, 97
  rate, 620
  reaction, 97
  reactive ion, 85
  selectivity, 96
  techniques, 93
  wet, 96
Eulerian cradles, 730
Eutectic, 57, 101
Evacuation, 78
Evaporation, 77–87
  electron beam, 83
  on ceramics, 155
  sequential, 84
  thermal, 83
Evaporator systems, 82
Exaltants, 91
Expansion
  mismatch, 6
  thermal, 39, 40
Exposure time, 95
Extenders for polymers, 50

F
Fablink, 482
Fabrication yield loss, 20
Face-centered cubic structure, 728
Face-up, 593
Failure analysis, 660
Failure mechanisms, 655
  corrosion, 656
  electrostatic discharge, 660
  mechanical stress, 659
  thermal, 250

Faraday constant, 92
Faraday's laws of electrolysis, 91
Fatigue, 321-323
Feedthroughs, 618
Ferric chloride, 96, 97
Ferroelectrics, 365–367
Fiducial cues, 595
Field-programmable gate array (FPGA), 443, 586, 596, 608–615
Figure of merit, 444
Filament, 83
    basket, 83
    boat, 83
    dimple, 83
    electron-emitting, 84
    spiral, 83
    wire, 83
Fillers for polymers, 52
Filters, 519
Fine-line
    processes, 90
    PWBs, 585
Fine-pitch solder ball array, 597
Finite element methods, 335-341, 524
Firing
    in ceramics, 164, 175
    processes, 592
First incident switching, 222
Flame retardants, 50, 52
Flammable, 89
Flap, 607
Flexible leads, 100
Flexible substrate, 599
Flip-chip, 417, 597, 604
    back-to-back schemes, 601
    bonding, 101
    C4 technology, 77
    copper pillar bump, 421
    face-to-face, 605
    gold stud bump, 421
    plated solder bump, 420
    polymer bump, 421
    printed solder bump, 420
Floorplan, 586
    MCM, 586
    overhead, 609
Flow-through systems, 614
Fluorine-based plasmas, 98
Fluorocarbon, 47
Fluoropolymer, 47, 68
Flux, 422
    brush, 423
    cleaning, 424
    density, 195
    dip, 422
    dispense, 423
    spray, 423
Forced convection, 268

Forecasting, 703
Footprint, 597
    substrate, 603
Formaldehyde, 52
Forward traveling wave, 228
Fosterite, 70
Fracture, 320
Free convection, 276
Frequency range, 487–489
FR4, 46, 67
    dielectric properties, 127
    processing, 131–134
Functional block, 443
Fusion welding, 98
Fuzz button, 602

G
Galvanostatically, 92
Gang bonding, 101, 597
Gases, 89
Gas phase, 77
Gate count, 610
Gerber, 106
Glass, 5, 71
    borosilicate, 42
    fabric, 108
    metal-filled, 5
    silver-filled, 57
    spodumene, 42
    vitreous sealing, 9
Glass beads, 590
Glass bell jar chamber, 82
Glass-ceramic, 35, 42, 71
    cofired, 42
Glass-epoxy laminate, 94
Glass transition temperature, 24, 33, 34, 44–50
    in PCB's, 121
Glazing, 39
Glob-top, 43, 50, 55
Global I/O count, 612
Glow discharge, 85
Glue lines, 617
Gold, 96
Gold chloride radicals, 96
Goniometer, 729
Grain size, 86, 93
Graphitic carbon, 736
Green sheet, 36, 39, 42
    ceramic, 36
Grid array assemblies, 598
Grit, 620
Growth mechanisms, 90

H
Half-cell potentials, 97
Hardeners, 50
Hard soldering, 101
HAST, 663

Hazard rate, 672
Heat, *see also* Thermal
  dissipation, 3
  embedded resistors, 361
  pipes, 286
  radiation, 254
  sinks, 266, 441
  sources, 247
  spreaders, 39, 565, 614
  transfer coefficients, 253
  transport, 251, 589
Heated substrate, 88
Helmholtz
  double layer, 92
  equation, 497
Hermetic
  digital mirror device (DMD) packaging, 646
  enclosures, 30
  packaging, 658
  sealing, 590
Hermeticity, 8, 24
Hermetic seal, 98
Hermetic sealed packages, 101
Heterogeneous reduction, 91
High-aspect-ratio through holes, 92
High-density interconnect (HDI)
  bumpless build-up layer (BBUL), 596
  chips first, 596
  edge-metal, 619
  modules, 15, 29, 66
  patterned overlay, 621
  PCB's, 134–140
  process, 593, 596, 616, 617
High-density plasmas, 89
High-end test equipment, 586
High-melting-point materials, 83
High-power amplifiers, 584
High-purity metals, 84
High-rate deposition, 84, 89
High-selectivity wet etchants, 96
High speed bath, 92
High-temperature
  electronics, 576
  furnace processes, 88
High temperature cofired ceramics (HTCC), 163, 180, 591
High-throw bath, 92
High vacuum, 78, 80
  pumps, 80
High vapor pressure components, 79
Highly integrated packaging and processing (HIPP), 606, 607
Highly-oriented pyrolytic graphics (HOPG), 736
Hitaceram, 41
Homopolymers, 44, 45
Hooke's law, 307
Horizontal reactor, 88
Hot air solder leveling (HASL), 123
Hot pressing, 40, 41

Hot pressed sintered disks, 85
Hot-wall reactor, 88, 89
Hybrid
  ceramic, 584
  chip and wire, 583
  microcircuits, 37, 70, 584, 585
Hydrocarbon, 47
Hydrocarbon contamination, 80
Hydrogen
  evolution, 92
  peroxide/sulfuric acid, 95, 96
  torch, 99
Hydrolysis, 40
Hydrolytic molecular breakdown, 49
Hydrostatic stress, 319

I

Image recognition, 100
Imide functionality, 53
Imidization, 53
Immersion cooling, 250
Impedance, 373, 490
Incident particle energy, 81
Inductance
  capacitors, 376
  mutual, 21, 211
  parasitic, 21, 199
  self and mutual, 194
  signal path, 199
Inductors, 31, 371. *See also* Passives and Embedded passives
  electrical performance, 373, 377, 516
Inert atmosphere, 40
Inert fillers, 50, 52
Inner lead bond pattern, 596
Input/Output (I/O)
  area, 600
  border, 600
  bound, 596
  configurations, 449, 600
  impedance, 218
  perimeter, 600
  terminals, 600
Insertion loss, 219
Inspection, 124
Institute for Printed Circuits (IPC), 141
Insulating epoxy-glass laminates, 90
Integrated circuit (IC), 1
Integrated passives, *see* Embedded passives
Intellectual property, 615
Intensity, 95
Interconnected mesh power system (IMPS), 230, 451
Interconnection
  capacitance, 201
  chip-to-chip, 587
  delay, 208
  densities, 71, 595
  gate-to-gate, 5

HDI, 619
layers, 617
levels, 4
manifold, 585, 595, 621
second-level, 7, 8
technologies, 588
3D high-density, 601
Interdiffusion, 101
Interfacial metallurgy, 101
Intermetallic
  compounds, 99
  formation, 101
International Technology Roadmap for Semiconductors (ITRS), 613
Interplanar spacing, 728
Interposer, 17, 602, 605, 606
  elastomeric, 602
Intrastack bonding, 603
Intrinsic stress, 86, 87
Ion, 84, 92
Ion
  beam, 98
  beam-assisted deposition (IBAD), 87
  bombardment, 86, 89
  guns, 759, 779
Ionization efficiency, 86
IPC, 141
Isotropic, 96, 97

J
JEDEC, 663
Joint testability action group (JTAG), 615
Junction temperature, 248

K
Kevlar, 67
Kinetic energy, 81, 84
Kirchhoff's law, 213, 228
Known good die (KGD), 14, 588, 608, 615
Kovar, 64, 102, 590

L
LaB6, 360
Laminar flow, 269
Laminate
  ceramics, 173
  double-sided, 95
  materials, 109
  substrate, 67, 584, 594
Land grid array, 606
Large-format MCM packages, 585
Large-scale join configurations, 599
  edge-edge, 599, 600
  edge-plane, 599, 600, 601
  mezzanine, 599, 600
  motherboard-to-daughtercard, 599
  plane-plane, 599, 600
Laser direct imaging, 113

Laser vias
  ceramics, 167
  FR4, 134–135
Lateral thermal shunts, 614
Latency of interconnections, 613
Lattice parameter, 729
Layout tools, 189
Layup, 118
Lead bend fatigue, 64
Lead frame, 5, 7, 55–57, 64, 98
  alloy 42, 64
  base materials, 65
  chemical etching, 64
  materials, 65
  metal, 5
  paddles, 590
  stamping, 64
Learning curve, 698
Least squares regression, 712
Level of packaging, 3
Lid
  sealing, 64
  brazing, 64
  cold welding, 64
  resistance welding, 64
  seam welding, 64
  solder, 64
Linear strains, 301
Line-of-sight areas, 81
Line-of-sight deposition, 78, 81
Linewidth, 96
Liquid cooling, 282
Liquid nitrogen trap, 82
Liquid photoresists, 94
Liquid recirculation, 96
Lithography
  adaptive, 594, 595, 621
  direct-write, 595, 621
Loading
  metal, 5
  factors, 592
Load-lock chamber, 82
Long-chain polymers, 93
Loss, 37
  dielectric, 37, 500, 510
  factor, 66
  related to frequency, 490
  tangent, 24, 33, 40, 68, 510, 593
Low noise amplifier, 490
Low-pressure CVD, 89
Low Temperature Cofired Ceramics (LTCC), 163, 592
  transmission circuit, 507
Lubricant, 79
Lumped circuit analysis, 209, 515

M
Magnesium aluminum silicate, 42
Magnesium oxide, 35–37

Magnetic
  fields, 86
  flux density, 196
  susceptibility, 31
  stirring, 96
Magnetron sources, 86
Magnetron sputtering, 85, 86
Maleability, 24
Manual grinding, 620
Manufacturing
  cycle time, 22
  tolerance, 111
  yield, 22
Manufacturability, 20–21
Manufacturing-induced defects, 22
Many-layer systems, 614
Marking, 430
Mask-based patterning, 595
Masks, 95
Mass analyzers, 779
  double-focusing magnetic sector, 780
  time-of-flight, 781
Mass of bombarding ions, 85
Mass spectroscopy, 785
Mass-to-charge ratio, 776
Master mask, 95
Matching network, 86
Materials, 23, 31, 86, 93
  conducting, 86
  delamination, 31
  for packaging, 23
  mean time to failure (MTTF), 34
  nonconducting, 86
  photosensitive, 93
  semiconducting, 86
MCM-C (ceramic) technology, 95
MCM-D (deposited) technology, 95
Mean free path, 78
Mechanical
  ablation, 620
  design, 299
  pump, 79
  spring array, 602
  stability, 20, 307
  stress, 659
Mechanical properties
  ceramic substrates, 153
  digital mirror devices (DMD), 644
  shrinkage in ceramics, 179
  thermal conductivity of ceramics, 180
  mechanical properties, 339
Megaframe imager, 619
Melting point, 24
Melt surface, 81
MEMS, 432
Mentor graphics, 460, 469–477
Metal
  bumps, 77
  cap, 98
  evaporation, 101
  lid joining, 101
  Matrix composites (MMC), 552
  pads, 77
  traces, 77, 93
Metal films
  Coefficient of thermal expansion (CTE), 659
  electrical properties, 360–361
  skin depth, 225
  thermal properties, 292
  thermodynamic stability, 656
  thick film, 161
Metallization, 93
  additive, 93
  subtractive, 93
Metal-organic
  compounds, 77
  CVD (MOCVD), 77
Metal-to-mask selectivities, 96
Metal-to-metal
  contacts, 98
  joining technique, 77
Mezzanine
  based MCM stacking, 601
  style construction, 600
Micro-Z ball stack technology, 604
Microchannel cooling, 288
Microelectromechanical systems (MEMS), 19, 609, 625
Micromanipulators, 95, 99
Micromirror devices, 636
Microoptoelectromechanical systems (MOEMS), 625
Microsectional analysis, 125
Microstrip circuitry, 40, 200, 215, 239
  transmission lines, 500
Microvias, 134, 448–450, 595
Microwave devices, 39
MIL-PRF, 121
MIL-STD, 663
Military
  aerospace systems, 588
  platforms, 586
Miller indices, 728
Misalignments, 595
Mixed signal systems, 626
Moat regions, 595
Modeling
  cost, 691
  packaging structures, 187, 523
  passive components, 372
  thermal, 289
Modulus, 22
MOEMS, 432
Moisture
  absorption, 19, 50
  diffusivity, 30
  penetration, 31, 657–662

Moisture sensitivity level (MSL), 392, 393, 407
Mold release agent, 50, 52
Molding, 55–56
Molecular
  architecture, 43
  beam, 81
  weight, 43
Molybdenum, 71
Molymanganese, 39
Moment, 313
Monolithic
  IC, 583
  wafer-scale integration (WSI), 584
Monomer, 43
Moore's law, 610
Moresco formula, 450
Motherboard, 443
Mounted components, 589
Moving average, 705
Moving probe tester, 457
Mullite, 71
Multi-chip modules (MCMs), 78, 90, 583–598
  ceramic (MCM-C), 584, 591, 592, 606
  ceramic/deposited (MCM-C/D), 613
  deposited (MCM-D), 584, 593
  deposited/laminate (MCM-D/L), 595
  folded flex-based, 607
  laminate (MCM-L), 584, 595
  packaging, 15, 16, 57, 89, 583–588, 598
  paper-thin, 617
  patterned overlay, 587, 620
  planar, 593
  power, 548
  stacks, 605
  thin-film, 586, 589
Multicomponent integration, 583
Multi-component package (MCP), 590, 603
Multifilament system, 82
Multilayer
  ceramic package, 98
  ceramic (MLC) PGAs, 101
  ceramic (MLC) substrates, 77
  circuit board fabrication, 90
  printed circuit boards (PCB), 77, 94, 95
  printed wiring structures, 606
  stackup, 591
  structure, 82
Multiple
  hearth e-beam source, 82
  layer buildups, 593
Mutual
  capacitance, 188, 226
  inductance, 194, 211, 226
Mylar, 57, 61, 68

N
Nanodevices, 19
Nanotechnology, 434

Natural convection, 276
Negative image, 95
Negative resist developer, 95
NEMI, 248, 353
Neopackaging, 601
Neo-stacking, 605
Net topology, 224
Network analyzer, 524
Newton's law of cooling, 253
NiCr, 357, 360
NiP, 357, 360, 362
Nipkow disk, 751
NIST, 584
Nitride, 35
  aluminum, 35
  boron, 35
  passivation layers, 88
  silicon, 35, 43
Noise margin, 21, 188, 453
Nomex, 68
Nonconducting
  materials, 86
  tiebars, 592
Non-demountable systems, 602
Nonmonotonic pin-to-pin relationship, 612
Nonrefractory materials, 592
Normal
  distribution, 677
  stress, 304
Nucleation, 82, 90, 91
Nuclei, 81, 85
Nusselt number, 275

O
Oil reservoir, 79
Operating pressures, 86
Optoelectronics, 19
Organic
  amines, 91
  solvents, 96
Orientation, 81
  preferential, 81
  random, 81
Orthoquinone diazides, 93
Outer lead bond pattern, 596
Outgassing, 58, 95, 632
  MEMS packaging, 632
Overpotential, 90, 92
Oxidation
  processing, 88
  reaction, 90, 97
Oxide, 35
  aluminum, 35
  beryllium, 35
  calcium, 37
  magnesium, 35, 37
  silicon, 35, 37

804 Index

Oxide
  passivation layers, 88
  on metals, 657
Oxidizers, 95
Oxygen, 78

P
Package
  ball grid array (BGA), 12, 598
  base, 64
  brazed ring-to-metal cap, 98
  ceramic ball grid array (CBGA), 13
  ceramic column grid array (CCBGA), 13
  ceramic leadless chip carrier (CLCC), 11
  CerDIP, 9
  discrete component, 551
  dual-in-line (DIP), 9, 36, 55, 598
  flatpack, 9, 55, 64
  level processes, 429
  lid, 598
  low-profile, 620
  metal, 63
  metal ball grid array (MBGA), 13
  metal transistor outline (TO), 8
  microball grid array, 14
  micro-SMT, 14
  miniball grid array, 14
  molding, 50
  multichip, 3, 4, 57
  multichip module, 15, 16, 57
  multicomponent, 590, 603
  pads, 98
  paper-thin, 17, 621
  pin grid array (PGA), 598
  pins, 90, 102
  quad-flat, 598
  stacking, 17
  surface-mount, 598
  thin small outline package (TSOP), 620
Packaging
  chip-scale, 14, 586, 597
  density, 29
  efficiency, 15, 71, 438
  few chip, 17
  first level, 5
  hermetic, 34
  hierarchy, 589, 615
  MEMS, 625
  molded plastic, 34
  multichip, 13, 15, 17, 66
  multichip module, 15, 16, 57
  radial spread coating, 56
  single-chip, 13
  thin-film, 57
  3D, 599
  ultradense, 613
  wafer-level, 17, 597
Pad-to-pad connectors, 90

Pads
  chip bonding, 5
Panalization, 108
Paper-thin
  HDI MCM, 617
  package, 621
Paraelectrics, 188, 365–367
  capacitance, 21
Parafocusing, 730
Parallel-plate electrode system, 89, 98
Parameter extraction, 202
Parasitic
  capacitance, 585
  inductance, 585
  latency reduction, 584
Partitioning, 442, 464
Pascal, 78
Pass-through
  connections, 600, 601
  contacts, 600
Passive
  arrays, 355
  capacitors, 363
  costing, 351
  decoupling, 381
  definitions, 354
  devices, 349
  dielectrics, 369
  electrical characterization, 373
  inductors, 371
  network, 355
  paraelectrics and ferroelectrics, 365
  parasitic properties, 361
  resistors, 358
  sizing embedded, 360, 367
  thick films, 363, 370
  value range in systems, 354
Passivation on chips, 50
Pastes, thick-film, 70
Patterned
  metal deposit, 93
  overlay, 594–616
  substrate, 96, 598
Patterning, 77, 93
PCB fabrication, 90
Peeling stress, 330
Performance metrics, 439
Permeability, 196, 491
Permittivity, 593
  free space, 200
  effective, 501
Personal digital assistant (PDA), 17
pH, 91
Phase
  change, 614
  velocity, 214
Phenolic polymer, 51
  flame-retardant, 67

Photo-definable material, 94
Photoengraving, 96, 100
Photolithography, 49, 77, 93, 593
Photonic devices, 636
Photoresist, 93, 50
   cross-linked negative, 95
   dry-film, 94, 95
   liquid, 94
   stripping, 95
Photosensitive materials, 54, 55, 93
Physical
   reactions, 88
   vapor deposition (PVD), 77, 78, 87
Physisorption, 81
Pin grid array (PGA), 12, 55, 585
Pin terminals, 590
Pincount, 592
   assemblies, 619
   overhead, 618
Pinhole defects, 593
Pins, 64
Pitch, 596, 609
Planar
   assembly, 599
   fabrication methods, 591
   inductance, 198
   magnetron sources, 86
Planarity, 593
Planarization, 24, 89
   degree of, 49
Planarized substrate, 596
Plane-plane (face-to-face) connections, 600
Plane stress and strain, 311-312
Plasma-enhanced CVD, 88, 89
Plasma surface treatment, 415
Plastic, *see also* Polymers
   ball grid array (PBGA), 13
   encapsulated microcircuits, 590
   leaded chip carrier (PLCC), 11, 55
   mechanical behavior, 309–310
Plasticizers, 40
Plated
   conductor foils, 77
   through holes (PTHs), 9, 55, 77, 90, 95, 118–122
Platen, 620
Plating, 77, 78, 89
   constant current, 92
   electro, 77, 89, 90, 94
   electroless, 77, 89, 90, 91, 94
   pulse, 92
Points of triple functionality, 44
Point-to-point connections, 602
Poisson's ratio, 308
Polyamide, 47
Polycrystalline, 81
   materials, 728
   silicon, 89
Poly(dimethylsiloxame) (PDMS), 751

Polyester, 56, 57, 61
   films, 100
   support sheet, 95
   unsaturated, 56
Polyimide, 7, 30, 39, 53, 61, 67, 72
   commercial, 72
   cyclic chain, 53
   films, 100
Polymer, 5, 29, 50, 93, 94, 100
   branched, 43–45
   chains, 43–45
   conductive, 63
   cross-linked, 43–46
   dielectrics, 593
   fluorocarbon, 30
   glassy, 47
   homo, 44, 45
   linear, 43, 45
   metal-filled, 5
   PVS, 43
   thermoplastic, 34, 46
   thermoset, 34, 45, 49, 50, 56, 94
Polymer thick film, *see* Thick film
Polymeric film-based chip carrier (PFBCC), 57
   films, 60
   quadpack, 10, 11, 12
   quad flatpack, 10, 12, 55
   single chip, 3
   slightly larger than IC carrier, 14
   small outline, 10, 11, 55
   small outline integrated circuit (SOIC), 10, 11
   solid logic technology (SLT), 37
   supporting film, 90
   surface-mount, 10
   tape ball grid array (TBGA), 13
   transistor outline (TO), 8, 36, 64
   ultrathin, 17
   wafer-level/chip-scale (WS-CSP), 18
Polymerization, 43, 46, 95
Polyolefin cover sheet, 95
Polyphenylquinoxaline (PPQ), 71
Polysilicon deposition rate, 89
Polyvinyl alcohol, 47
Polyvinylchloride (PVC), 43, 68
Positively charged argon ions, 85
Positive resist developer, 95
Postetch
   annealing, 98
   cleaning, 98
Potential gradient, 84
Potentiostatically, 92
Power
   consumption, 20
   cooling, 556
   density, 613
   diodes, 540
   dissipation, 21
   distribution, 3, 187, 231, 453

Power (*Continued*)
  dividers, 520
  electronics, 180, 537
  packages, 547
  plane, 450
  rail holdup, 595
  switches, 538
  transistors, 542
  wirebonding, 570
Prebuilt overlays, 596
Precipitation, 91
Preferential orientation, 81
Prepreg, 110
Pressure, 78, 85, 99
  base, 85
  sputtering, 85
Pressureless sintering, 40
Pretesting, 597
Primary sequencing, 45
Principal stress, 319
Printed circuit board (PCB), 4, 55, 67, 105, 586–591
  data formats, 106
  design, 477
  dimensional stability, 109
  electrical test, 124
  fine line, 586
  manufacturing tolerance, 111
  microvia, 595
  process flow, 112
  vias, 618
Printed wiring board (PWB), *see* Printed circuit board
Printing, 36, 37, 95
  contact, 95
  proximity, 95
  thick film, 36
  screen, 36, 37
Probability of ionization, 86
Process control
  variables, 592
  yield, 583
Processing, 88, 102. *See also* Substrate processing
Product cycles, 460
Profit model, 463
Programmable logic device (PLD), 443
Propagation
  constant, 491
  delay, 207, 613
  velocity, 229
Properties, 24
  ceramic materials, 38
  chemical, 24, 34
  electrical, 24, 31
  high-alumina content ceramics, 39
  material, 24, 29
  mechanical, 24, 30, 48
  physical
  thermal, 24, 33
Prototype, 585, 615

Proximity printing, 95
pSPICE, 237
Pulse
  design wavelength, 204
  sputtering, 87
  plating, 92
Pump, 79
  cryogenic, 80, 81
  diffusion, 80
  direct-drive, 79
  direct-drive, dual stage, oil-sealed, 79
  dual-stage, 79
  high vacuum, 80
  mechanical, 79
  mechanical rotary vane, 79
  oil vapors, 82
  rotary vane, 79
  speed, 79, 82
  turbomolecular, 80, 81
Punched vias, 167
Purple plague, 100
Physical vapor deposition (PVD), 84

Q

Quad flatpack, 10–12, 55
Quadruple flat packages, 585
Qualification test, 588
  individual, 588
  single, 588
Quality factor
  circuits, 491
  inductors, 371
Quantum numbers, 753

R

Radiation
  induced defects, 84
  thermal, 254
Radicals. 97
Radio frequency (RF)
  diode sputtering, 86
  generators, 86
Raleigh scattering, 735
Raman spectroscopy, 726, 734–740
  diamond, 736
  resonance, 739
  surface-enhanced, 739
Random
  motion, 78
  orientation, 81
Rate of diffusion, 85
RC delay, 207
Reactant gases, 89
Reactions, 88
  chemical, 88
  electrochemical, 90
  etching, 97
  oxidation, 90, 97

physical, 88
reduction, 90, 91, 97
Reactive
  gas, 81, 84, 88
  ion etching (RIE), 54, 85, 97, 98
  ion milling (RIM), 54
  metal layers, 49
  sputtering, 84, 87
Reactor, 88
  cold-wall, 88, 89
  continuous-feed, 88
  CVD, 88
  horizontal, 88, 89
  hot-wall, 88, 89
  vacuum, 88
  vertical, 89
Recessed patterned substrate, 595, 596
Reconfigurable logic, 608
Redox potentials, 90
Reducing agent, 90, 91
Reduction reaction, 90, 91, 97
Redundancy, 608
Reevaporation, 78
Reflow
  solder, 423, 398
  temperature, 393, 395, 396, 398, 407, 415, 417
Refractory
  materials, 591
  metal filament heater, 83
  metals, 82, 84
Reflection
  coefficient, 219
  noise, 225
Reflectometry, 733
Registration pins, 95
Relative die bonding, 603, 604
Reliability, 20, 95, 587, 613, 651,
  accelerated testing, 661–667. *See also* Accelerated testing
  definitions, 651
  failure analysis, 660
  failure mechanisms, 655. *See also* Failure mechanisms
  metrology, 668. *See also* Reliability metrology
  modeling, 668
  series and parallel systems, 681
Reliability metrology, 668
  hazard rate, 672
  normal distribution, 677
  reliability functions, 668
  Weibull distribution, 674
Rent's
  exponent, 610
  rule, 446, 610–612
Replacement analysis, 716
Residual gas molecules, 78
Resist, 93, 94, 95
  cross-linked, 95
  masking layer, 96

sensitivity, 95
stripping, 95
Resistance, 21, 32, 189
  contact, 192
  mechanisms, 190
  thermal, 252, 553
Resistive heating element, 88
Resistivity of materials, 24, 358, 360, 593
  parasitic, 21
Resistors, 358. *See also* Passives and Embedded passives
  design equations,, 189 358
  electrical performance, 373, 379
  materials, 361
  parasitic properties, 361
  sheet resistance, 359
  temperature coefficient of resistance (TCR), 163, 190, 360–361
  thick film, 162
  value range in systems, 354
Resolution, 95
Resonance, 232
Return circuit, 197
Rework, 48, 407
Reworkability, 98
RF
  modules, 431
  sputtering, 87
Ribbon bonds, 509
Risetime, 613
Rochow process, 53
Rocking curve, 732
Rotary vane pump, 79
Routing connections, 589
Ruggedized electronics, 586
Run-out errors, 592

S

Sacrificial substrate material, 617
Safety protocols, 89
Sample charging, 760
Sapphire, 37
Scalability, 609
Scalar network analyzer, 524
Scanning
  Auger microscopy (SAM), 753, 761
  electrochemical microscopy (SECM), 744
  electron microscopy (SEM), 726, 744–750
  force microscopy, 740
  parameters, 492
  probe microscopy (SPM), 740–744
  thermal microscopy, 744, 750
  tunneling microscopy (STM), 727, 740
Scattering parameters, 492
Schematics, 471
Screen-printing, 36, 156–159, 171
  green stack, 102
  ink and paste formulations, 159
  thick films, 77

808 Index

Scribing, 403
Seal ring, 591
Sealing 7, 425. *See also* Encapsulation
  hermetic, 7, 426
  package lid, 7
  pin, 7
Secondary
  drilling, 123
  electrons, 745
  ion mass spectrometry, 727, 752, 775–790
  ions, 776
Seed layer, 78, 90
Self inductance, 194
Self resonant frequency, 377, 515
Semiconducting materials, 86
Semiconductor Industry Association (SIA), 248
Sensitivity analysis, 717
Sensitizers, 93
Sensors, 627
Sequential deposition, 82
Sequential evaporation, 84
Serviceability, 20, 23, 608
S-glass, 67
S parameters, 492
Shadow mask, 101
Shear, 30
  force, 313
  modulus, 308
  strain, 302
  strength, 24
  stress, 305
  ultimate, 30
Sheet resistance, 191, 359–361
  ceramics, 154
  resistor materials, 359
  thick film materials, 161
Shrinkage, 24, 50
  on cure, 50
  polymer, 50
Side branching, 45
Siemens, 359
Signal, 3, 5
  delay, 5
  distortion, 33
  distribution, 3, 187
  integrity, 20, 202
  pins, 590
  plane, 450
  velocity, 207
Silane coupling agents, 49
Silicon, 15, 71, 89
  density, 15
  efficiency, 17
  on insulator (SOI), 620
  on-sapphire (SOS), 736
  polycrystalline, 89
Silicon carbide, 35, 41, 71, 151, 153
  semiconductors, 543
Silicon dioxide, 35, 37, 71, 593, 620
Silicon nitride, 35, 43, 71
Silicone, 7
Single-hearth electron beam evaporation source, 84
Single-chip packages, 585, 595
Singulation, 430
Skin depth, 223, 491
Slightly larger than IC carrier (SLICC), 14
Slip casting, 37
Small scale integration (SSI)
Sodium bisulfate/sulfuric acid mixture, 96
Solder, 5, 70, 407
  array pitches, 13
  bump attach, 102
  bumping, 77, 101, 420,
  power electronics, 556
  printing, 395
  reflow, 10, 398, 597
  systems, 586
Soldering, 77, 98, 101
  alloy, 7
  wave, 9
Soldermask, 123
Solids content, 50
Solid-state interdiffusion, 98
Soluble, 96
Solvent, 93, 95, 96
  acetone, 96
  organic, 96
  xylene-based, 95
Sources, 81
  electron beam, 81, 82
  magnetron, 85
  point, 81
  single-hearth electron beam evaporation, 84
  thermal evaporation, 81, 82
Space
  systems, 586
  transforming, 445
Spacer, 602, 604
Spatial efficiency, 598
Spanning vias, 618
Specific capacitance, 367
Specifications
  PCB, 106, 141
Spin-coating, 94
Spin speed, 94
Spiral filament, 83
Spiral paths, 86
Spodumene glass, 42
Spring-to-pin connectors, 90
Sputtered neutral mass spectroscopy (SNMS), 785
Sputtering, 77, 84, 87, 759
  cleaning, 759
  on ceramics, 155
  DC, 87
  dc diode, 85

Index 809

deposition, 85, 87
diode, 86
direct current, 84
magnetron, 85, 86
pressure, 85
pulsed, 87
radio frequency diode, 86
rate, 85
reactive, 84
RF, 87
system, 85
triode, 87
yield, 85
Stability, 24
dimensional, 24
thermal, 24, 33, 48
Stabilizers, 91
Stacked chips, 15–17, 430, 621
Interconnected, 15
Stacked 3D assemblies, 600
chip-on-chip, 601
die/package, 602
mezzanine-based MCM, 601
neopackaging, 601
sugar-cube, 601
thickness agnostic, 618
ultrathin, 671
Stainless steel chamber, 82
Standard electronic packages, 2
Standard Hydrogen Electrode (SHE), 92
Static Random Access Memory (SRAM), 15
Static Induction Transistors (SIT), 543
Steatite, 70
Step coverage, 77, 84, 87, 88
Sticking coefficient, 78
Stiction, 630
Stokes emission, 735
Strain, 30, 299–303
elastic, 30
energy, 318
epitaxial films, 732
plastic, 30
thermal, 303
ultimate, 30
Stray capacitance, 201
Strength, 24
adhesion, 24, 49
adhesive, 49
breakdown, 33
cohesive, 48, 49
flexural, 24, 37, 50
mechanical, 66
peel, 24
physical, 37
tensile, 24, 50
ultimate, 48
yield, 48

Stress, 24, 30, 86, 303-305
intensity factor, 321
intrinsic, 86, 87
measurement, 736
mechanical, 47, 48
MEMS, 632
normal, 30
printed circuit boards, 110
thermal, 86
Stress screening, 52
Stripline, 499
Stripper, 95
nophenol-based, 95
Sublimation, 620
Substrate, 55, 67, 70, 77
accessory, 604
ceramic, 70, 73, 584, 591
circuitized, 55
co-fired ceramic, 592
connection, 603
copper-plated, 94
deposited thin-film, 71, 73
dissipative, 614
double-sided, 601
efficiency, 609
electrical, 591
flexible, 599
heated, 88
interconnected, 594
interconnecting, 588, 589, 590, 592, 595
laminate, 67, 73
mechanical, 589, 593, 595, 603
multilayer ceramic (MLC), 77
planarized, 596
patterned, 96, 595, 598
patterned overlay, 595, 596, 597, 598
power electronics, 563
recessed patterned, 595
rigid-flex, 68
ultrathin, 617
Substrate
heater, 82
holder, 88
pads, 77
rotation, 82
Substrate materials
ceramic, 70, 149
Substrate processing
active metal brazing on ceramics, 182
ceramics, 151–171, 357, 362
cleanroom environments, 166
direct bonded copper on ceramic, 181
double-sided, 601
vias, 154, 167–171
Subtractive processes, 90–94
Sugar-cube stacking, 601
Sulfonic acid, 96
Sulfuric acid, 96

Supply rails, 613
Surface
  contamination, 85
  diffusion, 87
  energy, 31, 81
  finishes, 133
  mobility, 81
  morphology, 95
  mounting, 90
  mount pads, 98
  resistance, 491
  roughness, 81, 740
  tension, 83
  topology, 84, 621
Surface mount technology (SMT), 395
Susceptor, 88
Switching noise, 233, 453
System
  constraints, 440
  in a package (SIP), 17, 615, 622
  on a chip (SOC), 17, 615
  on a package (SOP), 17
  partitioning, 442
  requirements and design, 437
System-level
  constraints, 583
  contacts, 58

T
T junction, 506
TAB tape, 77, 93
Tailless bond, 99
TaN, 360
Tape-automated bonding (TAB), 77, 90–100, 414, 596, 604
  flip, 597
  inductance, 454
  3D Plus, 604
Tape for ceramics, 159, 164–166
  casting, 37, 40
  flexible polymer, 7
Target
  material. 85
  power supply, 86
Taxonomy, 602
  die-stacking, 602
Technology, 8
  curing, 24
  eutectic, 24
  flip-chip, 12
  glass transition, 24, 33, 34, 44, 45, 48, 50
  silicon planar, 8
  surface-mount, 10
  temperature, 24
Temperature, 99
  coefficient of capacitance (TCC), 365
  coefficient of resistance (TCR), 24, 360–361
  cycling and shock, 666
  window, 52

Tensile modulus, 48
Terminal
  pitch, 11
  pins, 591
Termination
  open and short, 239
  parallel and series, 221
  reflection, 217
Test code, 615
Testability, 22, 608, 615, 618
Testing, 22
  accelerated life, 34
  hermetic packages, 426
  in-process, 22, 23
  shock, 35, 66
  stress, 22
  vibration, 35
  wirebonding, 412
Thermal
  coefficient of expansion (TCE), 21, 24, 40
  conduction module, (TCM), 456, 584
  conductivity, 21, 34, 50, 66, 257, 555, 589, 614
  degradation, 33
  design, 573
  drift, 614
  expansion, 39
  expansion coefficient, 5, 591, 614
  management, 247, 601, 606, 613, 621
  mismatch, 30
  properties of materials, 292
  resistance, 248, 614, 554
  shock, 66
  stability, 24, 33, 48, 58, 66
  storage, 614
  strain, 303
  stress cycling, 35
  system considerations, 441
Thermal
  energy, 81
  evaporation sources, 81–83
  expansion coefficient, 86
  stress, 86
Thermocompression bonding, 99, 100
Thermodynamically favorable, 97
Thermoelectric coolers, 287, 614
Thermoplastic polymers, 45, 48, 406, 602
Thermoset polymers, 49, 50, 94, 406
Thick film
  capacitors, 370
  compared to LTCC, 164
  metallization, 161
  on ceramics, 156
  resistors, 357, 360, 363
  resistance equation, 358
  screen fabrication, 94
  screen printing, 156–159
Thickness uniformity, 82

Thin films
    capacitors, 363, 368, 370
    deposition, 84, 85
    embedded passives, 356–357
    metallization, 77
    on ceramics, 155
    resistance equation, 358
    resistors, 360
Thinned IC, 617, 620
Three-dimensional
    assemblies, 591
    chip stacks, 585
    demountable, 601
    folding, 607
    many-layered, 600
    stacking, 15
Three-dimensional die bond-out schemes, 603
    absolute or direct substrate, 603
    relative or intrastack, 603
Three-dimensional die layer configurations, 602, 603
    homogeneous, 602, 603
    staggered, 602, 603
    telescopic, 602, 603
Three-dimensional high density interconnect (HDI), 601, 606
Three-dimensional IC stacking methods based on wirebonding, 604
    spacer, 604
    staggered bond-out, 604
    staggered staircase, 604
    telescoping, 604
Three-dimensional packaging, 599
    folding, 599
    stacking, 17, 599
    stacking and folding, 599
Three-dimensional systems, 599, 606
    highly integrated packaging and processing (HIPP), 606, 607
Threshold energy, 85
Through hole, 92, 94, 595
    contacts, 18, 595
    high aspect-ratio, 92
Thyristors, 541
Tight-pitch assemblies, 598
Time constant, 593
Time delay, 32, 33
    signal propagation, 32, 33
Time domain measurements, 527
Time domain reflectometry (TDR), 191, 238
Time-of-flight (TOF), 584, 585, 613
    delay, 222
    signal, 584
Tolerances
    PCB's, 111
    repair, 608
Tooling for PCB's, 107
Topside oxide, 621
Torr, 78

Transistor, 8
    gain degradation, 8
    high-power, 39
Translation/rotation misalignments, 595
Transmission lines, 494
    coaxial, 498
    coplanar, 502
    microstrip, 500
    model, 209, 212, 226
    modes, 495
    planar, 499
    variations, 504
Transmit-receive antenna module (TRAM), 616
Transverse electromagnetic wave (TEM), 216
Triazine, 71
Trim, 430
Triode sputtering, 87
Tubes
    vacuum, 36
    electron, 37
Tube furnace, 88
Tungsten, 39, 71
Tungsten carbide, 35
Turbomolecular pump, 80, 81
Turbulent flow, 270

U

Ultradense packaging system, 613
Ultrasonic, 98
Ultrasonic agitation, 100
Ultrasonic source, 100
Ultrasonic wedge bonding, 100
Ultrathin silicon, 621
Ultrathin stacks, 617
Ultrathin substrate, 617
Ultraviolet light, 93
Underfill, 424
    flowable, 424
    no-flow, 425
Unifying package structure, 583
UV lamp, 95

V

Vacuum pumping system, 88
Vacuum reactors, 88
Van der Waals forces, 620, 741
Vaporization, 33, 34
Vapor-phase deposition, 77, 90
Vertical chambers, 88
Vertical nearest neighbor, 603
Vertical reactor, 88
Very large scale integration (VLSI), 19
Via-drilled, glass-epoxy laminates, 93
Via opening, 87
Vias, 18, 87
    barrel, 617
    densities, 593
    desmear, 119

Vias (*Continued*)
  drilling, 118
  filled, 593
  in ceramics, 154, 167–171
  in organic boards, 118, 618
  interlayer, 593
  laser, 167
  microvias, 134
  occlusion, 609
  PCB, 618
  spanning, 618
  staircase, 593
  thermal, 21
  through-wafer, 18
Via side walls, 87
Vinyl chloride, 43
Viscosity, 24, 94
  polymer, 50
Voids, 39, 40
Volatile, 96
Voltage (VCR), 163
Voltage
  reflection coefficient, 217
  standing wave ratio, 490
Volume energy, 81
Von Mises, 318, 320

W

Wafer backgrinding, 401. *See also* Thinning
  bumping, 419
  mounting, 401
  mounting equipment, 405
  preparation, 399
  probing, 399
  sawing, 402
  thinning, 401
Wafer level
  MEMS, 631
  packaging, 597
  redistribution, 18
  stacking (WLS), 18
Warpage, 19
Water
  absorption, 49
  cooled crucible, 84
  diffusivity, 30
  vapor, 78
Waveguides, 496
  coplanar, 502

Wax, 51
  palm tree, 51
  synthetic ester, 51
Wedge-shaped tool, 100
Weibull distribution, 674
Welding, 7, 590
Wet etching, 96
Wettability, 62
Windowing, 52
Windows in packages, 644
Winter's method, 710
Wire, 187
Wireability, 593
Wire bonders, 98
Wire bonding, 5, 98, 101, 409, 412, 451, 597, 603
  aluminum, 36
  ball bonding, 410
  density, 447
  gold-beryllium, 58
  gold-copper, 58
  Mg-Al, 58
  power electronics, 570
  ribbon bonding, 410
  Si-Al, 58
  testing, 412
  thermocompression, 59, 409
  thermosonic, 59, 410
  ultrasonic, 58, 409
  wedge bonding, 411
Wire filaments, 83
Wirepull, 413
Wiring density, 609, 616
Work function, 621, 767

Y

Yield, 601, 608
Young's modulus, 24, 30, 48
Yttrium-stabilized zirconium oxide, 732

X

X-ray, 84
  diffraction, 725–734
  photoelectron spectroscopy, 727, 752, 766–775
  sources, 729, 770

Z

Zero ohm resistor, 357
Zone model, 86